Accelerated Testing

Accelerated Testing

Statistical Models, Test Plans, and Data Analyses

WAYNE NELSON

Consultant, Schenectady, NY

A JOHN WILEY & SONS, INC., PUBLICATION

Library of Congress Cataloging-in-Publication is available.

10 9 8

This book is gratefully dedicated to the many clients whose fruitful collaboration and challenging applications stimulated my interest and developments in accelerated testing, and to the many colleagues who kindly provided me with examples, references, suggestions, and encouragement.

Contents

Preface, xi

1. **Introduction and Background, 1**

    1. Survey of Methodology and Applications, 3
    2. Types of Data, 12
    3. Types of Acceleration and Stress Loading, 15
    4. Engineering Considerations, 22
    5. Common Accelerated Tests, 37
    6. Statistical Considerations, 43
        Problems, 49

2. **Models for Life Tests with Constant Stress, 51**

    1. Introduction, 51
    2. Basic Concepts and the Exponential Distribution, 53
    3. Normal Distribution, 58
    4. Lognormal Distribution, 60
    5. Weibull Distribution, 63
    6. Extreme Value Distribution, 65
    7. Other Distributions, 68
    8. Life-Stress Relationships, 71
    9. Arrhenius Life-Temperature Relationship, 75
    10. Inverse Power Relationship, 85
    11. Endurance (Fatigue) Limit Relationships and Distributions, 92
    12. Other Single Stress Relationships, 95
    13. Multivariable Relationships, 98
    14. Spread in Log Life Depends on Stress, 105
        Problems, 107

3. **Graphical Data Analysis, 113**

    1. Introduction, 113
    2. Complete Data and Arrhenius-Lognormal Model, 114
    3. Complete Data and Power-Weibull Model, 128
    4. Singly Censored Data, 134

5. Multiply Censored Data, 139
6. Interval (Read-Out) Data, 145
   Problems, 154

4. Complete Data and Least Squares Analyses, 167

1. Introduction, 167
2. Least-Squares Methods for Lognormal Life, 170
3. Checks on the Linear-Lognormal Model and Data, 182
4. Least-Squares Methods for Weibull and Exponential Life, 189
5. Checks on the Linear-Weibull Model and Data, 203
6. Multivariable Relationships, 210
   Problems, 229

5. Censored Data and Maximum Likelihood Methods, 233

1. Introduction to Maximum Likelihood, 234
2. Fit the Simple Model to Right Censored Data, 242
3. Assess the Simple Model and Right Censored Data, 255
4. Other Models and Types of Data, 265
5. Maximum Likelihood Calculations, 284
   Problems, 302

6. Test Plans, 317

1. Plans for the Simple Model and Complete Data, 317
2. Plans for the Simple Model and Singly Censored Data, 328
3. Evaluation of a Test Plan by Simulation, 349
4. Survey of Test Plans, 361
5. ML Theory for Test Plans, 364
   Problems, 371

7. Competing Failure Modes and Size Effect, 377

1. Series-System Model, 378
2. Series Systems of Identical Parts, 383
3. Size Effect, 385
4. Nonuniform Stress, 387
5. Graphical Analysis, 392
6. ML Analysis for Competing Failure Modes, 407
7. ML Theory for Competing Modes, 413
   Problems, 417

8. Least-Squares Comparisons for Complete Data, 425

1. Hypothesis Tests and Confidence Intervals, 426
2. Graphical Comparisons, 429
3. Compare Log Standard Deviations, 434
4. Compare (Log) Means, 437
5. Compare Simple Relationships, 441

6. Compare Multivariable Relationships, 445
    Problems, 448

9. Maximum Likelihood Comparisons for Censored and Other Data, 451

1. Introduction, 451
2. One-Sample Comparisons, 452
3. Two-Sample Comparisons, 458
4. K-Sample Comparisons, 465
5. Theory for LR and Related Tests, 470
    Problems, 488

10. Models and Data Analyses for Step and Varying Stress, 493

1. Survey of Theory for Tests with Varying Stress, 494
2. Step-Stress Model and Data Analyses, 495
3. Varying-Stress Model and Data Analyses, 506
    Problems, 513

11. Accelerated Degradation, 521

1. Survey of Applications, 521
2. Degradation Models, 523
3. Arrhenius Analysis, 534
    Problems, 544

Appendix A.   Statistical Tables, 549

A1. Standard Normal Cumulative Distribution Function $\Phi(u)$, 550
A2. Standard Normal Percentiles z_P, 552
A3. Standard Normal Two-Sided Factors K_P, 552
A4. t-Distribution Percentiles $t(P;v)$, 553
A5. Chi-Square Percentiles $\chi^2(P;v)$, 554
A6a. F-Distribution 95% Points $F(0.95;v_1,v_2)$, 556
A6b. F-Distribution 99% Points $F(0.99;v_1,v_2)$, 558
A7. Probability Plotting Positions $F_i = 100(i-0.5)/n$, 560

References, 561

Index, 579

Preface to the Paperback Edition

First published in 1990, this book remains the most comprehensive presentation of statistical models and methods for accelerated test data. It is gratifying that it has been widely used and praised by practitioners and researchers in statistics and engineering. This paperback edition is available at a bargain price thanks to the fine work of Mr. Steve Quigley, Ms. Susanne Steitz, and the Wiley staff.

For subsequent advances in accelerated testing, the reader may wish to consult:

- Meeker, W.Q. and Escobar, L.A. (1998), *Statistical Methods for Reliability Data,* Wiley, New York, www.wiley.com. In particular, their degradation models and corresponding statistical methods and Chapter 11 of this book overlap little and together comprise a basic introduction to accelerated degradation.
- Nelson, Wayne (2004), "A Bibliography of Accelerated Test Plans," over 100 references, available from the author, WNconsult@aol.com.

Since 1990, commercial software for analysis of accelerated test data has continued to advance. To reflect these advances, Table 1.1 of Chapter 5 and corresponding text have been updated. Note that confidence limits using a normal approximation to the sampling distribution of a maximum likelihood estimator and its asymptotic standard error are not current best practice. Instead, one should use software that calculates confidence limits using the likelihood ratio, as described in Section 5.8 of Chapter 5, as these intervals are now known to be a better approximation in virtually all applications.

WAYNE B. NELSON
Consulting and Training
WNconsult@aol.com

Schenectady, New York
June 2004

Preface

Product reliability contributes much to quality and competitiveness. Many manufacturers yearly spend millions of dollars on product reliability. Much management and engineering effort goes into evaluating reliability, assessing new designs and design and manufacturing changes, identifying causes of failure, and comparing designs, vendors, materials, manufacturing methods, and the like. Major decisions are based on life test data, often from a few units. Moreover, many products last so long that life testing at design conditions is impractical. Many products can be life tested at high stress conditions to yield failures quickly. Analyses of data from such an accelerated test yield needed information on product life at design conditions (low stress). Such testing saves much time and money. This book presents practical, modern statistical methods for accelerated testing. Up-to-date, it provides accelerated test models, data analyses, and test plans. In recent years, much useful methodology has been developed, and this book makes it available to practitioners. This book will contribute to more efficient accelerated testing and to valid and more accurate information.

This book is written for practicing engineers, statisticians, and others who use accelerated testing in design, development, testing, manufacturing, quality control, and procurement. It will aid workers in other fields concerned with regression models for survival, for example, in medicine, biology, and actuarial science. Also, this book is a useful supplement for statistics and engineering courses, as it presents many stimulating real examples, emphasizes practical data analysis (employing graphical methods and computer programs), and shows how to use versatile maximum likelihood methods for censored data.

This book is organized to serve practitioners. The simplest and most useful material appears first. The book starts with basic models and graphical data analyses, and it progresses through advanced maximum likelihood methods. Available computer programs are used. Each topic is self-contained for easy reference, although this results in some repetition. Thus this book serves as a reference or textbook. Derivations are generally omitted unless they provide insight. Such derivations appear in advanced sections for those seeking deeper understanding or developing new statistical models, data analyses, and computer programs. Ample references to the literature will aid those seeking mathematical proofs.

Readers of this book need a previous statistics course for Chapter 4 and beyond. Chapters 1, 2, and 3 do not require a previous course. For advanced material, readers need facility in calculus through partial differentiation and the basics of matrix algebra.

There is a vast and growing literature on statistical methods for accelerated testing. However, this book has been limited to the most basic and widely used methods, as I did not wish to complete it posthumously. Topics not given in detail in this book are referenced. While I included my previously unpublished methods developed for clients, there are gaps in methodology, which are noted to encourage others to fill them. For advanced innovations and complex applications beyond the basics in this book, one can consult the literature and experts.

Chapter 1 introduces accelerated testing – basic ideas, terminology, and practical engineering considerations. Chapter 2 presents models for accelerated testing – basic life distributions and life-stress relationships for products. Chapter 3 explains simple graphical analyses to estimate product life. Requiring little statistical background, these data plots are easy and very informative. Chapter 4 covers least squares estimates and confidence limits for product life from complete data (all test specimens run to failure). Chapter 5 shows how to use maximum likelihood estimates and confidence limits for product life from censored data (some specimens not run to failure). Chapter 6 shows how to choose a test plan, that is, the stress levels and corresponding numbers of specimens. Chapter 7 treats data with competing failure modes – models, graphical analyses, and maximum likelihood analyses. Chapters 8 and 9 present comparisons (hypothesis tests) with least squares and maximum likelihood methods. Chapter 10 treats step-stress testing and cumulative damage models. Chapter 11 introduces aging-degradation testing and models.

The real data in all examples come mostly from my consulting for General Electric and other companies. Many data sets are not textbook examples; they are messy – not fully understood and full of pimples and warts. Proprietary data are protected by generically naming the product and multiplying the data by a factor. I am grateful to the many clients and colleagues who kindly provided their data for examples.

I am most grateful to people who contributed to this book. Dr. Gerald J. Hahn, above all others, encouraged my work on accelerated testing and is a valued, knowledgeable, and stimulating co-worker. Moreover, he helped me obtain support for this book from the General Electric Co. I am deeply indebted for support from my management at General Electric Co. Corporate Research and Development – Dr. Roland Schmitt (now President of Rensselaer Polytechnic Inst.), Dr. Walter Robb, Dr. Mike Jefferies, Dr. Art

Chen, Dr. Jim Comly, and Dr. Gerry Hahn. Professor Josef Schmee, when Director of the Graduate Management Institute of Union College, kindly provided an office where I worked on this book. He also gave me an opportunity to teach a course from the book manuscript and thereby improve the book.

Friends generously read the manuscript and offered their suggestions. I am particularly grateful for major contributions from Prof. Bill Meeker, Dr. Necip Doganaksoy, Dr. Ralph A. Evans, Mr. D. Stewart Peck, Dr. Agnes Zaludova, Mr. Don Erdman, Mr. John McCool, Mr. Walter Young, Mr. Dev Raheja, and Prof. Tom Boardman. My interest in and contributions to accelerated testing owe much to the stimulating applications of and many collaborations with Mr. Del Crawford, Mr. Don Erdman, Mr. Joe Kuzawinski, and Dr. Niko Gjaja, among others. Many experts on engineering topics and statistics provided key references and other contributions.

The illustrations are the superb work of Mr. James Wyanski (Scotia, NY) and Mr. Dave Miller. The manuscript benefited much from the skillful word processing of Mr. John Stuart (Desktop Works, Schenectady, NY) and Ms. Rita Wojnar.

Authors who wish to use examples, data, and other material from this book in journal publications may do so to the extent permitted by copyright law with suitable acknowledgement of the source. *Any* other use of such material requires the written permission of the publisher: Permissions Dept., John Wiley & Sons, 605 Third Ave., New York, NY 10158-0012.

I would welcome correspondence on suggestions on key references and improvements for the book.

WAYNE NELSON

Schenectady, New York, August 1989

About the Author

Dr. Wayne Nelson is a leading expert on statistical analysis of reliability and accelerated test data. He currently privately consults with companies on diverse engineering and scientific applications of statistics and develops new statistical methods and computer programs. He presents courses and seminars for companies, universities, and professional societies. He also works as an expert witness. An employee of General Electric Corp. Research & Development for 25 years, he consulted across GE on applications.

For his contributions to reliability, accelerated testing, and reliability education, he was elected a Fellow of the American Statistical Association (1973), the American Society for Quality (1983), and the Inst. for Electrical and Electronics Engineers (1988). GE Corp. R&D presented him the 1981 Dushman Award for outstanding developments and applications of statistical methods for product reliability and accelerated test data. The American Society for Quality awarded him the 2003 Shewhart Medal for outstanding technical leadership and innovative developments.

He has authored over 120 literature publications on statistical methods, mostly for engineering applications. For publications, he was awarded the 1969 Brumbaugh Award, the 1970 Youden Prize, and the 1972 Wilcoxon Prize, all of the American Society for Quality. The ASA has awarded him eight Outstanding Presentation Awards for papers at the national Joint Statistical Meetings.

In 1990, he was awarded the first NIST/ASA/NSF Senior Research Fellowship at the National Inst. of Standards and Technology to collaborate on modeling electromigration failure of microelectronics. On a Fulbright Award in 2001, he did research and lecturing on reliability data analysis in Argentina.

Dr. Nelson authored the book *Applied Life Data Analysis,* published by Wiley in 1982; it was translated into Japanese in 1988 by the Japanese Union of Scientists and Engineers. In 1990, Wiley published his landmark book *Accelerated Testing: Statistical Models, Test Plans, and Data Analyses.* In 2003, ASA-SIAM published his book *Recurrent Events Data Analysis for Product Repairs, Disease Episodes, and Other Applications.* He has authored various book chapters and tutorial booklets, and he contributed to standards of engineering societies.

Further details of his work appear at www.members.aol.com/WNconsult. He can be reached at WNconsult@aol.com, 739 Huntingdon Dr., Schenectady, NY 12309.

1

Introduction and Background

HOW TO USE THIS BOOK

This section describes this book's contents, organization, and how to use this book. This book presents statistical models, test plans, and data analyses for estimating product reliability from accelerated tests.

Chapter overview. This chapter presents an introduction to accelerated testing methods. Section 1 surveys common applications of accelerated testing and sources of information. Section 2 describes types of accelerated test data. Section 3 describes types of acceleration and types of stress loading. Section 4 discusses engineering considerations in planning and running an accelerated test. Section 5 describes common accelerated tests. Section 6 outlines the statistical steps and considerations in data collection and analysis. This chapter is background for the rest of this book. To profit from this chapter, readers need only a general engineering background. Of course, previous acquaintance with accelerated testing helps. Those lacking such acquaintance may benefit from reading Chapters 2 and 3 before Chapter 1.

Book overview. Chapter 1 gives an overview of the book and presents needed background. Chapter 2 describes accelerated life test models, consisting of a life distribution and a life-stress relationship. Chapter 3 presents simple probability and relationship plots for analyzing complete and censored data. Briefly stated, when all specimens have run to failure, the data are **complete**. When some specimens are unfailed at the time of the data analysis, the data are **censored**. The plots yield estimates for model parameters, product life (distribution percentiles, reliabilities, failure rates), and other quantities. Chapter 4 presents least-squares analyses of complete data; these analyses yield such estimates and corresponding confidence limits. Chapter 5 gives maximum likelihood methods for censored data; these methods yield estimates and confidence limits. Chapter 6 presents test plans. Chapter 7 presents models and graphical and maximum likelihood analyses for data with a mix of failure modes. Chapters 8 and 9 present comparisons (hypothesis tests) for complete and censored data. Chapter 10 treats step-stress testing, cumulative damage models, and data analyses. Chapter 11

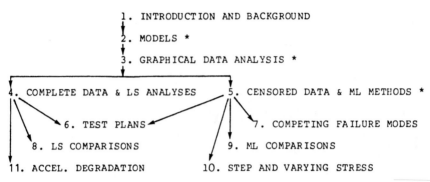

Figure 1.1.   Book organization (* basic chapter).

introduces accelerated degradation testing, models, and data analyses. Nelson (1990) briefly covers the most basic applied essentials of these topics.

Organization. Figure 1.1 shows this book's chapters. They are organized by type of data (complete, censored, and competing failure modes) and by statistical method (graphical, least squares, and maximum likelihood). The chapters are in order of difficulty. The arrows in Figure 1.1 show which chapters are background for later chapters. Also, each chapter introduction refers to needed background and describes the level of the chapter. The first three chapters are simple and basic reading for all. Chapter 2 on models is background for all else. Chapter 3 on graphical data analysis is most useful. The more advanced Chapters 4 through 6 are in order of difficulty. Chapter 6 (test plans) follows Chapter 5 (maximum likelihood analysis of censored data) in the logical development of the subject, but it can profitably be read after Chapter 2. Many readers who plan to use a particular model can selectively read just the material on data analysis with that model, skipping other material in data analysis chapters. Maximum likelihood methods (Chapter 5) are essential. They are versatile and apply to most models and types of data. Also, they have good statistical properties. If time is limited, read key Chapters 2, 3, and 5 for basics to solve most problems. Chapters 7 through 11 treat special topics and may be read in any order.

Numbering. The book numbers sections, equations, figures, and tables as follows. Within each chapter, the sections are numbered simply 1, 2, 3, etc.; subsections are numbered 4.1, 4.2, etc. Equation numbers give the (sub)section number and equation number; for example, (2.3) is the third numbered equation in Section 2. Figure and table numbers include the section number; Figure 2.3 is the third figure in Section 2. Such numbers do not include the chapter number. Unless another chapter is stated, any referenced equation, figure, or table is in the same chapter.

Problems. There are two types of problems at the end of a chapter. One type involves an analysis of data with the methods in that chapter; the other involves extending the results of the chapter to other problems. An asterisk (*) marks more laborious or difficult problems.

Citations. The book cites references by means of the Harvard system. A citation includes the author's name, year of publication, and his publications in that year. For example, "Nelson (1972b)" refers to Nelson's second referenced publication in 1972. All references are listed near the end of the book. Coauthored references follow all singly authored references by the first named coauthor. For example, Nelson and Hahn (1972) follows all references authored solely by Nelson.

Tables. Basic statistical tables are in Appendix A near the end of the book. Other tables must be obtained from the literature and are referenced.

Index. The index of the book is detailed. It will be an aid to those who wish to use the book as a reference for selected methods. Also, to aid users, some sections are self-contained, thus repeating some material.

Derivations. The book omits most derivations. Reasons for this are: (1) readers can properly apply most methods, knowing assumptions but not knowing derivations, (2) many derivations are easy for a reader or instructor to supply, and (3) more time can be spent on methods useful in practice. Many derivations appear in Mann, Schafer, and Singpurwalla (1974), Bain (1978), and particularly Lawless (1982) and Viertl (1988).

Terminology. This book uses common *statistical* terminology, whose meaning often differs from engineering and everyday meanings. Terms such as "normal," "independent," "dependent," and "confidence" are examples. Moreover, there are many instances of a single concept with several names. For example, independent, explanatory, and predictor variables are equivalent terms. Thus those not familiar with statistical terminology need to pay special attention to words in italics, boldface, and quotation marks. *Caveat lector.*

1. SURVEY OF METHODOLOGY AND APPLICATIONS

Overview. This section briefly surveys statistical and engineering methodology and the vast literature for accelerated testing. Also, this section briefly describes applications to indicate the wide use of accelerated testing. No doubt important references are lacking.

1.1. Methodology

Accelerated testing. Briefly stated, accelerated testing consists of a variety of test methods for shortening the life of products or hastening the degradation of their performance. The aim of such testing is to quickly obtain data which, properly modeled and analyzed, yield desired information on product life or performance under normal use. Such testing saves much time and money. The aim of this book is to provide practitioners with basic, practical statistical models, test plans, and data analyses for accelerated tests.

Statistical methodology. In recent years statisticians have developed much statistical methodology for accelerated testing applications. Indeed they solved most of the statistical problems listed by Yurkowski, Schafer, and Finkelstein (1967). For example, statisticians solved the big bugaboo of accelerated testing, namely, proper analysis of data with a mix of failure modes (Chapter 7). Recent books with chapters on statistical methodology for accelerated tests include Lawless (1982), Mann, Schafer, and Singpurwalla (1974), Jensen and Petersen (1982), Lipson and Sheth (1973), Tobias and Trindade (1986), Kalbfleisch and Prentice (1980), Cox and Oakes (1984), and Little and Jebe (1975). Viertl (1988) surveys statistical theory for accelerated testing. Nelson (1990) briefly presents the most basic essentials of applied statistical methods and models for accelerated testing. The present book provides applied statistical models and methods for accelerated testing. Statistical methodology is improving rapidly. Thus books over 5 years old lack important developments, and books over 10 years old are seriously out of date. This book is no exception.

Surveys of the statistical literature on accelerated testing include Viertl (1988), Nelson (1974), Ahmad and Sheikh (1983), Meeker's (1980) bibliography, Singpurwalla (1975), and Yurkowski, Schafer, and Finkelstein (1967). Peck and Trapp (1978) present simple graphical methods for semiconductor data. Peck and Zierdt (1974) survey semiconductor applications.

Journals. Journals with articles on accelerated testing may be found in the **References** at the back of this book. Journals with statistical methodology for accelerated testing include:
- *American Soc. for Quality Control Annual Quality Congress Transactions*
- *Annals of Reliability and Maintainability*
- *Applied Statistics*
- *IEEE Transactions on Reliability*
- *J. of Quality Technology*
- *J. of the American Statistical Assoc.*
- *J. of the Operations Research Soc. of America*
- *J. of Statistical Planning and Inference*
- *Naval Research Logistics Quarterly*
- *Proceedings of the Annual Reliability and Maintainability Symposium*
- *The Q R Journal – Theory and Practice, Methods and Management*
- *Quality and Reliability Engineering International*
- *Reliability Review of the American Soc. for Quality Control*
- *Technometrics*

Engineering methodology. Engineers have long used accelerated testing for diverse products. Governments and professional societies publish lists of standards and handbooks for testing methodology and data analysis. The engineering literature contains many papers on accelerated testing theory and applications. The bibliographies on accelerated testing of Meeker (1980) and Yurkowski and others (1967) with 524 older references show the scope of this literature. Also, various engineering books devote some space to this topic. A sample of references appears in the applications below.

Data banks/handbooks. This book covers **statistical methods** for collection and analysis of accelerated test data. It lacks data banks and handbooks for specific materials and products. The following brief list of sources may serve as a starting point in a search for such data. The US Department of Defense (DoD) (1981,1985) maintains Information Analysis Centers (IACs):
- Concrete Technology IAC, (601) 634-3269.
- DoD Nuclear IAC, (805) 963-6400.
- Infrared IAC, (313) 994-1200 ext. 214.
- Metals and Ceramics IC, (614) 424-5000. See the publication list of Metals and Ceramics IAC (1984)
- Metal Matrix Composites IAC, (805) 963-6452.
- Plastics Technical Evaluation Center, (201) 724-3189.
- Pavement and Soils Trafficability IAC, (601) 634-2209.
- Reliability AC, (315) 330-4151.
- Thermophysical and Electronic Properties IAC, (317) 494-6300.

Other sources of information include:
- Standards in many fields, American National Standards Inst. Catalog, 1430 Broadway, New York, NY 10018.
- National Nuclear Data Center, (516) 282-2103.
- Computerized references and data bases, STN International (Chemical Abstracts Service), PO Box 3012, Colombus, OH 43210-9989.
- Index to IEEE Publications (1988), (201)981-1393. Also, *Quick Reference to IEEE Standards.*
- *Ulrich's International Periodicals Directory*, R.R. Bowker Co., New York.
- *Science Citation Index* for locating more recent papers citing known papers on a topic.
- GIDEP, Government-Industry Data Exchange Program, for failure rates of electronic and mechanical components, (714)736-4677.
- CINDAS − Center for Information and Numerical Data Analysis and Synthesis, Purdue Univ., Dr. C. Y. Ho, (317)494-6300. Maintains data bases on dielectrics and other materials.

Omissions. This book omits various engineering aspects of accelerated testing. Omissions include:
- *Failure analysis.* Sources of information include the International Symposium for Testing and Failure Analysis, the Symposium on Mechanical Properties, Performance, and Failure Modes of Coatings (NBS/NIST), the Failure Analysis Special Interest Group of the Society of Plastics Engineers, and Chapter 13 of Ireson and Coombs (1988).
- *Test equipment and labs.* Sources of information include *Quality Progress*

(1988), *Evaluation Engineering* magazine (813) 966-9521, the Amer. Assoc. for Laboratory Accreditation, P.O. Box 200, Farifax Station, VA 22039, and various standards of professional societies.
- *Measurements/metrology and test methods.* Standards of engineering societies treat this in detail. Meetings include the Instrumentation and Measurement Technology Conference (IEEE). References include Heymen (1988).

Applications. For convenience, the following applications appear under three headings: (1) Materials, (2) Products, and (3) Degradation Mechanisms. These brief discussions are intended only to suggest the widespread use of accelerated testing. Those acquainted with a particular application will find the discussion rudimentary. Each discussion briefly describes applications, typical products, accelerating stresses, professional societies, journals, and meetings. Some references are included. Applications appear in the accelerated testing bibliographies of Meeker (1980) and Carey (1988), and older applications appear in the survey of Yurkowski and others (1967). Most applications involve time – either as time to failure or as time over which a performance property of a product degrades.

1.2. Materials

The following paragraphs briefly survey accelerated testing of materials. These include metals, plastics, dielectrics and insulations, ceramics, adhesives, rubber and elastics, food and drugs, lubricants, protective coatings and paints, concrete and cement, building materials, and nuclear reactor materials.

Metals. Accelerated testing is used with metals, including test coupons and actual parts, as well as composites, welds, brazements, bonds, and other joints. Performance includes fatigue life, creep, creep-rupture, crack initiation and propagation, wear, corrosion, oxidation, and rusting. Accelerating stresses include mechanical stress, temperature, specimen geometry and surface finish. Chemical acceleration factors include humidity, salt, corrosives, and acids. Societies include the American Society for Testing and Materials (ASTM), the American Society for Mechanical Engineers (ASME), American Powder Metallurgy Institute, ASM International (formerly the American Society for Metals), Institute of Metals, Society of Automotive Engineers (SAE), and the Society for Experimental Mechanics (SEM). References include ASTM STP 91-A, 744, and E739-80, Little and Jebe (1975), Graham (1968), Dieter (1961), Shelton (1982), Metals and Ceramics Information Center (1984), SAE Handbook AE-4 (1968), and Carter (1985).

Plastics. Accelerated testing is used with many plastics including building materials, insulation (electrical and thermal), mechanical components, and coatings. Materials include polymers, polyvinyl chloride (PVC), urethane foams, and polyesters. Performance includes fatigue life, wear, mechanical properties, and color fastness. Accelerating stresses include mechanical load (including vibration and shock), temperature (including cycling and shock), and weathering (ultraviolet radiation and humidity). Societies include the Plastics Institute of America, Plastics and Rubber Institute (PRI), and Society of Plastics Engineers (particularly its Failure Analysis Special Interest Group). Meetings include the International Conference on Fatigue in Polymers. Publications include *Polymer Engineering and Science* and *J. of Applied Polymer Science*. References include Mark (1985), Brostow and Corneliussen (1986), Hawkins (1984,1971), Underwriter Labs (1975), and Clark and Slater (1969).

Dielectrics and insulations. Accelerated testing is used with many dielectrics and electrical insulations including solids (polyethylene, epoxy), liquids (transformer oil), gases, and composites (oil-paper, epoxy-mica). Products include capacitors, cables, transformers, motors, generators, and other electrical apparatus. Performance includes time to failure and other properties (breakdown voltage, elongation, ultimate mechanical strength). Accelerating stresses include temperature, voltage stress, thermal and electrical cycling and shock, vibration, mechanical stress, radiation, and moisture. Societies include the Institute of Electrical and Electronics Engineers (IEEE), American Society for Testing and Materials (ASTM), and International Electrotechnical Commission (IEC). Publications include the *IEEE Trans. on Electrical Insulation* and *IEEE Electrical Insulation Magazine*. Meetings include the IEEE Annual Conference on Electrical Insulation and Dielectric Phenomena, IEEE Biannual International Symposium on Electrical Insulation, and Electrical/Electronics Insulation Conference. References include Sillars (1973), IEEE Standard 101 (1986), IEEE Standard 930 (1987), Goba (1969), IEEE Index (1988), Vincent (1987), Simoni (1974,1983), Vlkova and Rychtera (1978), and Bartnikas (1987, Chap. 5).

Ceramics. Applications are concerned with fatigue life, wear, and degradation of mechanical and electrical properties. References include Metals and Ceramics Information Center (1984). Societies include the United States Advanced Ceramics Association and American Ceramics Society. Publications include the *J. of the American Ceramics Soc.* Meetings include the World Materials Congress (ASM) and CERAMTEC Conference and Exposition (ASM/ESD). See Frieman (1980) and references for Metals.

Adhesives. Accelerated testing is used with adhesive and bonding materials such as epoxies. Performance includes life and strength. Accelerating stresses include mechanical stress, cycling rate, mode of loading, humidity, and temperature. References include Beckwith (1979,1980), Ballado-Perez (1986,1987), Millet (1975), Gillespie (1965), and Rivers and others (1981).

Rubber and elastics. Accelerated testing is used with rubbers and elastic materials (e.g., polymers). Products include tires and industrial belts. Performance includes fatigue life and wear. Accelerating stresses include mechanical load, temperature, pavement texture, and weathering (solar radiation, humidity, and ozone). Societies include the Plastics and Rubber Institute (PRI). References include Winspear's (1968) *Vanderbilt Rubber Handbook* and Morton (1987).

Foods and drugs. Accelerated testing is used with foods (e.g., browning of white wines), drugs, pharmaceuticals, and many other chemicals. Performance is usually shelf (or storage) life, usually in terms of amount of an active ingredient that degrades. Performance variables include taste, pH, moisture loss or gain, microbial growth, color, and specific chemical reactions. Accelerating variables include temperature, humidity, chemicals, pH, oxygen, and solar radiation. Societies include the American Society of Test Methods, US Pharmacopoeia, and Pharmaceutical Manufacturers Association. Major meetings include the Annual Meeting of Interplex. Kulshreshtha (1976) gives 462 references on storage of pharmaceuticals. References include Carstensen (1972), Connors et al. (1979), Bentley (1970), US FDA Center for Drugs and Biologics (1987), Young (1988), Labuza (1982), Beal and Sheiner (1985), and Grimm (1987).

Lubricants. Accelerated testing is used with solid (graphite, molybdenum disulphide, and teflon), oil, grease, and other lubricants. Performance includes oxidation, evaporation, and contamination. Accelerating stresses include speed, temperature, and contaminants (water, copper, steel, and dirt). Societies include the Society of Tribologists and Lubrication Engineers, STLE (formerly the American Society of Lubrication Engineers, ASLE). National Lubricating Grease Institute (NLGI), American Society for Testing and Materials (ASTM), and Society for Automotive Engineers (SAE). Elsevier Sequoia, S.A. (Switzerland) publishes *WEAR*, an international journal on the science and technology of friction, lubrication, and wear.

Protective coatings and paints. Accelerated testing is used for weathering of paints (liquid and powder), polymers, antioxidants, anodized aluminum, and electroplating. Performance includes color, gloss, and physical integrity (e.g., wear, cracking, and blistering). Accelerating stresses include weathering variables – temperature, humidity, solar radiation (wavelength and intensity) – and mechanical load. Societies include the American Electroplaters and Surface Finishers Society. Meetings include the World Materials Congress (ASM), and the Symposium on Mechanical Properties, Performance, and Failure Modes of Coatings (NBS/NIST).

Concrete and cement. Accelerated testing is used with concrete and cement to predict performance – the strength after 28 days of curing. The accelerating stress is high temperature applied for a few hours. Meetings include the Cement Industry Technical Conference.

Building materials. Accelerated testing is used with wood, particle board, plastics, composites, glass, and other building materials. Performance includes abrasion resistance, color fastness, strength, and other mechanical properties. Accelerating stresses include load and weathering (solar radiation, temperature, humidity). References include Clark and Slater (1969).

Nuclear reactor materials. Accelerated testing is used with nuclear reactor materials, for example, fuel rod cladding. Performance includes strength, creep, and creep-rupture. Accelerating stresses include temperature, mechanical stress, contaminants, and nuclear radiation (type, energy, and flux). Societies include the Institute of Environmental Sciences (1988) and American Nuclear Society. Journals include the *IEEE Trans. on Nuclear Science* and *Radiation Research*. DePaul (1957) surveys such work.

1.3. Products

The following paragraphs describe accelerated testing of certain products. Such products range from simple components through complex assemblies.

Semiconductors and microelectronics. Accelerated testing is used for many types of semiconductor devices including transistors such as gallium arsenide field emission transistors (GaAs FETs), insulated gate field emission transistors (IGFETs), Gunn and light emitting diodes (LEDs), MOS and CMOS devices, random access memories (RAMs), and their bonds, connections, and plastic encapsulants. They are tested singly and in assemblies such as circuit boards, integrated circuits (LSI and VLSI), and microcircuits. Performance is life and certain operating characteristics. Accelerating variables include temperature (constant, cycled, and shock), current, voltage (bias), power, vibration and mechanical shock, humidity, pressure, and nuclear radiation. Societies include the Institute for Electrical and Electronics Engineers (IEEE), American Electronics Association (AEA), Society for the Advancement of Material and Process Engineering (PO Box 2459, Covina, CA 91722). Major professional meetings include the International Reliability Physics Symposium, Annual Reliability and Maintainability (RAM) Symposium, International Symposium for Testing and Failure Analysis, Electronic Materials and Processing Congress (ASM), Annual Conference on Electronic Packaging and Corrosion in Microelectronics, and Gallium Arsinide Integrated Circuits Symposium (IEEE). References include Peck and Trapp (1978), Peck and Zierdt (1974), Reynolds (1977), IEEE Index (1988), and Howes and Morgan (1981). Publications include proceedings of the symposia above, *Microelectronics and Reliability, IEEE Trans. on Reliability, IEEE Journal of Solid-State Circuits, IEEE Trans. on Consumer Electronics, IEEE Circuits and Devices Magazine, IEEE Trans. on Circuits and Systems, IEEE Trans. on Electron Devices, IEEE Trans. on Power Electronics, Proceedings of the International SAMPE Electronics Materials Conference, IEEE J. of Quantum Electronics*, and *IEE Proceedings* (England).

Capacitors. Accelerated testing is used with most types of capacitors, including electrolytic, polypropylene, thin film, and tantalum capacitors. Performance is usually life. Accelerating variables include temperature, voltage, and vibration. Professional societies that publish standards and journal articles on accelerated test methods and applications include the Institute of Electrical and Electronics Engineers (IEEE) and the American Electronics Association (AEA). Also, see resources for semiconductor applications.

Resistors. Accelerated testing is used with thin and thick film, metal oxide, pyrolytic, and carbon film resistors. Performance is life. Accelerating variables include temperature, current, voltage, power, vibration, electrochemical attack (humidity), and nuclear radiation. References include Krause (1974). Also, see resources for semiconductor applications.

Other electronics. Accelerated testing is used with other electronic components such as optoelectronics (opto couplers and photo conductive cells), lasers, liquid crystal displays, and electric bonds and connections. The performance, accelerating stresses, professional societies, and references are much the same as those for semiconductors. Publications include *IEE Proceedings* (England), *IEEE Trans. on Power Electronics, IEEE Journal of Electronic Materials*, and *IEEE Trans. on Electron Devices*. Meetings include the Electronic Components Conference (IEEE) and International Electron Devices Meeting (IEEE).

Electrical contacts. Accelerated testing is used for electrical contacts in switches, circuit breakers, and relays. Performance includes corrosion and life. Metal fatigue, rupture, and welding are common failure mechanisms. Accelerating stresses include high cycling rate, temperature, contaminants (humidity), and current. References include the IEEE Index (1988). Meetings include the Holm Conference on Electrical Contacts (IEEE).

Cells and batteries. Accelerated testing is used with rechargable, nonrechargable, and solar cells. Performance includes life, self discharge, current, and depth of discharge. Accelerating variables include temperature, current density, and rate of charge and discharge. Societies include the Electrochemical Society (609) 737-1902. Publications include the *Journal of the Electrochemical Society, Solar Cells* (Switzerland), and *Proceedings of the Symposium on Lithium Batteries.* References include Sidik and others (1980), McCallum and others (1973), Linden (1984), and Gobano (1983). Meetings include the Battery Workshop (NASA), Annual Battery Conference on Applications and Advances (IEEE and California State University), and International Power Sources Symposium.

Lamps. Accelerated testing is used with incandescent (filament), fluorescent (including ballasts), mercury vapor, and flash lamps. Performance includes life, efficiency, and light output. Accelerating variables include voltage, temperature, vibration, and mechanical and electrical shock. Societies include the International Electrotechnical Commission (IEC). References include EG&G Electro-Optics (1984), IEC Publ. 64 (1973), and IEC Publ. 82 (1980).

Electrical devices. Accelerated testing is used with various electrical devices including motors, heating elements, and thermoelectric converters. References include the IEEE Index (1988). Motor and generator failures are almost always due to insulation or bearing failure. Thus their life distribution is inferred from that of their insulation and bearings (Chapter 7).

Bearings. Accelerated testing is used with roller, ball, and sliding (oil film) bearings. Performance includes life and wear (weight loss). Materials include steels and silicon nitride for rolling bearings and porous (sintered) metals, bronzes, babbitt, aluminum alloys, and plastics for sliding bearings. Accelerating stresses include overspeed, mechanical load, and contaminants. Societies include the Anti-Friction Bearing Manufacturers Association (AFB-MA), International Standards Organization (ISO), American Society for Testing and Materials (ASTM), Society of Automotive Engineers (SAE), and ASM International (formerly the American Society for Metals). References include Harris (1984), SKF (1981), and Lieblein and Zelen (1956).

Mechanical components. Accelerated testing is used with mechanical components and assemblies such as automobile parts, hydraulic components, tools, and gears. Performance includes life and wear. Accelerating stresses include mechanical load, vibration, temperature and other environmental factors, and combinations of such stresses. Societies include the American Society for Testing and Materials (ASTM), Society for Automotive Engineers (SAE), and American Society for Mechanical Engineers (ASME). Meetings include the International Machinery Monitoring and Diagnostic Conference (sponsored by Union College, Schenectady, NY). References include Collins (1981), Zalud (1971), and Boothroyd (1975). See resources for Metals.

1.4. Degradation Mechanisms

The following paragraphs describe common mechanisms for degradation of product *performance*. Such mechanisms are utilized or studied in accelerated tests. For more detail, refer back to discussions of materials and products. Meetings include the International Machinery Monitoring and Diagnostic Conference (sponsored by Union College, Schenectady, NY).

Fatigue. Materials eventually fail by fatigue if subjected to repeated mechanical loading and unloading, including vibration. Well studied are the fatigue of metals, plastics, glass, ceramics, and other structural and mechanical materials (see references on these). Fatigue is a major failure mechanism of mechanical parts including bearings and electrical contacts. The usual accelerating stress is load. Other stresses are temperature and chemicals (water, hydrogen, oxygen, etc.). References include Tustin and Mercado (1984), ASTM STP 648 (1978), ASTM STP 744 (1981), ASTM STP 748 (1981), ASTM STP 738 (1981), Frieman (1980), and Skelton (1982).

Creep. Creep, the slow plastic deformation of materials under constant mechanical load, may interfere with product function or cause rupture or fracture. Accelerating variables are typically temperature and mechanical load, load cycling, and chemical contaminants (for example, water, hydrogen, and fluorine). References include Goldhoff and Hahn (1968), Hahn (1979), and Skelton (1982). See resources for Metals and Plastics.

Cracking. Metals, plastics, glass, ceramics, and other materials crack. People study crack initiation and growth. Accelerating stresses include mechanical stress, temperature, and chemicals (humidity, hydrogen, alkalis, and acids). See resources for Metals and Plastics.

Wear. In applications, many materials are subjected to friction that removes the material. For example, rubber tires lose tread, house paints wash off, gears, bearings, and machine tools wear away. Accelerating stresses include speed, load (magnitude and type), temperature, lubrication, and chemicals (humidity). References include Rabinowicz (1988) and Peterson and Winer (1980). DePaul (1957) surveys nuclear applications. Boothroyd (1975) treats machine tool wear. Elsevier Sequoia, S.A. (Switzerland) publishes *WEAR*, an international journal on the science and technology of friction, lubrication, and wear.

Corrosion/oxidation. Most metals and many foods, pharmaceuticals, etc., deteriorate by chemically reacting with oxygen (oxidation and rusting), fluorine, chlorine, sulphur, acids, alkalis, salt, hydrogen peroxide, and water. Accelerating stresses include concentration of the chemical, activators, temperature, voltage, and mechanical load (stress-corrosion). Meetings include the Annual Conference on Electronic Packaging and Corrosion in Microelectronics. Professional societies include the National Assoc. of Corrosion Engineers (NACE). The publications list of the Metals and Ceramics Information Center (1984) includes work on corrosion. References include DePaul (1957) on nuclear applications, Rychtera (1985), and Uhlig and Revie (1985).

Weathering. This concerns the effects of weather on materials in outdoor applications. Such materials include metals, protective coatings (paint, electroplating, and anodizing), plastics, and rubbers. Accelerating stresses include solar radiation (wavelength and intensity) and chemicals (humidity, salt, sulphur, and ozone). The degradation generally involves corrosion, oxidation (rust), tarnishing, or other chemical reaction. Professional societies include Institute of Environmental Sciences (1988). Publications include the *Journal of Environmental Sciences*.

2. TYPES OF DATA

This section presents background on accelerated test data. Accelerated test data can be divided into two types. Namely, the product characteristic of interest is 1) **life** or is 2) some other measure of **performance**, such as tensile strength or ductility. Such data are described below. This background is essential for the rest of this book.

Performance data. One may be interested in how product performance degrades with age. In such performance testing, specimens are aged under high stress, and their performance measured at different ages. Such performance data are analyzed by fitting a degradation model to the data to estimate the relationship between performance, age, and stress. Chapter 11 discusses such data in detail and presents such models and data analyses. Such a degradation test has been used, for example, for temperature aging of electrical insulation and pharmaceuticals. Goba (1969) references such testing of electrical insulation.

Life data. The proper analysis of life data depends on the type of data. The following paragraphs describe the common types of life data from a single test or design condition.

Complete. Complete data consist of the *exact* life (failure age) of each sample unit. Figure 2.1A depicts a complete sample from a single test condition. There the length of a line corresponds to the length of life of a test unit. Chapters 3, 4, and 8 treat such data. Much life data are incomplete. That is, the exact failure times of some units are unknown, and there is only partial information on their failure times. Examples follow.

Censored. Often when life data are analyzed, some units are unfailed, and their failure times are known only to be beyond their present running times. Such data are said to be **censored on the right.** In older literature, such data or tests are called **truncated.** Unfailed units are called run-outs, survivors, removals, and suspensions. Such censored data arise when some units are (1) removed from test or service before they fail, (2) still running at the time of the data analysis, or (3) removed from test or service because they failed from an extraneous cause such as test equipment failure. Similarly, a failure time known only to be before a certain time is said to be **censored on the left.** If all unfailed units have a common running time and all failure times are earlier, the data are said to be **singly censored** on the right. Singly censored data arise when units are started together at a test condition and the data are analyzed before all units fail. Such data are singly **time censored** if the censoring time is fixed; then the number of failures in that fixed time is random. Figure 2.1B depicts such a sample. There the line for an unfailed unit shows how long it ran without failure, and the arrow pointing to the right indicates that the unit's failure time is later. Time censored data are also called **Type I censored.** Data are singly **failure censored** if the test is stopped when a specified number of failures occurs. The time to that fixed number of failures is random. Figure 2.1C depicts such a sample. Time censoring is more common in practice. Failure censoring is more common in the theoretical literature, as it is mathematically more tractable. Chapters 3 and 5 present analyses for singly censored data.

A. COMPLETE

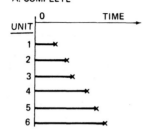

E. MULTIPLY FAILURE CENSORED (II)

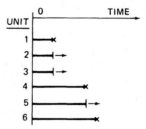

B. SINGLY TIME CENSORED (I)

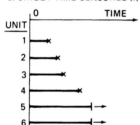

F. COMPETING FAILURE MODES (A,B,C)

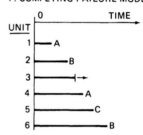

C. SINGLY FAILURE CENSORED (II)

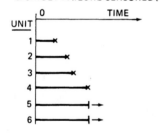

G. QUANTAL – RESPONSE

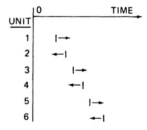

D. MULTIPLY TIME CENSORED (I)

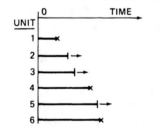

H. INTERVAL (GROUPED)

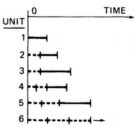

Figure 2.1.   Types of data (failure time ×, running time |→, failure occurred earlier ←|).

Multiply censored. Much data censored on the right have differing running times intermixed with the failure times. Such data are called **multiply censored** (also progressively, hyper-, and arbitrarily censored). Figure 2.1D depicts such a sample. Multiply censored data arise when units go on test at different times. Thus they have different running times when the data are recorded. Such data may be time censored (running times differ from failure times, as shown in Figure 2.1D) or failure censored (running times equal failure times, as shown in Figure 2.1E). Chapters 3, 5, and 7 treat such data.

Competing modes. A mix of **competing failure modes** occurs when sample units fail from different causes. Figure 2.1F depicts such a sample, where A, B, and C denote different failure modes. Data on a particular failure mode consist of the failure times of units failing by that mode. Such data for a mode are multiply censored. Chapters 3 and 7 treat such data.

Quantal-response. Sometimes one knows only whether the failure time of a unit is before or after a certain time. Each observation is either censored on the right or else on the left. Such life data arise if each unit is inspected **once** to see if it has already failed or not. Such inspection data are called **quantal-response** data, also called sensitivity, probit, binary, and all-or-nothing response data. Figure 2.1G depicts such a sample. There the arrow for each unit shows whether the unit failed before its inspection or will fail later. Chapter 5 treats such data.

Interval. When each unit is inspected for failure **more than once**, one knows only that a unit failed in an interval between inspections. So-called **interval**, grouped, or read-out data are depicted in Figure 2.1H. There a solid line shows the interval where a unit failed, and a dotted line shows an inspection interval where it did not fail. Such data can also contain right and left censored observations. Chapter 5 treats such data.

Mixture. Data may also consist of a mixture of the above types of data.

Purpose. Analyses of such censored and interval data have much the same purposes as analyses of complete data, for example, estimation of model parameters and the product life distribution and prediction of future observations.

3. TYPES OF ACCELERATION AND STRESS LOADING

This section describes common types of acceleration of tests (high usage rate, overstress, censoring, degradation, and specimen design) and stress loading. Test purposes (Section 4) might profitably be read first.

High Usage Rate

A simple way to accelerate the life of many products is to run the product more – at a **higher usage rate.** The following are two common ways of doing such **compressed time testing.**

Faster. One way to accelerate is to run the product faster. For example, in many life tests, rolling bearings run at about three times their normal speed. High usage rate may also be used in combination with overstress testing. For example, such bearings are also tested under higher than normal mechanical load. Another example of high usage rate involves a voltage endurance test of an electrical insulation by Johnston and others (1979). The AC voltage in the test was cycled at 412 Hz instead of the normal 60 Hz, and test was shorter by a factor of $412/60 = 6.87$.

Reduced off time. Many products are off much of the time in actual use. Such products can be accelerated by running them a greater fraction of the time. For example, in most homes, a major appliance (say, washer or dryer) runs an hour or two a day; on test it runs 24 hours a day. In use, a refrigerator compressor runs about 15 hours a day; on test it runs 24. A small appliance (say, toaster or coffee maker) runs a few cycles a day; on test it cycles many times a day.

Purpose. The purpose of such testing is to estimate the product life distribution at normal usage rates. It is assumed that the number of cycles, revolutions, hours, etc., to failure on test is the same that would be observed at the normal usage rate. For example, it is assumed that a bearing that runs 6.2 million revolutions to failure at high rpm would run 6.2 million revolutions at normal rpm. The data are treated as a sample from actual use. Then standard life data analyses provide estimates of the percentage failing on warranty, the median life, etc. They also provide comparisons of designs, manufacturing methods, materials, vendors, etc. Such analyses are explained by Nelson (1983c,1982, pp. 567-569) and by reliability books he references.

The assumption. It is not automatically true that the number of cycles to failure at high and normal usage rates is the same. Usually the test must be run with special care to assure that product operation and stress remain normal in all regards except usage rate. For example, high rate usage usually raises the temperature of the product. That usually results in fewer cycles to failure. It may even produce failure modes not seen at normal temperature and usage rate. Thus many such tests involve cooling the product to keep the temperature at a normal level. In contrast, products sensitive to thermal cycling may last longer if run continuously without thermal cycling. For this reason, toasters on test are force cooled by a fan between cycles.

Overstress Testing

Overstress testing consists of running a product at higher than normal levels of some accelerating stress(es) to shorten product life or to degrade product performance faster. Typical accelerating stresses are temperature, voltage, mechanical load, thermal cycling, humidity, and vibration. Later, overstress tests are described according to purpose and the nature of the test.

Overstress testing is the most common form of accelerated testing and is a main subject of this book.

Censoring

Modern practice includes accelerating tests through *censoring* (Chapters 5 and 6). That is, tests are terminated before all specimens run to failure. This shortens test time. For higher reliability products, one is usually interested in just the lower tail of the life distribution. Then usually little information is gained from data from the upper tail. However, sometimes an important failure mode is active at the design stress level and does not occur in the lower tail at the test stress levels. But it occurs in the upper tail. Then data from the upper tail is useful, and terminating the test early would miss that important failure mode (Chapter 7).

Degradation

Method. Accelerated degradation testing involves overstress testing. Instead of life, product *performance* is observed as it degrades over time. For example, the breakdown voltage of insulation specimens at high temperature is measured at various ages. A model for performance degradation is fitted to such performance data and used to extrapolate performance and time of failure. Thus failure and the life distribution can be predicted before any specimen fails. This accelerates the test. Failure is assumed to occur when a specimen performance degrades below a specified value. For example, an insulation specimen fails when its breakdown voltage degrades below the design voltage. Chapter 11 presents models and data analyses for accelerated degradation.

Specimen Design

Life of some products can be accelerated through the size, geometry, and finish of specimens.

Size. Generally large specimens fail sooner than small ones. For example, high capacitance capacitors fail sooner than low capacitance ones of the same design. The large capacitors merely have more dielectric area. Similarly, long cable specimens fail sooner than short ones. Time to breakdown of an insulating fluid is shorter for electrodes with a three-inch diameter than for electrodes with a one-inch diameter. Large (diameter or length) metal fatigue specimens fail sooner than short ones. Creep-rupture specimens have been linked end-to-end in a group to increase the amount of metal on a test machine. They are tested until the first specimen of a group fails; this is called **sudden-death testing.** Usually one wants to estimate the life of a smaller or larger standard size of the product. Such an estimate requires a model that takes specimen size into account (Chapter 7). Harter (1977) surveys such models.

Geometry. Specimen geometry may affect specimen life. For example, some metal fatigue, crack propagation, and creep-rupture specimens are notched. Such notches produce local high stress and early failure. One might argue that such specimens are locally overstressed or should be discussed under overstress testing. On the other hand, the average stress in such a specimen may be the normal design stress. Also, surface finish (roughness) and residual stresses of metal specimens affect fatigue life.

Stress Loading

The stress loading in an accelerated test can be applied various ways. Descriptions of common loadings follow. They include constant, cyclic, step, progressive, and random stress loading. The discussion is limited to a single stress variable, but the concepts extend to multiple stress variables.

Constant stress. The most common stress loading is **constant stress.** Each specimen is run at a constant stress level. Figure 3.1 depicts a constant stress test with three stress levels. There the history of a specimen is depicted as moving along a horizontal line until it fails at a time shown by an ×. An unfailed specimen has its age shown by an arrow. At the highest level, all four specimens ran to failure. At the middle level, four ran to failure, and one was unfailed. At the lowest level, four ran to failure, and four were unfailed. In use, most products run at constant stress. Then a constant stress test mimics actual use. Moreover, such testing is simple and has advantages. First, in most tests, it is easier to maintain a constant stress level. Second, accelerated test models for constant stress are better developed and empirically verified for some materials and products. Third, data analyses for reliability estimation are well developed and computerized. Chapter 6 presents test plans for constant stress tests. Such plans consist of the "best" test stress levels and number of specimens at each level.

Step stress. In **step-stress** loading, a specimen is subjected to successively higher levels of stress. A specimen is first subjected to a specified constant stress for a specified length of time. If it does not fail, it is subjected to a higher stress level for a specified time. The stress on a specimen is thus in-

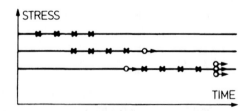

Figure 3.1.   Constant stress test (× failure, ○→ runout).

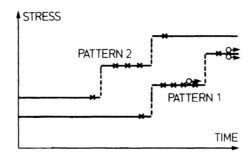

Figure 3.2.   Step-stress test ($\times$ failure, $\circ\!\!\to$ runout).

creased step by step until it fails. Usually all specimens go through the same specified pattern of stress levels and test times. Sometimes different patterns are applied to different specimens. Figure 3.2 depicts two such patterns. Such data may be censored. Pattern 1 has six failures and three runouts.

Advantages. The main advantage of a step-stress test is that it quickly yields failures. The increasing stress levels ensure this. Statisticians are happy to have failures, because they yield estimates of the model and of the product life. Engineers are happier when there are no failures, which suggests (perhaps incorrectly) that the product is reliable. Quick failures do not guarantee more accurate estimates. A constant stress test with a few specimen failures usually yields greater accuracy than a shorter step-stress test where all specimens fail. Roughly speaking, the total time on test (summed over all specimens) determines accuracy – not the number of failures.

Disadvantages. There is a major disadvantage of step-stress tests for reliability estimation. Most products run at constant stress – not step stress. Thus the model must properly take into account the cumulative effect of exposure at successive stresses. Moreover, the model must also provide an estimate of life under constant stress. Such a model is more complex than one for a constant stress test. Such **cumulative exposure models** (also called **cumulative damage models**) are like the weather. Everybody talks about them, but nobody does anything about them. Many models appear in the literature, few have been fitted to data, and even fewer assessed for adequacy of fit. Moreover, fitting such a model to data requires a sophisticated special computer program. Thus, constant stress tests are generally recommended over step-stress tests for reliability estimation. Another disadvantage of a step-stress test is that failure modes occurring at high stress levels (in later steps) may differ from those at use conditions. Some engineers who run elephant tests (Section 5.1) fail to note this. Chapter 10 presents step-stress models and data analyses. Step-stress data with a mix of failure modes can be properly analyzed with the methods of Chapters 7 and 10, provided one has an adequate cumulative exposure model.

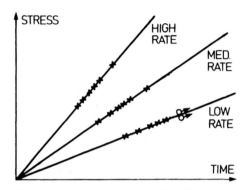

Figure 3.3.   Progressive stress test (× failure, ○→ runout).

Examples. Nelson (1980) describes a step-stress model and data analysis. Schatzoff and Lane (1987) use this model to optimize planning of a step-stress test with read-out data. Goba (1969) references such work on temperature accelerated testing of electrical insulation. ASTM Special Technical Publication No. 91-A (1963) provides methods for analyzing such data from metal fatigue tests. Yurkowsi and others (1967) survey early work.

Progressive stress. In **progressive stress** loading, a specimen undergoes a continuously increasing level of stress. Different groups of specimens may undergo different progressive stress patterns. Figure 3.3 depicts such a test with three patterns – each a linearly increasing stress. As shown in Figure 3.3, under a low rate of rise of stress, specimens tend to live longer and to fail at lower stress. Such life data may be censored. In metal fatigue, such a test with a linearly increasing mechanical load is called a **Prot test.**

Disadvantages. Progressive stress tests have the same disadvantages as step-stress tests. Moreover, it may be difficult to control the progressive stress accurately enough. Thus constant stress tests are generally recommended over progressive stress tests for reliability estimation. Chapter 10 presents progressive-stress models and data analyses.

Examples. Endicott and others (1961,1961,1965) used progressive testing on capacitors. Nelson and Hendrickson (1972) analyze such data on dielectric breakdown of insulating fluid. Prot (1948) introduced such testing into fatigue studies of metal. ASTM Special Technical Publication No. 91-A (1963) provides methods for analyzing such data from metal fatigue tests. Goba (1969) references work on such temperature accelerated testing of electrical insulation. ASTM Standard D2631-68 (1970) describes how to carry out progressive stress tests on capacitors.

Cyclic stress. In use, some products repeatedly undergo a cyclic stress loading. For example, insulation under ac voltage sees a sinusoidal stress. Also, for example, many metal components repeatedly undergo a mechanical

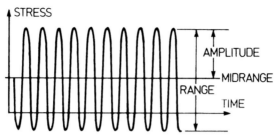

Figure 3.4.   Cyclic-stress loading.

stress cycle. A **cyclic stress** test for such a product repeatedly loads a specimen with the same stress pattern at high stress levels. Figure 3.4 depicts a cyclic stress test. For many products, a cycle is sinusoidal. For others, the duty (or test) cycle repeats but is not sinusoidal. The (usually high) number of cycles to failure is the specimen life. Such life data may be censored.

Insulation. For insulation tests, the stress level is the *amplitude* of the ac voltage sinusoid, which alternates from positive to negative voltage. So the level is characterized by a single number. For purposes of modeling and data analysis, such cyclic stress is regarded as a constant, and it can be depicted as in Figure 3.1, where the vertical axis shows the voltage amplitude.

Metals. In metal fatigue tests, usually a specimen undergoes (nearly) sinusoidal loading. But the sinusoid need not have a mean stress of zero. Figure 3.4 shows such a sinusoid with a positive mean. Tensile stress is positive, and compressive stress is negative in the figure. Thus, according to the figure, the specimen is under tension for most of a cycle and under compression for a small part of a cycle. Such sinusoidal loading is characterized by two numbers, say, the stress range and the mean stress. Frequency often has negligible effect. Thus, fatigue life can be regarded as a function of these two "constant" stress variables. In place of the stress range, metallurgists use the **A-ratio**; it is the stress amplitude (half the range) divided by the mean stress. For example, suppose a specimen is cycled from 0 psi to 80,000 psi compression and back to 0 psi. The mean stress is 40,000 psi, and the A-ratio is $0.5(80,000-0)/40,000 = 1$. The A-ratio for ac voltage cycling of insulation is infinity, since the mean voltage is zero. Usually, on a fatigue test, all specimens see the same A-ratio as does an actual part in use, but different groups of specimens run at different mean stress levels. Such a test is then regarded as a constant stress test. The test can then be depicted as in Figure 3.1 where the vertical axis shows the mean stress. Moreover, then specimen life is modeled with a constant stress model, and the data are analyzed accordingly.

Assumed. In most such tests, the frequency and length of a stress cycle are the same as in actual product use. For some products, they differ but are assumed to have negligible effect on life, and they are disregarded. For other

Figure 3.5.   Random stress loading.

products, the frequency and length of a cycle affect life; so they are included in the model as stress variables.

Random stress. Some products in use undergo randomly changing levels of stress, as depicted in Figure 3.5. For example, bridge members and airplane structural components undergo wind buffeting. Also, environmental stress screening (Section 5.2) uses random vibration. Then an accelerated test typically employs random stresses with the same distribution as actual random stresses but at higher levels. Like cyclic stress tests, random stress models employ some characteristics of the stress distribution as stress variables (say, the mean, standard deviation, correlation function, and power spectral density). Then such a test is regarded as a constant stress test; this, of course, is simplistic but useful. The test can then be depicted as in Figure 3.1 where the horizontal line shows the mean stress. Moreover, specimen life is modeled with a constant stress model, and the data are analyzed accordingly. Such life data may be censored.

4. ENGINEERING CONSIDERATIONS

Many considerations impact the validity and accuracy of information from an accelerated test. This section surveys certain engineering and management considerations involved in scientifically planning and carrying out a test. This section emphasizes accelerated tests for estimating product life at design conditions. However, most considerations apply to tests of performance degradation and to engineering experiments in general.

Overview. The following topics highlight certain management and engineering decisions on a test. Generally such decisions involve the collaboration of manager, designers, production and test engineers, statisticians, and others. The topics include

• Test Purpose	• Other Variables
• Product Performance	• Measurement Errors
• Realistic Test Specimens	• The Model
• Realistic Test Conditions	• Test Allocation
• Accelerating Stresses	• Planning Ahead

This section emphasizes the value of thinking these through in advance.

Test Purpose

The following describes engineering and statistical purposes for accelerated tests of products.

Purposes. Accelerated life tests and performance degradation tests serve various purposes. Common purposes include:

1. *Identify design failures.* Eliminate or reduce them through redundancy, better design, components, etc.
2. *Comparisons.* Choose among designs, components, suppliers, rated operating conditions, test procedures, etc.
3. *Identify manufacturing defects.* Eliminate them through better manufacturing, components, burn-in, etc. Estimate the reliability improvement from eliminating or reducing certain failure modes.
4. *Burn-in.* A manufacturing step to eliminate early failures. Determine burn-in time and conditions.
5. *Quality control.* Monitor product reliability and take corrective action as needed, for example, when a new failure mode arises, or life or performance degrades.
6. *Evaluate other variables.* Assess how much design, manufacturing, materials, operating, and other variables affect reliability. Optimize reliability with respect to them. Decide which need to be controlled.
7. *Acceptance sampling.* Assess production or incoming lots.
8. *Qualify* design and manufacturing changes, components, vendors, etc.
9. *Measure reliability.* Assess whether to release a design to manufacturing or product to a customer. Estimate warranty and service costs, failure rates, mean time to failure (MTTF), degradation rates, etc. Satisfy a customer requirement for such measurement. Use as marketing information.
10. *Demonstrate reliability.* Show that product reliability surpasses customer specifications.
11. *Validate the test.* Show that the accelerated test is consistent with itself over time, other tests (including those in other labs), and field data. Determine a range of test stresses. Develop a new test.
12. *Assess model.* Determine whether the engineering relationship and statistical distribution are adequate. Develop such a model.
13. *Operating conditions.* Develop relationships between reliability (or degradation) and operating conditions. Choose design operating conditions.
14. *Service policy.* Decide when to inspect, service, or replace and how many spares and replacements to manufacture and stock. Units may be taken out of service and tested under accelerated conditions when an unexpected problem shows up in service.

A test may have one or more such purposes, which concern design, manufacture, quality control, test, application, marketing, and field service.

Clear purpose needed. The most essential part of any test is a clear statement of the test purposes. Usually such **engineering purposes** concern decisions to be made by engineering and management. For example, they need (1) to decide whether an insulation will last long enough in service, (2) to decide which of several competing designs will last longest in service, or (3) to judge which manufacturing and service variables to control to prolong the service life of a product. These typical engineering and management decisions often must ultimately be made on the basis of test *and* other information. Statistical methods, of course, do not provide answers nor make decisions. They merely provide **numerical information** on product performance. Thus, if statistical methods are to be used, the engineering purpose must also be expressed as a **statistical purpose** in terms of needed *numerical information* on the product.

Statistical purpose. If an engineer (or manager) has difficulty specifying such needed numbers, the following may help. Imagine that one has all possible data on the entire population through all future time. Clearly the thousands or millions of individual data values are not used to aid a decision. A few simple useful numbers that describe the population need to be calculated from the mass of data. The engineer must decide which few numbers will help. Statistical analysis merely yields estimates of such population numbers from limited sample data. In addition, statistical test planning helps make those estimates more accurate, that is, closer to the true population values. Management and engineering must also specify how accurate such estimates must be for decision purposes. Thus, reiterating, engineers need to express their statistical purposes in terms of all data on the population and the useful population values to extract from that mass of data.

Examples. Such population values include the median life at design conditions, the percentage failing on warranty, the numerical relationship between product life and certain manufacturing variables, etc. This book presents statistical models, data analyses, and test plans for efficiently obtaining accurate estimates of such population values.

Performance

Choice of performance. Product performance is often measured in terms of life or physical properties. For example, cable insulation is measured with respect to mechanical properties (such as ultimate strength and percent ultimate elongation) and electrical properties (such as breakdown strength and dielectric loss). For a particular product, there usually are standard performance properties and methods for measuring them. They are outside the scope of this book, and they appear in engineering texts, standards, and handbooks. Certain properties degrade with product age, and such degradation may be accelerated by high stress levels. Such accelerated testing for performance degradation appears in Chapter 11. In statistical terminology, such a performance variable is called a **dependent** or **response variable.**

Failure. In accelerated life testing, time to failure is the performance characteristic or dependent variable. Thus "failure" must be precisely defined in practice. Agreement on a definition in advance is essential between producer and consumer to minimize misunderstandings and disputes. A discussion of "failure" follows.

Catastrophic failure. Many products undergo **catastrophic failure**; that is, they clearly and suddenly stop working. Incandescent light bulbs, for example, fail catastrophically.

Defined failure. For other products, performance degrades slowly over time, and there is no clear end of life. Then one can use a **defined failure** which occurs when performance first degrades below a specified value. For example, jet engine parts have four definitions of failure. (1) Critical parts "fail" when a crack initiates. In practice, this means when the crack is first detectable. (2) Less critical parts "fail" when the crack first reaches a specified size. (3) Any part "fails" when removed from service, usually the result of an inspection that shows that the part exceeds definition (1) or (2). This definition certainly is meaningful to accountants and top management. (4) Any jet engine part "fails" when there is a "part separation" – a marvelous engineering euphemism for catastrophic failure.

Coffee maker. Modern electric coffee makers heat water by passing it through heated tubing. In use, mineral build-up inside the tubing constricts the opening. This build-up gradually increases the time to make a pot of coffee. Engineers developed an accelerated test to study this problem – the sludge test. It was used to compare tubing designs, including competitors' designs. The test involved repeatedly running a coffee maker with a mineral-rich slurry that quickly produced build-up. By definition, a coffee maker "failed" on the first pot that took longer than 20 minutes. Of course, 20 minutes is a reasonable but arbitrary time; it owes much to the number of fingers we have. 17.2 minutes would be as good.

Customer-defined failure. One other definition is worth noting. The product "fails" when the customer says it fails. Marketing and top management take this definition seriously. Many failures as defined by engineering are not severe enough to concern customers.

The right definition. Which of several definitions is the right one? All are. Each has value. In practice, one can use several and analyze the data separately according to each. Each yields information. The definition of failure is an engineering or management decision.

Usage/exposure. Time to failure is only one possible measure of usage or exposure. For some products, other measures are used. Ball bearing usage is the number of revolutions. For non-rechargable battery cells, energy output is the usage. Locomotive usage is miles. Usage of many products is the number of duty cycles. Examples include toasters, dishwashers, rechargable battery cells, switches, and circuit breakers. Many such "cycled" products have customer warranties for a specified calendar time (for example, one year), although cycle usage differs greatly from customer to customer. Throughout this book "time" means any measure of usage. Even if using time, one must decide whether to measure it starting on the date of product installation or manufacture. Sometimes installation dates are not known. Also, if the product runs only part of the time, one must decide whether to measure the actual running time or merely use calendar days in service. Calendar days is usually easier and cheaper to determine. The choice of exposure and how to measure it are engineering and management decisions.

Realistic Test Specimens

Specimen versus product. Many test specimens differ from the actual product. Engineers realize that specimen life may differ from product life. Then they assume that specimen life is comparable to or below product life. Some differences between specimens and product are obvious, and others are subtle. Examples follow. It generally is worth the effort to use the actual product to get more accurate information. Statisticians will only acknowledge that specimen tests yield estimates of specimen life – not product life. Engineers must make the leap of faith from specimen life to product life.

Motor insulation. Specimens of motor insulation are usually motorettes – not motors. Motorettes contain phase, ground, and turn insulation, as do motors. The amount and geometry of such insulations differ from those in motors. Moreover, a motorette has no moving parts; it merely contains distinct samples of three insulations on substrates similar to those in a motor.

Metal fatigue. Metal fatigue specimens are often small cylinders. Metal parts are generally larger, have complex geometry with sharp corners that locally raise stress, have different surface treatment, etc. Engineers in airframe, jet engine, and other businesses partly compensate for these differences by running parts only one-third as long as the specimens would survive. For example, if 1 specimen in 1,000 fails by 30,000 cycles, the part is removed from service at $30,000/3 = 10,000$ cycles.

Cable insulation. Most specimens of cable insulation range in length from one-half to several feet. Usually, but not always, the conductor cross-section and insulation geometry and thickness are the same as in actual cable. Being much shorter, specimens have much longer lives than actual cable, typically many miles long. The cable can be regarded as a large number of specimens (end to end), and the cable fails when the first specimen fails. Statistical models for the effect of size appear in Chapter 7.

Sampling bias. Even when test specimens are the actual product, they usually are not a random sample. They are usually a biased sample and do not adequately represent the population. For example, an accelerated test of a new product typically uses prototype units. They may be fabricated in a lab or model shop. Even if made on the production line, they may differ from later production units. The design or the manufacturing process always changes during production. Some changes are obvious and intentional – new materials, cost savings, new vendors, etc. Others are subtle and often unrecognized – a vendor's process changes, plant humidity changes, employees start cleaning their bowling balls in the ultrasonic cleaning equipment for the product, etc. Quality control engineers spend much time hunting such unknown production changes that degrade the product. Engineers generally recognize that prototypes differ from production product.

Representative sample. A representative sample is best for estimating the reliability of a product population. Such a sample is heterogeneous and includes specimens from many production periods or batches. Ideally it is a truly *random* sample from the population. For other purposes, a homogeneous sample may serve better. For example, to compare two methods of manufacture, a *homogeneous* sample of specimens is randomly divided between the two methods. Due to the smaller variability between homogeneous specimens, the comparison will detect smaller differences between the two methods. Similarly, test variables such as humidity, temperature, etc., may be held constant over all specimens to minimize variability in a comparison. For still other purposes, sample specimens with *extreme* values of observed covariates may be chosen for a test. Such a wide range of covariate values provides more accurate estimates of their effects. Taguchi (1987), among others, advocates this in experiments with manufacturing processes.

What is the specimen? The following example suggests that the definition of "specimen" may need to be clarified. Much metal fatigue testing employs cylindrical specimens. At each end, a cylindrical specimen flares to a greater diameter, which the test machine grips. Failure can occur in the cylindrical portion, in the flare, or in the grips. Often just the cylindrical portion is regarded as the specimen. Then if the flare or grip portion fails, the cylindrical portion is treated as a run-out with that number of cycles. Another viewpoint was well expressed by a test technician – you can't ignore a failure in the grips, there's information in it. In practice, one benefits from analyzing the data with each definition of specimen.

Realistic Test Conditions

Test versus real use. Many accelerated tests attempt to simulate actual use conditions, except for the high level of the overstress variable. Ideally test conditions should exactly reproduce use conditions. Indeed much engineering thought and effort goes into making tests realistic. Professional societies issue standards for conducting tests. This is done to assure that the data accurately reflect product reliability in actual use. Nevertheless, many tests differ much from actual use but may still be useful. Engineers *assume* (based on experience) that a product that performs well on such a test will perform well in actual use. Also, they *assume* that a design, material, vendor, or manufacturing method that performs better than another (design, material, etc.) on test will perform better in actual use. Such tests may be regarded as **engineering tests** or **index tests**. They give a rough but useful measure of performance in actual use. Few accelerated tests exactly simulate actual use. There is only a question of how well a test simulates actual use. Engineers have always had to make this leap of faith from test to field performance. Statisticians acknowledge only that they estimate specimen reliability under test conditions. The following examples include obvious and subtle differences between test and use conditions. Thus it is useful to distinguish *test reliability* from *field reliability*.

Specifics of test methodology are outside the scope of this book. Refer to the many standards on test methodology published by engineering societies.

Motor insulation test. IEEE Standard 117 (1974) specifies how to run a temperature-accelerated life test of motor insulation. Specimens go into an oven, which is then raised to their test temperature. In the oven, no voltage is applied to the insulation. After a specified time at temperature, the specimens are removed from the oven and cooled to room temperature. The specimens are sprayed with water, and then a specified voltage is applied to the insulation, which breaks down (failure) or survives. Survivors go back into the oven for the next cycle at temperature. In actual use, motor insulation constantly runs under voltage, is not sprayed with water, and in many applications sees no thermal cycling.

Electric cord test. Electric cord for small appliances is subjected to a flex test of durability. A test machine repeatedly flexes each cord specimen until it fails electrically. The number of test hours to failure measures performance. A week on test is comparable to 10 to 20 years in service. To make the test reflect actual use, various versions and adjustments of the test machine were tried until it produced the same mix of failures (shorts, opens, etc.) seen in service. This "index" test is used to monitor quality, to compare proposed design and manufacturing changes (including cost savings), and to qualify new designs and vendors. This test ignores chemical degradation.

Toaster test. In a life test, toasters are repeatedly cycled. That is, the handle is pushed down, the toaster heats, and finally pops. One cycle immediately follows another (reduced off-time). To speed the test and make it more realistic, fans cool the toaster between cycles. Lack of such cooling may shorten the life of some components and prolong the life of others (those sensitive to mechanical stressing from temperature changes). Moreover, one version of the test runs by hand with bread (realistic), and another runs automatically with a slice of asbestos (a cost savings). Experience showed that bread and asbestos have the same effect on life.

Metal fatigue tests. Metal fatigue tests of specimens or actual parts generally employ a sinusoidal load. Each specimen runs at its own constant stress amplitude, frequency, and average stress; this yields data for a fatigue curve. In service, most parts do not see such simple loading. Many see other cyclic patterns of loads, whereas others see random loads (that is, with a spectrum of amplitudes and frequencies). More realistic fatigue tests mimic such complex loads.

Population and sample. The preceding discussions of specimens and test conditions deal with a basic statistical principle; namely, the sample must be **representative** of the population. In statistical theory, this means that the sample comes truly randomly from the entire population. Moreover, random numbers and simple random sampling should be used. In practice, this is seldom possible. Statistically speaking, a **sample** is any set of units from a population. In engineering work, "sample" also means a specimen or a test unit. Implicit in these comments is that the population must be clearly defined. Usually the **target** (ideal) population is all product of a particular design that will be made. In practice, the **sampled population** is often some small subset of the population. For example, the sample may consist of prototype units or early production units tested in a lab. Only engineering judgment can assess how representative of the population a particular sample is (test specimens and conditions).

Accelerating Stresses

In practice, one must decide how to accelerate a test. Should one use high temperature, mechanical load, voltage, current, vibration, humidity, or whatever? Should one use a combination of stresses?

Standard stresses. For many products there are standard test methods and accelerating stresses. For example, high temperature and voltage are usually used to accelerate life tests of electrical insulation and electronics. Such standard methods and stresses are usually based on much engineering experience. Moreover, many are documented in engineering standards.

No standard stresses. For other products there may be no standard stresses. Then the responsible engineers need to determine suitable accelerating stresses. Experimental work to determine appropriate stresses may be required. Such stresses should accelerate the failure modes of interest. Also, they should avoid accelerating failure modes that do not occur at design conditions. If there are extraneous failure modes, the data can still be properly analyzed (Chapter 7). The choice of such accelerating stresses is an engineering matter outside the scope of this book.

Multiple stresses. More than one accelerating stress may be used. This is done for various reasons. First, one may wish to know how product life depends on several stresses operating simultaneously. For example, derating curves for various electronic devices include two or more stresses, such as temperature, voltage, and current. Such stresses may interact and require more elaborate models, such as the generalized Eyring model (Chap. 2).

Second, one stress cannot be increased to a high enough level to yield failures quickly enough. So a second stress is used for added acceleration. The first stress may not go high enough, because the test equipment or product cannot go that high. For example, the product may melt, go through a phase change, anneal, etc. Third, one stress may accelerate only some of the failure modes observed at use conditions. So another stress can be used to accelerate other modes in a separate test of additional specimens. Also, both stresses could be used in a single experiment with a number of combinations of levels.

Single stress is simplest. For simplicity and validity of a test, it is generally best to use a single accelerating stress. There is less experience with products under multiple stresses, and fewer suitable models have been verified. For example, the IEEE formed a committee on multi-stress testing of electrical insulation and dielectrics, because of the controversy and lack of information on the subject.

The real stress. It is important to recognize which stress really accelerates product life. For example, one opinion was that the wear (and life) of brushes of electric motors is accelerated by the current through the brush. Of course, higher current produces higher brush temperature. So another opinion was that temperature accelerates brush life. A designed experiment with controlled combinations of temperature and current was run. It showed that only temperature determines brush life. This example indicates that temperature should be included in the model. Or temperature should be controlled and the same for all specimens when it is not the accelerating stress. High voltage, current, speed, etc., increase temperature, and specimens may need cooling.

Stress loading. One must decide how to apply the stress to the specimens. Over time, the stress level on a specimen can be constant, vary cyclically, vary randomly (with some distribution of frequency and amplitude), or increase continuously or in steps. The choice of such **stress loading** (Section 3) depends on how the product is loaded in service and on practical and theoretical limitations. For example, theories for the effect of varying stress on product life are crude and mostly unverified.

Constant stress is preferable. In service, many products run under a constant level of the accelerating stress. For example, most insulations run under constant temperature and voltage, most bearings run under constant load, most electronic components run under constant voltage and current, etc. This suggests that stress levels in accelerated tests to measure reliability should be constant to mimic actual use. Moreover, such constant-stress tests are easiest to run and model for reliability estimation.

Step-stress difficulties. For some products that run under constant stress, engineers sometimes choose progressive or step-stress testing. Such testing assures failures in a short time. There is no well verified theory for using such data to estimate product life under constant design stress. Moreover, although specimens of one design last longer on test than those of another design, they may not last longer at a lower constant design stress. This is especially so for products with a mix of failure modes. The high stress steps may produce failures that do not occur at design stress. Thus testing with varying stress is generally not recommended for reliability estimation. It may be useful for elephant testing (Section 5.1) to identify failure modes.

Products with varying stress. Some products run under varying stress in actual use. Then it is simplest to use the same stress loading but at a higher level in a test. The constant-stress models may adequately represent such a test. Some engineers choose a constant-stress test, because it is easy to run, even though in service the product sees varying stress. There is little established theory (Chapter 10) for using constant stress data to estimate life under varying stress. Then constant-stress testing must be regarded as an index test. That is, a product that performs well under constant stress is assumed to perform well under varying stress. Best and worst life distributions for such a product can be obtained from a constant-stress model. Use the model with the lowest and highest stress levels that the product sees. Even this contains implicit assumptions. Evaluating the model at the average or maximum stress level in use is misleading, since the dependence of product life on stress is highly nonlinear.

Test stress levels. Test stress levels should not be so high as to produce mostly other failure modes that rarely occur at the design stress. Yet levels should be high enough to yield enough of the failures active at the design stress. Also, an accelerated test model is valid over a limited range of stress. So levels of stress need to be in that range. This is so, for example, for high temperature which can melt, unbond, anneal, or cause a phase change in materials. Another example involved compressors tested under extreme high pressure to detect failure of the lubricant. A new compressor design could not withstand that pressure. So the standard test could not be used. Test plans that specify test stress levels appear in Chapter 6.

Other Variables

Types. Besides the accelerating stress, many accelerated tests involve other engineering variables (or factors) that affect product life. Such so-called **independent** or **explanatory variables** include product design, material, manufacturing, operation or test, and environment variables. For the following discussion, they are grouped according to whether the variable (1) is investigated at several values to learn its effect, (2) is held fixed at one value, (3) varies uncontrolled but is observed to learn its effect, and (4) varies uncontrolled and unobserved.

Continuous and categorical. Engineering variables can be continuous or categorical. A **continuous variable** can take on any numerical value in a continuous range. For example, binder content of an insulation can (theoretically) have any value from 0 to 100%. Similarly, insulation thickness (theoretically) can range from zero to infinity. A **categorical variable** can take on only distinct (discrete) values. For example, a categorical variable is the production shift that makes a specimen; it has three values – day, evening, and night. For material made and bought in distinct batches, production lot is a categorical variable. If three distinct designs are compared in a test, design is a categorical variable.

Experimental principles. Experimental design books discuss experimental designs and methods for investigating such variables. See, for example, Cox (1958), Little and Jebe (1975), Diamond (1981), Daniel (1976), and Box, Hunter, and Hunter (1978). These books present experimental principles such as randomization and blocking. Such principles help one get clear, correct results, as well as more accurate information. A few basic principles appear briefly below. Many so-called experimental design books only lightly touch on design principles. Instead they emphasize mathematical models and derivations and sophisticated data analysis. These are often not needed, as most well-designed experiments yield clear information from simple calculations and data plots.

Experimental variables. In some accelerated tests, **experimental** (controlled) **variables** are investigated. Different specimens are tested at different values of such variables. An example was an accelerated test of an insulation in the form of tape. There were two experimental variables: (1) the amount the tape overlaps itself as it is wound on the conductor and (2) the amount the next layer of tape is shifted relative to the previous layer of tape. The experiment involved three levels (amounts) of overlap and three levels of shift. Such specimens with various combinations of overlap and shift were accelerated at various voltage stresses. Another experiment with similar tape insulation investigated one experimental variable – total insulation thickness (tape can be laid in any number of layers). There were specimens with each of five thicknesses. Specimens of each thickness were accelerated at various voltage stresses.

Constant variables. In most accelerated tests, certain **constant variables** are each held at a single fixed value. For example, in the insulation tests of the previous paragraph, the tape came from a single homogeneous lot from the manufacturer. This avoids complications arising from differences between lots. Similarly, specimens were all made by the same technician, and all specimens went through each processing step as a batch; for example, all specimens were cured together. In contrast, certain variables in the fabrication of the Class-H insulation specimens in a later example were not held at a single fixed value. The 260° specimens were made separately from the rest, and they may have differed with respect to material lots and fabrication.

Uncontrolled observed variables. In some accelerated tests, certain variables vary uncontrolled but are observed or measured for each specimen.

Usually one intends to evaluate the effect of such **uncontrolled observed variables** ("covariates") on product life. If they are related to life, they may be included in the model fitted to the data. The tape insulation tests involved such variables, for example, binder content and dissipation factor. If binder content has negligible effect on life, then manufacturing specifications on it can be relaxed. If dissipation factor is related to life, then it can be used to decide which insulation goes into high or low stress applications. To more accurately estimate the effect of such variables, one can select and test specimens with extreme values of such variables, as advocated by Taguchi (1987).

Correlation. Data on such an observed, uncontrolled covariate may be "correlated" with life (or performance). As used by engineers, "correlated" means that life and the covariate have some relationship, usually assumed a physical cause and effect relationship. In particular, specimen data on (log) life crossplotted against such a covariate show a relationship, say, a trend to higher life for higher values of the covariate. Such a relationship in a plot does *not* necessarily mean that maintaining the covariate at a high level will produce a corresponding increase in life as suggested by the plot. That is, there may *not* be a cause and effect relationship between life and the covariate. Instead, there may be a more basic variable that simultaneously affects the covariate and life. Statisticians then say life and the covariate are (statistically) "correlated." When there is a cause and effect relationship, the covariate can be used to *control* life. Binder content above is such a covariate. When there is no cause and effect relationship but just statistical correlation, the covariate can be used to *predict* life but not to control it. Production shift below is such a covariate. Whether a covariate has a cause and effect relationship with life can be determined only with a designed experiment where the covariate is a controlled experimental variable.

In other tests of such tape insulation, production specimens were taped by the day, evening, and night shifts. There was no effort to make specimens by a single shift. Instead, test engineers just recorded whichever shift did the taping in the normal course of production. The data analyses indicated insulation taped by the night shift failed sooner. This is useful information for predicting life. Such insulation can be put into low-stress applications to reduce the risk of service failure. Similarly, production specimens were made from tape from many production lots; the tape lot for each specimen was recorded and analyzed.

Uncontrolled unobserved variables. In most accelerated tests, there are many **uncontrolled unobserved variables.** Engineers are aware of some of such variables and unaware of others. When aware, engineers may not measure them, because their effect is assumed negligible or they are difficult to measure. Such variables may include ambient humidity and temperature in fabrication and test. They may include various material and processing variables, such as specimen position in a curing oven.

For example, suppose that humidity during fabrication is such a variable. Also, suppose the first third of the specimens were made under high humidity and have short life, and the rest were made under low humidity and have long life. Suppose the first third are tested at high stress and the rest at lower stress. Then the estimate of life at design stress is biased low. Moreover, the effect of stress is "confounded" with (not separable from) the effect of humidity. Randomization avoids this.

Randomization. The effect of such biases from such variables can be reduced through randomization. **Randomization** involves using random numbers to determine the order in which specimens go through each step of fabrication, handling, test, measurement, etc. Order is not determined haphazardly. For example, randomly assign specimens to test stress levels and test equipment. Also, randomly assign specimens to positions in ovens, curing tanks, etc. Such randomization reduces the chance of bias. For example, the high and low humidity specimens would tend to be spread among the test stress levels, and their difference would tend to average out. Moreover, it is useful to record the random order (position or whatever) of each specimen through each step. Residual plots (Chapter 4) against each such order may be informative and identify problems in fabrication and testing.

Standard specimens. Another means of dealing with such variables is to regularly test standard specimens to monitor consistency of the test equipment, technique, and conditions. This requires a homogeneous sample of specimens that are archived and used as needed. If the life or degradation of such specimens changes from that seen before, then a cause needs to be identified and the test brought back into control. Also, one can use the life or degradation of standard specimens to "adjust" estimates for other specimens tested in the same period.

Choice of type of variable. In practice, one may be able to choose whether a variable will be experimental, controlled, or uncontrolled but observed. The choice depends on the purpose of the test. For example, suppose one wants to estimate the life of actual tape insulation. Then many tape lots should be employed in the test; that will produce realistic scatter in the data. In contrast, suppose one wants to discern the effect of tape overlap and shift with the greatest accuracy. Then one tape lot should be used; it will minimize scatter of the data and yield more accurate estimates of the effect of overlap and shift.

Measurement Errors

Engineers decide which stresses and other variables are included in a test. Also, they must decide how to measure them. Test results may be sensitive to measurement errors in life, stress, and other variables.

Life. Usually life on test is measured accurately enough. An exception may arise when specimens are periodically inspected for failure. If the first inspection is late, many specimens may fail before it. Then the data contain little information on the lower tail of the life distribution. That is, estimates of low percentiles are crude. So early inspection periods should be short.

Stress. Stress measurements are usually accurate enough. An exception arises, for example, in voltage-endurance tests of insulation. Then stress is the voltage across the insulation divided by the insulation thickness in mils (milli-inches). This so-called **voltage stress** has the dimensions of volts per mil. Such insulation may be thin, and a 1% measurement error for thickness typically yields a 6 to 15% error in life with an inverse power relationship. Fuller (1987) treats fitting of regression models to data with appreciable random errors in the independent variable measurements. This book does not address this complication. Moreover, nonuniform thickness complicates the meaning of voltage stress, as it differs from point to point on the insulation. Section 4 of Chapter 7 treats applications with nonuniform stress.

Corrected stresses. If observed to be different from its intended value, the actual value of a test stress should be used in the data analyses. For example, ambient temperature or humidity may be wrong. Suppose a specimen is hotter than ambient due to electrical or mechanical power dissipated in it. Then one must determine the correct specimen temperature or humidity. This may be done through direct measurement of specimen temperature or through a calculation of a correction based on physical theory.

Other variables. Other independent variables need to be measured with suitable accuracy. Usually one needs a random measurement error that is enough smaller than the differences between specimens. More precisely, the standard deviation of random measurement error should be a fraction of that between specimens.

The Model

Choice of model. Analysis of data from an accelerated life test employs a model. Such a model consists of a life distribution that describes the scatter in product life times and a relationship between typical life and the accelerating stress and other variables. Such a model is ideally based on engineering theory and experience. For many products, there are well-established models (Chapter 2) that are satisfactory over the desired ranges for stress and other variables. For example, for temperature accelerated life tests, the Arrhenius model is often satisfactory.

Needed models. For other products, engineers need to develop such models. The life-stress relationships ideally are based on physical failure mechanisms. Statisticians generally lack the physical insight needed to develop such relationships. However, they can help engineers efficiently collect and accurately analyze data to estimate product life using such a model. Modern statistical theory is versatile and capable of estimating product life, employing standard or new models. Lack of adequate physical models hinders the use of accelerated testing with some products.

Choose ahead. Such a model must be specified while planning a test. A concrete model in mind helps one plan a test that will yield the desired information. In particular, one needs to be sure of two things. First, the data can be used to estimate the model. Second, the estimate of the model yields the desired numerical information on product life. A test run without a model in mind may yield useless data. For example, the test may lack data on important variables. A statistician performing a post-mortem on such data usually can do little to extract desired information. Statistical consultation is most beneficial during planning. Of course, after seeing the data, one may discard the original model and use another.

Choice of the Number of Specimens, Stress Levels, and Test Length

Choice. A test plan includes the test levels of stress and the number of specimens to run at each level. If there are other variables, one must choose their test values and the number of specimens to run at each combination of test values. For example, in the tape insulation experiment with shift and overlap, specimens were made with different combinations of shift and overlap. Specimens of each combination were tested at different voltage levels.

Traditional plans. Traditional engineering plans have three or four test levels of a single stress and the same number of specimens at each level. This practice yields **less** accurate estimates of product life at low stress. As shown in Chapter 6, more accurate estimates are possible. One should run more specimens at low test stresses than at high test stresses. Also, allocating the specimens to four stress levels usually yields less accurate estimates than allocating them to three or two stress levels.

Number of specimens. *Sample size*, the number of test specimens, must be decided. Often it is the available number of specimens. In a development program the number is usually limited. Fabrication and test costs may limit the number. Ideally, the sample size is chosen to yield estimates with a desired accuracy. Of course, larger samples yield greater accuracy. Chapter 6 gives relationships between sample size and accuracy of estimates.

Constraints. Various aspects of a test, including test stress levels and numbers of specimens, are affected by constraints. For example, a deadline for terminating a test can affect the choice of stress levels and numbers of specimens. Limited oven space or test machines or personnel affect the choice. If the model is inadequate above a certain stress levels, test stresses must not exceed it. Even the available computer programs for data analysis may impact how the test is run.

Test length. Another aspect of test planning is the length of the test. Some tests run until all specimens fail – usually a poor practice. Other tests are terminated at deadlines or to free test equipment. Chapter 6 presents trade-off between accuracy of estimates and length of test.

Planning Ahead

Experience. "Experience is the name that people give their mistakes." Through experience we know the truth of the adages "Test in haste, repent at leisure" and "An ounce of planning is worth a pound of data analysis."

Aids to planning. Many accelerated tests are routine and get little forethought. However, this book can improve even such routine tests. New tests, especially, require forethought about the topics above. Ideally, as an aid to planning, one should write the final report before running the test. Blank spaces in the report can be filled in when data are collected. Also, it is very useful to run a **pilot test** with a small number of specimens. The pilot test should involve all steps in fabrication and testing. This will reveal problems that can be corrected before the main test. Also, analyze pilot or simulated data. This is a check that the planned analyses of the final data will yield the desired information accurately enough. Of course, the actual data may suggest other useful analyses. Section 6 gives an overview of the statistical aspects of such data analyses.

5. COMMON ACCELERATED TESTS

This background section briefly describes some common accelerated tests: elephant tests, environmental stress screening, a single test condition, a number of test conditions, and burn-in. Readers may skip this section, as it is not essential to the rest of this book.

5.1. Elephant Tests

Elephant tests go by many names including killer tests, design limits tests, design margin tests, design qualification tests, torture tests, and shake and bake. Such a test steps on the product with an elephant, figuratively speaking. (Luggage manufacturers use a gorilla on TV.) If the product survives, it passes the test, and the responsible engineers feel more faith in it. If the product fails, the engineers take appropriate action and usually redesign it or improve manufacturing to eliminate the cause of failure.

Test procedure. Such an elephant test generally involves one specimen (or a few). The specimen may be subjected to a single severe level of a stress (for example, temperature). It may be subjected to varying stress (for example, thermal cycling). It may be subjected to a number of stresses – either simultaneously or sequentially. An elephant test may not produce certain important failures that the product sees in service. So different elephant tests may be used to reveal different failure modes. For example, high voltage tests reveal electrical failures, and vibration tests reveal mechanical failures.

Cookware. For example, a manufacturer of ceramic cookware uses an elephant test to monitor production quality. Production is sampled regularly. Each item is heated to a specified temperature and plunged into ice water. The cycle of heating and thermal shock is repeated until the item fails. The number of cycles to failure is observed.

Standard elephants. For many products, companies have standard elephant tests. Professional societies and the government write standards for elephant tests. For example, there is a test standard for power cable insulation. It requires that a specified length of a new cable design go through a specified complex series of voltages and temperatures over a specified length of time. If the cable specimen survives, the new design qualifies for use. MIL-STD-883C describes such tests for qualifying microelectronics.

Purposes. In design and manufacturing *development* work, elephant tests reveal failure modes. Engineers then change the product or manufacturing process to overcome such failures. This important use of elephant testing is a standard engineering practice that often improves a product. It sometimes does not, as shown below by the TV transformer example. For *quality control*, elephant testing of samples from production can reveal product changes. Test failures indicate a process change that degraded the product. For example, compressors were sampled from production and subjected to a high pressure test for quick detection of failure of the lubricant. The test could not be used on a new design whose metal parts could not withstand the test pressure. Elephant tests are also used to qualitatively *compare* different designs, vendors, methods of manufacture, etc. Elephant tests for any purpose may mislead due to their limitations explained below.

Good elephant. Everyone asks: what is a good elephant test? The answer is easy: one that produces the same failures and in the same proportions that will occur in service. The hard question is: how does one devise such a test, especially for a new design, possibly with new failure modes? What kind of elephant (African or Asian)? What color (gray, white, or pink)? What sex and age? Should the elephant step on the product or do the boogaloo on it? These are harder questions. Should more than one elephant be used? If so, simultaneously or sequentially? A meaner elephant is not necessarily a better elephant, as the following example shows.

TV transformer. Hundreds of thousands of a type of transformer had recently been manufactured and had gone into service. An engineer devised a new elephant test, and it revealed a new failure mode. A new design overcame that failure mode, and production was changed over to the new design. Years later no transformer of the old design had failed from that mode. The redesign was unnecessary. Most companies have had such experiences with some products. Engineers agree that devising a good elephant test is a black art that requires engineering knowledge, experience, insight, and luck. Elephant tests that work are valuable. This book merely notes that they may be grossly misleading, especially for new designs, materials, and suppliers.

Limitations. Elephant tests provide only qualitative information on whether a product is good or bad. For the applications above, such information suffices. Moreover, elephant testing applies to complex products that may be assemblies of many diverse components. Many reliability applications require an estimate of a failure rate, percentage failing on warranty, typical life, etc. Elephant tests do not provide quantitative reliability information. This is a consequence of the few specimens. Also, more importantly, the data from a single test condition often cannot be extrapolated to actual use conditions (Section 5.3).

5.2. Environmental Stress Screening

ESS. This section briefly surveys Environmental Stress Screening. Environmental stress screening involves accelerated testing of products under a combination of random vibration and thermal cycling and shock – shake and bake. It has two major purposes. First, as an elephant test during development, its purpose is to reveal design and manufacturing problems. Second, in manufacturing, it is an accelerated burn-in to improve reliability. It is widely used for military, industrial, and consumer electronics – components and assemblies. With its roots in elephant testing and burn-in, ESS is a discipline less than 10 years old and is an engineering science or black art, opinion differing among experts.

Standards. Except for RADC TR-86-139, the following military standards are available from Naval Publications and Forms Center, 5801 Tabor Ave., Philadelphia, PA 19120, (215)697-2000.

MIL-HDBK-344 (20 Oct. 1986), "Environmental Stress Screening – Electronic Equipment,"
MIL-STD-810D (19 July 1983), "Environmental Test Methods and Engineering Guidelines,"
MIL-STD-883 (25 August 1983), "Test Methods and Procedures for Microelectronics,"
MIL-STD-2164 (5 April 1985), "Environmental Stress Screening Process for Electronic Equipments,"
RADC TR-86-139 (Aug. 1986), "RADC Guide to Environmental Stress Screening," RADC, Griffiss AFB, NY 13441.
U.S. Navy Document P-9492 (May 1979), "Navy Manufacturing Screening Program."

References. Tustin (1986) briefly surveys ESS. The Catalog of the Institute of Environmental Sciences (below) lists many books, standards, handbooks, and conference proceedings. Books include

- Schlagheck, J.G. (1988), *Methodology and Techniques of Environmental Stress Screening*, Tustin Technical Inst. (below).
- Tustin, W. and Mercado, R. (1984), *Random Vibration in Perspective*, Tustin Technical Inst. (below), 200 pp., $100.

Resources. Sources of expertise, conferences, courses, and literature on ESS include:

- Institute of Environmental Sciences, 940 E. Northwest Hwy., Mt. Prospect, IL 60056, (708)255-1561. Technical Publications Catalog.
- Tustin Technical Institute, Inc., 22 E. Los Olivos St., Santa Barbara, CA 93105, (805)682-7171, Dr. Wayne Tustin.
- Technology Associates, 51 Hillbrook Dr., Portola Valley, CA, (415)941-8276, Dr. O.D. "Bud" Trapp.
- Hobbs Engineering Corp., 23232 Peralta Dr., Suite 221, Laguna Hills, CA 92653, (714)581-9255, Dr. Gregg K. Hobbs.

5.3. A Single Test Condition

Some overstress testing for reliability *estimation* involves a single test condition. Then the estimate of the product life distribution at a use condition depends on certain assumptions. Often the assumptions are poorly satisfied, and the estimates may be quite crude. Two models for such tests are described below – the acceleration factor and partially known dependence of life on stress. In addition, such tests are used for demonstration testing and to compare designs, materials, manufacturing methods, etc.

Assumption. Such a test rests on an assumption. Namely, the test accelerates and reveals all the important failure modes at the use condition. Data with such a mix of failure modes (including ones not observed at use conditions) can properly be analyzed with the methods of Chapter 7.

Acceleration Factor

Definition. A certain test runs a diesel engine at 102% of rated power and is assumed to have an acceleration factor of 3. Roughly speaking, this means that if an engine ran 400 hours to failure on test, one assumes that it would have run $3 \times 400 = 1200$ hours to failure in service. Another engine test is assumed to have an acceleration factor of 5. Similarly, if an engine ran 300 hours without failure on the second test, one assumes that it would have run $5 \times 300 = 1500$ hours without failure in service. Two or more such tests usually are run to produce different failure modes seen in service. To analyze such accelerated data, one converts them to service hours. Then one analyzes the service hour data using standard statistical methods for a single sample to estimate the service life distribution. Such methods appear in reliability books referenced by Nelson (1983c, 1982, pp. 567-569).

Assumptions. This method involves various mathematical assumptions. Such assumptions may be marginally valid in practice.

1) Known factor. It is usually assumed that the acceleration factor is "known." In practice, often its value is a company tradition with unknown origins. Sometimes it is **estimated** from test and field data on a previous product design. The conversion of test time to service time and subsequent data analyses (particularly confidence intervals) do not take into account the uncertainty in the factor. Such uncertainty is due to the randomness in the statistical samples of the previous test and service data. More important, the factor is based on previous designs. So this method assumes (often in error) that the new design has the same acceleration factor.

2) Same shape. There is a subtle assumption in multiplying test data by the factor. Namely, this implies that the service life distribution is simply a multiple of the test life distribution. Consequently the two distributions are assumed to have the same shape, but the service life distribution is stretched out toward higher life by an amount equal to the factor. In other words, the scale parameter of the distribution of service life is a multiple of that of the test distribution. This may not be so.

3) Failure modes. Often there is more than one failure mode in the data. Then, if one ignores the different modes, one is assuming that they all have the same acceleration factor. Typically, different failure modes have different acceleration factors (Chapter 7). This is especially so for complex products consisting of assemblies of diverse components.

Performance. The discussion above applies to product life. Such a single accelerated test condition can also be used to accelerate the change or degradation in product performance over time. For example, the concrete industry uses the strength of concrete after 28 days of curing. Accelerated curing (say, 3 hours at high temperature) is used to predict 28-day strength.

Partly Known Relationship

Method. For some products and tests with a single stress, one may be able to use a partly known relationship between life and stress. For example, suppose an accelerated life test of insulation runs specimens at a single high temperature. Also, suppose that an Arrhenius relationship (Chap. 2, Sec. 9.1) describes the relationship between life and temperature. In particular, suppose life increases by a factor of 2 for every 10°C decrease in temperature; this is the assumed partial knowledge. This assumption is wrong (Problem 2.11), but it *crudely* allows one to derive an acceleration factor for converting data from any high test temperature to service data at any low temperature. For the insulation example, suppose that the test temperature is 40°C (= 4 × 10°) above the service temperature. Then life at the service temperature is $2^4 = 16$ times as great as at the test temperature. The test times are multiplied by this acceleration factor of 16 to crudely convert them to service times. Those service times are analyzed using standard statistical methods for a single sample to estimate the service life distribution. Such methods appear in reliability books, which Nelson (1983c, 1982, pp. 567-569) references. The resulting confidence limits are much too narrow since they do not reflect the error in the assumed factor of 2. Examples with acceleration factors include microprocessors (Chap. 3, Sec. 6.3) and lubricating oil (Chap. 5, Sec. 2.3).

Assumptions. This method makes various assumptions. First, it makes all of the assumptions for an acceleration factor. Moreover, this method employs an assumed form of the life-stress relationship (such as the Arrhenius relationship) and an assumed numerical value for the relationship (such as life increases by a factor of 2 for each 10°C decrease in temperature). This power law approximation may not adequately represent the true Arrhenius relationship. Even if it is adequate, the true factor may be 2.4 – not 2. Round numbers like 2, rather than 2.4, are automatically suspect. In practice, one might try different plausible factors to see how the corresponding estimates and conclusions differ. Hopefully, the conclusions and appropriate actions are not sensitive to the factor value. If they are, a more accurate factor must be obtained. Accelerated tests with more than one test stress level (Section 5.4) provide a statistical estimate of a factor. This is preferable to assuming a value, except when the statistical estimate has great uncertainty.

Advantages. In return for the added assumptions about the relationship, this method yields an advantage over the simple acceleration factor. Namely, the method works for any test temperature and any service temperature. In contrast, the simple factor applies only to the particular test and service conditions. However, such a relationship generally is satisfactory only for a material or a very simple product. It is unreliable for complex products consisting of diverse components, such as a circuit board or diesel engine.

Comparisons

Method. A single overstress condition is sometimes used to compare two or more different designs, materials, vendors, manufacturing methods, etc. A number of specimens of each design (or vendor, etc.) all undergo the same test condition. Such a condition may be high level of a single stress or a complex loading of one or more stresses. The life data on the different designs are compared. The flex test of appliance cord is such a test. It compares cord designs and monitors production quality.

Assumption. The design with the best life distribution while overstressed is **assumed** to have the best life distribution under normal conditions. Of course, it is possible that a design that is better at high stress is poorer under normal conditions, because different designs and failure modes usually have different acceleration factors. Accelerated tests with more than one test stress level (Section 5.4) surmount this difficulty. Then the data yield separate estimates of the acceleration factor for each design and failure mode.

5.4. A Number of Test Conditions

Description. Most overstress testing for reliability *estimation* employs a number of stress levels. A group of specimens is run at each level. A model for specimen life is fitted to the life data from the stress levels, and the fitted model is used to estimate the product life distribution at some design level of stress. Such testing can involve more than one stress variable. Then groups of specimens are each run at a different test condition, consisting of a combination of a level of each stress. A more complex model involving all stresses is fitted to the resulting life data, and it is used to estimate the product life distribution at some combination of a low level of each stress.

Validity. This type of testing, the accelerated test models, data analyses, and test plans are the main topics of this book. The ideas for such testing apply to other types of overstress tests, including elephant tests and especially tests for reliability estimation from a single test condition. Such tests and models have generally been most successful for materials and simple products. There are few successful applications to complex products consisting of assemblies of diverse components, such as a circuit board.

Comparisons. Such tests are suitable for comparisons of different designs (vendors, materials, etc.) at normal stress levels. A separate model is fitted to the data from each design, and each fitted model is used to estimate and compare the life distributions of the designs at normal stress levels.

5.5. Burn-In

Burn-in consists of running units under design or accelerated conditions for a suitable length of time. Burn-in is a manufacturing operation that is intended to fail short-lived units (defectives, sometimes called **freaks**). If burn-in works, the surviving units that go into service have few early failures. Units that fail early typically have manufacturing defects. Burn-in is primarily used for electronic components and assemblies. Jensen and Petersen's (1982) book surveys burn-in and includes a chapter on accelerated burn-in. Environmental stress screening (Section 5.2) includes burn-in as one of its purposes. Those interested in accelerated burn-in will find this book useful, as it provides more detail. In particular, Chapter 7 on competing failure modes pertains to accelerated burn-in.

6. STATISTICAL CONSIDERATIONS

Statistical considerations useful for the rest of this book are briefly presented here. The topics are statistical models, population and sample, valid data, nature of data analysis, estimates and confidence intervals, hypothesis tests, practical and statistical significance, numerical calculations, and notation. Many of these considerations reflect engineering considerations in Section 4.

Statistical models. Nominally identical units, made and used under the same conditions, usually have different values of performance, dimensions, life, etc. Such variability is inherent in all products, and it can be described by a statistical model or distribution. Chapter 2 presents such models.

Population and sample. A statistical model describes some **population.** A manufacturer of fluorescent lamps is concerned with the future production of a certain lamp design – an essentially infinite population. A manufacturer of locomotives is concerned with a small population of locomotives. A metallurgist is concerned with a future production of a new alloy – an essentially infinite population. A manufacturer is concerned with the performance of a small population of generators to be manufactured next year. The *target population* of interest should be clearly specified at the outset, as it affects the choice of sample specimens and other matters. To obtain information, we use a **sample** (a set of units) from the population. We analyze the sample data to get information on the underlying population distribution or to predict future data from the population.

Valid data. There are many practical aspects to the collection of valid and meaningful data. Some are briefly described below. Throughout, this book assumes that such aspects are properly handled. For example, measurements must be meaningful and correct. Also, one needs to avoid blunders in handling data. Bad data can be unknowingly processed by computers and by hand.

Population. Most statistical work assumes that the sample is from the target population. A sample from another population or a subset of the target population can give misleading information. For example, appliance failures on a service contract often overestimate failure rates for appliances not on contract. Also, laboratory test data may differ greatly from field data. Data on units made last year may not adequately predict this year's units. In practice, it is often necessary to use such data. Then engineering judgment must determine how well such data represent the population of interest and how much one can rely on the information.

Random sample. Most statistical theory assumes that the specimens are obtained by **simple random sampling** from the population of interest. Such sampling gives each possible set of n population units the same chance of being the chosen sample. Only the use of random numbers ensures random selection. In practice, other statistical sampling methods are sometimes used, the most common methods being haphazard sampling, stratified sampling, and two-stage sampling. Data analyses must take into account the sampling method. Like all others, this book assumes throughout that simple random sampling is used. Some samples are taken haphazardly, that is, without probability sampling. Other samples may be the available prototype units. Such samples may be quite misleading.

Experimental design. Many engineering experiments do not employ good experimental design principles. Use of such principles enhances the validity and clarity of experimental results. Moreover, a well designed and executed experiment is much easier to analyze and interpret. Also, such an experiment is easier to explain to others who must be convinced of the results. Such design principles include choosing suitable test conditions and specimens, using randomization throughout all steps of the experiment, using a statistical design, blocking, etc. Section 4 discusses some of these principles. Most statistical books on experimental design emphasize statistical models and data analyses. Few devote enough attention to good experimental principles, which most statisticians and engineers learn on the job.

Nature of data analysis. The following briefly describes statistical aspects of data analysis. It outlines how to define a statistical problem, select a mathematical model, fit the model to data, and interpret the results.

The statistical solution of a real problem involving data analysis has seven basic steps.

1. Clearly state the real problem and the purpose of the data analysis. In particular, specify the *numerical* information needed in order to draw conclusions and make decisions.
2. Formulate the problem in terms of a model.
3. Plan both collection and analyses of data that will yield the desired numerical information. Use experimental design principles.
4. Obtain appropriate data for estimating the parameters of the model.
5. Fit the model to the data, and obtain the needed information from the fitted model.
6. Check the validity of the model and data. As needed, change the model, omit some data, or collect more (often overlooked by statisticians), and redo steps 5 and 6.
7. Interpret the information provided by the fitted model to aid in drawing conclusions and making decisions for the engineering problem.

This book gives methods for steps 3, 5, and 6. The other steps involve the judgment of engineers, managers, scientists, etc. Those steps are discussed in Section 4 on engineering considerations, many of which have statistical consequences. In the planning stage, the data analyst can contribute by drawing attention to those considerations. Each of the steps is discussed further below, but full understanding of these steps comes only with experience. Data analysis is an iterative and exploratory process, and one usually subjects a data set to many analyses to gain insight. Thus, many examples in this book involve different analyses of the same set of data.

1. A clear statement of a real problem and the purpose of a data analysis is half of the solution. Having that, one can usually specify the numerical information needed to draw practical conclusions and make decisions. Of

course, statistical analysis provides no decisions – only numerical information for people who make them. If one has difficulty specifying the numerical information needed, the following may help. Imagine that any desired amount of data is available (say, the entire population). Then decide what values calculated from the data would be useful. Statistical analysis estimates such values from limited sample data. If such thinking does not clarify the problem, one does not understand it. Sometimes there is a place for exploratory data analyses that do not have clear purposes but that may reveal useful information. Data plots are particularly useful for such analyses.

2. One views a problem in terms of a model. Often the model is a simple and obvious one, widely used in practice. An example is the Arrhenius-lognormal model for time to insulation failure. Ideally a tentative model is chosen before the data are collected. After data collection, when a suitable model is not obvious, display the data various ways, say, on different probability and relationship papers. Such plots often suggest a suitable model. Indeed, a plot often reveals needed information and can serve as a model itself. Another approach is to use a very general model that is likely to include a suitable one as a special case. After fitting the general model to the data, one often sees which special case is suitable. Still another pragmatic approach is to try various plausible models and to select, as a working model, one that fits the data well. The chosen model should, of course, provide the desired information. Examples of these approaches appear in later chapters.

3. The test and data collection are planned, using experimental design principles. This insures that the model parameters and other quantities can be accurately estimated from the data. Sometimes, when data are collected before a model and data analyses are determined, it may not be possible to fit a desired model, and a less realistic model must be used. A statistician usually makes the greatest contribution at this test planning step. Chapter 6 discusses test plans.

4. Practical aspects of data collection and handling need much forethought and care. For instance, data may not be collected from the population of interest. For example, data may be from prototype units rather than from production units. Many companies go to great expense collecting test data, often ending with inadequate data owing to lack of forethought.

5. To fit a chosen model to the data, one has a variety of methods. This step is straightforward. It involves using methods described in this book to obtain estimates and confidence intervals. Confidence intervals are essential. Often wide, they help engineers take heed that the estimates have great statistical uncertainty. Much of the labor can (and often must) be performed by computer programs.

6. Of course, one can mechanically fit an unsuitable model just as readily as a suitable one. Computer programs do this well. An unsuitable model

may yield information leading to wrong conclusions and decisions. Before using information from a fitted model, one should check that the model adequately fits the data. Such checks usually employ graphical displays; they allow one to examine the model and the data for consistency. The model may also be checked against new data. Sometimes various models fit a set of data within the range of the data. However, they can give very different results outside that range. Then a model must be chosen on other considerations, for example, simplest or most conservative. Also, it may be possible to test more specimens to resolve such problems. Engineers will think to run more tests as needed; this important option is often overlooked by inexperienced statisticians and statistics texts.

7. Interpretation of results from the fitted model is usually easy when the above steps are done properly, as practical conclusions and decisions are usually apparent. A possible difficulty is that the information may not be accurate or conclusive enough for practical purposes. Then one needs more data for the analysis, and may do further testing. Also, one may have to make do with less accurate information. Most models and data are inaccurate to some degree. So the uncertainty in any estimate or prediction is greater than is indicated by the corresponding confidence or prediction interval.

Data analysis methods. Some specific data analysis methods are discussed below – estimates, confidence intervals, and hypothesis tests. These methods are treated in detail in later chapters. Nelson (1983c,1982, Chap. 6, Sec. 1) provides more detail. Nelson (1990) briefly presents the basics of analysis of accelerated test data. Older statistics books present individual data analysis techniques. They seem to imply that a data set is analyzed by calculating a t statistic or a hypothesis test. Modern books on data analysis emphasize exploratory data analysis involving many different analyses of a data set, especially graphical analyses. This book advocates multiple analyses of a data set. To illustrate this, many examples here appear in different chapters with different analyses to attain greater insight into the data.

Estimates and confidence intervals. Using sample data, this book provides estimates and confidence intervals for the model parameters and other quantities of interest. The estimates approximate the true parameter values. By their width, confidence intervals for parameters indicate the uncertainty in estimates. If an interval is too wide for practical purposes, a larger sample may yield one with the desired width. A wide confidence interval warns that the estimate may be too uncertain for practical purposes. Hahn and Meeker (1990) survey such confidence intervals.

Statistical comparisons. A statistical hypothesis test (Chapters 8 and 9) compares sample data with a hypothesis about the model. A common hypothesis is that a parameter equals a specified value; for example, a Weibull shape parameter equals 1 (that is, the distribution is exponential). Another common hypothesis is that corresponding parameters of two or more populations are equal; for example, the standard two-sample t-test compares two population means for equality. If there is a statistically significant difference between the data and the hypothesized model, then there is convincing evidence that the hypothesis is inadequate ("false" in statistical jargon). Otherwise, the hypothesized model is a satisfactory working assumption. Of course, data sets with few failures may be consistent with a physically inadequate model. Also, a test of fit or a test for outliers may result in rejection of the model or data.

Practical and statistical significance. Confidence intervals indicate how (im)precise estimates are; they reflect the random scatter in the data. Hypothesis tests indicate whether observed differences are statistically significant. That is, they indicate whether a difference between a sample of data and a hypothesized model (or whether the difference between a number of samples) is large relative to the random scatter in the data. **Statistically significant** differences are ones large enough to be convincing. In contrast, **practically significant** differences are ones big enough to be important in practice. Although results of an analysis may be practically significant (big), one should not rely on them unless they are also statistically significant, that is, convincing. Statistical significance assures that results are real rather than mere random sampling variation.

A confidence interval for such differences is usually easier to interpret and more informative than a statistical hypothesis test. The interval width allows one to judge whether the results are accurate enough to identify differences that are important in practice. Chapters 8 and 9 give such confidence intervals and their application to comparisons.

Numerical calculations. Numerical examples here are generally calculated with care. That is, extra figures are used in numbers substituted into formulas and in intermediate calculations. Only the final result is rounded to a suitable number of figures. This good practice helps assure that a result is accurate to the final number of figures shown. Common practice rounds all numbers to the number of significant figures desired in the final result. This practice yields fewer than the desired number of significant figures. For most practical purposes, two or three significant final figures suffice. A reasonable practice is to give estimates and confidence limits to enough figures so that they differ in just the last one or two places. For example, the estimate $\mu^* = 2.76$ and upper confidence limit $\tilde{\mu} = 2.92$ (for a parameter μ) differ in the last two places.

Many calculations for examples and problems can easily be done with an electronic pocket calculator. However, some calculations, particularly maximum likelihood calculations, require computer programs. Readers can develop their own programs from the descriptions given here or preferably use standard programs.

PROBLEMS

1.1. Search. Do a literature search on some area of application in Section 1 and compile a bibliography. Annotate it.

1.2. Extension. Extend the discussion of any topic in Sections 4, 5, and 6. Specify your audience.

2

Models for Life Tests
with Constant Stress

1. INTRODUCTION

Purpose. This chapter presents mathematical models for accelerated life tests with constant stress. These models are essential background for subsequent chapters. All planning and data analyses for accelerated tests are based on such models. Moreover, those who run elephant tests and tests with a single stress condition will find this chapter useful background, because this chapter provides concrete models and concepts. Such background will make apparent and reduce the vague nature of the meaning and interpretation of such tests. A model depends on the product, the test method, the accelerating stress, the form of the specimen, and other factors. Previous exposure to statistical life distributions and life-stress relationships is helpful but not essential background.

Model. A statistical **model** for an accelerated life test consists of 1) a life *distribution* that represents the scatter in product life and 2) a *relationship* between "life" and stress. Usually the mean (and sometimes the standard deviation) of the life distribution is expressed as a function of the accelerating stress. Sections 2 through 6 present the commonly used life distributions – the exponential, normal, lognormal, Weibull, and extreme value distributions. Section 2 also presents basic concepts for life distributions, including the reliability function and hazard function (instantaneous failure rate). Section 7 briefly presents a number of other distributions that are useful but less commonly used; these include mixture distributions and the log gamma distribution, which includes the Weibull and lognormal distributions as special cases. Sections 8 through 14 present life-stress relationships. Such relationships express a distribution parameter (such as a mean, percentile, or standard deviation) as a function of the accelerating stress and possibly of other variables. The most widely used basic relationships are 1) the Arrhenius relationship (Section 9) for temperature-accelerated tests and 2) the inverse power relationship (Section 10). Singpurwalla (1975) surveys a number of models. Nelson (1990) briefly presents the most basic applied models.

Standard Models. For many products, there are standard accelerating variables and models. For example, life testing of motor insulation is usually accelerated with high temperature, and the data are analyzed with the Arrhenius-lognormal model of Section 9.2. For some products with a standard accelerating variable, the form of the life distribution or the life-stress relationship may be in question. For example, for certain voltage-endurance testing of an insulation, the Weibull and lognormal distributions and the inverse power and other relationships were fitted to the data; the various distributions and relationships were compared to assess which fitted significantly better. Such standard accelerating variables and models appear in the product literature. For example, Meeker (1979) and Carey (1985) have run computerized literature searches on accelerated testing of a great variety of products. This chapter presents standard models for such products. Later chapters present data analyses using such models and test plans.

New models. For still other products, one may need to choose an accelerating variable and to develop and verify an appropriate model. This book does not deal explicitly with this difficult subject. Such work involves long-term effort of product experts, perhaps abetted by a skilled statistician. The book by Box and Draper (1987) is useful for such work.

Single failure cause. The models here are best suited to products that have only one cause of failure. However, they also adequately describe many products that have a number of causes. Chapter 7 presents the series-system model for products with a number of causes of failure.

Multiple tests. For some products there are two or more accelerated tests with different accelerating variables. Each test is run on a different set of units to accelerate different failure modes. For example, certain failure modes may be accelerated by high temperature, while others are accelerated by high voltage or vibration.

Other models. Models for more complex situations are not presented here in Chapter 2. These include models for step-stress testing (Chapter 10), for aging degradation of performance (Chapter 11), for the effect of specimen size (Chapter 7), for analysis of variance, and for components of variance. The models here do not apply to most repairable systems which can fail and be repaired any number of times. Ascher and Feingold (1984) and Nelson (1988) present models and data analyses for repairable systems. There is little theory for accelerated testing of repairable systems.

Notation. This book mostly follows common statistical notation. Population parameters and model coefficients are usually written as Greek letters; e.g., α, β, γ, and σ. Such numbers are usually unknown constants to be estimated from data. Estimates of such constants are random quantities and are denoted by the Greek letter with a caret $\hat{}$ or by a corresponding Latin letter, e.g., $\hat{\alpha}$, a, $\hat{\beta}$, b, $\hat{\gamma}$, c, $\hat{\sigma}$, and s. Notations for a true (unknown) population value σ and for an estimate $\hat{\sigma}$ are distinct; this attempts to avoid a common mistake of thinking that an estimate is the true population value. Random quantities, such as time to failure, are usually denoted by capital Latin letters, e.g., T and Y. A specific numerical outcome of such a random quantity is denoted by the corresponding lower case Latin letter, e.g., t and y. Latin letters (upper and lower case) are used for engineering variables, generally following standard engineering notation. A common exception to such notation is that Latin letters may be coefficients in engineering relationships.

2. BASIC CONCEPTS AND THE EXPONENTIAL DISTRIBUTION

This section presents basic concepts for product life distributions. The exponential distribution illustrates them.

Cumulative Distribution Function

Definition. A cumulative **distribution function** $F(t)$ represents the population fraction failing by age t. Any such continuous $F(t)$ has the mathematical properties:
a)   it is a continuous function for all t,
b) $\lim_{t \to -\infty} F(t) = 0$ and $\lim_{t \to \infty} F(t) = 1$, and
c) $F(t) \leq F(t')$ for all $t < t'$.
The range of t for most life distributions is from 0 to ∞, but some useful distributions have a range from $-\infty$ to ∞.

Exponential cumulative distribution function. The population fraction failing by age t is

$$F(t) = 1 - e^{-t/\theta}, \quad t \geq 0. \tag{2.1}$$

$\theta > 0$ is the *mean time to failure (MTTF)*. θ is in the same measurement units as t, for example, hours, months, cycles, etc. Figure 2.1 shows this cumulative distribution function. Its **failure rate** is defined as

$$\lambda \equiv 1/\theta \tag{2.2}$$

and is a constant. This relationship between the constant failure rate λ and the MTTF θ holds *only* for the exponential distribution. λ is expressed in failures per million hours, percent per month, and percent per thousand hours. For high reliability electronics, λ is often expressed in *FITs*, failures per billion hours. In terms of λ,

$$F(t) = 1 - e^{-\lambda t}, \quad t \geq 0. \tag{2.3}$$

The exponential distribution describes the life of insulating oils and fluids (dielectrics) and certain materials and products. It is often badly misused for products better described with the Weibull or another distribution. Some elementary reliability books mistakenly suggest that the exponential distribution describes many products. In the author's experience, it adequately describes only 10 to 15% of products in the lower tail of the distribution.

Engine fan example. The exponential distribution with a mean of θ = 28,700 hours was used to describe the hours to failure of a fan on diesel engines. The 28,700 hours was estimated from data and has great statistical uncertainty. It and other numbers are treated in this chapter as if exact. The corresponding failure rate is λ = 1/28,700 = 35 failures per million hours. For the engine fans, the population fraction failing on an 8,000 hour warranty is calculated from (2.1) as $F(8,000)$ = 1 $-$ exp($-8,000/28,700$) = 0.24. Thinking 24% too high, management decided to use a better fan design.

Reliability Function

Definition. The **reliability function** $R(t)$ for a life distribution is the probability of survival beyond age t, namely,

$$R(t) \equiv 1 - F(t). \tag{2.4}$$

This is also called the *survivor* or *survivorship function.*

Exponential reliability. The population fraction surviving age t is

$$R(t) = e^{-t/\theta}, \quad t \geq 0. \tag{2.5}$$

Figure 2.2 shows this reliability function. It is the cumulative distribution function (Figure 2.1) "turned over."

For the engine fans, reliability for 8,000 hours is $R(8,000)$ = exp($-8,000$ /28,700) = 0.76. That is, 76% of such fans survive warranty.

Percentile

Definition. The 100Pth **percentile** of a distribution $F(\ )$ is the age τ_P by which a proportion P of the population fails. It is the solution of

$$P = F(\tau_P). \tag{2.6}$$

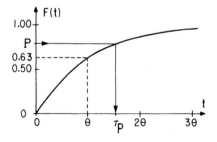

Figure 2.1.   Exponential cumulative distribution.

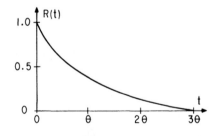

Figure 2.2.   Exponential reliability function.

In life data work, one often wants to know low percentiles such as the 1% and 10% points, which correspond to early failure. The 50% point is called the **median** and is commonly used as a "typical" life. τ_P can be obtained as shown in Figure 2.1. Enter the figure on the vertical axis at the value P, go horizontally to the curve for $F(t)$, and go down to the time axis to read τ_P.

Exponential percentile. The 100Pth percentile is

$$\tau_P = -\theta \ln(1-P). \tag{2.7}$$

For example, the mean θ is roughly the 63rd percentile of the exponential distribution. For the diesel engine fans, median life is $\tau_{.50} = -28,700\ln(1-0.50) = 19,900$ hours. The 1st percentile is $\tau_{.01} = -28,700\ln(1-0.01) = 288$ hours.

Probability Density

Definition. The *probability density* is the derivative

$$f(t) \equiv \frac{dF(t)}{dt}, \tag{2.8}$$

which must exist mathematically. It corresponds to a histogram of the population life times. Equivalently, the population fraction failing by age t is the integral of (2.8), namely,

$$F(t) = \int_{-\infty}^{t} f(u)\, du. \tag{2.9}$$

If the lower limit of a distribution range is 0, the integral ranges from 0 to t. Similarly,

$$R(t) = \int_{t}^{\infty} f(t)dt.$$

Exponential probability density. Differentiation of (2.1) yields

$$f(t) = (1/\theta)\, e^{-t/\theta}, \quad t \geq 0. \tag{2.10}$$

Figure 2.3 depicts this probability density. Also,

$$f(t) = \lambda e^{-\lambda t}, \quad t \geq 0. \tag{2.11}$$

Mean

Definition. The *mean* or *expectation* $E(T)$ of a distribution for random time to failure T with probability density $f(t)$ is the value of the integral

$$E(T) \equiv \int_{-\infty}^{\infty} t\, f(t)\, dt. \tag{2.12}$$

The integral runs over the range of the distribution (usually 0 to ∞ or $-\infty$ to ∞). The mean is also called the *average* or *expected life*. It corresponds to the arithmetic average of the lives of all units in a population. Like the median, it is used as still another "typical" life. Here the terms "expectation"

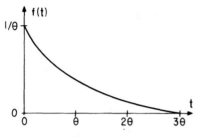

Figure 2.3.   Exponential probability density.

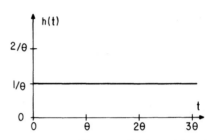

Figure 2.4.   Exponential hazard function.

and "expected life" have the precise statistical meaning in (2.12) and do not mean "anticipated life."

Exponential mean. The mean is

$$E(T) = \int_0^\infty t \,(1/\theta)\, e^{-t/\theta}\, dt = \theta. \tag{2.13}$$

This shows why θ is called the **mean time to failure** (MTTF). Also $E(T) = 1/\lambda$; as noted above, this relationship holds *only* for the exponential distribution. For the diesel engine fans, $E(T) = 28{,}700$ hours.

Variance and Standard Deviation

Definition. The *variance* of a distribution with a probability density $f(t)$ is

$$\mathrm{Var}(T) \equiv \int_{-\infty}^\infty [t - E(T)]^2 f(t) dt. \tag{2.14}$$

The integral runs over the range of the distribution. The variance is a measure of the spread of the distribution. An equivalent formula is

$$\mathrm{Var}(T) = \int_{-\infty}^\infty t^2 f(t) dt - [E(T)]^2. \tag{2.15}$$

$\mathrm{Var}(T)$ has the units of time squared, for example, hours squared. Used by statisticians, variance is equivalent to standard deviation below, which is easier to interpret.

Exponential variance. For an exponential distribution,

$$\mathrm{Var}(T) = \int_0^\infty t^2 (1/\theta)\exp(-t/\theta) dt - \theta^2 = \theta^2. \tag{2.16}$$

This is the square of the mean. For the diesel engine fans, the variance of the time to failure is $\mathrm{Var}(T) = (28{,}700)^2 = 8.24 \times 10^8$ hours^2.

Definition. The *standard deviation* $\sigma(T)$ of a life distribution is

$$\sigma(T) = [\mathrm{Var}(T)]^{1/2}. \tag{2.17}$$

This has the units of life, for example, hours. The standard deviation is a more commonly used measure of distribution spread than the variance, because it has the same dimensions as life.

Exponential standard deviation. For an exponential distribution,

$$\sigma(T) = (\theta^2)^{1/2} = \theta. \tag{2.18}$$

This equals the mean. For the fans, $\sigma(T) = (8.24 \times 10^8)^{1/2} = 28,700$ hours.

Hazard Function (Instantaneous Failure Rate)

Definition. The *hazard function* $h(t)$ of a distribution is defined as

$$h(t) \equiv f(t)/[1 - F(t)] = f(t)/R(t). \tag{2.19}$$

It is the **instantaneous failure rate** at age t. That is, in the short time Δ from t to $t + \Delta$, a proportion $\Delta \cdot h(t)$ of the population that reached age t fails. $h(t)$ is a measure of proneness to failure as a function of age. It is also called the *hazard rate* and the *force of mortality*. In many applications, one wants to know whether the failure rate of a population increases or decreases with product age, that is, whether service failures will increase (or decrease) with product age.

The exponential hazard function is

$$h(t) = [(1/\theta) e^{-t/\theta}]/e^{-t/\theta} = 1/\theta = \lambda, \quad t \ge 0. \tag{2.20}$$

Figure 2.4 shows this constant hazard function. Also, $h(t) = \lambda$, $t \ge 0$; this explains why λ is called the failure rate. *Only* the exponential distribution has a constant failure rate, a key characteristic. That is, for this distribution only, an old unit and a new unit have the same chance of failing over a future time interval Δ. Such products are said to lack memory and, like some people, do not remember how old they are. Also, products with a constant failure rate are said to have *random failures,* sometimes implying due to shocks and external events. The term "random" is misleading, as many such products fail from wear-out or manufacturing defects. For example, engine fans of any age failed at a constant rate of 35 failures per million hours from fatigue.

Cumulative hazard. The *cumulative hazard function* is

$$H(t) \equiv \int_{-\infty}^{t} h(u)du. \tag{2.21}$$

Here the lower limit is the lower end of the distribution range. This function is employed in hazard plotting (Chapter 3). For the exponential distribution,

$$H(t) = \int_{0}^{t} \lambda \, du = \lambda t, \quad t \ge 0.$$

For any distribution, $H(t) = -\ln[R(t)]$. Equivalently,

$$R(t) = \exp[-H(t)]. \tag{2.22}$$

This basic relationship is employed on hazard plotting papers (Chapter 3).

Infant mortality. A decreasing hazard function during the early life of a product is said to correspond to *infant mortality.* Figure 7.4 shows this near time zero. Such a failure rate often indicates that the product has design or manufacturing defects. Some products, such as capacitors and some semiconductor devices, have a decreasing failure rate over their observed life.

Wear–out. A hazard function that increases without limit during later life of a product is said to correspond to *wear-out* failure. This often indicates that failures are due to the product wearing out. Figure 7.4 shows this feature in the later part of the hazard curve. Many products have an increasing failure rate over the entire range of life. If the failure rate increases, preventive replacement of units in service can prevent costly service failures.

3. NORMAL DISTRIBUTION

This section presents the normal (or Gaussian) distribution. Its hazard function increases without limit. Thus it may describe products with wear-out failure. It has been used to describe the life of incandescent lamp (light bulb) filaments and of electrical insulations. It is also used as the distribution for product properties such as strength (electrical or mechanical), elongation, and impact resistance in accelerated tests. It is important to understanding and using the lognormal distribution (Section 4), which is widely used to interpret accelerated test data. Also, the sampling distribution of many estimators is approximately normal. This fact yields approximate confidence intervals in later chapters. Thus knowledge of the normal distribution is essential. Books on it include Schneider (1986) and Johnson and Kotz (1970).

Normal cumulative distribution function. The population fraction failing by age y is

$$F(y) = \int_{-\infty}^{y} (2\pi\sigma^2)^{-1/2} \exp\left[-\frac{1}{2}\left(\frac{x-\mu}{\sigma}\right)^2\right] dx, \quad -\infty < y < \infty. \quad (3.1)$$

Figure 3.1 depicts this function. μ is the population mean and may have any value. σ is the population standard deviation and must be positive. μ and σ are in the same measurement units as y, for example, hours, months, cycles, etc. (3.1) can be expressed in terms of the standard normal cumulative distribution function $\Phi(\ )$ as

$$F(y) = \Phi[(y-\mu)/\sigma], \quad -\infty < y < \infty. \quad (3.2)$$

$\Phi(\ )$ is (3.1) evaluated at $\mu = 0$ and $\sigma = 1$; it is tabulated in Appendix A1. Many tables give $\Phi(z)$ only for $z \geq 0$; then use $\Phi(-z) = 1 - \Phi(z)$. $z = (y-\mu)/\sigma$ is called the (standardized) *normal deviate*.

The range of y is from $-\infty$ to $+\infty$. Life must, of course, be positive. Thus the distribution fraction below zero must be small for this distribution to be a satisfactory approximation in practice.

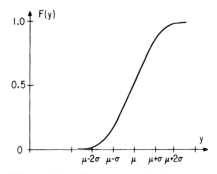

Figure 3.1.   Normal cumulative distribution.

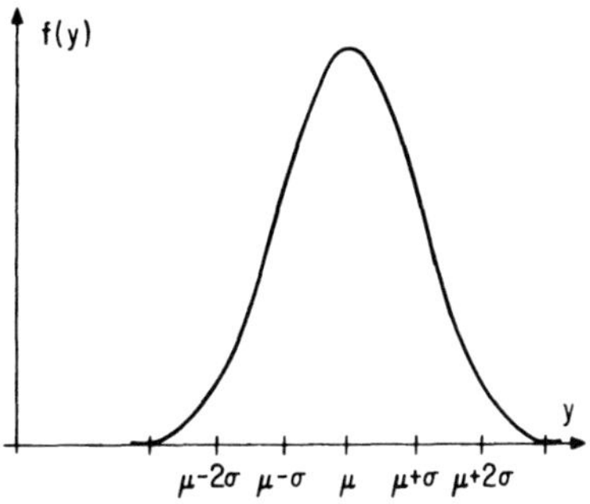

Figure 3.2.   Normal probability density.

Insulation example. Nelson (1981) approximates the life of a type of insulation specimen with a normal distribution with $\mu = 6{,}250$ and $\sigma = 600$ years. The distribution fraction with negative life is $F(0) = \Phi[(0-6{,}250)/600] = \Phi[-10.42] \approx 1.0 \times 10^{-25}$, which is negligible.

Normal probability density. The probability density is

$$f(y) = (2\pi\sigma^2)^{-1/2}\exp[-(y-\mu)^2/(2\sigma^2)], \quad -\infty < y < \infty. \tag{3.3}$$

Figure 3.2 depicts this probability density, which is symmetric about the mean μ. The figure shows that μ is the median and σ determines the spread.

Normal percentile. The 100Pth percentile is

$$\eta_P = \mu + z_P\sigma; \tag{3.4}$$

z_P is the 100Pth standard normal percentile and is tabled briefly below and in Appendix A2. The *median* (50th percentile) of the normal distribution is $\eta_{.50} = \mu$, since $z_{.50} = 0$. Some standard percentiles are:

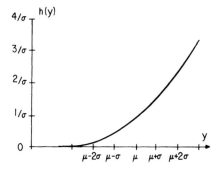

Figure 3.3.   Normal hazard function.

$100P\%$:	0.1	1	2.5	5	10	50	90	95	97.5	99
z_P:	-3.090	-2.326	-1.960	-1.645	-1.282	0	1.282	1.645	1.960	2.326

For the insulation specimens, median life is $\eta_{.50}$ = 6,250 years. The 1st percentile is $\eta_{.01}$ = 6,250 + (-2.326)600 = 4,830 years.

Normal mean and standard deviation. For the normal distribution,

$$E(Y) = \mu \quad \text{and} \quad \sigma(Y) = \sigma \tag{3.5}$$

are the distribution parameters. For the insulation specimens, $E(Y)$ = 6,250 years and $\sigma(Y)$ = 600 years.

Normal hazard function. The normal hazard function appears in Figure 3.3, which shows that the normal distribution has an *increasing failure rate with age* (wear-out behavior). Thus, the insulation above has an increasing failure rate. This suggests that older units with the insulation are more failure prone. In a preventive replacement program, older units should be replaced first to minimize the number of service failures.

4. LOGNORMAL DISTRIBUTION

The lognormal distribution is widely used for life data, including metal fatigue, solid state components (semiconductors, diodes, GaAs FETs, etc.), and electrical insulation. The lognormal and normal distributions are related; this fact is used to analyze lognormal data with methods for normal data. Books on the distribution include Crow and Shimizu (1988), Schneider (1986), Johnson and Kotz (1970), and Aitchinson and Brown (1957).

Lognormal cumulative distribution. The population fraction failing by age t is

$$F(t) = \Phi\{[\log(t) - \mu]/\sigma\}, \quad t > 0. \tag{4.1}$$

Figure 4.1 shows lognormal cumulative distribution functions. μ is the mean of the *log* of life – not of life. μ is called the *log mean* and may have any value from $-\infty$ to ∞. σ is the standard deviation of the *log* of life – not of life. σ is called the *log standard deviation* and must be positive. μ and σ are not "times" like t; instead they are unitless pure numbers. Here log( ) denotes the common (base 10) logarithm. Some authors use the natural (base e) logarithm, denoted by ln( ) in this book. $\Phi($) is the standard normal cumulative distribution function; it is tabulated in Appendix A1. (4.1) is like the normal cumulative distribution function (3.1); however, $\log(t)$ appears in place of t. The cumulative distribution can also be written as

$$F(t) = \Phi\{[\log(t/\tau_{.50})]/\sigma\} = \Phi\{\log[(t/\tau_{.50})^{1/\sigma}]\}; \tag{4.1'}$$

here $\tau_{.50}$ = antilog(μ) is the median. (4.1') is similar to the Weibull cumula-

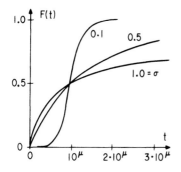

Figure 4.1.   Lognormal cumulative distribution.

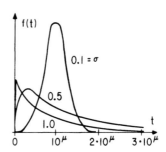

Figure 4.2.   Lognormal probability densities.

tive distribution (5.1). (4.1′) shows that $\tau_{.50}$ is a scale parameter, and σ is a shape parameter.

Lognormal probability density. For a lognormal distribution,

$$f(t) = \{0.4343/[(2\pi)^{1/2}t\sigma]\} \exp\{-[\log(t)-\mu]^2/(2\sigma^2)\}, \quad t > 0. \quad (4.2)$$

Here $0.4343 \approx 1/\ln(10)$. Figure 4.2 shows probability densities, which have a variety of shapes. The value of σ determines the shape of the distribution, and the value of μ determines the 50% point and the spread in life t.

Class-H insulation example. The life of specimens of a proposed Class-H insulation at design temperature was described with a lognormal distribution with $\mu = 4.062$ and $\sigma = 0.1053$. The population fraction failing during the 20,000-hour design life is $F(20,000) = \Phi\{[\log(20,000)-4.062]/0.1053\} = \Phi[2.270] = 0.988$. Thus most would fail, and the insulation was abandoned.

Percentile. The 100Pth lognormal percentile is

$$\tau_P = \text{antilog}[\mu+z_P\sigma] = 10^{\mu+z_P\sigma}; \quad (4.3)$$

here z_P is the 100Pth standard normal percentile. The *median* (50th percentile) is $\tau_{.50} = \text{antilog}[\mu]$.

For the Class-H insulation, the median is $\tau_{.50} = \text{antilog}(4.062) \approx 11,500$ hours. The 1% life is $\tau_{.01} = \text{antilog}[4.062+(-2.326)0.1053] \approx 6,600$ hours.

Lognormal reliability function. The population fraction surviving age t is

$$R(t) = 1 - \Phi\{[\log(t)-\mu]/\sigma\} = \Phi\{-[\log(t)-\mu]/\sigma\}. \quad (4.4)$$

Lognormal mean and standard deviation. The mean and standard deviation of time t are little used in accelerated testing. Formulas appear, for example, in Nelson (1982).

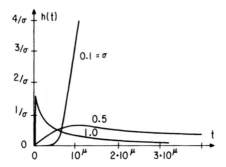

Figure 4.3.   Lognormal hazard functions.

Lognormal hazard functions. Hazard functions appear in Figure 4.3. For $\sigma \approx 0.5$, $h(t)$ is roughly constant. For $\sigma \leq 0.2$, $h(t)$ increases and is much like that of a normal distribution but eventually decreases off scale; also, then the lognormal cumulative distribution and probability density are close to normal ones. For $\sigma > 0.8$, $h(t)$ increases quickly and decreases slowly. This flexibility makes the lognormal distribution popular and suitable for many products. However, the lognormal hazard function has a property seldom seen in products. It is zero at time zero, increases to a maximum, and then decreases to zero with increasing age. Nonetheless, over most of its range and especially over the lower tail, the lognormal distribution fits life data of many products. Often one uses only the lower tail in applications.

For the Class-H insulation, $\sigma = 0.1053 \leq 0.2$. Thus, its failure rate increases with age (except in the far upper tail), according to the fitted lognormal distribution. The failure rate of most insulations, metals, and other materials strictly increases. Sometimes a single lognormal distribution is fitted to pooled data from different distributions (batches, test conditions, etc.), then the resulting estimate of σ tends to be above those of the individual distributions. Consequently, the actual failure rate when such differing data are pooled is less than that suggested by the separate σ estimates.

Relationship with the normal distribution. The following helps one understand the lognormal distribution in terms of the simpler normal distribution. Suppose life t has a lognormal distribution with parameters μ and σ. Then the (base 10) log of life $y = \log(t)$ has a normal distribution with mean μ and standard deviation σ. Thus the analysis methods for normal data can be used for the logarithms of lognormal data.

Base e lognormal. Much engineering work now employs the base e lognormal distribution. The base e lognormal cumulative distribution is

$$F(t) = \Phi\{[\ln(t) - \mu_e]/\sigma_e\}. \tag{4.5}$$

Its parameters are related to the parameters $\mu_{10} = \log(\tau_{.50})$ and σ_{10} of the corresponding base 10 lognormal distribution; namely,

$$\mu_e = \ln(\tau_{.50}) = \mu_{10} \times \ln(10), \quad \sigma_e = \sigma_{10} \times \ln(10). \tag{4.6}$$

5. WEIBULL DISTRIBUTION

The Weibull distribution is often used for product life, because it models either increasing or decreasing failure rates simply. It is also used as the distribution for product properties such as strength (electrical or mechanical), elongation, resistance, etc., in accelerated tests. It is used to describe the life of roller bearings, electronic components, ceramics, capacitors, and dielectrics in accelerated tests. According to extreme value theory, it may describe a "weakest link" product. Such a product consists of many parts from the same life distribution, and the product fails with the first part failure. For example, the life of a cable or capacitor is determined by the shortest lived portion of its dielectric. Statistical theory for such cable or extreme phenomena appears in Galambos (1978) and Gumbel (1958).

Weibull cumulative distribution. The population fraction failing by age t is

$$F(t) = 1 - \exp[-(t/\alpha)^\beta], \quad t > 0. \tag{5.1}$$

The *shape parameter* β and the *scale parameter* α are positive. α is also called the *characteristic life*. It is always the 63.2th percentile. α has the same units as t, for example, hours, months, cycles, etc. β is a unitless pure number; it is also called the β *parameter* and *"slope" parameter* when estimated from a Weibull plot (Chapter 3). For most products and materials, β is in the range 0.5 to 5. Figure 5.1 shows Weibull cumulative distribution functions.

Capacitor example. The life of a type of capacitor is represented with a Weibull distribution with $\alpha = 100,000$ hours and $\beta = 0.5$. The population probability of failure in the first year of service is $F(8,760) = 1 - \exp[-(8,760/100,000)^{0.5}] = 0.26$ or 26%.

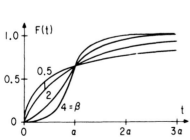

Figure 5.1.   Weibull cumulative distributions.

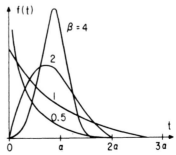

Figure 5.2.   Weibull probability densities.

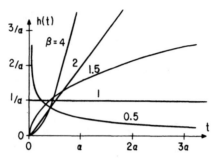

Figure 5.3.   Weibull hazard functions.

Weibull probability density. For a Weibull distribution,

$$f(t) = (\beta/\alpha^{\beta}) \, t^{\beta-1} \exp[-(t/\alpha)^{\beta}], \quad t > 0. \tag{5.2}$$

The Weibull probability densities in Figure 5.2 show that β determines the shape of the distribution, and α determines the spread. β determines the spread in *log* life; high (low) β corresponds to small (great) spread. For $\beta = 1$, the Weibull distribution is the exponential distribution. For much life data, the Weibull distribution fits better than the exponential, normal, and lognormal distributions.

Weibull reliability function. The population fraction surviving age t is

$$R(t) = \exp[-(t/\alpha)^{\beta}], \quad t > 0. \tag{5.3}$$

For the capacitors, the population reliability for one year is $R(8760) = \exp[-(8760/100,000)^{0.5}] = 0.74$ or 74%.

Weibull percentile. The 100Pth percentile of a Weibull distribution is

$$\tau_P = \alpha[-\ln(1-P)]^{1/\beta}; \tag{5.4}$$

ln( ) is the natural log. For example, $\tau_{.632} \approx \alpha$ for any Weibull distribution. This may be seen in Figure 5.1. For the capacitors, the 1st percentile is $\tau_{.01} = 100,000[-\ln(1-0.01)]^{1/0.5} = 10$ hours, a measure of early life.

Weibull mean and standard deviation. These are little used in accelerated testing. Formulas appear, for example, in Nelson (1982) and many reliability texts.

Weibull hazard function. For a Weibull distribution,

$$h(t) = (\beta/\alpha) \, (t/\alpha)^{\beta-1}, \quad t > 0. \tag{5.5}$$

Figure 5.3 shows Weibull hazard functions. A power function of time, $h(t)$ increases for $\beta > 1$ and decreases for $\beta < 1$. For $\beta = 1$ (the exponential distribution), the failure rate is constant. With a simple increasing or decreasing failure rate, the Weibull distribution flexibly describes product life. β tells the nature of the failure rate. This is often important information, particularly regarding whether to use preventive replacement.

For the capacitors, $\beta = 0.5 < 1$. Thus their failure rate decreases with age, called infant mortality behavior. This is consistent with $\tau_{.01} = 10$ hours although $\alpha = 100,000$ hours is large. A decreasing failure rate is typical of capacitors with solid dielectrics.

Relationship to the Exponential Distribution. The following relation is used later to analyze Weibull data in terms of the simpler exponential distribution. Suppose time T to failure has a Weibull distribution with parameters α and β. Then $T' = T^{\beta}$ has an exponential distribution with mean $\theta = \alpha^{\beta}$. In such analyses, one assumes β is known, and α and other quantities are estimated from the data.

Three-parameter Weibull. Nelson (1982) and Lawless (1982), among others, present the three-parameter Weibull distribution. It is seldom used for accelerated testing.

Weibull versus lognormal. In many applications, the Weibull and lognormal distributions (and others) may fit a set of data equally well, especially over the middle of the distribution. When both are fitted to a data set, the Weibull distribution has an earlier lower tail than the corresponding lognormal distribution. That is, a low Weibull percentile is below the corresponding lognormal one. Then the Weibull distribution is more pessimistic. Chapters 4 and 5 present methods for assessing which distribution fits a set of data better.

6. EXTREME VALUE DISTRIBUTION

The (smallest) extreme value distribution is needed background for analytic methods for Weibull data. Indeed the (base e) log of time to failure for a Weibull distribution has an extreme value distribution. The extreme value distribution also describes certain extreme phenomena; these include electrical strength of materials and certain types of life data. Like the Weibull distribution, the smallest extreme value distribution may be suitable for a "weakest link" product. In other words, suppose a product consists of many nominally identical parts from the same strength (life) distribution (unbounded below) and the product strength (life) is that of the weakest (first) part to fail. Then the smallest extreme value distribution may describe the strength (life) of units (Galambos (1978)). An example is the life or electrical strength of cable insulation. That is, a cable may be regarded as consisting of many segments, and the cable fails when the first segment fails.

Extreme value cumulative distribution. The population fraction below y is

$$F(y) = 1 - \exp\{-\exp[(y-\xi)/\delta]\}, \quad -\infty < y < \infty. \tag{6.1}$$

The *location parameter* ξ may have any value from $-\infty$ to ∞. ξ is the 63.2 percentile. The *scale parameter* δ is positive, and it determines the spread of the distribution. ξ and δ are in the same units as y, for example, hours, cycles, etc. Figure 6.1 depicts this function. The distribution range is $-\infty$ to $+\infty$. Lifetimes must, of course, be positive. Thus the fraction below zero must be small for this to be a satisfactory life distribution.

Material strength. Weibull (1951) describes the ultimate strength of a material with an extreme value distribution with $\xi = 108 \text{ kg/cm}^2$ and $\delta = 9.27 \text{ kg/cm}^2$. The proportion of such specimens with strength below 80 kg/cm^2 is $F(80) = 1 - \exp\{-\exp[(80-108)/9.27]\} = 0.048$ or 4.8%.

Extreme value density. For an extreme value distribution,

$$f(y) = (1/\delta)\exp[(y-\xi)/\delta]\cdot\exp\{-\exp[(y-\xi)/\delta]\}, \quad -\infty < y < \infty. \tag{6.2}$$

Figure 6.2 shows the probability density, which is asymmetric.

Extreme value reliability function. For an extreme value distribution,

$$R(y) = \exp\{-\exp[(y-\xi)/\delta]\}, \quad -\infty < y < \infty. \tag{6.3}$$

For the material, reliability for a stress of 80 kg/cm^2 is $R(80) = 1 - F(80) = 0.952$. That is, 95.2% of such specimens withstand 80 kg/cm^2.

Extreme value percentile. The $100P$th percentile is

$$\eta_P = \xi + u_P\delta; \tag{6.4}$$

here

$$u_P \equiv \ln[-\ln(1-P)] \tag{6.5}$$

is the $100P$th *standard extreme value percentile* ($\xi = 0$ and $\delta = 1$). For example, $\eta_{.632} \approx \xi$, the location parameter, since $u_{.632} \approx 0$. Standard percentiles are

$100P\%$:	0.1	1	5	10	50	63.2	90	99
u_P:	-6.907	-4.600	-2.970	-2.250	-0.367	0	0.834	1.527

For the material, $\eta_{.50} = 108 + (-0.367)9.27 = 104.6 \text{ kg/cm}^2$.

Extreme value mean and standard deviation. For an extreme value distribution,

$$E(Y) = \xi - 0.5772\delta, \quad \sigma(Y) = 1.283\delta; \tag{6.6}$$

here $0.5772 \cdots$ is Euler's constant and $1.283 \cdots = \pi/\sqrt{6}$. The mean is the 42.8% point of the distribution. For any extreme value distribution, mean <

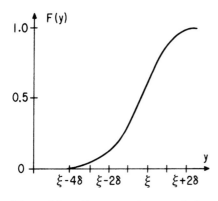

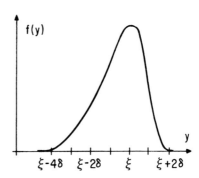

Figure 6.1. Extreme value cumulative distribution.

Figure 6.2. Extreme value density.

median $< \xi$. For the material, $E(Y) = 108 - 0.5772(9.27) = 102.6 \text{ kg/cm}^2$. Also $\sigma(Y) = 1.283(9.27) = 11.9 \text{ kg/cm}^2$.

Extreme value hazard function. For an extreme value distribution,

$$h(y) = (1/\delta)\exp[(y-\xi)/\delta], \quad -\infty < y < \infty. \tag{6.7}$$

Figure 6.3 shows that $h(y)$ increases exponentially with age (wear out).

Relationship to the Weibull distribution. The extreme value distribution is used to analyze Weibull data. The following relationships are used to analyze the base e logs of Weibull data. The log data are easier to analyze with the simpler extreme value distribution, because it has a single shape and simple location and scale parameters, similar to the normal distribution. Sup-

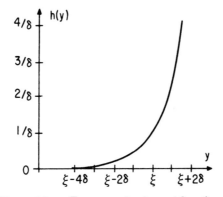

Figure 6.3. Extreme value hazard function.

pose a Weibull life distribution has shape and scale parameters β and α. The (base e) logarithm $y = \ln(t)$ of life t has an extreme value distribution with

$$\xi = \ln(\alpha), \quad \delta = 1/\beta. \tag{6.8}$$

The last equation shows that the spread in ln life is the reciprocal of β. Thus low β corresponds to high spread of log life, and high β to low spread. The Weibull parameters can be expressed as

$$\alpha = \exp(\xi), \quad \beta = 1/\delta. \tag{6.9}$$

Similarly, the Weibull parameters in terms of the standard deviation $\sigma(Y)$ and mean $E(Y)$ of the extreme value distribution are

$$\beta = 1.283/\sigma(Y), \quad \alpha = \exp[E(Y)+0.4501\,\sigma(Y)]. \tag{6.10}$$

These relationships are used in Chapter 4 to analyze Weibull data.

7. OTHER DISTRIBUTIONS

The basic distributions above (Sections 2 through 6) are commonly used for accelerated tests. The following other distributions and ideas may be useful. They include failure at time zero, eternal survivors, a mixture of distributions, the generalized gamma distribution, and nonparametric analysis. Chapter 7 is devoted to the important series-system model for the life distribution of a product with a mix of competing failure modes.

Distributions with failure at time zero. A fraction of a population may already be failed at time zero or fail soon after. For example, consumers may purchase a product that does not work when installed. The model for this consists of the proportion p failed at time zero and a continuous life distribution for the rest. Such a cumulative distribution appears in Figure 7.1. The sample proportion failed at time zero is used to estimate p, and the other sample failure times are used to estimate the continuous distribution.

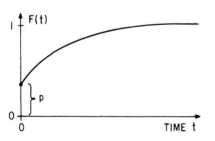

Figure 7.1.   A cumulative distribution with failures at time zero.

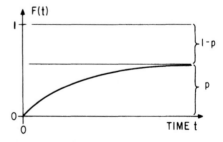

Figure 7.2.   A cumulative distribution with eternal survivors.

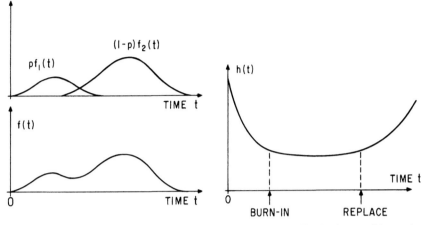

Figure 7.3.   A mixture of distributions.   **Figure 7.4.**   "Bathtub curve" hazard function.

Distributions with eternal survivors. Some units may never fail. This applies to (1) the time to death from a disease when some individuals are immune, (2) the time to redemption of trading stamps (some stamps are lost and never redeemed), (3) the time to product failure from a particular defect when some units lack that defect, and (4) time to warranty claim on a product whose warranty applies only to original owners, some of which sell the product before failure. Figure 7.2 depicts such a cumulative distribution. Meeker (1985,1987) presents an application to integrated circuits.

Mixtures of distributions. A population may consist of two or more subpopulations. Figure 7.3 depicts two subpopulations comprising proportions p and $1-p$ of the population. Units from different production periods may have different life distributions due to differences in design, raw materials, environment, usage, etc. It is often important to identify such a situation and the production period, customers, environment, etc., with poor units. Then suitable action may be taken on that portion of the population. If two subpopulations have cumulative distribution functions $F_1(t)$ and $F_2(t)$, then the population has the cumulative distribution function

$$F(t) = pF_1(t) + (1-p)F_2(t). \tag{7.1}$$

Such a *mixture* should be distinguished from competing failure modes, described in Chapter 7. Everitt and Hand (1981), McLachlan and Basford (1987), and Titterington, Smith, and Makor (1986) treat mixture distributions in detail. Hahn and Meeker (1982) offer practical advice on mixtures in analyzing product life data. Peck and Trapp (1978) divide certain semiconductors into two subpopulations; they call the early failures "freaks," which may be as much as 20 or 30% of a product starting development and typically 1 to 2% of a mature product. Vaupel and Yashin (1985) show how one may badly misinterpret life data when unaware that the population is a mixture.

The bathtub curve. Some products have a decreasing failure rate in the early life and an increasing failure rate in later life. Figure 7.4 shows such a hazard function, called a "bathtub curve." However, most products have a failure rate that just decreases throughout their observed life or else just increases. Thus, for many products, the bathtub curve does not hold water. It commonly appears in reliability books but describes only 10 to 15% of the author's applications, usually products with competing failure modes (Chapter 7). Hahn and Meeker (1982) carefully distinguish between models for a mixture of subpopulations and those for competing failure modes. Both situations can have a bathtub failure rate for the population.

Burn-in. Some products, such as high-reliability capacitors and semiconductor devices have a decreasing failure rate and are subjected to a burn-in. This weeds out early failures before units are put into service. Such burn-in is most effective if the population is a mixture of a small subpopulation of defectives (from manufacturing problems) that all fail early and a main population with satisfactory life. Peck and Trapp (1978) and Jensen and Petersen (1982) comprehensively treat planning and analysis of burn-in procedures, including the economics and accelerated burn-in. Such burn-in to weed out early failures is one of the purposes of environmental stress screening (ESS), also called shake and bake. Tustin (1986) surveys ESS. Also, some other products may be removed from service before wear-out starts. Thus units are in service only in the low failure rate portion of their life. This increases their reliability in service.

Generalized gamma distribution. Farewell and Prentice (1977), Kalbfleisch and Prentice (1980), Cohen and Whitten (1988), and Lawless (1982) present the generalized gamma distribution. It includes the lognormal and Weibull distributions as special cases. The distribution (of log life) has three parameters (location, scale, and shape). When the distribution is fitted to data, the estimate of the shape parameter is used to compare the Weibull and lognormal fits. Such a comparison is useful if experience does not suggest either distribution. Farewell and Prentice have a computer program that fits this distribution to censored data from an accelerated test where the location parameter is a linear function of (possibly transformed) stress. Bowman and Shenton (1987) survey the simpler gamma distribution.

Birnbaum-Saunders distribution. Birnbaum and Saunders (1969) proposed this distribution to describe metal fatigue. They mathematically derive their distribution from a model for crack propagation (Chap. 11, Sec. 2.4). Its cumulative distribution function is

$$F(t) = \Phi\{[(t/\beta)^{1/2} - (\beta/t)^{1/2}]/\alpha\}, \quad t > 0; \tag{7.2}$$

here $\Phi\{\ \}$ is the standard normal cumulative distribution function, $\beta > 0$ is the median, and $\alpha > 0$ determines the distribution shape.

It is an alternative to the lognormal distribution, which is widely used to describe metal fatigue life. It is comparable to the lognormal distribution in important respects. This cumulative distribution is close to a lognormal one for small α (usually found in practice), say, $\alpha < 0.3$. Then the corresponding (base e) lognormal parameters are approximately $\mu_e \approx \ln(\beta)$ and $\sigma_e \approx \alpha$. Its hazard function is zero at $t = 0$, increases to a maximum with age, and then finally decreases to a constant value; the lognormal hazard function is similar but finally decreases to zero. The distribution has a shorter lower tail than the corresponding lognormal one. That is, its 0.1% point is above the corresponding lognormal 0.1% point. This assumes that the distributions are matched by equating two percentiles above 0.1% (say, for 1% and 50%) or matched by some other reasonable means.

Nonparametric analysis. Nonparametric analysis of data does not involve an assumed (parametric) form of the distribution; that is, fitting is distribution free. Widely used for biomedical life data, nonparametric estimates are rarely used for engineering data. First, nonparametric estimates are not as accurate as parametric ones, provided the assumed parametric distribution is adequate. Second, nonparametric estimates of percentiles or fraction failed outside the range of the sample data do not exist; that is, one cannot extrapolate nonparametrically into the lower or upper tail of the distribution. Nonparametric fitting of distributions and regression models to censored life data appears in various biomedical books. These include (from basic to advanced) Lee (1980), Miller (1981), Kalbfleisch and Prentice (1980), Cox and Oakes (1984), Viertl (1988), and Lawless (1982). They all present the widely used Cox model, also called the proportional hazards model. All such regression models employ parametric relationships between life and stress or other variables; only the life distribution does not have an assumed parametric form. Little used for accelerated tests, such models are not presented in this book.

8. LIFE-STRESS RELATIONSHIPS

This section motivates the life-stress relationships and models in following sections. These relationships are for constant-stress tests. Indeed, many products run at nominally constant stress in actual use and at constant stress in accelerated tests. Some readers may wish to skip to Section 9.

Relationships

Typical life data from a constant-stress test are plotted as $\times$'s against stress in Figure 8.1A. The figure has linear scales for life and for stress. Life times at low stress tend to be longer than those at high stress. Also, the scatter in life is greater at low stress than at high stress. The smooth curve through the data represents "life" as a function of stress. Engineering theory for some curves does not specify exactly what "life" means; it is some "nomi-

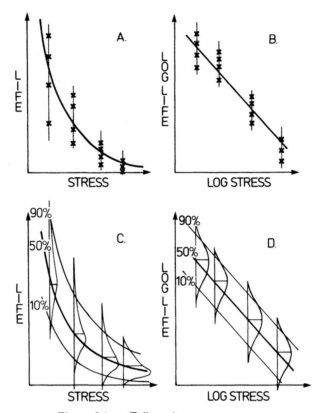

Figure 8.1.   Failure time versus stress.

nal" life, which is not made precise. In this book, such "nominal" life is a specific characteristic of the life distribution – usually the mean, median, or other distribution percentile. MIL-HDBK-217E (1986) presents a great variety of such relationships. They serve as derating curves for many types of electronic components.

Data can be conceptually simpler when plotted on paper with logarithmic or other suitable scales. On suitable paper the plotted points tend to follow a straight line, as in Figure 8.1B. Then a straight line through the data represents the **life-stress relationship** between product "life" and stress. A straight line is easier to fit to the data than a curve. Moreover, it is mathematically easy to extrapolate the straight line to a low stress to estimate the nominal life there, assuming the straight line is adequate. On the other hand, it is difficult to extrapolate a curve like that in Figure 8.1A. Use of a straight line on a special plotting paper is equivalent to using a particular equation to represent life versus stress. The Arrhenius relationship (Section 9) and the inverse power relationship (Section 10) are such equations.

The preceding simple view of accelerated test data has long been used by engineers. It is adequate for some work. However, most applications benefit from more refined models which follow.

Applications

Life-stress relationships for many products and materials have appeared in the engineering literature. The computerized literature searches by Meeker (1979) and Carey (1985) contain many applications. Relationships for various applications are surveyed in the following references:
- Electrical insulation: Goba (1969)
- Electronic components: Grange (1971), MIL-HDBK-217E (1986)
- Metal fatigue: ASTM STP 91-A (1963) and STP 744 (1979), Gertsbakh and Kordonskiy (1969).

Models with Distributions and Relationships

A simple relationship does not describe the scatter in the life of the test units. For each stress level, the units have some statistical distribution of life. A more refined model employs a statistical distribution to describe the scatter in life. Figure 8.1C depicts such statistical distributions. The curve for the probability density (histogram) of life at a stress would be perpendicular to the page, but it has been drawn flat on the figure. A heavy curve passes through the 50 percent point of the distribution at each stress. Lighter curves pass through the 10 and 90 percent points. Such a curve can be imagined for any percentile. Thus the *model* here consists of a combination of a life distribution and a life-stress relationship. The percentile curves depict the model.

Many such models are simpler on plotting paper (with logarithmic or other suitable scales) where the relationship between life and stress is a straight line. The model in Figure 8.1C is depicted in Figure 8.1D on paper on which the relationship is a straight line. The relationships for other percentiles of the life distributions plot as parallel straight lines for many models, as shown in the figure. For example, percentiles of the Arrhenius model (Section 9) and the inverse power law model (Section 10) plot as parallel straight lines on suitable paper. Such models incorporating a distribution are more realistic than a simple relationship. For a particular stress level, such a model gives the age at which, for example, 1 percent or any other percentage of the units fail. More general models need not have percentiles that plot as parallel straight lines on such paper.

Commonly used theoretical life distributions include the exponential, lognormal, and Weibull distributions, described in previous sections.

Distribution Plots

In Figure 8.1, life data are plotted against test stress. Another plot of such data is useful. Figure 8.2A shows a plot of the cumulative percentage of the sample that has failed as a function of time. The plotted points are the

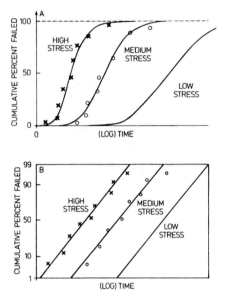

Figure 8.2.   Cumulative percentage failed versus time.

sample failure times, and the smooth curve depicts the population cumulative percentage failing as a function of time. The plot shows sample data for two stresses and the corresponding population distributions. Figure 8.2A also shows the population cumulative distribution for a low design stress.

Such a plot is simpler on probability paper. Such paper has a suitable data scale (usually logarithmic) for time and a probability scale for the cumulative percentage failed. One uses a paper on which the plotted points tend to follow a straight line, as in Figure 8.2B. Then a straight line represents the population cumulative percentage failing as a function of time at that stress. There are probability plotting papers for the exponential, normal, lognormal, Weibull, extreme value, and other distributions. A straight line on a probability paper is a cumulative distribution function for its distribution. The straight lines for the various stresses are parallel in Figure 8.2B. The Arrhenius model (Section 9) and the inverse power law model (Section 10) have distributions that are parallel lines on suitable probability paper. More general models need not have distributions that plot as parallel straight lines on such paper. Instead the distributions plot as curves that do not cross.

Overview

The preceding paragraphs provide background for the following sections. Section 9 presents the Arrhenius relationship for temperature-accelerated tests. Section 10 presents the inverse power relationship. These two basic relationships are widely used, and many other relationships are generalizations of them. Section 11 introduces endurance (or fatigue) limit relationships for products that do not fail when stressed below the endurance (fatigue) limit. Section 12 surveys a number of other relationships involving a single stress. Section 13 covers multivariable relationships for more than one accelerating stress or other variables, such as design, manufacturing, and operating variables. Section 14 gives relationships for the spread of a life distribution as a function of stress and other variables.

Such relationships usually apply to a single failure mode. Such relationships may not be suitable for a product that fails from a number of causes. Models for such products appear in Chapter 7. Also, special models for the effect of size of the test specimen appear in Chapter 7.

9. ARRHENIUS LIFE-TEMPERATURE RELATIONSHIP

Applications. The Arrhenius life relationship is widely used to model product life as a function of temperature. Applications include
- electrical insulations and dielectrics, as surveyed by Goba (1969)
- solid state and semiconductor devices, Peck and Trapp (1978)
- battery cells   • corrosion
- lubricants and greases   • electro migration   failure
- plastics   in microcircuits
- incandescent lamp filaments, IEC Publ. 64 (1974).
Based on the Arrhenius Law for simple chemical-reaction rates (Chapter 11), the relationship is used to describe many products that fail as a result of degradation due to chemical reactions or metal diffusion. The relationship is adequate over some range of temperature.

Overview. Section 9.1 presents the Arrhenius law for reaction rates and motivates the Arrhenius life relationship. Then, the distribution of life around the relationship is modeled with the lognormal (Section 9.2), Weibull (Section 9.3), and exponential (Section 9.4) distributions.

Analyses of various types of data with this relationship appear as follows:
- complete data (Chapter 4, Section 2),
- censored data (Chapter 5, Section 2),
- data with a mix of failure modes (Chapter 7).
The Arrhenius life relationship below describes the life of products and test specimens that run under constant temperature. Chapter 10 presents models for life where temperature is not constant. Chapter 11 presents a related model for degradation of performance of a product as a function of temperature and age.

9.1. The Relationship

This section motivates the Arrhenius life relationship. Readers may wish to skip to the Arrhenius life relationship in Section 9.2. Also, Chapter 11 provides other motivation.

Arrhenius law. According to the *Arrhenius rate law,* the rate of a simple (first-order) chemical reaction depends on temperature as follows

$$\text{rate} = A' \exp[-E/(kT)]; \tag{9.1.1}$$

E is the *activation energy* of the reaction, usually in electron-volts.

k is Boltzmann's constant, 8.6171×10^{-5} electron-volts per °C.

T is the absolute *Kelvin temperature;* it equals the Centigrade temperature plus 273.16 degrees; the absolute Rankine temperature equals the Fahrenheit temperature plus 459.7 Fahrenheit degrees.

A' is a constant that is characteristic of the product failure mechanism and test conditions.

The rate of metal diffusion is described by the same equation. Thus the following Arrhenius life relationship based on (9.1.1) may describe failures due to diffusion in solid state devices and certain other products made of metal, if geometry of the distinct metals is not an important factor. The effect of geometry has largely been ignored in the literature.

Motivation. The following relationship is based on a simple view of failure due to such a chemical reaction (or diffusion). The product is assumed to fail when some critical amount of the chemical has reacted (or diffused); a simple view of this is

$$(\text{critical amount}) = (\text{rate}) \times (\text{time to failure}).$$

Equivalently,

$$(\text{time to failure}) = (\text{critical amount}) / (\text{rate}).$$

While naive, this suggests that nominal time τ to failure ("life") is inversely proportional to the rate (9.1.1). This yields the **Arrhenius life relationship**

$$\tau = A \exp[E/(kT)]. \tag{9.1.2}$$

Here A is a constant that depends on product geometry, specimen size and fabrication, test method, and other factors. Products with more than one failure mode have different A and E values for each mode. Adequacy of (9.1.2) has been confirmed experimentally for certain products and failure modes. In fact, in certain applications (e.g., motor insulation), if the Arrhenius life relationship (9.1.2) does not adequately fit the data, the data are suspect rather than the relationship. So the usual handwaving between (9.1.1) and (9.1.2) is not worth faulting. This connection between (9.1.1) and (9.1.2) has value. It suggests a mechanism for failure when the relationship (9.1.2) holds, namely, degradation due to a single chemical reaction (or to metal diffusion).

Linearized relationship. The (base 10) logarithm of (9.1.2) is

$$\log(\tau) = \gamma_0 + (\gamma_1/T) \tag{9.1.3}$$

where

$$\gamma_1 = \log(e)(E/k) \approx 0.4343E/k. \tag{9.1.4}$$

Thus the log of "nominal life," $\log(\tau)$, is a linear function of inverse absolute temperature $x = 1/T$. For example, Sillars (1973, p. 29) presents (9.1.3) for insulation life. "Life" τ is usually taken to be a specified percentile or the mean of the (log) life distribution. Common choices are the 50th, 63.2th, and 10th percentiles. (9.1.4) can be expressed as

$$E = 2.303 \, k \, \gamma_1. \tag{9.1.5}$$

For most diodes, transistors, and other solid state devices, E is in the range 0.3 to 1.5 electron-volts. Moreover, E varies from one failure mode to another, even for the same device.

Class-H insulation example. An Arrhenius relationship used to describe life (in hours) of a new Class-H motor insulation has $\gamma_0 = -3.16319$ and $\gamma_1 = 3,273.67$. In practice, such parameters are estimated from data, and they often have great statistical uncertainty. They and other numbers are treated in this chapter as if exact. The log life at the design temperature of 180°C (453.16°K) is $-3.16319 + (3,273.67/453.16) = 4.0611$. The antilog is about 11,500 hours. The activation energy is $E = 2.303 \times 8.6171 \times 10^{-5} \times 3,273.67 \approx 0.65$ eV. Dakin (1948) proposed the Arrhenius relationship to describe temperature-accelerated life tests of electrical insulations and dielectrics. The relationship is now widely used for such tests and other products.

Arrhenius acceleration factor. By (9.1.2), the Arrhenius acceleration factor between life τ at temperature T and life τ' at reference temperature T' is

$$K = \tau/\tau' = \exp\{(E/k)[(1/T)-(1/T')]\}. \tag{9.1.6}$$

For the Class-H insulation, the acceleration factor between $T = 453.16°$K (180°C) and $T' = 533.16°$K (260°C) is

$$K = \exp\{(0.65/8.6171 \times 10^{-5})[(1/453.16)-(1/533.16)]\} = 12.$$

Thus specimens run 12 times longer at 180°C than at 260°C.

Larsen-Miller relationship. The effect of temperature on time τ to creep (to a specified % elongation) or to rupture of metals under load is discussed, for example, by Dieter (1961). The Larsen-Miller relationship for an absolute temperature T is (9.1.3) written as

$$T\,[-\gamma_0 + \log(\tau)] = \gamma_1. \tag{9.1.7}$$

γ_1 is called the *Larsen-Miller parameter*; it depends only on load (stress in psi) and not on τ or T. Dieter (1961) presents other such relationships which are fitted to creep-rupture data at high temperatures to estimate life at lower design temperatures. Usually τ is taken to be the median life, and the scatter in life is often ignored in metallurgical studies. Of course, in high reliability applications where failure is to be avoided, the lower tail of the life distribution must be modeled with a life distribution.

Arrhenius paper. Figure 9.1 shows Arrhenius paper which has a log scale for life and a nonlinear (Centigrade) temperature scale, which is linear in inverse absolute temperature. The linear scale for inverse absolute temperature was added to the figure to show its relation to the temperature scale. On such paper, the Arrhenius (life-temperature) relationship (9.1.2) plots as a straight line. The value of A (or equivalently γ_0) determines the intercept of the line (at $T = \infty$ or $1/T = 0$). The value of E (or equivalently γ_1) deter-

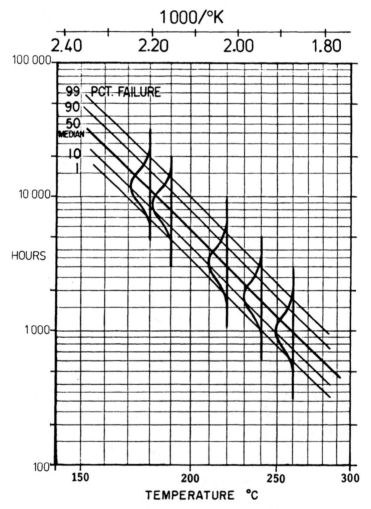

Figure 9.1.   Arrhenius relationship (and lognormal percentile lines) on Arrhenius paper.

mines the slope. The previously calculated 11,500-hour "life" of Class-H insulation at 180°C appears as the median (50%) in Figure 9.1.

9.2. Arrhenius-Lognormal Model

The life of many products and materials in a temperature-accelerated test is described with a lognormal distribution. IEEE Standard 101 (1988) uses the lognormal distribution for motor insulation. Peck and Trapp (1978) use it for semiconductors and solid state devices. Described below is the Arrhenius-lognormal model. It combines a lognormal life distribution with an Arrhenius dependence of life on temperature.

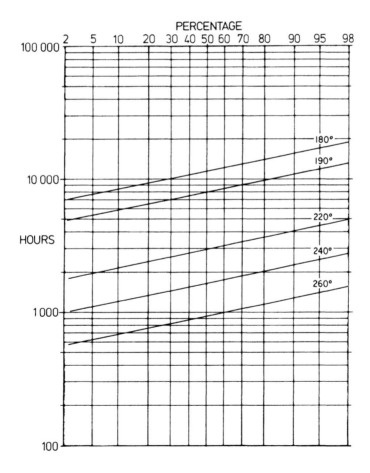

Figure 9.2.   Cumulative distributions on lognormal probability paper – Arrhenius-lognormal model.

Assumptions. The assumptions of the **Arrhenius-lognormal model** are:

1. At absolute temperature T, product life has a lognormal distribution. Equivalently, the log (base 10) of life has a normal distribution.
2. The standard deviation, σ, of *log* life is a constant, i.e., independent of temperature. Section 14 extends the model to nonconstant σ.
3. The log of median life $\tau_{.50}$ is a linear function of the inverse of the absolute temperature T; that is,

$$\log[\tau_{.50}(T)] = \gamma_0 + (\gamma_1'/T), \qquad (9.2.1)$$

which is called the *Arrhenius life relationship*. Parameters γ_0, γ_1', and σ are characteristic of the product and test method; they are estimated from data. An example of (9.2.1) is depicted on Arrhenius paper in Figure 9.1. Equivalently, the mean $\mu(x)$ of log life is a linear function of $x = 1000/T$:

$$\mu(x) = \gamma_0 + \gamma_1 x. \qquad (9.2.2)$$

Here and elsewhere, 1000 is used to scale inverse temperatures, and $\gamma_1 = \gamma_1'/1000$. Resulting numbers are more convenient. These assumptions yield the following cumulative distribution of life and its percentiles.

Base e. If the base e log is used, then (9.2.1) and (9.2.2) become

$$\ln[\tau_{.50}(T)] = \gamma_0^e + (\gamma_1'^e/T), \qquad (9.2.1')$$

$$\mu'(x) = \gamma_0^e + \gamma_1^e x; \qquad (9.2.2')$$

here $\gamma_0^e = 2.30\,\gamma_0$, $\gamma_1^e = 2.30\,\gamma_1$, and $\sigma^e = 2.30\,\sigma$ where $\ln(10) \approx 2.30$. Then ln replaces log in formulas below.

Fraction failed. At absolute temperature T, the cumulative distribution function (population fraction failed) at age t is

$$F(t,T) = \Phi\{[\log(t) - \mu(x)]/\sigma\}; \qquad (9.2.3)$$

$\Phi\{\ \}$ is the standard normal cumulative distribution function (Appendix A1). This fraction failed plots as a straight line versus t on lognormal probability paper in Figure 9.2. The value of σ determines the slope of such lines. A high (low) σ value corresponds to a high (low) slope and to a wide (narrow) distribution of log life. In Figure 9.2, the distribution lines are parallel. This reflects assumption 2, which is needed for the following reason. Different σ values at different temperatures result in distribution lines with different slopes. Such lines cross, resulting in a lower fraction failed for higher temperature beyond the age where the lines cross. Such crossing is physically implausible. Thus a common (constant) value for σ is assumed.

The model for the Class-H insulation provides an example of evaluating (9.2.3). The assumed value of σ is 0.10533. At 180°C, the fraction failed by 10,000 hours is calculated as follows. The absolute temperature is $T = 180 + 273.16 = 453.16°K$, and $x = 1000/453.16 = 2.2067$. The log mean is

$\mu(2.2067) = -3.16319 + (3.27367) 2.2067 = 4.0611$. The fraction failed is

$F(10,000,453.16°K) = \Phi\{[\log(10,000) - 4.0611]/0.10533\} = \Phi(-0.580) = 0.281$.

This 28% is a point on the 180° line at 10,000 hours in Figure 9.2. To plot the entire distribution, calculate the fraction failed at another age, plot that point on the lognormal paper, and draw a straight line through the two points.

Percentiles. At temperature T, the $100P$th percentile (P fractile) of life is

$$\tau_P(T) = \text{antilog}[\mu(x) + z_P\, \sigma] = \text{antilog}[\gamma_0 + \gamma_1(1000/T) + z_P\sigma]; \quad (9.2.4)$$

here z_P is the standard normal percentile (Appendix A2). For a fixed P, $\tau_P(T)$ plotted against Centigrade temperature on Arrhenius paper is a straight line. Figure 9.1 shows such lines for several percentiles; the value z_P determines the vertical position of the corresponding line. The corresponding percentile of *log* life is

$$\eta_P(x) = \log[\tau_P(x)] = \mu(x) + z_P\sigma. \quad (9.2.5)$$

Thus, one can think in terms of the percentile of life or of log life. The median (50th percentile) is a special case; namely,

$$\tau_{.50}(T) = \text{antilog}[\mu(x)] = \text{antilog}[\gamma_0 + \gamma_1(1000/T)],$$

$$\eta_{.50}(x) = \mu(x) = \gamma_0 + \gamma_1 x. \quad (9.2.6)$$

For the Class-H insulation at 180°C, the 10th percentile of log life is

$$\eta_{.10}(2.2067) = 4.0611 + (-1.282)0.10533 = 3.9261.$$

The 10th percentile of life is

$$\tau_{.10}(453.16°K) = \text{antilog}(3.9261) \equiv 10^{3.9261} = 8,435 \text{ hours}.$$

This is a point on the 10% line in Figure 9.1. Also, it is a corresponding point on the 180° line in Figure 9.2.

Design temperature. In some applications, one must choose a design temperature that yields a desired "life." Such a desired life is usually a specified value $\tau_P{}^*$ of a percentile. For the Arrhenius-lognormal model, the desired absolute temperature is

$$T^* = 1000\gamma_1/[\log(\tau_P{}^*) - \gamma_0 - z_P\sigma]. \quad (9.2.7)$$

For example, the desired median life of the Class-H insulation is $\tau_{.50}{}^* = 20,000$ hours. This life is achieved at a design temperature of

$$T^* = 1000(3.27367)/[\log(20,000) - (-3.16319) - 0(0.10533)] = 438.58°K.$$

This is $(438.58 - 273.16) \approx 165°C$. This can also be obtained from Figure 9.1. Enter the figure on the time scale at 20,000 hours; go horizontally to the 50% line; then go down to the temperature scale to read 165°C.

9.3. Arrhenius-Weibull Model

The life of some products and materials in a temperature-accelerated test is described with a Weibull distribution. For example, the author has used it for capacitor dielectric and for insulating tape. Described below is the Arrhenius-Weibull model. It combines a Weibull life distribution with an Arrhenius dependence of life on temperature.

Assumptions. The assumptions of the **Arrhenius-Weibull model** are:

1. At absolute temperature T, product life has a Weibull distribution; equivalently, the natural log of life has an extreme value distribution.
2. The Weibull shape parameter β is a constant (independent of temperature); equivalently, the extreme value distribution of the natural log of life has a constant scale parameter $\delta = 1/\beta$. Section 14 extends the model to nonconstant β.
3. The natural log of the Weibull characteristic life α is a linear function of the inverse of T:

$$\ln[\alpha(T)] = \gamma_0 + (\gamma_1'/T). \tag{9.3.1}$$

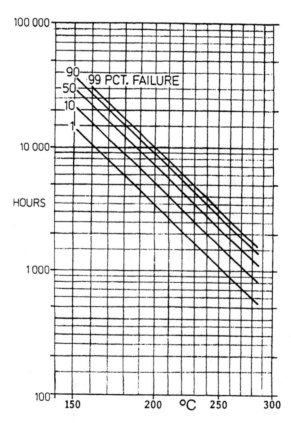

Figure 9.3.   Arrhenius relationship and Weibull percentile lines.

The parameters γ_0, γ_1', and β are characteristic of the product and test method; they are estimated from data. $\alpha(T)$ plots as a straight line on Arrhenius paper (Figure 9.3). Equivalently,

3′. The extreme value location parameter of the distribution of natural log life is a linear function of $x = 1000/T$; that is,

$$\xi(x) = \ln[\alpha(T)] = \gamma_0 + \gamma_1 x. \qquad (9.3.2)$$

These assumptions yield the cumulative distribution of life and its percentiles.

Fraction failed. At absolute temperature T, the cumulative distribution function (population fraction failed) at age t is

$$F(t;T) = 1 - \exp\{-[t/\alpha(T)]^\beta\} = 1 - \exp\{-[t\exp[-\gamma_0 - (\gamma_1'/T)]]^\beta\}. \qquad (9.3.3)$$

For a specific temperature T, this fraction failed plots as a straight line versus t on Weibull probability paper. Such Weibull distribution lines appear in Figure 10.2 for another relationship. The value of β determines the slope of such lines on Weibull paper. Thus β is also unfortunately called the *slope parameter*, not to be confused with the relationship slope γ_1. A high β value corresponds to a narrow distribution of ln life; a low β value corresponds to a wide distribution of ln life. In Figure 10.2, the distribution lines are parallel, a result of assumption 2.

Percentile. At temperature T, the $100P$th percentile (P fractile) is

$$\tau_P(T) = \alpha(T) \left[-\ln(1-P)\right]^{1/\beta} = \exp[\gamma_0 + \gamma_1(1000/T)][-\ln(1-P)]^{1/\beta}. \qquad (9.3.4)$$

For a fixed P, $\tau_P(T)$ plotted against T on Arrhenius paper is a straight line, as in Figure 9.3. However, the spacing of these parallel Weibull percentile lines differs from that of the lognormal percentile lines in Figure 9.1. The corresponding percentile of log life is

$$\eta_P(x) = \xi(x) + u_P\delta; \qquad (9.3.5)$$

here $x = 1000/T$, and $u_P = \ln[-\ln(1-P)]$ is the standard extreme value percentile. Then

$$\tau_P(T) = \exp[\eta_P(x)]. \qquad (9.3.6)$$

The 63.2 percentile is the special case $\tau_{.632}(T) = \alpha(T)$, $\eta_{.632}(x) = \gamma_0 + \gamma_1 x$.

Design temperature. Suppose a desired life is specified as a percentile value $\tau_P{}^*$. For the Arrhenius-Weibull model, the absolute temperature that yields this life is

$$T^* = 1000\gamma_1/\ln\{\tau_P{}^*/[-\ln(1-P)]^{1/\beta}\}. \qquad (9.3.7)$$

9.4. Arrhenius-Exponential Model

The life of semiconductor and solid state devices and other electronic components is often (incorrectly) represented with an exponential distribution. For example, MIL-HDBK-217E (1986) does so. The exponential distribution is often a reasonable approximation for the distribution of times between failure for a complex electronic system. However, it is often a poor or misleading approximation to the life distribution of a (nonrepaired) component or material. Then engineers incorrectly use the exponential distribution for various reasons. Some do not know better. Knowing better, some follow common practice and use the exponential distribution, because it is easy to use and handbooks quote only constant failure rates. Moreover, a crude reliability estimate is better than no estimate. Some would use the Weibull distribution, which is better, but lack appropriate test or field data or handbook information; so they are unable to estimate Weibull parameters. Of course, the Arrhenius-exponential model is a special case of the Arrhenius-Weibull model with $\beta = 1$. Some other suitable assumed β value will usually yield better results.

Assumptions. The assumptions of the **Arrhenius-exponential model** are:

1. At any absolute temperature T, life has an exponential distribution.
2. The natural log of the mean life θ is a linear function of the inverse of T:

$$\ln[\theta(T)] = \gamma_0 + (\gamma_1'/T). \qquad (9.4.1)$$

Model parameters γ_0 and γ_1' are characteristic of the product and test method; they are estimated from data. $\theta(T)$ plots as a straight line on Arrhenius paper. Equivalently,

$2'$. The natural log of the (constant) failure rate $\lambda = 1/\theta$ is

$$\ln[\lambda(T)] = -\gamma_0 - (\gamma_1'/T). \qquad (9.4.2)$$

Also, $\lambda(T)$ plots as straight line on Arrhenius paper. This model is presented, for example, by Evans (1969). These assumptions yield the cumulative distribution function of life and percentiles below.

Temperature derating curve. MIL-HDBK-217E (1986) presents failure rate information on electronics. For example, for a Metal Oxide Semiconductor (MOS), the failure rate (failures per million hours) is (9.4.2) expressed as

$$\lambda(T) = 1.08 \times 10^8 \, e^{-6373/T}.$$

For a design application at 55°C ($T = 273.16 + 55 = 328.16$), $\lambda(328.16) = 1.08 \times 10^8 \, e^{-6373/328.16} = 0.39$ failures per million hours. This failure-rate relationship plots as a straight line on Arrhenius paper.

Fraction failed. At absolute temperature T, the cumulative distribution function (population fraction failed by age t) is

$$F(t; T) = 1 - \exp[-t/\theta(T)] = 1 - \exp[-t\lambda(T)]$$
$$= 1 - \exp\{-t \exp[-\gamma_0 - (\gamma_1'/T)]\}. \qquad (9.4.3)$$

For a temperature T, this fraction failed plots as a straight line versus t on

Weibull probability paper. Such distribution lines are parallel, as in Figure 9.2, but have different spacings for the exponential distribution.

Percentiles. At absolute temperature T, the $100P$th percentile (P fractile) is

$$\tau_P(T) = \theta(T)\,[-\ln(1-P)] = \exp[\gamma_0 + \gamma_1(1000/T)]\,[-\ln(1-P)]. \qquad (9.4.4)$$

For fixed P, $\tau_P(T)$ plotted against T on Arrhenius paper is a straight line. Such percentile lines for different P are parallel. Their spacing is determined by (9.4.4). The 63.2th percentile is, of course, the mean $\theta(T)$.

10. INVERSE POWER RELATIONSHIP

Applications. The inverse power relationship is widely used to model product life as a function of an accelerating stress. Applications include:
- Electrical insulations and dielectrics in voltage-endurance tests. Examples include Cramp (1959), Kaufman and Meador (1968), Zelen (1959), Simoni (1974), and IEEE Standard 930 (1987).
- Ball and roller bearings, for example, Lieblein and Zelen (1956), Harris (1984), and the SKF catalog (1981).
- Incandescent lamps (light bulb filaments), IEC Publ. 64 (1974).
- Flash lamps, EG&G Electro-Optics (1984).
- Simple metal fatigue due to mechanical loading, for example, Prot (1948) and Weibull (1961), and due to thermal cycling, for example, Coffin (1954,1974) and Manson (1953,1966).

The relationship is sometimes called the **inverse power law** or simply the **power law**. The term "law" suggests it is universally valid, which it is not. However, while usually not based on theory, the relationship is empirically adequate for many products.

Overview. This section first presents the (inverse) power relationship. Then the distribution of product life around the relationship is modeled with the lognormal, Weibull, and exponential distributions.

Analysis of various types of data with this relationship appear as follows:
- complete data (Chapter 4, Section 4)
- censored data (Chapter 5, Section 2)
- data with a mix of failure modes (Chapter 7)
- step-stress test data (Chapter 10)

The power relationship below describes the life of products and test specimens that run under constant stress. Chapter 10 presents models for life where stress varies.

10.1. The Relationship

Definition. Suppose that the accelerating stress variable V is positive. The **inverse power relationship** (or law) between "nominal" life τ of a product and V is

$$\tau(V) = A/V^{\gamma_1} ; \qquad (10.1.1)$$

here A and γ_1 are parameters characteristic of the product, specimen geometry and fabrication, the test method, etc. Equivalent forms are

$$\tau(V) = (A'/V)^{\gamma_1} \text{ and } \tau(V) = A''(V_0/V)^{\gamma_1} ;$$

here V_0 is a specified (standard) level of stress. The parameter γ_1 is called the *power* or *exponent*.

Transformer oil example. The inverse power relationship was used to describe the life (in minutes) of a transformer oil tested under certain conditions. Assumed parameter values are $A = 1.2284 \times 10^{26}$ and $\gamma_1 = 16.3909$. The voltage stress V is in kilovolts (kV). Thus $\tau(V) = 1.2284 \times 10^{26}/V^{16.3909}$. At $V = 15$ kV, life is $\tau(15) = 1.2284 \times 10^{26}/15^{16.3909} = 6.45 \times 10^6$ minutes.

Coffin-Manson relationship. The inverse power law is used to model fatigue failure of metals subjected to thermal cycling. The "typical" number N of cycles to failure as a function of the temperature *range* ΔT of the thermal cycle is

$$N = A/(\Delta T)^B . \qquad (10.1.2)$$

Here A and B are constants characteristic of the metal and test method and cycle. Then (10.1.2) is called the *Coffin-Manson relationship*. Coffin (1954,1974) and Manson (1953,1966) present it and applications. The relationship has been used for mechanical and electronic components. In electronics it is used for solder and other connections. For metals, the fatigue life is often modeled with the lognormal distribution, which is combined with the inverse power relationship in Section 10.2. For metals, the B is near 2. For plastic encapsulants for microelectronics, B is near 5. Problem 3.15 applies this relationship. Nachlas (1986) proposes a general life relationship for thermal cycling. Nishimura and others (1987) show that life of plastic packaging for electronics also depends on the minimum cycle temperature.

Palmgren's equation. Life tests of roller and ball bearings employ high mechanical load. In practice, life (in millions of revolutions) as a function of load is represented with *Palmgren's equation for* the 10th percentile B_{10} of the life distribution, namely,

$$B_{10} = (C/P)^p ; \qquad (10.1.3)$$

C is a constant called the *bearing capacity* and p is the power. B_{10} is called the "B ten" bearing life. P is the (equivalent radial) load in pounds. Bearing life is usually modeled with the Weibull distribution, which is combined with the inverse power relationship in Section 10.3. For rolling steel bearings, the Weibull shape parameter β is typically in the range 1.1 to 1.3 in actual use and in the range 1.3 to 1.5 in laboratory tests. For steel ball bearings, the power $p = 3$ is used, and, for steel roller bearings, $p = 10/3$ is used. This model is presented by Harris (1984) and the catalog of SKF (1981).

Taylor's model. Boothroyd (1975) gives *Taylor's model* for the median life τ of cutting tools, namely,

$$\tau = A/V^m \; ;$$

here V is the cutting velocity (feet/sec), and A and m are constants depending on the tool material, geometry, etc. For high strength steels $m \approx 8$, for carbides $m \approx 4$, and for ceramics $m \approx 2$.

Linearized relationship. The natural logarithm of (10.1.1) is

$$\ln(\tau) = \gamma_0 + \gamma_1 [-\ln(V)]. \tag{10.1.4}$$

Thus the log of "typical life", $\ln(\tau)$, is a linear function of the transformed stress $x = -\ln(V)$. "Life" τ is usually taken to be a specified percentile of the life distribution. Common choices are the 50th, 63.2th, and 10th percentiles.

Log-log paper. (10.1.4) shows that the inverse power relationship (10.1.1) is a straight line on log-log paper. Figure 10.1 shows such a straight line on such paper. The previously calculated nominal life of transformer oil of 6.45 $\times$ 10^6 minutes at 15 kV appears off scale on the line labeled 63.2%. Special log-log papers are needed. Namely, the length of the log cycle for time is much shorter than that for voltage as in Figure 10.1. Ordinary log-log paper

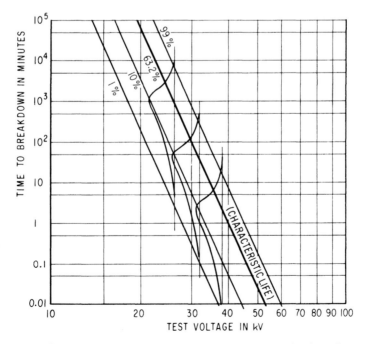

Figure 10.1.   Inverse power relationship (Weibull percentile lines) on log-log paper.

has the same cycle length on each axis, and it is not suitable for most accelerated testing work. Some organizations have developed their own suitable log-log paper, similar to that in Figure 10.1. Current graphics packages on computers readily custom make such paper with plotters. Figure 10.1 shows time on the vertical axis. Some papers have time on the horizontal axis and stress on the vertical axis; this is common in applications with metal fatigue and insulation endurance.

Power acceleration factor. By (10.1.1), the power acceleration factor between life τ at stress V and life τ' at reference stress V' is

$$K = \tau/\tau' = (V'/V)^{\gamma_1} . \tag{10.1.5}$$

For the transformer oil, the acceleration factor between $V = 15$ kV and $V' = 38$ kV is $K = (38/15)^{16.3909} = 4.1 \times 10^6$. Thus such oil runs 4.1 million times longer at 15 kV than at 38 kV, *if* the relationship is valid.

10.2. Power-Lognormal Model

The life of certain products is described with a lognormal life distribution whose median is an inverse power function of stress. This is the simplest model used for metal fatigue; it appears, for example, in ASTM STP 744 (1979) and ASTM STP 91-A (1963). The author has used this model for voltage-endurance of insulating tape. A description of the model follows.

Assumptions. The assumptions of the **power-lognormal model** are:

1. At any stress level V, product life has a lognormal distribution.
2. The standard deviation, σ, of log life is a constant (independent of V). Section 14 extends the model to nonconstant σ.
3. The median life, $\tau_{.50}$, is an inverse power function of stress; that is,

$$\tau_{.50}(V) = 10^{\gamma_0}/V^{\gamma_1} . \tag{10.2.1}$$

Parameters γ_0, γ_1, and σ are characteristic of the product and test method. An example of (10.2.1) is plotted in Figure 10.1. Equivalently, the mean $\mu(t)$ of (base 10) log life is a linear function of transformed stress $x = -\log(V)$:

$$\mu(x) = \gamma_0 + \gamma_1 x . \tag{10.2.2}$$

These assumptions yield the following equations for the cumulative distribution of life and its percentiles.

Fraction failed. At stress level V, the cumulative distribution function (population fraction failed by age t) is

$$F(t;V) = \Phi\{[\log(t) - \mu(x)]/\sigma\} ; \tag{10.2.3}$$

here $\Phi\{\ \}$ is the standard normal cumulative distribution function (Appendix A1). This fraction plots as a straight line versus t on lognormal probability paper. For example, see Figure 9.2.

Percentiles. At stress level V, the $100P$th percentile (P fractile) is

$$\tau_P(V) = \text{antilog}[\mu(x) + z_P \sigma] = (10^{\gamma_0}/V^{\gamma_1}) \times \text{antilog}(z_P\sigma); \qquad (10.2.4)$$

here z_P is the standard normal percentile (Appendix A2). For fixed P, $\tau_P(V)$ plots against V on log-log paper as a straight line. Figure 10.1 shows such lines, but their vertical spacings correspond to a Weibull distribution. The corresponding percentile of log life is

$$\eta_P(x) = \log[\tau_P(V)] = \mu(x) + z_P \sigma. \qquad (10.2.5)$$

The median life or median log life is a special case with $z_{.50} = 0$.

Design stress level. In some applications, one must choose a stress level that gives a desired "life." Such a desired life is usually a specified value $\tau_P{}^*$ of a percentile. For example, at what stress level will the 0.1th percentile of a fatigue life distribution of a metal be 12,000 cycles. For the power-lognormal model, the corresponding stress level V^* is

$$V^* = (1/\gamma_1) \, \text{antilog}[\gamma_0 + z_P\sigma - \log(\tau_P{}^*)]. \qquad (10.2.6)$$

In airplane frame and engine design, such a stress level is used, and the part is removed from service at $(\tau_P{}^*/3)$ cycles. This practice seeks to avoid what the industry terms a "part separation." The safety factor of 3 helps compensate for the uncertainties in estimating fatigue life due to differences between the geometry of specimens and that of actual parts and to differences between test and actual stress, environment, etc.

10.3. Power-Weibull Model

The life of certain products is described with a Weibull life distribution whose characteristic life is a power function of stress. Applications include:

- Electrical insulation and dielectrics, for example, IEEE Publication P930 (1987), Nelson (1970), and Simoni (1974). Voltage stress is the accelerating variable.
- Ball and roller bearings, for example, Lieblein and Zelen (1956), Harris (1984), and SKF (1981). Load is the accelerating variable.
- Metal fatigue, for example, Weibull (1961). Mechanical stress (pounds per square inch) is the accelerating variable.

The assumptions and properties of the model follow.

Assumptions. The assumptions of the **power-Weibull model** are:

1. At stress level V, product life has a Weibull distribution.
2. The Weibull shape parameter β is a constant (independent of V). Section 14 extends the model to nonconstant β.
3. The Weibull characteristic life α is an inverse power function of V:

$$\alpha(V) = e^{\gamma_0}/V^{\gamma_1}. \qquad (10.3.1)$$

The parameters γ_0, γ_1, and β are characteristic of the product and test method. $\alpha(V)$ plots as a straight line versus V on log-log paper as shown in Figure 10.1. Equivalently,

1′. The natural log of product life has an extreme value distribution.
2′. The extreme value scale parameter $\sigma = 1/\beta$ is a constant.
3′. The extreme value location parameter $\xi = \ln(\alpha)$ is a linear function of $x = -\ln(V)$; that is,

$$\xi(x) = \gamma_0 + \gamma_1 x. \tag{10.3.2}$$

These assumptions yield the cumulative life distribution and its percentiles.

Fraction failed. At stress level V, the cumulative distribution function (population fraction failed by age t) is

$$F(t;V) = 1 - \exp\{-[t/\alpha(V)]^\beta\} = 1 - \exp\{-[t\,e^{-\gamma_0}\,V^{\gamma_1}]^\beta\}. \tag{10.3.3}$$

For a specific stress level V, $F(t;V)$ plots as a straight line versus t on Weibull probability paper. Such distribution lines appear in Figure 10.2. A high (low) β corresponds to a narrow (wide) distribution of ln life.

Transformer oil. The voltage endurance (time to breakdown) of a certain transformer oil was described with the power-Weibull model. Assumed parameter values were $\gamma_0 = 60.1611$, $\gamma_1 = 16.3909$, and $\beta = 0.8084$. Time t is in minutes, and voltage stress V is in kilovolts (kV). Usually voltage stress is in volts/mil. The gap between the test electrodes was constant throughout such tests; the gap is not stated here for proprietary reasons. A factory test of a transformer runs 10 minutes at 20 kV. The probability of oil failure on test is $F(10;20) = 1 - \exp\{-[10e^{-60.1611}\,20^{16.3909}]^{0.8084}\} = 0.0008$ or 0.08%.

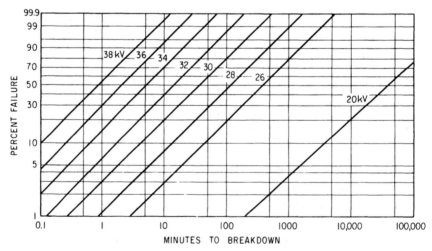

Figure 10.2. Cumulative distributions on Weibull probability paper – power-Weibull model.

Percentiles. At stress level V, the $100P$th percentile (P fractile) is

$$\tau_P(V) = \alpha(V)\,[-\ln(1-P)]^{1/\beta} = [e^{\gamma_0}/V^{\gamma_1}]\,[-\ln(1-P)]^{1/\beta}. \quad (10.3.4)$$

This equation shows that any $\tau_P(V)$ plotted against V on log-log paper is a straight line, as in Figure 10.1. The spacing of such parallel percentile lines depends on the Weibull distribution through the factor $[-\ln(1-P)]^{1/\beta}$.

For the transformer oil, the 0.1 percentile at 20 kV is $\tau_{.001} = [e^{60.1611}/20^{16.3909}]\,[-\ln(1-0.001)]^{1/0.8084} = 12.4$ minutes.

The corresponding percentile of ln life is

$$\eta_P(x) = \ln[\tau_P(V)] = \xi(x) + u_P\,\delta; \quad (10.3.5)$$

here $u_P = \ln[-\ln(1-P)]$ and $\delta = 1/\beta$. The 63.2 percentile is the special case

$$\tau_{.632}(V) = \alpha(V) = e^{\gamma_0}/V^{\gamma_1}, \quad \eta_{.632}(x) = \xi(x) = \gamma_0 + \gamma_1 x. \quad (10.3.6)$$

Design stress level. Suppose a desired life is specified as a percentile value $\tau_P{}^*$. For the power-Weibull model, the stress level that yields this life is

$$V^* = \{e^{\gamma_0}\,[-\ln(1-P)]^{1/\beta}/\tau_P{}^*\}^{1/\gamma_1}. \quad (10.3.7)$$

10.4. Power-Exponential Model

The life of semiconductor and solid state devices and other electronic components is often (incorrectly) represented with an exponential distribution. For example, MIL-HDBK-217E (1986) does so. The following simple power-exponential model is presented only because it is commonly used. Generally the power-Weibull model provides a much better representation of the life of electronic components. On the other hand, the power-exponential model adequately fits voltage-endurance data on transformer oil, as described by Cramp (1959) and Nelson (1970) and certain semiconductors after burn-in. The power-exponential model is a special case of the power-Weibull model with $\beta=1$.

Assumptions. The assumptions of the **power-exponential model** are:

1. At any stress level V, life has an exponential distribution.
2. The mean life θ is an inverse power function of V:

$$\theta(V) = e^{\gamma_0}/V^{\gamma_1}. \quad (10.4.1)$$

Model parameters γ_0 and γ_1 are characteristic of the product and test method. $\theta(V)$ plots as a straight line on log-log paper. Equivalently,

2′. The failure rate $\lambda = 1/\theta$ is a power function of V:

$$\lambda(V) = e^{-\gamma_0}V^{\gamma_1}. \quad (10.4.2)$$

Also, $\lambda(V)$ plots as a straight line on log-log paper.

It follows that the natural log of life has an extreme value distribution with scale parameter $\delta = 1$ and location parameter

$$\xi(x) = \ln[\theta(x)] = \gamma_0 + \gamma_1 x \qquad (10.4.3)$$

where $x = -\ln(V)$.

Transformer oil. The voltage endurance (time to breakdown) of the transformer oil described above was also described with the power-exponential model. Assumed parameter values are $\gamma_0 = 60.1611$ and $\gamma_1 = 16.3909$. The mean time to failure at 20 kV is $\theta(20) = e^{60.1611}/20^{16.3909} \approx 63,500$ minutes. The corresponding failure rate is $\lambda(20) = 1/63,500 \approx 16$ failures per million minutes.

Fraction failed. At stress level V, the cumulative distribution function (population fraction failed by age t) is

$$F(t;V) = 1-\exp[-t/\theta(V)] = 1-\exp[-t\lambda(V)] = 1-\exp[-te^{-\gamma_0}V^{\gamma_1}]. \quad (10.4.4)$$

For a specific V, this fraction failed plots as a straight line versus t on Weibull probability paper. Such distribution lines are parallel as in Figure 10.2.

The probability of failure of the transformer oil during a 10-minute factory test at 20 kV is $F(10;20) = 1 - \exp[-10/63,500] \approx 0.00016$ or 0.016%. The Weibull distribution with $\beta = 0.8084$ gives this as 0.08%.

Percentiles. At stress level V, the $100P$th percentile (P fractile) is

$$\tau_P(V) = \theta(V)[-\ln(1-P)] = [e^{\gamma_0}/V^{\gamma_1}]\cdot[-\ln(1-P)]. \qquad (10.4.5)$$

For fixed P, $\tau_P(V)$ plots against V on log-log paper as a straight line, as in Figure 10.1. Such percentile lines for different P are parallel. Their spacing is determined by the exponential distribution through (10.4.5).

The 0.1 percentile of the transformer oil at 20 kV is $\tau_{.001}(20) = 63,500 \cdot [-\ln(1-0.001)] = 63.5$ minutes.

Design stress level. Suppose that a desired life is specified as a mean time to failure θ^*. For the power-exponential model, the stress level that yields this life is

$$V^* = (e^{\gamma_0}/\theta^*)^{1/\gamma_1}. \qquad (10.4.6)$$

In terms of a specified failure rate λ^*, $V^* = (e^{\gamma_0}\lambda^*)^{1/\gamma_1}$.

11. ENDURANCE (OR FATIGUE) LIMIT RELATIONSHIPS AND DISTRIBUTIONS

Fatigue data on certain steels suggest that specimens tested below a certain stress run virtually indefinitely without failure. That stress is called the *fatigue limit*. Graham (1968) and Bolotin (1969) discuss this phenomenon, and ASTM STP 744 (1979, p. 92) gives a number of proposed life-stress relationships with a fatigue limit for steels. For many components, such low design stress is uneconomical, and such components are designed for finite life and removal before failure.

Similarly, voltage endurance data on certain dielectrics and insulations suggest that specimens tested below a certain voltage stress (electric field strength in volts per mil) run virtually indefinitely without failure. That voltage stress is called the *endurance limit*. While the existence of such a limit stress may be in doubt, it would allow designers to put critical components under a low enough stress to prevent failures during design life.

The simple relationship below may be useful even if there is no physical endurance limit. It has three coefficients and may be a better fit to non-linear data over the range of interest than the quadratic relationship (12.4), which also has three coefficients.

Power-Type Relationship

A commonly used relationship for "nominal" time τ to failure with an endurance (or fatigue) limit $V_0 > 0$ is

$$\tau = \begin{cases} \gamma_0/(V-V_0)^{\gamma_1}, & V>V_0, \\ \infty, & V\leq V_0. \end{cases} \tag{11.1}$$

Here V is the positive stress and γ_0 and γ_1 are product parameters. V_0, γ_0, and γ_1 are estimated from data. (11.1) reduces to the inverse power law if $V_0 = 0$. Product life is infinite for stresses below V_0. This relationship appears, for example, in Bolotin (1969) and ASTM STP 91-A and ASTM STP 744 (1979, p. 92) for metal fatigue; then V is mechanical stress in psi. For dielectric endurance, V is voltage stress in volts per mil.

Straight Line

Another simple relationship with endurance (or fatigue) limit $V_0 > 0$ is

$$\log(\tau) = \begin{cases} \gamma_0 + \gamma_1\log(V), & V>V_0, \\ \infty, & V\leq V_0. \end{cases} \tag{11.2}$$

Here V is the stress, and γ_0 and γ_1 are product parameters. (11.2) is a straight line on log-log paper for $V > V_0$. It is used to represent fatigue data on certain steels as described in ASTM STP 744 (1979, pp. 92 and 111) and by Dieter (1961, p. 304). (11.2) is the inverse power law if $V_0 = 0$.

Fatigue-Limit (or Strength) Distribution

The preceding simple relationships involve a sharp fatigue limit. Stressed below that limit, all specimens last forever. Stressed above that limit, all specimens have finite life. A more plausible model assumes that each speci-

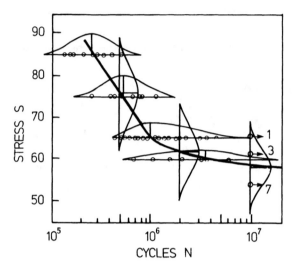

Figure 11.1.   Fatigue life distributions and strength distributions ($\bigcirc\!\!\rightarrow$ is a runout).

men has its own distinct fatigue limit. Thus there is a distribution of fatigue limits of such specimens. This distribution is also called the **strength distribution**; it is depicted as a vertical distribution in Figure 11.1. Of course, fatigue specimens cannot be tested an infinite time. Thus such fatigue tests of steel typically run for 10^7 cycles. For most applications, 10^7 cycles well exceeds design life. A designer then chooses a design stress such that (almost) all product units survive 10^7 cycles. The desired design stress is typically the 0.001 fractile (0.1 percentile) or 10^{-6} fractile (1 in 10^6 fails) of the fatigue limit distribution. In some applications, design life is short, and the design stress is well above the fatigue limit. Then one uses the strength distribution at the design life. The following paragraphs survey such fatigue limit (strength) distributions, the form of the test data, and data analyses.

Fatigue limit (or strength) distributions have been represented with the normal, lognormal, Weibull, extreme value, logistic, and other distributions; for example, see ASTM STP 744 (1979, p. 174), Little and Jebe (1975, Chaps. 10-12), and Serensen and others (1967, Chap. 3). The normal distribution provides a simple and concrete example. Suppose the entire population runs at stress level S. Then the population fraction that fails before N cycles is

$$F(S;N) = \Phi[(S - \mu_N)/\sigma_N]. \tag{11.3}$$

Here $\Phi[\ ]$ is the standard normal cumulative distribution function, and μ_N and σ_N are the mean and standard deviation of *strength* at N cycles. The vertical strength distributions in Figure 11.1 are normal distributions. For a design life of N^* cycles with a small percentage $100P$ failing, the design stress from (11.3) is

$$S^* = \mu_{N^*} + z_P \, \sigma_{N^*} \, ; \tag{11.4}$$

here z_P is the standard normal $100P$th percentile. In practice, one estimates μ_{N^*} and σ_{N^*} from data.

Fatigue data used to estimate a strength distribution generally have the form depicted in Figure 11.1. A number of specimens is run at each of a small number of test stresses. Testing is stopped at, say, 10^7 cycles. The runouts (nonfailures) are depicted with arrows in Figure 11.1. For a particular design life (say, 10^7 cycles), the data at each test stress consist of the number of specimens that failed before that life and the number that reached that life without failure. The exact failure ages are ignored. Such binary data (failed before or survived the design life) are called **quantal-response data.** Such data are sometimes obtained from **up-down** testing. Such testing involves a number of stress levels and running one specimen at a time. If a specimen fails before (after) the design life, the next specimen is run at the next lower (higher) stress level. Thus most specimens are run near the middle of the strength distribution.

Methods and tables for fitting a strength distribution to such quantal-response data are given, for example, by Little (1981) and Little and Jebe (1975). Also, many computer programs (Chapter 5, Section 1) fit distributions to such quantal-response data. Typically such methods and up-down testing, which were developed for biological applications, yield efficient estimates and confidence limits for the median of the strength distribution. In fatigue design, estimates of low percentiles of a strength distribution are usually more important. Efficient test plans for estimating low percentiles have not been used in practice. Meeker and Hahn (1977) present some optimum plans for a logistic strength distribution. Use of quantal-response data and the strength distribution at a design life ignores the relationship between fatigue life and stress. This simplification has a drawback; namely, run-outs below the design life cannot be included in a quantal-response analysis. A more complex model (including the fatigue curve and life distribution) is required to properly include such early run-outs in an analysis.

12. OTHER SINGLE STRESS RELATIONSHIPS

The Arrhenius and inverse power relationships are the most commonly used life-stress relationships. People, of course, use a great variety of other relationships. This section briefly surveys a number of such relationships with a single stress variable. A statistical distribution of life about such relationships is omitted here. One can combine any such distribution with a relationship as done in Sections 9 and 10.

This section first presents simple relationships with just two coefficients. Then it proceeds to relationships with three or more coefficients. It also presents special purpose relationships (for example, the elastic-plastic fatigue relationship). MIL-HDBK-217E (1986) presents a variety of derating curves and relationships for failure rates of electronic components. The relationships are typically a function of temperature, voltage, or current.

Exponential Relationship

The *exponential relationship* for "life" τ as a function of stress V is

$$\tau = \exp(\gamma_0 - \gamma_1 V). \tag{12.1}$$

This has been used, for example, for the life of dielectrics, according to Simoni (1974). Simoni notes that usually there are not enough test data to assess whether this or the inverse power relationship (10.1.1) fits the data better. MIL-HDBK-217E (1986) also uses it for various electronic components. The ln of (12.1) is

$$\ln(\tau) = \gamma_0 - \gamma_1 V. \tag{12.2}$$

This shows that (12.1) plots as a straight line on semi-log paper. Then life τ appears on the log scale, and stress V appears on the linear scale.

Exponential-Power Relationship

The *exponential-power relationship* for "nominal" life τ as a function of (possibly transformed) stress x is

$$\tau = \exp(\gamma_0 - \gamma_1 x^{\gamma_2}). \tag{12.3}$$

This is used, for example, in MIL-HDBK-217E (1986), where x is often voltage or inverse of absolute temperature. This relationship has three parameters γ_0, γ_1, and γ_2; thus it is not linear on any plotting paper.

Quadratic and Polynomial Relationships

The *quadratic relationship* for the log of nominal life τ as a function of (possibly transformed) stress x is

$$\log(\tau) = \gamma_0 + \gamma_1 x + \gamma_2 x^2. \tag{12.4}$$

This relationship is sometimes used when a linear relationship $(\gamma_0 + \gamma_1 x)$ does not adequately fit the data. For example, the linearized form of the Arrhenius relationship (9.2.2) or of the power law (10.1.4) may be inadequate. Then (12.4) is sometimes used, and its curve on the corresponding plotting paper is a quadratic. For example, Nelson (1984) applies (12.4) to metal fatigue data. A quadratic relationship is often adequate over the range of the test data, but it can err much if extrapolated much outside of that range. It is best to regard (12.4) as a curve fitted to data rather than as a physical relationship based on theory.

For accurate calculation, (12.4) is often expressed as the equivalent

$$\log(\tau) = \gamma_0' + \gamma_1'(x - x_0) + \gamma_2(x - x_0)^2,\qquad (12.5)$$

where x_0 is a chosen stress value near the center of the data or test range.

A *polynomial relationship* for the log of "nominal" life τ as a function of (possible transformed) stress x is

$$\log(\tau) = \gamma_0 + \gamma_1 x + \gamma_2 x^2 + \cdots + \gamma_K x^K.\qquad (12.6)$$

Such relationships are used for metal fatigue data over the stress range of the data. Such a polynomial for $K \geq 3$ is virtually worthless for extrapolation, even short extrapolation.

Elastic-Plastic Relationship for Metal Fatigue

Used for metal fatigue over a wide range of stress, the *elastic-plastic relationship* between "life" N (number of cycles) and constant strain (or pseudo-stress) amplitude S is

$$S = AN^{-a} + BN^{-b}.\qquad (12.7)$$

This relationship (S-N curve) has four parameters A, a, B, and b, which must be estimated from data. Here N cannot be written as an explicit function of S. Consequently, some data analysts incorrectly treat S as the dependent (and random) variable in least-squares fitting of this and other fatigue relationships to data. Hahn (1979) discusses the consequences of incorrectly treating stress as the dependent (and random) variable. Least-squares theory, of course, assumes that the random variable (life) is the dependent variable. (12.7) is sometimes incorrectly called the Coffin-Manson relationship, which is (10.1.2) according to Dr. Coffin.

A metallurgical interpretation of (12.7) is given by Graham (1968, p. 25). For example, the *elastic term* AN^{-a} usually has an a value in the range 0.10 to 0.20 and a typical a value of 0.12. The *plastic term* BN^{-b} usually has a b value in the range 0.5 to 0.7 and a typical b value of 0.6. Fatigue life of metals is complex, and no one S-N curve is universally applicable, even one with four parameters such as (12.7). ASTM STP 744 (1981, p. 92) lists some of the proposed curves. Some applications involving temperature and other variables employ polynomial fatigue curves. Such polynomial curves merely smooth the data. They have no physical basis.

Eyring Relationship for Temperature Acceleration

An alternative to the Arrhenius relationship (Section 9) for temperature acceleration is the Eyring relationship. Based on quantum mechanics, it is presented as a reaction rate equation for chemical degradation by Glasstone, Laidler, and Eyring (1941).

The Eyring relationship for "nominal" life τ as a function of absolute temperature T is

$$\tau = (A/T)\exp[B/(kT)];\tag{12.8}$$

here A and B are constants characteristic of the product and test method, and k is Boltzmann's constant. For the small range of absolute temperature in most applications, (A/T) is essentially constant, and (12.8) is close to the Arrhenius relationship (9.1.2). For most applications, both fit the data well.

The methods for fitting the Eyring model to data are the same as those for fitting the Arrhenius model, described in later chapters. One just analyzes transformed times $t' = t \cdot T$ as if they come from an Arrhenius model.

In the Eyring-lognormal model, (12.8) is the relationship for median life. Also, the standard deviation σ of log life t is assumed to be a constant. Then the standard deviation of the log of transformed life is the constant value σ.

13. MULTIVARIABLE RELATIONSHIPS

The life-stress relationships in Sections 8 through 12 involve a single accelerating stress. Such relationships are appropriate for many accelerated tests. However, some accelerated tests involve more than one accelerating stress or an accelerating stress and other engineering variables. For example, an accelerated life test of capacitors employed both high temperature and high voltage – two accelerating variables. The effect of both stresses on life was sought. Also, for example, in an accelerated life test of tape insulation, the accelerating variable was voltage stress, but the dissipation factor of specimens was included in the relationship. The factor was related to life and could be used to determine which insulation batches to install at higher voltage. Also, for example, MIL-HDBK-217E (1986) uses multivariable derating curves for failure rates of electronic components. The curves are functions of temperature, voltage, current, vibration, and other variables.

This section presents relationship between life and two or more variables, which may be stress or other predictor variables. For simplicity, accelerating stresses and other variables are all called variables.

It is useful to divide nonaccelerating variables into two groups. One group consists of variables that are experimentally varied. That is, the value of such an *experimental variable* is chosen for each specimen. For example, a taping experiment involved insulating tape wound on a conductor. The amount of overlap of successive layers of tape was varied – different specimens with different amounts of overlap. The other group of variables are uncontrolled, and such *uncontrolled variables* are only measured. For example, the same tape specimens were each measured for dissipation factor, a property of the specimen, because it is related to life. In experimental design books (e.g., Box, Hunter, and Hunter (1978)), such uncontrolled variables which are observed and included in a relationship are called *covariates*.

This section first presents a general multivariable relationship – the log-linear relationship; it includes the generalized Eyring relationship and indicator variables. The section then briefly presents nonlinear relationships and the Cox (proportional hazards) model.

Log-Linear Relationship

General relationship. A general, simple relationship for "nominal" life τ (say, a percentile) is the *log-linear relationship*

$$\ln(\tau) = \gamma_0 + \gamma_1 x_1 + \cdots + \gamma_J x_J. \tag{13.1}$$

Here $\gamma_0, \gamma_1, \cdots, \gamma_J$ are coefficients characteristic of the product and test method; they are usually estimated from data. $x_1, x_2, \cdots, x_J$ are (possibly transformed) variables. Any x_j may be a function (transformation) of one or any number of basic engineering (predictor or independent) variables. (13.1) is used in parametric analyses with an assumed form of the life distribution. It is also used in nonparametric analyses without an assumed form of the life distribution. For example, the Cox proportional hazards model below is nonparametric and uses (13.1).

(13.1) is a linear function of each of the coefficients $\gamma_0, \gamma_1, \cdots, \gamma_J$. It is not necessarily linear in the original variables used to calculate $x_1, x_2, \cdots, x_J$. Relationships that are linear in the coefficients are used mostly because they are mathematically convenient and physically adequate rather than "correct." The log-linear relationship includes a number of special cases below. Also, the relationship can be used to represent the spread in life (lognormal σ or Weibull β) as a function of a number of variables (Section 14).

Taping experiment. An experiment with insulating tape sought to evaluate the effect on life of the amount w that tape overlaps itself when wound on a conductor. The effect was modeled with (13.1) and sinusoidal terms

$$x_1 = \sin(2\pi w/W), \quad x_2 = \cos(2\pi w/W),$$

where W is the tape width. Also, the life test was voltage accelerated; so the effect of voltage stress was modeled with $x_3 = -\ln(V)$, that is, with the inverse power law.

Battery cells. In a battery cell application, Sidik and others (1980) use a quadratic relationship in five variables. The terms of (13.1) corresponding to linear, quadratic, and cross terms of the quadratic relationship. They sought to maximize the quadratic function of life by optimizing the five design and operating variables.

Generalized Eyring Relationship

Relationship. The *generalized Eyring relationship* has been used to describe accelerated life tests with temperature and one other variable. Glasstone, Laidler, and Eyring (1941) present it as a reaction rate equation. Rewritten to express "nominal" product life τ as a function of absolute temperature T and a (possibly transformed) variable V, it is

$$\tau = (A/T)\exp[B/(kT)] \times \exp\{V[C+(D/kT)]\}. \tag{13.2}$$

Here A, B, C, and D are coefficients to be estimated from data, and k is Boltzmann's constant. Most engineering applications of (13.2) use V or the transformation $\ln(V)$ in its place; see Peck's relationship below. In some applications the first $1/T$ is omitted. (13.2) is equivalent to (13.1) where

$$\tau' = \tau \cdot T \text{ is "life"}, \quad \gamma_0 = \ln(A), \quad \gamma_1 = B/k, \quad x_1 = 1/T,$$

$$\gamma_2 = C, \quad x_2 = V, \quad \gamma_3 = D/k, \quad x_3 = V/T.$$

$x_3 = x_1 x_2 = V(1/T)$ is an "interaction term" for $x_1 = (1/T)$ and $x_2 = V$. Applications follow. (13.2) like all other relationships needs to be supported in applications by data and experience, since theory is merely theory.

Capacitors. The author used (13.2) without the first $1/T$ as the Weibull characteristic life for an accelerated life test of capacitors with V equal to the ln of voltage. It was assumed that the *interaction term* between temperature and voltage was zero ($D = 0$). If $D \neq 0$, then in an Arrhenius plot, the straight line for (13.2) has a different slope for each voltage level. Because V is the ln of voltage, there was an assumed inverse power relationship between life and voltage. Figure 13.1 depicts this relationship as a *plane* in three dimensions. In the figure, temperature is on a reciprocal absolute scale, and life and voltage are on log scales. Figure 13.2 depicts contours of constant

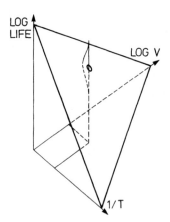

Figure 13.1. Generalized Eyring relationship with a power relationship in V.

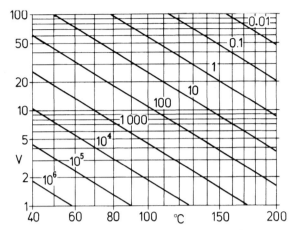

Figure 13.2.   Contours of constant life τ for a generalized Eyring relationship with no interaction ($D = 0$).

life (height on the relationship plane) projected onto the temperature-voltage plane of Figure 13.1. That plane is scaled like Arrhenius paper in Figure 13.2. The contours are straight lines since $D = 0$. Montanari and Cacciari (1984) extend the model by assuming that the Weibull shape parameter is a linear function of temperature. They fit their model to accelerated life test data on low density polyethylene.

Electromigration. Aluminum conductors in VLSI and other microelectronics fail from electromigration, which d'Heurle and Ho (1978) and Ghate (1982) survey. High current densities in such conductors promote movement of aluminum atoms, resulting in voids (opens) or extrusions (shorts). Accelerated tests of this phenomenon employ elevated temperature T and current density J (amps per square centimeter). Black's (1969a,b) formula for median life τ of such conductors is the Erying relationship; namely,

$$\tau = AJ^{-n} \exp[E/(kT)] ;$$

here $V = -\ln(J)$ and $D = 0$ (no interaction term). Figures 13.1 and 13.2 depict Black's formula for median life if voltage V is replaced by current density J. Using a debated physical model and argument, Black derives $n = 2$. Much data support this value. Recently Shatzkes and Lloyd (1986) propose physical mechanisms and argue that

$$\tau = A(T/J)^2 \exp[E/(kT)].$$

They state that this fits two data sets as well as Black's formula does. Many use the lognormal distribution for such life data.

Temperature-humidity tests. Many accelerated life tests of epoxy packaging for electronics employ high temperature and humidity. For example,

85°C and 85% relative humidity (RH) is a common test condition. Peck (1986) surveys such testing and proposes an Eyring relationship for life

$$\tau = A(RH)^{-n} \exp[E/(kT)],$$

called *Peck's relationship.* Data he uses to support it yield estimates $n = 2.7$ and $E = 0.79$ eV. Figures 13.1 and 13.2 depict this relationship if relative humidity RH replaces voltage V and its axis is linear instead of logarithmic. Intel (1988) uses another Eyring relationship

$$\tau = A \exp(-B \cdot RH) \exp[E/(kT)].$$

Intel notes that this differs little from Peck's relationship relative to uncertainties in such data.

Rupture of solids. *Zhurkov's* (1965) *relationship* for "time" τ to rupture of solids at absolute temperature T and tensile stress S is

$$\tau = A \exp[(B/kT) - D(S/kT)].$$

This is the Eyring relationship with $C = 0$ and a minus sign for D. Also, it is a form of the Larsen-Miller relationship (9.1.7). Zhurkov motivates this relationship with chemical kinetic theory, and he presents data on many materials to support it. He interprets B as the energy to rupture molecular bonds and D as a measure of the disorientation of the molecular structure. Ballado-Perez (1986,1987) proposes this relationship for life of bonded wood composites; he extends it to include indicator variables (below) for type of wood, adhesive, and other factors.

Indicator Variables

Category variables. Most variables in relationships are numerical, and they can mathematically take on any value in an interval. For example, absolute temperature can mathematically have any value from 0 to infinity. On the other hand, some variables can take on only a finite number of discrete values or categories. Examples include 1) insulation made on three *shifts*, 2) insulation coated on two different conductor *metals*, and 3) material from two *vendors*. Relationships for such categorical variables are expressed as follows in terms of indicator variables.

Shifts example. For concreteness, suppose that insulation is made on three production shifts, denoted by 0, 1, and 2. Also, suppose that insulation life in a voltage-endurance test is modeled with the power-Weibull model. Also, suppose that the insulations from the shifts have the same power in the power law but different constant coefficients (intercepts). Then the characteristic life α_j for shift j is

$$\ln[\alpha_0(V)] = \gamma_0 + \gamma_3 \ln(V), \quad \ln[\alpha_1(V)] = \gamma_1 + \gamma_3 \ln(V),$$
$$\ln[\alpha_2(V)] = \gamma_2 + \gamma_3 \ln(V); \tag{13.4}$$

here V is the voltage stress, γ_3 is the power coefficient and is negative, and γ_j is the intercept coefficient for shift $j = 0, 1, 2$. The straight lines for the three shifts are parallel on log-log paper. The power γ_3 is assumed to be the same for all three shifts, as it is regarded as a physical property of the insulating material. The intercept is assumed to depend on shift, as skills of the workers differ. Of course, both assumptions were assessed using data.

Definition. Define the *indicator variable* $z_j = 1$ if the corresponding test specimen is from shift j; otherwise, $z_j = 0$ if the specimen is from another shift. For example, a specimen made on shift 1 has values $z_0 = 0, z_1 = 1$, and $z_2 = 0$ for the three indicator variables, also called *dummy variables*. An indicator variable takes on only the values 0 and 1. For that reason it is also called *0-1 variable*. Let $z_3 = \ln(V)$. The relationships (13.4) can then be written in a single equation

$$\ln[\alpha(V)] = \gamma_0 z_0 + \gamma_1 z_1 + \gamma_2 z_2 + \gamma_3 z_3. \tag{13.5}$$

This equation has four variables (z_0, z_1, z_2, z_3) and four coefficients $(\gamma_0, \gamma_1, \gamma_2, \gamma_3)$; it has no intercept coefficient, that is, a coefficient without a variable. Most computer programs that fit linear relationships to data require that the relationship have an intercept coefficient. This can be achieved by awkwardly rewriting (13.5) as

$$\ln[\alpha(V)] = \delta_0 + \delta_1 z_1 + \delta_2 z_2 + \gamma_3 z_3; \tag{13.6}$$

here $\delta_0 = \gamma_0$ is the intercept, $\delta_0 + \delta_1 = \gamma_1$, and $\delta_0 + \delta_2 = \gamma_2$. Equivalently, $\delta_1 = \gamma_1 - \gamma_0$ and $\delta_2 = \gamma_2 - \gamma_0$. The δ_j coefficients are not as natural or simple a way of representing the relationship. Yet this representation is better suited to most computer programs, which require an intercept term.

Some relationships have more than one categorical variable. For example, the three shifts each make two types of insulation. Shift is a categorical variable requiring two indicator variables $(z_1$ and $z_2)$, and insulation is another requiring one (z_3). Then

$$\ln[\alpha(V)] = \delta_0 + \delta_1 z_1 + \delta_2 z_2 + \delta_3 z_3 + \gamma_3 \ln(V).$$

A relationship that is just a linear function of the indicator variables for two or more categorical variables is call a *main-effects* relationship. More complex relationships involve *interaction terms*; they appear in books on analysis of variance, for example, Box, Hunter, and Hunter (1978). Zelen (1959) presents an application with interaction terms for life of glass capacitors over a range of temperature and of voltage.

Logistic Regression Relationship

The logistic regression relationship is widely used in biomedical applications where the dependent variable is binary; that is, it is in one of two mutually exclusive categories, for example, dead or alive. The *logistic relationship* for the proportion p in a particular category (say, "failed") as a function of J independent variables $x_1, \cdots, x_J$ is

$$\ln[(1-p)/p] = \gamma_0 + \gamma_1 x_1 + \cdots + \gamma_J x_J ; \tag{13.7}$$

here $\gamma_0, \gamma_1, \cdots, \gamma_J$ are unknown coefficients to be estimated from data.

In accelerated testing, (13.7) might be used when the life data are quantal-response data; that is, each specimen is inspected once to determine whether it has failed by its inspection age. Then p is the fraction failed, and one of the independent variables is (log) age at inspection.

Introductions to the logistic relationship and fitting it to data are given by Neter, Wasserman, and Kutner (1983) and Miller, Efron, and others (1980). A comprehensive presentation of logistic regression is given by Breslow and Day (1980). The BMDP Statistical Software of Dixon (1985) is one of the many statistical packages that fit the model to data.

Nonlinear Relationships

The log-linear relationship (13.1) is linear in the unknown coefficients. Engineering theory may suggest relationships that are nonlinear in the coefficients. However, most computer packages fit only linear relationships. Nonlinear relationships usually must be programmed into certain packages.

Nelson and Hendrickson (1972, p. 5-3-4) give an example of such a nonlinear relationship. A test involved time to breakdown of an insulating fluid between parallel disk electrodes. The voltage across the electrodes was increased linearly with time at different rates R (volts per second). Electrodes of various areas A were employed. The assumed distribution for time to breakdown is Weibull with parameters

$$\alpha(R,A) = \{\gamma_1 R/[A \exp(\gamma_0)]\}^{1/\gamma_1} , \quad \beta = \gamma_1 .$$

The $\ln(\alpha)$ relationship is nonlinear in γ_1.

Cox (Proportional Hazards) Model

Used in biomedical applications, the *Cox* (or *proportional hazards*) *model* can be used as an accelerated life testing model. It does not assume a form for the distribution – possibly an attractive feature. It can be used to extrapolate in stress but not in time, because it is distribution free ("nonparametric"). It cannot be used to extrapolate the distribution to early time in the lower tail or later time in the upper tail outside the range of the life data. So it is useful only for estimating the observed range of the life distribution at actual use conditions. Such extrapolation in stress is desired in many applications. A brief description of the model follows.

Let $x_1, \cdots, x_J$ denote the (possibly transformed and centered) variables, and let $h_0(t)$ denote the hazard function of the unknown life distribution at

$x_1 = x_2 = \cdots = x_J = 0$. The **Cox model** for the hazard function of the distribution at variable values $x_1, \cdots, x_J$ is

$$h(t;x_1, \cdots, x_J) = h_0(t) \cdot \exp(\gamma_1 x_1 + \cdots + \gamma_J x_J). \qquad (13.8)$$

The base hazard function $h_0(t)$ and the coefficients $\gamma_1, \cdots, \gamma_J$ are estimated from data. Note the similarity of (13.8) to the log-linear relationship (13.1). (13.8) does not have an intercept coefficient γ_0, as $h_0(t)$ takes the place of γ_0. The corresponding reliability functions are

$$R(t;x_1, \cdots, x_J) = [R_0(t)]^{\exp(\gamma_1 x_1 + \cdots + \gamma_J x_J)}. \qquad (13.9)$$

The reliability function at $x_1 = \cdots = x_J = 0$ is

$$R_0(t) = \exp[-\int_0^t h_0(t)dt]. \qquad (13.10)$$

The life distribution (13.9) is complex. Its "typical" life (say, a percentile) depends on $x_1, \cdots, x_J$ in a complex way. Moreover, the spread and shape of the distribution of *log* life generally depend on $x_1, \cdots x_J$ in a complex way. Previous parametric models, such as the Arrhenius-lognormal and linear-Weibull, have simpler form. The Weibull distribution with a multivariable log linear relationship (13.1) for characteristic life is a special case of (13.9).

Kalbfleisch and Prentice (1980), Lee (1980), Miller (1981), and Cox and Oakes (1984) among others present the model in detail. They also give methods and computer programs for fitting it to data and evaluating the fit.

14. SPREAD IN LOG LIFE DEPENDS ON STRESS

In many accelerated life test models, the spread in log life is assumed to be the same at all stress levels of interest. For example, the standard deviation σ of log life in the Arrhenius-lognormal model is assumed to be a constant. Similarly, the shape parameter β in the power-Weibull model is assumed to be a constant. A constant spread is assumed for two reasons. First, the data or experience with such data suggests that a constant spread adequately models log life. Second, the analyst uses a model with constant spread because it is traditional or easy to use. For example, almost all model fitting programs (especially least-squares programs) assume that the data spread is constant.

On the other hand, experience with certain products indicates that the spread of log life is a function of stress. For example, for metal fatigue and roller bearing life, the spread is greater at lower stress. For some electrical insulations, the spread is smaller at lower stress. The following paragraphs present some simple "heteroscedastic" relationships for spread as a function of stress. For concreteness, the standard deviation σ of log life with a lognormal distribution is used. Equivalently, one could use the shape parameter β of a Weibull distribution or any measure of spread of a life distribution.

Log-Linear Relationship for Spread

The simplest relationship for a spread parameter σ is the log-linear relationship

$$\ln[\sigma(x)] = \delta_0 + \delta_1 x. \qquad (14.1)$$

Here δ_0 and δ_1 are parameters characteristic of the product; they are estimated from data. x is the (possibly transformed) stress. Equivalently,

$$\sigma(x) = \exp[\delta_0 + \delta_1 x]. \qquad (14.2)$$

Figure 14.1A depicts (14.2) where $\mu(x)$ is a linear function of $x = \log(\text{stress})$. Nelson (1984) gives an example of fitting (14.1) to metal fatigue data with runouts. Of course, σ must be positive. The logarithmic form of (14.1) assures that is so. Glaser (1984) assumes σ is simply a linear function (no logarithm) of x. His function yields an incorrect negative σ for extreme values of x. Thus it is satisfactory over only a limited range of x.

Other mathematically plausible relationships could be used, for example, the log-quadratic relationship

$$\ln[\sigma(x)] = \delta_0 + \delta_1 x + \delta_2 x^2. \qquad (14.3)$$

A model with a lognormal (or Weibull) distribution where $\sigma(x)$ is a func-

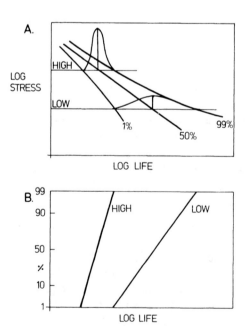

Figure 14.1.   (A) Log spread as a function of stress, (B) Probability plot of crossing distributions.

tion of stress has the following drawback. Figure 14.1B depicts on lognormal (or Weibull) paper a life distribution for low stress and another for high stress. If the straight lines are extended into the lower tail, the distributions cross there. This is physically implausible and undesirable if the lower tail is important. Then such a model with $\sigma(x)$ is inaccurate low enough in the tails. This drawback is not apparent in Figure 14.1A. There is a need for more sophisticated models that do not have distributions that cross.

The spread may depend on a number of accelerating and other variables. This dependence can be represented with the log-linear and other multivariable relationships of Section 13.

Components of Variance

Metals and many other products are made in batches. For metals, the fatigue life distribution may differ appreciably from batch to batch. Thus the life distribution of the entire product population is a mixture of the distributions of the batches. Such a situation with many batches is modeled with a components-of-variance model. Such models for a single stress level appear in most books on analysis of variance, for example, Box, Hunter, and Hunter (1978). Some recent research is extending such models to regression situations. However, such extensions do not yet apply to censored data or to standard deviations that depend on the accelerating variable. Such models need to be developed for metal fatigue and other applications.

PROBLEMS (* denotes difficult or laborious)

2.1. Exponential. For the exponential distribution function $F(t) = 1 - \exp(-\lambda t)$, $t > 0$, derive the following.
(a) Probability density.
(b) Hazard function.
(c) 100Pth percentile.
An exponential distribution used to model time to breakdown of an insulating fluid has $\lambda = 0.946$ failures/thousand-hours. Calculate the following.
(d) The fraction failing in 20 minutes.
(e) The mean time to failure.
(f) The median life.
(g) Plot this life distribution on Weibull probability paper.

2.2. Weibull. For the Weibull cumulative distribution function, $F(t) = 1 - \exp[-(t/\alpha)^{\beta}]$, $t > 0$, derive the following.
(a) Probability density.
(b) Hazard function.
(c) 100Pth percentile.
A Weibull distribution used to model time to breakdown of an insulating fluid has $\alpha = 63,400$ minutes and $\beta = 0.8084$. Calculate the following.

(d) Median life.
(e) Fraction failing on at 20-minute test.
(f) Plot this distribution on Weibull probability paper.

    2.3. Log_{10} normal. For the lognormal cumulative distribution function, $F(t) = \Phi\{[\log_{10}(t) - \mu]/\sigma\}$, $t > 0$, derive the following.
(a) Probability density.
(b) Hazard function.
(c) 100Pth percentile.
A lognormal distribution for life (in hours) of a Class-B insulation at 130°C (design temperature) has $\mu = 4.6698$ and $\sigma = 0.2596$. Calculate:
(d) Median life.
(e) 1st percentile.
(f) Fraction failing by 40,000 hours.
(g) Plot this distribution on lognormal probability paper.

    2.4. Log_e normal. The lognormal distribution with base e logs has the cumulative distribution function

$$F(t) = \Phi\{[\ln(t) - \mu_e]/\sigma_e\}, \quad 0 < t < \infty;$$

here μ_e is the mean and σ_e is the standard deviation of ln life. Derive the:
(a) 100Pth percentile.
(b) Probability density in terms of the standard normal density $\phi(\ )$.
(c) Hazard function.
(d) Relationship between μ_e and σ_e and the μ and σ of the same lognormal distribution expressed with base 10 logs. Evaluate μ_e and σ_e for the Class-B insulation in the previous problem.

    2.5.* Mixture of exponentials. Suppose that a population contains a proportion p of units from an exponential life distribution with mean θ_1 and the remaining proportion $1-p$ from an exponential life distribution with mean θ_2. Proschan (1963) treats such a problem.
(a) Derive the hazard function of the mixture distribution.
(b) Show that the failure rate of the mixture distribution decreases with age.
(c) Extend this result to a mixture of two Weibull distributions with characteristic lives α_1 and α_2 and a common shape parameter β.
These results indicate that the failure rate of a mixture has a lower β than do the individual Weibull subpopulations. For example, bearings under (homogeneous) laboratory test conditions have a Weibull life distribution with β in the range 1.3 to 1.5. Under a mix of diverse field conditions, the same type of bearing has β in the range 1.1 to 1.3.

    2.6. Class-B insulation (lognormal). The life (in hours) of a Class-B motor insulation was approximated with an Arrhenius-lognormal model (base 10) with $\sigma = 0.2596$, $\gamma_0 = -6.0098$, and $\gamma_1 = 4.3055$ where $x = 1000/T$ and absolute temperature T is in Centigrade degrees.

(a) Calculate the 1st, 10th, 50th, 90th, and 99th percentiles of life at 220, 190, 170, 150, and 130°C (the design temperature).
(b) Plot these percentiles on Arrhenius paper, and draw the percentile lines.
(c) Plot these percentiles on lognormal probability paper, and draw the straight distribution lines.
(d) Comment on the behavior of the lognormal hazard function.

2.7. Class-B insulation (Weibull). A Weibull distribution for the life of the Class-B insulation can be fitted as follows. Assume that the 50th and 10th percentile calculated above are correct. This choice of percentiles (10th and 50th) is reasonable but arbitrary.
(a) Calculate the coefficients of the relationship for the log (base e) of median life.
(b) Using the 10th and 50th percentiles at a particular temperature, calculate the value of the Weibull shape parameter. Comment on the nature of the failure rate (increasing or decreasing?).
(c) Calculate the Weibull 1st, 90th, and 99th percentiles at 220, 190, 170, 150, and 130°C (the design temperature).
(d) Plot these percentiles on Arrhenius paper, and draw the percentile lines.
(e) Plot these percentiles on Weibull probability paper, and draw the straight distribution lines.
(f) Plot the lognormal percentiles on the Weibull plot, and draw the curved distribution lines through them. Comment on how the Weibull and lognormal distributions compare i) below the 10th percentile, ii) between the 10th and 50th percentiles, and iii) above the 50th percentile.

2.8. Eyring model. For the Eyring relationship (12.8) and a lognormal life distribution, do the following.
(a) Assuming (12.8) is the median, derive the equation for mean log life.
(b) Express the equation from (a) in a form that is linear in coefficients γ_0 and γ_1 (which are functions of A and B) and $x = 1/T$.
(c) Write the reliability function in terms of temperature T and age t.
(d) Write the log of the 100Pth percentile in terms of x and σ.
(e) Write the 100Pth percentile in terms of x and σ and in terms of T and σ.
(f) The Class-H insulation example of Section 9.2 employs the Arrhenius model with $\sigma = 0.10533$, $\tau_{.50}(260°C) = 940$ hours, and $\tau_{.50}(190°C) = 8,030$ hours. (Calculate the corresponding parameter values γ_0 and γ_1 for the Eyring-lognormal model.
(g) Calculate the 10th and 50th percentiles of the Eyring-lognormal model for temperatures of 300, 260, 225, 190, 180, and 130°C.
(h) Plot the percentiles from (g) on Arrhenius paper, and draw smooth curves through them. Plot the corresponding percentiles from the Arrhenius-lognormal model, and draw straight lines through them. How do the two models compare at "design" stresses of 180 and 130°C (small and medium extrapolation)? Which is more optimistic?

2.9. Superalloy fatigue. Nelson (1984) describes fatigue life (in cycles) of specimens of a nickel base superalloy with a lognormal life distribution with

$$\mu(S) = 4.573919 - 5.004997[\log(S) - 2.002631],$$

$$\sigma(S) = \exp\{-1.262398 - 4.921923[\log(S) - 2.002631]\};$$

here S is the stress in thousands of pounds per square inch (ksi).
(a) Calculate $\mu(S)$ and $\sigma(S)$ for S = 75, 85, 100, 120, and 150 ksi.
(b) Calculate the 0.1th, 10th, and 50th percentiles at each of those stresses.
(c) Plot the percentiles on log-log paper (stress vertical versus life horizontal, engineering practice), and draw percentile curves.
(d) Plot the percentiles on lognormal probability paper, and draw straight distribution lines through them.
(e) In view of the plots, comment on peculiarities or inadequacies of the model. For this alloy and others, design engineers usually use the 0.1 percentile divided by 3 as a design life.

2.10. Relay failure rate. MIL-HDBK-217E (1986) gives the failure rate (failures per million hours) of a certain class of resistively loaded relays rated for 85°C ambient. MIL-HDBK-217E assumes that the failure rate is constant. Consequently the life distribution is assumed to be exponential. As a function of absolute temperature T and "stress" S = (operating load current)/(rated resistive load current),

$$\lambda(T,S) = 5.55\times10^{-3}\exp[(T/352.0)^{15.7}]\times\exp[(S/0.8)^{2.0}].$$

(a) For 85°C, calculate λ for S = 0.2, 0.4, 0.6, 0.8, and 1.0.
(b) Repeat (a) for 25, 50, and 125°C.
(c) Plot the preceding four curves (25, 50, 85, and 125°C) on appropriate relationship plotting paper for λ versus S.
(d) For a "stress" of S = 0.8, calculate the operating temperature that will result in a failure rate of 0.018 failures per million hours.
(e) For λ = 0.010 failures per million hours, calculate T as a function of S for S = 0.0, 0.2, 0.4, 0.6, 0.8, and 1.0. T versus S is a trade-off curve.
(f) Repeat (e) for λ = 0.050 and 0.100.
(g) Plot the three trade-off curves from (e) and (f) on paper with appropriate scales for T and S. This is a contour plot of constant λ.

2.11. Life doubles for each 10°C. There is a widely used rule of thumb based on an inverse power approximation to the Arrhenius relationship; namely, product life doubles for every 10°C drop in temperature. The following problem shows that the rule can be very crude and misleading.
(a) For a given activation energy E, show that this is true at just one temperature T_2. Give the formula for T_2 in terms of E.
(b) Evaluate T_2 for E = 0.50 electron-volts.
(c) Evaluate T_2 for E = 1.00 electron-volts and for E = 1.50 electron-volts.

(d) Show how much the factor could differ from 2 for a 10°C drop at a temperature different from T_2.

(e) Comment on the validity of the rule of thumb.

2.12. Generalized Eyring acceleration factor. For the generalized Eyring relationship (13.2),

(a) Give the acceleration factor between conditions (T,V) and (T',V').

(b) Show that the acceleration factor is the product of the separate T and V acceleration factors only if $D = 0$.

Use Peck's relationship (Section 13) with his values $n = 2.7$ and $E = 0.79$ eV.

(c) On Arrhenius paper, make a contour plot of lines of constant acceleration factor K. Use 85°C and 85% relative humidity (RH) as the reference condition, a common test condition for electronics in MIL-STD-883. Use $K = .001, .01, .1, 1, 10, 100, 1000$. Include RH from 10 to 100% and temperature from 40 to 200°C. This plot shows how much the reference condition accelerates life compared to any design temperature and humidity.

2.13. Zhurkov's relationship. Sketch contours of constant life for Zhurkov's relationship on suitable plotting paper.

2.14. MOS. Use the failure rate relationship of the MOS in Section 9.4.

(a) Calculate and plot the failure rate on Arrhenius paper.

(b) Calculate and plot the life distributions on Weibull paper for 25, 50, 100, and 150°C.

3

Graphical Data Analysis

1. INTRODUCTION

Purpose. This basic chapter presents simple data plots for analysis of accelerated life test data. These plots provide desired estimates of a product life distribution at design stress and of model parameters. Also, the data plots are used to assess the validity of the model and data. In addition, the plots help one understand the complex numerical methods of later chapters.

Background. Needed background appears in Chapter 2, particularly, simple models. These include lognormal, Weibull, and exponential life distributions and Arrhenius and inverse power relationships. Unlike analytic methods, graphical methods do not require a knowledge of statistical theory.

Advantages. Graphical methods for data analysis have advantages and disadvantages relative to analytic methods. Graphical methods are multipurpose, and their advantages include:
- They are simple – quick to make and easy to interpret. At age 11, the author's son could make and interpret such data plots. Moreover, they do not require special computer programs, as do analytic methods. However, computer programs (see Chap. 5, Sec. 1) readily make such plots. Moreover, well-made computer plots are convincing and authoritative.
- They provide estimates of the product life distribution (percentiles, percentage failed) at any stress and estimates of model parameters.
- They allow one to assess how well a model fits the data and how valid the data are. Such assessments are also needed before using analytic methods.
- Most important they help convince others of conclusions based on plots or analytic results, which others accept more readily after seeing such plots.
- They reveal unsought insights into the data. Such a discovery in the Class-H insulation example below yields $1,000,000 yearly. Analytic methods reveal less. They have tunnel vision and rarely reveal anything not specifically calculated.

Disadvantages. Analytic methods have certain advantages over graphical methods. These include:
- The statistical uncertainty of analytic estimates can be given objectively by means of confidence intervals. This is important because inexperienced

data analysts tend to think estimates are more accurate than they really are. On the other hand, it is often apparent that graphical estimates are (or are not) accurate enough for practical purposes.

* Comparisons (say, of two products) can be made objectively with a statistical confidence interval or hypothesis test. Such a test indicates whether an observed difference is statistically significant, that is, convincing. Of course, graphical comparisons can be convincing, namely, when observed differences are large compared to the scatter in the data. Subjective judgments of what is convincing can differ. Viewing the same data plots, three data analysts can have six different opinions on what they see. Thus use subjective judgement cautiously, aided by objective analytic methods.
* Appropriate sample sizes can be determined, as well as, optimum or good test plans, based on analytic methods.

For most work, it is essential to use both graphical and analytic methods. Each provides certain information not provided by the other. A proper analysis of data always requires many different analyses. Examples repeatedly appear in different chapters to show the wide variety of analyses that should be applied to a data set.

Method. The graphical method employs the simple model (Chapter 2) and involves two data plots. The first employs probability paper for the assumed life distribution (for example, lognormal, Weibull, and exponential). The second employs paper which linearizes the assumed relationship (for example, the Arrhenius and inverse power laws). The simple model is usually suitable only for data with **a single cause of failure.** Models and analyses for data with more than one cause of failure appear in Chapter 7. Chapters 4 and 5 provide corresponding analytic methods for complete and censored data. Models and data analyses for products whose performance degrades with age appear in Chapter 11. Nelson (1990) briefly presents basic applications of graphical analyses for accelerated test data. General presentations of data plotting are given by Cleveland (1985), Tufte (1983), and Chambers and others (1983).

Overview. In this chapter, Section 2 presents the basic probability and relationship plots in detail; the presentation treats complete data, the lognormal probability plot, and the Arrhenius (relationship) plot. Section 3 briefly does the same for complete data, the Weibull probability plot, and the inverse power relationship. Section 4 presents probability and relationship plots for singly censored data (the most efficient type of test plan). Section 5 presents such plots for multiply censored data. Section 6 presents plots for interval (read-out) data.

2. COMPLETE DATA AND ARRHENIUS-LOGNORMAL MODEL

Introduction. This section presents simple graphical methods for analyzing complete data, using the simple model of Chapter 2. The methods are illustrated with the lognormal distribution and the Arrhenius relationship presented in Chapter 2. The methods yield estimates of model parameters and the product life distribution at any stress. The plots are also used to assess the validity of the model and data.

Corresponding numerical methods in Chapter 4 provide confidence intervals as well as estimates. The simpler graphical methods fill most practical needs. Moreover, they provide information that numerical methods do not, including checks on the validity of the data and the model. On the other hand, graphical estimates have unknown accuracy whereas numerical confidence intervals indicate the accuracy of estimates. It is best to use a combination of graphical and numerical methods.

Overview. Section 2.1 describes illustrative data. Section 2.2 shows how to make and use a probability plot of the data. Section 2.3 presents a data plot of the relationship between life and stress. Section 2.4 shows how to estimate the model parameters and life distribution at any stress. Section 2.5 concludes with assessments of the model and data.

2.1. Data (Class-H Insulation)

The complete data in Table 2.1 are hours to failure of 40 motorettes with a new Class-H insulation run at 190, 220, 240, and 260°C. These data were the first clue to a discovery that is reducing a business's insulation costs one million dollars yearly at 1989 prices. For each test temperature, the 10 motorettes were periodically examined for insulation failure, and the given failure time is midway between the inspection time when the failure was found and the time of the previous inspection. The test purpose was to estimate the median life of such insulation at its design temperature of 180°C. A median of 20,000 hours was desired. The Arrhenius-lognormal model is used, based on engineering experience.

Table 2.1. Class-H Insulation Life Data and Plotting Positions

Hours to Failure				Rank	Plotting Positions	
190°C	220°C	240°C	260°C	i	F_i	F_i'
7228	1764	1175	600	1	5	9.1
7228	2436	1175	744	2	15	18.2
7228	2436	1521	744	3	25	27.3
8448	2436	1569	744	4	35	36.4
9167	2436	1617	912	5	45	45.5
9167	2436	1665	1128	6	55	54.5
9167	3108	1665	1320	7	65	63.6
9167	3108	1713	1464	8	75	72.7
10511	3108	1761	1608	9	85	81.8
10511	3108	1953	1896	10	95	90.9

There are four test temperatures each with an *equal* number of specimens. This is a traditional engineering test plan, but it is very *inefficient*.
Better test plans involve two or three stress levels and run more specimens at
the low stress. Better plans, described in Chapter 6, provide more accurate
estimates of the life distribution at the design stress.

2.2. Lognormal Probability Plot

The life data for each test stress level are plotted as follows on probability
paper. The plot provides estimates of the model parameters and life distribution percentiles. The data plot resembles the model on probability paper
in Figure 9.2 of Chapter 2. On such probability paper, the theoretical cumulative distribution function of life is a straight line. How to make a probability plot follows. Estimates for distribution percentiles and model parameters
are presented later. Section 2.5 explains how to use the plots to assess the
validity of the data and model.

Plotting Positions. Order the n failure times at a test stress from smallest
to largest as shown in Table 2.1. Give the earliest failure rank 1, the second
earliest failure rank 2, etc. Calculate the probability plotting position for
each failure from its rank i as

$$F_i = 100(i - 0.5)/n. \tag{2.1}$$

F_i values appear in Table 2.1. These *midpoint plotting positions* approximate
the percentage of the population below the ith failure. People also use the
expected plotting position

$$F_i' = 100i/(n + 1). \tag{2.2}$$

F_i' values also appear in Table 2.1. People also use the *median plotting position* approximated as

$$F_i'' \approx 100(i - 0.3)/(n + 0.4). \tag{2.3}$$

Consistently use any plotting position for all stresses. The F_i are tabulated in
Appendix A7, the F_i' by King (1971), and the F_i'' by Johnson (1964). Plotting
positions differ little compared to the usual random variation in the data.
Some authors strongly argue for a particular plotting position. This is as
fruitless as arguing religions; they all get you to heaven.

Probability plot. Use probability paper for the model distribution.
Choose probability and data scales with the smallest range that encloses the
data and any distributions and percentiles of interest. This spreads out the
plotted data and reveals details. Two or more plotting papers can be joined
to obtain more log cycles on the data scale. Label the data scale to span the
data and any distribution of interest, say, at the design temperature. Plot
each failure time against its plotting position on the probability scale.
Figure 2.1A is a plot of the Class-H data on lognormal probability paper.

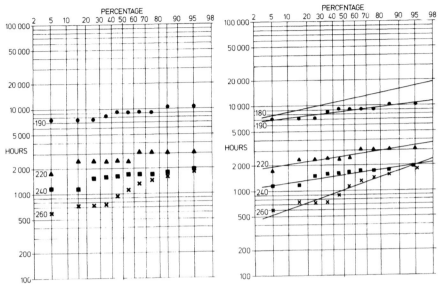

Figure 2.1A.   Lognormal plot of
Class-H data.

Figure 2.1B.   Separate fits to
Class-H data.

Assess distribution. If the plotted points for a stress level tend to follow a straight line, then the distribution appears to describe the data adequately. Sample sizes are typically small (below 20 specimens) at each stress level in accelerated life tests, and the plots may appear erratic. A more sensitive assessment of a distribution is in Chapter 4 and uses residuals.

Distribution lines. Draw a separate straight distribution line through the data for each test stress as in Figure 2.1B. The vertical (time) deviations between the line and the plotted points should be as small as possible. Such fitted lines are not necessarily parallel due to random variation in the data or unequal true slopes. However, the Arrhenius and other simple models assume that the life distributions at different stresses have a common slope, as depicted in Figure 9.2 of Chapter 2. The slope of a line for a lognormal distribution corresponds to the log standard deviation σ. The 260° data have a slope different from the others. Note that the fitted lines distract the eye from the data and can distort the data. Thus it is best to examine plots both with and without fitted lines.

Parallel lines. *If* parallel lines are appropriate, a more refined method fits parallel lines to the data as in Figure 2.2. Guided by the separately fitted lines, fit parallel lines with a compromise common slope to the data for each stress. The numbers of specimens may differ from stress to stress. Then try to choose the common slope by weighting the slope at a stress proportional to its number of specimens. This parallel fit to the data looks like the plot of the model in Figure 9.2 of Chapter 2. Estimates are obtained from the paral-

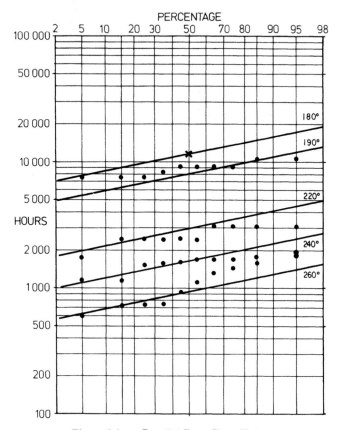

Figure 2.2.   Parallel fit to Class-H data.

lel lines as described below. Analytic methods for fitting parallel lines appear in Chapter 4. Again note in Figure 2.2 how the fitted parallel lines distract and incorrectly suggest that the data plots are parallel.

Percent failed at a test stress. Either type of fitted distribution line provides an estimate of the percentage of units failing by a given age. Enter the probability plot at that age on the time scale, go sideways to the fitted line for that stress, and then go up to the probability scale to read the percentage. For example, for 190°C in Figure 2.1B, the estimate of the percentage failing by 7,000 hours is 7 percent. Similarly from Figure 2.2, the estimate is 27%. Estimates from the parallel distribution lines tend to be more accurate, provided the model is correct.

Percentile estimate. The following provides an estimate of a percentile at a test stress. Enter the probability plot at the desired percentage on the probability scale, go vertically to the fitted distribution line for the test stress, and go horizontally to the time scale to read the estimate. For example, this estimate of the median life at 190°C is 8,700 hours from Figure 2.1B. These estimates of the medians are ×'s in Figure 2.3 on Arrhenius paper.

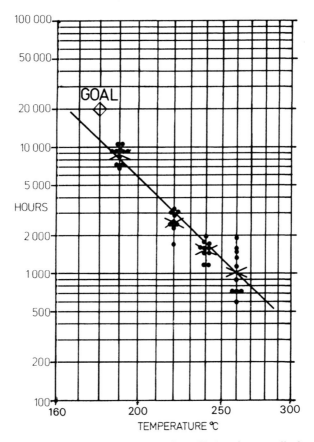

Figure 2.3.   Arrhenius plot of Class-H data (× = median).

Lognormal papers. Lognormal probability plotting papers are available from the following:

1) TEAM (Technical and Engineering Aids for Management), Box 25, Tamworth, NH 03886, (603)323-8843. The wide selection in the TEAM (1988) catalog includes:

3112	1-99% (11" horizontal)	3 log cycles (8 1/2" vertical)
3212	0.01-99.99% (11" horizontal)	3 log cycles (8 1/2" vertical)
3312	0.0001-99.9999% (11" horizontal)	3 log cycles (8 1/2" vertical)
311	0.01-99.99% (8 1/2" horizontal)	3 log cycles (11" vertical)
313	0.01-99.99% (8 1/2" horizontal)	5 log cycles (11" vertical)
314	0.01-99.99% (8 1/2" horizontal)	7 log cycles (11" vertical)
315	0.01-99.99% (8 1/2" horizontal)	10 log cycles (11" vertical)
30101	2-98% (8 1/2" horizontal)	2 log cycles (11" vertical)
SA-5	Assortment of 311, 313, 314, 315, 3112, 3212, 3312.	

2) CODEX Book Co., 74 Broadway, Norwood, MA 02062, (617)769-1050.

Y3 210 0.01-99.99% (11" horizontal)   2 log cycles (8½" vertical)
Y3 213 0.1-99.9% (8½" vertical)   3 log cycles (11" horizontal)
Y4 211 0.01-99.99% (11" horizontal)   4 log cycles (16½" vertical)

3) K + E (Keuffel & Esser Co.), 20 Whippany Rd., Morristown, NJ 07960, (800)538-3355.

46 8040 0.01-99.99% (11" horizontal)   2 log cycles (8½" vertical)
46 8080 2-98% (8½" horizontal)   3 log cycles (11" vertical)

4) Craver (1980) gives reproducible copies of

pg. 220 0.01-99.99% (11" horizontal)   1 log cycle (8½" vertical)
pg. 221 0.01-99.99% (11" horizontal)   2 log cycles (8½" vertical)
pg. 222 0.01-99.99% (11" horizontal)   3 log cycles (8½" vertical)
pg. 225 2-98% (8½" horizontal)   3 log cycles (11" vertical)
                                         is K + E 46 8080
pg. 229 0.1-99.9% (8½" vertical)   3 log cycles (11" horizontal)

5) Technology Associates, 51 Hillbrook Dr., Portola Valley, CA 94025, (415)941-8272.

5001 0.1-99.5% (8½" horizontal)   6 log cycles (11" vertical)

Symbols. Plots with professional quality plotting symbols and lettering can be achieved with transfer lettering (press type). The catalogs of Chartpak (1988) and Letraset (1986) list sheets of such symbols and lettering. Also, some computer packages (Chap. 5, Sec. 1) produce quality plots.

2.3. Relationship Plot (Arrhenius)

Data plot. The following *relationship plot* of data resembles the relationship plot of the model in Figure 9.1 of Chapter 2. The plot graphically estimates the relationship between life and stress. The method employs plotting paper on which the relationship is a straight line. For example, Arrhenius paper (Figure 2.3) has a log time scale and a reciprocal scale for absolute temperature. Plot each failure time against its temperature as in Figure 2.3, which shows the Class-H data of Table 2.1. Draw a line through the data to estimate "life" as a function of temperature. Note that Figure 2.3 also shows that the 260° data has greater scatter than data from other test temperatures. This was noted earlier from the lognormal plot, Figure 2.2, which shows this more clearly. Probability plots often reveal more. Figures 2.2 and 2.3 are side by side and have identical time scales. A particular failure time is plotted at the same height on both figures. This pair of figures depicts that the probability and relationship plots are two views of the same model and data.

Life estimate at any stress. The "life" line is used to estimate life at any stress level. Enter the plot at the stress level, go up to the fitted life line, and then go horizontally to the time scale to read the estimate. For the Class-H insulation, the estimate of life at the design temperature of 180°C is 11,500 hours from Figure 2.3. This is well below the desired 20,000 hours. "Life" as used here is a vague "typical" life and is common engineering usage. The uncertainty of a graphical estimate can be gauged subjectively. Wiggle the fitted line (clear straight edge) to change the estimate while still passing the line through the data well. The largest and smallest such estimates for which the line fits well enough indicate the uncertainty. For example, a line through 20,000 hours at 180°C does not pass through the data well enough. This is convincing that the insulation does not meet the 20,000 hour requirement. Chapter 4 presents confidence intervals, a precise means of evaluating uncertainties of numerical estimates.

Percentile lines. The "life" line has more specific meaning if fitted as follows. On the relationship paper, plot the estimate of a chosen percentile at each stress. Such percentile estimates from the distribution line on a probability plot are described above. The sample percentile and the geometric mean are other useful estimates; they are described below. Fit a line to the percentile estimates. Try to weight each estimate proportional to the number of test units at its stress. This line graphically estimates the relationship between the percentile and stress. Such a line is fitted to medians (×s) in Figure 2.3. Plot all points to display the data and to help spot peculiar data.

Analytic methods for estimating the median and other percentile lines appear in Chapters 4 and 5. It is useful to plot such fitted lines with the data as in Figure 2.3. The line helps check the fit. Also, such a plot is a good means of presenting results to others, regardless whether analytic or graphical.

Sample percentile. The sample percentile is another useful percentile estimate. Suppose that the chosen percentage is a plotting position, say, 25%. Then the **sample percentile** is the corresponding observation. For example, the sample 25th percentile at 240°C is 1521 hours from Table 2.1. Suppose the chosen percentage is not a plotting position, say, 50%. On the relationship plot, one graphically interpolates appropriately between the observations with plotting positions just above and below the chosen percentage. For the Class-H data, the sample 50th percentile is graphically midway between the 5th and 6th largest observations, which have plotting positions of 45 and 55%. Mark such sample percentiles with ×s in relationship plots.

The sample percentile is robust. That is, it is a valid estimate of a distribution percentile even when the assumed distribution is not valid over the entire actual distribution. This is so because the estimate does not use the distribution line on the probability plot. If the assumed distribution is valid, then the sample percentile is not as accurate as the previous estimate. This is so because the sample percentile uses one or two observations, and the previous estimate uses the entire sample at that stress. Usually the percentile is chosen in the middle or lower tail of the distribution, near the percentiles of interest at the design stress.

Geometric mean. Another useful estimate of a median is the geometric mean. For a stress, take the $\log_{10}$ of each time, sum those logs, and divide the sum by the number of observations in the sum. The antilog of this result is the geometric mean. For example, for the Class-H data at 190°C, the geometric mean is 8,701 hours. This estimate is valid only for the lognormal distribution and when there are no peculiar data. Then it is the most accurate estimate of the median. If the lognormal distribution is in doubt, the previous two estimates are likely more accurate.

Arrhenius papers. Arrhenius plotting papers are available from

1) TEAM, Box 25, Tamworth, NH 03886, (603)323-8843.

   8112   − 100 to + 200°C (11" horizontal)   3 log cycles (8 1/2" vertical)
   8212   20 to   400°C (11" horizontal)   3 log cycles (8 1/2" vertical)
   8312   20 to   1600°C (11" horizontal)   3 log cycles (8 1/2" vertical)

2) K + E (Keuffel & Esser Co.), 20 Whippany Rd., Morristown, NJ 07960, (800)538-3355.

   46 8200   40 to 300°C (8 1/2" horizontal)   4 log cycles (11" vertical)
   46 8242   100 to 300°C (8 1/2" horizontal)   4 log cycles (11" vertical)

3) Technology Associates, 51 Hillbrook Dr., Portola Valley, CA 94025, (415)941-8272. For a convenient scale for *activation energy* (Figure 6.3):

   5001   25 to 400°C (8 1/2" vertical)   7 log cycles (11" horizontal)

2.4. Graphical Estimates

Graphical estimates of the life distribution at any stress (such as a design stress), of the model parameters, and of a design stress are given below.

Distribution for a stress. The distribution line for any stress, such as a design stress, is fitted as follows. First, use the percentile line on the relationship plot to estimate the percentile at the chosen stress level. For example, the estimate of median life at the design temperature of 180°C is 11,500 hours from Figure 2.3. Plot that estimate on the probability plot. This point is shown as an × on Figure 2.2. Draw a distribution line through that point parallel to the distribution lines for the test stresses. Such a 180° line with a compromise ("average") slope appears in Figure 2.2. This line provides estimates of percentiles and fraction failed for the 180° life distribution.

If the plot or other knowledge suggests that the distribution lines are not parallel, fit a line with an appropriate slope, for example, with the slope of the data at the nearest test stress. In Figure 2.1B, one would use the slope of the 190° data, not the compromise 180° line in the plot. Information given later suggests that the 260° data should be ignored in drawing the 180° line.

Percent failed at a (design) stress. As follows, estimate the percent failed by a given age at a specified stress, such as a design stress. Use the preceding distribution line for that stress. Enter the probability plot on the time scale at the given age, go horizontally to the distribution line, and then go vertically to the probability scale to read the percentage failed. For the Class-H insulation, the estimate of the percentage failed by 10,000 hours at 180°C is 27% from Figure 2.2.

Percentile at a (design) stress. A percentile at any specified stress (such as design stress) is estimated from the fitted distribution line as follows. Enter the probability plot at that percentage, go to the fitted line, and then go to the time scale to read the estimate. For example, in Figure 2.2, the estimate of median life at the design temperature of 180°C is 11,500 hours. This is much below the desired 20,000 hour median. Also, the estimate of the 1st percentile (off scale) at 180°C is 6600 hours.

Other percentile lines. Other percentile lines are drawn on the relationship plot as follows. As described in a previous paragraph, obtain a percentile estimate at the design stress. For example, the estimate of the 1st percentile at 180°C is 6600 hours. Plot this estimate on the relationship plot. Then through this point draw a line that is parallel to the original relationship line as in Figure 9.1 of Chapter 2. This added line estimates the desired percentile line. It is used as described next to estimate the design stress that yields a specified life. Such parallel lines for a number of percentiles may be added. Then the plot looks like the model in Figure 9.1 of Chapter 2.

Design stress. One may need to estimate a design stress that yields a specified life. To do this, enter the relationship plot at the specified life on the time scale, go to the fitted line for the desired percentile, and go to the stress scale to read the estimate. For example, a median life of 20,000 hours results from an estimated temperature of 165°C from Figure 2.3. The analytic estimate and confidence limits for this stress are in Chapter 4.

Relationship parameters (activation energy). Sometimes one wishes to estimate the coefficients γ_0 and γ_1 of the simple linear relationship. For the Arrhenius relationship, γ_1 is related to the activation energy E of the chemical degradation that produces failure; thus γ_1 has a physical interpretation. Choose two widely spaced temperatures $T < T'$. Obtain graphical estimates of the median lives t^*_{50} and $t^{*'}_{50}$ from the relationship plot. Then the estimates of γ_1, γ_0, and the activation energy E are

$$\gamma_1^* = [TT'/(T'-T)]\log(t^*_{50}/t^{*'}_{50}), \quad \gamma_0^* = \log(t^*_{50}) - (\gamma_1^*/T'),$$

$$E^* = 2.303k\gamma_1^*; \tag{1.3}$$

here $k = 0.8617 \times 10^{-4}$ is Boltzmann's constant in electron volts per degree

Kelvin. Analytic estimates and confidence limits for γ_1, γ_0, and E are in Chapter 4. Peck and Trapp's (1978) Arrhenius paper has a special scale that estimates activation energy directly. Other papers listed above lack this scale.

For example, from Figure 2.3, the graphical estimates of the medians at $T = 453.2°K$ (180°C) and $T' = 533.2°K$ (260°C) are $t^*_{.50} = 11,500$ and $t^{*'}_{.50} = 950$ hours. Thus

$$\gamma^*_1 = [(453.2)533.2/(533.2-453.2)]\log(11,500/950) = 3271,$$

$$\gamma^*_0 = \log(950) - (3271/533.2) = -3.17,$$

$$E^* = 2.303 \times 0.8617 \times 10^{-4}(3271) = 0.65 \text{ eV}.$$

Log standard deviation. Estimate σ from the slope of a fitted line in the lognormal probability plot as follows. Enter the plot on the probability scale at the 50% point, go down to one of the parallel fitted lines, and then go sideways to the time scale to read the median estimate $t^*_{.50}$. Similarly, obtain the estimate $t^*_{.16}$ of the 16th percentile from the same line. The estimate of σ is

$$\sigma^* = \log(t^*_{.50}/t^*_{.16}). \qquad (1.4)$$

Analytic estimates and confidence limits for σ are in Chapter 4. Another graphical estimate of σ uses residuals (Chapter 4).

For example, for the 180° line in Figure 2.1B, $t^*_{.50} = 11,500$ hours and $t^*_{.16} = 9,000$ hours. Then $\sigma^* = \log(11,500/9,000) = 0.11$. This small estimate indicates that such insulation has a failure rate that increases over most of the distribution. The line for a test temperature could also be used to estimate σ, say, the 190° line.

2.5. Assess the Model and Data

The validity of the graphical analyses and estimates above depend on how valid the assumptions of the model are. The methods below check the assumptions of the simple (Arrhenius-lognormal) model and the validity of the data. In particular, the methods check that the assumed distribution fits the data, that the distribution spread is the same for all stresses, and that the (transformed) life-stress relationship is linear. Also, the methods check the data for outliers and other peculiarities due to blunders or faulty testing.

The methods below are applied to the Class-H insulation data of Section 2.1 and the Arrhenius-lognormal model. However, the methods apply to data from any simple model. They apply, for example, to the power-Weibull model of Section 3.

Analytic methods for assessing the model and complete data appear in Chapter 4. That chapter also includes graphical analyses of residuals. Use graphical and analytic methods to get the most information.

Lognormal distribution. A probability plot allows one to assess how well the theoretical distribution fits the data. Relatively straight lognormal plots suggest the distribution adequately fits the data. To judge straightness sensitively, hold the plot at eye level and sight along the line of plotted points. For example, the four data plots in Figure 2.1A are relatively straight and indicate that the lognormal distribution adequately fits at the four temperatures.

Curved distribution plots. Plots curved the same way suggest that another distribution, such as the Weibull distribution, may fit the data better. Another analysis of curved plots involves drawing smooth curves through the points. Such curves (nonparametric fits) can be used as distribution lines. Estimates of percentiles and fraction failed are read from them as explained above. Such percentile estimates can be plotted on a relationship plot and percentile lines fitted to them.

Sometimes the lower tails of curved plots are straight. If interested only in the lower tail, one can treat the data above some point in each lower tail as censored. The censored data can then be analyzed with the methods of Section 4 or 5. Then the fitted model and estimates describe only the lower tail of the fitted distribution. Hahn, Morgan, and Nelson (1985) present methods for such censoring and analytically fitting to just the early failure data.

Outliers. Individual points out of line with the rest may indicate unusual or mishandled specimens, rather than poor fit of the distribution. Such "outliers" usually are failures that are too early relative to others at the same stress. Peculiar data are discussed in detail below.

Overinterpretation. Those inexperienced in analyzing data tend to overinterpret plots. They assume that any noticeable gap, flat spot, curvature, etc., of a data plot has physical meaning and is a property of the population. But only pronounced features of a plot should be assumed to be properties of the population. Hahn and Shapiro (1967) and Daniel and Wood (1971) display probability plots of Monte Carlo data from a true normal distribution. Their plots are sobering. Their samples of 20 and even 50 observations appear erratic, having peculiarities such as curvature, gaps, and outlying points. Most people's subjective notions of randomness are stringent and orderly; they incorrectly expect points in a probability plot to fall on a straight line. Some authors have promoted this fallacy by creating illustrative data by drawing data points on a straight line on a probability plot. *Caveat lector.* This book presents only real data, complete with pimples and warts.

Constant standard deviation. The Arrhenius-lognormal model assumes that the standard deviation σ of log life is a constant. If σ does depend on stress, then the estimates for percentiles at a stress may be inaccurate. A σ that depends on stress may be a property of the product, may result from a faulty test, or may be due to competing failure modes (Chapter 7).

The following method assesses whether the spread σ is independent of stress. If the spread is constant, probability plots (Section 2.3) of the data should yield parallel lines. Most samples at each test stress are small (less than 20 specimens). So the slopes of the distribution lines may randomly vary a lot when the true population spreads are the same. Thus, only extreme differences in the slopes are convincing. The slope changing systematically with stress may indicate that the spread depends on stress or there are competing failure modes. If the slope for one test stress differs greatly from the others, the data or the test at that stress may be faulty.

Clue to $1,000,000. In Figure 2.1A, the separate lines for 190, 220, and 240° lines are parallel, and the 260° line is steeper. The slope of a distribution line corresponds to the log standard deviation σ. So the plot suggests that the scatter in (log) life at 260° is greater; this is contrary to the assumed constant scatter (log standard deviation) of the simple model. In insulation work, the validity of the Arrhenius model is well established, and departures from the model are usually due to faulty experimental technique or competing failure modes (Chapter 7). The nonparallel lines were the first clue leading to a yearly savings of one million dollars (1989 prices) for the product department. After being shown the nonparallel plots, the responsible insulation engineer suggested two possible reasons.

First, the 10 motorette specimens at 260° were made after the other 30. The 30 had not failed fast enough (few failures yield little information). So the engineer decided to make 10 more and test them at 260°C to get failures quickly. When questioned, the engineer did not know if the 10 specimens were made from the same lots of raw materials or by the same shop people as the previous 30. This suggests that the nonparallel plots are possibly due to lack of quality control in the materials or in making the specimens.

Second, the engineer then revealed that the motorettes failed from three causes. A green statistician at the time, the author did not know enough to ask the engineer at the outset if the data contained more than one cause of failure. Riper now, the author asks this right away, even when just planning a test. Autopsy of each failed motorette revealed the cause as a 1) Turn failure, 2) Ground failure, or 3) Phase failure. Each corresponds to failure of a distinct part of the insulation system. Moreover, most of the 260° failures were Ground failures, and most of the failures at the other three temperatures were Turn failures. So the scatter in log life of Ground insulation may just be greater than that of Turn insulation. Clearly the Ground failure data at 260° should not be used to estimate life at 180° where motorettes mostly fail from Turn failures. Section 5 presents valid graphical analyses for such data and further insight leading to the $1,000,000 yearly saving. Chapter 7 presents numerical analyses of data with competing failure modes.

The nonparallel distribution lines and competing failure modes are ignored in what follows. Ignoring these complications is done only to present the graphical methods.

So there is a question of whether or not to use the 260° data in the analyses. In practice, it is best to do two analyses, one with and the other without the suspect data. If the results of both are the same for practical purposes, one need not choose an analysis. Otherwise, one must decide which analysis is better, say, more correct or more conservative.

Linear (Arrhenius) relationship. The Arrhenius relationship is a linear relationship between transformed life and temperature. This assumed linearity is important when one extrapolates the fitted relationship to estimate life at low stress. The relationship plot may be nonlinear for various reasons. The life test may not have been carried out properly. The data may contain a number of competing failure modes, each with a different linear relationship. Also, the (transformed) true relationship just may not be linear.

Subjective assessment of the linearity comes from examining the relationship plot of the data (or better the percentile estimates). The data and the medians of the Class-H insulation in Figure 2.3 are close to a straight line compared to the scatter in the data. This suggests that the Arrhenius relationship adequately fits the data. Chapter 4 presents analytic methods for checking linearity.

Nonlinear relationship plot. If nonlinearity is convincing, examine the relationship plot of the percentile estimates to determine how the relationship departs from linearity. For competing failure modes (Chapter 7), the resulting relationship is concave downwards in Figure 2.3. If a sample percentile is out of line with the others, the data for that stress may be in error. Look for erroneous data, and determine if those data should be used or not. After examining the relationship plot, one may do any of the following:

- If the plot shows that a smooth curve describes the relationship, fit a curve to the data. Be sure that the apparent curvature is not a result of erroneous data. Extrapolating such a curve to the design stress may be difficult to justify and is likely to be inaccurate.
- Analyze the data with a linear fit but subjectively take into account the nonlinearity in interpreting the data and coming to conclusions. The linear fit may ignore certain data or weight some data more or less.

Valid data. The probability and relationship plots may reveal peculiarities of the data. In a sense, the data are always valid, and our assumptions are often invalid if the data differ from the model. Sometimes peculiar data arise from blunders in recording or transcribing the data. More often such data arise from specimens that are mismade or mistested or an inaccurate model. Methods above check the adequacy of the model, namely, the distribution, a common standard deviation of log life, and a linear relationship.

Peculiar data. If the model does not fit the data well, it is important to determine the reason. Some people speak backwards and say the data do not fit the model. The data are almost always right, and understanding the reason for peculiar data is often more important than the good data. For example, the greater scatter in the 260° Class-H data led to yearly savings of

$1,000,000. Thus it is essential to determine the cause of peculiar data. It is less important to decide whether to include or exclude peculiar data from an analysis. Usually it is best to do two or more analyses, excluding some or all of the peculiar data. Often the practical conclusions are the same for such analyses (for example, Problem 3.7). When the conclusions differ, then one must chose an analysis, say, the most conservative or most realistic analysis.

Outliers. Sometimes there is one or a few data points that stand out in a plot. Such points are usually failures that are too early relative to the rest of the data. As noted above, it is usually most informative to determine the cause of such "outliers." Otherwise, one can do analyses with and without such outliers. Section 4 presents data with outliers.

3. COMPLETE DATA AND POWER-WEIBULL MODEL

Introduction. This section presents simple graphical methods for complete life test data, using Weibull and exponential life distributions and the (inverse) power relationship. Exactly like those in Section 2, the methods estimate the model parameters and the product life distribution at any stress. The plots also are used to assess the validity of the model and data. Corresponding least-squares methods appear in Chapter 4. To use the methods, one needs background on the Weibull and exponential distributions and the inverse power law in Chapter 2. While the example employs the inverse power law, the methods apply to other relationships between life and stress. These methods are the same as those of Section 2, which is needed background. Thus this section aims only to acquaint readers with Weibull probability paper and the inverse power law, since both are widely used for accelerated life test data.

Overview. Section 3.1 describes data that illustrate the graphical methods. Section 3.2 explains how to make and use Weibull probability plots of the data. Section 3.3 presents a log-log plot of the relationship between life and stress. Section 3.4 presents estimates of the model parameters and other quantities. Section 3.5 shows how to assess the model and data.

3.1. Data (Insulating Fluid)

The data in Table 3.1 illustrate the graphical methods for complete data, using the power-Weibull model. The data are the times to oil breakdown under high test voltages. High voltages quickly yield breakdown data. At design voltages, time to breakdown runs thousands of years. The tests employed two parallel plate electrodes of a certain area and gap. The electrical stress is given as a voltage, since the electrode geometry was constant.

The main purpose was to estimate the relationship between time to breakdown and voltage. This involves fitting the model to the data. The model is used to estimate the probability of product failure during a factory

Table 3.1. Times to Breakdown of an Insulating Fluid

26 kV Min-utes	Plotting Position	28 kV Min-utes	Plotting Position	30 kV Min-utes	Plotting Position	32 kV Min-utes	Plotting Position
5.79	16.3	68.85	10.0	7.74	4.5	0.27	3.3
1579.52	50.0	108.29	30.0	17.05	13.6	0.40	10.0
2323.70	83.3	110.29	50.0	20.46	22.7	0.69	16.7
		426.07	70.0	21.02	31.8	0.79	23.3
		1067.60	90.0	22.66	40.9	2.75	30.0
				43.40	50.0	3.91	36.7
				47.30	59.1	9.88	43.3
				139.07	68.2	13.95	50.0
				144.12	77.3	15.93	56.7
				175.88	86.4	27.80	63.3
				194.90	95.5	53.24	70.0
						82.85	76.7
						89.29	83.3
						100.58	90.0
						215.10	96.7

34 kV Min-utes	Plotting Position	36 kV Min-utes	Plotting Position	38 kV Min-utes	Plotting Position
0.19	2.6				
0.78	7.9				
0.96	13.2	0.35	3.3		
1.31	18.4	0.59	10.0		
2.78	23.7	0.96	16.7		
3.16	28.9	0.99	23.3		
4.15	34.2	1.69	30.0		
4.67	39.5	1.97	36.7		
4.85	44.7	2.07	43.3	0.09	6.2
6.50	50.0	2.58	50.0	0.39	18.7
7.35	55.3	2.71	56.7	0.47	31.2
8.01	60.5	2.90	63.3	0.73	43.7
8.27	65.8	3.67	70.0	0.74	56.2
12.06	71.1	3.99	76.7	1.13	68.7
31.75	76.3	5.35	83.3	1.40	81.2
32.52	81.6	13.77	90.0	2.38	93.7
33.91	86.8	25.50	96.7		
36.71	92.1				
72.89	97.4				

test at 20 kV. Another purpose was to assess whether the distribution of time to breakdown is exponential.

3.2. Weibull Probability Plot

Weibull plot. A probability plot is made and interpreted as described in Section 2.2. However, Weibull probability paper is used (Figure 3.1). Such paper has a log scale for data (time) and a Weibull scale for probability. Data from all of the stresses are plotted on in Figure 3.1 and clutter the plot. There is a separate straight line through the data for each test stress. The

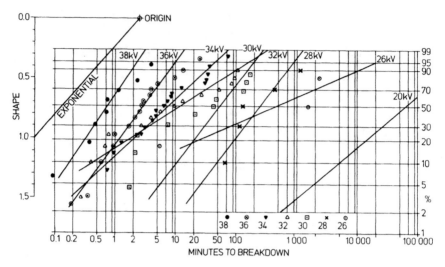

Figure 3.1.   Weibull plot of insulating fluid data.

lines estimate the cumulative distribution functions at those stresses, that is, the percentage failed versus time.

For the model for the example, the Weibull life distribution has the same shape parameter value at any stress. This means that the distribution lines are all parallel. Such parallel lines are fitted in Figure 3.2; there data from half of the stresses are plotted to avoid clutter.

Shape parameter estimate. The Weibull shape parameter indicates the

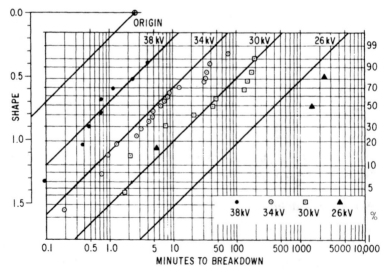

Figure 3.2.   Parallel fits to insulating fluid data.

behavior of the failure rate with time. It is estimated from the Weibull probability plot (Figure 3.2). Through the point labeled "origin," draw a line parallel to the common slope of the fitted lines. The line intersects the shape parameter scale at a value which is the graphical estimate. Figure 3.2 gives an estimate of 0.81. The value less than 1 indicates that the failure rate decreases with time. Also, the life distribution is close to exponential. It is not clear that this estimate convincingly differs from 1. Also, uncontrolled test conditions (voltage etc.) may have created greater scatter in the data and thereby lowered the observed shape value below 1. Chapter 4 gives an analytic method that assesses whether the sample shape parameter significantly differs from 1, that is, an exponential life distribution.

Specified Weibull shape. Sometimes the value of the Weibull shape parameter is specified. In the example, the life distribution is assumed to be exponential – a shape parameter of 1. One can fit distributions with that shape parameter value to the data as follows. On the Weibull paper, draw a line from the point labeled "origin" through the shape parameter scale at the specified value. Such a line in Figure 3.1 passes through the value 1 (an exponential distribution). Through the data for a stress draw a line parallel to the shape parameter line. That fitted distribution has the specified Weibull shape parameter value.

Weibull papers. Weibull probability papers are available from:

1) TEAM, Box 25, Tamworth, NH 03886, (603)323-8843.

118	0.0001-99.9% (8½" vertical)	3 cycles (11" horizontal)
218	0.0001-99.9% (8½" vertical)	5 cycles (11" horizontal)
318	0.0001-99.9% (8½" vertical)	7 cycles (11" horizontal)
118-2	0.01-99.9% (8½" vertical)	3 cycles (11" horizontal)
218-2	0.01-99.9% (8½" vertical)	5 cycles (11" horizontal)
318-2	0.01-99.9% (8½" vertical)	7 cycles (11" horizontal)
118-3	1.0-99.9% (8½" vertical)	3 cycles (11" horizontal)
218-3	1.0-99.9% (8½" vertical)	5 cycles (11" horizontal)
318-3	1.0-99.9% (8½" vertical)	7 cycles (11" horizontal)
228	0.0001-99.9% (11" vertical)	7 cycles (14" horizontal)
112*	0.01-99.99% (11" horizontal)	3 cycles (8½" vertical)
113*	0.01-99.99% (11" horizontal)	5 cycles (8½" vertical)
122*	0.0001-99.9999% (14" horizontal)	3 cycles (11" vertical)
123*	0.0001-99.9999% (14" horizontal)	5 cycles (11" vertical)
10/18-3	1-99.9% (11" horizontal)	1 cycle (8½" vertical)
20/18-3	1-99.9% (11" horizontal)	2 cycles (8½" vertical)
40/18-2	0.01-99.9% (11" horizontal)	4 cycles (8½" vertical)
SA-6	Assortment of 118, 118-2, 118-3, 218, 218-2, 218-3, 318, 318-2, 318-3, 518-2.	

* The catalog calls these extreme value probability papers. Their data

scale is logarithmic, and they are actually Weibull papers but lack a shape parameter scale.

2) CODEX Book Co., 74 Broadway, Norwood, MA 02062, (617)769-1050:

Y4 280*   0.1-99.9% (11" vertical)   4 cycles (16½" horizontal)

3) Craver (1980) gives reproducible copies of

pg. 228   0.1-99.9% (11" vertical)   2 log cycles (8½" horizontal)
pg. 230   0.001-99.9% (11" vertical)   3 log cycles (8½" horizontal)

Exponential paper. Use Weibull paper for an exponential fit. Exponential probability paper badly compresses data in the lower tail, which is usually of greatest interest.

3.3. Relationship Plot (Inverse Power)

Relationship line. For the inverse power law, plot the data on log-log paper (Figure 3.3). Then fit a straight line by eye to pass through the data. This is best done by fitting the line to an estimate of a percentile at each stress. Such estimates of the characteristic life (63.2 percentile) are shown as ×'s in Figure 3.3. The fitted line in Figure 3.3 graphically estimates the inverse power law relationship between the characteristic life and stress. Compare Figure 3.3 with the plotted model in Figure 10.1 of Chapter 2. Here the

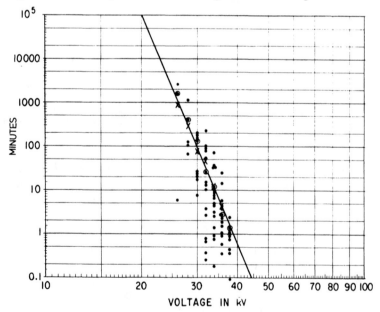

Figure 3.3.   Log-log plot of insulating fluid data (× is a graphical estimate of α and O is the sample 63rd percentile).

relationship plot (Figure 3.3) and the corresponding probability plot (Figure 3.1 or 3.2) do not appear side by side with identical time scales, as do Figures 2.2 and 2.3, and one is turned 90° from the other. Log-log and Weibull papers with identical time scales show the correspondence between the two data plots. However, such pairs of papers have not been developed.

Percentile estimates. Two estimates of a percentile at a stress can be used. One is obtained graphically from the fitted distribution line as described in Section 2.2. Such estimates appear as $\times$'s in Figure 3.3. The other is the sample percentile of Section 2.3. These observations (63.2 percentiles) are circles in Figure 3.3. In practice, only one such estimate is marked at each test stress. The two percentile estimates differ with respect to simplicity and statistical accuracy. The sample percentile is easier to use and less accurate. For most purposes, the graphical estimate is recommended.

Log-log papers. Most commercial log-log paper has cycles of the same length on both axes. Such paper is often not suitable for accelerated test data. Such paper is available through CODEX (1988), Craver (1980), Dietzen (1988), Keuffel & Esser (1988), and TEAM (1988) among others. Better paper (Figure 3.3) has one or two large cycles for stress and many small cycles for life; some engineering departments have developed such papers. Semi-log paper with one or two log cycles can be used. Mark the linear scale with powers of 10 and treat it as a log scale.

3.4. Graphical Estimates

Graphical estimates are obtained as described in Section 2. Examples of such estimates follow and include the Weibull characteristic life, percentiles, relationship parameters, and the design stress that yields a specified life.

Characteristic life. The estimate of the characteristic life at any stress can be read directly from the fitted line in the relationship plot. Enter the plot at that stress on the horizontal scale. Go up to the fitted line, and then go sideways to the time scale to read the estimate. For example, in Figure 3.3, the estimate of the characteristic life at 20 kV is 105,000 min. The estimate may be in error if the relationship is nonlinear on log-log paper. Analytic methods for checking linearity appear in Chapter 4.

Percentile lines. Section 2 shows how to estimate percentile lines in a relationship plot. Such lines appear in Figure 10.1 of Chapter 2. The lines estimate those percentiles at any stress. For example, the 1st percentile at 20 kV is 210 min. from Figure 10.1.

Relationship parameters (power). The power γ_1 in the inverse power law $\alpha = e^{\gamma_0}/V^{\gamma_1}$ is often of interest. The larger its (absolute) value, the more sharply life decreases as stress increases. For two stresses $V < V'$, graphically estimate their characteristic lives α^* and α'^*. Then the graphical estimates of γ_1 and γ_0 are

$$\gamma_1^* = \ln(\alpha^*/\alpha'^*)/\ln(V'/V), \quad \gamma_0^* = \ln(\alpha^* V'^{\hat{n}}).$$ (3.1)

For example, in Figure 3.3 $\alpha^* = 105,000$ min at $V = 20$ kV and $\alpha'^* = 0.60$ min at $V' = 40$ kV. Thus

$$\gamma_1^* = \ln(105,000/0.60)/\ln(40/20) = 17.4, \quad \gamma_0^* = \ln(105,000 \cdot 20^{17.4}) = 63.7.$$

A power of 17.4 is unusually large, even for insulation life.

Choice of design stress. Sometimes one needs to estimate a design stress with a desired characteristic life. Enter the log-log plot at that life on the time scale, go sideways to the fitted line, and go down to the stress scale to read the estimate. For example, in Figure 3.3, a characteristic life of 1,000,000 min (offscale) is provided by an estimated voltage of 17.6 kV. This method can also be used with percentile lines to estimate a design stress with a specified low percentile. For example, if the 1st percentile must be 10,000 min., then the stress must be 15.8 kV from Figure 10.1 of Chapter 2.

Distribution line at any (design) stress. The distribution line at any stress is estimated on the probability paper as described in Section 2.4. Such a line for the insulating fluid at the factory test voltage of 20 kV appears in Figure 3.1. Percentiles and percentage failing are estimated from the line as described in Section 2.4. For example, the estimate of the 1st percentile at 20 kV is 210 min. from Figure 3.1. Also, the estimate of the percentage failing by 10 min. at 36 kV is 95%.

3.5. Assess the Model and Data

Graphical methods of Section 2.5 are used to assess the (Weibull) distribution, the (power law) relationship, and the data. Corresponding analytic methods appear in Chapter 4. Use both graphical and analytic methods to get the most information from the data.

The straight Weibull plots (Figure 3.1) indicate that the Weibull distribution fits adequately. The relatively parallel Weibull plots (Figure 3.1) suggest a common shape parameter is reasonable. The straight relationship plot (Figure 3.3) indicates that the power law fits over the range of the data. In both plots, a low outlier at 26 kV is evident. No reason for it was found. Eliminating it has little effect on estimates, since the sample is relatively large (76 times to breakdown) and the Weibull distribution has a long lower tail.

4. SINGLY CENSORED DATA

Introduction. Often life data are not complete. When the data are analyzed, some units may still be running. Such data are **singly censored** when the failure times of unfailed units are known only to be beyond their current common (single) running time. A censored life clearly cannot be discarded or treated as a failure as this ignores and arbitrarily changes such data.

This section presents simple graphical methods for estimating from singly censored data the model and the life distribution at any stress. The methods are like those for complete data in Sections 2 and 3 and include 1) a probability plot of singly censored data from each test stress and 2) a relationship plot of life against stress.

Also, a distribution may adequately fit only data in the lower tail at each stress. When the upper tail is not of interest, then the influence of data in the upper tail can be removed. One treats all upper tail data as censored at some time in the lower tail. Hahn, Morgan, and Nelson (1985) present such *artificial censoring*.

Overview. Section 4.1 describes data used to illustrate the methods here. Section 4.2 presents probability plotting of such data, and Section 4.3 presents relationship plotting. Section 4.4 briefly reviews graphical estimates. Section 4.5 discusses checks on the model and data. Chapter 5 provides advanced analytic methods for such data. A combination of graphical and analytic analyses of such data is most effective.

4.1. Data (Class-B Insulation)

Table 4.1 displays censored data from a temperature-accelerated life test of a Class-B insulation for electric motors (Crawford 1970). Ten motorettes were tested at each of four temperatures (150°C, 170°C, 190°C, 220°C). The test purpose was to estimate the life distribution (in particular, its median and 10% point) at the design temperature of 130°C. At the time of the analysis, 7 motorettes at 170°C had failed, five each at 190°C and 220°C had failed, and none at 150°C had failed. The + in Table 4.1 indicates a running motorette at that number of hours. The motorettes were periodically checked for failure, and a failure time in Table 4.1 is the upper endpoint of the period in which the failure occurred. It is better to use the midpoint.

Table 4.1. Class-B Insulation Life Data and Plotting Positions

150 °C		170°C		190°C		220°C	
Hours	F_i	Hours	F_i	Hours	F_i	Hours	F_i
8064+	-	1764	5%	408	5%	408	5%
8064+	-	2772	15	408	15	408	15
8064+	-	3444	25	1344	25	504	25
8064+	-	3542	35	1344	35	504	35
8064+	-	3780	45	1440	45	504	45
8064+	-	4860	55	1680+	-	528+	-
8064+	-	5196	65	1680+	-	528+	-
8064+	-	5448+	-	1680+	-	528+	-
8064+	-	5448+	-	1680+	-	528+	-
8064+	-	5448+	-	1680+	-	528+	-

The lognormal distribution and Arrhenius relationship are used to analyze these data. The methods also apply to other simple models, using the Weibull distribution and the inverse power law and other relationships.

4.2. Probability Plot (Lognormal)

A probability plot of censored data at each test stress is made as follows. Use probability plotting paper for the model distribution (lognormal here). Suppose, at a test stress, a sample of n units has r failure times. Order the times from smallest to largest, and assign rank i to the ith ordered failure time. As before, the plotting position of the ith failure is

$$F_i = 100(i - 0.5)/n, \quad i = 1, 2, \cdots, r.$$

These plotting positions are in Table 4.1. Nonfailure times are not assigned plotting positions. Other plotting positions of Section 2.2 can be used.

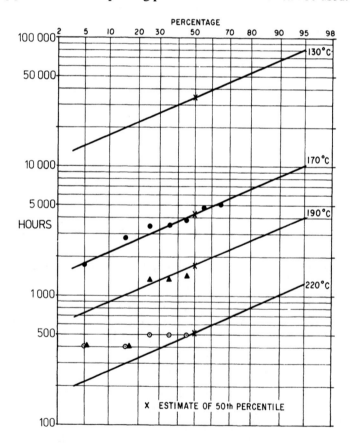

Figure 4.1.   Lognormal plot of censored Class-B data.

On the probability paper, plot each failure time against its plotting position. Nonfailure times are not plotted. Figure 4.1 shows a lognormal probability plot of the Class-B data. By eye fit a straight line (or, if necessary, a curve) to the plotted points as shown in Figure 4.1. Fitted lines can be parallel, since the model has parallel lines. As before, the fitted lines obscure the data. So it is best to also have a plot without lines.

The probability plot of singly censored data yields the same information and is interpreted the same as a plot of complete data (Sections 2.2-2.5 and 3.2-3.5). Such information includes estimates of distribution percentiles and fraction failing by a given age. For example, Figure 4.1 yields median estimates of 4300 hours at 170°, 1650 hours at 190°, and 510 hours at 220°. For 190 or 220°, estimating the median involves extrapolating from data in the lower tail of the distribution to the middle of the distribution. The lognormal (cumulative) distribution is used to extrapolate over time. This is similar in spirit to using the Arrhenius relationship to extrapolate over temperature.

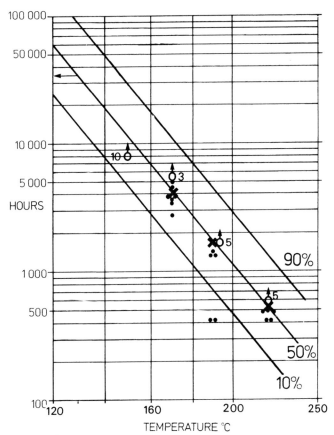

Figure 4.2.   Arrhenius plot of Class-B data (• observed failure, × estimate of median, Ô# unfailed).

4.3. Relationship Plot (Arrhenius)

A relationship plot of life against stress is made as described in Section 2.3 or 3.3. In particular, estimate a specific percentile at each test stress from the probability plot. Then plot each estimate against stress on paper where the relationship between "life" and stress is linear. Finally draw a line through the plotted estimates to estimate the relationship.

For the Class-B insulation, each median estimate is plotted with a cross against its test temperature on Arrhenius paper in Figure 4.2. A straight line is fitted by eye to the crosses in Figure 4.2. The failure and nonfailure times are also plotted to display the data. Nonfailures make such a plot more difficult to grasp. This is one reason the line is fitted to percentile estimates rather than directly to the data.

The fitted relationship is used as described in Sections 2.3, 2.4, 3.3 and 3.4 to estimate model parameters and life at a given stress. For example, the estimate of median life at the design temperature of 130°C is 35,000 hours. This key estimate indicates that insulation life is satisfactory.

Analytic fitting of a relationship to censored data appears in Chapter 5. Such fitting provides confidence limits, as well as estimates.

4.4. Graphical Estimates

The probability and relationship plots yield estimates as described in Sections 2.4 and 3.4. An example follows.

To estimate the life distribution line at 130°C, mark the median life of 35,000 hours on the lognormal probability paper in Figure 4.1. Then draw a straight line through this median. Choose the slope of this 130° line as the visual average of the slopes for the test temperatures. The slope (log standard deviation) is assumed to be the same for all temperatures. This line estimates the 130° life distribution. For example, the estimate of the 10th percentile at 130°C is 17,300 hours.

4.5. Assess the Model and Data

Graphical methods of Sections 2.5 and 3.5 are used to assess the (lognormal) distribution, the (Arrhenius) relationship, and the data. Censored data require greater care in interpretation; for example, the nonfailures in a relationship plot must be visually assessed differently from the failures. Thus it is best to use estimates of percentiles in such a plot to assess linearity.

The lognormal plots (Figure 4.1) of the 170 and 220° data are relatively straight. However, their slopes differ some. When informed of this, the insulation engineer revealed that post mortem of failures had identified different dominant failure modes for these two temperatures. For more details on a mixture of failure modes, see Chapter 7.

At 190°C, two failures are much too early. Review of the data and test method did not reveal their cause. The data were reanalyzed without them. The estimate of median life at 130° changed little. Of course, the estimate of the log standard deviation decreased, and the estimate of the 10% point at 130°C increased.

Examination of the plot of the medians suggests that over the range 170°C, 190°C, and 220°C the linear relationship is adequate.

5. MULTIPLY CENSORED DATA

Introduction. In some accelerated life tests, data at a stress level are multiply censored. Such data contain running and failure times that are intermixed. Such data result from (1) analysis of the data while specimens are still running, (2) removal of specimens from test at various times, (3) starting specimens on test at various times, and (4) loss of specimens through failure modes not of interest or through extraneous causes such as test equipment failure. This section describes graphical estimates of the accelerated test model and of the product life distribution at any stress. Graphical analysis involves (1) a hazard plot of the multiply censored data at each test stress (like a probability plot) and (2) a relationship plot of life versus stress. Analytic methods for such data appear in Chapter 5.

The example employs the lognormal distribution and Arrhenius relationship. Of course, the methods apply to other distributions and other (transformed) linear relationships.

Overview. Section 5.1 presents illustrative data. Section 5.2 explains hazard plotting for multiply censored data. Section 5.3 presents the relationship plot. Section 5.4 describes graphical estimates. Section 5.5 describes checks for the validity of the model and data.

5.1. Data (Turn Failures)

Data in Table 5.1 illustrate the graphical methods here. The data are hours to Turn failure of a new Class-H insulation system tested in motorettes at high temperatures of 190, 220, 240, and 260°C. A purpose was to estimate the median time to Turn failure at the design temperature of 180°C. A median life over 20,000 hours was desired.

Ten motorettes were run at each temperature and periodically inspected for failure. The time in Table 5.1 is midway between the time when the failure was found and the time of the previous inspection. The times between checks are short, and using the midpoint has little effect on the plots. The times between checks (called cycle lengths) were nominally 7, 4, 2, and 2 days for 190°, 220°, 240°, and 260°C, respectively.

Table 5.1. Turn Failure Data in Hours

190°C	220°C	240°C	260°C
7228	1764	1175	1128
7228	2436	1521	1464
7228	2436	1569	1512
8448	2436+	1617	1608
9167	2436	1665	1632+
9167	2436	1665	1632+
9167	3108	1713	1632+
9167	3108	1761	1632+
10511	3108	1881+	1632+
10511	3108	1953	1896

The data on Turn failures are not complete, since some motorettes were removed from test before having a Turn failure. Each running time is marked with a + in Table 5.1. Failure times are unmarked. Such multiply (or progressively) censored data and must be analyzed with special methods like those below.

5.2. Hazard Plot (Lognormal)

A hazard plot of the multiply censored data from each test stress is made to estimate each life distribution. A hazard plot is a probability plot and is used and interpreted like one. Other methods for plotting multiply censored data are given by Kaplan and Meier (1958), Herd (1960), Johnson (1964), and Nelson (1982, p. 147). They employ probability paper. Hazard plotting also applies to singly censored and complete data; for such data, probability plotting is usually used, as it is better known and understood by others. The hazard plotting method is explained with the 220° data in Table 5.2.

Hazard calculations. For a test stress, suppose there are n tests units ($n = 10$ for the 220° data). Order the n times from smallest to largest as shown in Table 5.2 (ignore whether they are running or failure times). Then label the times with reverse ranks k; that is, label the first with n, the second with $n - 1$, $\cdots$, and the nth with 1 as in Table 5.2.

Calculate a hazard value for each **failure** time as $100/k$, where k is its reverse rank. The hazard values for the Turn failures are shown in Table 5.2. For example, the failure at 2436 hours with reverse rank 8 has a hazard value of $100/8 = 12.5\%$. Hazard values are not calculated for running times.

Calculate the cumulative hazard value for each **failure** as the sum of its hazard value and the cumulative hazard value of the preceding failure. For example, for the same failure at 2436 hours, the cumulative hazard value is $33.6 = 12.5 + 21.1$. The cumulative hazard values of the Turn failures are

Table 5.2. Hazard Calculations for 220° Turn Data

220°C Hours	Reverse Rank k	(100/k)% Hazard	% Cum. Hazard	Modified Cum. Haz.
1764	10	10.0	10.0	5.0
2436	9	11.1	21.1	15.6
2436	8	12.5	33.6	27.4
2436+	7			
2436	6	16.7	50.3	42.0
2436	5	20.0	70.3	60.3
3108	4	25.0	95.3	82.8
3108	3	33.3	128.6	112.0
3108	2	50.0	178.6	153.6
3108	1	100.0	278.6	228.6

+ censoring time

shown in Table 5.2. Cumulative hazard values have no physical meaning and can be larger than 100%. They are just proper plotting positions.

Modified values. Modified cumulative hazard values may be better for plotting small samples. The modified value for a failure is the average of its cumulative hazard value and that of the preceding failure. The modified cumulative hazard value of the first failure is half of its cumulative hazard value. Such modified values appear in Table 5.2.

Hazard paper. Choose the hazard paper of a theoretical distribution. There are hazard papers for the exponential, Weibull, extreme value, normal, and lognormal distributions. These papers are available from TEAM, Box 25, Tamworth, NH 03886. Lognormal hazard paper is used for the insulation life data (Figure 5.1). Suitably label the vertical (data) scale.

Hazard plot. On the hazard paper, plot each failure time vertically against its cumulative hazard value on the horizontal axis. Nonfailure times are not plotted. Such a plot is made with the data for each stress as shown in Figure 5.1. Fit parallel straight lines to the data at each stress if desired. Each line is a graphical estimate of the cumulative distribution at that stress.

How to use a hazard plot. The probability (percentage) scale on a hazard paper for a distribution is exactly the same as that on the corresponding probability paper. Thus, a hazard plot is used the same way as a probability plot, as described in Sections 2.2, 2.4, and 2.5. The hazard scale is only a convenience for plotting multiply censored data. Examples follow.

Percentile estimates. An estimate of a distribution percentile comes from a hazard plot in the same way as from a probability plot. Enter the hazard plot on the probability scale at the desired percentage. Go down to the fitted line for the stress, and then go sideways to the time scale to read the percen-

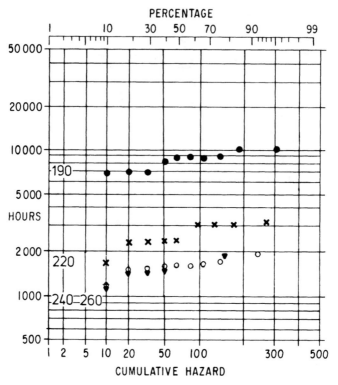

Figure 5.1.   Lognormal hazard plot of Turn data.

tile estimate. For example, the estimate of the median time (50th percentile) to Turn failure at 220°C is 2900 hours from Figure 5.1. Estimates of the medians at the test temperatures are plotted as crosses in Figure 5.2.

Estimate of a percentage failing. A hazard plot is used like a probability plot to estimate the percentage of units that fail by a given age at a stress. Enter the hazard plot on the time scale at that age. Go sideways to the fitted line for the stress, and then go up to the probability scale to read the percentage. For example, in Figure 5.1, the estimate of the percentage that fail by 3,000 hours at 220° is 55%.

5.3. Relationship Plot (Arrhenius)

The relationship between life (some distribution percentile) and stress is estimated with the same method described in Sections 2.3 and 3.3. Namely, use paper where the relationship is a straight line, and plot the estimate of the chosen percentile for each test stress against the stress. The Turn failure medians are plotted on Arrhenius paper in Figure 5.2. The Turn failure times and running times could be plotted to display the data. However, the plot of failure and running times is cluttered and difficult to interpret.

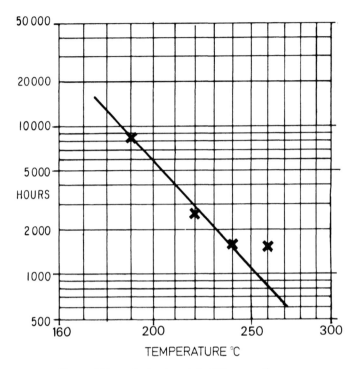

Figure 5.2.   Arrhenius plot of Turn medians ×.

Finally fit a straight line by eye to the percentile estimates. This line graphically estimates the relationship between life (the percentile) and stress. For reasons given later, the 260° data were not used to estimate the line in Figure 5.2. The median life at any temperature is estimated from this line. In particular, the estimate of median life of Turn insulation at the design temperature of 180°C is 12,300 hours. This is well below the desired 20,000 hours. Other percentile lines can be estimated on the relationship plot as described in Sections 2.4 and 3.4.

5.4. Graphical Estimates

The hazard and relationship plots yield estimates as described in Sections 2.4 and 3.4. Some examples follow.

Estimates of μ and σ. The hazard plot yields estimates of the mean log life μ and the log standard deviation σ. As before, the estimate of the mean log life μ at a stress is just the log of the median there. For example, the graphical estimate of the median at 220°C is 2900 hours from Figure 5.1. The corresponding mean log life is log(2900) = 3.462. the estimate of the log standard deviation σ is the difference between the logs of the 50th and 16th percentiles at a stress. For 220°C, the estimate of the 16th percentile is 2400 hours. The estimate of the log standard deviation is log(2900) − log(2400) = 0.08. This small value indicates an increasing failure rate.

Median at design temperature. The relationship plot yields an estimate of the median at any temperature. For example, the estimate for 180°C, the design temperature, is 12,300 hours.

5.5. Assess the Model and Data

The graphical methods of Section 2.5 are used to assess the distribution (lognormal here), the relationship (Arrhenius here), and the data. Censored observations in a relationship plot are difficult to interpret. Thus one uses estimates of percentiles in such a plot. Some remarks on the plots follow.

Common σ. Figure 5.1 shows noteworthy features. The plots at the four temperatures are parallel. This indicates that σ has the same value at all test temperatures. This is consistent with the Arrhenius-lognormal model. In contrast, Figure 2.1A does not have parallel plots because the Class-H data there contain a mix of failure modes. One expects the model to fit data on a single failure mode better than it fits data with a mix of failure modes.

Relationship not Arrhenius. The 260° data coincide with the 240° data in Figure 5.1. Consequently the 260° median is the same as the 240° median in Figure 5.2. However, insulation life should be less at 260° than at 240°. One possible reason for the peculiar 260° data is that the 260° motorettes were not made at the same time as the others. So they may differ with respect to materials or fabrication and consequently life.

$1,000,000 insight. Another possible reason for the peculiar 260° data is that the test method based on IEEE Standard 117 may be misleading. The Standard recommends how long motorettes be held at temperature in an oven between inspections. Following the Standard, the test used 7 days between inspections at 190°, 4 days at 220°, and 2 days at 240°, but did not use 1 day at 260°. Instead the test used 2 days at 260°, the same as at 240°. The inspection involves removing the motorettes from the oven, cooling them to room temperature, and applying the design voltage to see if the insulation withstands it. Unfailed motorettes are put back into the oven and heated to the test temperature. Thus the insulation is thermally cycled, and the resulting mechanical stressing may degrade life. According to this theory, if the 260° motorettes had been cycled every day, instead of every two days, they would have failed sooner and the data would have looked "right." A subsequent designed experiment (in Problem 3.9) involved combinations of temperature and cycle length. The experiment showed that thermal cycling has an important effect on the insulation life. The insulation engineer knew that such a motor is used one of two ways: (1) continuously or (2) frequently on and off. The engineer saw that continuously running motors were not thermally cycled, and a cheaper insulation would suffice for them. This insight annually saves $1,000,000 at 1989 prices.

6. INTERVAL (READ-OUT) DATA

Introduction. This section describes how to graphically analyze read-out (interval) data. It shows how to estimate the product life distribution under (accelerated) test conditions and under design conditions. Topics include
- 6.1. Interval (read-out) data and Microprocessor example,
- 6.2. Probability plot and confidence limits,
- 6.3. Relationship plot and acceleration factors.

Tobias and Trindade (1986) present some of these topics with electronics applications. Analytic methods for such data appear in Chapter 5.

6.1. Read-Out (Inspection) Data and Microprocessor Example

Overview. This section describes read-out data and how it arises, a Microprocessor example, removals and censoring, and assumptions.

Description. Some life tests yield *read-out (interval) data* on time to failure of specimens. In such tests, sample specimens start on test together (at test time 0), and they are inspected periodically for failure. Finding a specimen failed on inspection i at read-out time t_i, one knows only that it failed between the previous read-out time t_{i-1} and t_i ($t_0 = 0$). The exact failure time is not observed, because it is difficult or costly to instrument each specimen to observe failure.

Example. Table 6.1 shows typical read-out data on a sample of Microprocessors tested at 125°C and 7.0 V. The inspection (read-out) times are 6, 12, 24, 48, 168, 500, 1000, and 2000 test hours. For inspection interval i, the data consist of the number n_i of devices that started through the interval at time t_{i-1} and the number f_i that failed by the interval end at time t_i. These numbers appear Table 6.1 as f_i/n_i. For example, interval 2 runs from 6 to 12 hours, and the data 2/1417 means that 1417 devices entered the interval at 6 test hours, and 2 were found failed on inspection at 12 test hours. Such data are also called *inspection* and *interval data*.

Purpose. A purpose of the analyses below is to estimate the life distribution of such devices at the *test* condition (125°C and 7.0 V) and at the *design*

Table 6.1. Microprocessor Read-Out Data

Interval i:	1	2	3	4	5	6	7	8
Hours t_i:	6	12	24	48	168	500	1000	2000
f_i/n_i:	6/1423	2/1417	0/1415	2/1414	1/573	1/422	2/272	1/123

condition (55°C and 5.25 V). Also, the device failure rate is to be compared with a goal of 200 FITs, which is equivalent to 0.02% per 1000 hours, assuming an exponential distribution.

Removals and censoring. In such testing, some unfailed devices may be censored after any inspection time. For example, Table 6.1 shows that 1414 devices entered interval 4 at 24 hours and 2 failed. Of the $1414 - 2 = 1412$ that survived to 48 hours, 573 devices entered interval 5 at 48 hours. Thus $1412 - 573 = 839$ unfailed devices were censored at 48 hours. Such censoring arises various ways. Some unfailed devices may be removed from test at various read-outs. Such *removals* free the test equipment for other tests and reduce test cost. Of course, such removals result in less accurate estimates of the life distribution at later inspection times. However, it is often best to run more devices through early inspection times to accurately estimate the lower tail of the life distribution. Below it is assumed that removals occur only at inspection times. Also, censoring may result from loss of devices, say, from failure of the test equipment or other extraneous causes, for example, a failure mode not of interest. Also, censoring results from having several tests in progress, each having run a different length of time when the data are analyzed. Analysis of read-out data without (intermediate) censoring is much simpler; see, for example, Nelson (1982, Chapter 9).

Assumptions. (1) In some tests, specimens are run at an accelerated temperature and are inspected at room temperature. Such thermal cycling at each inspection may affect device life. This possibility is often overlooked. (2) A common inspection scheduled is assumed throughout. In Table 6.1, all devices were inspected on the same schedule. Consequently, the data plots below are simple. Otherwise, use Peto's (1973) and Turnbull's (1976) plot.

6.2. Probability Plot and Confidence Limits

Overview. This section presents a probability plot (nonparametric estimate and confidence limits) for the life distribution for read-out data with removals or censoring.

Estimate and Plot

Estimate. For read-out data, the life distribution can be nonparametrically estimated only at the inspection times. The following method yields a nonparametric estimate F_i and plot of the population fraction failed at each inspection time t_i. The random quantities in the read-out data are the numbers f_i of devices failing in each inspection period. The actual random failure times are not observed. Thus the usual probability plotting methods of previous sections for observed failure times do not apply. The following steps yield the Kaplan-Meier (1958) estimate adapted to read-out data.

Steps. Table 6.2 shows the steps in the calculation of the reliability esti-

Table 6.2. Calculation of Microprocessor Reliability Estimates

(1) i	(2) t_i	(3) f_i/n_i	(4) $R'_i =$ $1-(f_i/n_i)$	(5) $R_i =$ $R'_1 R'_2 \cdots R'_i$	(6) $F_i =$ $1-R_i$	(7) 95% conf.
1	6	6/1423	$.99578_{35}$	$.99578_{35}$	0.42%	±.34%
2	12	2/1417	$.99858_{85}$	$.99437_{80}$	0.56%	±.39%
3	24	0/1415	1.00000_{00}	$.99437_{80}$	0.56%	±.39%
4	48	2/1414	$.99858_{55}$	$.99297_{16}$	0.70%	±.44%
5	168	1/573	$.99825_{48}$	$.99123_{86}$	0.88%	±.56%
6	500	1/422	$.99763_{03}$	$.98888_{97}$	1.11%	±.73%
7	1000	2/272	$.99264_{70}$	$.98161_{85}$	1.84%	±1.26%
8	2000	1/123	$.99186_{99}$	$.97363_{78}$	2.64%	±2.02%

mate at each inspection time t_i. The Microprocessor data in Table 6.2 come from Table 6.1.

1. Column (1) shows the number i of each inspection, $i = 1, 2, \cdots , 8$.
2. Column (2) shows the inspection time t_i in hours. For example, inspection 2 was at 12 test hours, and the inspection interval ran from 6 to 12 test hours.
3. Column (3) shows the data f_i/n_i. f_i is the number of devices failing in interval i, and n_i is the number on test at the interval start at time t_{i-1}.
4. Column (4) shows the calculation of the estimate of the *conditional* reliability for interval i as $R'_i = 1-(f_i/n_i)$. This estimates the fraction of the population unfailed at time t_{i-1} that also reach time t_i unfailed. f_i/n_i estimates the conditional fraction failed in interval i. For example, for interval 2, $R'_2 = 1-(2/1417) = 0.9985885$. For confidence limits later, these and all other calculations employ at least seven significant figures.
5. Column (5) shows the recursive calculation of the estimate of the (*unconditional*) reliability at time t_i as $R_i = R'_i \times R_{i-1}$ where $R_0 = 1$. For example, for interval 2, $R_2 = R'_2 \times R_1 = 0.9985885 \times 0.9957835 = 0.9943780$. The flow of the recursive calculations is shown by arrows in Table 6.2.
6. Column (6) shows the estimate of the (*unconditional*) population fraction failed by time t_i, namely, $F_i = 1-R_i$, expressed as a percentage.
7. Column (7) shows the ± uncertainty (95% confidence) in the estimate F_i of the fraction failed. Its calculation appears below.

Plot. On probability paper, plot each estimate F_i of the population fraction failed against its time t_i. These estimates for the Microprocessor data appear as ×s on Weibull paper in Figure 6.1. In previous probability plots, each point corresponds to one specimen failure at its actual time of failure. However, in a plot of read-out data, a point may correspond to one or more failures. Moreover, the point appears at the inspection time − after the actu-

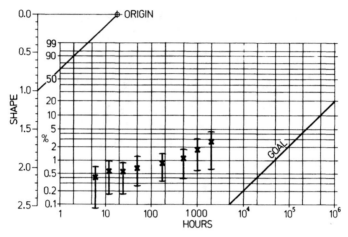

Figure 6.1.   Weibull plot of Microprocessor estimates and 95% limits.

al failure time. To interpret such a plot, one must take into account such differences. To help do this, one can write the actual number of failures in each interval in place of the point. Or, better, plot confidence limits (Figure 6.1) about each plotted F_i to indicate its accuracy. No failures occurred in interval 3. It is difficult to say whether to plot the estimate F_3, since the point does not correspond to any failures.

   Interpretation. Interpret the plot of the F_i like other probability plots as follows. But take into account the interval nature of the data.

- *Distribution fit.* If the plot is relatively straight, then the distribution adequately fits the data over the observed inspection times. For example, the Weibull plot of the Microprocessor data (Figure 6.1) is relatively straight. Thus the Weibull distribution appears to adequately fit the data. To compare how well different distributions fit, plot the data on various distribution papers, and choose the distribution with the straightest plot. The confidence limits below help make this choice.

- *Failure rate.* The graphical estimate of the Weibull shape parameter indicates the nature of the failure rate (increasing, decreasing, or constant as the population ages). For the Microprocessor data, this estimate is 0.3. This indicates a decreasing failure rate. Thus such devices would benefit from burn-in – *if the failure mode at the test condition is the dominant one at the design condition.*

- *Goal.* A goal is a failure rate of 200 FITs. The corresponding exponential distribution appears in Figure 6.1 as a straight line with a shape parameter (slope) of 1. The distribution estimate is well below (worse than) the goal. Of course, this estimate is for an accelerated condition. Section 6.3 shows how to estimate the distribution at a design condition.

Confidence Limits

Overview. This section presents nonparametric confidence limits for the population reliability at the read-out times where the data are (1) uncensored and (2) censored. These confidence limits help one judge the accuracy of the plotted points and draw a line through them. Chapter 5 presents parametric confidence limits.

(1) Uncensored data. Suppose n devices are tested and are *uncensored* before inspection i. The following *simple* estimate and confidence limits apply to all inspection times through t_i. For example, for the Microprocessor data in Table 6.1, the limits apply through $t_3 = 24$ hours. Also, to a good approximation they apply through $t_4 = 48$ hours, since only one device out of 1415 is censored at 24 hours.

Estimate. Suppose C_i is the *cumulative* number failed by read-out time t_i. Then the simple estimate of the population fraction failed by time t_i is the sample fraction $F_i = C_i/n$. The reliability estimate is $R_i = 1-(C_i/n)$. The estimate above for censored data reduces to this simple one when there is no intermediate censoring. Plot these F_i on probability paper.

Exact limits. If there is no censoring before t_i, the cumulative number of failures C_i has a binomial distribution, and exact binomial confidence limits apply. These limits appear in most statistics texts, for example, in Nelson (1982, p. 205). Simple approximate limits follow.

Poisson approximation. For few failures (say, $C_i < 10$ and $n > 10C_i$), use the two-sided $100P\%$ confidence limits (Poisson approximation):

$$\underline{F_i} \cong (0.5/n)\chi^2[(1-P)/2;2C_i], \quad \tilde{F}_i \cong (0.5/n)\chi^2[(1+P)/2;2C_i+2]. \quad (6.1)$$

Here $\chi^2[P';D]$ is the $100P'$th percentile of the chi-square distribution with D degrees of freedom. Note that the two limits have different degrees of freedom: $2C_i$ and $2C_i+2$. The confidence limits for reliability are $\underline{R}_i = 1-\tilde{F}_i$ and $\tilde{R}_i = 1-\underline{F}_i$. The one-sided upper $100P\%$ confidence limit is

$$\tilde{F}_i \cong (0.5/n)\chi^2(P;2C_i+2).$$

With $100P\%$ confidence, the population fraction failing by time t_i is no worse than this $\tilde{F}_i$.

Example. $n = 1423$ for the Microprocessor data (Table 6.1). The simple estimate at $t_1 = 6$ hours is $F_1 = 6/1423 = 0.0042$ or 0.42%. The 95% confidence limits are

$$\underline{F}_1 = (0.5/1423)\chi^2[(1-0.95)/2;2(6)] = (0.5/1423)4.404 = 0.0015 \text{ or } 0.15\%,$$

$$\tilde{F}_1 = (0.5/1423)\chi^2[(1+0.95)/2;2(6)+2] = (0.5/1423)26.12 = 0.0092 \text{ or } 0.92\%.$$

Normal approximation. For many failures (say, $10 < C_i < n - 10$), use the two-sided $100P\%$ limits (normal approximation):

$$\underset{\sim}{F_i} \cong F_i - K_P[F_i(1-F_i)/n]^{1/2}, \quad \tilde{F_i} = F_i + K_P[F_i(1-F_i)/n]^{1/2}. \quad (6.2)$$

Here K_P is the standard normal $100(1+P)/2$ percentile. For example, $K_{.95} = 1.96 \cong 2$. The one-sided upper $100P\%$ confidence limit is

$$\tilde{F_i} = F_i + z_P[F_i(1-F_i)/n]^{1/2};$$

here z_P is the standard normal $100P$th percentile. For example, $z_{.95} = 1.645$. Confidence limits for reliability are $\underset{\sim}{R_i} = 1 - \tilde{F_i}$ and $\tilde{R_i} = 1 - \underset{\sim}{F_i}$. Thomas and Grunkemeier (1975) investigate better approximations.

(2) Censored data. For *censored* read-out data, the following are approximate confidence limits for the population fraction failed by read-out time t_i. Based on a normal approximation to the distribution of the estimate F_i, they are adequate when $C_i > 10$, and can be used even for $C_i > 5$. For smaller C_i, use the binomial confidence limits for uncensored data if feasible. Plot these confidence limits on the probability paper with the estimates F_i. Figure 6.1 shows such limits on the Weibull plot for the Microprocessor data.

Limits. The following limits employ an estimate $v(F_i)$ of the variance of the estimate F_i. Two-sided approximate $100P\%$ confidence limits are

$$\underset{\sim}{F_i} \cong F_i - K_P[v(F_i)]^{1/2}, \quad \tilde{F_i} \cong F_i + K_P[v(F_i)]^{1/2}. \quad (6.3)$$

Here K_P is the standard normal $100(1+P)/2$ percentile (Appendix A3).

Approximate variance. For F_i small (say, $F_i < 0.10$), a simple approximation is

$$v(F_i) \cong R_i^2\{[F_1'/(n_1 R_1')] + [F_2'/(n_2 R_2')] + \cdots + [F_i'/(n_i R_i')]\}. \quad (6.4)$$

The notation follows that in Table 6.2. Note that both primed and unprimed estimates appear here, and $F_i' = f_i/n_i = 1 - R_i'$. For example, for the Microprocessor data, $v(F_8) \cong 0.000100$ at 2000 hours. The approximate 95% confidence limits are

$$\underset{\sim}{F_8} = 0.0264 - 2(0.000100)^{1/2} = 0.0064 \text{ or } 0.64\%,$$

$$\tilde{F_8} = 0.0264 + 2(0.000100)^{1/2} = 0.0464 \text{ or } 4.64\%.$$

These limits are plotted at 2000 hours in Figure 6.1.

Exact variance. The exact variance estimate entails the calculations in Table 6.3. The calculations in columns (1) through (5) of Table 6.3 are exactly the same as those in Table 6.2. However, in column (3) of Table 6.2, use $n_i' = n_i - 1$ in place of n_i. In Table 6.3, r_i' denotes the estimate of the conditional reliability, and r_i denotes the estimate of the reliability at time t_i. Then the exact variance estimate is

$$v(F_i) = R_i(R_i - r_i). \quad (6.5)$$

Table 6.3. Exact Calculation of $v(F_i)$

(1) i	(2) t_i	(3) f_i/n_i'	(4) $r_i' = 1-[f_i/n_i']$	(5) $r_i = r_1'r_2'\cdots r_i'$	(6) $v(F_i) = R_i(R_i-r_i)$	(7) $K_{.95}[v(F_i)]^{1/2} = 200[v(F_i)]^{1/2}$
1	6	6/1422	$.995780_6$	$.995780_6$	$.0000028$	$\pm.34\%$
2	12	2/1416	$.998585_{75}$	$.994374_1$	$.0000038$	$\pm.39\%$
3	24	0/1414	1.000000_{00}	$.994374_1$	$.0000038$	$\pm.39\%$
4	48	2/1413	$.998584_5$	$.992966_6$	$.0000048$	$\pm.44\%$
5	168	1/572	$.998252_{17}$	$.991230_7$	$.0000078$	$\pm.56\%$
6	500	1/421	$.997624_7$	$.988876_2$	$.0000132$	$\pm.73\%$
7	1000	2/271	$.992619_9$	$.981578_2$	$.0000394$	$\pm1.26\%$
8	2000	1/122	$.991803_2$	$.973532_5$	$.0001024$	$\pm2.02\%$

Here R_i comes from Table 6.2, and r_i comes from Table 6.3. Note that $(R_i - r_i)$ is a small difference of nearly equal numbers. Thus R_i and r_i must be accurate to seven significant figures to assure that $v(F_i)$ is accurate to two figures. The $v(F_i)$ appear in column (6) of Table 6.3. Column (7) shows $2 \times 100 \times [v(F_i)]^{1/2}$ where $2 \cong K_{.95}$, and 100 yields a percentage. This $v(F_i)$ is Greenwood's (1926) variance for the Kaplan-Meier estimate extended for read-out data.

Computer packages. The procedure LIFETEST of SAS Inst (1985) and other computer packages calculate the Kaplan-Meier estimate (1958) and such confidence limits for multiply censored data with *observed* failure times. Such routines can be used for censored *interval* data if all specimens have the same inspection schedule. Then one must take into account the interval nature of the data in inputting the data and using the output. If groups of specimens have different inspection schedules, one must use the more complex confidence limits for the Peto (1973) and Turnbull (1976) estimates. STAR of Buswell and others (1984) performs the complex Peto calculations and calculates confidence limits.

6.3. Relationship Plot and Acceleration Factors

Overview. This section first shows how to use an acceleration factor (defined below) to analyze data from a test at a *single* accelerated stress level. Such analyses yield an estimate of the life distribution at a design condition for interval and other types of data. Interval data from *two or more* stress levels are plotted on relationship and probability plots and analyzed with methods in Sections 4.3 and 5.3, including assessing the model and data.

One condition. In some accelerated tests, specimens are run at just *one* accelerated test condition. Moreover, that condition may result from accelerating a number of variables, such as temperature, temperature cycling, humidity, vibration, etc. Such testing of electronics is common, and MIL-STD-883 specifies standard tests. Run during development, such a test is usually intended to identify failure modes so they can be corrected. Also, they are used as demonstration tests (MIL-STD-883) to assess whether a device has satisfactory reliability. One can estimate device life at a design condition *only if* one knows the acceleration factor between life at the accelerated and design conditions.

Acceleration factor. Suppose that "typical life" of a *failure mode* is t at a design condition and is t' at an accelerated test condition. Then the *acceleration factor K* for those two conditions is

$$t = K \cdot t'. \tag{6.6}$$

For example, if $K = 500$, the failure mode lasts 500 times as long at the design condition as at the accelerated condition. Also, loosely speaking, one hour at the accelerated condition equals K hours at the design condition. Equivalently, a read out time of 6 hours under acceleration corresponds to $500 \cdot 6 = 3000$ hours at the design condition. An acceleration factor is calculated as follows from a *known* life-stress relationship. Each failure mode has a separate relationship and acceleration factor (Chapter 7).

Arrhenius factor. The Arrhenius relationship is often used to describe temperature-accelerated tests where product failure is due to chemical degradation or intermetallic diffusion. Suppose that T is the design temperature, and T' is the test temperature, both in degrees Kelvin. Kelvin = Centigrade + 273.16. Then the Arrhenius acceleration factor (Section 9.1 of Chapter 2) for a failure mode is

$$K = \exp\{(E/k)[(1/T)-(1/T')]\}. \tag{6.7}$$

Here E is the activation energy (in eV) of the failure mode, and $k = 8.6171 \times 10^{-5}$ is Boltzmann's constant in eV per Kelvin degree. E corresponds to the slope of an Arrhenius relationship on an Arrhenius plot. To evaluate the factor, one must know E or assume a value for it. There are such factors for other life-stress relationships.

Example. For the Microprocessor, the test temperature is $T' = 125 + 273.16 = 398.16$, and the design temperature is $T = 55 + 273.16 = 328.16$. For some types of failure modes, the activation energy is assumed to be $E = 1.0$ eV. The corresponding acceleration factor is

$$K = \exp\{(1.0/8.6171 \times 10^{-5}) [(1/328.16)-(1/398.16)]\} = 501.$$

Thus such a failure mode lasts about 500 times longer at the design temperature. For such microprocessors, other failure modes are also *assumed* to have activation energies of 0.8 and 0.3 eV.

Design life. The following provides an estimate of the life distribution at a design condition. For each read-out time t_i', use the acceleration factor K to calculate the equivalent time $t_i = K t_i'$ at the design condition. Use the

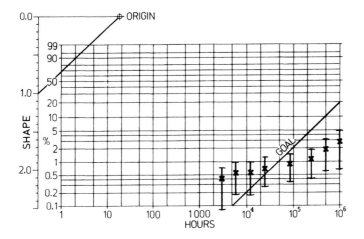

Figure 6.2.   Weibull plot of 55°C estimate and 95% limits.

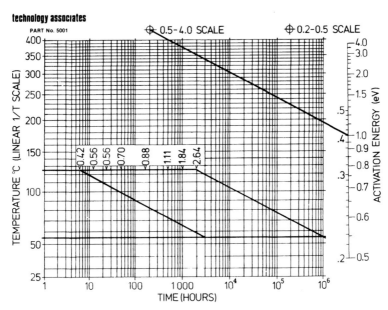

Figure 6.3.   Arrhenius plot of Microprocessor data, slope of 1.0 eV. (Paper provided by courtesy of D. Stewart Peck.)

equivalent times to estimate the life distribution at the design condition, using the methods above. Equivalently, move the estimate of the accelerated distribution toward longer life by a factor K. For example, Figure 6.2 shows a Weibull plot of this estimate of the Microprocessor life distribution at 55°C. The distribution at 55°C is higher than that at 125°C (Figure 6.1) by a factor $K = 500$. For example, the inspection at 6 hours at 125°C corresponds to an inspection at 500·6 = 3000 hours at 55°C. Confidence limits and parametric estimates similarly are at higher times by a factor K as in Figure 6.2.

Multiple acceleration. There can be several accelerating variables, each with an accelerating factor. The product of those factors is the combined accelerating factor. This assumes that the variables do not interact. The generalized Eyring model (Chapter 2) represents such interactions.

Uncertainty. Note that the width of a confidence interval is the same at the accelerated and design conditions. This results from assuming that the acceleration factor is correct. In practice, the factor is approximate. Thus the uncertainty of the estimate of life at the design condition really exceeds the confidence limits. A better analysis would include the uncertainty in the acceleration factor.

Arrhenius plot. The acceleration factor can be graphically evaluated with Arrhenius paper as follow. The Arrhenius paper in Figure 6.3 shows each read-out time and sample cumulative percent failed at 125°C, the accelerated temperature. In Figure 6.3, a line passes through an origin and the Activation Energy scale at 1.0 eV. Its slope corresponds to $E = 1.0$ eV. Draw lines parallel to that line from the read-out times t_i' at 125°C to get the corresponding times t_i at 55°C. The estimates F_i correspond to these new t_i. The times scales in Figures 6.2 and 6.3 coincide, showing the relationship between the Weibull and Arrhenius plots.

PROBLEMS (* denotes difficult or laborious)

3.1. Three insulations. Specimens lives (in hours) of three electrical insulations at three test temperatures appear below. Failure times are the midpoints of inspection intervals. The main aim is to compare the three insulations at the design temperature of 200°C and at 225° and 250°, occasional operating temperatures. Use the Arrhenius-lognormal model.

Insulation 1			Insulation 2			Insulation 3		
200°C	225°C	250°C	200°C	225°C	250°C	200°C	225°C	250°C
1176	624	204	2520	816	300	3528	720	252
1512	624	228	2856	912	324	3528	1296	300
1512	624	252	3192	1296	372	3528	1488	324
1512	816	300	3192	1392	372			
3528	1296	324	3528	1488	444			

(a) On separate lognormal probability paper for each insulation, plot the data from the three test temperatures.

(b) Are there any pronounced peculiarities in the data?

(c) How do the three insulations compare at 200°C, the usual operating temperature, with respect to the median and spread in (log) life? Which differences are convincing? Why?

(d) How do the three insulations compare at 225 and 250°C, occasional operating temperatures, with respect to the median and spread in (log) life? Which differences are convincing? Why?

(e) How do the three compare overall, and are any differences convincing to you? Write a short recommendation to management.

(f) Estimate σ for each insulation. Gauge the amount of uncertainty in each estimate. Do the slopes (σ's) for the insulations differ convincingly? Comment on the nature of the failure rate (increasing or decreasing?).

(g) On separate Arrhenius papers, plot the data for each insulation.

(h) On each Arrhenius plot, mark the sample medians and fit a straight line to them. Judge the linearity of each plot.

(i) Use the Arrhenius line to estimate the median life at 200°C of each insulation. Gauge the amount of uncertainty in each estimate. Do you prefer these estimates or those from (c)? Why?

(j) Suggest further analyses and other models.

(k) Carry out (j).

3.2. Plotting positions. Use the data on Insulation 2 at 225°C above. On lognormal probability paper make three plots of the data at 225°C, using the (1) midpoint, (2) mean, and (3) median plotting positions.

(a) How do the graphical estimates of the distribution median, 1% point, and σ compare for the three plotting positions?

(b) Which yields the most conservative (pessimistic) and optimistic estimates for reliability purposes?

(c) Do the differences in the estimates look large compared to the uncertainties in the estimates?

3.3. Heater data. Complete life data (in hours) from a temperature-accelerated test of sheathed tubular heaters appear below. The main aim was to estimate the median and 1% point of life at the design temperature of 1100°F. Absolute (Rankine) temperature is the Fahrenheit temperature plus 460°F. Life was assumed to have a lognormal distribution and to follow an Arrhenius relationship.

Temp. °F	Hours					
1708:	511	651	651	652	688	729
1660:	651	837	848	1038	1361	1543
1620:	1190	1286	1550	2125	2557	2845
1520:	1953	2135	2471	4727	6134	6314

(a) Plot the data on lognormal probability paper. Comment whether the lognormal distribution adequately fits.

(b) From separately fitted lines, estimate the log standard deviation and median at each test temperature. Gauge the amount of uncertainty in each estimate. Do the slopes of the lines differ convincingly?

(c) From parallel fitted lines, estimate a common log standard deviation. Gauge the amount of uncertainty in this estimate.

(d) Plot the data and sample medians on Arrhenius paper. First make such paper from semi-log paper, as none is available for 1100 to 1708°F.

(e) Fit a straight line to the sample medians and comment on linearity.

(f) Use the fitted line to estimate the median at 1100°F. Gauge the amount of uncertainty in this estimate.

(g) On the lognormal paper, plot the estimate of the 1100° median, and draw the line for the 1100° life distribution.

(h) Estimate the 1% point of life at 1100°F. Gauge its uncertainty.

(i) Comment on convincing lack of fit of the model or peculiar data.

(j) Suggest further analyses and other models.

(k) Carry out (j).

3.4. Class-H without 260° data. As follows, reanalyze the Class-H insulation data of Table 2.1 without the 260° data. Use Weibull probability paper rather than lognormal paper. As before, the main aim is to estimate the 50th percentile at 180°C.

(a) Plot the data on Weibull probability paper.

(b) Comment on the adequacy of the Weibull fit to the data?

(c) Is the Weibull or lognormal fit convincingly better? Why?

(d) From separately fitted lines, estimate the shape parameter and 50th percentile at the three test temperatures. Gauge the amount of uncertainty in each estimate. Do the slopes of the lines differ convincingly?

(e) From parallel fitted lines, estimate a common shape parameter. Gauge the amount of uncertainty in this estimate. Is the estimate convincingly different from 1? Comment on the behavior of the failure rate with age.

(f) Plot the data on Arrhenius paper.

(g) On the Arrhenius plot, mark the sample medians. Then fit a line to the sample medians. Comment on the linearity of the relationship.

(h) Use the fitted line to estimate the median at the design temperature of 180°C. Gauge the amount of uncertainty in this estimate.

(i) On the Weibull paper, plot the estimate of the median at 180° and draw the line for the 180° life distribution.

(j) Comment on any convincing lack of fit of the model or peculiar data.

(k) Suggest further analyses and other models.

(l) Carry out (k).

3.5. Bearing data. Courtesy of Mr. John McCool, the complete life data below are from a load accelerated test of rolling bearings. Forty bearings were tested, ten at each of four test loads. The usual assumed model consists of a Weibull life distribution and an inverse power relationship (Palmgren's equation) between life and load. The failure labeled * is an obvious outlier.

Load	Life (10^6 revolutions)									
0.87	1.67	2.20	2.51	3.00	3.90	4.70	7.53	14.70	27.76	37.4
0.99	.80	1.00	1.37	2.25	2.95	3.70	6.07	6.65	7.05	7.37
1.09	.012*	.18	.20	.24	.26	.32	.32	.42	.44	.88
1.18	.073	.098	.117	.135	.175	.262	.270	.350	.386	.456

(a) Make a Weibull probability plot of the four data sets.
(b) Comment on the outlier and the adequacy of the Weibull distribution.
(c) Separately fit distribution lines, and estimate the shape parameter for each test load and the 10th percentile (B_{10} life) for each test load. Gauge the amount of uncertainty in each estimate. Is it reasonable to use a common shape parameter?
(d) Fit parallel lines, and estimate a common shape parameter. Gauge the amount of uncertainty in this estimate. Is the estimate convincingly different from 1? Does the failure rate increases or decreases with age?
(e) Make a relationship plot of the data on log-log paper.
(f) On the relationship plot, mark the sample 10th percentile estimates (midway between the 1st and 2nd smallest observations) for each test load. Fit a line to those estimates, and judge whether it fits adequately.
(g) Use the fitted line to estimate the 10th percentile at load of 0.75. Gauge the amount of uncertainty in this estimate.
(h) On the Weibull plot, plot the estimate of the 10th percentile and draw the line for the life distribution at the design load 0.75.
(i) Comment on convincing lack of fit of the model or peculiar data.
(j) Suggest further analyses and other models.
(k) Carry out (j).

3.6. Censored bearing data. Redo Problem 3.5 (a)-(i), plotting only the first four failures at each load. Delete the outlier. Note answers differing much from those in 3.5.

3.7. Censored Class-B insulation life. Delete the two low outliers from the data of Table 4.1, and reanalyze the data as follows. As before, the main aim is to estimate median life at 130°C.
(a) Plot the data on a lognormal probability paper.
(b) Comment on the adequacy of a lognormal fit, and compare this plot with Figure 4.1.
(c) From separately fitted lines, estimate the log standard deviation and 50th percentile at each test temperature. Gauge the amount of uncertainty in each estimate. Do the slopes of the lines differ convincingly?
(d) From parallel fitted lines, estimate a common log standard deviation. Gauge the amount of uncertainty in this estimate.
(e) Plot the data and estimates of the 50th percentiles on Arrhenius paper.
(f) Fit a straight line to the sample medians and comment on the linearity of the relationship.
(g) Use the fitted line to estimate the 50th percentile at 130°C, the design temperature. Gauge the amount of uncertainty in this estimate.

(h) On the lognormal paper, plot the estimate of the 130° median, and draw the line for the 130° life distribution.

(i) Comment on convincing lack of fit of the model or peculiar data.

(j) Compare the estimates obtained here with those in Section 4. Which analysis (with or without outliers) do you prefer and why? State any compelling reasons that you prefer either analysis.

(k) Suggest further analyses and other models.

(l) Carry out (k).

3.8. Later Class-B insulation data. The test of the Class-B insulation of Table 4.1 continued running. The test was terminated with the data below.

150°C	170°C	190°C	220°C
9429+	1764	408	408
9429+	2772	408	408
9429+	3444	1344	504
9429+	3542	1344	504
9429+	3780	1440	504
9429+	4860	1920	600
9429+	5196	2256	600
9429+	6206	2352	648
9429+	6792+	2596	648
9429+	6792+	3120+	696

Do (a) through (j) from Problem 3.7. Ignore the interval nature of the data.

(k) Do the additional failures here yield appreciably different estimates or conclusions from those of Problem 3.7?

3.9.* $1,000,000 experiment. The analysis of the data on Turn failures of Class-H insulation in Section 5.5 suggested that the length of the oven cycle may affect insulation life. A subsequent experiment with a Class-H insulation with various combinations of temperature and cycle length was run. The resulting data (in hours) on Turn failures appear below. The aims of the analyses are to estimate Turn life at the design temperature of 180°C and to evaluate the effect of cycle length. Note that there are different numbers of the 43 motorettes at the test conditions. This yields greater accuracy in extrapolating to 180°C, the design temperature, as described in Chapter 6.

200 °C/7 days:	9 survived 7392+
215 °C/28 days:	9072, 7 survived 11424+
215 °C/2 days:	6 survived 2784+
230 °C/7 days:	4286, 4452, 2 at 4620, 5746, 6216
245 °C/28 days:	2352, 3 at 3024, 4363, 7056
245 °C/2 days:	1660, 1708, 1996, 3008
260 °C/7 days:	3 at 1088, 1764

(a) Make a lognormal plot of the data from the four test combinations (230°, 7 days), (245°, 28 days), (245°, 2 days), and (260°, 7 days). Comment on (i) adequacy of the lognormal fit and (ii) parallelness of the four plots (common σ).

(b) Estimate a common σ value and gauge its amount of uncertainty. Comment on the nature of the failure rate.

(c) Compare the lognormal plots of the 245° data with cycle lengths of 2 and 28 days. Is there a convincing difference between the two sets of data? Do the data support the theory that shorter cycle length (more frequent temperature cycling) reduces insulation life?

(d) IEEE Std 117 suggests using test conditions (215°, 28 days), (230°, 7 days), and (245°, 2 days). Make a lognormal plot of these three data sets. Comment on the appearance of the plot.

(e) Plot the data (including nonfailures) and the sample medians from (d) on Arrhenius paper, and fit the Arrhenius line. Estimate the median at 180°C, and gauge its amount of uncertainty.

(f) Another analysis employs data with just one cycle length, 7 days: (200°, 7 days), (230°, 7 days), and (260°, 7 days). Make a lognormal plot of these three data sets. Comment on the appearance of the plot.

(g) Repeat (e) for the 7-day data on another Arrhenius paper.

(h) Do the estimates from (e) and (g) differ convincingly? Which do you prefer and why?

(i) Are the two estimates of median life at 180°C sensitive to assuming a lognormal life distribution?

(j) Suggest further analyses and other models.

(k) Carry out (j).

(l) Write a brief report for engineers to summarize your findings. Include appropriate plots and output.

3.10.* Left and right censored data. The data below came from a voltage-endurance test of an insulating fluid. Some breakdowns occurred within one second while the voltage was being raised to 45 kV; they are denoted by 1− below. Such data are **censored on the left**; that is, failure occurred before the given time. Include them in determining plotting positions of later failures, but do not plot them. The aim is to estimate the 1% life at 15 kV, a design voltage.

Time to Breakdown (seconds)

45 kV	40 kV	35 kV	30 kV	25 kV	Plotting Position
1−	1	30	50	521	4.2
1−	1	33	134	2,517	12.5
1−	2	41	187	4,056	20.8
2	3	87	882	12,553	29.2
2	12	93	1,448	40,290	37.5
3	25	98	1,468	50,560+	45.8
9	46	116	2,290	52,900+	54.2
13	56	258	2,932	67,270+	62.5

47	68	461	4,138	83,990*	70.8
50	109	1182	15,750	85,500+	79.2
55	323	1350	29,180+	85,700+	87.5
71	417	1495	86,100+	86,420+	95.8

– denotes left censored (failure occurred earlier).
+ denotes right censored (unfailed). * unplotted failure.

(a) Make a Weibull plot of the five data sets. Comment on the adequacy of the Weibull distribution.
(b) Separately fit distribution lines, and separately estimate the same percentile and the shape parameter for each test voltage. Gauge the amount of uncertainty in each estimate. Is it reasonable to use a common shape parameter?
(c) Fit parallel lines, and estimate a common shape parameter. Gauge the amount of uncertainty in this estimate.
(d) Engineering theory says that the life distribution of such fluid is exponential. Does the shape estimate convincingly differ from 1? Comment whether the failure rate increases or decreases.
(e) Make a relationship plot on log-log paper, showing censored data.
(f) Fit a line to percentile estimates. Judge whether the line fits adequately.
(g) Use the fitted line to estimate the percentile at 15 kV. Gauge the amount of uncertainty in this estimate. On Weibull paper, draw the distribution line through this estimate.
(h) Comment on convincing lack of fit of the model or peculiar data.
(i) Suggest further analyses and other models.
(j) Carry out (i).

3.11. Multiply censored relay data. Below are test data on a production relay and on a proposed design change. Engineering experience suggested that life has a Weibull distribution and life is an exponential function of current. Engineering sought to compare the production and proposed designs over the range of test currents.

Production	Thousands of cycles									
16 amps:	38+	77+	138	168+	188	228	252	273	283+	288
	291	299	317	374	527	529	559	567	656	873
26 amps:	103	110	131	219	226+					
28 amps:	84	92	121	138	191	206	254	267	308	313
Proposed										
26 amps:	110	138	249	288	297					
28 amps:	8+	51+	118+	144	219	236+	236+	252	252+	

(a) Make separate Weibull hazard plots for the two designs.
(b) Comment on the adequacy of the Weibull distribution.
(c) Estimate the shape and scale parameter for each of the five data sets. Gauge the amount of the uncertainty in each estimate.

(d) Estimate a common shape parameter for each design separately and for both designs together. Gauge the amount of the uncertainty in each estimate. Is a common value reasonable? Is the common estimate convincingly different from 1? Does the failure rate increase or decrease?

(e) Make a separate relationship plot for each design on semilog paper, using the log scale for cycles and the linear scale for current. Plot runouts.

(f) On each relationship plot, mark sample percentile estimates (your choice) for each test current. Fit a line to the percentile estimates, and comment on the linearity of the two relationships.

(g) Comment on any lack of fit of the model or peculiar data.

(h) Is one design convincingly better than the other?

(i) Suggest further analyses and other models.

(j) Carry out (i).

3.12. Transformer turn data. Transformer life testing at high voltage (rms) resulted in the multiply censored data below. All failures were turn-to-turn failures of the primary insulation. The Weibull distribution and inverse power law were assumed to fit such data. The main aim was to estimate the 1st percentile at 15.8 kV, 110% of design voltage 14.4 kV.

Voltage	Hours									
35.4 kV:	40.1	59.4	71.2	166.5	204.7	229.7	308.3	537.9	1002.3+	1002.3+
42.4 kV:	0.6	13.4	15.2	19.9	25.0	30.2	32.8	44.4	50.2+	56.2
46.7 kV:	3.1	8.3	8.9	9.0	13.6	14.9	16.1	16.9	21.3	48.1+

(a) Plot the data on Weibull hazard paper. Comment on the Weibull fit.

(a′) Repeat (a) using the modified hazard plotting positions. Compare plots (a) and (a′).

(b) From separately fitted lines, estimate the shape parameter and a chosen percentile at each test voltage. Gauge the amount of uncertainty in each estimate. Do the slopes of the lines differ convincingly?

(c) From parallel fitted lines, estimate a common shape parameter. Gauge the amount of uncertainty in this estimate. Does this estimate differ convincingly from 1? Does the failure rate increase or decrease?

(d) Plot the data (including nonfailures) and percentile estimates on log-log paper. Fit a line to the percentile estimates. Comment on the linearity of the relationship.

(e) Estimate the power of the inverse power law. Gauge the amount of uncertainty in this estimate.

(f) Use the fitted line to estimate the chosen percentile at 15.8 kV. Gauge its amount of uncertainty. Plot this estimate on Weibull paper.

(g) Draw a distribution line through the estimate (f) on Weibull paper. Estimate the 1st percentile at 15.8 kV and convert it from hours to years. Gauge the amount of uncertainty in this estimate.

(h) Comment on any convincing lack of fit or peculiar data. Suggest further analyses.

(i)   Comment on the advantages and disadvantages (including ease and accuracy) of the following. Treat the failure at 56.2 hours as if it were censored at 50.2 hours, and analyze the data as singly censored, using Weibull probability paper.

(j)   Using lognormal paper, do (a), (b), and (c) and assess whether the Weibull or lognormal fits better. Is one convincingly better?

(k)   Suggest further analyses and other models.

(l)   Carry out (k).

(m)   Write a brief report for management to summarize your findings. Include appropriate plots.

3.13. Permalloy corrosion. Thirty permalloy specimens were tested in a corrosive atmosphere at six relative humidities (%RH). The weight change of each specimen over a specified time is tabled below.

%RH	Corrosion (weight change)							
30	0.0144	0.0153	0.0092	0.0120	0.0111	0.0163	0.0193	0.0244
40	0.0221	0.0280	0.0287	0.0301	0.0301	0.0330		
50	0.0744	0.0684						
60	0.1050	0.1110	0.1160	0.1185				
70	0.1665	0.2065	0.2220	0.2540	0.2930	0.3008	0.3408	0.3807
78	0.6549	0.6660						

The main purpose was to assess whether the weight change at 10% and 20% relative humidity is below the specified 0.0050.

(a)   Plot the rate data on lognormal and Weibull probability paper. Does one distribution fit significantly (convincingly) better the the other?

(b)   Are the samples for the different humidities adequately described with parallel lines? Explain.

(c)   Plot the weight change data versus humidity on log-log and semi-log paper. On semi-log paper, plot weight change on the log scale. Are the plots straight? Which paper straightens the plot better? Is it convincingly better?

(d)   Fit a line to your chosen relationship plot, and extend it to 10% humidity. What percentile does your line estimate? Estimate the percentile at 10 and 20% humidity.

(e)   Plot the two percentile estimates on your chosen probability paper, and draw the distribution lines. How do the two distributions of weight change compare with the specification? Low weight change is desired.

(f)   Viewing the relationship and probability plots, discuss the adequacy of the models with respect to conclusions in (e). Also, note any pecularities of the data.

(g)   Write your chosen relationship in algebraic form with symbols for unknown parameters. Estimate the parameters of the relationship from your relationship plot. Write the relationship in terms of the numerical parameter estimates.

(h)   From the probability plot, estimate the distribution "spread" parameter. Write the equation for the cumulative distribution of weight loss at a

given humidity – both with symbols and numbers for parameters. Do this for the equation for a percentile as a function of humidity.
(i) Suggest further analyses and other methods.
(j) Carry out (i).

3.14. Eyring relationship. Repeat the graphical analyses of the Class-H data, but use the Eyring relationship instead of the Arrhenius relationship. In particular, plot the transformed times $t_i' = t_i T_i$ where t_i is the failure time and T_i the absolute temperature for specimen i.

3.15. Thermal cycling. Eighteen specimens of encapsulant were thermally cycled – six at each of the temperature changes. Inspection after 12, 50, 100, and 200 cycles detected cracked specimens and yielded the following interval data. The aim is to estimate life under a 40° cycle.

Temp.	Cycles				Survived
Change	0-12	13-50	51-100	101-200	200
190°C	1	1	2	1	1
140°C	-	-	2	1	3
100°C	-	-	-	-	6

(a) Plot the data on lognormal probability paper. For a single failure in an interval, plot it at the interval middle. For two failures, plot them equally spaced in the interval. Comment on whether the plots are parallel.
(b) Plot the data on log-log paper, including nonfailures. Theory for the Coffin-Manson relationship suggests life goes inversely as the fifth power of the temperature change. Fit to the data a line with the corresponding slope. Comment on how well the line fits the data.
(c) Use both plots to estimate the life distribution for a temperature change of 40°C. Plot the estimate of distribution line on the lognormal paper.
(d) Comment on the validity of the data and model.
(e) In a test of a new encapsulant, 42 specimens were cycled through a temperature change of 190° and all survived 200 cycles. How does the new encapsulant compare with the old under a 190° cycle? Under 40°? Are differences convincing? What assumptions are used? Are they valid?

3.16. Wire varnish. Twisted wire pairs coated with a varnish (electrical insulation) were subjected to a temperature-accelerated test. The midpoints of the inspection intervals in hours and the numbers of failures (in parentheses) follow.

Temp.	Hours (No. of failures)				
220°C	1092 (1),	2184 (2),	2436 (4),	2604 (1)	
240°C	528 (5)				
260°C	108 (2),	132 (4),	156 (3),	180 (2),	204 (5)

(a) Do a complete graphical analysis of the data.
(b) Estimate median life at 180°C, the design temperature.

(c) Why were unequal numbers of specimens run at the test temperatures?

3.17. CMOS RAMs. A temperature-accelerated test of CMOS RAMs yielded following numbers of failures on inspection (read-out data). The purpose of the test was to estimate the life at the design temperature of 55°C and at the worst service temperature of 85°C. A device "failed" when its leakage current reached a specified value, a result of mobile ion contamination. The test ended at 336 hours. Temperatures are ambient. Devices ran under a reverse bias but not under power. There are no specimen removals.

125°C (106 RAMs)	150°C (48 RAMs)	175°C (24 RAMs)
No. at hours (days)	No. at hours (days)	No. at hours (days)
2 at 72 (3)	3 at 24 (1)	2 at 24 (1)
1 at 168 (7)	1 at 136 (5.7)	3 at 72 (3)
2 at 336 (14)	2 at 168 (7)	3 at 336 (14)
101 survived 336 hours	2 at 336 (14)	16 survived 336 hours
	40 survived 336 hours	

(a) Make a Weibull plot of the data, taking into account the inspection intervals by spreading the individual failures over their intervals. Does the Weibull distribution adequately fit the data? Explain. Repeat, plotting the fraction failed at inspection times. Compare the plots.

(b) Estimate the shape parameter. Does the failure rate increase or decrease?

(c) Does the shape parameter depend on temperature? Explain.

(d) Estimate the same low percentile at each temperature. Plot those estimates on Arrhenius paper. Draw a line through them. Also, plot the individual failures and indicate numbers of survivors. Comment on adequacy of the Arrhenius relationship.

(e) Estimate the activation energy from (d), and compare the estimate with 1.05 eV, the traditional value for the failure mechanism.

(f) Use (d) to plot estimates of the life distributions at 55 and 85°C on the Weibull paper. In particular, estimate the 1st and 50th percentiles.

(g) The devices all came from the same wafer. Comment on how this affects previous estimates and interpretation of the plots.

(h) Comment on how to improve the schedule of inspection times.

(i) Comment on the advantages and disadvantages of the unequal numbers of specimens at the three test temperatures.

(j) Calculate an optimum test plan (Chapter 6), and compare it with that above.

(k) Fit the Arrhenius-Weibull model to the (interval) data by maximum likelihood (Chapter 5). Plot the fitted model and confidence limits on such papers without the data.

(l) Repeat the preceeding with the Arrhenius-lognormal model.

3.18.* Microprocessor. The following read-out data are from an accelerated test of the same Microprocessor whose data appear in Table 6.1.

The accelerated temperature is the same (125°C) but here the voltage is 5.25 V, the design voltage.

Interval i	1	2	3	4	5
Hours t_i	48	168	500	1000	2000
f_i/n_i	1/1413	3/1411	1/316	2/315	1/165

(a) Using the format of Table 6.2, calculate the estimates F_i.

(b) Plot the distribution estimate on Weibull paper. Does the Weibull distribution adequately fit the data?

(c) Estimate the Weibull shape parameter, and comment on the failure rate at that test condition. Would burn-in improve production? Why?

(d) Do the shape parameters at the two voltages differ convincingly? Suggest explanations for the difference?

(e) Calculate two-sided 95% confidence limits for the fraction failed, and plot them. Do the two distributions differ convincingly?

(f) Calculate the acceleration factor between 125°C and the design temperature 55°C, using (i) $E = 0.8$ eV and (ii) $E = 0.3$ eV. On Arrhenius paper, plot slopes for 0.3, 0.8 and 1.0 eV, and plot the 5.25 V data. Which activation energy is most pessimistic at the design temperature?

(g) Move the estimate and confidence limits from 125°C to 55°C on the Weibull plot, assuming $E = 1.0$ eV. Compare the 5.25 and 7.0 V distributions to each other and to the goal. What differences are convincing?

(h) On another Weibull paper, do (g) for both distributions, assuming $E = 0.8$ eV. Do the two distributions at design temperature differ much from a practical point of view?

(i) Do (b) and (g) on lognormal paper, assuming $E = 1.0$ eV.

(j) Criticize the preceding analyses.

4

Complete Data and Least Squares Analyses

1. INTRODUCTION

Contents. This chapter presents least squares analyses for complete life data (all specimens run to failure). The analyses provide estimates and confidence limits for product life, namely, for model parameters, mean (log) life, percentiles, the stress yielding a desired life, and the fraction failing by a given age. The methods are used here with normal, lognormal, Weibull, and exponential life distributions. The methods apply to data on a measure of product performance which degrades with time (Chapter 11). This chapter presents checks on the data and assumed model. These checks help one assess the validity of the estimates and confidence limits. More important, such checks often yield insight into improvements of a product.

Background. Needed background for this chapter includes the distributions and linear life-stress relationships of Chapter 2. Knowledge of the graphical methods of Chapter 3 is useful. Also needed is basic knowledge of statistical estimates (sampling distributions and standard errors), confidence limits, and hypothesis tests; for example, Nelson (1982, Chaps. 6 and 10) covers these basics. Knowledge of least squares regression methods is helpful but not essential; regression texts are referenced below. This chapter is intended for those who know only the basics of statistical methods. Those acquainted with least squares methods may choose to read selectively.

Chapter overview. Section 2 presents least squares methods for estimates and confidence limits for a (log) normal life distribution and a linear life-stress relationship. Section 3 describes data analyses and plots for checking the data and a model with a (log) normal life distribution. Section 4 extends least squares methods for estimates and confidence limits to Weibull and exponential life distributions. Section 5 describes data analyses and plots for checking the data and a model with a Weibull (or exponential) life distribution. Section 6 presents least squares fitting of multivariable models; such models contain two or more accelerating and engineering variables.

Advantages and disadvantages. Analytic methods like least squares have both advantages and disadvantages compared with graphical methods. Analytic methods are objective; that is, if two people use the same analytic methods on a set of data, they get exactly the same results. This is not true of graphical methods, but two people will usually arrive at the same conclusions from graphical analysis. Also, analytic methods indicate the accuracy of estimates by means of standard errors and confidence intervals. Statistical uncertainties in estimates of product life are usually large and startling. The accuracy is important if graphical methods do not clearly indicate that the information is accurate enough for practical purposes. A disadvantage of analytic methods is that they do not readily reveal certain information in data, whereas graphical methods do. Chapter 3 and Sections 3 and 5 below give examples of this. Also, the computations for most analytic methods are too laborious; so they require special computer programs. In contrast, graphical methods are easy to use by hand, and there are computer programs that do graphical analyses. Moreover, graphical methods help one present results to others. Seeing is believing.

It is usually best to use graphical methods and analytic methods together. The graphical methods help one assess whether the data and analytic methods are reasonable. Moreover, each provides certain information not provided by the other. Understanding a set of data requires many different analyses. Many modern statistics books expound such iterative and exploratory data analysis. Older books naively "solve" a problem by simply calculating a t statistic, a confidence interval, or a significance level.

Why this chapter? Methods for complete data deserve a chapter for several reasons. First, many accelerated tests are run until all specimens fail. This is usually inefficient and not recommended for reasons given in Chapter 5. However, this is a common practice for many products. Second, the least squares methods (calculations) for such data are well known and relatively simple. For example, IEEE Standard 101, ASTM Special Technical Publication STP 313, and ASTM Standard Practice E 739-80 present least squares methods. Moreover, regression books present the calculations and theory; examples include Draper and Smith (1981), Weisberg (1985), and Neter, Wasserman, and Kutner (1983,1985). Theory for the methods of this chapter appears in such books. Third, least squares computer programs that do the calculations are widely available. Even some pocket calculators do them. Fourth, for a lognormal life distribution, least squares methods yield the "best" estimates and exact confidence limits. In contrast, least squares methods for a Weibull or exponential life distribution are not statistically efficient (most accurate). Fifth, many readers are acquainted with regression methods for complete data. So this chapter is a simple introduction to Chapter 5, which extends regression methods to censored data and is more complex.

Avoid complete data. Running an accelerated test until all specimens fail is generally inefficient; that is, it wastes time and money, It is generally better to stop a test before all specimens fail. Then one analyzes the censored data with the maximum likelihood methods in Chapter 5. Developed in the 1960s, those methods are yet unknown to some engineers. Moreover, the calculations for censored data are complex and require special computer programs. Also, many major companies do not yet have such programs, as they are not part of many well-known statistical packages. Consequently, engineers will continue to collect complete data and use least squares methods.

Model error. Besides efficiency, there is another reason to stop a test before all specimens fail, namely, accuracy of results. Suppose one wants to estimate a low percentile of the life distribution at some stress level. Also, suppose that the assumed life distribution does not fit the data over the entire range of the distribution; that is, the model is inaccurate. Then it is usually better to fit the distribution only to the early failures at each test stress. Then the later failures do not bias the estimate of a low percentile. Figure 1.1 shows the reason for such bias on probability paper. Hahn, Morgan, and Nelson (1985) explain in detail the use of artificial censoring to reduce such model bias. In particular, they recommend treating failures in the upper tail as if they were censored at some earlier time.

Computer programs. Least-squares computer programs are widely available. Most statistical packages with such programs also provide probability plots and crossplots, useful for checking the model and data. Major packages used in industry include:
- BMD, edited by Dixon (1974).
- BMDP, Dixon (1983). This has a friendlier user language than BMD does.
- SAS, offered by the SAS Institute, Inc. (1982), (919)467-8000.
- SPSS^x, offered by SPSS, Inc. (1986).
- Minitab, offered by Minitab Project, 215 Pond Laboratory, University Park, PA 16802, (814)238-3280 or 865-1595.

These all run on main frame and minicomputers. Some are available on PCs.

Most least squares programs lack certain output useful for accelerated testing. This includes estimates and confidence limits for percentiles, for a

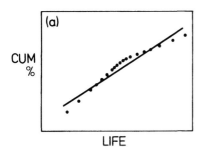
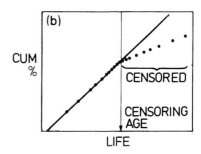

Figure 1.1.   (a) Biased fit to all data, (b) better fit to lower tail.

fraction failing by a given age, and for a design stress that yields a specified life. Such programs assume life has a (log) normal distribution, and most lack the necessary modifications for Weibull and exponential distributions. These modifications appear in Sections 4 and 5. Most readers will use such programs. However, some readers may wish to write their own programs. The detailed calculations presented here make this easy. Be aware though that such personally written programs usually suffer from round-off error, over- and under-flow, crude approximations to t percentiles and other statistical functions, etc. Most mature standard statistical packages perform acceptably in these regards, as most use sophisticated algorithms. Moreover, they calculate and plot residuals, as described in Sections 3 and 5.

Section 1 of Chapter 5 lists computer packages that analyze censored (and complete) life data with maximum likelihood methods. For complete data and a lognormal life distribution, maximum likelihood estimates are the same as least squares estimates. Moreover, maximum likelihood programs give estimates and approximate confidence limits for percentiles and for a fraction failing by a given age, whereas least squares programs do not.

Uncertainties are greater. Practical considerations for data collection and analysis are discussed in Chapter 1. For example, data ideally should be a random sample from the population of interest, that is, from production units tested under actual use conditions. Also, the model should adequately represent the product life as a function of stress. For example, the methods of this chapter may mislead if the data contain more than one failure mode; Chapter 7 treats this complication. Of course, in practice, the test units and conditions only approximate the population and service conditions, and the model is inaccurate to some degree. Thus the estimates below have greater uncertainty than indicated by standard errors and confidence limits.

Related material. Subsequent chapters extend the methods here. Chapter 5 presents maximum likelihood methods for analysis of censored and interval data. Chapter 7 presents graphical and maximum likelihood methods for data with a mix of failure modes. Also, Chapter 3 provides graphical methods for complete and censored data. Chapter 6 gives guidance on the choice of a test plan and the number of test specimens.

2. LEAST-SQUARES METHODS FOR LOGNORMAL LIFE

Purpose. This section describes least squares methods for estimates and confidence limits for parameters, percentiles, reliabilities, and other quantities. These methods apply when product life has a lognormal or normal distribution and the life-stress relationship is linear. The methods also apply to data on product performance that degrades with time (Chapter 11).

Background. Needed background for this section includes the normal and lognormal distributions and the simple linear life-stress relationship

(Chapter 2). The widely used Arrhenius-lognormal model is a special case. Also needed is basic knowledge of statistical estimates, confidence limits, and hypothesis tests, and the graphical methods of Chapter 3.

Overview. Section 2.1 describes the example data, model, and computer output used to illustrate the least squares methods. Section 2.2 presents estimates for model parameters, mean (log) life, percentiles, a design stress, and the fraction failing by a given age. Section 2.3 presents confidence limits for the same quantities. Section 3 provides checks on the data and model.

2.1. Example Data, Model, and Computer Output

Example data. The least squares methods for a normal or lognormal life distribution are illustrated with the Class-H insulation data. Chapter 3 presents these complete data and graphical analyses of them. Table 2.1 contains the (base 10) log times to failure of the 40 specimens. For the assumed lognormal distribution, one analyzes the log times. As in Chapter 3, we use the middle of an inspection interval as the failure time.

Purpose. The main purpose of the least squares analysis is to estimate the median life of such insulation at the design temperature of 180°C. This includes comparing the estimate with a desired life of 20,000 hours by means of a confidence interval. Also, the validity of the data and assumed model must be assessed with methods in Section 3; they led to a discovery of how to save $1,000,000 annually.

Model assumptions and notation. For complete (log) normal data, least squares analysis involves the following assumptions. A random sample of n units is put on test, and all run to failure. There are J test stress levels. n_j units are tested at (transformed) stress level x_j, $j = 1, 2, \cdots, J$. For example, for the Arrhenius model, $x_j = 1000/T_j$ is the reciprocal of absolute temperature T_j. For the inverse power law, $x_j = \log(V_j)$ is the log of voltage V_j. y_{ij} denotes the (log) time to failure of test unit i at test stress x_j. Such log times for the Class-H insulation data appear in Table 2.1. The total number of test units is $n = n_1 + \cdots + n_J$.

190 C	220 C	240 C	260 C
3.8590	3.2465	3.0700	2.7782
3.8590	3.3867	3.0700	2.8716
3.8590	3.3867	3.1821	2.8716
3.9268	3.3867	3.1956	2.8716
3.9622	3.3867	3.2087	2.9600
3.9622	3.3867	3.2214	3.0523
3.9622	3.4925	3.2214	3.1206
3.9622	3.4925	3.2338	3.1655
4.0216	3.4925	3.2458	3.2063
4.0216	3.4925	3.2907	3.2778

Table 2.1. Log Failure Times of Class-H Insulation

The model for the (log) failure time y_{ij} is

$$y_{ij} = \mu(x_j) + e_{ij} \qquad (2.1.1)$$

for $i = 1, \cdots, n_j$ and $j = 1, \cdots, J$. The *random variation* (or *random error*) e_{ij} of the (log) life y_{ij} has a normal distribution with a mean of zero and an unknown standard deviation σ. The linear life-stress relationship

$$\mu(x_j) = \gamma_0 + \gamma_1 x_j \qquad (2.1.2)$$

gives the mean log life. (2.1.2) includes the Arrhenius and inverse power relationships. Also, the random variations e_{ij} are all assumed statistically independent. These are the usual assumptions for least squares regression theory; see Draper and Smith (1981) or Neter, Wasserman, and Kutner (1983,1985). This section presents least squares estimates and confidence intervals for the model parameters γ_0, γ_1, and σ and for other quantities.

For some purposes, (2.1.2) is used in the form

$$\mu(x_j) = \gamma_0' + \gamma_1(x_j - x') . \qquad (2.1.2')$$

Here x' is a chosen value, usually near the center of the data or a key value such as the design stress. Then $\mu(x') = \gamma_0'$ and $\gamma_0 = \gamma_0' - \gamma_1 x'$. Here the stress x is "centered" on x' or "coded."

Output. Least squares fitting of the model above can be carried out with most statistical packages. Figure 2.1 displays selected output from such fitting to the Class-H data with the SAS package. The SAS calculations employ data accurate to 7 figures; Table 2.1 is accurate to 4 or 5. Most such least squares packages lack certain output useful for accelerated test data. Sections 2.2 and 2.3 describe the underlying calculations and define all notation and terminology. Equation numbers for such calculations appear below. The variables are LOGLIFE (the base 10 log of insulation life) and INVTEMP (1000 divided by the absolute temperature). The following comments refer to numbered lines in the output.

- Line 1 and following show descriptive statistics. These include the sums, means (2.2.4 and 2.2.5), and sums of squares (2.2.6, 2.2.7, and 2.2.8) for the variables LOGLIFE and INVTEMP.
- Line 2 shows the estimate 0.1110659 of the log standard deviation σ (2.2.12). Here it is labeled the Root Mean Square Error. SAS does not provide confidence limits (2.3.6) for σ.
- Line 3 shows the least squares estimate -3.16328 for the intercept coefficient γ_0 (2.2.10). It also shows the standard error (2.3.8) for that estimate. SAS output lacks confidence limits (2.3.9) for this coefficient.
- Line 4 shows the least squares estimate 3.27284166 for the slope coefficient for INVTEMP (2.2.9). It also shows the standard error (2.3.11) for that estimate. SAS output lacks confidence limits (2.3.12) for this coefficient.
- Line 5 and following show the variances (2.3.7 and 2.3.10) and covariance of the coefficient estimates.

1 DESCRIPTIVE STATISTICS

VARIABLE	SUM	MEAN	UNCORRECTED SS
LOGLIFE	135.6608131	3.391520328	465.2941647
INVTEMP	80.1114010	2.002785025	160.8874031
INTERCEP	40.0000000	1.000000000	40.0000000

VARIABLE	VARIANCE	STD DEVIATION
LOGLIFE	0.1332758829	0.3650696959
INVTEMP	0.0113202295	0.1063965671
INTERCEP	0.0000000000	0.0000000000

SUMS OF SQUARES AND CROSSPRODUCTS

SSCP	LOGLIFE	INVTEMP	INTERCEP
LOGLIFE	465.2942	273.1444	135.6608
INVTEMP	273.1444	160.8874	80.1114
INTERCEP	135.6608	80.1114	40

DEP VARIABLE: LOGLIFE
ANALYSIS OF VARIANCE

SOURCE	DF	SUM OF SQUARES	MEAN SQUARE	F VALUE	PROB>F
MODEL	1	4.72900560	4.72900560	383.362	0.0001
ERROR	38	0.46875383	0.01233563		
C TOTAL	39	5.19775943			

2
ROOT MSE	0.1110659	R-SQUARE	0.9098	
DEP MEAN	3.39152	ADJ R-SQ	0.9074	
C.V.	3.274811			

PARAMETER ESTIMATES

VARIABLE	DF	PARAMETER ESTIMATE	STANDARD ERROR	T FOR HO: PARAMETER=0	PROB > |T|
3 INTERCEP	1	-3.16328	0.33523683	-9.436	0.0001
4 INVTEMP	1	3.27284166	0.16715551	19.580	0.0001

5 COVARIANCE OF ESTIMATES

COVB	INTERCEP	INVTEMP
INTERCEP	0.1123837	-0.0559597
INVTEMP	-0.0559597	0.02794096

TEMP	MEAN	STD ERR	LOWER95%	UPPER95%	MEDIAN	LMEDIAN	UMEDIAN
6 180 deg.	4.05899	0.0383472	3.98136	4.13654	11454.8	9579.83	13696.7
190 deg.	3.90305	0.0314793	3.83933	3.96678	7999.3	6907.59	9263.6
220 deg.	3.47319	0.0180497	3.43665	3.50973	2973.0	2733.08	3233.9
240 deg.	3.21454	0.0197508	3.17456	3.25452	1638.9	1494.71	1796.9
260 deg.	2.97530	0.0275735	2.91948	3.03111	944.7	830.76	1074.3

Figure 2.1.   SAS least squares output for Class-H data.

- Line 6 and following show estimates and symmetric two-sided 95% confidence limits (2.2.3) for the log mean life at the test and design temperatures. They also show estimates and two-sided 95% confidence limits for the median life in hours at the test and design temperatures. These limits for 180°C are below 20,000 hours. This is convincing evidence that the insulation does not meet the 20,000-hour goal. SAS does not automatically calculate such results.

SAS provides other standard output, but it has little value for accelerated tests. Most readers will analyze data with such a standard package. They can skip the theory in Sections 2.2 and 2.3.

2.2. Estimates for the Model

This section presents formulas for the least squares estimates for the model parameters and other quantities of interest. Section 2.3 presents corresponding confidence intervals. The estimates and confidence intervals are standard ones from least squares regression theory; see Draper and Smith (1981) or Neter, Wasserman, and Kutner (1983,1985). They have good statistical properties and suit practical work.

Preliminary calculations. Preliminary calculations with the data follow. Calculate the *sample average* $\bar{y}_j$ and *standard deviation* s_j for each test stress:

$$\bar{y}_j = (y_{1j} + y_{2j} + \cdots + y_{n_j j})/n_j, \tag{2.2.1}$$

$$s_j = \{[(y_{1j} - \bar{y}_j)^2 + \cdots + (y_{n_j j} - \bar{y}_j)^2]/(n_j - 1)\}^{1/2}$$
$$= \{[(y_{1j}^2 + \cdots + y_{n_j j}^2) - n_j \bar{y}_j^2]/(n_j - 1)\}^{1/2}; \tag{2.2.2}$$

here the sums run over all observations y_{ij} at test stress j. Antilog($\bar{y}_j$) is called the *sample geometric average* or *mean*. The first formula for s_j has less round off error. The *number of degrees of freedom* of s_j is

$$\nu_j = n_j - 1. \tag{2.2.3}$$

If $n_j = 1$, s_j is not calculated. Table 2.2 shows these calculations for the example. All calculations use six-figure accuracy; this assures 3- or 4-figure accuracy in final results. It would also be best to use six-figure data.

Calculate the *grand averages* of all data

$$\bar{x} = (n_1 \bar{x}_1 + \cdots + n_J \bar{x}_J)/n, \tag{2.2.4}$$

Table 2.2. Least Squares Calculations for the Class-H Insulation

(2.2.1) $\bar{y}_1 = 3.93958$, $\bar{y}_2 = 3.41500$, $\bar{y}_3 = 3.19395$, $\bar{y}_4 = 3.01755$

Antilog($\bar{y}_j$): 8,701   2,600   1,563   1,041

(2.2.2) $s_1 = \{[(3.8590 - 3.93958)^2 + \cdots + (4.0216 - 3.93958)^2]/(10-1)\}^{1/2} = 0.0624891$

(2.2.2) $s_2 = \{[(3.2465 - 3.41500)^2 + \cdots + (3.4925 - 3.41500)^2]/(10-1)\}^{1/2} = 0.0791775$

(2.2.2) $s_3 = \{[(3.0700 - 3.19395)^2 + \cdots + (3.2907 - 3.19395)^2]/(10-1)\}^{1/2} = 0.0716720$

(2.2.2) $s_4 = \{[(2.7782 - 3.01755)^2 + \cdots + (3.2778 - 3.01755)^2]/(10-1)\}^{1/2} = 0.170482$

(2.2.4) $\bar{x} = [10(2.159) + 10(2.026) + 10(1.949) + 10(1.875)]/40 = 2.00225$

(2.2.5) $\bar{y} = [10(3.93958) + 10(3.41500) + 10(3.19395) + 10(3.01755)]/40 = 3.39152$

(2.2.6) $S_{yy} = (3.8590 - 3.39152)^2 + \cdots + (3.2778 - 3.39152)^2 = 5.19746$

(2.2.7) $S_{xx} = 10(2.159 - 2.00225)^2 + 10(2.026 - 2.00225)^2 + 10(1.949 - 2.00225)^2$
   $= 0.441627.$

(2.2.8) $S_{xy} = 10(2.159 - 2.00225)3.93958 + \cdots + 10(1.875 - 2.00225)3.01755 = 1.44574$

(2.2.9) $c_1 = 1.44574/0.441627 = 3.27367$

(2.2.10) $c_0 = 3.39152 - 3.27367(2.00225) = -3.16319$

(2.2.11) $s = \{[9(0.0624891)^2 + 9(0.0791775)^2 + 9(0.0716720)^2 + 9(0.170482)^2]/(40-4)\}^{1/2}$
    $= 0.105327$

(2.2.12) $s' = \{[5.19746 - 3.27367(1.44574)]/(40-2)\}^{1/2} = 0.110569$

$$\bar{y} = (n_1\bar{y}_1 + \cdots + n_J\bar{y}_J)/n . \tag{2.2.5}$$

Each is also the sum over the entire sample divided by n. Calculate the *sums of squares*

$$S_{yy} = \sum_{j=1}^{J}\sum_{i=1}^{nj}(y_{ij}-\bar{y})^2 = \sum_{j=1}^{J}\sum_{i=1}^{nj}y_{ij}^2 - n\bar{y}^2 , \tag{2.2.6}$$

$$S_{xx} = n_1(x_1-\bar{x})^2 + \cdots + n_J(x_J-\bar{x})^2 = n_1x_1^2 + \cdots + n_Jx_J^2 - n\bar{x}^2 , \tag{2.2.7}$$

$$S_{xy} = n_1(x_1-\bar{x})\bar{y}_1 + \cdots + n_J(x_J-\bar{x})\bar{y}_J = n_1x_1\bar{y}_1 + \cdots + n_Jx_J\bar{y}_J - n\bar{x}\bar{y}; \tag{2.2.8}$$

here the double sum of S_{yy} runs over all n observations in the sample. Table 2.2 shows the sums for the example. Least-squares computer programs perform the calculations for equations (2.2.1) through (2.2.12). Results in Table 2.2 differ from those in Figure 2.1 due mostly to the fewer number of significant figures in the data in Table 2.1.

Some data sets have a different x_j value for each or most specimens; that is, $n_j = 1$. Then the sums in (2.2.5) through (2.2.8) must run over all n specimens. All subsequent calculations are the same. However, then use the estimate s' (2.2.12) rather than s (2.2.11) for σ.

Coefficient estimates. The least squares estimates of γ_1 and γ_0 are

$$c_1 = S_{xy}/S_{xx} , \tag{2.2.9}$$

$$c_0 = \bar{y} - c_1\bar{x} . \tag{2.2.10}$$

For the example in Table 2.2, $c_1 = 3.27367$ and $c_0 = -3.16319$. Confidence limits for the true γ_1 and γ_0 appear in Section 2.3. Graphical estimates appear in Chapter 3. The estimate of activation energy (in electron-volts) is

$$E^* = 2303 k \, c_1$$

where Boltzmann's constant is $k = 8.6171 \times 10^{-5}$ electron-volts per Centigrade degree. For the example in Table 2.2, $E^* = 2303 \, (8.6171 \times 10^{-5}) \, 3.27367 = 0.65$ electron-volts (rounded to a physically useful number of figures).

σ **estimate.** The *pooled estimate of the (log) standard deviation* σ is

$$s = [(\nu_1 s_1^2 + \cdots + \nu_J s_J^2)/\nu]^{1/2}; \tag{2.2.11}$$

here $\nu = \nu_1 + \cdots + \nu_J = n - J$ is its number of degrees of freedom. This is called the *estimate of the standard deviation based on replication* (or *on pure error*). For the example in Table 2.2, $s = 0.105327$. This small value indicates that the failure rate increases with insulation age. Another pooled estimate is

$$s' = [(S_{yy} - c_1 S_{xy})/(n-2)]^{1/2} . \tag{2.2.12}$$

This is called the *estimate based on lack of fit* about the equation. It has $\nu' = (n-2)$ degrees of freedom. Most regression programs give this estimate. For the example in Table 2.2, $s' = 0.110569$. For practical purposes, this differs little from $s = 0.105327$.

Either σ estimate can be used below, but one must use the corresponding

ν or ν'. s is recommended and used below. s' tends to overestimate σ if the true (transformed) life-stress relationship is not linear. However, s' may sometimes be preferable; it is conservative, tending to overestimate σ and to make the product look worse. Confidence limits for σ appear in Section 2.3. Graphical estimates of σ appear in Chapter 3.

Estimate for mean (log) life. For any (transformed) stress level x_0, the least squares estimate of the (log) mean $\mu(x_0) = \gamma_0 + \gamma_1 x_0$ is

$$m(x_0) = c_0 + c_1 x_0. \qquad (2.2.13)$$

This estimate assumes that the linear relationship (2.2.2) holds; otherwise, the estimate may be inaccurate. Methods for assessing linearity appear in Section 3.1. Confidence limits for $\mu(x_0)$ appear in Section 2.3. The graphical estimate appears in Chapter 3. For the lognormal distribution, the estimate of the median life is $t_{.50}(x_0) = \text{antilog}[m(x_0)]$.

For the insulation at absolute temperature T_0 ($x_0 = 1000/T_0$),

$$m(x_0) = -3.1632 + 3.2737x_0 = -3.1632 + (3273.7/T_0);$$

here c_0 and c_1 come from Table 2.2. The coefficient estimates in this equation require five or more significant figures; this assures that any final calculation is accurate enough. This line is shown on Arrhenius paper in Figure 2.2. At the design temperature of 180°C ($x_0 = 2.207$), $m(2.207) = -3.1632 + 3.2737(2.207) = 4.062$. The estimate of the median life at 180°C is

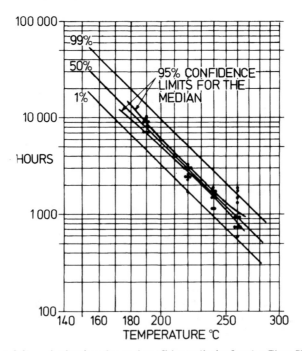

Figure 2.2. Arrhenius plot and confidence limits for the Class-H data.

$t_{.50}(2.207)$ = antilog(4.062) = 11,500 hours. This point is on the median line in Figure 2.2.

Estimates for percentiles. For any (transformed) stress level x_0, the estimate for the normal $100P$th percentile $\eta_P(x_0) = \mu(x_0) + z_P \sigma$ is

$$y_P(x_0) = m(x_0) + z_P s = c_0 + c_1 x_0 + z_P s; \qquad (2.2.14)$$

here z_P is the $100P$th standard normal percentile. This estimates the (log) mean if $P = 0.50$. The estimate of the lognormal $100P$th percentile is $t_P(x_0) = $ antilog$[y_P(x_0)]$.

For the example, the estimate of the 1st percentile of log life at 180°C is $y_{.01}(2.207) = 4.062 + (-2.2363)0.1053 = 3.827$ where $z_{.01} = -2.2363$. This estimate of life is $t_{.01}(2.207) = $ antilog(3.827) = 6,710 hours. The 1st percentile line in Figure 2.2 passes through this estimate.

Lower confidence limits for such percentiles appear in Section 2.3. Graphical estimates appear in Chapter 3.

Estimate for a stress with desired life. One may need to estimate the (transformed) stress level such that a given percentage $100R$ of the population survives a specified (log) time η^*. The estimate x^* of the (transformed) stress comes from (2.2.14) written as

$$x^* = (\eta^* - c_0 + z_R s)/c_1. \qquad (2.2.15)$$

For the example, suppose 99 percent of the population is to survive 10,000 hours ($\eta^* = \log(10,000) = 4.000$). The estimate of the transformed stress that produces this life is

$$x^* = [4.000 - (-3.163) + (2.2363)0.1053]/3.2737 = 2.260.$$

Converted, this is $(1000/2.260) - 273.16 \approx 169$°C.

Confidence limits for such a stress appear in Section 2.3. A graphical estimate appears in Chapter 3.

Estimate for a fraction failing. For any (transformed) stress x_0, the estimate for the fraction failing by a given (log) age η_0 is

$$F^*(\eta_0; x_0) = \Phi\{[\eta_0 - m(x_0)]/s\} = \Phi[(\eta_0 - c_0 - c_1 x_0)/s]; \qquad (2.2.16)$$

$\Phi\{\ \}$ is the standard normal cumulative distribution function (Appendix A1).

For Class-H insulation at 190°C, $x_0 = 1000/(190 + 273.16) = 2.159$. The estimate of the fraction failing by 7000 hours ($\eta_0 = \log(7000) = 3.84510$) is

$$F^*(3.84510; 2.159) = \Phi\{[3.84510 - (-3.16319) - (3.27367)2.159]/0.105327\}$$

$$= \Phi(-0.566) = 0.286.$$

The graphical estimate appears in Chapter 3. Confidence limits for the true value of such a fraction appear in Section 2.3.

2.3. Confidence Intervals

The accuracy of least squares estimates is given by their standard errors and confidence intervals. These appear below for the estimates of the model parameters and the other quantities of interest. Many of these intervals are given by Draper and Smith (1981), Neter, Wasserman, and Kutner (1983, 1985), and Weisberg (1985).

Confidence limits for the (log) mean. $m(x_0) = c_0 + c_1x_0$ is the estimate for the (log) mean $\mu(x_0) = \gamma_0 + \gamma_1x_0$ at a stress x_0. It has a normal sampling distribution with a mean equal to $\mu(x_0)$. Thus $m(x_0)$ is an unbiased estimate for $\mu(x_0)$. The standard deviation of this sampling distribution is the *standard error* of $m(x_0)$ and is

$$\sigma[m(x_0)] = \{(1/n) + [(x_0 - \bar{x})^2/S_{xx}]\}^{1/2}\sigma; \qquad (2.3.1)$$

the notation is defined in Section 2.2. The size of this standard error depends on the test plan, that is, the choice of test stress levels x_j and the numbers n_j of test units run at them. The estimate of this standard error is

$$s[m(x_0)] = \{(1/n) + [(x_0 - \bar{x})^2/S_{xx}]\}^{1/2}s; \qquad (2.3.2)$$

here s is an estimate of σ and has ν degrees of freedom (Section 2.2).

A two-sided $100\gamma\%$ confidence interval for the true $\mu(x_0)$ has limits

$$\mu(x_0) = m(x_0) - t(\gamma';\nu) \cdot s[m(x_0)],$$
$$\tilde{\mu}(x_0) = m(x_0) + t(\gamma';\nu) \cdot s[m(x_0)]; \qquad (2.3.3)$$

here $t(\gamma';\nu)$ is the $100\gamma'$th t percentile with ν degrees of freedom where $\gamma' = (1+\gamma)/2$. These percentiles appear in Appendix A4. Confidence limits for the true median $\tau_{.50}(x_0)$ of a lognormal life distribution are

$$\underset{\sim}{\tau}_{.50}(x_0) = \text{antilog}[\mu(x_0)], \quad \tilde{\tau}_{.50}(x_0) = \text{antilog}[\tilde{\mu}(x_0)].$$

For the Class-H insulation at 180°C ($x_0 = 2.207$),

$$s\,[m(2.207)] = \{(1/40)+[(2.207\text{-}2.002)^2/0.4416]\}^{1/2}0.1053 = 0.0365;$$

here $s = 0.1053$ has 36 degrees of freedom. The 95% confidence limits are

$$\mu(2.207) = 4.062 - 2.029(0.0365) = 3.988,$$
$$\tilde{\mu}(2.207) = 4.062 + 2.029(0.0365) = 4.136;$$

here $t(0.975;36) = 2.029$ is the 97.5th t percentile with $\nu = 36$ degrees of freedom. The limits for the median life are $\underset{\sim}{\tau}_{.50}(2.207) = \text{antilog}(3.988) = 9,730$ and $\tilde{\tau}_{.50}(2.207) = \text{antilog}(4.136) = 13,700$ hours. These limits do not enclose the desired median life of 20,000 hours. So clearly the true median life is statistically significantly (convincingly) below 20,000 hours. This answers the main question. The 95% confidence limits for median lives at the design and test temperatures are tabulated below. These limits appear on curves in Figure 2.2. The limits are narrowest near the center of the data and are wider the farther from the center – a general characteristic of such limits.

Temperature (°C)	Lower Limit for Median (hours)	Upper Limit for Median (hours)
180	9,730	13,700
190	6,980	9,220
220	2,720	3,190
240	1,510	1,780
260	836	1,070

Often one seeks only a lower limit, since long life is desired. A lower one-sided $100\gamma\%$ confidence limit for $\mu(x_0)$ is

$$\mu(x_0) = m(x_0) - t(\gamma;\nu)\, s[m(x_0)]; \tag{2.3.4}$$

here $t(\gamma;\nu)$ is the 100γth t percentile with ν degrees of freedom. The lower limit for the median of a lognormal life distribution is the antilog of this.

For the Class-H insulation at 180°C ($x_0 = 2.207$), the lower 95% confidence limit for $\mu(2.207)$ is $\mu(2.207) = 4.062 - 1.689(0.0365) = 4.000$; here $t(0.95;36) = 1.689$ is the 95th t percentile with $\nu = 36$ degrees of freedom. For the median, $\underset{\sim}{\tau}_{.50}(2.207) = \text{antilog}(4.000) = 10{,}000$ hours.

A lower limit for a two-sided $100\gamma\%$ confidence interval is also a lower limit for a one-sided $100(1+\gamma)/2\%$ confidence interval. For example, the lower limit of a two-sided 95% confidence interval is a lower limit for a one-sided 97.5% confidence interval.

Confidence limits for a percentile. Confidence limits for the $100P$th percentile $\eta_P(x_0)$ of a (log) normal life distribution at a given stress x_0 follow. Usually one wants that no more than a proportion P of the population failing below the limit with 100γ percent confidence. An approximate limit is

$$\eta_P(x_0) \approx y_P(x_0) - z_\gamma\{[z_P^2/(2\nu)] + (1/n) + [(x_0 - \bar{x})^2/S_{xx}]\}^{1/2}s; \tag{2.3.5}$$

here the notation is the same as before. $\underset{\sim}{\tau}_P(x_0) = \text{antilog}[\eta_P(x_0)]$ is a lower confidence limit for the $100P$th lognormal percentile. $\underset{\sim}{\tau}_P(x_0)$ is also called a lower **tolerance limit** for $100(1-P)\%$ of the population.

An exact lower limit is given by Easterling (1969). No least squares regression programs give confidence limits for percentiles. The confidence limit is inaccurate if the population distribution is not (log) normal.

For the Class-H insulation at the design temperature of 180°C ($x_0 = 2.207$), the 95% confidence limit for the log 1st percentile is

$$\eta_{.01}(2.207) = 3.827 - 1.645\{[(-2.326)^2/(2 \times 36)]$$

$$+ (1/40) + [(2.207 - 2.002)^2/0.4416]\}^{1/2}0.1053 = 3.741.$$

The limit for the lognormal 1st percentile of life is $\underset{\sim}{\tau}_{.01}(2.207)$ = anti-log(3.741) = 5500 hours (two-figure accuracy).

Confidence limits for σ. A two-sided $100\gamma\%$ confidence interval for the true (log) normal standard deviation σ has limits

$$\underset{\sim}{\sigma} = s\cdot\{\nu/\chi^2[(1+\gamma)/2;\nu]\}^{1/2}, \quad \tilde{\sigma} = s\cdot\{\nu/\chi^2[(1-\gamma)/2;\nu]\}^{1/2} ; \qquad (2.3.6)$$

here $\chi^2(\delta;\nu)$ is the 100δth chi-square percentile with ν degrees of freedom, that of s (Section 2.2). These percentiles appear in Appendix A5. Some regression programs calculate these limits.

For the example, 95% confidence limits are $\underset{\sim}{\sigma}$ = $0.1053[36/51.0]^{1/2}$ = 0.0885 and $\tilde{\sigma}$ = $0.1053[36/23.3]^{1/2}$ = 0.1317; here $\chi^2(0.975;36)$ = 51.0 and $\chi^2(0.025;36)$ = 23.3. This could employ s', which has ν' degrees of freedom.

This confidence interval for the standard deviation of (log) life assumes that the life distribution is (log) normal. Otherwise, the interval is incorrect.

Confidence limits for γ_0. The estimate c_0 for the intercept coefficient γ_0 has a normal sampling distribution with a mean of γ_0. Thus c_0 is an unbiased estimate for γ_0. The standard deviation of this sampling distribution is the *standard error* of c_0 and is

$$\sigma(c_0) = [(1/n) + (\bar{x}^2/S_{xx})]^{1/2}\sigma. \qquad (2.3.7)$$

Its estimate is

$$s(c_0) = [(1/n) + (\bar{x}^2/S_{xx})]^{1/2}s; \qquad (2.3.8)$$

here s is an estimate for σ and has ν degrees of freedom (Section 2.2).

A two-sided $100\gamma\%$ confidence interval for γ_0 has limits

$$\underset{\sim}{\gamma}_0 = c_0 - t(\gamma';\nu)s(c_0), \quad \tilde{\gamma}_0 = c_0 + t(\gamma';\nu)s(c_0) ; \qquad (2.3.9)$$

here $t(\gamma';\nu)$ is the $100\gamma'$ = $100(1+\gamma)/2$th t percentile with ν degrees of freedom. These percentiles appear in Appendix A4. Most regression programs calculate these limits. This interval is seldom used in accelerated testing work, since γ_0 usually does not a have useful physical meaning.

For the example, $s(c_0)$ = $[(1/40)+(2.002^2/0.4416)]^{1/2}0.1053$ = 0.318; here s = 0.1053 has 36 degrees of freedom. The 95% limits are $\underset{\sim}{\gamma}_0$ = $-3.163-2.029(0.318)$ = -3.808 and $\tilde{\gamma}_0$ = $-3.163+2.029(0.318)$ = -2.518 where $t(0.975;36)$ = 2.029.

This interval is exact if the life distribution is lognormal. For large samples, the confidence of the interval is close to $100\gamma\%$ for other distributions.

Confidence limits for γ_1. The estimate c_1 for the slope coefficient γ_1 has a normal sampling distribution with a mean of γ_1. Thus, c_1 is an unbiased estimate for γ_1. The standard deviation of this sampling distribution is the *standard error* of c_1 and is

$$\sigma(c_1) = [1/S_{xx}]^{1/2}\sigma. \tag{2.3.10}$$

Its estimate is

$$s(c_1) = [1/S_{xx}]^{1/2}s; \tag{2.3.11}$$

here s is an estimate of σ and has ν degrees of freedom (Section 2.2).

A two-sided $100\gamma\%$ confidence interval for γ_1 has limits

$$\underline{\gamma}_1 = c_1 - t(\gamma';\nu)s(c_1), \quad \tilde{\gamma}_1 = c_1 + t(\gamma';\nu)s(c_1); \tag{2.3.12}$$

here $t(\gamma';\nu)$ is the $100\gamma' = 100(1+\gamma)/2$th t percentile with ν degrees of freedom. These percentiles appear in Appendix A4. Most regression programs give these limits. Limits for activation energy (electron-volts) are

$$\underline{E} = 2{,}303k\,\underline{\gamma}_1, \quad \tilde{E} = 2{,}303k\,\tilde{\gamma}_1$$

where Boltzmann's constant is $k = 8.6171{\times}10^{-5}$ electron-volts per $°C$.

For the example, $s(c_1) = [1/0.4416]^{1/2}0.1053 = 0.1585$. The 95% confidence limits are $\underline{\gamma}_1 = 3.2737 - 2.029(0.1585) = 2.952$ and $\tilde{\gamma}_1 = 3.2737 + 2.029 (0.1585) = 3.595$. For the activation energy, $\underline{E} = 2{,}303(8.6171{\times}10^{-5}) 2.952 = 0.59$ and $\tilde{E} = 2{,}303(8.6171 \times 10^{-5})3.595 = 0.71$ electron-volts.

This interval is exact if the life distribution is (log) normal. For large samples, its confidence is close to $100\gamma\%$ for other life distributions.

Confidence limits for a fraction failing. Two-sided approximate $100\gamma\%$ confidence limits for the fraction failing by age t_0 at (transformed) stress x_0 are calculated as follows. Calculate the standardized deviate

$$Z = [\eta_0 - m(x_0)]/s \tag{2.3.13}$$

where $\eta_0 = \log(t_0)$. Calculate its approximate variance as

$$\text{var}(Z) = (1/n) + [(x_0 - \bar{x})^2/S_{xx}] + [Z^2/(2\nu)]. \tag{2.3.14}$$

Calculate

$$\underline{Z} = Z - z_{\gamma'}[\text{var}(Z)]^{1/2}, \quad \tilde{Z} = Z + z_{\gamma'}[\text{var}(Z)]^{1/2}, \tag{2.3.15}$$

where $\gamma' = (1+\gamma)/2$. The approximate confidence limits are

$$\underline{F}(t_0;x_0) = \Phi(\underline{Z}), \quad \tilde{F}(t_0;x_0) = \Phi(\tilde{Z}); \tag{2.3.16}$$

here $\Phi(\ )$ is the standard normal cumulative distribution function (Appendix A1). For a one-sided $100\gamma\%$ confidence limit, replace γ' by γ. Usually one uses an upper limit. Confidence limits for reliability are

$$\underline{R}(t_0;x_0) = 1 - \tilde{F}(t_0;x_0), \quad \tilde{R}(t_0;x_0) = 1 - \underline{F}(t_0;x_0).$$

These limits tend to be too narrow. The approximation usually suffices if $\nu \geq 15$. Owen (1968) gives exact confidence limits based on the non-central t

distribution. Such limits are inaccurate if the life distribution is not (log) normal, no matter how large the sample.

Confidence limit for a design stress. Sometimes one seeks a design stress x such that the 100Pth percentile $\tau_P(x)$ equals τ_P^*, a specified life. The estimate x^* (2.2.15) randomly falls below the true x with about 50% probability. One may prefer a design stress that is on the low (safe) side of x with high probability γ. Then one uses the lower one-sided 100γ% confidence limit $\underline{x}$ for x as the design stress. An approximate limit $\underline{x}$ is the solution of

$$\eta_P^* = c_0 + c_1 \underline{x} + z_P s - z_\gamma \{[z_P^2/(2\nu)] + (1/n) + [(\underline{x} - \bar{x})^2/S_{xx}]\}^{1/2} s, \qquad (2.3.17)$$

where $\eta_P^* = \log(\tau_P^*)$. This equation assumes that the transformed stress x is an increasing function of stress, such as $x = \log(V)$. If not (e.g., $x = 1000/T$), replace $-z_\gamma$ by $+z_\gamma$ and solve (2.3.17). No least squares program calculates this limit.

Easterling (1969) and Owen (1968) give an exact limit, which employs the non-central t distribution. This limit is inaccurate if the population distribution is not (log) normal, no matter how large the sample.

3. CHECKS ON THE LINEAR-LOGNORMAL MODEL AND DATA

Least-squares analyses of data involve assumptions about the model and data. Thus, the accuracy of the estimates and confidence limits depends on how well the assumptions hold. Some estimates and confidence limits are accurate enough even when the assumptions are far from satisfied, and others may be quite sensitive to an inaccurate model or faulty data.

Sections present checks for (3.1) linearity of the life-stress relationship, (3.2) dependence of the (log) standard deviation on stress, (3.3) the (log) normal distribution, and (3.4) the data. Section 3.5 shows also how to use residuals to estimate σ graphically and to assess the effect of other variables on life. More important, such checks may reveal useful information on the product or test method.

3.1. Is the Life-Stress Relationship Linear?

A test for linearity of the (transformed) relationship between the mean (log) life and stress follows. It tests whether the sample means of log life for the test stresses are statistically significantly far from the fitted straight line. Sample means may depart significantly from the line for two main reasons:

1. The true relationship is not a straight line.

2. The true relationship is a straight line, but other variables or factors have produced data departing from the line. Examples include:

a. inaccurate stress levels (mismeasured or not held constant);
b. malfunctioning test equipment (for example, failure of a specimen in a rack of specimens produces failure of others in the rack because they are not electrically isolated);
c. differing test specimens due to differing raw materials, fabrication, handling, and personnel (for example, specimens made by the third shift failed sooner);
d. differing test conditions due to uncontrolled variables other than stress (for example, temperature increases with voltage used as the accelerating stress, and cycle time in the oven and temperature cycling can also affect life when temperature is the accelerating stress);
e. blunders in recording, transcribing, and analyzing the data;
f. the combined effect of two or more failure modes (Chapter 7).

Suppose there are n specimens among J test stress levels where $J > 2$. The notation follows that of Section 2. Calculate the F statistic for linearity

$$F = [(n-2)s'^2 - (n-J)s^2] / [(J-2)s^2];\qquad(3.1)$$

here s is the estimate of σ based on pure error (2.2.11), and s' is the estimate based on lack of fit (2.2.12). The numerator of F may be a small difference between two large numbers. So calculate with extra significant figures.

The test for linearity of the relationship is

1. If $F \leq F(1-\alpha;J-2,\nu)$, there is no evidence of nonlinearity at the $100\alpha\%$ level;
2. If $F > F(1-\alpha;J-2,\nu)$, there is statistically significant nonlinearity at the $100\alpha\%$ level.

Here $F(1-\alpha;J-2,\nu)$ is the $1-\alpha$ point of the F distribution with $(J-2)$ degrees of freedom in the numerator and $\nu = n-J$ in the denominator. Appendix A6 contains values of $F(1-\alpha;J-2,\nu)$. This test is exact for a (log) normal life distribution. Also, it is a useful approximation for other distributions. If there is statistically significant nonlinearity, examine the relationship plot of the data (Chapter 3) to understand the nonlinearity.

For the Class-H data in Table 2.1,

$$F = [(40-2)(0.110569)^2 - (40-4)(0.105327)^2] / [(4-2)(0.105327)^2] = 2.94.$$

The F distribution has $4-2 = 2$ degrees of freedom in the numerator and $\nu = 40-4 = 36$ in the denominator. Since $F = 2.94 < 3.27 = F(.95;2,36)$, there is no statistically significant (that is, no convincing) nonlinearity.

3.2. Is the (Log) Standard Deviation Constant?

An assumption is that σ is constant, that is, independent of stress. If σ does depend on stress, then the estimates and confidence intervals for percentiles at a stress are inaccurate. However, the estimates for γ_0 and γ_1 and the relationship $\mu(x_0)$ generally suffice, even when σ depends on stress. The following test objectively assesses whether σ depends on stress. Dependence of σ on stress may be an inherent characteristic of the product; it may result from a faulty test; or it may be due to different failure modes acting at different stress levels.

The following comparison of (log) standard deviations is *Bartlett's test*.

n_j denotes the number of specimens at the test stress j,
s_j denotes the sample (log) standard deviation for test stress j,
ν_j denotes its degrees of freedom ($\nu_j = n_j - 1$), and
J denotes the number of test stresses.

The pooled estimate (2.2.11) of a common (log) standard deviation is

$$s = [(\nu_1 s_1^2 + \cdots + \nu_J s_J^2)/\nu]^{1/2}, \tag{3.2}$$

and $\nu = \nu_1 + \cdots + \nu_J$ denotes its number of degrees of freedom.

Bartlett's test statistic is

$$Q = C\{\nu \cdot \log(s) - [\nu_1 \cdot \log(s_1) + \cdots + \nu_J \cdot \log(s_J)]\}, \tag{3.3}$$

(base 10 logs), and

$$C = 4.605 \Big/ \left\{ 1 + \frac{1}{3(J-1)}\left[\frac{1}{\nu_1} + \cdots + \frac{1}{\nu_J} - \frac{1}{\nu}\right]\right\}. \tag{3.4}$$

The approximate level α test for equality of the s_j is

1. if $Q \le \chi^2(1-\alpha;J-1)$, the s_j do not differ significantly at the $100\alpha\%$ level;
2. if $Q > \chi^2(1-\alpha;J-1)$, they differ significantly at the $100\alpha\%$ level;

here $\chi^2(1-\alpha;J-1)$ is the $100(1-\alpha)$th chi-square percentile with $(J-1)$ degrees of freedom. Appendix A5 contains these percentiles.

If the s_j differ significantly, then examine them to determine how they differ. For example, examine a plot of the estimates and confidences limits side by side. Figure 3.1 shows such a plot for the Class-H insulation. Most applied books state the value of plotting data, but few note the value of plotting estimates and confidence limits. Also, then take those differences into account in interpreting the data. For example, if the data from a stress level are suspect, one might discard them and analyze only the "good" data.

For the Class-H insulation data,

$$C = 4.605 \Big/ \left\{ 1 + \frac{1}{3(4-1)}\left[\frac{1}{9} + \frac{1}{9} + \frac{1}{9} + \frac{1}{9} - \frac{1}{36}\right]\right\} = 4.401,$$

$$Q = 4.401\{36 \cdot \log(0.1053) - [9 \cdot \log(0.0625)] + 9 \cdot \log(0.0792)$$

$$+ 9 \cdot \log(0.0717) + 9 \cdot \log(0.1705)]\} = 12.19.$$

The chi-square distribution for Q has $J - 1 = 3$ degrees of freedom. Since

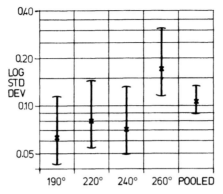

Figure 3.1.   Estimates × and confidence limits I for σ.

Q = 12.19 > 11.34 = $\chi^2(0.99; 3)$, the s_j differ very significantly (1% level). Figure 3.1 suggests that the log standard deviation at 260° is greater than the others. There are possible reasons for this. For example, the main failure mode at 260° differs from that at the other temperatures. Proper analyses for data with a mix of failure modes appear in Chapter 7. Also, other lots of raw materials were used in the 260° specimens. Should one use the 260° data? Insights gained from understanding the 260° data yield $1,000,000 annual savings. It is best to do two analyses, one with and the other without suspect data. If the results of both differ little, either can be used. If they differ much, decide which analysis is more accurate or more conservative.

Two other tests for comparing standard deviations are:

1. Cochran's test, presented by Draper and Smith (1981) and
2. the maximum F ratio, presented by Pearson and Hartley (1954, p. 60).

Tabulations for these apply only when the n_j are all equal. Nelson (1982, p. 481 ff) presents plots for simultaneous comparisons based on 2. These and Bartlett's test assume that the life distribution is (log) normal. If it is not, these comparisons are crude at best. Checks for normality appear next.

3.3. Is the Life Distribution (Log) Normal?

Analyses above assume that the life distribution is (log) normal at any stress level of interest. Estimates of percentiles and a fraction failing and certain confidence intervals are sensitive to this assumption. As indicated, other estimates and confidence limits are accurate enough with most life distributions.

A simple check on how well the (log) normal distribution fits the data employs the (log) normal probability plot in Section 2.2 of Chapter 3. Such plots appear in Figure 2.1 of Chapter 3. The plots should follow a straight line reasonably well. Pronounced curvature at most stresses indicates that the true distribution is not well described by the (log) normal. In judging curvature, people tend to expect straighter plots than random samples yield. The erratic nature of such plots may be gathered from plots of Monte Carlo samples from a normal distribution by Hahn and Shapiro (1967) and Daniel and Wood (1980).

A more sensitive check employs a probability plot of residuals as follows. Other uses of such residuals are described by Draper and Smith (1981) and Daniel and Wood (1980).

Calculate the *adjusted residuals about the (log) mean* at test stress j as

$$r_{ij} = (y_{ij} - \bar{y}_j)[n_j/(n_j - 1)]^{1/2}; \qquad (3.5)$$

here y_{ij} is (log) observation i, and $\bar{y}_j$ is the average of the n_j (log) observations at stress j. If the n_j are all equal, the *standardizing factor* $[n_j/(n_j - 1)]^{1/2}$ is not necessary. The simple differences $(y_{ij} - \bar{y}_j)$ are called the *raw residuals about the (log) mean* .

Most regression programs calculate a *raw residual about the fitted line* as $r'_{ij} = y_{ij} - m(x_j)$; this is the (log) observation minus the estimate of the (log) mean at that stress. Many statistical programs display such residuals in normal probability plots and in crossplots (which are described in Section 5.5). The methods below also apply to such residuals, but the adjustment factor $[n_j/(n_j - 1)]^{1/2}$ must be replaced by that in Section 5.3.

Pool all standardized residuals into a single sample. They should look like a sample from a normal distribution with mean of zero and a standard deviation of σ. Plot the pooled sample on normal probability paper as described in Chapter 3. The plot of the pooled sample reveals more than the separate plots of the data for each test stress. Assess whether the residuals reasonably follow a straight line. If not, then the (log) normal distribution does not adequately describe time to failure. The curvature may indicate how the true distribution differs from the (log) normal distribution. Also, lack of fit may indicate that some observations are in error; that is, the specimens, the test, or the data handling are faulty.

The standardized residuals for the Class-H insulation data are calculated in Table 3.1. Their normal plot appears in Figure 3.2. The plot is remarkably straight. This suggests that the lognormal distribution fits the data well. This plot does not reveal that the 260° data have a larger σ. This shows that no one plot reveals all information in the data. In view of the larger σ at 260°C, one might redo this plot without the 260° residuals.

If the plot clearly curves, plot the residuals on other probability papers to assess how well other distributions fit. In particular, try extreme-value paper. If it yields a straight plot, then the Weibull fits better than the lognormal distribution. The same residuals are plotted on extreme value paper in Figure 3.3. The curved plot suggests that the Weibull distribution is not suitable.

Table 3.1.   Calculation of Standardized Residuals of Class-H Data

	190°C			220°C	
Obs.	Mean	Std. Res.	Obs.	Mean	Std. Res.
(3.8590	- 3.9396) $\sqrt{10/9}$	= -0.085	(3.2465	- 3.4150)$\sqrt{10/9}$	= -0.177
(3.8590	- 3.9396) $\sqrt{10/9}$	= -0.085	(3.3867	- 3.4150)$\sqrt{10/9}$	= -0.030
(3.8590	- 3.9396) $\sqrt{10/9}$	= -0.085	(3.3867	- 3.4150)$\sqrt{10/9}$	= -0.030
(3.9268	- 3.9396) $\sqrt{10/9}$	= -0.013	(3.3867	- 3.4150)$\sqrt{10/9}$	= -0.030
(3.9622	- 3.9396) $\sqrt{10/9}$	= 0.024	(3.3867	- 3.4150)$\sqrt{10/9}$	= -0.030
(3.9622	- 3.9396) $\sqrt{10/9}$	= 0.024	(3.3867	- 3.4150)$\sqrt{10/9}$	= -0.030
(3.9622	- 3.9396) $\sqrt{10/9}$	= 0.024	(3.4925	- 3.4150)$\sqrt{10/9}$	= 0.082
(3.9622	- 3.9396) $\sqrt{10/9}$	= 0.024	(3.4925	- 3.4150)$\sqrt{10/9}$	= 0.082
(4.0216	- 3.9396) $\sqrt{10/9}$	= 0.086	(3.4925	- 3.4150)$\sqrt{10/9}$	= 0.082
(4.0216	- 3.9396) $\sqrt{10/9}$	= 0.086	(3.4925	- 3.4150)$\sqrt{10/9}$	= 0.082

	240°C			260°C	
Obs.	Mean	Std. Res.	Obs.	Mean	Std. Res.
(3.0700	- 3.1940)$\sqrt{10/9}$	= -0.131	(2.7782	- 3.0176)$\sqrt{10/9}$	= -0.252
(3.0700	- 3.1940)$\sqrt{10/9}$	= -0.131	(2.8716	- 3.0176)$\sqrt{10/9}$	= -0.154
(3.1821	- 3.1940)$\sqrt{10/9}$	= -0.012	(2.8716	- 3.0176)$\sqrt{10/9}$	= -0.154
(3.1956	- 3.1940)$\sqrt{10/9}$	= 0.002	(2.8716	- 3.0176)$\sqrt{10/9}$	= -0.154
(3.2087	- 3.1940)$\sqrt{10/9}$	= 0.016	(2.9600	- 3.0176)$\sqrt{10/9}$	= -0.061
(3.2214	- 3.1940)$\sqrt{10/9}$	= 0.029	(3.0523	- 3.0176)$\sqrt{10/9}$	= 0.037
(3.2214	- 3.1940)$\sqrt{10/9}$	= 0.029	(3.1206	- 3.0176)$\sqrt{10/9}$	= 0.109
(3.2338	- 3.1940)$\sqrt{10/9}$	= 0.042	(3.1655	- 3.0176)$\sqrt{10/9}$	= 0.156
(3.2458	- 3.1940)$\sqrt{10/9}$	= 0.055	(3.2063	- 3.0176)$\sqrt{10/9}$	= 0.199
(3.2907	- 3.1940)$\sqrt{10/9}$	= 0.101	(3.2778	- 3.0176)$\sqrt{10/9}$	= 0.274

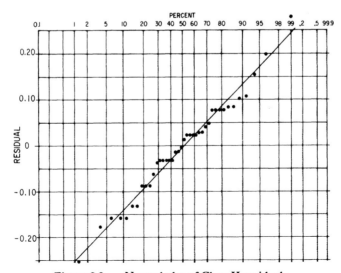

Figure 3.2.   Normal plot of Class-H residuals.

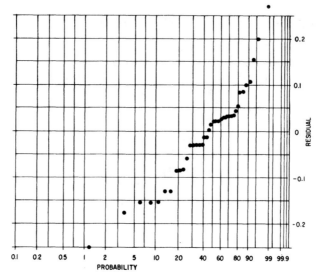

Figure 3.3.   Extreme value plot of Class-H residuals.

Analytic methods for testing data for normality are given, for example, by Wilk and Shapiro (1968). These methods are laborious but are sensitive to nonnormality. Such methods are correct only for the data from a single stress level. They are not correct for pooled residuals. Pooled residuals are "closer" to normal than a sample from a single stress level. Even after using such methods and finding nonnormality, one must look at the plot of residuals to see the nature of the nonnormality.

Normal probability papers are widely available. Sources include:

1. TEAM, Box 25, Tamworth, NH 03886, (603)323-8843.

3111	1–99% (11" horizontal)	100 divisions (8 1/2" vertical)
3211	0.01–99.99% (11" horizontal)	100 divisions (8 1/2" vertical)
3311	0.0001–99.9999% (11" horizontal)	100 divisions (8 1/2" vertical)
3111.5	0.2–99.8% (11" horizontal)	50 divisions (8 1/2" vertical)

2. CODEX Book Co., 74 Broadway, Norwood, MA 02062, (617)769-1050.

Y3 200	0.01–99.99% (11" horizontal)	40 divisions (8 1/2" vertical)
Y3 201	0.01–99.99% (11" horizontal)	100 divisions (8 1/2" vertical)
Y3 09	0.01–99.99% (11" horizontal)	90 divisions (8 1/2" vertical)

3. K + E (Keuffel & Esser Co.), 20 Whippany Rd., Morristown, NJ 07960, (800)538-3355.

46 8000	0.01–99.99% (11" horizontal)	90 divisions (8 1/2" vertical)
46 8003	is like 46 8000 but has orange grid instead of green.	

4. Craver (1980) provides reproducible copies:

pg. 216   0.01 – 99.99% (11" horizontal)   40 divisions (8 1/2" vertical)
pg. 217   0.01 – 99.99% (11" horizontal)   80 divisions (8 1/2" vertical)
pg. 218   0.01 – 99.99% (11" horizontal)   90 divisions (8 1/2" vertical)
pg. 219   0.01 – 99.99% (11" horizontal)   100 divisions (8 1/2" vertical)
pg. 227   0.01 – 99.99% (11" horizontal)   80 divisions (8 1/2" vertical)

Papers with a narrow probability range (1 – 99%) are best for residual plots, since most accelerated tests have fewer than 100 specimens.

3.4. Checks on the Data

Suspect data are data subsets or individual observations that deviate from the assumed model. Such data can arise from faulty test methods, an inadequate model, misrecording and handling the data, etc. The previous checks on the model also check the data. Nonlinearity of the life-stress relationship, dependence of σ on stress, and poor fit of the assumed distribution – all may indicate suspect data. Probability plots of the data for the different stress levels (Figure 2.1 of Chapter 3) .may reveal suspect data. A probability plot of the pooled standardized residuals may reveal individual wild points called *outliers;* such points do not fall along the straight line determined by the bulk of the data. The points in the tails of a plot vary much from random sampling; thus outliers should be very out of line with the other data before they are suspect or discarded. Preferably, one should find a cause for suspect data before omitting the data. Knowledge of the cause may help improve the product, test method, or model.

Formal numerical methods for identifying outliers in a sample from a single population are given by Beckman and Cook (1983), Barnett and Lewis (1984), and Daniel and Wood (1980). Flack and Flores (1989) extend such methods to residuals from fitting a regression model.

3.5. Effect of Other Variables

The effect of some other variable on life can be assessed by crossplotting residuals against the variable. Also, a plot of the residuals against the accelerating variable may yield information. Such crossplots are presented in detail by Draper and Smith (1981), Neter, Wasserman, and Kutner (1983), and Daniel and Wood (1980). Sections 6.3 and 6.4 give examples.

4. LEAST-SQUARES METHODS FOR WEIBULL AND EXPONENTIAL LIFE

This section describes least squares methods for fitting a model with a Weibull or exponential distribution to complete life data. The methods provide estimates and confidence limits for model parameters, percentiles, fraction failing, and design stress.

The methods and calculations are slight modifications of those for a (log) normal life distribution in Section 2. Chapter 5 presents more accurate maximum likelihood estimation for the Weibull and exponential distributions. Thus many readers may choose to skip this section. This section is intended to be self-contained for ease of use. The model and analyses are widely used. So they merit the separate detailed presentation here. Moreover, these modifications do not appear in regression texts nor in other sources.

Least-squares fitting for a Weibull life distribution yields estimates that are not as accurate as those from maximum likelihood fitting (Chapter 5). However, least squares regression programs are widely available. Moreover, the least squares calculations are simple. Thus readers can do the (simple but laborious) calculations on a pocket calculator, or they can write their own computer programs. Of course, most standard programs have better round-off error and acceptance by other people. However, most standard programs lack some output useful to accelerated testing; examples include estimates and confidence limits for percentiles and a fraction failing. Readers can write programs with such output. In contrast, maximum likelihood calculations for fitting a Weibull distribution are complex; they require a sophisticated computer program, which some readers lack.

4.1. Example Data and Weibull Model

This section presents 1) data used to illustrate least squares analyses and 2) the Weibull model fitted to the data. The model includes the power-Weibull and Arrhenius-Weibull models of Chapter 2.

Data. The analyses are illustrated with the insulating oil data of Cramp (1959) in Section 3.1 of Chapter 3. Table 4.1 shows the ln times (in minutes) to breakdown of 76 specimens tested at voltages from 26 to 38 kV. The main purpose is to estimate the relationship between time to breakdown and voltage. This involves fitting the power-Weibull model, used to estimate the probability of oil failure during a transformer test at 20 kV. Another purpose is to assess the exponential distribution for time to oil breakdown.

Model, assumptions, and notation. Least-squares analysis using a Weibull distribution for complete accelerated life test data involves the following assumptions. A random sample of n units is put on test, and all run to failure. There are J test stress levels. n_j units are tested at (transformed) stress level $x_j, j = 1, 2, \cdots, J$. For example, for the inverse power law, $x_j = \ln(V_j)$ where V_j is the (positive) stress. y_{ij} denotes the ln time to failure of test unit i at test stress x_j. Such ln times for the insulating fluid appear in Table 4.1. The total number of test units $n = n_1 + \cdots + n_J$.

The model for the ith ln failure time y_{ij} at (transformed) stress x_j is

$$y_{ij} = \mu(x_j) + e_{ij} \qquad (4.1.1)$$

Table 4.1. ln Times to Breakdown of an Insulating Fluid

26 kV	28 kV	30 kV	32 kV	34 kV	36 kV	38 kV
1.7561	4.2319	2.0464	-1.3094	-1.6608	-1.0499	-2.4080
7.3648	4.6848	2.8361	-0.9163	-0.2485	-0.5277	-0.9417
7.7509	4.7031	3.0184	-0.3711	-0.0409	-0.0409	-0.7551
	6.0516	2.0154	-0.2358	0.2700	-0.0101	-0.3148
	6.9731	3.1206	1.0116	1.0224	0.5247	-0.3012
		3.7704	1.3635	1.1505	0.6780	0.1222
		3.8565	2.2905	1.4231	0.7275	0.3364
		4.9349	2.6354	1.5411	0.9477	0.8671
		4.9706	2.7682	1.5789	0.9969	
		5.1698	3.3250	1.8718	1.0647	
		5.2724	3.9748	1.9947	1.3001	
			4.4170	2.0806	1.3837	
			4.4918	2.1126	1.6770	
			4.6109	2.4898	2.6224	
			5.3711	3.4578	3.2386	
				3.4818		
				3.5237		
				3.6030		
				4.2889		

for $i = 1, \cdots, n_j$ and $j = 1, \cdots, J$. The *random variation* or *error* e_{ij} of ln life y_{ij} has an extreme value distribution with a mean of zero and an unknown scale parameter δ. The linear life-stress relationship

$$\mu(x_j) = \gamma_0 + \gamma_1 x_j \qquad (4.1.2)$$

gives the mean ln life. The Arrhenius and inverse power relationships are special cases of (4.1.2). The extreme value location parameter $\xi(x_j)$ is related to the mean; namely,

$$\xi(x_j) = \mu(x_j) + 0.5772\delta = \gamma_0 + \gamma_1 x_j + 0.5772\delta. \qquad (4.1.3)$$

This is the ln of the Weibull scale parameter $\alpha(x_j)$, and 0.5772 is Euler's constant. In Chapter 2, the equation for $\xi(x_j)$ is written differently as $\xi(x_j) = \gamma_0' + \gamma_1 x$ where $\gamma_0' = 0.5772\delta + \gamma_0$. The form (4.1.2) and (4.1.3) is more convenient for the purposes of Chapter 4. The random variations e_{ij} have the same standard deviation

$$\sigma = 1.283\delta = 1.283/\beta, \qquad (4.1.4)$$

where β is the Weibull shape parameter. Also, the e_{ij} are all assumed statistically independent. Except for the assumed extreme value distribution, these are the usual assumption for least squares regression theory; see Draper and Smith (1981) or Neter, Wasserman, and Kutner (1983,1985). This section presents least squares estimates and confidence intervals for the model parameters γ_0, γ_1, and δ and for other quantities.

4.2. Estimates for the Parameters

The least squares estimates for the model parameters and other quantities of interest are presented below. Corresponding confidence intervals are in Section 4.3. The estimates and confidence intervals are modifications of those from standard regression theory (for example, in Draper and Smith (1981) or Neter, Wasserman, and Kutner (1983,1985)). Computer programs for these calculations are widely available.

Preliminary calculations. First calculate the sample ln averages and standard deviations for each test stress. For test stress level j, they are

$$\bar{y}_j = (y_{1j} + y_{2j} + \cdots + y_{njj})/n_j, \tag{4.2.1}$$

$$s_j = \{[(y_{1j}-\bar{y}_j)^2 + \cdots + (y_{njj}-\bar{y}_j)^2]/(n_j-1)\}^{1/2}$$

$$= \{[(y_{1j}^2 + \cdots + y_{njj}^2) - (n_j\bar{y}_j^2)]/(n_j-1)\}^{1/2}; \tag{4.2.2}$$

here the summations run over all n_j observations at stress level j. $\exp(\bar{y}_j)$ is called the *sample geometric average* or *mean*. The second expression for s_j is commonly used in computer programs. The first is recommended for greater accuracy. s_j has $\nu_j = n_j-1$ degrees of freedom. If $n_j = 1$, s_j is not calculated. These calculations are shown in Table 4.2 for the insulating fluid data. Intermediate calculations have six or more figure accuracy to assure three or four figure accuracy in final results.

For a single distribution, Menon (1963) presents Weibull parameter estimates based on $\bar{y}_j$ and s_j. He discusses their properties and gives their large sample variances. His estimates are closely related to the ones below.

Next calculate the *grand averages* of all the data

$$\bar{x} = (n_1\bar{x}_1 + \cdots + n_J\bar{x}_J)/n, \tag{4.2.3}$$

$$\bar{y} = (n_1\bar{y}_1 + \cdots + n_J\bar{y}_J)/n. \tag{4.2.4}$$

Each of these is the sum over the entire sample divided by n. Calculate the *sums of squares*

$$S_{yy} = \sum_{j=1}^{J}\sum_{i=1}^{nj}(y_{ij}-\bar{y})^2 = \sum_{j=1}^{J}\sum_{i=1}^{nj}y_{ij}^2 - n\bar{y}^2, \tag{4.2.5}$$

$$S_{xx} = n_1(x_1-\bar{x})^2 + \cdots + n_J(x_J-\bar{x})^2 = n_1x_1^2 + \cdots + n_Jx_J^2 - n\bar{x}^2, \tag{4.2.6}$$

$$S_{xy} = n_1(x_1-\bar{x})\bar{y}_1 + \cdots + n_J(x_J-\bar{x})\bar{y}_J = n_1x_1\bar{y}_1 + \cdots + n_Jx_J\bar{y}_J - n\bar{x}\bar{y}. \tag{4.2.7}$$

The double sum for S_{yy} runs over all n observations in the sample. These calculation are shown in Table 4.2 for the example. Regression programs automatically do these calculations.

Coefficient estimates. The least squares estimates of γ_1 and γ_0 are

$$c_1 = S_{xy}/S_{xx}, \tag{4.2.8}$$

$$c_0 = \bar{y} - c_1\bar{x}. \tag{4.2.9}$$

Table 4.2. Least Squares Calculations for an Insulating Fluid

	26 kV	28 kV	30 kV	32 kV	34 kV	36 kV	38 kV
Total:	16.8718	26.6475	42.0415	33.4272	33.9405	13.5327	−3.3951
Avg. $\bar{y}_j$:	5.62393	5.32950	3.82195	2.22848	1.78634	0.902180	−0.424388
$\exp(\bar{y}_j)$:	276.977	206.335	45.6934	9.28574	5.96758	2.46497	0.65417
No. n_j:	$n_1 = 3$	$n_2 = 5$	$n_3 = 11$	$n_4 = 15$	$n_5 = 19$	$n_6 = 15$	$n_7 = 8$

$$n = 3 + 5 + 11 + 15 + 19 + 15 + 8 = 76$$

(4.2.2) $s_1 = \{[(1.7561-5.62393)^2 + \cdots + (7.7509-5.62393)^2]/(3-1)\}^{1/2} = 3.35520$

(4.2.2) $s_2 = \{[(4.2319-5.32950)^2 + \cdots + (6.9731-5.32950)^2]/(5-1)\}^{1/2} = 1.14455$

(4.2.2) $s_3 = \{[(2.0460-3.82195)^2 + \cdots + (5.2724-3.82195)^2]/(11-1)\}^{1/2} = 1.11119$

(4.2.2) $s_4 = \{[(-1.3094-9.28574)^2 + \cdots + (5.3711-9.28574)^2]/(15-1)\}^{1/2} = 2.19809$

(4.2.2) $s_5 = \{[(-1.6608-1.78634)^2 + \cdots + (4.2889-1.78634)^2]/(19-1)\}^{1/2} = 1.52521$

(4.2.2) $s_6 = \{[(-1.0499-0.90218)^2 + \cdots + (3.2386-0.90218)^2]/(15-1)\}^{1/2} = 1.10989$

(4.2.2) $s_7 = \{[(-2.4080+0.424388)^2 + \cdots + (0.8671+0.424388)^2]/(8-1)\}^{1/2} = 0.991707$

(4.2.3) $\bar{x} = [3(3.2581) + 5(3.33221) + 11(3.4012) + 15(3.46754)$
$+ 19(3.52637) + 15(3.58352) + 8(3.63759)]/76 = 3.49591$

(4.2.4) $\bar{y} = [3(5.62393) + 5(5.32950) + 11(3.82195) + 15(2.22848)$
$+ 19(1.78634) + 15(0.90218) + 8(-0.424389)]/76 = 2.14561$

(4.2.5) $S_{yy} = (1.7561-2.14561)^2 + (7.3648-2.14561)^2 + \cdots + (0.8671-2.14561)^2$
$= 370.228.$

(4.2.6) $S_{xx} = 3(3.2581-349591)^2 + 5(3.3322-3.49591)^2 + \cdots + 8(3.63759-3.49591)^2$
$= 0.709319$

(4.2.7) $S_{xy} = 3(3.2581-349591)5.62393 + 5[3.33221(-3.49591)]5.32950 + \cdots$
$+ 8(3.63759-3.49591)(-0.424388) = -11.6264$

(4.2.8) $c_1 = -11.6264/0.709319 = -16.3909$

(4.2.9) $c_0 = 2.14561 - (16.3909)(-3.49591) = 59.4468$

(4.2.10) $s = \{[3(3.35520)^2 + 5(1.14455)^2 + \cdots + 8(0.991707)^2]/(76-7)\}^{1/2} = 1.587$
$\nu = 76-7 = 69$

(4.2.11) $s' = \{[0.709319 - (-16.3909)(-11.6264)]/(76-2)\}^{1/2} = 1.558$
$\nu' = 76-2 = 74$

(4.2.12) $d = (0.7797)1.587 = 1.237$

(4.2.13) $b = 1.283/1.587 = 0.808$

Their calculation appears in Table 4.2.

σ **estimate.** The *pooled estimate of the standard deviation* σ is

$$s = [(\nu_1 s_1^2 + \cdots + \nu_J s_J^2)/\nu]^{1/2}; \qquad (4.2.10)$$

here $\nu = \nu_1 + \cdots + \nu_J = n - J$ is its number of degrees of freedom. This is called the estimate of the ln standard deviation based on *replication* or *pure error*. Another estimate of σ is

$$s' = [(S_{yy} - c_1 S_{xy})/(n-2)]^{1/2}. \qquad (4.2.11)$$

This is called the estimate of the standard deviation based on *lack of fit* (scatter of the data) about the fitted line. It is also called the *standard error of estimate*. It has $\nu' = (n-2)$ degrees of freedom. Most regression programs give the s' estimate.

Either estimate of σ can be used in later calculations. However, one must use the corresponding number of degrees of freedom. For analysis of accelerated life test data, the estimate s based on pure error is recommended; s is used throughout the example. s is recommended because s' tends to overestimate σ, if the relationship between the mean ln life and the stress x is not linear. However, s' may sometimes be preferable, since it is conservative in the sense that it tends to make the product spread look worse. Also, most computer programs give s', not s.

The estimate of the extreme value scale parameter δ is

$$d = (0.7797)\,s.\tag{4.2.12}$$

The estimate of the Weibull shape parameter β is

$$b = 1/d = 1.283/s.\tag{4.2.13}$$

The calculations for these estimates are in Table 4.2 for the example. The estimate of the Weibull shape parameter is $b = 0.808$. This value near 1 indicates that the failure rate is near constant (decreases slightly with age). Thus the life distribution is close to exponential, which is suggested by engineering theory. Table 4.2 is used in examples throughout this section.

Estimate of the mean ln life. For any stress level x_0, the least squares estimate $m(x_0)$ of the ln mean $\mu(x_0) = \gamma_0 + \gamma_1 x_0$ is

$$m(x_0) = c_0 + c_1 x_0.\tag{4.2.14}$$

For example, x_0 may be the design level of stress. The estimate of the 42.8th Weibull percentile is $t_{.428}(x_0) = \exp[m(x_0)]$. (4.2.14) assumes that the life-stress relationship is linear. If not, the estimate may be inaccurate. Methods for assessing linearity appear in Section 5. The estimate for the Weibull scale parameter is

$$\alpha^*(x_0) = \exp[m(x_0)+0.5772d].$$

For the example, the estimate of the relationship is

$$m(x_0) = 59.4468 - 16.3909x_0 = 59.4468 - 16.3909 \cdot \ln(V_0);$$

the values of c_0 and c_1 come from Table 4.2. This equation is the center line in Figure 4.1. The estimate of the ln mean at 20 kV ($x_0 = 2.99573$) is $m(2.99573) = 59.4468 - 16.3909(2.99573) = 10.3439$. The estimate of the 42.8th Weibull percentile at 20 kV is $t_{.428}(2.99573) = \exp(10.3439) = 31{,}000$ minutes. The estimates of the ln means and 42.8th percentiles at the design and test voltages appear in Table 4.3. The estimate of the Weibull scale parameter at 20 kV is $\alpha^*(2.99573) = \exp[10.3439 + 0.5772\,(1.237)] \approx 63{,}400$ minutes (3-figure accuracy). The fitted line for the ln mean in Figure 4.1 passes through these estimates.

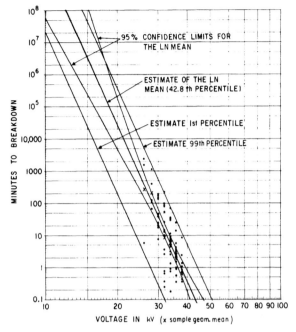

Figure 4.1.   Log-log plot of the fitted model for insulating fluid.

Estimate of a fraction failing. For a (transformed) stress x_0, the estimate of the fraction failing by a given age t_0 is

$$F^*(t_0;x_0) = 1 - \exp\{-[t_0/\alpha^*(x_0)]^b\}. \qquad (4.2.15)$$

For the example,

$$F^*(t_0;V_0) = 1 - \exp\{-[t_0/(e^{60.1608}/V_0^{16.3909})]^{0.808}\}.$$

Table 4.3. Estimates of the Mean ln Life vs. Voltage

Voltage (kV)	Estimate of ln Mean	Estimate of 42.8th Percentile (minutes)
38	−0.1767	0.838
36	0.7096	2.03
34	1.6464	5.19
32	2.6401	14.0
30	3.6980	40.4
28	4.8288	125.
26	6.0435	421.
20	10.3439	31,000
15	15.0592	3.45×10^6
10	21.7052	2.67×10^9
5	33.0666	2.29×10^{14}

The estimate of the fraction failing on a t_0 = 10 minute test at V_0 = 20 kV is

$$F^*(10;20) = 1 - \exp\{-[10/63,400]^{0.808}\} = 8.5 \times 10^{-4}.$$

Chapter 3 presents a graphical estimate of such a fraction failing.

Estimates of percentiles. For stress level x_0, the estimate of the 100Pth percentile $\eta_P(x_0)$ of ln life is

$$y_P(x_0) = m(x_0) + [0.5772 + u(P)]\,0.7797\,s$$
$$= c_0 + c_1 x_0 + [0.5772 + u(P)]\,0.7797\,s; \qquad (4.2.16)$$

$u(P) = \ln[-\ln(1-P)]$ is the 100Pth standard extreme value percentile. The estimate of the 100Pth percentile of the Weibull life distribution is $t_P(x_0) = \exp[y_P(x_0)]$. The estimate (4.2.14) of the ln mean is a special case of (4.2.16), namely, the 42.8th percentile where $u(0.428) = -0.5772$.

The estimate of the extreme value location parameter $\xi(x_0)$ is a special case of (4.2.16), namely,

$$\xi^*(x_0) = m(x_0) + 0.4501s, \qquad (4.2.17)$$

since the location parameter is the 63.2th percentile and $u(0.632) = 0$. The estimate of Weibull characteristic life (63.2th percentile) at x_0 is

$$\alpha^*(x_0) = \exp[\xi^*(x_0)] = \exp[c_0 + c_1 x_0 + (0.4501)s].$$

For the insulating fluid at 20 kV, the estimate of the 1st percentile is

$$y_{.01}(2.99573) = 10.3439 + [0.5772 + (-4.6002)]0.7797(1.587) = 5.3664;$$

here $-4.6002 = u(0.01)$. The estimate of the 1st percentile of the Weibull life distribution is $t_{.01}(2.99573) = \exp(5.3664) = 214$ minutes. This time can also be obtained graphically from Figure 4.1. Similarly, the estimate of the location parameter at 20 kV is

$$\xi^*(2.99573) = 10.3439 + 0.4501(1.587) = 11.0579.$$

Then $\alpha^*(2.99573) = \exp(11.0579) = 63,400$ minutes. The estimates of selected percentiles at the design and test stresses appear in Table 4.4.

Estimate of design stress. In some applications, one seeks a design stress level such that a specified small percentage 100P of the population will fail by a specified time τ_P^*. The estimate x^* of that stress level from (4.2.16) is

$$x^* = \{\eta_P^* - c_0 - [0.5772 + u(P)]0.7797s\}/c_1 \qquad (4.2.18)$$

where $\eta_P^* = \ln(\tau_P^*)$.

For the insulating fluid, suppose only 1% of the population is to fail before $\tau_{.01}^* = 10,000$ minutes ($\eta_{.01}^* = 9.2103$). The stress estimate is

$$x^* = \{9.2103 - 59.4468 - [0.5772 + (-4.6002)]0.7797(1.587)\}/(-16.3909)$$

$$= 2.7612.$$

Table 4.4. Percentile Estimates vs. Voltage

Voltage	Percentiles (in minutes)		
(kV)	1st	63.2th	99th
38	0.00058	1.71	11.3
36	0.00140	4.15	27.5
34	0.0358	10.6	70.1
32	0.0966	28.6	189.
30	0.278	82.4	545.
28	0.862	255.	1690.
26	2.90	861.	5690.
20	214.	63,400	420,000
15	23.9×10^3	7.08×10^6	46.9×10^6
10	18.4×10^6	5.45×10^9	36.1×10^9
5	1.58×10^{12}	0.469×10^{15}	3.10×10^{15}

The corresponding estimate of voltage is $V^* = \exp(2.7612) = 15.8$ kV. This estimate can also be obtained graphically from Figure 4.1. Enter the graph on the time scale at τ_P^*; go horizontally to the $100P\%$ line and then down to the stress scale to read x^* (or the untransformed stress V^* here).

4.3. Confidence Intervals

The accuracy of an estimate is given by its standard error and confidence interval. These are given below for the least squares estimates of the model parameters and other quantities of interest. Some of these approximate intervals are given by Draper and Smith (1981) and by Neter, Wasserman, and Kutner (1983,1985). This subsection presents only approximate confidence limits, as theory for exact limits is too complex. McCool (1980) presents simulation methods for better approximate limits; his work must be modified for least squares fitting. Lawless (1976) gives exact limits requiring a special computer program.

Confidence limits for a ln mean. For stress level x_0, $m(x_0) = c_0 + c_1x_0$ is the estimate for the ln mean $\mu(x_0) = \gamma_0 + \gamma_1x_0$. Its sampling distribution is approximately normal when n is large and has a mean equal to the true $\mu(x_0)$. Thus, $m(x_0)$ is an unbiased estimate for $\mu(x_0)$. The accuracy of $m(x_0)$ is given by its standard error

$$\sigma[m(x_0)] = \{(1/n) + [(x_0 - \bar{x})^2/S_{xx}]\}^{1/2} \sigma. \qquad (4.3.1)$$

Its estimate is

$$s[m(x_0)] = \{(1/n) + [(x_0 - \bar{x})^2/S_{xx}]\}^{1/2}s; \qquad (4.3.2)$$

here s is an estimate of σ and has ν degrees of freedom. See equations (4.2.10) and (4.2.11) for estimates of σ.

A two-sided approximate $100\gamma\%$ confidence interval for $\mu(x_0)$ has limits

$$\mu(x_0) = m(x_0) - t(\gamma';\nu)\, s[m(x_0)], \quad \tilde{\mu}(x_0) = m(x_0) + t(\gamma';\nu)\, s[m(x_0)];$$

here $\tau(\gamma';\nu)$ is the $100\gamma' = 100(1+\gamma)/2$th t percentile with ν degrees of freedom. Such limits for the 42.8th percentile of the life distribution are

$$\underline{\tau}_{.428}(x_0) = \exp[\mu(x_0)] \text{ and } \tilde{\tau}_{.428}(x_0) = \exp[\tilde{\mu}(x_0)]. \tag{4.3.4}$$

This and other confidence intervals in this section assume that the sampling distribution of the estimate is approximately normal. This is usually so if the sample size is moderately large.

For the insulating fluid at 20 kV ($x_0 = 2.99573$),

$$s[m(2.99573)] = \{(1/76) + [(2.99573 - 3.49591)^2/0.709319]\}^{1/2} 1.587 = 0.9598.$$

The estimate $s = 1.587$ has $\nu = 69$ degress of freedom. The limits of the 95% confidence interval for $\mu(2.99573)$ are

$$\mu(2.99573) = 10.3439 - 1.9949(0.9598) = 8.4292,$$

$$\tilde{\mu}(2.99573) = 10.3439 + 1.9949(0.9598) = 12.2586;$$

here $1.9949 = t(0.975;69)$. The corresponding confidence limits for the 42.8th percentile at 20 kV are $\underline{\tau}_{.428}(2.99573) = \exp(8.4292) = 4{,}580$ and $\tilde{\tau}_{.428}(2.99573) = \exp(12.2586) = 211{,}000$ minutes. The 95% confidence limits for the 42.8th percentiles at design and test voltages appear in Table 4.5. These limits are on the curves in Figure 4.1.

The preceeding confidence intervals for the ln mean are two-sided. Sometimes one wants only a lower limit for the ln mean, since long life is desired. A lower one-sided approximate $100\gamma\%$ confidence limit for $\mu(x_0)$ is

$$\mu(x_0) = m(x_0) - t(\gamma;\nu)\, s[m(x_0)]; \tag{4.3.5}$$

here $t(\gamma;\nu)$ is the 100γth t percentile with ν degrees of freedom. This limit for the 42.8th percentile of the life distribution is $\underline{\tau}_{.428}(x_0) = \exp[\mu(x_0)]$.

For the insulating fluid at 20 kV ($x_0 = 2.99573$), the lower one-sided 95% confidence limit for $\mu(2.99573)$ is

$$\mu(2.99573) = 10.3439 - 1.6672(0.9598) = 8.7437;$$

here $1.6672 = t(0.95;69)$. The lower confidence limit for the 42.8th percentile at 20 kV is $\underline{\tau}_{.428}(2.99573) = \exp(8.7437) = 6270$ minutes.

A lower limit for a two-sided $100\gamma\%$ confidence interval is also a lower limit for a one-sided $100(1+\gamma)/2\%$ confidence interval. Thus, in Table 4.5 the lower limits of two-sided 95% confidence intervals are lower one-sided 97.5% confidence limits.

Confidence limits for a percentile. Approximate two-sided $100\gamma\%$ confidence limits for the 100Pth ln percentile $\eta_P(x_0)$ at a stress x_0 are

Table 4.5. Estimates and Confidence Limits for 42.8th Percentiles

	Percentiles (in minutes)		
Voltage	Estimate of 42.8th	Two-Sided 95% Confidence Limits	
(kV)	Percentile (minutes)	Lower (minutes)	Upper (minutes)
38	0.838	0.440	1.60
36	2.03	1.25	3.32
34	5.19	3.55	7.59
32	14.0	9.58	20.5
30	40.4	24.3	67.1
28	125.	61.2	256.
26	421.	161.	1106.
20	31.1×10^3	4.58×10^3	$210. \times 10^3$
15	3.47×10^6	0.176×10^6	68.5×10^6
10	2.67×10^9	0.0297×10^9	$240. \times 10^9$
5	0.229×10^{15}	0.189×10^{12}	0.278×10^{18}

$$\eta_P(x_0) = y_P(x_0) - t(\gamma';\nu) s[m(x_0)] = \mu(x_0) + [u(P) + 0.5772]0.7797s ,$$
$$\tilde{\eta}_P(x_0) = y_P(x_0) + t(\gamma';\nu) s[m(x_0)] = \tilde{\mu}(x_0) + [u(P) + 0.5772]0.7797s ; \qquad (4.3.6)$$

here the notation has the usual meaning. For the Weibull percentiles,

$$\tau_P(x_0) = \exp[\eta_P(x_0)] \text{ and } \tilde{\tau}_P(x_0) = \exp[\tilde{\eta}_P(x_0)]. \qquad (4.3.7)$$

These limits ignore the random variation in the estimate s. This is relatively small if x_0 is far from the test stresses.

Often a lower limit is desired in practice. One then wants at most a proportion P of the units at that stress to fail before the limit with $100\gamma\%$ confidence. Replace γ' by γ to get such a limit from (4.3.7). A one-sided lower confidence limit for a percentile is also known as a statistical *lower tolerance limit*. $\tau_P(x_0) = \exp[\eta_P(x_0)]$ is a lower approximate $100\gamma\%$ confidence limit for the 100Pth percentile of the Weibull life distribution.

For the insulating fluid, assume that $\beta = 1/\delta = 1$; that is, the life distribution is exponential. The approximate lower 95% confidence limit for the ln 1st percentile at 20 kV ($x_0 = 2.99573$) is

$$\eta_{.01}(2.99573) = 8.7437 + [-4.6002 + 0.5772]1 = 4.7207.$$

The approximate lower 95% confidence limit for the 1st percentile of the Weibull life distribution is $\tau_{.01}(2.99573) = \exp(4.7207) = 112$ minutes.

Confidence limits for a design stress. Sometimes one wants a design stress x^* such that its 100Pth percentile of life has a specified value $\tilde{\tau}_P^*$. Two-sided approximate $100\gamma\%$ confidence limits $(\underset{\sim}{x}, \tilde{x})$ for x^* are the solutions of

$$\overset{*}{\eta_P} = c_0 + c_1 \, x + [0.5772 + u(P)]0.7797s - t(\gamma'; \nu)\{(1/n) + [(x - \bar{x})^2/S_{xx}]\}^{1/2}s,$$
$$(4.3.8)$$
$$\overset{*}{\eta_P} = c_0 + c_1 \, \tilde{x} + [0.5772 + u(P)]0.7797s + t(\gamma'; \nu)\{(1/n) + [(\tilde{x} - \bar{x})^2/S_{xx}]\}^{1/2}s;$$

here $\overset{*}{\eta_P} = \ln(\overset{*}{\tau_P})$, and $t(\gamma'; \nu)$ is the $100\gamma' = 100(1 + \gamma)/2$th percentile of the t distribution with ν degrees of freedom. These limits assume that the transformed stress x is an increasing function of the actual stress. If not, interchange x and $\tilde{x}$ and solve (4.3.8). These limits are crude and ignore the random variation in s. If x^* is far from the test stresses, the error is smaller.

One usually uses a one-sided $100\gamma\%$ limit x as a design stress. Then the true percentile $\eta_P(x)$ at x is above $\overset{*}{\eta_P}$ with probability γ. For a one-sided $100\gamma\%$ confidence limit, replace γ' by γ in (4.3.8).

The limits $(x, \tilde{x})$ can be obtained graphically from a graph of $\eta_P(x)$ and $\tilde{\eta}_P(x)$ versus x. Such curves for $\eta_{.428}(x)$ and $\tilde{\eta}_{.428}(x)$ appear in Figure 4.1. Enter the graph on the time axis at τ_P; go horizontally to each curve and then down to the stress scale to read the corresponding x and $\tilde{x}$ values. Figure 4.1 directly yields voltage $V = 12.1$ and $\tilde{V} = 18.6$ kV with 95% confidence for $\overset{*}{\tau}_{.428} = 10^6$ minutes.

Confidence limits for the shape parameter. Least-squares regression theory for confidence limits for the Weibull shape parameter does not exist yet. Confidence limits based on maximum likelihood theory appear in Chapter 5. Lawless (1976,1982) gives other exact confidence limits.

Confidence limits for the intercept γ_0. The least squares estimate c_0 for the intercept coefficient γ_0 has a sampling distribution that is approximately normal and has a mean equal to the true value γ_0. Thus, c_0 is an unbiased estimate for γ_0. The true standard error of c_0 is

$$\sigma(c_0) = [(1/n) + (\bar{x}^2/S_{xx})]^{1/2}\sigma. \qquad (4.3.9)$$

Its estimate is

$$s(c_0) = [(1/n) + (\bar{x}^2/S_{xx})]^{1/2}s; \qquad (4.3.10)$$

here s is an estimate for σ and has ν degrees of freedom. Two-sided approximate $100\gamma\%$ confidence limits for γ_0 are

$$\gamma_0 = c_0 - t(\gamma'; \nu)\, s(c_0), \quad \tilde{\gamma}_0 = c_0 + t(\gamma'; \nu)\, s(c_0); \qquad (4.3.11)$$

here $t(\gamma'; \nu)$ is the $100\gamma' = 100(1 + \gamma)/2$th t percentile with ν degrees of freedom. This interval is robust in the sense that its true confidence is close to $100\gamma\%$, if the sample size is moderately large. This interval is rarely used in accelerated testing work, because γ_0 usually lacks a useful physical meaning.

For the insulating fluid,

$$s(c_0) = \{(1/76) + [(3.49591)^2/0.709319]\}^{1/2}\, 1.587 = 6.589;$$

the estimate $s = 1.587$ has 69 degrees of freedom. Approximate 95%

confidence limits for γ_0 are

$$\gamma_0 = 59.4468 - 1.9949(6.589) = 46.30,$$

$$\tilde{\gamma}_0 = 59.4468 + 1.9949(6.589) = 72.59,$$

where $1.9949 = t(0.975;69)$.

Confidence limits for the slope γ_1. The estimate c_1 for the slope coefficient γ_1 has a sampling distribution that is approximately normal, and its mean equals the true value γ_1. Thus, c_1 is an unbiased estimate for γ_1. The standard error of c_1 is

$$\sigma(c_1) = [1/S_{xx}]^{1/2}\sigma. \tag{4.3.12}$$

Its estimate is

$$s(c_1) = [1/S_{xx}]^{1/2}s. \tag{4.3.13}$$

Two-sided approximate $100\gamma\%$ confidence limits for γ_1 are

$$\gamma_1 = c_1 - t(\gamma';\nu)\,s(c_1), \quad \tilde{\gamma}_1 = c_1 + t(\gamma';\nu)\,s(c_1); \tag{4.3.14}$$

$t(\gamma';\nu)$ is the $100\gamma' = 100(1+\gamma)/2$th t percentile with ν degrees of freedom.

For the insulating fluid, $s(c_1) = [1/0.709319]^{1/2}\,1.587 = 1.884$. Approximate 95% confidence limits for the power γ_1 are

$$\gamma_1 = -16.3909 - 1.9949(1.884) = -20.149,$$

$$\tilde{\gamma}_1 = -16.3909 + 1.9949(1.884) = -12.632;$$

here $1.9949 = t(0.975;69)$. These limits could be rounded to one decimal place in view of the large uncertainty.

Confidence limits for a fraction failing. Such approximate limits are not given here. Use the maximum likelihood ones in Chapter 5.

Exponential Fit

The exponential distribution is the Weibull distribution with a shape parameter value of $\beta = 1$. Thus the model, estimates, and confidence limits of Subsections 4.1, 4.2, and 4.3 apply. However, the estimate b (and d) is replaced by the assumed value $\beta = 1$. The affected equations are listed below. Their equation numbers are the same as in those subsections. But they are distinguished here by a prime. These methods with suitable modification can be used with any specified value of β; this is useful when the distribution is not exponential.

Model. Because $\beta = 1/\delta = 1$, model equations simplify as follows:

$$\ln[\theta(x_j)] = \xi(x_j) = \mu(x_j) + 0.5772, \quad \sigma = 1.283,$$

$$\ln[\tau_P(x_j)] = \mu(x_j) + [0.5772 + u(P)]. \tag{4.1.3'}$$

Estimates. Modified estimates for an exponential distribution follow.

Other estimates are not modified. The estimate of a fraction failing is

$$F^*(t_0 ; x_0) = 1 - \exp\{-t_0/\exp[m(x_0)+0.5772]\}. \quad (4.2.15')$$

The estimate of a (ln) percentile is

$$\ln[t_P(x_0)] = m(x_0)+[0.5772+u(P)] = c_0 + c_1 x_0 + [0.5772+u(P)]. \quad (4.2.16')$$

The estimate of the mean life is the special case $P = 0.632$; namely,

$$\theta^*(x_0) = \exp(0.5772+c_0+c_1 x_0).$$

The estimate of the failure rate is

$$\lambda^*(x_0) = 1/\theta^*(x_0) = 1/\exp(0.5772+c_0+c_1 x_0).$$

The estimate of a design stress that yields a given percentile τ_P^* is

$$x^* = \{\ln(\tau_P^*)-c_0-[0.5772+u(P)]\}/c_1 . \quad (4.2.18')$$

Confidence limits. Modified $100\gamma\%$ confidence limits for an exponential distribution $(\beta = 1)$ appear below. All limits are modified. The normal percentile $z_{\gamma'}$ replaces the t percentile $t(\gamma';\nu)$ throughout, where $\gamma' = (1+\gamma)/2$. For a one-sided $100\gamma\%$ limit, use γ in place of γ'.

Two-sided approximate $100\gamma\%$ confidence limits for the mean ln life are

$$\underline{\mu}(x_0) = m(x_0) - z_{\gamma'}\sigma[m(x_0)], \quad \tilde{\mu}(x_0) = m(x_0) + z_{\gamma'}\sigma[m(x_0)] \quad (4.3.3')$$

where

$$\sigma[m(x_0)] = \{(1/n)+[(x_0-\bar{x})^2/S_{xx}]\}^{1/2} 1.283. \quad (4.3.1')$$

Two-sided approximate $100\gamma\%$ confidence limits for a $100P$th percentile are

$$\underline{\tau}_P(x_0) = \exp\{\underline{\mu}(x_0)+[0.5772+u(P)]\},$$
$$\tilde{\tau}_P(x_0) = \exp\{\tilde{\mu}(x_0)+[0.5772+u(P)]\}. \quad (4.3.7')$$

Confidence limits for the exponential mean are the special case $P = 0.632$:

$$\underline{\theta}(x_0) = \exp[\underline{\mu}(x_0) + 0.5772], \quad \tilde{\theta}(x_0) = \exp[\tilde{\mu}(x_0) + 0.5772].$$

Two-sided approximate $100\gamma\%$ confidence limits for the failure rate are

$$\underline{\lambda}(x_0) = 1/\tilde{\theta}(x_0) \text{ and } \tilde{\lambda}(x_0) = 1/\underline{\theta}(x_0).$$

Two-sided approximate $100\gamma\%$ confidence limits $(\underline{x},\tilde{x})$ for a design stress are the solutions of

$$\eta_P^* = c_0+c_1\underline{x}+[0.5772+u(P)]-z_{\gamma'}\{(1/n)+[(\underline{x}-\bar{x})^2/S_{xx}]\}^{1/2}1.283,$$
$$\eta_P^* = c_0+c_1\tilde{x}+[0.5772+u(P)]+z_{\gamma'}\{(1/n)+[(\tilde{x}-\bar{x})^2/S_{xx}]\}^{1/2}1.283. \quad (4.3.8')$$

These limits assume that the transformed stress x is an increasing function of the actual stress V. If not, interchange $\underline{x}$ and $\tilde{x}$ and solve (4.3.8'). Then V

corresponds to $\tilde{x}$ (and $\tilde{V}$ to $\underline{x}$).

Two-sided approximate $100\gamma\%$ confidence limits for the intercept γ_0 are

$$\underline{\gamma}_0 = c_0 - z_\gamma' \, \sigma(c_0), \quad \tilde{\gamma}_0 = c_0 + z_\gamma' \, \sigma(c_0), \tag{4.3.11'}$$

where

$$\sigma(c_0) = [(1/n) + (\bar{x}^2/S_{xx})]^{1/2} 1.283. \tag{4.3.9'}$$

Two-sided approximate $100\gamma\%$ confidence limits for the slope γ_1 are

$$\underline{\gamma}_1 = c_1 - z_\gamma' [1/S_{xx}]^{1/2} 1.283, \quad \tilde{\gamma}_1 = c_1 + z_\gamma' [1/S_{xx}]^{1/2} 1.283. \tag{4.3.15'}$$

Two-sided approximate $100\gamma\%$ confidence limits for a fraction failing by age t_0 are

$$\underline{F}(t_0 ; x_0) = 1 - \exp[-t_0/\tilde{\theta}(x_0)], \quad \tilde{F}(t_0 ; x_0) = 1 - \exp[-t_0/\underline{\theta}(x_0)].$$

5. CHECKS ON THE LINEAR-WEIBULL MODEL AND DATA

The preceding estimates and confidence limits are based on certain assumptions. Thus, their accuracy depends on how well the assumptions are met. This section presents methods for checking the assumptions. From such checks one can judge how much to rely on various estimates and confidence limits. Of course, some are reliable even when the assumptions are far from satisfied, and others are quite sensitive to departures from the assumptions. Also, presented here are checks on the validity of the data. These checks reveal suspect data points that may result from faulty testing or blunders in handling the data. The checks examine
- linearity of the (transformed) life-stress relationship,
- dependence of the Weibull shape parameter on stress,
- that life has a Weibull distribution,
- the effect of other variables,
- deviant data.

These checks also appear in Section 3. The effect of such departures from the assumptions on the estimates and confidence limits is discussed below.

5.1. Is the Life-Stress Relationship Linear?

The model assumes that the transformed life-stress relationship (4.1.2) is linear. Linearity is important if one extrapolates the fitted line outside the range of the test stresses. The data may be nonlinear for various reasons:

1. The true relationship may be linear, but the observed relationship may be nonlinear, because the life test was run improperly. For example, specimens at some stress levels may differ in their fabrication or handling from those at other levels, or the actual levels may differ from intended ones.

2. The relationship may be nonlinear due to several failure modes acting, each with a different linear relationship.
3. Also, the relationship may just be inherently nonlinear, that is, not described by the chosen life-stress relationship.

A subjective assessment of the linearity of the relationship may be obtained from examination of a plot of the data on paper for the relationship. This is described in Section 3.5 of Chapter 3. For the insulating fluid, such a log-log plot appears as Figure 3.3 in Chapter 3 and is reasonably linear.

The F test for linearity (Section 3.1) applies, but it is approximate for a Weibull life distribution. For the insulating fluid, the F statistic is

$$F = [(76-2)(1.55815)^2 - (76-7)(1.58685)^2]/[(7-2)(1.58685)^2] = 0.47.$$

Here the F distribution has $J-2=7-2=5$ degrees of freedom in the numerator and $n-J=76-7=69$ in the denominator. Since $F = 0.47 < 2.35 = F(0.95;5,69)$, there is no statistically significant (convincing) evidence of nonlinearity. The observed F value is so small that the F approximation may be quite crude and still suffice. Examination of the log-log plot of the data in Figure 3.3 of Chapter 3 supports this conclusion.

If statistically significant nonlinearity is found, examine the relationship plot to determine how the relationship departs from linearity. Nonlinearity may result from curvature in the relationship, defective data, or different failure modes. If there are several failure modes, the relationship would be concave downwards (life on the vertical axis). If the data from a stress are out of line with data from the other stresses, those data may be in error. Reasons for that error must be examined to determine whether those data are valid and should be used or not. After examining the relationship plot, do one or more of the following.

1. If the plot indicates that a smooth curve describes the relationship, fit a curve to the data or transform the data differently. A quadratic equation in the (transformed) stress may suffice. Before doing this, be sure that the apparent curvature is not a result of erroneous data.
2. If the plot contains data points that are not consistent with the rest, set them aside and analyze the apparently valid data. Before doing this, one should try to justify rejecting those data, say, due to some flaw in the experimental work. It is most important to determine the cause of such data, as that can lead to improving the product or test method.
3. Use the estimates and confidence limits as is, but subjectively take into account the nonlinearity in interpreting them and coming to conclusions.

5.2. Is the Weibull Shape Parameter Constant?

The model assumes that the shape parameter β has the same true value for all stress levels. β may depend on stress, or different β values may result from faulty testing. If so, the percentile estimates should not be calculated with the pooled estimate of the shape parameter. Instead, an estimate of a percentile at a stress level should employ an estimate of β for that level. If no data have been collected at a particular level, then one must express β as some function of the stress. That function would have to be estimated from the data. The following graphical method compares shape parameters at different stress levels. Analytic comparisons with maximum likelihood estimates appear in Chapter 5. One can subjectively assess whether the shape parameter is constant as described in Section 2.5 of Chapter 3 using Weibull paper. Weibull probability plots of the data should yield parallel lines with the same slope. When the samples are small as in Figure 3.1 of Chapter 3, one should expect much random variation in the plotted slopes. Only extreme or systematic differences in the plotted slopes are evidence that the shape parameter is not constant. In particular, a systematic change in the slope over successive stress levels suggests that the shape parameter depends on stress. A single slope that differs greatly from the other slopes indicates possibly faulty testing at that stress level.

For the insulating fluid, the Weibull plot in Figure 3.1 of Chapter 3 suggests that the Weibull shape parameter is not the same at all test voltages. In particular, the slope at 32 kV is smaller than at the other test voltages. This may indicate that the test for the 32 kV data was faulty. The slope at 26 kV is also small; however, that sample is too small to warrant any conclusions.

If the Weibull plot is not convincing, use an objective statistical test for equality of the shape parameters (Chapter 5). For the insulating fluid, such a test shows that the shape parameters estimates at different test voltages do not differ statistically significantly.

5.3. Is the Life Distribution Weibull?

Certain estimates and confidence limits are sensitive to the assumption that the life distribution is Weibull. These include estimates of percentiles and a fraction failing and certain confidence intervals. Also, only if the Weibull distribution fits, is the fitted life-stress line an estimate of the 42.8th percentile. Otherwise, the line is some vague nominal life. The following are subjective checks that the Weibull distribution is adequate.

Weibull plot. As described in Section 3.2 of Chapter 3, use Weibull plots of the data from the different test stress levels. Examine them to assess if the plots are reasonably straight. A pronounced systematic curvature at most stresses suggests that the Weibull distribution is not adequate. In judging, inexperienced people tend to be too critical of such plots; they expect much better linearity than is usually observed. Subjective notions of how a plot should look tend to be too stringent and orderly. The great normal erraticness of such plots may be seen in plots of Monte Carlo samples in Hahn and Shapiro (1967) and Daniel and Wood (1980). Figure 3.2 of Chapter 3 looks erratic to the inexperienced eye.

Residuals plot. A more sensitive check on the Weibull assumption employs a plot of the *adjusted ln residuals about the fitted line.* Such a residual of the ith ln observation y_{ij} at the test stress x_j is

$$r_{ij} = [y_{ij} - m(x_j)] / \{1 - (1/n) - [(x_j - \bar{x})^2 / S_{xx}]\}^{1/2} ;$$

here $m(x_j)$ is the least squares estimate (4.2.14) of the ln mean at stress x_j. The notion is defined in Section 4.2. The differences $r'_{ij} = y_{ij} - m(x_j)$ are called the *raw residuals about the fitted line*; they are easier to calculate and most regression programs calculate them. The raw residuals have unequal standard deviations, all slightly less than σ; so they are not from distributions with the same standard deviation. The adjusted residuals are better, because each has a standard deviation equal to σ. But the raw residuals suffice for a moderately large sample, such as for the insulating fluid.

All adjusted residuals are pooled as a single sample. They should look like a sample from an extreme value distribution with a mean of zero and a standard deviation equal to σ. Plot the pooled sample on (smallest) extreme value probability paper. Extreme value probability paper are available from TEAM, Box 25, Tamworth, NH 03886, (603)323 – 8843:

#111	0.01 – 99.99% (11" horizontal)	100 divisions (8 1/2" vertical)
#112	0.0001 – 99.9999% (14" horizontal)	100 divisions (11" vertical)

These and most other extreme value papers are for the *largest* extreme value distribution. For the smallest extreme value distribution, change each percentage $100F$ on such probability papers to $100(1 - F)$.

Examine the plot to assess whether the data follow a straight line. The best way to judge straightness is to hold the paper horizontal at eye level and sight along the points. If the plot is not reasonably straight, then the Weibull distribution is inadequate. The nature of the curvature may indicate how the true distribution differs from the Weibull. Also, a non-straight plot may indicate that some observations are in error (that is, a faulty test) or that the life-stress relationship is inadequate. In view of this, one should plot these residuals only after checking the relationship plot.

For the insulating fluid, the adjusted residuals about the fitted line are calculated in Table 5.1. Note that the raw and adjusted residuals differ little, since the adjusted factors are close to 1 (1.007 to 1.050). Thus, the raw residuals would do. The extreme value plot of the adjusted residuals appears in Figure 5.1. Note that the probability scale of the *largest* extreme value paper has been relabeled. Each percentage $100F$ was changed to $100(1 - F)$. The plot is relatively straight and shows no peculiarities. Thus, the Weibull distribution appears adequate.

5.4. Estimate the Weibull Shape Parameter

Shape parameter estimate. The plot of the pooled adjusted residuals

Table 5.1. Calculation of Adjusted Residuals

26 kV

$$1 / \sqrt{1 - \frac{1}{72} - \frac{(3.25810 - 3.49591)^2}{0.709319}} = 1.050$$

$(1.7561 - 6.0435)1.050 = -4.50$
$(7.3648 - 6.0435)1.050 = 1.39$
$(7.7509 - 6.0435)1.050 = 1.79$

28 kV

$$1 / \sqrt{1 - \frac{1}{72} - \frac{(3.33221 - 3.49591)^2}{0.709319}} = 1.026$$

$(4.2319 - 4.8288)1.026 = -0.61$
$(4.6848 - 4.8288)1.026 = -0.15$
$(4.7031 - 4.8288)1.026 = -0.13$
$(6.0546 - 4.8288)1.026 = 1.26$
$(6.9731 - 4.8288)1.026 = 2.20$

30 kV

$$1 / \sqrt{1 - \frac{1}{72} - \frac{(3.40120 - 3.49591)^2}{0.709319}} = 1.013$$

$(2.0464 - 3.698)1.013 = -1.67$
$(2.8361 - 3.698)1.013 = -0.87$
$(3.0184 - 3.698)1.013 = -0.69$
$(3.0454 - 3.698)1.013 = -0.66$
$(3.1206 - 3.698)1.013 = -0.58$
$(3.7704 - 3.698)1.013 = 0.07$
$(3.8565 - 3.698)1.013 = 0.16$
$(4.9349 - 3.698)1.013 = 1.25$
$(4.9706 - 3.698)1.013 = 1.29$
$(5.1698 - 3.698)1.013 = 1.49$
$(5.2724 - 3.698)1.013 = 1.60$

32 kV

$$1 / \sqrt{1 - \frac{1}{72} - \frac{(3.46754 - 3.49591)^2}{0.709319}} = 1.007$$

$(-1.3094 - 2.64013)1.007 = -3.98$
$(-0.9163 - 2.64013)1.007 = -3.58$
$(-0.3711 - 2.64013)1.007 = -3.03$
$(-0.2358 - 2.64013)1.007 = -2.90$
$(1.0116 - 2.64013)1.007 = -1.64$
$(1.3635 - 2.64013)1.007 = -1.29$
$(2.2905 - 2.64013)1.007 = -0.35$
$(2.6354 - 2.64013)1.007 = -0.00$
$(2.7682 - 2.64013)1.007 = 0.13$
$(3.3250 - 2.64013)1.007 = 0.69$
$(3.9748 - 2.64013)1.007 = 1.34$
$(4.4170 - 2.64013)1.007 = 1.79$
$(4.4918 - 2.64013)1.007 = 1.87$
$(4.6109 - 2.64013)1.007 = 1.99$
$(5.3711 - 2.64013)1.007 = 2.75$

34 kV

$$1 / \sqrt{1 - \frac{1}{72} - \frac{(3.52637 - 3.49591)^2}{0.709319}} = 1.007$$

$(-1.6608 - 1.64635)1.007 = -3.33$
$(-0.2485 - 1.64635)1.007 = -1.91$
$(-0.0409 - 1.64635)1.007 = -1.70$
$(0.2700 - 1.64635)1.007 = -1.39$
$(1.0224 - 1.64635)1.007 = -0.63$
$(1.1505 - 1.64635)1.007 = -0.50$
$(1.4231 - 1.64635)1.007 = -0.22$
$(1.5411 - 1.64635)1.007 = -0.11$
$(1.5789 - 1.64635)1.007 = -0.07$
$(1.8718 - 1.64635)1.007 = 0.23$
$(1.9947 - 1.64635)1.007 = 0.35$
$(2.0806 - 1.64635)1.007 = 0.44$
$(2.1126 - 1.64635)1.007 = 0.47$
$(2.4898 - 1.64635)1.007 = 0.85$
$(3.4578 - 1.64635)1.007 = 1.82$
$(3.4818 - 1.64635)1.007 = 1.85$
$(3.5237 - 1.64635)1.007 = 1.89$
$(3.6030 - 1.64635)1.007 = 1.97$
$(4.2889 - 1.64635)1.007 = 2.66$

36 kV

$$1 / \sqrt{1 - \frac{1}{72} - \frac{(3.58352 - 3.49591)^2}{0.709319}} = 1.012$$

$(-1.0499 - 0.709603)1.012 = -1.78$
$(-0.5277 - 0.709603)1.012 = -1.25$
$(-0.0409 - 0.709603)1.012 = -0.76$
$(-0.0101 - 0.709603)1.012 = -0.73$
$(0.5247 - 0.709603)1.012 = -0.19$
$(0.6780 - 0.709603)1.012 = -0.03$
$(0.7275 - 0.709603)1.012 = 0.02$
$(0.9477 - 0.709603)1.012 = 0.24$
$(0.9969 - 0.709603)1.012 = 0.29$
$(1.0647 - 0.709603)1.012 = 0.36$
$(1.3001 - 0.709603)1.012 = 0.60$
$(1.3837 - 0.709603)1.012 = 0.68$
$(1.6770 - 0.709603)1.012 = 0.98$
$(2.6224 - 0.709603)1.012 = 1.94$
$(3.2386 - 0.709603)1.012 = 2.56$

38 kV

$$1 / \sqrt{1 - \frac{1}{72} - \frac{(3.63759 - 3.49591)^2}{0.709319}} = 1.021$$

$(-2.4080 - 0.176655)1.021 = -2.28$
$(-0.9417 - 0.176655)1.021 = -0.78$
$(-0.7551 - 0.176655)1.021 = -0.59$
$(-0.3148 - 0.176655)1.021 = -0.14$
$(-0.3012 - 0.176655)1.021 = -0.13$
$(0.1222 - 0.176655)1.021 = 0.31$
$(0.3364 - 0.176655)1.021 = 0.52$
$(0.8671 - 0.176655)1.021 = 1.07$

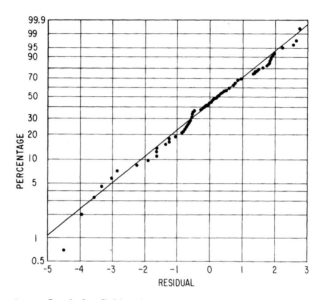

Figure 5.1.   Insulating fluid residuals on extreme value probability paper.

yields a pooled graphical estimate of the Weibull shape parameter. Visually fit a straight line to the plot so its 42.8% point is zero. The 78.3th percentile of that line is an estimate of the extreme value scale parameter δ. Its reciprocal is the shape parameter estimate. This is more accuracte than the graphical estimate of Chapter 3. Such a fitted line in Figure 5.1 has a 78.3th percentile of 1.24. The pooled estimate is $\beta^* = 1/1.24 = 0.81$.

   Check. Using the plot, one can subjectively assess whether the data are consistent with a specified β_0 value. For example, theory for the insulating fluid says $\beta_0 = 1$. On the plot, mark the value $\delta_0 = 1/\beta_0$ at 78.3%. Draw a line through this point and through zero at 42.8%. Compare the slope of this line with that of the data. A convincing difference indicates that the data are not consistent with β_0. For the insulating fluid, the line for $\beta_0 = 1$ can be added to Figure 5.1. The slope of that line borders on differing convincingly from that of the data. When the subjective assessment is marginal, an objective assessment (Chapter 5) is needed. The observed $\beta^* = 0.81$ is below the theorized $\beta_0 = 1$. Such a lower estimate would result if the test conditions lacked control and varied slightly, causing greater scatter in (ln) life.

5.5. The Effect of Other Variables

   Sometimes there are other variables whose values are observed for each test specimen. Such variables include, for example, the shift (personnel) that made the specimens, the length of the test cycle (see Problem 3.9), and dissipation factor (of insulation). One wishes to know if and how such variables affect product life. The effect of such variables on life may be examined (Section 6) through fitting multivariable relationships and through crossplots.

A variable not in a model can be examined in a crossplot of the residuals (from the fitted model) against that variable. If the variable affects life, the plot will have a trend. Figure 6.7 of Section 6 is such a plot of residuals against binder content. The residuals come from fitting the power-Weibull model to voltage endurance data from an insulating tape. The plot indicates that life decreases with increasing binder content, a measure of energy loss in the tape dielectric.

Binder content (a percentage) is a *continuous variable*; that is, it can have any (positive) numerical value. Some variables have only discrete category values, for example, year of manufacture. Figure 6.11 of Section 6 is a crossplot of the tape residuals against year of manufacture. The crossplot slightly suggests that the recent years produced specimens with shorter life.

Regression books explain in detail how to make and interpret such crossplots of residuals. See, for example, Draper and Smith (1981), Neter, Wasserman, and Kutner (1983), or Daniel and Wood (1980). A trend in such a crossplot by itself does not prove cause and effect of the variable. Other variables could be producing the effect on life and on the variable.

Most regression packages calculate the raw residuals about the fitted relationship. Also, most crossplot residuals against other variables.

5.6. Checks on the Data

Suspect data are data subsets or individual observations that deviate from the assumed model or the bulk of the data. Such data can arise from faulty testing, errors in handling the data, the effect of other variables, etc. The previous checks on the model also are checks on the data. Nonlinearity of the life-stress relationship, dependence of the Weibull shape parameter on stress, inadequate fit of the Weibull distribution – each may indicate suspect data. In addition to these methods, one can examine the Weibull plots of the data and the extreme value plot of the residuals for individual extremely high or low points, called *outliers*. Such points may be erroneous; determine their cause and decide whether to use them in the analyses. Points in the extreme tails of a plot vary just from random sampling much more than most people expect. Thus, points must be much out of line before they are labeled outliers or discarded. Preferably, one should find the cause of suspect data. Then one can improve the testing and data handling. In a sense, suspect data are always right; that is, they reflect something real happening. Only the model or our understanding is inadequate.

Formal numerical methods for identifying outliers are given by Beckman and Cook (1983), Barnett and Lewis (1984), and Daniel and Wood (1980).

6. MULTIVARIABLE RELATIONSHIPS

6.1. Introduction

Overview. This section presents least squares fitting of multivariable relationships to data. This section explains only how to interpret computer output on such fitting. Theory for such fitting is complex and is omitted here. Such theory appears in the references below. Needed background for this section includes multivariable relationships (Section 13 of Chapter 2) and Sections 2 and 3 of the present chapter. Section contents include 1) example data, 2) fitting a specified relationship, 3) assessing the fit and the data, and 4) stepwise fitting of a general relationship.

Why multivariable? As explained in Section 13 of Chapter 2, multivariable relationships are used when there is more than one accelerating variable or there are other variables that may affect product life. For example, such relationships may serve as derating curves for the combined effects of a number of accelerating stresses, a common practice for electronic devices. Also, such relationships may yield insights that improve a product's manufacture, operation, or testing. Even tests with just an accelerating stress and presumably no other variable often involve other variables. Examples of such variables include: 1) order of a specimen in going through each step of fabrication and testing, 2) different technicians who carry out each step, and 3) position on the test rack. Good experimental procedure has the specimens go through each fabrication and testing step in a separate random order.

Test planning. For principles of experimental design, refer to Box, Hunter, and Hunter (1978), one of many introductions to the subject. Chapter 1 discusses test design in detail. Few accelerated tests are statistically designed. This usually results in less accurate information and sometimes no information. A statistical consultant is most helpful at the test planning stage. Most books, including this one, overemphasize data analysis. Few devote enough attention to test planning, which is more important. For most well planned tests, the conclusions are apparent without fancy data analyses. And the fanciest data analyses often cannot salvage a poorly planned test.

References. There are many books that present least squares fitting of multivariable relationships to data. They explain how to interpret output of standard computer programs, and they derive the underlying theory. Examples include Draper and Smith (1981) and Neter, Wasserman, and Kutner (1983,1985). Also, there are many computer programs that do the least squares calculations. Section 1 references some of these programs; consult them for more detail. This section can only briefly survey some basics of this vast subject. Also, note that such books and programs generally assume that the life distribution is (log) normal; many confidence limits are sensitive to that assumption.

Linear relationship. For most applications, the multivariable relationship is linear *in the unknown coefficients*, which are estimated from the data. Such relationships may be nonlinear in the independent variables. Section 13 of Chapter 2 presents such relationships, and the following examples employ them. Such "linear" relationships are employed for two reasons. First, they are easy to fit to data with standard computer programs. Second, they usually adequately represent data. Often they do not have a theoretical basis but are adequate empirical models in practice. This book does not address the difficult engineering problem of identifying important variables and the form of their relationship with life. Box and Draper (1987) treat this problem. This book treats the simpler problems of fitting chosen relationships to data and assessing a relationship and the data.

Assumptions. Most computer programs assume that the standard deviation of (log) life is a constant. Of course, some products have a standard deviation that is a function of the accelerating stress or other variables; Chapter 5 presents fitting of such models. Also, in some applications, the multivariable relationship is a nonlinear function of the coefficients. Such relationships can be fitted by the nonlinear least squares method as described in the references in Section 1 or by maximum likelihood (Chapter 5).

Degradation. Least squares can be used to fit performance degradation relationships to multivariable data. This is described in Chapter 11. Then performance is the statistically dependent variable, and time is one of the independent variables.

6.2. Example Data

Test purposes. This subsection describes insulation data that illustrate least squares fitting of a multivariable relationship. Each insulation specimen is production insulation on a conductor taken from a different electrical machine. The data consist of design, manufacturing, and test variables in Table 6.1. One hundred and six specimens underwent an accelerated voltage-endurance test. Purposes of data collection and testing include

1. Estimate life (a low percentile) of insulation at its (low) design voltage.
2. Evaluate the effect of manufacturing and design variables on life.
3. Retrospectively assess quality control; that is, assess whether the life distribution was stable over the manufacturing period of several years. If not stable, review of past data on manufacturing variables may reveal assignable causes or variables that may be profitably controlled in future manufacture.
4. Assess whether a Weibull or lognormal distribution fits the data better. Also, assess whether either fits adequately over the range of the data before extrapolating far into the lower tail. This is related to purpose 1.

The life data are complete. This is unusual (and usually undesirable) in ac-

Table 6.1. List of Variables

ELNO	Electrode (specimen) number.
INSLOT	The lot of insulating material used to make the electrode.
TAPLOT	The lot of tape used to make the electrode.
TPDATE	The date the electrode was taped – year, month, day.
INSWTH	The width in centimeters of the insulated electrode.
INSHGT	The height in centimeters of the insulated electrode.
INSTHK	The insulation thickness in mm.
LAYERS	The number of layers of tape applied to the electrode.
MMPERL	mm per layer = INSTHK/LAYERS.
MATDAM	A measure of material damage during tape application.
DENSTY	Insulation density.
VOLTLE	Percentage volatile content of the insulation.
BINDER	Percentage organic binder in the insulation.
EXTRAC	Percentage of unreacted organic binder in the insulation.
BKNKV	Breakdown voltage in kV of insulation after binder is removed.
KVSLOT	Breakdown voltage in kV of the slot portion of the electrode.
KVBEND	Breakdown voltage in kV of the bend portion of the electrode.
DFRTXV	Dissipation factor at room temperature and 10 V/mm.
DFCDXV	Dissipation factor at 100°C and 10 V/mm.
DFRTTU	Dissipation factor tip up at room temperature, 10 to 100 V/mm.
DFCDTU	Dissipation factor tip up at 100°, 10 to 100 V/mm.
HOURS	Life in hours of the insulation.
VPM	The test stress in volts/mm applied to the electrode.
STRINS	The type of strand insulation used.
LOGHRS	Base 10 log of HOURS.
LOGVPM	Base 10 log of VPM.
ASPECT	= INSHGT/INSWTH.

celerated testing. One usually cannot wait until all specimens fail. A listing of the data is omitted, as the listing is not essential to the purposes here.

Purpose of analyses. The analyses below are intended only to illustrate such least squares fitting. They are incomplete and address only purpose 2 above. Thorough analysis would involve many such analyses.

Details. It is useful to remark that the conductors have rectangular cross sections. Their heights and widths differ due to design differences among the machines. The electrical stress (VPM) is the applied voltage across the insulation divided by its thickness (INSTHK) in millimeters.

Test plan. The test levels of voltage stress (the accelerating variable) were not chosen so the test plan yielded accurate estimates of life at the low design stress. The voltage stresses of the specimens were spread roughly uniformly over the test range. This is inefficient for extrapolation. As shown in Chapter 6, one can get more accurate estimates at a design stress with a better choice of stress levels. This results from putting more specimens at the extremes of the test range and a few in the middle. Moreover, more specimens should be tested at low stress than at high.

6.3. Fitting a Specified Model

Purpose. This section shows the fitting of a specified multivariable relationship to data. The relationship expresses the mean of log hours to failure (LOGHRS) as a function of log voltage stress in volts per mm (LOGVPM) and of dissipation factor (DFRTTU), a measure of electrical loss. All logarithms are base 10.

Model. The specified linear relationship for mean log life μ is

$$\mu = \gamma_1 + \gamma_2 \text{ LOGVPM} + \gamma_3 \text{ DFRTTU}. \tag{6.1}$$

The coefficients are estimated from the data by least squares and denoted by C1, C2, and C3 below. The standard deviation σ of log life is assumed to be constant. Also, for estimating low percentiles, the life distribution is assumed to be lognormal. Interest centers on low percentiles of the life distribution. Because σ is assumed constant, any percentile equations are parallel to (6.1). Thus one need only work with (6.1). Those who use a relationship like (6.1) when interested in low percentiles need to be aware they implicitly assume σ is constant.

Variables. DFRTTU or its log could be used here. Engineering knowledge does not suggest which is better, and both adequately fit the data over the observed range of DFRTTU. Also, engineering opinion is divided on whether LOGVPM or VPM is better in (6.1). Here (6.1) is the inverse power relationship for life versus VPM.

Picture. The following mental picture of the data and the fitted equation aids understanding. For fitting this model, each specimen is regarded to have a numerical data value for LOGHRS, LOGVPM, and DFRTTU. Each specimen can then be plotted as a point in a three-dimensional space. The three perpendicular axes of the space correspond to LOGHRS (the vertical axis), LOGVPM, and DFRTTU. The points form a cloud in that space, where (6.1) is a plane passing through the cloud. Figure 6.1 shows three projections of the data cloud onto a plane determined by a pair of axes. The high point in Figure 6.1A is influential in determining the estimate C3 for γ_3.

Fitting. Fitting (6.1) to the data amounts to determining a "best" plane through the cloud. The least squares "best" plane minimizes the sum of the squared vertical LOGHRS distances between the plane and the data points. Each difference (vertical distance on the LOGHRS scale) between a point and the plane is the residual for that specimen. A constant σ corresponds to the points having the same vertical scatter around the plane everywhere.

Significant variable. A variable in (6.1) is statistically significant if the estimate of its coefficient (C2 or C3) differs significantly from zero; this is so

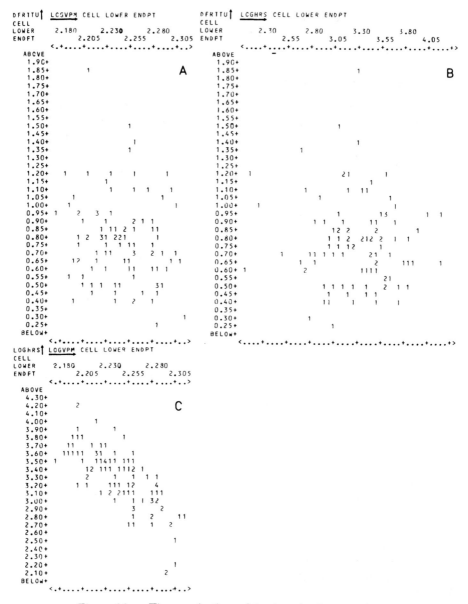

Figure 6.1. Three projections of the data cloud onto a plane.

if the confidence interval for the coefficient does not enclose zero. This is equivalent to the data cloud having a convincing slope along the axis of that variable. "Convincing" means relative to the scatter of the data points about the fitted plane.

Centering. For various purposes, it can be better to write (6.1) as

$$\mu = \gamma_1' + \gamma_2(\text{LOGVPM} - \text{LOGVPM}') + \gamma_3(\text{DFRTTU} - \text{DFRTTU}'). (6.1')$$

Here LOGVPM´ and DFRTTU´ are chosen values. Usually such a value is (1) the mean value of the independent variable for the sample specimens, (2) some other value near the center of the data values of the variable, or (3) some other meaningful value such as the design value. Then $\mu = \gamma_1'$ for LOGVPM = LOGVPM´ and DFRTTU = DFRTTU´. So γ_1' is the mean log life at those variable values – a simple, useful interpretation of that intercept coefficient. Such *centering* (or *coding*) of the independent variables is discussed further in Chapter 5. There such centering aids numerical computation, in addition to interpretation. Of course,

$$\gamma_1 = \gamma_1' - \gamma_2 \text{LOGVPM}' - \gamma_3 \text{DFRTTU}'.$$

Output. Figure 6.2 shows output from fitting (6.1´) to the data. Key lines are numbered. Their meanings and interpretations follow. Consult regres-

1⟩ REGRESSION ALL(LOGHRS LOGVPM DFRTTU)

```
 * SUMMARY STATISTICS

 CASES
  106
```

2⟩
```
 VARIABLE   AVERAGE          VARIANCE         STD DEV

   LOGVPM   2.248131       0.1077328E-02    0.3282267E-01
   DFRTTU  0.7980566       0.6769291E-01    0.2601786
   LOGHRS   3.323656       0.1511960        0.3888394
```

```
 CORRELATION MATRIX OF VARIABLES

 VARIABLE      LOGVPM           DFRTTU          LOGHRS

 LOGVPM    1.0000000
 DFRTTU   -0.1461116        1.0000000
 LOGHRS   -0.7387652       -0.6100866E-01    1.000000
```

```
 COVARIANCE MATRIX OF VARIABLES

 VARIABLE      LOGVPM           DFRTTU          LOGHRS

 LOGVPM    0.1077328E-02
 DFRTTU   -0.1247757E-02     0.6769291E-01
 LOGHRS   -0.9431225E-02    -0.6172105E-02    0.1511960
```

```
 * LEAST SQUARES ESTIMATE OF THE FITTED EQUATION
```

3⟩
```
 MEAN =   23.88226
          + ( -9.053152    )*LOGVPM
          + (-0.2580513    )*DFRTTU
```

4⟩ STD DEV = 0.2558672

```
 LEAST SQUARES ESTIMATES OF COEFFICIENTS WITH 95% LIMITS

 VAR     COEFF     ESTIMATE       LOWER LIMIT      UPPER LIMIT
```

5⟩
```
 INTR    C00000    23.88226        20.42740        27.33713
 LOGVPM  C00001   -9.053152       -10.57830        -7.528002
 DFRTTU  C00002   -0.2580513      -0.4504556       -0.6564692E-01
```

Figure 6.2.   Output for model (6.1).

sion texts for interpretation of other output. Output is from STATPAC of Nelson and others (1983).

- Line 1 is the command for least squares fitting; LOGHRS is the dependent variable, and LOGVPM and DFRTTU are independent variables.
- Line 2 and following lines show summary statistics on the sample data.
- Line 3 shows the fitted equation (6.1). The coefficient estimates for LOGVPM and DFRTTU are both negative; this indicates that life decreases as either variable increases, which is consistent with engineering experience. The coefficient estimate for LOGVPM is -9.05; thus insulation life decreases as the inverse of the 9.05 power of voltage stress (VPM). DFRTTU is an uncontrolled but measured covariate. Its physical effect on life is gauged from the size of (its coefficient)×(twice its standard deviation), namely, $(-0.258)×(2×0.260) = -0.134$. Then life is reduced by a factor antilog$(-0.134) = 0.73$, because DFRTTU is not controlled and not at the lower end of its range. Better manufacture or materials may increase life 27% or so. This is so only if there is a cause and effect relationship between DFRTTU and life. Whether 27% improvement warrants the effort of controlling DFRTTU depends on engineering judgment. Also, engineering would have to know or learn how to control DFRTTU.
- Line 4 shows the estimate of the standard deviation σ about the equation. This corresponds to the estimate s' in previous sections. Its number of degrees of freedom is 103, the sample size (106 here) minus the number of coefficients (3 here) in the fitted equation.
- Line 5 and following lines show the coefficient estimates and corresponding two-sided 95% confidence limits. The confidence limits for the DFRTTU coefficient are -0.450 and -0.0656. These limits do not enclose 0. Thus the coefficient estimate -0.258 is statistically significant (5% level). That is, there is convincing evidence that DFRTTU is related to life. Of course, this does not necessarily mean that there is a cause and effect relationship. Cause and effect should be supported by physical theory, a designed experiment, or experience. DFRTTU may depend on more basic manufacturing or material variables which also affect life.

The other output is less useful, but it may be worthwhile in some applications. Computer packages provide it mostly as a matter of tradition. Explaining such output usually causes more confusion than enlightenment.

Specified coefficient. Sometimes it is useful to specify the value of a coefficient in a relationship – rather than estimate it from the data. Then the other coefficients are fitted to the data using the specified coefficient value. Section 4.4 of Chapter 5 describes why and how to do this.

Test fit of the relationship. The following hypothesis test assesses how well the assumed relationship fits the data. It generalizes the test of linearity of Section 3.1. Suppose that the relationship contains P estimated coefficients, including the intercept. Also, suppose that there are n specimens among J distinct test conditions. Here J must be less than n; that is, there must be replicates at some test conditions. Also, P must be less that J;

otherwise, the P coefficients cannot all be estimated. Fit the relationship to the data (by computer) to get the estimate s' of the standard deviation based on lack of fit, the multivariate extension of (2.2.12). Also, calculate the estimate s based on replication, the multivariate extension of (2.2.11). Roughly speaking, if s' is much greater than s, then there is lack of fit. Section 3.1 lists possible causes of lack of fit.

F test of fit. The hypothesis test employs the F *statistic for lack of fit*

$$F = [(n-P)s'^2 - (n-J)s^2]/[(J-P)s^2]. \tag{6.2}$$

The numerator may be a small difference between two large numbers. So carry extra significant figures in the calculation. The F *test of fit* is

- If $F \leq F(1-\alpha; J-P, n-J)$, there is no evidence of lack of fit at the $100\alpha\%$ significance level.
- If $F > F(1-\alpha; J-P, n-J)$, there is statistically significant lack of fit at the $100\alpha\%$ level.

Here $F(1-\alpha; J-P, n-J)$ is the $1-\alpha$ point (upper α point) of the F distribution with $J-P$ degrees of freedom in the numerator and $n-J$ in the denominator. This test is exact for a (log) normal life distribution. Also, it is a useful approximation for other distributions. If there is lack of fit, examine data and residual plots to determine why (Section 3.1). A more general relationship with other functions of the same or additional variables may yield a better fit. A relationship may show no lack of fit and yet not be satisfactory. This happens if other important variables are not in the relationship. Also, a relationship may show lack of fit and be adequate for practical purposes.

Compare relationships. The following hypothesis test assesses whether one linear relationship fits the data better than another. This test applies *only if* one relationship *contains* the other. Suppose that the general relationship has Q estimated coefficients (including the intercept), and the simpler "contained" one has P. For example, a quadratic relationship in two variables has $Q=6$ coefficients. It contains a simpler linear relation in the two variables, which has $P=3$ coefficients. In particular, if the quadratic and interaction coefficients equal zero, the linear equation results. The contained relationship can result from setting certain coefficients in the general one equal to each other (Problem 4.10) or to zero (example below) or to some other specified constant. (In the most general theory, the contained relationship can result from setting $Q-P$ specified distinct linear functions of the Q coefficients equal to constants.) Also, suppose that there are n specimens among J distinct test conditions, $J < n$.

Incremental F test. For the general relationship, calculate s'_Q, the estimate of the standard deviation based on lack of fit. Also, for the contained relationship, calculate s'_P. Roughly speaking, if s'_Q is enough less than s'_P, then the general relationship fits significantly better. The exact hypothesis test employs the *incremental F statistic*

$$F = [(n-P)s_P'^2 - (n-Q)s_Q'^2]/[(Q-P)s_Q'^2]. \qquad (6.3)$$

Carry extra significant figures in this calculation. The *incremental F test* to compare the two relationships is

- If $F \leq F(1-\alpha;Q-P,n-Q)$, the general relationship does not fit the data significantly better than the contained one.
- If $F > F(1-\alpha;Q-P,n-Q)$, the general relationship fits the data statistically significant better at the $100\alpha\%$ level.

Here $F(1-\alpha;Q-P,n-Q)$ is the $1-\alpha$ point of the F distribution with $Q-P$ degrees of freedom in the numerator and $n-Q$ in the denominator.

Discussion. s is the estimate of the standard deviation based on replication; suppose its number of degrees of freedom $n-J$ is large (say, over 20). Then use s in place of s_Q' in the denominator of (6.3), and use its degrees of freedom $n-J$ in place of $n-Q$ in the F percentile. This usually yields a more sensitive comparison. If the general relationship is statistically significantly better, assess whether it yields any practical benefits over the simpler one in terms of insights to improve the product. Often s_Q' is comparable to s_P' for practical purposes. Then the general relationship is little better for estimating product life. This incremental F test is also used in Section 6.4 for step-wise fitting; there coefficients are successively added to or deleted from a candidate relationship one at a time. This test is exact for a (log) normal life distribution. Also, it is a useful approximation for other distributions.

Example. Relationship (6.1) has $P=3$ coefficients for the intercept and variables LOGVPM and DFRTTU. In Section 6.4, a more general "final" relationship has $Q=5$ coefficients for the intercept and variables LOGVPM, BINDER, and INSTHK, and VOLTLE. These two relationships cannot be compared as described above for two reasons. First, neither relationship contains the other. Second, the relationships were fitted to different subsets of the data, due to missing values of some variables for some specimens.

Residuals. The following are examples of residual plots to assess the model and data. A residual for a specimen is its (log) life minus the fitted equation evaluated at the variable values of the specimen. Numerical assessments for a model and data appear in the references of Section 3.

- Figure 6.3 is a normal probability plot of the residuals. It is relatively straight; this suggests that the lognormal distribution adequately describes the data over its range. This plot merely suggests that the lognormal distribution may adequately extrapolate into the lower tail. The residuals can also be plotted on extreme value paper to assess a Weibull fit. However, least squares fitting makes the residuals "more normal."
- Figure 6.4 is a crossplot of the residuals against LOGVPM, a variable in the relationship. It is good practice to plot the residuals against all variables in and out of the relationship. The plot has two noteworthy features. There are five isolated points in the upper left corner that merit investiga-

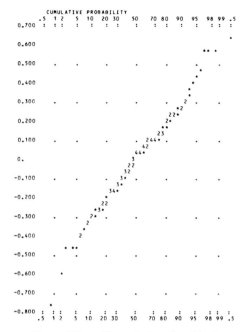

Figure 6.3. Normal plot of residuals.

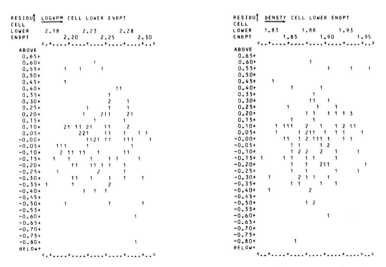

Figure 6.4. Crossplot residuals vs. LOGVPM.

Figure 6.5. Crossplot residuals vs. DENSTY.

tion. There is a slight curvature; the peak is near the middle. Check this by fitting the relationship (6.1) plus a quadratic term in LOGVPM.

- Figure 6.5 is a crossplot of the residuals against DENSTY, a variable not in the relationship. The plot is horizontal; this suggests that DENSTY is not related to life (LOGHRS). Here and in other plots, the lowest residual appears a bit too low. So it may be suspect and left out of some analyses. It is important to determine why it is low.

- Figure 6.6 is a crossplot of the residuals against INSTHK. The trend suggests that life decreases with increasing insulation thickness with constant voltage stress. This is a well-known insulation phenomenon. This suggests that the relationship would benefit from an INSTHK term. Such a term improves reliability prediction in the design process. The two points in the lower left corner seem outliers, which merit investigation.

- Figure 6.7 is a crossplot of the residuals against BINDER. The trend suggests that insulation life decreases as binder content increases. This is a well-known phenomenon. This suggests including a BINDER term in the relationship. Also, this suggests that putting insulation with low binder content into high voltage applications will improve reliability.

- Figure 6.8 is a crossplot of the residuals against VOLTLE. There is a slight suggestion of a trend. However, cover the point in the lower right corner, and there is no trend. Conclusions should generally not be based on one point. This shows the value of covering influential points with a finger while viewing a plot.

- The vertical scatter of the points in Figures 6.4 through 6.8 is constant versus each variable. This supports the assumption that σ is constant with respect to those variables over their observed ranges. Bartlett's test and the maximum F ratio cannot be used here to test the assumption of constant σ. These tests require replicate specimens with the exact same values of all variables.

To help detect a relationship in such crossplots, one can calculate and plot the smoothed locally weighed regression for a scatterplot, as described by Chambers and others (1983, Sec. 4.6).

Conclusions. The most important conclusion from the plots is that adding other variables to equation (6.1) likely improves the fit and understanding. In particular, BINDER and INSTHK appear related to life. Otherwise, the model looks reasonable. Also, it would be useful to investigate outliers for assignable causes that may lead to better product. This example points out the general need for many analyses of a data set, especially data plots.

6.4. Stepwise Fitting of a General Model

Purpose. This subsection presents stepwise fitting of a general model. In some tests, there is data on a large number of engineering variables, and it is useful to know which of these "candidate" variables are related to product life. Stepwise fitting of a general relationship provides an answer. Stepwise

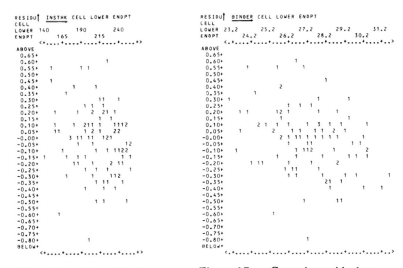

Figure 6.6.   Crossplot residuals vs. INSTHK.

Figure 6.7.   Crossplot residuals vs. BINDER.

fitting selects statistically significant variables and fits the relationship with those variables. The "selected" variables are merely related to life. Whether there is a cause and effect relationship must be decided from engineering considerations. A designed experiment is best for identifying cause and effect.

Example data. The insulation data above are used to illustrate stepwise fitting. Data were collected on a number of design, manufacturing, and test variables (Table 6.1). The relationship between them and life may yield in-

Figure 6.8.   Crossplot residuals vs. VOLTLE.

sights to improving the product. Nine such variables are used in the following analyses. Eighty-three specimens have the values of all nine variables. They are analyzed here, ignoring the other $106 - 83 = 23$ specimens with some values missing. Methods for analyzing data that are "incomplete" or "missing" are presented by Little and Rubin (1987). References to such methods appear in the yearly Current Index of the American Statistical Association (1987). There are no stepwise regression programs for missing data.

General model. Suppose the data contain P variables $x_1, x_2, \cdots, x_P$. These variables may be functions of more basic variables; examples include squares, crossproducts, and logarithms. The general linear relationship for mean log life is

$$\mu = \gamma_0 + \gamma_1 x_1 + \cdots + \gamma_P x_P. \tag{6.4}$$

The coefficients are estimated from data. The standard deviation σ of log life is assumed to be constant. It is assumed that the form (or shape) of the log life distribution is constant. In stepwise fitting, certain coefficients are set equal to zero, because the corresponding variables do not improve the fit of (6.4) to the data. The resulting equation then contains only the statistically significant variables. For stepwise fitting of (6.4), it is recommended that there be at least $5P$ data cases (specimens).

Picture. The following mental picture of the example data and the fitted equation aids understanding. The data for each specimen can be plotted as as a point in a 10-dimensional space. The 10 orthogonal axes correspond to life (LOGHRS) on the "vertical" axis and the 9 predictor variables. The points form a cloud in the space, where (6.4) is a hyperplane. Fitting of (6.4) amounts to finding the "best" plane through the cloud. Suppose a coefficient of (6.4) is zero (say, γ_1). Then (6.4) projected onto the plane of LOGHRS and x_1 is a horizontal line; that is, life is the same for any value of x_1. Finding a "best" plane involves assigning coefficient values of zero to certain variables, because the slope of the data cloud with respect to each such variable is close to zero. "Close to zero" means relative to the scatter in the data. The final equation of selected variables contains only the statistically significant variables; that is, inclusion or deletion of any such variable significantly affects the fit.

Stepwise logic. Various schemes are used to select variables in the final equation. Draper and Smith (1981) review such schemes, including Mallow's C_p. In their Section 6.4, they present in detail the stepwise logic of most programs. It is briefly stated here. The program can start with all or no variables in the equation; the user makes that choice. "All" is recommended when possible. At a general step, there is a number of variables in the current equation. The program adds or deletes a variable from the current equation as follows.

- Add step. For a variable not in the current equation, the program fits a new equation adding that variable to the current variables. After fitting such an equation for each such variable, the program identifies the variable whose addition yields the smallest sample standard deviation about its fitted equation. The incremental F test (Section 6.3) determines if that standard deviation is statistically significantly smaller than the one for the current equation. If so, that variable is added to obtain the new current equation. Then the program tries to delete a variable from the new current equation as follows.

- Delete step. For each variable in the current equation, the program fits a new equation with the current variables except that one. After doing this for each current variable, the program identifies the one whose deletion yields the largest sample standard deviation about the resulting equation. The incremental F test is used to determine if that standard deviation is statistically significantly bigger than the one for the current equation. If it is not, that variable is deleted from the current equation. Then the program tries to add a variable as described above.

The program continues to add and delete until it fails both to add or delete in adjoining steps.

Final equation. Starting with all or no variables in the equation can yield different final equations. Changing the significance levels for adding and deleting variables can yield different final equations. Thus the final equation is not unique, and one must judge which is best. Some programs can "force" user-specified variables to be in the equation at every step. Certain variables are forced, for example, if engineering experience or theory suggests them.

Categorical variables. Most programs enter or delete a categorical variable as follows. The coefficients for the indicator functions for that variable are all added or deleted as a group – not one at a time.

Output. Figure 6.9 shows output on stepwise fitting to the data.
- The variables in the general equation appear in line 1.
- Line 2 and following show summary statistics on the 83 specimens; they are used in the stepwise calculations, particularly the correlation matrix.
- Line 3 and following show step 1 where LOGVPM is added; line 4 shows the fitted equation, and line 5 shows the sample standard deviation about that equation. LOGVPM is the variable most strongly related to (log) life.
- Line 6 and following show step 2 where BINDER is added; line 7 shows the fitted equation, and line 8 shows the sample standard deviation s'.
- Line 9 and following show step 3 where INSTHK is added; line 10 shows the fitted equation, and line 11 shows the sample standard deviation.
- Line 12 and following show step 4 with an unsuccessful attempt to delete a variable. Line 13 and following show VOLTLE is added; line 14 shows the fitted equation, and line 15 shows the sample standard deviation.
- Line 16 and following show step 5 with an unsuccessful attempt to delete a variable. Line 17 and following show an unsuccessful attempt to add a variable. The stepwise fitting stops here.
- Lines 18 and 19 show the final equation and sample standard deviation about it. Following lines are statistics on the fit of the final equation.

Comments. Note how the sample standard deviation (of LOGHRS) decreases as variables are added:

line 1: 0.392,
line 5: 0.262 with LOGVPM,
line 8: 0.242 with BINDER added,
line 11: 0.235 with INSTHK added, and
line 15: 0.227 with VOLTLE added.

Most of the decrease occurs with the first added variable LOGVPM and little thereafter. Thus, for prediction purposes, little statistical accuracy results from additional variables. However, controlling such additional variables

1) REGRESSION STEP ALL(LOGHRS LOGVPM BINDER ASPECT DENSTY VOLTLE EXTRAC

   DFRTTU INSTHK DFCDTU)

2) * SUMMARY STATISTICS

CASES
  83

VARIABLE	AVERAGE	VARIANCE	STD DEV
LOGVPM	2.248549	0.1075133E-02	0.3278923E-01
BINDER	27.28711	3.920270	1.979967
ASPECT	1.913630	0.5203001	0.7213183
DENSTY	1.891712	0.1054115E-02	0.3246714E-01
VOLTLE	0.8452530	0.8201226E-01	0.2363778
EXTRAC	14.91687	13.16459	3.628304
DFRTTU	0.7716145	0.5288268E-01	0.2299623
INSTHK	212.5422	847.9708	29.11994
DFCDTU	0.8154458	0.6688810E-01	0.2586273
LOGHRS	3.337746	0.1540612	0.3925063

CORRELATION MATRIX OF VARIABLES

VARIABLE	LOGVPM	BINDER	ASPECT	DENSTY
LOGVPM	1.0000000			
BINDER	-0.7127754E-01	1.000000		
ASPECT	0.4152845E-01	0.2417110	1.0000000	
DENSTY	-0.4571351E-01	-0.6546124	-0.6878039E-01	1.000000
VOLTLE	0.1855866	-0.2745033	0.1764015	-0.5147923E-01
EXTRAC	0.4996372E-01	-0.3981858	0.1764736	0.2495507
DFRTTU	-0.1055449	0.6099511E-01	0.2523931E-01	-0.1007431
INSTHK	-0.5495333	0.1242588	-0.3872157	-0.7165637E-01
DFCDTU	-0.2060215	0.8406925E-02	0.1867475	-0.1138306
LOGHRS	-0.7479180	-0.2097922	-0.1064552	0.2180162

VARIABLE	VOLTLE	EXTRAC	DFRTTU	INSTHK
VOLTLE	1.0000000			
EXTRAC	0.6151215	1.000000		
DFRTTU	0.2609756E-01	0.4854147E-01	1.0000000	
INSTHK	-0.3833542	-0.3040712	0.7373877E-01	1.000000
DFCDTU	0.4236557	0.3325855	0.5584080	-0.4472697E-01
LOGHRS	-0.1615158	0.2196694E-01	-0.4989165E-01	0.2556711

VARIABLE	DFCDTU	LOGHRS
DFCDTU	1.000000	
LOGHRS	0.7189336E-01	1.000000

Figure 6.9.   Output on stepwise fitting.

3〉 *STEP NO.   1

TEST TO ENTER NEW VARIABLE

OLD ERROR MEAN SQUARE: 0.1540612

VAR	CORR	INCREMENTAL F-RATIO	% PT. OF F-DIST	NEW ERROR MEAN SQUARE	INCREMENTAL ERROR MEAN SQ
LOGVPM	0.7479	102.8	100.	0.6372030E-01	0.8534091E-01
BINDER	0.2098	3.729	94.3	0.1490988	0.4962385E-02
ASPECT	0.1065	0.9285	66.2	0.1541957	-0.1345016E-03
DENSTY	0.2180	4.042	95.2	0.1485501	0.5511109E-02
VOLTLE	0.1616	2.172	85.6	0.1518895	0.2171716E-02
EXTRAC	0.2197E-01	0.3911E-01	15.6	0.1559879	-0.1826730E-02
DFRTTU	0.4989E-01	0.2021	34.6	0.1555750	-0.1513770E-02
INSTHK	0.2557	5.665	98.0	0.1457682	0.8292971E-02
DFCDTU	0.7189E-01	0.4208	48.2	0.1551571	-0.1095869E-02

*ENTERED: LOGVPM 4.0000 IS THE SPECIFIED VALUE FOR ENTERING A VARIABLE

* LEAST SQUARES ESTIMATE OF THE FITTED EQUATION

4〉 MEAN =   23.46925
            + (-8.953018   )*LOGVPM

5〉 STD DEV =   0.2621456

6〉 *STEP NO.   2

TEST TO ENTER NEW VARIABLE
*ENTERED: BINDER 4.0000 IS THE SPECIFIED VALUE FOR ENTERING A VARIABLE

* LEAST SQUARES ESTIMATE OF THE FITTED EQUATION

7〉 MEAN =   25.40708
            + (-9.178652   )*LOGVPM
            + (-0.5242336E-01)*BINDER

8〉 STD DEV =   0.2420586

9〉 *STEP NO.   3

TEST TO REMOVE OLD VARIABLE
*REMOVED   NONE   3.0000 IS THE SPECIFIED VALUE FOR DELETING A VARIABLE

TEST TO ENTER NEW VARIABLE
*ENTERED: INSTHK 4.0000 IS THE SPECIFIED VALUE FOR ENTERING A VARIABLE

* LEAST SQUARES ESTIMATE OF THE FITTED EQUATION

10〉 MEAN =   28.68199
            + (-10.42966   )*LOGVPM
            + (-0.4916296E-01)*BINDER
            + (-0.2592094E-02)*INSTHK

11〉 STD DEV =   0.2350513

Figure 6.9.   Continued

12⟩ *STEP NO.   4

TEST TO REMOVE OLD VARIABLE
*REMOVED   NONE   3.0000 IS THE SPECIFIED VALUE FOR DELETING A VARIABLE

TEST TO ENTER NEW VARIABLE
13⟩ *ENTERED: VOLTLE   4.0000 IS THE SPECIFIED VALUE FOR ENTERING A VARIABLE

* LEAST SQUARES ESTIMATE OF THE FITTED EQUATION
14⟩ MEAN =   29.47465
        + (-10.50796   )*LOGVPM
        + (-0.5726074E-01)*BINDER
        + (-0.3487699E-02)*INSTHK
        + (-0.2428519   )*VOLTLE

15⟩ STD DEV =   0.2277949

16⟩ *STEP NO.   5

TEST TO REMOVE OLD VARIABLE
*REMOVED   NONE   3.0000 IS THE SPECIFIED VALUE FOR DELETING A VARIABLE

17⟩ TEST TO ENTER NEW VARIABLE
*NO VARIABLE IS SIGNIFICANT
        4.0000 IS THE SPECIFIED VALUE FOR ENTERING A VARIABLE

VARIABLES IN THE EQUATION:
LOGVPM BINDER INSTHK VOLTLE

PARTIAL CORRELATIONS

VARIABLE   LOGVPM   BINDER   INSTHK   VOLTLE
LOGHRS -0.7914638   -0.4404170   -0.3376917   -0.2695911

* LEAST SQUARES ESTIMATE OF THE FITTED EQUATION
18⟩ MEAN =   29.47465
        + (-10.50796   )*LOGVPM
        + (-0.5726074E-01)*BINDER
        + (-0.3487699E-02)*INSTHK
        + (-0.2428519   )*VOLTLE

19⟩ STD DEV =   0.2277949

LEAST SQUARES ESTIMATES OF COEFFICIENTS WITH 95% CONFIDENCE LIMITS

VAR   COEFF   ESTIMATE   LOWER LIMIT   UPPER LIMIT   STANDARD ERROR

INTR C00000   29.47465   25.01163   33.93768   2.241773
LOGVPMC00001 -10.50796   -12.33721   -8.678714   0.9188296
BINDERC00002-0.5726074E-01 -0.8357311E-01 -0.3094333E-01   0.1321667E-01
INSTHKC00008-0.3487699E-02 -0.5679076E-02 -0.1296521E-02   0.1100726E-02
VOLTLEC00005-0.2428519   -0.4383946   -0.4730927E-01   0.9822084E-01

Figure 6.9.   Continued

may result in an increase in life that has practical value. The coefficient estimate for a variable changes when other variables are added or deleted. Designed experiments with "orthogonal" controlled variables avoid this.

Interpretation. The practical effect of a variable on life is judged by the product of its coefficient in the final equation (in line 18) times twice its standard deviation (in line 1). For example, the product for BINDER is

.05726074(2×1.979967) = 0.22675, and its antilog is 1.686. The antilogs of these products are the factors: 4.8 for LOGVPM, 1.7 for BINDER, 1.6 for INSTHK, and 1.4 for VOLTLE. Life would be improved by the factor if such a variable could be controlled at the low end of its range. In particular, BINDER and VOLTLE are controlled in manufacture and might be better controlled. Also, batches of conductors with low (high) BINDER and VOLTLE can be assembled into high (low) stress locations in machines. The negative coefficients are consistent with experience with insulations. It also is reassuring that the four selected variables are design and manufacturing variables that are controlled. If an uncontrolled variable like DRFTTU were selected, it could not be used to improve life. However, it might be used to decide which insulation to put into high stress applications. The "final" equation should be regarded as merely a working equation that may be revised with further data collection, understanding, analyses with other relationships, etc. In particular, interaction terms are worth including in such a general relationship.

Assess fit and data. Various plots of residuals of the final equation help one assess the fit and data.
- Figure 6.10A is a normal probability plot of the residuals, and Figure 6.10B is an extreme value plot. Both plots have low points in the lower tail, suggesting possible outliers. The normal plot has less curvature. Thus the lognormal distribution looks slightly better than the Weibull.
- Figure 6.11 is a crossplot of the residuals against TPDATE, year of manufacture. There is a slight suggestion that more recent insulation has lower life. This was not a candidate variable in the stepwise fitting above.
- Figure 6.12 is a crossplot of the residuals against ASPECT, the ratio of height to width of the conductor cross section. There is no trend of the points. This is consistent with the program not selecting this candidate variable to be in the equation.
- Figure 6.13 is a crossplot of the residuals against BINDER, a selected variable. The lack of curvature of the plot suggests that a linear BINDER term is adequate.
- Figure 6.14 is a crossplot of the residuals against MATDAM, a measure of damage to the insulation, measured after manufacture. The outlier is contrary. That specimen has the greatest damage. Yet it has the longest life relative to the equation. This contradiction merits investigation. The other points suggest little dependence of life on MATDAM.
- Crossplots of the residuals against each variable in the data set were made. This is always worth doing and usually informative. Only those plots presented above had any noteworthy features.

Concluding remarks. The preceding analyses are not "the answer," and likely none exists. However, they do provide understanding and insight. More important, they suggest further analyses. It is useful to do many such exploratory analyses to develop understanding of the data. Many analyses will be uninformative and ignored. Computer programs make this easy.

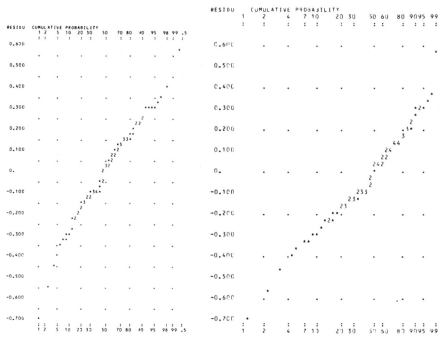

Figure 6.10A. Normal plot of residuals.   **Figure 6.10B.** Extreme value plot of residuals.

Figure 6.11. Residuals vs. TPDATE.   **Figure 6.12.** Residuals vs. ASPECT.

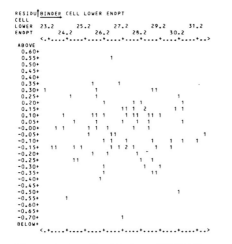

Figure 6.13.   Residuals vs. BINDER.   **Figure 6.14.**   Residuals vs. MATDAM.

PROBLEMS (* denotes difficult or laborious)

4.1. Three insulations. For each insulation in Problem 3.1, do the following based on the Arrhenius-lognormal model. Use a regression program.

(a) Calculate the mean and standard deviation of the log data for each temperature. Calculate the sums of squares.

(b) Calculate estimates and two-sided 95% confidence limits for the coefficients and log standard deviation (use both estimates s and s' and both corresponding limits).

(c) Plot corresponding estimates and confidence limits side by side for the three insulations. Judging by the confidence limits, are there any significant differences among the coefficients or log standard deviations?

(d) Calculate the estimate and 95% confidence limits for the mean log life and median life at the design temperature of 200°C. Plot them as in (c).

(e) Judging by the confidence limits, are there any significant differences among the three median lives?

(f) Repeat (d) and (e) for 225 and 250°C, occasional operating temperatures.

(g) Calculate standardized residuals and make a normal probability plot. Assess (log) normality and the data.

(h) Assess linearity with the F test.

(i) Use Bartlett's test to compare the log standard deviations at the three test temperatures. Also, calculate two-sided 95% confidence limits for them at each temperature and plot them side by side.

(j) Use Bartlett's test to compare the pooled log standard deviations of the three insulations.

(k) Plot the estimates and confidence limits on a separate Arrhenius paper for each insulation.

(l) Write a short management report describing the findings on how the insulations compare. Do not describe the statistical methods. Refer to plots where appropriate.

4.2. Use of s'. Repeat the calculations of the Class-H example using s' (and its degrees of freedom) in place of s. Comment on any important differences in estimates or confidence limits.

4.3. Heater data. Use the data of Problem 3.3 and the Arrhenius-lognormal model. Use a regression program if you wish. Absolute (Rankin) temperature is the Fahrenheit temperature plus 459.7 Fahrenheit degrees.
(a) Calculate the means and standard deviations of the log data at each test temperature. Calculate the sums of squares.
(b) Calculate estimates and two-sided 95% confidence limits for the model coefficients and log standard deviation. Do the same for the activation energy in electron-volts.
(c) Calculate the estimate and two-sided confidence limits for the log mean and the median at the test temperatures and design temperature 1100°F.
(d) Repeat (c) for the 1% point.
(e) Plot the estimates, confidence limits, and data on Arrhenius paper.
(f) Calculate two-sided 95% confidence limits for the log standard deviations for the four test temperatures, and plot them side by side. Compare them with Bartlett's test.
(g) Test for linearity with the F statistic.
(h) Calculate standardized residuals and plot them on normal and extreme value probability paper. Comment on the fit of the lognormal and Weibull distributions. Are there any peculiar data points?
(i) Plot the residuals against temperature. Does the plot yield insights?
(j) Suggest and carry out further analyses if useful.

4.4. Eyring relationship. Fit the Eyring-lognormal model to the Class-H insulation data. In place of specimen life t_i, use $t_i' = t_i \cdot T_i$ as the dependent variable, where T_i is absolute temperature. Do the analyses of Problem 4.3. Comment whether the Eyring relationship fits better than the Arrhenius relationship. Is the difference in fit convincing? Explain.

4.5. Class-H without 260° data. For the Class-H insulation data of Table 2.1 without the 260° data, perform the analyses listed in Problem 4.3. The design temperature is 180°C. Use a regression program if you wish.

4.6. Bearing data. Use the data in Problem 3.5, and omit the outlier. Use the power-Weibull model. Use a regression program if you wish.
(a) For each test load, calculate the mean and standard deviation of the ln data. Calculate the sums of squares.
(b) Calculate estimates and two-sided approximate 95% confidence limits for the model coefficients and the shape parameter. Is the power coefficient estimate consistent with the standard value of 3?
(c) Calculate estimates and two-sided approximate 95% confidence limits for the mean ln life at the test and design loads.

(d) Repeat (c) for the 10th percentile.
(e) Assess linearity with the approximate F test.
(f) Calculate standardized residuals and plot them on extreme value and normal probability papers. Assess the distribution fit and the data.
(g) Plot the fitted 42.8 and 10th percentile lines and corresponding confidence limits on log-log paper. Also, plot the data.

4.7. Weibull fit to Class-H data. Repeat the calculations of the Class-H example, using the Weibull distribution in place of the lognormal distribution. Also, repeat the analyses of standardized residuals.

4.8. Exponential fit. For the insulating fluid data, redo all calculations in the example, assuming an exponential distribution.

4.9. Permalloy corrosion data. Analyze the Permalloy corrosion data of Problem 3.13 as follows. Assume a lognormal distribution for weight change and an exponential relationship. Use a regression program if you wish.
(a) Fit the relationship to the data by least squares.
(b) Calculate 95% confidence limits for all model parameters.
(c) Use Bartlett's test for constant standard deviation. State conclusions.
(d) Use the F test for fit of the assumed relationship. State conclusions.
(e) Calculate residuals and plot them on normal and extreme value paper. Assess which distribution fits better and whether the fit is adequate.
(f) Estimate the population percentage above the specification for 10% and 20% humidity.
(g) Calculate 95% confidence limits for the two percentages in (f).
(h) Comment on the additional insights (or lack thereof) provided by these numerical analyses over those provided by the graphical analyses of Problem 3.13. Is there any question whether the Permalloy meets the corrosion specification?

4.10. Oil. An accelerated test employed a pair of parallel disk electrodes immersed in an insulating oil. Voltage V across the pair was increased linearly with time t at a specified rate R, and the voltage at oil breakdown was recorded. Since $V = Rt$, time to breakdown equivalently could have been recorded. Below are 60 such breakdown voltages measured at each combination of three rates of rise, R, and two electrode areas, A. The test purpose was to verify the engineering model for such data. Under the model, breakdown voltage (and time to breakdown) has a Weibull distribution with the same shape parameter value at each of the six test conditions. Two proposed relationships for mean ln breakdown voltage are

$$\mu(R,A) = \gamma_0 + \gamma_1 \ln(R) + \gamma_2 \ln(A), \tag{1}$$

$$\mu(R,A) = \gamma_0 + \gamma_1 \ln(R/A). \tag{2}$$

(a) Plot the three samples from the electrodes with $A = 1$ in.^{2} on a Weibull paper, and plot the other three samples with $A = 9$ in.^{2} on another. Comment on the fit of the Weibull distribution, whether the shape parameter is constant, and any peculiar data.

1 Sq. Inch Electrodes 10 volts/sec Rate of Rise	9 Sq. Inch Electrodes 10 volts/sec Rate of Rise
41 43 42 43 44 40 38 47 43 45	33 37 38 38 38 37 27 42 39 38
38 44 49 42 42 51 39 34 41 41	42 32 42 40 32 38 36 42 20 37
35 44 46 39 41 40 52 40 35 40	43 40 38 43 39 41 35 41 40 32
39 46 47 44 41 46 46 42 45 42	38 40 37 29 31 41 38 36 35 40
44 41 44 38 36 44 50 47 49 46	37 41 36 39 43 42 43 43 41 44
34 47 49 43 43 48 34 38 47 35	37 43 38 40 40 38 33 40 35 41

1 Sq. Inch Electrodes 100 volts/sec Rate of Rise	9 Sq. Inch Electrodes 100 volts/sec Rate of Rise
46 50 39 36 47 55 49 58 50 48	43 42 45 48 38 44 37 44 43 42
53 54 55 37 53 52 53 50 52 50	43 49 44 45 50 44 44 45 41 48
45 48 53 50 43 50 42 45 47 34	45 48 43 49 50 45 45 46 47 42
46 42 46 46 52 47 53 52 45 47	47 48 47 48 39 49 44 47 34 41
43 45 54 51 46 55 44 49 49 53	45 48 44 47 45 50 40 47 47 43
53 54 53 53 51 48 49 52 45 49	49 45 45 45 47 39 44 37 47 48

1 Sq. Inch Electrodes 1000 volts/sec Rate of Rise	9 Sq. Inch Electrodes 1000 volts/sec Rate of Rise
55 57 59 57 55 60 53 51 57 54	50 53 50 49 53 51 47 44 53 42
57 64 53 63 51 62 62 56 62 57	49 46 50 38 48 43 52 53 52 48
41 41 51 58 59 60 58 55 59 63	45 53 52 50 55 50 43 52 50 54
63 53 63 61 59 53 60 58 62 56	51 40 52 53 47 45 53 47 54 50
69 65 51 56 55 57 54 63 65 65	32 48 53 52 45 48 48 51 53 48
56 54 65 60 60 64 60 54 57 61	54 51 50 54 35 56 51 48 48 46

(b) Graphically estimate the six 42.8% points from (a) and plot them on suitable relationship paper. Fit two lines to the estimates, one for $A = 1$ and the other for $A = 9$. Are the two plots straight enough? Explain. Graphically estimate the coefficients in (1). (2) is a special case of (1) where $\gamma_1 = -\gamma_2$. Are the coefficient estimates for (1) consistent with this? Explain.

(c) Devise a suitable way to plot all observations on (b). Plot the data.

(d) By computer, calculate the six sample ln means and ln standard deviations. Suitably plot them on the Weibull and relationship plots.

(e) Use a least squares program to fit (1) to the ln data. Plot the estimates and confidence limits for γ_1 and $-\gamma_2$ side by side on suitable paper. Comment on whether the results are consistent with (2).

(f) Estimate the Weibull shape parameter.

(g) Use the F test for lack of fit of (1). State and explain your conclusions.

(h) Make suitable probability and crossplots of the (ln) residuals from (e). The observations are in the time order observed (going across rows at a test condition). Comment on what the plots reveal.

(i) Use a least squares program to fit (2) to the ln data.

(j) Use the incremental F test to compare the fit of (1) and (2). State and explain your conclusions.

(k) Do (h) to residuals from (i).

(l) Suggest further analyses.

(m) Carry out (k).

(n) Write a short report for the dielectrics engineers, explaining your findings. Include plots and output as appropriate.

4.11. Au-Al Bonds. Do complete graphical and least squares analyses of each data set in Problem 8.3.

5

Censored Data and Maximum Likelihood Methods

This important chapter presents maximum likelihood (ML) methods, which are basic for analyzing censored (and complete) data. These widely used methods are versatile; they apply to most models, types of data, and types of stress loading. This chapter pertains only to constant-stress tests and data with a single failure mode. Chapter 7 presents models and data analyses for a mix of failure modes. ML methods provide estimates and confidence limits for model parameters, a product life distribution at a design stress, and other quantities of interest. ML methods also provide checks on the validity of the model and data. In practice, it is most informative to use both ML methods and the data plots of Chapter 3. Although mathematically complex, ML methods are easy to apply in practice using special computer packages now widely available. Nelson (1990) explains the basics of ML methods for applied analysis of accelerated test data.

Overview. Section 1 briefly surveys properties of ML methods and ML computer packages. Section 2 presents ML fitting by computer of a simple linear model to right censored data – the most common application; in particular, Section 2.2 is most important. Section 3 describes methods for assessing the model and data; these include plots of censored residuals and likelihood ratio tests of model assumptions. Section 4 presents a variety of applications with various types of data and models. Section 5 presents theory underlying ML methods.

Maximum likelihood methods apply to more complex models and data in subsequent chapters. For example, Chapter 7 employs ML methods for data with a mix of failure modes. Chapter 9 presents ML comparisons (hypothesis tests). Also, Chapter 10 employs ML methods for step-stress data. Chapter 11 employs ML methods for accelerated degradation data.

Background. This chapter employs various models of Chapter 2. It particularly employs the simple linear model (for example, the Arrhenius-lognormal and power-Weibull models). This chapter has the same goals of, concepts for, and approaches to data analysis as does Chapter 4; the only differences are that this chapter applies to censored data and uses ML fitting. Thus, Chapter 4, which deals with the simpler and more familiar topics of complete data and least-squares fitting, is useful background. Also, this chapter employs probability and hazard plotting of Chapter 3. To follow Section 5 on ML theory, one needs to know partial differentiation, basic matrix algebra, and the multivariate normal distribution. Section 1 may be easier to read after Section 2, since Section 2 provides concrete examples of the concepts in Section 1.

1. INTRODUCTION TO MAXIMUM LIKELIHOOD

Overview. This section surveys maximum likelihood (ML) methods – the properties of ML estimators and confidence intervals. Topics include censoring, properties of ML methods, the method, asymptotic theory, invariance property, confidence intervals, computer packages (most important), other methods, and nonparametric methods. Background provided by this section is helpful for following sections on applications. Theoretical assumptions and justification for ML methods appear in Wilks (1962), Rao (1973), Hoadley (1971), Rao (1987), and Nelson (1982).

Censoring. ML methods apply to multiply **time** censored data (Type I); such data are common in practice and are treated in detail here. The methods also apply to multiply **failure** censored data (Type II); such data are common in the theoretical literature, as they are mathematically more tractable. Multiply censored data include as special cases singly censored data (Types I and II) and complete data. ML methods also apply to right and left censored data, quantal-response data, interval data, and any combination of such types of censored and failure data.

Value of censoring. In accelerated testing, tests are stopped or data are analyzed before all specimens fail. The estimates from the censored data are less accurate than those from complete data, *if* the model and data are valid. However, this is more than offset by the reduced test time and expense. Optimum (most accurate) censored test plans appear in Chapter 6. Engineers now recognize the value of stopping a test before all specimens fail; examples of such tests include Crawford (1970) and Brancato and others (1977).

Artificial censoring. In some applications, there may be benefits from artificially treating later failures as if they were censored at some earlier time. This is so if only the lower tail of the life distribution is of interest and the assumed life distribution fits adequately in the lower tail but does not at the same time fit the upper tail. Figure 1.1 of Chapter 4 depicts this situation. Hahn, Morgan, and Nelson (1985) discuss the choice of such censoring times and present a metal fatigue application.

Basic assumption. Like other methods (e.g., those in Chapter 3) for analysis of multiply right (or left) censored data, ML methods depend on a basic assumption. It is assumed that units censored at any specific time come from the same life distribution as the units that run beyond that time. This assumption does not hold, for example, if units are removed from service unfailed when they look like they are about to fail. Lagakos (1979) discusses in detail this assumption, which he calls **random** or **noninformative censoring,** and alternative assumptions about the censoring.

Properties of ML methods. ML methods are very important for analysis of accelerated test data and other data because they are very versatile. That is, they apply to most theoretical models and kinds of censored data. Also, there are sophisticated computer programs that do the difficult ML calculations. Moreover, most ML estimators have good statistical properties, as shown in the references in the **Overview** paragraph above. For example, under certain conditions (usually met in practice) on the model and data, ML estimators are best asymptotically normal (BAN). That is, for a "large" number of failures in the sample, the cumulative distribution function of the sampling distribution of a ML estimator is close to a normal cumulative distribution whose mean equals the quantity being estimated. Moreover, the standard deviation (called the *standard error* of the estimator) of that normal distribution is no greater than that of any other asymptotically normal estimator. A ML estimator also usually has good properties for samples with few failures. Also, for testing hypotheses, likelihood ratio tests based on ML theory (Chapter 9) have asymptotically optimum properties; for example, they are locally most powerful. In Section 3, such tests are used to assess the assumptions of a model.

The method. In principle, the ML method is simple. One first writes the sample likelihood (or its logarithm, the log likelihood) as shown in Section 5. It is a function of the assumed model (distribution and relationships), the model parameters (or coefficients), and the data (including the censoring or other form of the data). The ML estimates of the parameters are the parameter values that maximize the sample likelihood (or, equivalently, the log likelihood). The exact sampling distributions of many ML estimators, confidence limits, and test statistics are not known. However, the asymptotic (large-sample) theory gives approximate normal distributions for them. These distributions provide approximate confidence limits and hypothesis tests. The theory (Section 5) is mathematically and conceptually advanced. However, one need not understand the theory to use computer output for estimates, confidence intervals, and hypothesis test statistics.

Asymptotic theory. Section 5 presents the asymptotic theory for ML estimators, confidence limits, and hypothesis tests. For small samples, such intervals tend to be narrower than exact ones. Exact intervals from small samples are referenced; they have been developed for few distributions and only singly Type II (failure) censoring. For the asymptotic theory to be a good approximation, the number of **failures** in the sample should be **large.** How large depends on the distribution, what is being estimated, the confidence level of limits, etc. A rough rule of thumb is at least 20 failures, but 10 may suffice. Such asymptotic theory is also called **large-sample** theory. This terminology is misleading for multiply and singly censored data, since the number of *failures* needs to be large. The theory is crude for a large sample (many specimens) with few failures. In practice, the asymptotic theory is applied to small samples, since crude theory is better than no theory. Confidence limits are then much too short, but usually are wide enough to be sobering. Shenton and Bowman (1977) give theory for higher-order terms for greater accuracy of the asymptotic theory for ML sampling distributions.

Invariance property. The ML method provides estimates of the parameters of a model. In practice, one also wants estimates of functions of model parameters. Examples of such functions are (1) the reliability at some age, (2) a percentile of the life distribution at a design stress level, and (3) the design stress corresponding to a specified life (percentile). Due to the invariance of ML estimators, the estimate of such a function is simply the function evaluated at the ML estimates of the model parameters. The asymptotic theory (Sections 5.6 and 5.7) gives the variance of the ML estimator and approximate confidence limits for the true function value.

Confidence intervals. The asymptotic theory provides approximate confidence limits for a true population value. They are based on a simple *normal* approximation to the sampling distribution of the corresponding estimator. A better approximation involves likelihood ratio (LR) limits (Section 5.8). Exact confidence limits for certain models and singly censored data have been developed. For example, Lawless (1982) and McCoun and others (1987) present exact theory for confidence limits for the power-exponential model and singly censored data. Such limits require a special computer program. McCool (1980) uses Monte Carlo simulation to tabulate exact confidence limits for the power-Weibull model and singly failure censored data. These intervals can be extended to lognormal and other distributions. Bootstrapping methods for better approximate confidence limits for regression models may soon be extended to censored data. Recent work includes Freedman (1981), Shorack (1982), and Robinson (1983).

LR limits. The maximized log likelihood as a function of a parameter can also be used to obtain better approximate *likelihood ratio* (LR) confidence limits for the parameter, as described by Lawless (1982), Doganaksoy (1989a,b), and Vander Wiel and Meeker (1988). Advantages of these limits include: (1) each one-sided confidence level is usually closer to the specified confidence, even for samples with few failures, and (2) the limits are never outside the natural range of the parameter. For example, confidence limits for a fraction failing should be in the range 0 to 1. Such LR limits (Section 5.8) are more laborious to calculate, and few maximum likelihood programs readily do such calculations. Such confidence limits will, no doubt, be a feature of such programs in the future.

Computer packages. The following information and Table 1.1 were provided by the package developers.

- LIMDEP is described in Greene's (2000) manual. 8.0 runs on Windows PCs. Contact: Econometric Software, Inc., sales@limdep.com, www.limdep.com.
- ReliaSoft's RELIABILITY OFFICE SUITE includes WEIBULL++ for life data analysis, ALTA for accelerated life data analysis, repairable systems analysis, and degradation analysis. They can be purchased as a suite or individually. All run under Microsoft Windows (95, 98, NT, 2000 & XP). Download free evaluation copies from http://www.reliasoft.com/products.htm. Contact: ReliaSoft@ReliaSoft.com, www.ReliaSoft.com, (888)886-0410.
- The SAS RELIABILITY Procedure provides statistical modeling and analyses of life data, accelerated life test data, recurrence data, and regression models. It fits standard life distributions to data with interval, right, and left censoring. It makes graphs, including probability and percentile plots for survival data, and mean cumulative function plots for recurrence data. The SAS 9.1 System runs on MicroSoft Windows, workstations (UNIX and OpenVMS), OS/2, and mainframes. Contact: software@sas.com. For general information and documentation, visit www.sas.com/rnd/app/.
- JMP 5.1 from SAS Inst. is a general statistical package. Reliability features include fitting of life distributions, competing risk models, accelerated test models, the Cox model, user programmed models, and recurrence data analysis. Contact: jmpsales@jmp.com, www.jmp.com , (800)594-6567.
- SPLIDA is an add-on to S-PLUS, a general statistical package of Insightful, www.insightful.com, (800)569-0123. SPLIDA fits life distributions and regression relationships to censored and truncated data. It analyzes recurrence data and degradation data from repeated and destructive measurements. Much SPLIDA output appears in Meeker and Escobar (1998), *Statistical Methods for Reliability Data,* Wiley. SPLIDA runs on Windows 95/98 (and after) and on Windows NT4.0 (and after) using version 2000 (and after) of S-PLUS. SPLIDA functions, their interface, and documentation can be downloaded from www.public.iastate.edu/~splida/. Contact: wqmeeker@iastate.edu.
- SURVIVAL is described in the 200-page manual of Steinberg and Colla (1988). A module of SYSTAT, it runs on PCs and the Macintosh. Contact: info-usa@systat.com, www.systat.com.
- IMSL Numerical Libraries for Fortran and C and the JMSL Numerical Library (for Java Applications) are comprehensive math and statistical code libraries provided by Visual Numerics. Detailed contact information, product system compatibility, product features, and more can be found at www.vni.com.
- WinSMITH^{TM} comes bundled in SuperSMITH^{TM} for probability plotting, growth modeling, and accelerated test analysis. User guides and DEMO software can be downloaded at www.weibullnews.com. WinSMITH 4.0WH runs on Windows-based systems (3.1, 95, 98, 2000, NT, XT). Contact: Wes Fulton, (310)548-6358, wes33@pacbell.net; or Dr. Bob Abernethy, (561)842-4082, Weibull@worldnet.att.net.

Table 1.1. Features of Computer Packages for ML Fitting

Updated June 2004	LIMDEP 8.0 Greene	ReliaSoft ALTA	SAS 9.1 RELIABILITY	JMP 5.1	S-PLUS SPLIDA Meeker	SYSTAT 11 SURVIVAL	IMSL/ JMSL	Win-SMITH
DATA								
Observed & Rt. Censored	Yes	Yes	Yes	Yes	Yes	Yes	Yes	Yes
Left Censored	Yes	Yes	Yes	Yes	Yes	Yes	Yes	Yes
Interval	Yes	Yes	Yes	Yes	Yes	Yes	Yes	Yes
Transformations	Yes	Yes	In SAS	Yes	Yes	Yes	Yes	Yes
Subset Selection	Yes	Yes	Yes	Yes	Yes	Yes	Yes	Yes
Simulation	Yes	Yes	In SAS	Yes	Yes	No	Yes	Yes
DISTRIBUTIONS								
Exponential	Yes	Yes	Yes	Yes	Yes	Yes	Yes	Yes
Weibull	Yes	Yes	Yes	Yes	Yes	Yes	Yes	Yes
(Log)Normal	Yes	Yes	Yes	Yes	Yes	Yes	Yes	Yes
(Log)Logistic	Yes	Yes	Yes	Yes	Yes	Yes	Yes	No
Gamma	Yes	Yes	No	Yes	Yes	Yes	Yes	No
Extreme Value	No	No	Yes	Yes	Yes	Yes	Yes	Yes
Generalized Gamma	Yes	Yes	Yes	Yes	Yes	No	Yes	No
Other	Gompertz	Various	No	Yes	Yes	No	Yes	No
User Programmed Dist's	Yes	No	In SAS	Yes	w Effort	Yes	Yes	No
RELATIONSHIPS								
Linear for Location Param.	Yes	Yes	Yes	Yes	Yes	Yes	Yes	Yes
" without Intercept	Yes	Yes	Yes	Yes	Yes	No	Yes	Yes
Log Linear for Scale Param	Yes	Yes	Yes	No	Yes	No	Yes	Yes
Cox Proportional Hazards	Yes	Yes	In SAS	Yes	In S+	Yes	Yes	No
User Programmed Relation	Yes	No	In SAS	Yes	w Effort	Yes	Yes	No
Model Step/Varying Stress	No	Yes	No	No	Yes	No	No	Yes
ML FITTING								
Stepwise	No	Yes	No	No	In S+	Yes	Yes	Yes
Hold Coefs / Params Fixed	Yes	Yes	Some	No	Yes	Yes	Yes	Yes
Freq. Count Data (Weights)	Yes	Yes	Yes	Yes	Yes	Yes	Yes	Yes

Table 1.1. Continued

Updated June 2004	LIMDEP 8.0 Greene	ReliaSoft ALTA	SAS 9.1 RELIA-BILITY	JMP 5.1	S-PLUS SPLIDA Meeker	SYSTAT 11 SUR-VIVAL	IMSL/JMSL	Win-SMITH
FIT OUTPUT								
Ests & Normal Conf Limits:								
Parameters	Yes	Yes	Yes	Yes	Yes	Yes	Yes	Yes
Percentiles	Yes	Yes	Yes	Yes	Yes	Yes	Yes	Yes
Fraction Failing	Yes	Yes	Yes	Yes	Yes	Yes	Yes	Yes
User Function of Params	Yes	No	No	Yes	w Effort	Yes	No	No
Covariance Matrix of Ests	Yes	Yes	Yes	No	Yes	Yes	Yes	No
Maximum Log Likelihood	Yes	Yes	Yes	Yes	Yes	Yes	Yes	Yes
LR Confidence Limits	No	Yes	Yes	Yes	Some	No	No	Yes
Plot of Fitted Relationship	No	Yes	Yes	Yes	Yes	Yes	Yes	Yes
Plot of Fitted CDF	Yes	Yes	Yes	Yes	Yes	Yes	Yes	Yes
MODEL EVALUATION								
Residuals	No	Yes	Yes	No	Yes	No	No	Yes
Prob. Plots (Right Cens'd)	No	Yes	Yes	Yes	Yes	Yes	Yes	Yes
Exponential	No	Yes	Yes	Yes	Yes	Yes	Yes	Yes
Weibull	No	Yes	Yes	Yes	Yes	Yes	Yes	Yes
Extreme Value	No	No	Yes	Yes	Yes	Yes	Yes	Yes
Lognormal	No	Yes	Yes	Yes	Yes	Yes	Yes	Yes
Normal	No	Yes	Yes	Yes	Yes	Yes	Yes	Yes
Linear % and Data Scales	Yes	Yes	In SAS	Yes	Yes	Yes	Yes	Yes
Other	No	Logistic	Logistic	No	Yes	Logistic	Yes	Weibayes
Peto-Turnbull cdf estimate	No	No	Yes	Yes	Yes	Yes	Yes	No
LR Tests	Yes	Yes	Yes	Yes	Yes	Yes	Yes	Yes
Crossplots	Yes	Yes	In SAS	No	Yes	w Effort	No	Yes

Hitz, Hudec, and Müllner (1985b) and Harrell (1988) also survey such packages. Dallal (1988) warns about common deficiencies of packages. Many major companies unfortunately still lack such packages.

Some general statistical packages require programming by the user to fit accelerated testing models to censored data. These include BMDP, GLIM, and SPSS, which are included in the survey of Wagner and Meeker (1985), who also list a number of special purpose routines for such model fitting. For example, Aitkin and Clayton (1980) jury-rig GLIM to fit a linear-Weibull model to censored data. NAG (1984) provides the GLIM manual.

Numerical accuracy. Most packages print estimates, confidence limits, and other calculated quantities to six or seven digits. Such quantities typically are accurate to three or four digits, which suffices for most applications. Moreover, confidence limits often differ from the corresponding estimate in the second or even first digit. So the computational accuracy is usually satisfactory relative to the statistical uncertainty. Some packages achieve such accuracy of three or four digits through sophisticated algorithms and astute programming. It is best to use such mature programs. Those who write their own programs can expect to encounter numerical problems. Nelson (1982, Chap. 8, Sec. 6) discusses numerical techniques for ML calculations. Kennedy and Gentle (1980), Chambers (1977), Maindonald (1984), and especially Thisted (1988) present other numerical aspects of statistical computation.

Other estimation methods. Maximum likelihood methods are recommended for fitting parametric regression models to censored data. Other methods merit mention. Hahn and Nelson (1974) compare such methods. Schneider (1986) devotes a chapter to regression for the (log)normal distribution. Viertl (1988) surveys in detail the theory of various methods. Of course, most important are the graphical methods of Chapter 3 – essential to any data analysis. The methods below yield estimators that have accuracy (standard errors) comparable to that of ML estimators.

Linear. Estimates based on order statistics of failure (Type II) censored samples are presented by Nelson and Hahn (1972,1973). They provide best (minimum variance) linear unbiased estimators and simple (and less accurate) linear unbiased estimators and their standard errors and approximate confidence limits. Exact confidence limits based on linear estimates have not been developed for regression models. Monte Carlo simulation of such limits could be carried out, as has been done for ML estimates. The calculation of linear estimates is generally much less laborious than the calculation of ML estimates. However, the calculation is still laborious, and no such computer programs exist. Bugaighis (1988) ran simulations where ML and linear estimates have comparable mean squared errors; however, for very few failures, the ML estimate is better. Mann (1972) and Escobar and Meeker (1986) study optimum test plans for such linear estimators.

Iterative least squares. Schmee and Hahn (1979,1981) and Aitkin (1981) present an iterative least-squares fitting method and computer program. Their method replaces a censored observation by an "observed" value equal to its expected failure time conditional on how long the specimen ran without failure, assuming a (log)normal life distribution. Then the method fits the regression model by least squares to the failure times and such conditional expected times. At each iteration the method uses the previously fitted model to get the conditional expected failure times of censored observations. Neither exact nor asymptotic theory for the properties of the estimators has been developed. Monte Carlo simulation by Schmee and Hahn indicate that their estimators perform comparably to ML estimators. Morgan (1982) extends their method to an extreme value (Weibull) life distribution. The method applies to both time (Type I) and failure (Type II) censored data.

Weighted regression. Lawless (1982), Nelson (1970), and Lieblein and Zelen (1956) calculate linear or ML estimates of distribution parameters separately for each stress level. They then fit the relationship to the location parameter estimates using weighted least squares regression. The weight for an estimate is the inverse of its variance. Also, they estimate the common scale parameter with a weighted sum of the estimates of the scale parameter.

Bayesian analysis. Bayesian analysis involves expressing subjective knowledge or degree-of-belief about model parameter values as an a priori distribution for them. This distribution is then mathematically combined with observed data to yield the posterior distribution for the parameter values. The posterior distribution is narrower than the a priori one, thereby reflecting the added information from the data. The posterior yields a Bayesian estimate and probability limits for the true parameter values and functions of them. After analyzing data, some practioners do not agree with the posterior distribution. They revise the a priori distribution until they get a "satisfactory" posterior distribution. Bayesian analysis is somewhere between using assumed values for the parameters (corresponding to no spread in the a priori distribution) and using standard "classical" methods of this book (corresponding to a non-informative a priori distribution). As they often have subjective notions about model parameter values, engineers and others find Bayesian analysis philosophically attractive. The considerable theoretical literature reflects this interest. Statisticians talk about Bayesian analysis more often than the weather. However, it is rarcly applied, due often to the difficulty of specifying an a priori distribution. Efron and others (1986) discuss the philosophical pros and cons of Bayesian analysis. Clarotti and Lindley (1988), Proschan and Singpurwalla (1979), and Viertl (1987,1988) survey Bayesian theory for accelerated testing. Martz and Waller (1982) present Bayesian reliability analysis but do not include accelerated testing and regression models.

Nonparametric fitting. The method used to fit the Cox proportional hazards model to data is a nonparametric form of maximum likelihood fitting. Biomedical books on such fitting include Lee (1980), Cox and Oakes (1984), Kalbfleisch and Prentice (1980), and Lawless (1981) among others. Many statistical packages do such fitting. Other regression methods appear in the nonparametric literature. As nonparametric methods are rarely used for accelerated test data, they are not discussed here. References include Viertl (1988), Shaked, Zimmer, and Ball (1979), and Basu and Ebrahimi (1982).

2. FIT THE SIMPLE MODEL TO RIGHT CENSORED DATA

Purpose. This important section presents ML fitting of the simple model to right censored data – the most common model and type of data. So they are presented in detail. Such data may be complete or singly or multiply censored. The ML fitting yields estimates and approximate confidence limits for parameters, percentiles, reliabilities, and other quantities. This section explains only how to interpret computer output on such fitting. The underlying ML theory and calculations appear in Section 5.

Background. Needed background includes:
• Simple models of Chapter 2.
• An understanding of results of least-squares fitting of such models, as described in Chapter 4.
• Basic understanding of statistical estimates, their standard errors, and confidence limits. Statistics books provide this background.
• Understanding of data plots in Chapter 3.

Overview. Section 2.1 presents example data and the simple model. Section 2.2 explains how to interpret computer output on fitting a simple model to right censored data; this is the most important material, as most analyses use such output. Section 2.3 presents a number of special cases, namely,
• the exponential, Weibull, and lognormal distributions,
• complete and singly censored data,
• assumed ("known") value of the slope coefficient (acceleration factor) or Weibull shape parameter.
Section 3 provides checks on the model and data. Section 4 presents ML fitting of other models to other types of data. Section 5 presents general ML theory and calculations.

2.1. Example Data and Model

Data. Data that illustrate ML fitting appear in Table 4.1 of Chapter 3. The data come from a temperature accelerated life test of Class-B insulation. Ten specimens were tested at each of four test temperatures, and the data at each temperature were singly censored. The main purpose of the test was to estimate the median life of such insulation at the design temperature of

130°C. Periodic inspection of specimens determined when they failed. Below, each failure is treated as if it occurred at the midpoint of its interval. The effect of this is small, since the intervals are enough narrower than the life distribution at a temperature.

Model. The model fitted to the example data is the Arrhenius-lognormal model. It is a special case of the general simple model of Chapter 2. Briefly, the general model consists of:
- a distribution with a constant scale parameter σ and
- a location parameter μ that is a linear function of a single (possibly transformed) stress x, namely,

$$\mu(x) = \gamma_1 + \gamma_2 x. \tag{2.1}$$

In particular, for the exponential distribution, the mean θ is

$$\ln\theta(x) = \gamma_1 + \gamma_2 x. \tag{2.2}$$

For the Weibull distribution, the scale parameter α is

$$\ln\alpha(x) = \gamma_1 + \gamma_2 x. \tag{2.3}$$

For the lognormal distribution, the log mean μ is

$$\mu(x) = \gamma_1 + \gamma_2 x. \tag{2.4}$$

ML programs fit these models to data.

2.2. ML Output

Overview. Figure 2.1 shows STATPAC output on ML fitting of the Arrhenius-lognormal model to the Class-B data. Such output usually includes estimates and confidence limits for model coefficients and parameters and for percentiles and reliabilities at any stress level. These and other key output are discussed below. Most ML programs (Section 1) give such output.

Line 1 marks STATPAC commands that specify the Arrhenius-lognormal model. The stress variable is INTEMP, inverse (absolute) temperature. It equals 1000/(absolute temperature in degrees Kelvin). Degrees Kelvin equals the Centigrade temperature plus 273.16°C.

Log likelihood. Line 2 contains the maximum log likelihood -148.53734. It is used in various tests of fit of the model, described in Section 3. Its calculation appears in Section 5.

Matrices. Line 3 points out the Fisher matrix, and 5 and 6 point out the covariance and correlation matrices of the coefficient and parameter estimates. Their calculation and use are described with the theory in Section 5. Each matrix is symmetric about its main diagonal; so only the elements on and below the main diagonal are printed.

1) DISTRIBUTION LOGNORMAL(HOURS)

RELATION CENTER 1STORDER (INTEMP)

FIT

CASES
17 WITH OBSERVED VALUES.
23 WITH VALUES CENSORED ON THE RIGHT.
---
40 IN ALL

2) MAXIMUM LOG LIKELIHOOD = -148.553736

* FISHER MATRIX

COEFFICIENTS	C 1	C 2	= C 3
C 1	427.9488		
C 2	-10.00322	5.729874	
C 3	-65.34026	-1.238175	41.34756

3) * MAXIMUM LIKELIHOOD ESTIMATES FOR DIST. PARAMETERS
WITH APPROXIMATE 95% CONFIDENCE LIMITS

PARAMETERS	ESTIMATE	LOWER LIMIT	UPPER LIMIT	STANDARD ERROR
C 1	3.471803	3.358559	3.585048	0.5777776E-01
C 2	4.308286	3.453056	5.163516	0.4363418
SPREAD σ	0.2591552	0.1811721	0.3707053	0.4733279E-01

5) * COVARIANCE MATRIX

PARAMETERS	C 1	C 2	SPREAD
C 1	0.3333270E-02		
C 2	0.7013480E-02	0.1903941	
SPREAD σ	0.1421717E-02	0.4350154E-02	0.2240395E-02

6) * CORRELATION MATRIX

PARAMETERS	C 1	C 2	SPREAD
C 1	1.000000		
C 2	0.2781929	1.000000	
SPREAD	0.5193648	0.2106278	1.000000

7) MAXIMUM LIKELIHOOD ESTIMATE OF THE FITTED EQUATION

CENTER
$\mu = 3.471803$
+ (INTEMP - 2.201629) * 4.308286

SPREAD
$\sigma = 0.2591552$

PCTILES (2.4804048)

INTEMP 2.4804048 130°C

* MAXIMUM LIKELIHOOD ESTIMATES FOR DIST. PCTILES
WITH APPROXIMATE 95% CONFIDENCE LIMITS

PCT.	ESTIMATE	LOWER LIMIT	UPPER LIMIT	STANDARD ERROR
0.1	7446.657	3539.210	15665.58	2825.631
0.5	10120.60	5078.786	20167.53	3560.286
1	11744.99	6021.356	22909.26	4003.571
5	17639.50	9401.507	33095.95	5663.273
10	21912.13	11768.12	40800.17	6949.789
20	28495.97	15247.77	53255.00	9091.509
8) 50	47081.61	24089.76	92017.46	16096.53
80	77789.17	36522.75	165681.8	30007.06
90	101162.1	44852.02	228167.6	41980.00
95	125665.6	52885.30	298605.4	55491.54
99	188733.8	71366.06	499123.2	93646.19

PCTILES (ALL)

INTEMP 2.3631723 150°C

* MAXIMUM LIKELIHOOD ESTIMATES FOR DIST. PCTILES
WITH APPROXIMATE 95% CONFIDENCE LIMITS

PCT.	ESTIMATE	LOWER LIMIT	UPPER LIMIT	STANDARD ERROR
0.1	2327.116	1252.640	4323.989	735.5537
0.5	3163.263	1837.178	5446.522	876.9552
1	3670.578	2202.390	6118.843	956.9371
5	5513.345	3539.238	8588.563	1246.860
10	6848.786	4484.059	10460.58	1479.976
20	8906.609	5862.999	13530.22	1900.072
9) 50	14313.54	9262.188	23405.41	3484.114
80	24313.54	13803.60	42325.65	7022.442
90	31883.91	16782.27	59372.12	10218.59
95	39677.64	19627.87	78599.11	13901.58
99	53990.05	26118.72	133231.1	24520.50

Figure 2.1.   ML fit of Arrhenius-lognormal model to Class-B data.

Estimates and confidence limits. Line 4 points out the ML estimates and approximate "normal" confidence limits for the coefficients (γ_1, γ_2) and σ. For example, the ML estimate for σ (SPREAD) is $\hat{\sigma} = 0.259 \cdots$, and the approximate 95% confidence limits are $(0.181 \cdots, 0.370 \cdots)$. Such approximate normal confidence limits tend to be narrower than exact ones. Thus the uncertainty of such estimates is greater than normal confidence limits suggest. The more failures in the data, the closer the approximate limits are to exact ones, roughly speaking. Section 5 describes the calculation of such estimates and confidence limits. Likelihood ratio (LR) limits of Section 5.8 are generally a better approximation. In the output, C1 and C2 denote the coefficients in (2.5).

Standard errors. To the extreme right of line 4 are estimates of the standard errors of the ML estimators of the coefficients and σ. The standard error is the standard deviation of the sampling distribution of the estimator. For samples with many failures, the sampling distribution of most ML estimators is approximated well by a normal distribution. That normal distribution has a mean (asymptotically) equal to the true value of the quantity estimated, and it has a standard deviation equal to the true asymptotic standard error of the ML estimator. These estimates of the standard errors are calculated as described in Section 5. Also, they are used to calculate the approximate normal confidence limits as described in Section 5.

Relationship. Line 7 points out the estimate of the linear relationship for μ as a function of $x =$ INTEMP in an equivalent form:

$$\mu(x) = \gamma_1' + \gamma_2(x - \bar{x}); \qquad (2.5)$$

here $\bar{x}$ is the average transformed stress (INTEMP) of the data. Line 7 shows $\bar{x} = 2.201629$. This form of the relationship is more numerically robust and yields more accurate estimates; also, it improves the speed and sureness of convergence of the ML fitting. Many ML packages do not automatically subtract the mean (or some other value near the center of the x variable) from the data. Users then should do so to assure accurate fitting. Below line 7 is the ML estimate of σ (SPREAD). The estimate for γ_1' is labeled C1 in the output in line 4.

Percentiles. Lines 8 and 9 point out estimates, confidence limits, and standard errors of percentiles at a design and a test temperature. For example, the estimate of median (PCT = 50) life at 130°C (INTEMP = 2.4804048) is 47,081 hours. Corresponding approximate 95% confidence limits are (24,089, 92,017) hours. For reporting purposes, such values should be rounded to two (or at most three) significant figures. Most computer programs do not round appropriately. Thus the estimate is reported as 47,000 hours and the confidence limits as 24,000 and 92,000 hours. This information suggested that the insulation life would be satisfactory at 130°C, the design temperature.

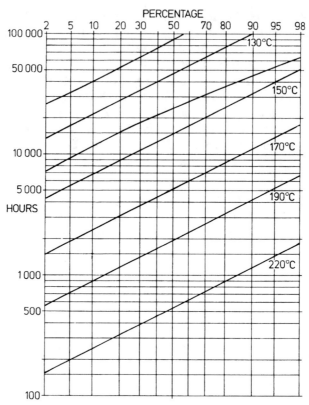

Figure 2.2.   Lognormal plot of fitted distributions and confidence limits.

Depiction. Figure 2.2 contains lognormal probability paper. There the fitted distributions (percentile estimates at a temperature) plot as straight lines. The 95% confidence limits for percentiles (Line 8 in Figure 2.1) plot as curves. Such limits are plotted at 130°C but could be plotted for any temperature. This plot usefully depicts the fitted model and the statistical uncertainty in the fit. Such a plot should also display the data, as in Chapter 3.

Similarly, Figure 2.3 depicts the fitted model on Arrhenius (relationship) paper. The fitted percentile lines are straight, and the 95% confidence limits for a line are curves. The figure shows such curves for the median line. It is more informative if such plots also show the data. Then such a plot summarizes much information well, but may be cluttered, as in Chapter 3.

Reliabilities. Many ML programs provide estimates and confidence limits for a reliability or fraction failed at a user-specified age and stress level. STATPAC does so, but Figure 2.1 does not show them. One can easily calculate such estimates as described below. Also, the probability plot (Figure 2.2) yields a reliability estimate and confidence limits. Enter the plot on the time scale at the desired age; go to the fitted distribution line for the stress level; and then go up to the probability scale to read the reliability esti-

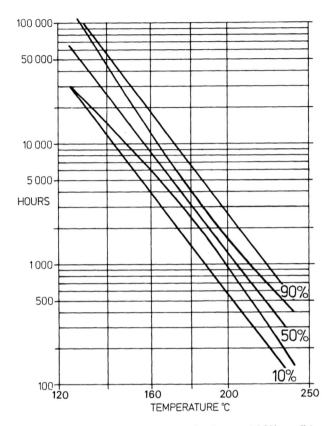

Figure 2.3.   Arrhenius plot of fitted percentile lines and 95% confidence limits.

mate. Repeat this with the confidence limit curves to get the confidence limits for that reliability. This relationship between estimates and confidence limits for percentiles and those for reliabilities applies to any model.

Other estimates. Percentiles and reliabilities are functions of population (true model) coefficients and parameters. Other such functions are of interest in some applications. Examples include 1) the failure rate versus age at a particular stress level and 2) the design stress that yields a specified life (say, a percentile must have a specified value). The corresponding ML estimate is the function evaluated at the ML estimates for the model coefficients and parameters. The corresponding confidence limits appear in Section 5. Some ML programs have such functions as standard features, and some others let the user program such functions. Also, one can calculate such an estimate by hand from the coefficient and parameter estimates. Use many significant figures in the coefficient and parameter estimates and in intermediate calculations. This assures that final estimates and confidence limits are accurate to two or three figures.

2.3. Special Cases

Purpose. This section briefly surveys some special cases of the simple model and right censored data. These include:
- the basic distributions (exponential, Weibull, and lognormal) and others,
- simpler data (singly censored or complete),
- known slope coefficient,
- known acceleration factors, and
- known Weibull shape parameter.

Exponential Distribution

Misuse. The exponential distribution has been widely used (often with crude results) with the simple linear model. For example, it is used in MIL-HDBK-217 for electronic components. Many reliability texts state as gospel that most electronic components "follow" an exponential life distribution. In the author's experience, only 10 to 15% of products are adequately described with an exponential distribution. There are reasons for the misuse of this distribution. First, it can sometimes yield crude but useful results. Second, it has been sanctified by long usage. Third, its simplicity appeals to many; namely, it is characterized by a single number – the mean time to failure (MTTF) or the failure rate, which is usually more suitable. Fourth, data analysis methods for it are simpler than those for other distributions. It is presented here, because it adequately describes some products and because even knowledgeable people continue to use it despite its inaccuracy. They do so to facilitate dealing with less sophisticated clients, management, and associates. Also, many contracts specify the use of government reliability standards based on the exponential distribution. A better alternative, described below, is the Weibull distribution with a specified shape parameter. Indeed, the exponential distribution is a Weibull distribution with a specified shape parameter of 1. Most products are described better with another value.

Literature. The simple model with an exponential distribution appears widely in the literature. References include Evans (1969), Fiegl and Zelen (1965), Glasser (1967), Lawless (1976), Zelen (1969), Mann, Schafer, and Singpurwalla (1974, Chap. 9), and many recent reliability texts. Most authors employ ML fitting to right censored data. Hamada (1988) compares asymptotic variances of ML estimates with true variances from simulations.

Simple censoring. ML fitting of such a model to right censored data is complex and laborious. It must be done by a computer program (Section 1). Such fitting is no less complex for singly censored or complete data.

Confidence intervals. There are no tables for exact confidence limits for this model, even for complete data. Lawless (1976, 1982, Sec. 6.3.2) presents theory and a computer program for exact limits. McCoun and others (1987) extend this theory to generalizations of the exponential distribution such as the Weibull distribution with a known shape parameter.

Two stress levels. The ML estimates simplify if there are just two test stresses x_1 and x_2. Suppose that the number of failures at x_i is r_i, and suppose that the total running time of the n_i specimens is T_i. The total includes all failed and censored specimens in the multiply censored sample at x_i. The ML estimate $\hat{\theta}_i$ of the mean at x_i is

$$\hat{\theta}_i = T_i/r_i. \tag{2.6}$$

The estimate does not exist if $r_i = 0$. The ML fit of the simple relationship (2.2) passes through both points $(x_i, \hat{\theta}_i)$. Thus the coefficient estimates are

$$\hat{\gamma}_2 = \{\ln(\hat{\theta}_2) - \ln(\hat{\theta}_1)\}/(x_2 - x_1), \quad \hat{\gamma}_1 = \ln(\hat{\theta}_1) - \hat{\gamma}_2 x_1. \tag{2.7}$$

The estimate of mean life at stress level x is

$$\hat{\theta}(x) = \exp(\hat{\gamma}_1 + \hat{\gamma}_2 x). \tag{2.8}$$

Approximate confidence limits for such quantities are calculated with methods of Section 5 and of Lawless (1976, 1982, Section 6.3.2). Tables for exact confidence limits have not been developed. Of course, the failure rate estimate is $\hat{\lambda}(x) = 1/\hat{\theta}(x)$.

Weibull Distribution

Use. The simple linear model with a Weibull distribution is widely used. For most products, it is more suitable than the exponential distribution.

Literature. References on ML fitting of the model to right censored data include McCool (1981,1984), Lawless (1982), and Singpurwalla and Al-Khayyal (1977). Such fitting is complex and requires special computer programs (Section 1).

Simple censoring. The fitting does not simplify for singly censored or complete data.

Confidence limits. Using methods of Section 5, computer programs calculate ML estimates and approximate confidence limits for model parameters and other quantities. Lawless (1982) gives exact limits calculated with a special computer program; such limits are exact only for failure censored data, including complete data. McCool (1980,1981) gives tables for exact limits for singly failure censored data, which includes complete data. McCool's tables apply only to equal sample sizes (inefficient for extrapolation), to equal numbers of failures at each test stress level, and to certain spacings of the stress levels.

Two levels. ML fitting does not simplify for data with just two stress levels, even for singly censored or complete data.

Lognormal Distribution

Use. The simple model with the lognormal distribution is widely used. For many products it is not clear if it or the Weibull distribution fits better.

Literature. References on ML fitting of the model to right censored data include Glasser (1965), Hahn and Miller (1968a,b), Schneider (1986), Aitken (1981), and Lawless (1982). Such fitting is complex and requires special computer programs (Section 1).

Simple censoring. The ML fitting is simpler only for complete data. Then ML fitting is equivalent to least-squares fitting (Chapter 4).

Confidence intervals. Using methods of Section 5, computer programs calculate ML estimates and approximate confidence limits for parameters and other quantities. Lawless (1982) presents theory for exact confidence limits, which require a special computer program. Schneider and Weisfeld (1987) present approximate normal limits based on other ML statistics. There are no tables for exact limits for singly censored data. For complete data, the confidence limits of Chapter 4 apply.

Two levels. Only for *complete* data does ML fitting simplify with two stress levels. Then ML estimates are the same as those above for the exponential distribution, Replace θ by the mean log life μ. For each stress level, replace each estimate of θ by the sample mean log life. Confidence limits in Chapter 4 apply.

Other Distributions

Literature. ML fitting of a simple linear relationship with other distributions to right censored data appears in the literature. References on the log gamma distribution include Farewell and Prentice (1977), who offer a computer program, and Lawless (1982). Lawless (1982) presents the gamma and logistic distributions. The nonparametric Cox (proportional hazards) model is treated by many, including Lee (1980), Kalbfleisch and Prentice (1980), Cox and Oakes (1984), and Lawless (1982).

Slope Coefficient Specified

Motivation. In some applications, one may specify (or assume) a value γ_2' for the slope coefficient γ_2. That is, γ_2 is not estimated from the data. This is done for various reasons including:
- The number of specimen failures is so small that the estimate of the slope is less accurate than the assumed value.
- Failures occurred at just one test stress. So the slope cannot be estimated.
The following paragraphs explain how to analyze data using a specified slope. The method applies to right censored and other types of data. Singpurwalla (1971) treats this problem but does not note that it reduces to a simple one with a single sample, as shown below.

Coefficient value. The slope coefficient corresponds, for example, to the power in the inverse power law and to the activation energy in the Arrhenius relationship. For certain products, experience suggests a coefficient value.

For example, the power in Palmgren's equation is sometimes taken to be 3 for steel ball bearings. The activation energy of semiconductor failures is sometimes taken to be 1.0 electron-volts or a value in the range 0.35 to 1.8 eV, depending on the failure cause. Assuming (incorrectly) that life doubles for every 10°C decrease in temperature is equivalent to specifying the slope coefficient. In the Coffin-Manson relationship, the power is often assumed to be 2. It is important to realize that the slope is a characteristic of a failure mode. If a product has more than one failure mode, it is usually appropriate to use a different slope for each mode, as in Chapter 7. The following applies to the data on a failure mode.

Equivalent data. The analysis proceeds as follows. Suppose one wishes to estimate some characteristic of the life distribution at stress level x', which may be a design level. Suppose that the log failure or censoring time of specimen i is y_i. Use the natural log for Weibull and exponential distributions, and use the base 10 log for lognormal distributions. Use either log for unspecified distributions. Also, suppose that test stress level of the specimen is x_i. Calculate each equivalent log life y_i' at x' as

$$y_i' = y_i + (x' - x_i)\gamma_2'. \tag{2.9}$$

This results in a single sample of equivalent log times all at x'. Figure 2.4 motivates this; there each data point is translated to stress x' by moving it along a line with slope γ_2'. One can also work with the original times $t_i = \exp(y_i)$, assuming natural logs. Then the equivalent times are

$$t_i' = t_i \exp[(x' - x_i)\gamma_2']. \tag{2.10}$$

Here exp[ ] is the "acceleration factor."

Analyses. Use standard graphical or numerical analyses for fitting a single distribution to such equivalent data, which are usually multiply censored.

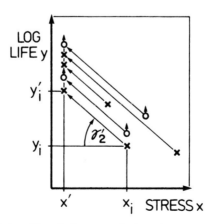

Figure 2.4.   Translation to equivalent times at stress x'.

Such analyses appear in, for example, Nelson (1982), Lawless (1982), Lee (1980), and others. Also, standard computer programs (Section 1) perform such analyses. Such analyses yield estimates and confidence limits for percentiles and other quantities. Also, hazard plot (Chapter 3) such data.

Other slopes. The specified γ_2' is seldom correct. So it is useful to repeat the analyses using other γ_2 values. This "sensitivity analysis" tells one how much other values affect results.

Lubricant example. Table 2.1 shows data from a temperature accelerated test for failure of lubricating oil in 50 compressors. The purpose of the test was to estimate the life distribution of the oil at the design temperature of 220°F. The equivalent data at 220°F were calculated by assuming that oil life doubles with every 18°F (10°C) *drop* in temperature, a naive rule of thumb from freshman chemistry. This widely used inverse power relationship approximates the more correct Arrhenius one. The equivalent time is

$$t_i' = t_i/2^{drop/18}. \tag{2.11}$$

Table 2.1 shows the calculation of these equivalent times. Figure 2.5 is a Weibull plot of the equivalent times and the Weibull distribution fitted by ML. The life distribution was marginal for the 1,110-day design life. Figure 2.5 also shows equivalent data from factors of 1.5 and 2.5 instead of 2. Here the percentage failing during design life changes greatly with the factor. This suggested it was important to determine the factor more accurately.

Check slope. The assumed value of the slope coefficient can be checked as follows. Suppose that the test was run at several stress levels or that roughly equal numbers of specimens are grouped into several stress intervals. Calculate the equivalent times, and treat the equivalent data from each group (stress level) as a separate sample. Compare these samples with graphical or numerical methods to see if they are from the same distribution – a test of homogeneity. Such comparisons appear in the references in the **Analyses** paragraph above. Significant differences between the groups may indicate the specified slope is in error. If the observed group distributions are in the same (or reversed) order as their stress levels, then the slope may be in error.

Table 2.1. Lubricant Data and Equivalent Times

Temp.	No.		Days	Equiv. Days at 220°F	
260°F:	30	unfailed at	88 +	$88 \times 2^{(260\text{-}220)/18}$ =	411 +
310°F:	1	failed at	25	$25 \times 2^{(310\text{-}220)/18}$ =	800
	1	failed at	43	43	1376
	1	failed at	75	75	2400
	1	failed at	87	87	2784
	1	failed at	88	88	2816
	15	unfailed at	88 +	88	2816 +

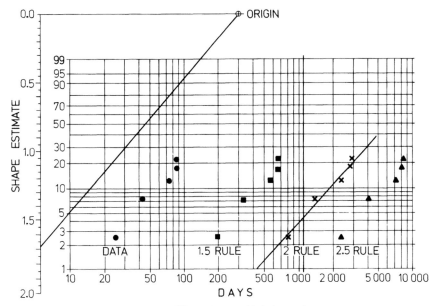

Figure 2.5.   Weibull plot of lubricant data.

If the observed group distributions are in a random order, then the model or data may be in error.

Acceleration Factors

Definition. In some applications, life t' under design use is expressed in terms of life t at an accelerated test condition as

$$t' = Kt. \qquad (2.12)$$

Here K is called the *acceleration factor* and is assumed known. As above, one can calculate equivalent times t' from accelerated times t and estimate the life distribution under design conditions.

Factor value. Such an analysis is only as accurate as the acceleration factor. The value of the factor can be arrived at in a number of ways.
- Suppose the product is described by the simple model with a known slope coefficient. Then the model and known slope can be used to calculate K. The preceding lubricant application is an example of this.
- Many companies run traditional accelerated tests. They may include any number of accelerating variables and may involve complex sequences of different stressings. The acceleration factors are often company tradition, and their origins may be long forgotten.
- Some acceleration factors are estimated from data. This involves estimating a typical accelerated life and a typical "use" life. The "use" life may come from field data or from test data from simulated use conditions. The K factor is the ratio of observed "use" life over accelerated life.

- Other acceleration factors appear in handbooks. For example, MIL-STD-217 gives such factors for electronic components.

Separate factors. Each failure mode of a product needs its own K factor. Moreover, there may be more than one accelerated test for a product. Then each test is designed to accelerate different failure modes, and each combination of test and mode has its own K factor. For example, a vibration test of an electronic system yields mechanical failures of solder and pin connections. A temperature-accelerated test of such systems yields thermal degradation failures of capacitors, semiconductors, and other system components; each component failure mode has its own K factor. In practice, some analysts apply a single factor to all modes of a component, to a group of components, or to the entire system. This is usually crude.

Several tests. When there are several tests, each with its own K factor, the data can be combined. Each set of test data is converted to equivalent data at use conditions, and the data are suitably analyzed, for example, as data with competing failure modes (Chapter 7). Also, the sets may be compared for equal distributions if appropriate.

Weibull Distribution – Shape Parameter Specified

Motivation. In some applications, there are few or no failures. Then one has no estimate or a crude one for the shape parameter of an assumed Weibull life distribution. Assuming a value for the shape parameter, one can then analyze such data with no failures as follows. Also, one can then get more accurate estimates from data with few failures. Assuming an exponential distribution is the same as assuming a Weibull distribution with a shape value of 1. Usually another shape value yields more accurate results.

Relationship. The analyses below employ the following relationship between the exponential and Weibull distributions. Suppose that T_i is a random observation from a Weibull distribution with parameter values β and α_i at stress level x_i. Then

$$T_i' = T_i^\beta \qquad (2.13)$$

is a random observation from an exponential distribution with mean $\theta_i = \alpha_i^\beta$.

Analyses. The analyses employ the transformed exponential data T_i', which are analyzed with the preceding methods for an exponential distribution. The fitted log-linear model for θ must be transformed back to

$$\alpha(x) = \{\theta(x)\}^{1/\beta} = \exp\{(\gamma_1 + \gamma_2 x)/\beta\} \qquad (2.14)$$

for the Weibull model. Nelson (1985) describes in detail this type of analysis for a single distribution. McCoun and others (1987) extend Lawless's (1976,1982) exact limits to this model.

Other shape values. The assumed shape value is, of course, in error some. It is useful to redo the analyses with both smaller and larger values. Hopefully, the practical conclusions remain the same. If not, one must choose among them, for example, the most conservative.

Assumed slope coefficient. Analyses may be based on assumed values of the shape parameter *and* slope coefficient. Then the data are converted to a single sample of exponential data at a selected stress level. The estimates for the exponential distribution must be converted back to those for a Weibull distribution with (2.14).

3. ASSESS THE SIMPLE MODEL AND RIGHT CENSORED DATA

Purpose. The analyses of Section 2 depend on assumptions about the simple model and right censored data. The accuracy of estimates and confidence limits depend on how well those assumptions are satisfied. This section presents methods for assessing:

3.1.   a model characteristic has a specified value,
3.2.   the scale parameter (Weibull shape β or lognormal σ) has the same value at all stress levels,
3.3.   the (transformed) life-stress relationship is linear,
3.4.   the assumed life distribution is adequate,
3.5.   the data are valid (identify outliers),
3.6.   the effect of other variables.

This section follows the pattern of Section 3 of Chapter 4 on complete data. That section is useful, but not essential, background. This section provides graphical and numerical methods suitable for right censored data. Numerical methods include likelihood ratio (LR) tests, which serve various purposes; theory for them appears in Chapter 9. Escobar and Meeker (1988) present further methods to assess assumptions and the influence of observations.

3.1. A Model Characteristic Has a Specified Value

Overview. In some applications, there is a standard, specified, or proposed value for a certain model coefficient, parameter, quantity, or characteristic (c.g., MTTF). Then one may want to assess whether the data are consistent with that value. Then the specified value, employed in further analyses, may yield more accurate results than those based on its estimate from the data. Examples include
- a Weibull shape parameter of 1.1 to 1.3 for life of rolling bearings,
- a power of 3 in Palmgren's equation for life of ball bearings,
- a power of 2 in the Coffin-Manson relationship for thermal fatigue,
- an activation energy of 1.0 electron-volts for the Arrhenius relationship for solid state electronics.

In demonstration tests, one wants to assess whether a population value exceeds a specified value. Examples include
- Does the median life of a proposed Class-H motor insulation exceed 20,000 hours at 180°C ?
- Does the MTTF of a power supply exceed a specified value?

Graphical and numerical methods for such problems follow.

Graphical analysis. Using the methods of Chapter 3, one can graphically estimate a quantity. The specified value can also be plotted and compared with the estimate and data. Either they are consistent (relative to the scatter in the data), or they differ convincingly. For example, in Chapter 3, the graphical estimate of median life of the Class-H insulation at 180°C is convincingly below the desired 20,000 hours. This can be seen in the Arrhenius plot (Figure 2.3 of Chapter 3). Another example is the insulating fluid in Chapter 3. In Figures 3.1 and 3.2 there, the data appear consistent with a theoretical value of $\beta = 1$ for the Weibull shape parameter. If a graphical analysis is not conclusive, a numerical analysis is helpful as follows.

Confidence limits. If confidence limits for a quantity enclose its specified value, then the data are consistent with that value. Otherwise, the data contradict that value. For example, in Chapter 4, the 95% confidence limits for the median life of Class-H insulation at 180°C are 9,730 and 13,700 hours. These limits are well below the specified 20,000 hours. Thus, the data convincingly indicate a true median below 20,000 hours. In reliability demonstration, the confidence limit(s) must exceed the specified value. For example, the lower confidence limit for a MTTF must exceed the specified MTTF.

Hypothesis tests. Much statistical theory concerns hypothesis tests. Such hypothesis tests simply answer yes or no to the question: are the data consistent with a specified value of a model (population) quantity? Confidence intervals answer the same question but are more informative. In particular, the length of a confidence interval indicates the accuracy of the estimate and how much one may rely on the yes-or-no conclusion. See Chapters 8 and 9 for such hypothesis tests. Likelihood ratio tests (Chapter 9) apply to censored data of all types.

Demonstration tests. A reliability demonstration test of a product is run to decide whether some observed measure of product reliability is convincingly *better* than a specified value. Long used for military products, such tests are now being used by more and more companies on commercial products. The language of a typical test is that "the (constant) product failure rate (or other measure of reliability) must meet a specified (constant) rate λ^* with C% confidence." Operationally this means that the product passes the test if the test data yield a *one-sided* upper C% confidence limit λ for the true (unknown constant) rate λ such that $\lambda < \lambda^*$. That is, the confidence limit must be better than the specification. Otherwise the product fails the test, and appropriate action is taken (redesign, new supplier, etc.). Some product designers misunderstand this language and mistakenly design the product to have the specified λ^*. Then λ falls above λ^* with a high probability $C\%$. That is, the product **fails** the test with high probability $C\%$. To ensure that the product passes with high probability, the designer must achieve a true λ *well below* the specified λ^*. The designer can use the OC curve for the equivalent hypothesis test to determine a suitable λ value – one that gives the product a desired high probability of passing the test. Unfortunately such OC curves have not been developed for accelerated tests. Nelson (1972c) presents an accelerated demonstration test. The preceding discussion could be in terms of the MTTF θ or some other measure of reliability. Then the product passes if $\theta > \theta^*$. The normal approximate confidence limits are inaccurate one-sided limits. The one-sided LR limits (Section 5.8) are more accurate and preferred for demonstration tests. Some standards for reliability demonstration tests assume a Weibull or other distribution with a nonconstant failure rate.

3.2. Constant Scale Parameter

Overview. The simple model assumes that the scale parameter of *log* life (Weibull $1/\beta$ or lognormal σ) is constant. That is, the scale parameter has the same value at all stress levels of interest. Estimates of low percentiles are sensitive to a nonconstant scale parameter. Also, most analyses in Sections 3.3 through 3.5 are based on this assumption. The following graphical and numerical methods assess this assumption. Each method yields different insights and is worth using.

Graphical analysis. As described in Chapter 3, a probability plot provides an assessment of a constant scale parameter. On probability paper, the plotted data from different stress levels should be close to parallel. In Figure 4.1 of Chapter 3, the Class-B data from the three test temperatures do not look parallel. This may result from possible outliers at 190°C.

Confidence limits. There is a crude comparison of scale parameters. It employs a separate estimate and confidence limits for the scale parameter at each stress level with at least two distinct failures. Thus one must fit a separate distribution to the data from each stress level. If an interval for one stress level overlaps the estimate for another level, the two estimates are comparable (do not differ statistically significantly). If two intervals do not overlap, the two estimates differ statistically significantly. An intermediate situation occurs if two intervals overlap, but neither overlaps the other estimate. Then one cannot draw a conclusion using this method. Then use the maximum ratio test below.

Class-B insulation. For the Class-B data, such ML estimates and approximate normal 95% confidence limits follow.

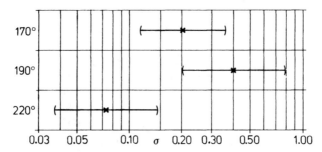

Figure 3.1.   Estimates and confidence limits for σ by temperature.

Temp.	σ Estimate	95% Limits
170	0.2028	(0.1158, 0.3549)
190	0.3996	(0.2013, 0.7931)
220	0.0728	(0.0366, 0.1449)

Figure 3.1 displays these estimates and limits on semi-log paper. The plot makes comparison easier. The estimates at 170° and 190° are consistent with each other, since the 190° interval overlaps the 170° estimate. The 190° and 220° estimates differ significantly since their intervals do not overlap. The 170° and 220° estimates are inconclusive, since their intervals overlap, but neither interval overlaps the other estimate. Also, note that there is no trend to the σ estimates as a function of temperature. Thus σ does not appear to be a simple and plausible function of temperature. Outliers at 190° seem to be a more likely cause of differing σ's.

LR test. The scale parameters can be compared for equality with the following LR test. It generalizes Bartlett's test for equality of (log)normal standard deviations with complete data (Chapter 4). It applies to any distribution with a scale parameter. For example, for Weibull and lognormal life, the distribution of log life has scale parameters $\delta = 1/\beta$ and σ, respectively. The test involves the following steps.

1. Fit a separate distribution to the data from each of the J test stresses with two or more distinct failure times. Obtain the corresponding J maximum log likelihoods $\hat{\mathcal{L}}_1, \cdots, \hat{\mathcal{L}}_J$. For example, for the Class-B data, there are $J=3$ such test temperatures, and $\hat{\mathcal{L}}_{170} = -64.270$, $\hat{\mathcal{L}}_{190} = -43.781$, and $\hat{\mathcal{L}}_{220} = -32.302$.
2. Fit a model with a common scale parameter and a separate location parameter for each of the J stress levels. Such fitting employs J indicator variables in a relationship for the location parameter. Use only the data from the J stress levels with two or more distinct failures. Obtain the corresponding maximum log likelihood $\hat{\mathcal{L}}_0$. For the Class-B data, $\hat{\mathcal{L}}_0 = -145.198$ for the data from $J=3$ temperatures with failures.
3. Calculate the LR test statistic as

$$T = 2(\hat{\mathcal{L}}_1 + \cdots + \hat{\mathcal{L}}_J - \hat{\mathcal{L}}_0). \tag{3.1}$$

For the Class-B data, $T = 2[-64.270-43.781-32.302-(-145.198)] = 9.69$.

4. If the true scale parameters are all equal, the distribution of T is approximately chi square with $J-1$ degrees of freedom. The approximation is better the larger the number of failures at each such stress, say, at least 5. If the true scale parameters differ, T tends to have larger values. Evaluate T as follows where $\chi^2(1-\alpha;J-1)$ is the $100(1-\alpha)$ percentile of the chi-square distribution with $J-1$ degrees of freedom.

   • If $T \leq \chi^2(1-\alpha;J-1)$, the J scale parameter estimates do not differ statistically significantly at the $100\alpha\%$ level.
   • If $T > \chi^2(1-\alpha;J-1)$, the scale parameters differ statistically significantly (convincing evidence) at the $100\alpha\%$ level.

   For the Class-B data, $T = 9.69 > 9.210 = \chi^2(0.99;2)$. Thus the σ's differ highly significantly (at the 1% level). That is, if the population σ's were equal, so large a T value would be observed less than 1% of the time. This unusually high T value suggests that the population σ's differ.

5. If T is significant, examine the estimates and confidence limits for the J scale parameters from step 1, and determine how they differ. Figure 3.1 helps do this. Change the model or delete data, and reanalyze as appropriate. For the Class-B data, the high σ estimate at 190° may be due to two low outliers at 408 hours. These outliers can be seen in Figures 4.1 and 4.2 of Chapter 3.

Maximum ratio. Another hypothesis test for equality of scale parameters employs a separate distribution fitted to each of the J stress levels with at least two distinct failures. The test statistic is the ratio of the largest to smallest ML estimates of the scale parameters. An example of this test is the maximum F-ratio for complete normal data in Chapter 4. Specifically, the test employs the maximum ratio statistic

$$R = \max(\hat{\sigma}_j/\hat{\sigma}_{j'}). \tag{3.2}$$

The test employs $R(1-\alpha;J)$, the $100(1-\alpha)\%$ point of the distribution of R:
   • If $R \leq R(1-\alpha;J)$, the estimates do not differ statistically significantly at the $100\alpha\%$ level.
   • If $R > R(1-\alpha;J)$, the estimates differ significantly at the $100\alpha\%$ level.

The distribution of R depends on the sizes and censorings of the J samples. For censored data, there is only one exact table of such R percentiles. McCool (1981,1974) gives such a table for comparing Weibull shape parameters; it applies only to equal sample sizes and single censoring with the same number of failures in each sample. For other distributions, approximate the percentiles of R with the Bonferroni inequality in Nelson (1982, p. 553).

Relationship. Constancy of a scale parameter can also be assessed by fitting the model plus a relationship for the scale parameter. This method has an advantage over previous ones; it applies when there are many stress levels, each with one or a few specimens. Also, it applies when some stress levels have no failures. The simplest such relationship in stress x is

$$\sigma(x) = \exp(\gamma_3 + \gamma_4 x). \tag{3.3}$$

This relationship is a monotone function of x. If the confidence interval for γ_4 does not contain zero, then there is convincing evidence that σ depends on stress. Otherwise, there is no convincing evidence.

The Arrhenius relationship and (3.3) were fitted to the Class-B data, including the 150° data with no failures. The ML estimate and approximate 95% confidence limits for γ_4 are -1.069 and $(-6.764, 4.626)$. Here $x = 1000/(\text{absolute temperature})$. This interval contains zero. Thus there is no significant trend to σ. Section 4 presents another example with this relationship. Fitting (3.3) is more sensitive means of detecting a trend of σ than are the previous tests, which are more sensitive to detecting erratic behavior of σ. Other relationships may be useful, for example, one including a quadratic term in x. This test may also be sensitive to an inadequate relationship for the location parameter μ; the previous ones are not.

3.3. Is the Life-Stress Relationship Linear?

Overview. The simple model includes a *linear* relationship for the (transformed) life-stress relationship. This assumption is critical for long extrapolations to low stress. Graphical and numerical assessments of this assumption follow. Each method yields different insights and is worth using.

Graphical. Chapter 3 explains how to use data plots to assess linearity of the life-stress relationship. For the Class-B data, the relationship appears adequate in Figure 4.2 of Chapter 3.

LR test. The following LR test checks linearity. It assumes that the scale parameter is a constant at all test stress levels. It generalizes the F test for linearity of Chapter 4. It employs only the data from the J stress levels each having at least one failure. For the Class-B insulation, the data at 150° are not used since they contain no failures. Perform the following steps.

1. Fit the simple model (Section 2.1), and obtain its maximum log likelihood $\hat{\mathcal{L}}_0$. For the Class-B data, $J=3$ and $\hat{\mathcal{L}}_0 = -145.867$ for those 3 temperatures.
2. Fit a model with a common scale parameter and a separate location parameter for each of the J stress levels, using indicator variables. Obtain its maximum log likelihood $\hat{\mathcal{L}}$. For the Class-B data, $\hat{\mathcal{L}} = -145.198$.
3. Calculate the LR test statistic

$$T = 2(\hat{\mathcal{L}} - \hat{\mathcal{L}}_0). \tag{3.4}$$

   For the Class-B data, $T = 2[-145.198 - (-145.867)] = 1.34$.
4. If the relationship is linear, the distribution of T is approximately chi square with $J-2$ degrees of freedom. If the relationship is not linear, T tends to have larger values. Evaluate T as follows where $\chi^2(1-\alpha; J-2)$ is the $100(1-\alpha)$th chi square percentile with $J-2$ degrees of freedom.

- If $T \leq \chi^2(1-\alpha;J-2)$, the data are consistent with a linear relationship at the $100\alpha\%$ level.
- If $T > \chi^2(1-\alpha;J-2)$, the data differ statistically significantly from a linear relationship at the $100\alpha\%$ level.

For the Class-B data, $T = 1.34 < 3.842 = \chi^2(0.95;1)$. Thus there is no evidence of nonlinearity at the 5% level.

5. If there is significant nonlinearity, the reason should be sought. For example, examine the data in a relationship plot. Fit another model or delete data, as appropriate, and reanalyze the data. For example, for the Turn data in Figures 5.1 and 5.2 of Chapter 3, the 260° data are not consistent with the rest. Later discovery of the cause of this yielded $1,000,000 yearly. Physical understanding of nonlinearity or other poor fit is usually more valuable than a reanalysis.

Nonlinear fit. Nonlinearity can also be assessed by fitting a more general relationship that includes the linear one. This method has an advantage over previous ones; it applies when there are many stress levels, each with one or a few specimens. Also, it applies when some stress levels have no failures. The simplest such relationship in stress x is

$$\mu(x) = \gamma_1 + \gamma_2(x - x') + \gamma_3(x - x')^2. \tag{3.5}$$

This relationship has a quadratic term. x' is some chosen value in the middle of the data; it "centers" the data and makes the ML fitting converge better. Some packages automatically center each independent variable. If not, users should do so to assure accurate fitting. If the confidence interval for γ_3 does not contain zero, then there is convincing evidence of nonlinearity. Otherwise, there is no convincing evidence. A quadratic was fitted to the Class-B data including the 150° data with no failures. The ML estimate and approximate normal 95% confidence limits for γ_3 are 5.02 and $(-13.8, 23.8)$. Here $x = 1000/(\text{absolute temperature})$, and $x' = 2.201629$ is the average of x over the data. This interval contains zero. Thus there is no significant nonlinearity. This test is more sensitive to detecting a curved relationship. Tests above are more sensitive than this one to detecting erratic behavior of the data. Also, a nonlinear life-stress relationship can be used with a relationship for the scale parameter, e.g., (3.3). Section 4 gives an example of this.

3.4. Assess the Life Distribution

Purpose. This section presents graphical and numerical methods for assessing the assumed life distribution. Estimates of low percentiles are sensitive to the assumed life distribution. Each method below yields different insights, and all are worth using.

Graphical. Chapter 3 presents probability and hazard plots as a means of assessing an assumed life distribution. For example, the Class-B data appear in Figure 4.1 of Chapter 3. The plotted data from the three test temperatures are not straight and parallel, suggesting that the lognormal is a poor fit or that the data have problems. Nelson (1973b) uses residuals for a more sensitive check on the distribution as follows.

Plot of residuals. A ("raw") **residual** $r_i = y_i - \hat{\mu}(x_i)$ is the difference between a log observation y_i and the estimate of its location parameter $\hat{\mu}(x_i)$ at its x_i value. Such a residual is *censored* if the observation is censored. Table 3.1 shows the calculation of such residuals for the Class-B data. The pooled sample of all such log residuals comes from the assumed (log) life distribution. That distribution has a location parameter equal to zero and scale parameter equal to σ, the population value. This pooled sample is usually multiply censored. Make a hazard plot of such log residuals to assess the distribution fit. Use normal paper to assess a lognormal fit. Use extreme value paper to assess a Weibull fit. Figure 3.2 is a normal hazard plot of the Class-B residuals. The plot is not straight. The early failures are low relative to the rest of the data. In general, this may indicate that the distribution is inadequate or that there are low outliers. Figure 4.1 of Chapter 3 and Figure 3.1 here suggest that low outliers are the more likely cause. This suggests redoing the analyses without the two low outliers at 190°. A cause of the

Table 3.1. Calculation of Class-B Residuals (+ Denotes Censored)

\multicolumn{3}{c}{150°C}			\multicolumn{3}{c}{190°C}		
y_i $-$ $\hat{\mu}(x_i)$ $=$ r_i			y_i $-$ $\hat{\mu}(x_i)$ $=$ r_i		
3.9066+	− 4.1684 =	−0.2618+	2.6107	− 3.2900 =	−0.6973
3.9066+	− 4.1684 =	−0.2618+	2.6107	− 3.2900 =	−0.6973
3.9066+	− 4.1684 =	−0.2618+	3.1284	− 3.2900 =	−0.1616
3.9066+	− 4.1684 =	−0.2618+	3.1284	− 3.2900 =	−0.1616
3.9066+	− 4.1684 =	−0.2618+	3.1584	− 3.2900 =	−0.1316
3.9066+	− 4.1684 =	−0.2618+	3.2253+	− 3.2900 =	−0.0647+
3.9066+	− 4.1684 =	−0.2618+	3.2253+	− 3.2900 =	−0.0647+
3.9066+	− 4.1684 =	−0.2618+	3.2253+	− 3.2900 =	−0.0647+
3.9066+	− 4.1684 =	−0.2618+	3.2253+	− 3.2900 =	−0.0647+
3.9066+	− 4.1684 =	−0.2618+	3.2253+	− 3.2900 =	−0.0647+

\multicolumn{3}{c}{170°C}			\multicolumn{3}{c}{220°C}		
3.2465	− 3.7077 =	−0.4612	2.6107	− 2.7217 =	−0.1110
3.4428	− 3.7077 =	−0.2649	2.6107	− 2.7217 =	−0.1110
3.5371	− 3.7077 =	−0.1706	2.7024	− 2.7217 =	−0.0193
3.5492	− 3.7077 =	−0.1585	2.7024	− 2.7217 =	−0.0193
3.5775	− 3.7077 =	−0.1302	2.7024	− 2.7217 =	−0.0193
3.6866	− 3.7077 =	−0.0211	2.7226+	− 2.7217 =	0.0009+
3.7157	− 3.7077 =	0.0080	2.7226+	− 2.7217 =	0.0009+
3.7362+	− 3.7077 =	0.0285+	2.7226+	− 2.7217 =	0.0009+
3.7362+	− 3.7077 =	0.0285+	2.7226+	− 2.7217 =	0.0009+
3.7362+	− 3.7077 =	0.0285+	2.7226+	− 2.7217 =	0.0009+

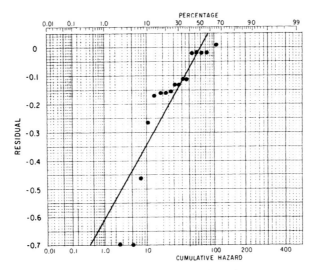

Figure 3.2.   Normal hazard plot of Class-B residuals.

suspected outliers was sought without success. Most computer packages for life data calculate and plot such residuals.

Other residuals. Other residuals can be plotted. These include
- Residuals from fitting a separate distribution to the data from each test stress level. Lack of fit of the linear relationship does not affect these log residuals. With lack of fit, raw residuals tend to be too large (absolute magnitude), and their distribution differs from the population distribution. Equation (3.5) of Chapter 4 gives this residual for complete data.
- Residuals adjusted for the fitting of the relationship to reduce the bias in the scale parameter estimate based on them. Cox and Snell (1968) derive such *adjusted residuals* for ML fitting. Section 5.3 of Chapter 4 gives the formula for such residuals for complete data and least-squares fitting.

It is worth plotting such residuals. Each type reveals different information.

Generalized gamma. The Weibull and lognormal often are candidate distributions. To choose between them, one can fit the simple relationship and the generalized gamma distribution to *log* life data as described by Farewell and Prentice (1973). This distribution has a shape parameter q, as well as a location and a scale parameter. For $q = 0$, the distribution is normal (lognormal), and for $q = 1$, it is extreme value (Weibull). If the maximum log likelihood is greater for $q = 0$ than for $q = 1$, the lognormal distribution fits better. Otherwise, the Weibull fits better. Calculate twice the absolute difference of the two maximum log likelihoods. If this exceeds $\chi^2(1-\alpha;1)$, then the distribution with the higher log likelihood fits significantly better than the other at the $100\alpha\%$ level.

For example, for Class-B insulation data, the two maximum log likelihoods are $\hat{\mathcal{L}}_0 = -148.54$ and $\hat{\mathcal{L}}_1 = -147.02$. The Weibull distribution fits better since $\hat{\mathcal{L}}_1 > \hat{\mathcal{L}}_0$. $T = 2(\hat{\mathcal{L}}_1 - \hat{\mathcal{L}}_0) = 3.04 < 3.842 = \chi^2(0.95;1)$. Thus the Weibull fits better but not significantly better. This test does not require a program for fitting a simple relationship and generalized gamma distribution. Merely use the maximum log likelihoods from separately fitting the models with Weibull and lognormal distributions. Of course, a computer package must calculate Weibull and lognormal likelihoods consistent with each other. Some omit constants from the likelihood. This test is sensitive to outliers.

Tests of fit. Standard tests of fit for various distributions are presented by D'Agostino and Stephens (1986). They apply only to a single sample from a single distribution. They do not apply to residuals from fitting a regression model. At best such standard tests are crude when applied to residuals. Moreover, such residuals, as a result of the model fitting, tend to resemble the fitted distribution. Also, such tests are sensitive to outliers, which do not come from the main population distribution. Outliers need to be investigated and understood and usually omitted from subsequent analyses. Moreover, if such a test is used and the assessed distribution has significant lack of fit, one must examine a probability plot to see why. Such tests give no indication of the nature of the lack of fit. Thus a probability plot of residuals is essential.

3.5. Checks on the Data

Purpose. This section briefly reviews graphical and numerical checks on the data. They mainly seek to identify outliers and other problems in the data. As always, determining the cause of such data is more important than deciding whether to include or exclude such data in analyses.

Graphical. All of the plots of the data (Chapter 3) and plots of the residuals (Section 3.4) can reveal outliers and other problems in the data. For example, Figure 5.1 of Chapter 3 shows that the 260° data on Turn failures are not consistent with the rest of the data. Also, for example, Figure 4.2 of Chapter 3 suggests two low outliers in the Class-B data. Usually the adjusted residuals show outliers more clearly than do raw ones.

Outlier tests. Standard outlier tests are presented by Barnett and Lewis (1984) and surveyed by Beckman and Cook (1983). Most of these tests apply only to a single sample from a single distribution. Few tests are suitable for regression models and complete data. Standard (single-sample) tests are crude for a sample of residuals, which tend to resemble the assumed distribution as a result of fitting the model. Flack and Flores (1989) give outlier tests for residuals. Few tests apply to censored data.

3.6. Effects of Other Variables

Purpose. This section briefly describes graphical and numerical methods for assessing the effect of other variables on product life.

Crossplotting. Crossplotting of residuals against such a variable is useful for complete data, as described in Chapter 4. Censored data in such crossplots are often difficult to interpret. One knows only that the residual for the eventual failure is somewhere above the censored residual. See Figures 4.7 and 4.8 (next section), which are such crossplots. There each censored residual appears as an A, and the residual if observed would be above its A. The residuals come from fitting a model for life as a function of stress to metal fatigue data. The residual for each specimen is plotted against the pseudo stress in Figure 4.7. By imagining each censored residual as higher and observed, one might see a very slight positive trend – likely not statistically significant. Lawless (1982, p. 281) recommends replacing each censored residual with its conditional expected failure time; this expected observed residual is crossplotted as an observed residual. The GRAFSTAT package of Schatzoff (1985) calculates and crossplots such residuals.

Hazard plots. The effect of other variables on product life can be assessed with hazard plots of residuals. Divide the residuals into two or more groups according ranges of the variable being examined. Make a separate hazard plot of each group. The plots can be put on the same or separate hazard papers. If the plots are comparable, the variable appears unrelated to product life. If there is a systematic trend in the distributions, the variable appears related to life. Nelson (1973) gives examples of such hazard plots.

Numerical. One can assess the effect of other variables on life by including them in the model fitted to the data. Fitting of such multivariable models appears in Section 4.3. It is a useful means of assessing such variables.

4. OTHER MODELS AND TYPES OF DATA

Introduction. This section extends the methods of Sections 2 and 3 on the simple model and right censored data. Section 4.1 extends the methods to other types of data, including left censored, quantal-response, and interval data. Section 4.2 extends the methods to more complex models with one accelerating variable; these include fatigue (or endurance) limit models, and models where the spread in log life depends on stress. Section 4.3 extends them to multivariable models. Sections 2 and 3 are needed background.

4.1. Fit the Simple Model to Other Types of Data

Purpose. This section presents ML fitting of the simple model to interval, quantal-response, observed, and right and left censored data, as well as a mix of such data. This section extends the methods of Sections 2 and 3 to other types of data. In particular, this section presents ML estimates and confidence limits for model coefficients, parameters, and other quantities.

Also, this section provides methods to assess the model, the data, and the effect of other variables. Theory for such ML fitting appears in Section 5.

Same information. The information one seeks from other data is the same as that from right censored data (Sections 2 and 3). Indeed, one fits the same simple model to other types of data, and one gets the estimates and confidence limits for the same coefficients, parameters, etc. Also, one uses the same methods to assess the model and data. Only the form of the data differs from that in Sections 2 and 3. Indeed the computer output for other types of data has the same appearance as that for right censored data.

Differences. Results of analyses of other types of data differ from results from right censored data in some respects. For censored, interval, and quantal-response data, standard errors of estimates tend to be greater and confidence intervals wider than if the data were observed. That is, there is less information (or accuracy) in other types of data. Also, the normal approximation for the sampling distribution of an estimator requires more failures to be adequate for other types of data. The type of data on a specimen determines the mathematical form of the specimen's likelihood; Section 5 presents such likelihoods.

Interval data. Table 4.1 of Chapter 3 shows the Class-B insulation data. The data resulted from repeated inspection of the specimens. That is, they are interval data, but each failure was previously treated as if it occurred at the middle of its interval. The 220° and 190° specimens were inspected every 48 hours, the 170° specimens every 96 hours, and the 150° specimens every 168 hours. The Arrhenius-lognormal model was fitted to the correct interval data. STATPAC output on the fit appears in Figure 4.1. Statistical packages listed in Section 1 fit models to such data.

Accuracy of estimates. In Figure 4.1, the normal confidence limits are close to those in Figure 2.1. However, those in Figure 4.1 are slightly wider. This results from interval data not being as informative as exactly observed failures. The inspection intervals for the Class-B data are narrow relative to the spread of the (log) life distribution. So the proper confidence limits in Figure 4.1 are only slightly wider than those in Figure 2.1. If the inspection intervals are wide compared to the distribution, the confidence intervals are much wider than those from data with exact failure times. Of course, all data are interval data since they are recorded to just a few figures accuracy. For most work, rounding is small and can be ignored. Problem 5.13 presents data where the inspection intervals are wide relative to the distribution.

Literature on interval data. Various references treat ML analyses of interval data for a single distribution, for example, Nelson (1982, Chap. 9). In addition to such analyses, the following references present optimum choices of the inspection times. Ehrenfeld (1962) and Nelson (1977) do so for the exponential distribution. Kulldorff (1961) does so for the (log)normal distribution. Meeker (1986) does so for the Weibull distribution.

```
MAXIMUM LOG LIKELIHOOD =   -73.957765

* MAXIMUM LIKELIHOOD ESTIMATES FOR MODEL COEFFICIENTS
  WITH APPROXIMATE 95% CONFIDENCE LIMITS

COEFFICIENTS  ESTIMATE      LOWER LIMIT      UPPER LIMIT   STANDARD ERROR

C00001        -6.016456     -7.907212        -4.125699       0.9646718
C00002         4.309538      3.437492         5.181584       0.4449215
C00003         0.2588536     0.1658819        0.3518253    4.743456E-02

* COVARIANCE MATRIX

COEFFICIENTS  C00001          C00002           C00003

C00001         0.9305916
C00002        -0.4284944      0.1979551
C00003        -8.728676E-03   4.613323E-03     2.250037E-03

* FISHER MATRIX

COEFFICIENTS  C00001          C00002           C00003

C00001         428.5899
C00002         933.5882       2038.920
C00003        -251.4781       -558.6667        614.2957

PERCENTILES(130.)

TEMP      130.

* MAXIMUM LIKELIHOOD ESTIMATES FOR  DIST.   PCTILES
  WITH APPROXIMATE 95% CONFIDENCE LIMITS

PCT.    ESTIMATE      LOWER LIMIT      UPPER LIMIT     STANDARD ERROR

0.1     7463.632      3529.896         15781.15         2851.313
0.5     10140.87      5057.636         20333.05         3599.356
1       11766.47      5991.561         23107.47         4051.641
5       17663.39      9334.361         33424.40         5747.740
10      21936.28      11670.98         41230.98         7062.543
20      28518.65      15103.33         53849.93         9248.815
50      47091.56      23818.50         93104.73         16377.08
80      77760.15      36069.61         167638.1         30476.36
90      101093.5      44276.43         230820.1         42582.69
95      125548.6      52189.44         302023.7         56228.65
99      188468.9      70393.80         504619.0         94702.79
```

Figure 4.1.   Output on Arrhenius-lognormal fit to Class-B interval data.

Quantal-response data. Regression analyses of quantal-response data appear often in the biomedical literature. Key references include Finney (1968), Breslow and Day (1980), Miller, Efron, and others (1980), and Nelson (1982, Chap. 9). Nelson (1979) fits the power-Weibull model to such data on time to cracking of turbine disks. ASTM STP 731 (1981) presents analyses of such data on metal fatigue to estimate a strength (fatigue limit) distribution. Meeker and Hahn (1977,1978) present optimum inspection times for such data from a logistic distribution. Statistical packages listed in Section 1 fit models to such data.

Left censored data. Left censored data arise when failures occur before the specimens are continuously monitored. For example, a test may start on

Friday and not be observed over the weekend. In some tests, failures occur so quickly that their times cannot be observed. Problem 3.10 presents such data. Such data yield less accurate estimates of lower percentiles of the life distribution. When such percentiles are sought, it is best to observe early and frequently to make such estimates more accurate. Statistical packages listed in Section 1 fit models to such data.

Constant scale parameter. Whether the scale parameter is constant is assessed with the methods in Section 3.2. The methods include the LR test, the maximum ratio, and expressing the scale parameter as a function of stress. All of these methods are appropriate for the other types of data.

Linearity. Adequacy of the simple linear relationship is assessed with the methods in Section 3.3. Namely, the LR test applies if there are a few test stresses, each with failures. Also, if there are many stress levels each with few specimens, then fitting a more general nonlinear relationship is suitable. Such a relationship must include the simple linear one. Examples include a quadratic relationship or a power relationship with an endurance limit.

Distribution. Adequacy of the assumed distribution is assessed with the methods in Section 3.4. The log gamma distribution can be used to compare the Weibull and (log)normal distributions. A plot of the residuals is more complex with the other types of data. Residuals can be left censored, right censored, and intervals, as well as observed. Often interval residuals are narrow relative to the distribution spread. Such residuals can be treated as observed residuals equally spaced over their intervals and plotted on hazard paper. The Class-B interval residuals can be handled this way. If the interval residuals are wide or if there are left censored residuals, then a Peto (1973) plot must be made. Turnbull (1976) develops the Peto plot further. Some computer packages provide such Peto plots.

Effect of other variables. Crossplots of interval and censored residuals against stress and other variables are difficult to interpret. It is best to fit relationships including such variables to the data to assess their effect. Section 4.3 presents fitting of such multivariable relationships.

4.2. Other Models with a Single Stress

Purpose. This section extends the methods for the simple model to other models with a single stress variable. In particular, this section presents ML fitting of other models to all types of data, and it presents methods for assessing the model and data. A variety of such models appear in Sections 11, 12, and 14 of Chapter 2. This section illustrates such fitting with two examples. The first employs a model with a fatigue limit, and the second employs a model with a log standard deviation that depends on the stress.

Fatigue or Endurance Limit Model

Overview. The following paragraphs present ML fitting of the fatigue (endurance) limit model of Section 11 of Chapter 2. The topics below include the example data, the model, computer output on the fit, assessments of the model and data, and extensions of the model.

Insulation example. Data that illustrate such fitting come from a voltage-endurance test of electrical insulation. The purpose of fitting such a model was to assess whether the insulation has an endurance limit (a voltage stress above zero). If so, the simple power-lognormal model, typically used for such insulation, would be inadequate for extrapolation to the design stress. Also, the design voltage stress might be chosen below the endurance limit to eliminate insulation failure. The data contain 110 specimens, of which 92 are failures, and the rest are censored on the right.

Model. The assumed model has a lognormal life distribution. The endurance-limit relationship for median life has the form

$$\tau_{.50} = \gamma_1/(V-\gamma_3)^{\gamma_2}; \qquad (4.2.1)$$

here V is the voltage stress, and γ_3 is the endurance limit, which is a voltage stress. They satisfy $V>\gamma_3>0$. This relationship has three coefficients; moreover, the relationship is nonlinear in them. Figure 11.1 of Chapter 2 depicts an endurance limit relationship. This is the (inverse) power relationship if $\gamma_3 = 0$. The simple (transformed power) relationship has two coefficients and is linear in them. The relationship above is not a standard one in any package in Section 1. It must be programmed into a package.

Model inadequacies. The preceding relationship assumes that there is a sharp endurance limit γ_3 which is the same for all specimens. Other models in Section 11 of Chapter 2 contain a strength distribution for the endurance limit. Such models seem more plausible; moreover, such a strength distribution is observed in steels in fatigue tests. The sharp endurance limit may cause problems in fitting the relationship when the test stresses are near the endurance limit, as in Problem 5.14. In the insulation problem, the test stresses are well above the endurance limit, and there are no problems in fitting. In this example, the relationship appears adequate.

Output. Figure 4.2 shows STATPAC output on the ML fit of the model. The ML estimate of the endurance limit is C3 $\approx$ 73 volts/mil. This is close enough to the design stress to interest the designers. Moreover, the approximate 95% confidence limits 36 and 110 do not enclose zero. This is convincing evidence of a positive endurance limit, provided that the model and data are valid. Even if the relationship is not valid outside the test stress range (116, 207), this is convincing evidence that the true relationship is not an inverse power law. The correlation coefficient for C2 and C3 is -0.9841222.

```
MAXIMUM LOG LIKELIHOOD =   -674.97635

* FISHER MATRIX

COEFFICIENTS  C    1            C    2          C    3         C    4

C    1     0.1343242
C    2     5.298513      223.8206
C    3     0.3771258      16.04091       1.153382
C    4    -0.9484622     -63.77421      -5.197079      2298.933

* MAXIMUM LIKELIHOOD ESTIMATES FOR MODEL COEFFICIENTS
  WITH APPROXIMATE 95% CONFIDENCE LIMITS

COEFFICIENTS  ESTIMATE      LOWER LIMIT     UPPER LIMIT   STANDARD ERROR

C    1     258.1657       234.8524        281.4789      11.89451
C    2     6.510000        3.611426         9.408574      1.478864
C    3    73.24000        36.78775       109.6923       18.59809
C    4 σ  0.2920000       0.2499631        0.3340368     0.2144737E-01

* COVARIANCE MATRIX

COEFFICIENTS  C    1            C    2          C    3         C    4

C    1     141.4793
C    2    -10.30933        2.187040
C    3     97.08243      -27.06735       345.8889
C    4    -0.8150106E-02  -0.4772817E-02    0.7111651E-01   0.4599897E-03

* CORRELATION MATRIX

COEFFICIENTS  C    1            C    2          C    3         C    4

C    1     1.000000
C    2    -0.5860783       1.000000
C    3     0.4388599      -0.9841222       1.0000000
C    4    -0.3194794E-01  -0.1504778        0.1782905      1.000000
```

Figure 4.2.   Output on ML fit of an endurance-limit relationship.

The value is close to -1, but not close enough to cause concern that the fitting is not numerically accurate enough. The numerical fitting of another form of the relationship may be more accurate.

LR test. An alternate test for $\gamma_3 = 0$ is the likelihood ratio test. Here let $\hat{\mathcal{L}}$ denote the maximum log likelihood with the endurance-limit model, and let $\hat{\mathcal{L}}'$ denote the maximum log likelihood with the power-lognormal model ($\gamma_3 = 0$). The LR test statistic is $T = 2(\hat{\mathcal{L}} - \hat{\mathcal{L}}')$. If $T > \chi^2(1-\alpha;1)$, then the endurance limit significantly differs from zero at the $100\alpha\%$ level. Otherwise, the estimate C3 is consistent with $\gamma_3 = 0$. For the insulation data, $T = 2[-674.976 - (-677.109)] = 4.266 > 3.841 = \chi^2(0.95;1)$. Thus the estimate C3 is significantly different from zero at the 5% level.

Assess relationship. The adequacy of the endurance-limit relationship can be assessed two ways. First, fit other relationships. Usually one tries a more general relationship, say, with four coefficients. The LR test can be used to assess if the general relationship fits significantly better. Also, one can fit an entirely different relationship, say, a quadratic one in $\log(V)$. If used for extrapolation, the relationship should be physically plausible. A

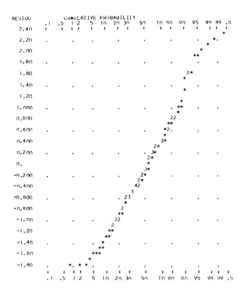

Figure 4.3.   Normal plot of the log residuals.

quadratic relationship may not be plausible outside the test range. Second, crossplot the residuals against stress, and examine the plot.

Constant scale parameter. The assumption of a constant scale parameter can be assessed with the methods of Section 3.

Assess distribution. Adequacy of the assumed distribution is assessed as described in Section 3. In particular, fit the log gamma distribution, and make a probability plot of the residuals. Figure 4.3 is a normal probability plot of the log residuals of the insulation. The plot is reasonably straight.

Other variables. The effect of other variables is assessed as described in Section 3. In particular, one can fit relationships with other variables, as described in Section 4.3, or one can plot the residuals against such variables. For example, the insulation specimens were made on the first, second, and third shifts. A probability plot of the log residuals from each shift was made, and the plots were compared. They all have comparable medians (near zero) and comparable standard deviations. Thus shift does not appear to significantly affect insulation life.

Model extensions. The model above can be made more realistic. For example, the model could contain a distribution for the endurance limit as described in Section 11 of Chapter 2. Also, the standard deviation of log life could be a function of stress. This effect is large in fatigue of steels and bearings, but it is smaller in dielectrics.

Optimum plans. Test plans for endurance-limit models have not been investigated. Thus there is no guidance on the choice of the test stress levels and the number of specimens at each.

Spread Depends on Stress

Overview. The following paragraphs present ML fitting of the log linear relationship for a scale parameter. In some applications, the usual assumption of a constant scale parameter is not adequate. This is particularly true for metal fatigue. Such relationships appear in Section 14 of Chapter 2. The topics below include the example data, the model, computer output on the fit, and assessments of the model and data.

Fatigue example. Low cycle fatigue data from a strain-controlled test appear in Table 4.1. For each of 26 specimens, the data consist of the pseudo stress (S, the specimen's Young's modulus times its strain), the number of test CYCLES, and the failure code (1 = failed, 2 = runout). In the analyses, failures in the uniform cross section of a specimen are treated as failures. Runouts and failures in the specimen radius, weld, or threads are treated as runouts (censored). Thus the uniform cross section of the specimen is treated as the actual specimen in the analyses. Other definitions of failure and specimen could also be used. The purpose of the analyses is to obtain fatigue curves for use by designers. In applications of the alloy, designers typically use the "-3σ" curve, roughly the 0.1% failure curve. The data appear in Nelson (1984).

Model. The assumed model has a lognormal life distribution. The fitted relationship for the mean log life is quadratic in LPS = log(S); namely,

$$\mu(S) = \gamma_1 + \gamma_2[LPS - A] + \gamma_3[LPS - A]^2; \qquad (4.2.2)$$

Table 4.1. Fatigue Data (1. = Failure, 2. = Runout)

S	CYCLES	CODE			
145.9	5733.00	1.	114.8	21300.0	1.
85.20	13949.0	1.	144.5	6705.00	1.
116.4	15616.0	1.	91.30	112002.	1.
87.20	56723.0	1.	142.5	11865.0	1.
100.1	12076.0	1.	100.5	13181.0	1.
85.80	152680.	1.	118.4	8489.00	1.
99.80	43331.0	1.	118.6	12434.0	1.
113.0	18067.0	1.	118.0	13030.0	1.
120.4	9750.00	1.	80.80	57923.0	2.
86.40	156725.	1.	87.30	121075.	1.
85.60	112968.	2.	80.60	200027.	1.
86.70	138114.	2.	80.30	211629.	1.
89.70	122372.	2.	84.30	155000.	1.

here A is the average log(S) over the test specimens. The log standard deviation σ is assumed to be a log linear function of stress; that is,

$$\ln[\sigma(S)] = \gamma_4 + \gamma_5(LPS - A). \tag{4.2.3}$$

Output. Figure 4.4 shows STATPAC output on the ML fit of the model. The approximate normal 95% confidence limits for γ_5 are -8.89 and -2.06. These limits do not enclose 0. Thus there is evidence that σ depends on stress. Figure 4.5 displays the data and fitted percentile curves. Runouts are denoted by A; B denotes a runout and a failure; D denotes a runout and three failures.

Other relationships. Various other relationships were fitted to the data:

Model:	1	2	3	4
$\mu(S)$:	linear	quadratic	linear	quadratic
$\ln[\sigma(S)]$:	constant	constant	linear	linear
max $\hat{\mathcal{L}}$	-252.64	-250.55	-250.70	-246.73

Likelihood ratio tests (Chapter 9) were used to compare the various models. Model 2 is significantly better than model 1, and model 4 is significantly better than model 3; both results indicate the quadratic term improves the fit. Model 3 is significantly better that model 1, and model 4 is significantly better than model 2; both results indicate that the log-linear σ is better than the constant σ. Thus model 4 was used.

Unitized residuals. Model 4 has a σ that depends on stress. Thus the usual log residuals about the fitted equation do not have a common standard deviation. The following generalized definition of a residual is needed. We assume the distribution has a location parameter $\mu(x)$ and a scale parameter $\sigma(x)$ which are functions of stress x (and possibly other variables). Let y_i denote the (log) life of specimen i, $\hat{\mu}_i$ the estimate of $\mu(x_i)$ at its stress level x_i, and $\hat{\sigma}_i = \exp(\hat{\gamma}_4 + \hat{\gamma}_5 x_i)$. The **unitized (log) residual** of specimen i is

$$r_i \equiv (y_i - \hat{\mu}_i)/\hat{\sigma}_i. \tag{4.2.4}$$

Such residuals are, roughly speaking, a sample from the assumed distribution with $\mu = 0$ and $\sigma = 1$, if the assumed distribution is correct. Such a residual is censored if y_i is censored.

Probability plot. Figure 4.6 is a normal plot of the unitized residuals (some of which are censored) from the example. The plot is curved in the middle of the distribution, and there is a high outlier. Thus the distribution appears inadequate, or the relationships are inadequate. Note that the lower tail of the plot is relatively straight. That is, the lognormal distribution adequately fits the lower tail. Hahn, Morgan, and Nelson (1985) suggest and explain how to censor observations in the upper tail to make the model fit

MAXIMUM LOG LIKELIHOOD = <u>-246.72675</u>

* MAXIMUM LIKELIHOOD ESTIMATES FOR DIST. PARAMETERS
  WITH APPROXIMATE 95% CONFIDENCE LIMITS

PARAMETERS		ESTIMATE	LOWER LIMIT	UPPER LIMIT	
C	1	4.482099	4.360507	4.603691	
C	2	-7.012116	-8.745509	-5.278724	⟨0
C	3	19.96210	0⟨ 6.936051	32.98814	
C	4	-1.411449	-1.712305	-1.110592	⟨0—
C	5	-5.482505	-8.897899	-2.067112	⟨0

* COVARIANCE MATRIX

PARAMETERS		C 1	C 2	C 3	C 4
C	1	0.3848529E-02			
C	2	-0.2416480E-02	0.7821349		
C	3	-0.1670752	-4.839053	44.16854	
C	4	0.7317652E-03	-0.1446559E-01	0.6022734E-01	0.2356174E-01
C	5	-0.1097854E-01	0.1467465E-01	0.5314569	-0.3835418E-01

PARAMETERS		C 5
C	5	3.036473

* MAXIMUM LIKELIHOOD ESTIMATE 'OF THE FITTED EQUATION

CENTER $\mu(S)$
    =   4.482099

    + (LPS   -   2.002631   )
      *(-7.012116
    +   19.96210   * (LPS   -   2.002631   ))

SPREAD $\ln(\sigma)$
    =   -1.411449

    + (LPS   -   2.002631   ) *   -5.482505

PCTILES(1.875061)

         LPS

    1.8750610   **75 ksi**

* MAXIMUM LIKELIHOOD ESTIMATES FOR   DIST.   PCTILES
  WITH APPROXIMATE 95% CONFIDENCE LIMITS

PCT. P	ESTIMATE N_P	LOWER LIMIT	UPPER LIMIT
0.1	15316.75	1831.747	128076.1
0.5	27383.94	4368.243	171666.3
1	36298.30	6624.217	198901.5
5	78395.90	20112.79	305572.7
10	118204.0	35473.36	393878.6
20	194378.0	67925.30	556240.6
50	502920.1	196105.8	1289756.
80	1301220.	426712.4	3967950.
90	2139762.	590307.8	7756263.
95	3226298.	752268.8	0.1383681E 08
99	6968052.	1140934.	0.4255616E 08

Figure 4.4.   Output on ML fit of the fatigue model.

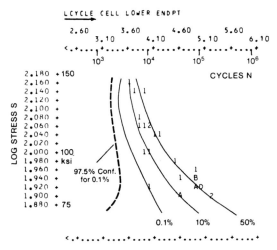

Figure 4.5.   Log-log plot of data and fitted percentile curves.

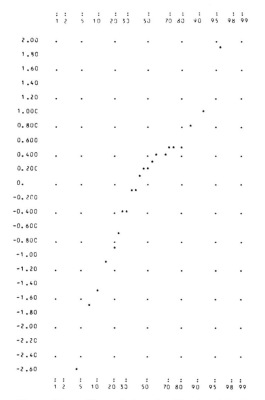

Figure 4.6.   Normal plot of unitized residuals.

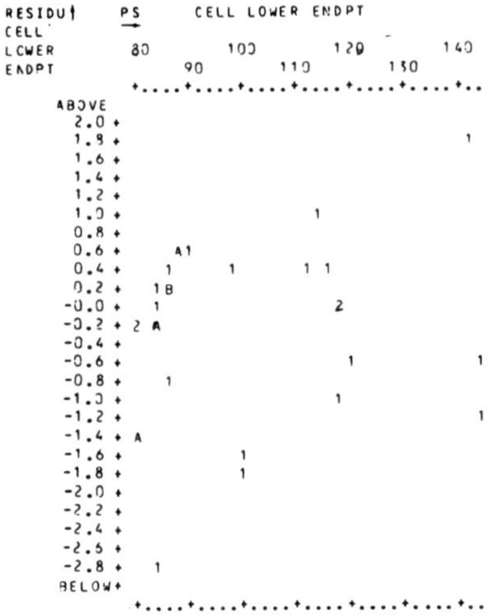

```
RESIDU†      PS       CELL LOWER ENDPT                    RESIDU†      BLEND CATEGORIES
CELL                  ————                                CELL         ————
LOWER        80        100        120      140            LOWER         172        177      185
ENDPT             90        110      150                  ENDPT            173      179
             +....+....+....+....+....+....+..                          ++++....++++....++++
  ABOVE                                                     ABOVE
   2.0 +                                                     2.0+
   1.8 +                            1                        1.8+                      1
   1.6 +                                                     1.5+
   1.4 +                                                     1.4+
   1.2 +                                                     1.2+
   1.0 +                       1                             1.0+   1
   0.8 +                                                     0.8+
   0.6 +        A1                                           0.6+  11
   0.4 +     1      1       1 1                              0.4+                3     1
   0.2 +    1 B                                              0.2+  10    1   1
  -0.0 +    1                  2                            -0.0+   1    2
  -0.2 +  2 A                                               -0.2+   1    1         10
  -0.4 +                                                    -0.4+
  -0.6 +                       1         1                  -0.6+   1          1
  -0.8 +    1                                               -0.8+              1
  -1.0 +                       1                            -1.0+                        1
  -1.2 +                                 1                  -1.2+                        1
  -1.4 + A                                                  -1.4+  10
  -1.6 +           1                                        -1.6+              1
  -1.8 +           1                                        -1.8+              1
  -2.0 +                                                    -2.0+
  -2.2 +                                                    -2.2+
  -2.4 +                                                    -2.4+
  -2.6 +                                                    -2.6+
  -2.8 +    1                                               -2.8+                        1
 BELOW+                                                    BELOW+
             +....+....+....+....+....+....+..                          ++++....++++....++++
```

Figure 4.7.   Residuals versus pseudo stress.

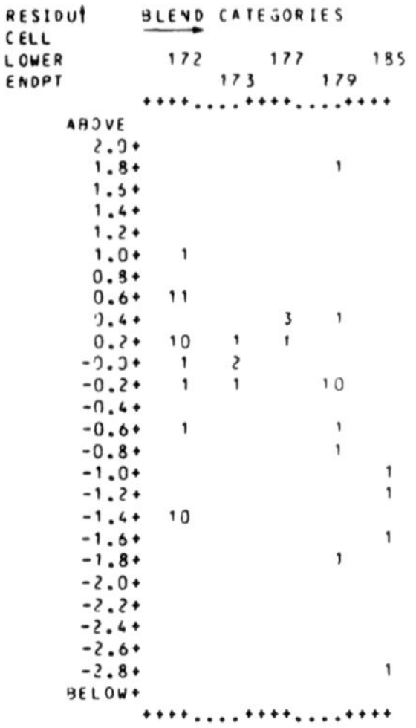

Figure 4.8.   Residuals versus blends.

better in the lower tail. In this application, only the lower tail is of interest; so their suggestion is useful. An interval observation yields an interval residual (two endpoints). Interval residuals can be plotted as if observed at their midpoints or plotted with the method of Peto (1973) and Turnbull (1976).

Crossplots. Crossplots of unitized residuals against stress or other variables can yield useful information. Figure 4.7 shows these residuals plotted against pseudo stress. Censored residuals are plotted as A's. B denotes an observed and a censored residual in that rectangle on the plot. If observed, the censored (unitized) residuals would be above their points in Figure 4.7. The crossplot shows no unusual features; thus the model and data appear satisfactory. Figure 4.8 shows these residuals plotted against blend. The metal is made in batches called blends. In the figure, 10 denotes a runout, and 11 denotes a runout and a failure. The plot suggests that blend 185 has a shorter fatigue life. The reason should be sought if it is statistically significantly shorter. So a formal comparison of the blends is needed first. This can be done by fitting a model that includes indicator variables for blends. Nelson (1984) plots these residuals against other variables.

Optimum plans. Test plans for models where σ is a function of stress were investigated by Meeter and Meeker (1989). They provide guidance on the choice of the test stress levels and the number of specimens at each.

4.3. Multivariable Relationships

Introduction. This section presents ML fitting of models with multivariable relationships for the location and scale parameters of the life distribution. This section also includes methods for assessing the model and data. Such fitting and assessments extend those in preceding sections to multivariable relationships. Thus preceding sections of this chapter are needed background. Also needed are Sections 13 and 14 of Chapter 2 on multivariable relationships. This section first describes the example data and model. Then it presents the ML output and discusses how to assess the model and data. These methods extend, as described in previous sections, to all types of data. The example presents right censored data, the most common type.

Two insulations. The example data come from two similar insulations, designated as 0 and 1. The data are from a voltage accelerated test with specimens of differing insulation thickness. The purpose of the analysis is to compare the insulations with respect to their dependence of life on voltage and thickness. All of the 106 specimens of insulation 0 are failed, and 92 of the 99 specimens of insulation 1 are failed.

Model. The fitted model has a lognormal distribution for life in hours. The mean log life is a function of the insulation SOURCE (the $0-1$ variable), LOGVPM (the log of voltage stress – volts per mm), and THICK (insulation thickness in cm); namely,

$$\mu = \gamma_1 + \gamma_2 SOURCE + \gamma_3 LOGVPM + \gamma_4 SOURCE \times LOGVPM$$

$$+ \gamma_5 THICK + \gamma_6 SOURCE \times THICK. \qquad (4.3.1)$$

This relationship uses the indicator variable SOURCE and its cross terms to compare the two insulations. Thus this relationship equivalent to two separate relationships for the two insulations; namely,

$$\mu_0 = \gamma_1 + \gamma_3 LOGVPM + \gamma_5 THICK, \qquad (4.3.2)$$

$$\mu_1 = (\gamma_1 + \gamma_2) + (\gamma_3 + \gamma_4) LOGVPM + (\gamma_5 + \gamma_6) THICK. \qquad (4.3.3)$$

Thus for an insulation, median life is an inverse power function of voltage stress and an exponential function of thickness. (4.3.3) shows that γ_2 is the difference between the intercepts of the two insulations. γ_4 is the difference between the powers in the power laws of the insulations. γ_6 is the difference between the thickness coefficients of the insulations. If an estimate of any of these three coefficients differs statistically significantly from zero, then the two insulations differ convincingly in that respect.

The standard deviation of log life is assumed to depend on the insulation. The assumed relationship is

$$\ln(\sigma) = \gamma_7 + \gamma_8 SOURCE. \qquad (4.3.4)$$

The use of the indicator variable *SOURCE* is equivalent to

$$\sigma_0 = \exp(\gamma_7) \quad \text{and} \quad \sigma_1 = \exp(\gamma_7 + \gamma_8). \qquad (4.3.5)$$

The model fitting program STATPAC "centers" all variables (including

SOURCE and the cross terms) in the relationships above. That is, the sample average of each variable is subtracted from each variable value before the variable is used in an equation. Thus the meaning and value of some coefficients in the output differs from that in the equations above. However, the key coefficients γ_2, γ_4, γ_6, and γ_8 are the same. Moreover, the relationships are the same but differently parameterized. It is better if a package lets the user decide how the relationship is parameterized. STATPAC uses its parameterization to make the iterative fitting converge more surely and rapidly and to insure more accurate calculations. When a ML package does not center each independent variable, the user should to assure accurate fitting.

More general models could use more general distributions and relationships. For example, the log gamma distribution may be useful when one needs to decide on the form of the distribution. Also, σ could depend on LOGVPM and THICK.

Output. STATPAC output from fitting the model appears in Figure 4.9. Line 1 shows the variables in the relationship for μ, where SRCLVM = SOURCE×LOGVPM and SCRBLD = SOURCE×THICK. Line 2 shows the variable SOURCE in the ln linear relationship for σ. Line 3 shows a maximum log likelihood of -1759.246; this can be used in various LR tests for the model. Line 4 shows the coefficient estimates and approximate 95% confidence limits. Conclusions are:

- The ML estimate of γ_2, the difference in intercepts, is C2 = 0.17, and the confidence limits are $(-8.51, 8.86)$. The interval encloses zero; thus there is no convincing difference in the intercepts of the two insulations.
- The ML estimate of γ_4, the difference in powers, is C4 = 0.01, and the confidence limits are $(-3.58, 3.61)$. The interval encloses zero; thus there is no convincing difference in the powers.
- The ML estimate of γ_6, the difference in thickness coefficients is C6 = -0.10, and the confidence limits are $(-4.98, 4.78)$. The interval encloses zero; thus there is no convincing difference in the thickness coefficients.

Each of these three estimates is very close to zero compared to the width of its confidence interval.

- The ML estimate of γ_8, the difference of the ln standard deviations, is C8 = 0.41, and the confidence limits are $(0.21, 0.60)$. The interval does not enclose zero; thus there is a convincing difference in the σ's.

In summary, only the insulation σ's differ convincingly.

Line 6 displays the form of the fitted relationship for μ, called CENTER here. Note that all variables, including cross terms, are centered. Line 7 displays the relationship for $\ln(\sigma)$, called SPREAD here.

Adequate σ relationship. Most models assume that the scale parameter (σ here) is constant. Then one tests whether it is constant. In models with a relationship for σ, assess the fit of the σ relationship as follows:

- Crossplot the **unitized** residuals against each variable in the relationship. Censored residuals should be identified in the plot. If there are many cen-

```
  DISTRIBUTION LOGNORMAL(HOURS)

1 RELATION CENTER 1STORDER(SOURCE LOGVPM SRCLVM THICK SCRBLD)

2 RELATION SPREAD 1STORDER(SOURCE)

  CASES
   198 WITH OBSERVED VALUES.
     7 WITH VALUES CENSORED ON THE RIGHT.
   ----
   205 IN ALL

3 MAXIMUM LOG LIKELIHOOD =  -1759.2460

  * MAXIMUM LIKELIHOOD ESTIMATES FOR MODEL COEFFICIENTS
    WITH APPROXIMATE 95% CONFIDENCE LIMITS

  PARAMETERS    ESTIMATE      LOWER LIMIT      UPPER LIMIT   STANDARD ERROR

4 C00001       3.404593        3.359651         3.449536    0.2292991E-01
  C00002       0.1765872      -8.510413         8.863587    4.432143
  C00003      -9.917424      -11.66295         -8.171894    0.8905768
  C00004       0.1156985E-01  -3.588896         3.612035    1.836972
  C00005      -2.220991       -4.203069        -0.2389132   1.011264
  C00006      -0.1013898      -4.984599         4.781819    2.491433
  C00007      -1.164480       -1.263478        -1.065483    0.5050898E-01
  C00008       0.4115606       0.2131610        0.6099603   0.1012243

  * CORRELATION MATRIX

  PARAMETERS   C00001          C00002           C00003          C00004

  C00001      1.0000000
  C00002      0.0163908      1.0000000
  C00003      0.0005551      0.4917794        1.0000000
5 C00004     -0.0124775     -0.9947363       -0.5000622      1.0000000
  C00005      0.0002882      0.2868598        0.5136944     -0.2552890
  C00006     -0.0045986     -0.6058051       -0.2153168      0.5215846
  C00007      0.0196192     -0.0374700       -0.0253823      0.0366767
  C00008      0.0197336     -0.0156666        0.0227263      0.0144723

  PARAMETERS   C00005          C00006           C00007          C00008

  C00005      1.0000000
  C00006     -0.4084182      1.0000000
  C00007     -0.0133711      0.0291553        1.0000000
  C00008      0.0120174      0.0199704        0.0435017       1.0000000

  * MAXIMUM LIKELIHOOD ESTIMATE OF THE FITTED EQUATION

6 CENTER  μ
    =   3.404593   +   (SOURCE -   0.4829268   ) * ( 0.1765872      )

               +   (LOGVPM -   2.248905.    ) * ( -9.917424        )

               +   (SRCLVM -   1.086457     ) * ( 0.1156985E-01)

               +   (THICK  -   0.2128927    ) * ( -2.220991        )

               +   (SCRBLD -   0.1025707    ) * ( -0.1013898       )

7 SPREAD  ln σ
    =  -1.164480   +   (SOURCE -   0.4829268   ) * ( 0.4115606      )
```

Figure 4.9.   STATPAC fit to two insulations.

sored residuals, they can be replaced by their conditional expected values, as described by Lawless (1982, p. 281). If the spread of the unitized residuals is not constant in a crossplot, then the effect of that variable is not adequately modeled.

- To help detect nonconstant σ and see how it depends on a variable, one can extend the scatterplot smoothing of residuals of Chambers and others (1983, Sect. 4.7) to censored residuals.
- Divide the residuals into groups according to a variable, for example, low, middle, and high values of the variable. Make a hazard plot of the unitized residuals for each group and compare their slopes. If the scale parameter estimate (slope) for any plot is not near 1, then the effect of that variable on σ is not adequately modeled.
- Fit to the data a more general relationship that includes the original relationship for σ, say, one that includes quadratic or cross terms. If some added coefficients are statistically significantly different from zero, then the original relationship is not adequate. One can then add those coefficients (terms) to the relationship.
- Use the LR test below for a more general relationship.

If the assumed relationship is not adequate, then these methods usually suggest how to improve it. Note that an inadequate relationship for the location parameter (μ here) can distort the fit of the relationship for σ.

Adequate μ relationship. For models with a multivariable relationship for the location parameter μ, one needs to assess the fit of the relationship. Methods for this appear in preceding sections. They include:

- Crossplot the residuals against each variable in the relationship. Plot censored residuals as described above. If such a plot is not a horizontal cloud of points, then the relationship needs to be improved with respect to that variable as suggested by the plot.
- Fit a more general relationship that includes the original relationship, say, one with quadratic or cross terms. If some added coefficients are statistically significantly different from zero, then the original relationship is not adequate. Then add those coefficients (terms) to the relationship.

If the assumed relationship for μ is not adequate, then these methods usually suggest how to improve it.

LR test for a relationship. A LR test can be used to compare the fit of a more general relationship and of the assumed one (for a scale or location parameter). The assumed relationship must be a special case of the general one. Usually the assumed relationship is the general one with certain coefficients equal to zero. Suppose that the number of distinct estimated coefficients in the general relationship is J, and the number in the assumed relationship is J'. Also, denote the maximum log likelihoods with the two relationships by $\hat{\mathcal{L}}$ (general) and $\hat{\mathcal{L}}'$ (assumed). Calculate the LR test statistic as $T = 2(\hat{\mathcal{L}} - \hat{\mathcal{L}}')$. If the true values of the extra $(J - J')$ coefficients in the general relationship are all zero, then the distribution of T is approximately

χ^2 with $(J-J')$ degrees of freedom. If any true values of those coefficients differ from zero, then T tends to have larger values. Thus the LR test is:

- If $T > \chi^2(1-\alpha; J-J')$, then the general relationship fits significantly better at the $100\alpha\%$ level. Such χ^2 percentiles appear is Appendix A5.
- If $T \le \chi^2(1-\alpha; J-J')$, then the general relationship does not fit significantly better.

If the general relationship fits significantly better, then one should determine which extra coefficients differ from zero. This LR test corresponds to the incremental F test in least-quares regression theory; there the test usually appears in an analysis-of-variance table. The test can also be simultaneously applied to location and scale parameter relationships, as shown next.

Example. An example of this test comes from the data on two insulations. Another model fitted to the data involved (1) a linear relationship in *LOGVPM* and *THICK* for μ and (2) a constant σ. That is, (4.3.2) was fitted to the pooled data. This model has $J' = 4$ estimated coefficients (or parameters) – γ_1, γ_3, γ_5, and σ. The maximum log likelihood is $\hat{\mathcal{L}}' = -1774.986$. The more general model above has $J = 8$ estimated coefficients, and $\hat{\mathcal{L}} = -1759.246$. The LR statistic is $T = 2[-1759.246 - (-1774.986)] = 31.48$ and has $(8-4) = 4$ degrees of freedom. Since $T = 31.48 > 18.47 = \chi^2(0.999; 4)$, the more general model fits very highly significantly better. The reason for this better fit can be seen from previous analyses; namely, the two σ's differ.

Stepwise fitting. Another means of arriving at an adequate model is stepwise fitting. Some packages listed in Section 1 do such fitting to censored data, employing the LR test above. Then fit a very general model with many terms. And the stepwise procedure selects the significant ones. The discussion of stepwise fitting in Chapter 4 is useful background. Peduzzi, Holford, and Hardy (1980) discuss stepwise procedures for ML fitting.

Assess the distribution. Adequacy of the assumed distribution can be assessed as described in Section 3.4. For example, a normal plot of the pooled unitized residuals of the two insulations appears in Figure 4.10. The plot is quite straight except for the two lowest points, possible outliers. They should be investigated to seek a cause. Otherwise, the lognormal distribution appears adequate. Also, the log gamma distribution can be fitted to assess whether the lognormal or Weibull distribution fits better.

Assess the data. The data can be assessed as described in Section 3.5. The chief methods are a probability plot and and crossplots of the residuals.

Effect of other variables. The effect of other variables can be assessed as in Section 3.6. The chief methods are crossplots of residuals and fitting of models with the other variables. Figures 4.7 and 4.8 are such crossplots. To help detect a relationship in a crossplot like Figure 4.7, one can extend to censored data the calculation and plot of the locally weighted regression scatterplot smoothing of Chambers and others (1983, Sec. 4.6).

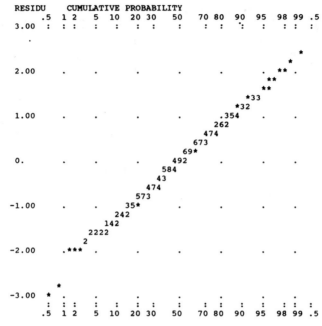

Figure 4.10.   Normal plot of unitized residuals for two insulations.

Applications. Multivariable relationships have been fitted to censored life data on various products. Examples include
- Sidik and others (1980) on battery cells,
- Zelen (1959) on capacitors,
- Montanari and Cacciari (1984) on electrical insulation.

The bibliography by Meeker (1980) contains multivariable applications.

Cox model. The Cox model, described in Chapter 2, usually involves a multivariable relationship. The model is nonparametric in the sense that no form of the life distribution is assumed. Engineering applications generally employ parametric models.

4.4. Some Coefficients Specified

Purpose. This section describes a method to estimate a multivariable relationship for life when the values of some coefficients are specified. There are two main reasons to specify the value of a coefficient. First, sometimes data are collected at just one level of the corresponding variable. The compressor oil test of Section 2.3 is an example; data were collected at just one test temperature. Then the coefficient for such a variable cannot be estimated from the data, and it is otherwise not possible to extrapolate to other levels of that variable. Indeed, many of the tests of MIL-STD-883 involve a single level of a accelerating stress. By assuming or specifying a value for such a coefficient, one can extrapolate as shown below. Second, if the data likely yield an inaccurate estimate of a coefficient, a specified value can yield more accurate results.

Coefficient value. The specified value of a coefficient may be
- An estimate based on data from a similar product.
- A handbook value (e.g., from MIL-HDBK-217).
- A traditional value (life doubles with each 10°C decrease in temperature or the activation energy is 1.0 eV).
- An educated guess.

Model. The following simple and concrete relationship illustrates the method. Assume that the typical (log) life is a linear function

$$\mu(x_1, x_2) = \gamma_0 + \gamma_1 x_1 + \gamma_2 x_2 \tag{4.4.1}$$

of two variables x_1 and x_2. Suppose that the specified value γ_2' is assigned to γ_2; for most models, this value is negative. Then the coefficients γ_0 and γ_1 are to be estimated from data. The method extends to nonlinear relationships, any number of variables, and any number of specified coefficients. Choose a reference value x_2' of variable x_2. x_2' may be the design value or some other value including zero. The relationship can be rewritten as

$$\mu(x_1, x_2) = \gamma_0' + \gamma_1 x_1 + \gamma_2'(x_2 - x_2'). \tag{4.4.2}$$

Here $\gamma_0' = \gamma_0 + \gamma_2' x_2'$. Then the relationship between typical (log) life and x_1 when $x_2 = x_2'$ (the reference value) is

$$\mu'(x_1) = \gamma_0' + \gamma_1 x_1. \tag{4.4.3}$$

This relationship is fitted to transformed data as follows.

Transformed data. Suppose that specimen i has variable values x_{1i} and x_{2i}. Also, suppose its observed (or censored) log life is y_i. Calculate the *transformed log life*

$$y_i' = y_i - \gamma_2'(x_{2i} - x_2'). \tag{4.4.4}$$

This is the log life that specimen i would have had if run at $x_{2i} = x_2'$. Such a transformed life may be observed or censored on the right or left. For interval data, calculate transformed lower and upper endpoints. Here $K_i = \exp[-\gamma_2'(x_{2i} - x_2')]$ is an *acceleration factor* that converts life $t_i = \exp(y_i)$ at stress level x_{2i} to life $t_i' = \exp(y_i')$ at stress level x_2'. That is, $t_i' = K_i t_i$.

Data fitting. Then fit the relationship (4.4.3) to the transformed data (4.4.4) by graphical or numerical means. This yields estimates and confidence limits for the unspecified coefficients and other quantities. For censored (and interval) data, use maximum likelihood fitting. For complete data, use ML or least squares fitting. The estimates are unbiased and the confidence limits are correct only if the specified coefficient values are correct. Thus the uncertainties in estimates are greater than such confidence limits indicate. Redo the fitting with other specified coefficient values to assess their effect on results. Also, one can assess the fit of the model and the residuals as described in Section 3.

5. MAXIMUM LIKELIHOOD CALCULATIONS

This advanced section presents maximum likelihood (ML) theory and calculations for fitting a model to observed, censored, and interval data. These calculations yield ML estimates and approximate confidence intervals. In principle the ML method is simple and consists of three steps:

1. $\mathcal{L}$. Write the sample log likelihood $\mathcal{L}$ (Section 5.3). It is (i) a function of the sample data (Section 5.1) and the form of the life data (observed, censored, or interval) and (ii) a function of the model (Section 5.2) and its J parameters and coefficients γ_j.

2. $\partial\mathcal{L}/\partial\gamma_j$. From the sample likelihood $\mathcal{L}$ or from the likelihood equations $\partial\mathcal{L}/\partial\gamma_j = 0$ (5.12), calculate the ML estimates $\hat{\gamma}_j$ of the J model coefficients γ_j (Section 5.4).

3. $\partial^2\mathcal{L}/\partial\gamma_j\partial\gamma_{j'}$. From the $J \times J$ matrix (5.13) of second partial derivatives of $\mathcal{L}$ with respect to the model coefficients γ_j, calculate approximate confidence limits for model coefficients and other quantities (Sections 5.4-5.7).

The presentation covers
- The form of the data,
- The model, consisting of a statistical distribution for the dependent variable and relationships for the distribution parameters in terms of the independent variables and unknown model coefficients,
- The sample log likelihood,
- Maximum likelihood estimates of model coefficients,
- Fisher and covariance matrices of the estimates,
- Estimate of a function of the coefficients and its variance,
- Approximate normal confidence intervals,
- Approximate LR confidence intervals.

These are illustrated with the Arrhenius-lognormal model applied to the Class-B insulation data. Readers need to know the basics of matrix algebra and partial differentiation. Further theory appears in Chapters 6 (true covariance matrix) and 9 (LR tests). Results and methods are presented without regularity conditions, proofs, and motivation, which are given by Wilks (1962), Rao (1973), Hoadley (1971), Rao (1987), and Nelson (1982).

5.1. Data

This section describes the general form of data from an accelerated test, particularly, the data matrix and the censoring.

Data matrix. The data are assumed to comprise a data matrix. Each of the n sample specimens has a value of a dependent variable y (usually life or log life) and a value for each of the J independent variables $x_1, \cdots, x_J$. For specimen i, these are denoted by $(y_i, x_{1i}, \cdots, x_{Ji})$, called a **data case**, where $i = 1, 2, \cdots, n$. The independent variables may be numerical or indicator variables. For the calculations, the data values of these variables are regarded as given and fixed. Also, the data include the censoring information.

Censoring. For interval data, the dependent variable consists of two values y_i and y_i', the lower and upper limits of the interval. Also, for censored or interval data, the form of the data is usually indicated by an additional variable. For example, STATPAC of Nelson and others (1983) employs a variable u that takes on the value 1 for an exact observed failure time, 2 for a right censored time, 4 for a left censored time, and 6 for an interval.

Example. The Arrhenius-lognormal model is used for the Class-B insulation data (Table 4.1 of Chapter 3). The data on specimen i consist of its log time y_i, its reciprocal absolute temperature $x_i = 1000/T_i$, and the form u_i (censoring) of its data. In this application, times are either observed or censored on the right.

5.2. The Model

A **model** fitted to data consists of a **statistical distribution** for the dependent variable (usually life) and **relationships** for the distribution parameters. The relationships express the distribution parameters as functions of the independent variables and unknown model coefficients. These are described next in general terms.

The statistical distribution. The dependent variable y is assumed to have a specified continuous cumulative distribution function

$$F(y; \theta_1, \cdots, \theta_Q); \tag{5.1}$$

here $\theta_1, \cdots, \theta_Q$ are the Q distribution parameters. The probability density is

$$f(y; \theta_1, \cdots, \theta_Q) \equiv dF(y; \theta_1, \cdots, \theta_Q)/dy. \tag{5.2}$$

More generally, the form of distribution can differ from specimen to specimen and would be denoted F_i or f_i. This generality is not needed here and is omitted. It is needed in Chapter 10 for different step stress patterns.

Example. For the Arrhenius-lognormal model, the cumulative distribution for y, the (base 10) log of life, is

$$F(y; \mu, \sigma) = \Phi[(y - \mu)/\sigma];$$

here $\Phi[\ ]$ is the standard normal cumulative distribution function, and μ and σ are the $Q = 2$ distribution parameters. The probability density is

$$f(y; \mu, \sigma) = (1/\sigma)\, \phi[(y - \mu)/\sigma];$$

here $\phi[z] = (2\pi)^{-1/2} \exp(-z^2/2)$ is the standard normal probability density.

Relationships for the distribution parameters. Each distribution parameter is expressed as an assumed function of the J independent variables $x_1, \cdots, x_J$ and the P distinct model coefficients $\gamma_1, \cdots, \gamma_P$; that is,

$$\theta_1 = \theta_1(x_1, \cdots, x_J; \gamma_1, \cdots, \gamma_P),$$

$$\cdot$$
$$\cdot \qquad\qquad\qquad (5.3)$$
$$\cdot$$

$$\theta_Q = \theta_Q(x_1, \cdots, x_J; \gamma_1, \cdots, \gamma_P).$$

Usually a particular coefficient appears in only one of the relationships. The notation in (5.3) does not mean that each coefficient appears in each relationship but only that a coefficient could do so. The functional form of each relationship is specified (assumed), but the values of the model coefficients are unknown. They are estimated from the data as described below.

Example. For the Arrhenius-lognormal model, the relationships are

$$\mu = \gamma_1 + \gamma_2 x, \quad \sigma = \gamma_3,$$

where x is the reciprocal absolute temperature. This agrees with earlier notation. The terms "parameter" and "coefficient" are sometimes equivalent when a distribution parameter is a constant, e.g., $\sigma = \gamma_3$. Here $P = 3$.

Other parameterizations. One can reparameterize the relationships (5.3). This is done for various reasons. For example, sometimes one uses $\sigma = \exp(\gamma_3')$. This form of σ allows γ_3' to range from $-\infty$ to $+\infty$ while σ remains positive; this allows unconstrained optimization and avoids negative trial values of $\hat{\sigma}$ as a program searches for the ML estimate. Also, the sampling distribution of the ML estimate of γ_3' may be closer to normal than that for γ_3. So the approximate normal confidence limits (Section 5.7) for γ_3' yield positive limits for σ that may be more accurate. The parameter σ could also be modeled as a function of x, for example, $\sigma = \exp(\gamma_3 + \gamma_4 x)$.

Notation. For specimen i, the values of the distribution parameters are

$$\theta_{1i} = \theta_1(x_{1i}, \cdots, x_{Ji}; \gamma_1, \cdots, \gamma_P),$$

$$\cdot$$
$$\cdot \qquad\qquad\qquad (5.4)$$
$$\cdot$$

$$\theta_{Qi} = \theta_Q(x_{1i}, \cdots, x_{Ji}; \gamma_1, \cdots, \gamma_P).$$

The distribution parameters for specimen i are denoted simply by $\theta_{1i}, \cdots, \theta_{Qi}$. Their dependence on $x_{1i}, \cdots, x_{Ji}$ and $\gamma_1, \cdots, \gamma_P$ is thus not explicit in the notation but should be understood. For the Arrhenius-lognormal model,

$$\mu_i = \gamma_1 + \gamma_2 x_i, \quad \sigma_i = \sigma, \quad \text{for } i = 1, \cdots, n.$$

For example, consider the Class-B insulation data of Section 2, appearing in Table 4.1 of Chapter 3. Specimen 25 at 190°C is a failure at 1440 hours, and $x_{25} = 1000/(190 + 273.16) = 2.1591$. Its log mean and standard deviation are $\mu_{25} = \gamma_1 + \gamma_2 2.1591$ and $\sigma_{25} = \sigma$.

5.3. Likelihood Function

The ML method for estimating the model coefficients is based on the sample (log) likelihood of a set of data. This section describes in detail its calculation. Its use is described below.

Specimen likelihood. The **likelihood** for a specimen (or data case) is, loosely speaking, the probability of its observed value of the dependent variable (usually life). The dependent (or response) variable can be observed to have an exact value (failure age) or be censored and in an interval. In this context, left and right censored data are regarded as interval data. The specimen likelihood for each type of data follows.

Observed. Suppose specimen i has an observed value y_i of the dependent variable. Then its likelihood is

$$L_i = f(y_i; \theta_{1i}, \cdots, \theta_{Qi}); \tag{5.5}$$

here $f(\ )$ is the probability density of the assumed distribution, and $\theta_{1i}, \cdots, \theta_{Qi}$ are the specimen's parameter values. This L_i is the "probability" of an observed failure at y_i. This L_i is a function of $y_i, x_{1i}, \cdots, x_{Ji}$ and $\gamma_1, \cdots, \gamma_P$. Class-B specimen 25 failed at log age $y_{25} = \log(1440) = 3.1584$, and

$$L_{25} = (\sqrt{2\pi}\,\sigma)^{-1} \exp[-(3.1584 - \gamma_1 - \gamma_2 2.1591)^2/(2\sigma^2)].$$

Right censored. Suppose specimen i has the dependent variable censored on the right at y_i; that is, its value is above y_i. Its likelihood is

$$L_i = 1 - F(y_i; \theta_{1i}, \cdots, \theta_{Qi}). \tag{5.6}$$

This L_i is the probability that the specimen's life is above y_i. The distribution parameters $\theta_{1i}, \cdots, \theta_{Qi}$ contain $x_{1i}, \cdots, x_{Ji}$ and $\gamma_1, \cdots, \gamma_P$. For example, Class-B specimen 26 survived log age $y_{26} = \log(1680) = 3.2253$, and

$$L_{26} = 1 - \Phi[(3.2253 - \gamma_1 - \gamma_2 2.1591)/\sigma].$$

Left censored. Suppose specimen i has the dependent variable censored on the left at y_i; that is, its value is below y_i. Its likelihood is

$$L_i = F(y_i; \theta_{1i}, \cdots, \theta_{Qi}). \tag{5.7}$$

This L_i is the probability that the specimen's life is below y_i. As before, the parameters $\theta_{1i}, \cdots, \theta_{Qi}$ contain $x_{1i}, \cdots, x_{Ji}$ and $\gamma_1, \cdots, \gamma_P$.

Interval. Suppose specimen i has a dependent variable value that is known only to be in an interval with endpoints $y_i < y_i'$. Then its likelihood is

$$L_i = F(y_i'; \theta_{1i}, \cdots, \theta_{Qi}) - F(y_i; \theta_{1i}, \cdots, \theta_{Qi}). \tag{5.8}$$

This L_i is the probability that the specimen's life is in the interval (y_i, y_i'). As before, the parameters $\theta_{1i}, \cdots, \theta_{Qi}$ contain $x_{1i}, \cdots, x_{Ji}$ and $\gamma_1, \cdots, \gamma_P$. This likelihood reduces to (5.7) if the lower endpoint is $-\infty$ or zero and to (5.6) if the upper endpoint is $+\infty$.

Sample likelihood. It is assumed that the n sample specimens have **statistically independent** random variations in their values of the dependent variable. Or, briefly stated, their lifetimes are statistically independent. Then, by independence, the likelihood (probability) L of the sample outcome is the product of likelihoods (probabilities) of the specimens:

$$L \equiv L_1 \times L_2 \times \cdots \times L_n. \tag{5.9}$$

This **sample likelihood** is the joint probability of the n dependent variable outcomes. For example, for the Class-B data, the sample likelihood is the product of $n = 40$ specimen likelihoods like the two above.

Log likelihood. The **log likelihood** of specimen i is

$$\mathcal{L}_i \equiv \ln(L_i). \tag{5.10}$$

Note that this is the base e log and should be called the ln likelihood if consistent with notation in this book. The **sample log likelihood** is

$$\mathcal{L} \equiv \ln(L) = \mathcal{L}_1 + \mathcal{L}_2 + \cdots + \mathcal{L}_n. \tag{5.11}$$

In what follows, $\mathcal{L}$ and $\mathcal{L}_i$ are regarded as functions of $\gamma_1, \cdots, \gamma_P$:

$$\mathcal{L} = \mathcal{L}(\gamma_1, \cdots, \gamma_P), \quad \mathcal{L}_i = \mathcal{L}_i(\gamma_1, \cdots, \gamma_P), \quad i = 1, \cdots, n.$$

Figure 5.1 depicts a sample likelihood that is a function of two coefficients. Such likelihoods depend implicitly on the assumed statistical distribution, on the assumed relationships for the distribution parameters (which are functions of the independent variables and model coefficients), on the form of the data, and on $x_{1i}, \cdots, x_{Ji}$. The Class-B sample log likelihood is a function of γ_1, γ_2, and σ and is the sum of 40 specimen log likelihoods like the two above.

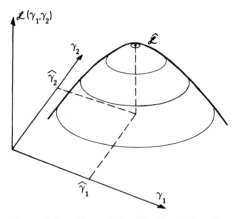

Figure 5.1.   Sample log likelihood function.

Example. For the Arrhenius-lognormal model and an observed log life y_i, the specimen log likelihood is

$$\mathcal{L}_i(\gamma_1,\gamma_{2,}\sigma) = -(1/2)\ln(2\pi) - \ln(\sigma) - (1/2)[(y_i - \gamma_1 - \gamma_2 x_i)/\sigma]^2.$$

Here $z_i = (y_i - \gamma_1 - \gamma_2 x_i)/\sigma$ is a **standardized deviate**. For a specimen life censored on the right at log age y_i, the specimen log likelihood is

$$\mathcal{L}_i(\gamma_1,\gamma_2,\sigma) = \ln\{1 - \Phi[(y_i - \gamma_1 - \gamma_2 x_i)/\sigma]\}.$$

The sample log likelihood is a function of γ_1, γ_2, and σ. Namely,

$$\mathcal{L}(\gamma_1,\gamma_2,\sigma) = \sum_i' \{-(1/2)\ln(2\pi) - \ln(\sigma) - (1/2)[(y_i - \gamma_1 - \gamma_2 x_i)/\sigma]^2\}$$
$$+ \sum_i'' \ln\{1 - \Phi[(y_i - \gamma_1 - \gamma_2 x_i)/\sigma]\};$$

here the first sum $\sum_i'$ runs over the observed failures, and the second sum $\sum_i''$ runs over the nonfailures (censored on the right). The Class-B sample log likelihood has 17 such failure terms and 23 such survivor terms.

Basic likelihoods. Table 5.1 displays the log likelihoods for a specimen for the basic distributions and the various types of data. The notation for the distributions and their parameters is that in Chapter 2. The tabled log likelihoods are expressed in terms of standardized deviates z_i. Their use simplifies the formulas and the calculations. Each z_i may contain the specimen's $x_{1i}, \cdots, x_{Ji}$ and the $\gamma_1, \cdots, \gamma_P$. These log likelihoods for data cases are used to calculate the sample log likelihood in (5.11).

Table 5.1. Log Likelihood $\mathcal{L}_i$ for Types of Data and Basic Distributions

Distribution	Standardized Deviate	Observed Value y_i
Normal (μ,σ)	$z_i = (y_i - \mu_i)/\sigma_i$	$-(1/2)\ln(2\pi) - \ln(\sigma_i) - (1/2)\, z_i^2$
Log$_{10}$normal (μ,σ)	$z_i = [\log(t_i) - \mu_i]/\sigma_i$	$-(1/2)\ln(2\pi\sigma_i^2) + \ln[\log(e)] - (1/2)\, z_i^2 - \ln(t_i)$
Extr. Value (ξ,δ)	$z_i = (y_i - \xi_i)/\delta_i$	$-\ln(\delta_i) - \exp(z_i) + z_i$
Weibull (α,β)	$z_i = [\ln(t_i) - \ln(\alpha_i)]\beta_i$	$\ln(\beta_i) - \exp(z_i) + z_i - \ln(t_i)$
Exponential (θ)	$z_i = t_i/\theta_i$	$-\ln(\theta_i) - z_i$

Distribution	Value above y_i	Value below y_i	Interval Value Between y_i and y_i'
Normal (μ,σ)	$\ln[1 - \Phi(z_i)]$	$\ln[\Phi(z_i)]$	$\ln[\Phi(z_i') - \Phi(z_i)]$
Log$_{10}$normal (μ,σ)	$\ln[1 - \Phi(z_i)]$	$\ln[\Phi(z_i)]$	$\ln[\Phi(z_i') - \Phi(z_i)]$
Extr. Value (ξ,δ)	$-\exp(z_i)$	$\ln\{1 - \exp[-e^{z_i}]\}$	$\ln\{\exp[-e^{z_i}] - \exp[-e^{z_i'}]\}$
Weibull (α,β)	$-\exp(z_i)$	$\ln\{1 - \exp[-e^{z_i}]\}$	$\ln\{\exp[-e^{z_i}] - \exp[-e^{z_i'}]\}$
Exponential (θ)	$-z_i$	$\ln[1 - \exp(-z_i)]$	$\ln[\exp(-z_i) - \exp(-z_i')]$

Other likelihoods. The preceding formulas also yield the log likelihoods for any other distribution. The log likelihoods are then used as described below to fit the model to data.

5.4. ML Estimates of Model Coefficients

Estimates. The **maximum likelihood estimates** $\hat{\gamma}_1, \cdots, \hat{\gamma}_P$ of $\gamma_1, \cdots, \gamma_P$ are the coefficient values that maximize the sample log likelihood (5.11) over the allowed ranges of $\gamma_1, \cdots, \gamma_P$. Figure 5.1 depicts a sample log likelihood and the corresponding ML estimates. Thus, the estimates are the coefficient values that maximize the probability (likelihood) of the sample data. This intuitively plausible criterion for fitting yields estimators with good properties, as theory shows. Usually the estimates are obtained by iterative numerical optimization methods described by Nelson (1982, Chap. 8), Kennedy and Gentle (1980), and especially Ross (1990) and Thisted (1987). For some models each ML estimate $\hat{\gamma}_p$ is a function of all the sample values $(y_i, x_{1i}, \cdots, x_{Ji})$, $i = 1, \cdots, n$. An estimate can in theory be written as $\hat{\gamma}_p = \hat{\gamma}_p(y_1, x_{11}, \cdots, x_{J1}; \cdots; y_n, x_{1n}, \cdots, x_{Jn})$. In practice, however, one usually cannot write $\hat{\gamma}_p$ as an explicit function of the data. The *maximum log likelihood* value is $\hat{\mathcal{L}} = \mathcal{L}(\hat{\gamma}_1, \cdots, \hat{\gamma}_P)$. For the Class-B example in Figure 2.1, line 2 shows $\hat{\mathcal{L}}$, and line 4 shows the ML estimates.

Likelihood equations. The values $\hat{\gamma}_1, \cdots, \hat{\gamma}_P$ that maximize $\mathcal{L}(\gamma_1, \cdots, \gamma_P)$ sometimes can be found by the usual calculus method. Namely, set equal to zero the P derivatives of $\mathcal{L}(\gamma_1, \cdots, \gamma_P)$ with respect to $\gamma_1, \cdots, \gamma_P$, and solve the following **likelihood equations** for $\hat{\gamma}_1, \cdots, \hat{\gamma}_P$:

$$\partial \mathcal{L}(\gamma_1, \cdots, \gamma_P)/\partial \gamma_1 = 0,$$
$$\vdots \qquad\qquad (5.12)$$
$$\partial \mathcal{L}(\gamma_1, \cdots, \gamma_P)/\partial \gamma_P = 0.$$

Usually these nonlinear equations in $\gamma_1, \cdots, \gamma_P$ cannot be solved algebraically. Then they must be solved with numerical methods. Alternatively, many computer programs directly optimize $\mathcal{L}(\ )$ by numerical search. Nelson (1982, Chap. 8) and his references discuss numerical aspects of such optimization methods. After solving (5.12), one needs to check that the solution is at the global maximum – not at a local maximum nor saddle point. For example, the solution is at an optimum if all eigenvalues of the local Fisher matrix are positive. Such optimization converges faster and more accurately if each independent variable is "centered" by subtracting from it some value near the center of the data on that variable, such as the mean.

Example. For the Arrhenius-lognormal model and complete data, the likelihood equations can be solved explicitly. The equations are

$$0 = \partial \mathcal{L}/\partial \gamma_1 = \sum_{i=1}^{n}(y_i - \gamma_1 - \gamma_2 x_i)/\sigma^2, \quad 0 = \partial \mathcal{L}/\partial \gamma_2 = \sum_{i=1}^{n} x_i (y_i - \gamma_1 - \gamma_2 x_i)/\sigma^2,$$

$$0 = \partial \mathcal{L}/\partial \sigma = \sum_{i=1}^{n} \{(-1/\sigma) + [(y_i - \gamma_1 - \gamma_2 x_i)^2/\sigma^3]\}.$$

Let $\bar{y} = \sum_{i=1}^{n} y_i/n$ and $\bar{x} = \sum_{i=1}^{n} x_i/n$. Then the explicit and unique solutions are

$$\hat{\gamma}_2 = \left[\sum_{i=1}^{n}(y_i - \bar{y})(x_i - \bar{y})\right] \Big/ \left[\sum_{i=1}^{n}(x_i - \bar{x})^2\right], \quad \hat{\gamma}_1 = \bar{y} - \hat{\gamma}_2 \bar{x},$$

$$\hat{\sigma} = \left\{\sum_{i=1}^{n}(y_i - \hat{\gamma}_1 - \hat{\gamma}_2 x_i)^2/n\right\}^{1/2}.$$

These are the same as the least-squares estimates of Chapter 4. However, in Chapter 4, the estimate s' for σ contains $(n-1)$ in place of n. That is, $\hat{\sigma} = s'\sqrt{(n-1)/n}$.

Existence and uniqueness. For some models and data, the estimates may not exist; for example, they may have physically unacceptable values such as infinity or zero (say, for a standard deviation). For most models and data, the ML estimates are unique. But log likelihood functions for some models and data can have any number of local maxima. Of course, the ML estimates correspond to the global maximum. Then special numerical optimization techniques are needed to locate the global maximum.

Asymptotic theory. The following asymptotic theory for ML estimates applies to samples with **many failures** and models that satisfy certain regularity conditions referenced in Section 1. Most models used in practice satisfy such conditions. Then the joint sampling distribution of $\hat{\gamma}_1, \cdots, \hat{\gamma}_P$ is close to a multivariate normal cumulative distribution with means $\gamma_1, \cdots, \gamma_P$ and the covariance matrix estimated below. That is, the cumulative sampling distribution of $\hat{\gamma}_1, \cdots, \hat{\gamma}_P$ converges "in law" ("in distribution") to a joint normal cumulative distribution. This does not necessarily mean that the true means, variances, and covariances of $\hat{\gamma}_1, \cdots, \hat{\gamma}_P$ converge to those of the asymptotic normal distribution. However, the asymptotic (approximate) normal distribution is valid for calculating the usual approximate normal confidence limits (Section 5.7) for $\gamma_1, \cdots, \gamma_P$. Such limits employ the (normal) probability that a ML estimate is within a specified multiple of its standard error from its true value. Chapter 6 presents the true theoretical asymptotic covariance matrices. The asymptotic normality depends on central limit theorems for independent but not identically distributed random variables; see, for example, Hoadley (1971). The LR confidence limits (Section 5.8) generally are closer to the specified confidence level in each tail. However, the normal approximate limits are calculated by most ML packages.

5.5. Fisher and Covariance Matrices

After obtaining $\hat{\gamma}_1, \cdots, \hat{\gamma}_P$, one estimates their covariance matrix as described next. This matrix, calculated from the Fisher matrix, later yields approximate confidence intervals for various quantities.

Fisher matrix. The **local Fisher information matrix** is the $P{\times}P$ symmetric matrix of negative second partial derivatives

$$
\mathbf{F} = \begin{bmatrix}
-\partial^2\hat{\mathcal{L}}/\partial\gamma_1^2 & -\partial^2\hat{\mathcal{L}}/\partial\gamma_1\partial\gamma_2 & \cdots & -\partial^2\hat{\mathcal{L}}/\partial\gamma_1\partial\gamma_P \\
-\partial^2\hat{\mathcal{L}}/\partial\gamma_2\partial\gamma_1 & -\partial^2\hat{\mathcal{L}}/\partial\gamma_2^2 & \cdots & -\partial^2\hat{\mathcal{L}}/\partial\gamma_2\partial\gamma_P \\
\vdots & \vdots & \ddots & \vdots \\
-\partial^2\hat{\mathcal{L}}/\partial\gamma_P\partial\gamma_1 & -\partial^2\hat{\mathcal{L}}/\partial\gamma_P\partial\gamma_2 & \cdots & -\partial^2\hat{\mathcal{L}}/\partial\gamma_P^2
\end{bmatrix}. \tag{5.13}
$$

The caret $\hat{}$ indicates that the derivative is evaluated at $\gamma_1 = \hat{\gamma}_1, \cdots, \gamma_P = \hat{\gamma}_P$. $\mathbf{F}$ estimates the true (asymptotic) Fisher matrix, which appears in Chapter 6 with the asymptotic theory. For the Class-B example, line 3 of Figure 2.1 shows the local Fisher matrix. However, $C3 = \ln(\sigma)$ is the parameter there.

Some computer packages evaluate $\mathbf{F}$ from analytic expressions for the derivatives. Others numerically approximate the derivatives by a perturbation calculation of second differences. Then, for accuracy, the package must first calculate the second difference for each data case and then sum the second differences over the sample.

Example. For the Arrhenius-lognormal example with complete data, the second partial derivatives are

$$
\partial^2\mathcal{L}/\partial\gamma_1^2 = \sum_{i=1}^{n}(-1/\sigma^2), \quad \partial^2\mathcal{L}/\partial\gamma_1\partial\gamma_2 = \sum_{i=1}^{n}(-x_i/\sigma^2),
$$

$$
\partial^2\mathcal{L}/\partial\gamma_1\partial\sigma = -2\sum_{i=1}^{n}(y_i-\gamma_1-\gamma_2 x_i)/\sigma^3, \quad \partial^2\mathcal{L}/\partial\gamma_2^2 = \sum_{i=1}^{n}(-x_i^2/\sigma^2),
$$

$$
\partial^2\mathcal{L}/\partial\gamma_2\partial\sigma = -2\sum_{i=1}^{n}x_i(y_i-\gamma_1-\gamma_2 x_i)/\sigma^3,
$$

$$
\partial^2\mathcal{L}/\partial\sigma^2 = \sum_{i=1}^{n}\{(1/\sigma^2)-3[(y_i-\gamma_1-\gamma_2 x_i)^2/\sigma^4]\}.
$$

Evaluated at $\gamma_1 = \hat{\gamma}_1$, $\gamma_2 = \hat{\gamma}_2$, and $\sigma = \hat{\sigma}$,

$$
\partial^2\hat{\mathcal{L}}/\partial\gamma_1^2 = -n/\hat{\sigma}^2, \quad \partial^2\hat{\mathcal{L}}/\partial\gamma_1\partial\gamma_2 = -n\bar{x}/\hat{\sigma}^2, \quad \partial^2\hat{\mathcal{L}}/\partial\gamma_2\partial\sigma = 0,
$$

$$
\partial^2\hat{\mathcal{L}}/\partial\gamma_1\partial\sigma = 0, \quad \partial^2\hat{\mathcal{L}}/\partial\gamma_2^2 = -\left(\sum_{i=1}^{n}x_i^2\right)/\hat{\sigma}^2, \quad \partial^2\mathcal{L}/\partial\sigma^2 = -2n/\hat{\sigma}^2.
$$

Thus the local Fisher information matrix is

$$\mathbf{F} = \begin{bmatrix} n/\hat{\sigma}^2 & n\bar{x}/\hat{\sigma}^2 & 0 \\ n\bar{x}/\hat{\sigma}^2 & \sum_{i=1}^{n} x_i^2/\hat{\sigma}^2 & 0 \\ 0 & 0 & 2n/\hat{\sigma}^2 \end{bmatrix}.$$

ML programs give the numerical value of the matrix.

Covariance matrix. The inverse of $\mathbf{F}$ is the *local estimate* $\mathbf{V}$ *of the* (asymptotic) covariance matrix for $\hat{\gamma}_1, \cdots, \hat{\gamma}_P$. That is,

$$\mathbf{V} = \mathbf{F}^{-1} = \begin{bmatrix} \text{var}(\hat{\gamma}_1) & \text{cov}(\hat{\gamma}_1,\hat{\gamma}_2) & \cdots & \text{cov}(\hat{\gamma}_1,\hat{\gamma}_P) \\ \text{cov}(\hat{\gamma}_2,\hat{\gamma}_1) & \text{var}(\hat{\gamma}_2) & \cdots & \text{cov}(\hat{\gamma}_2,\hat{\gamma}_P) \\ \vdots & \vdots & \ddots & \vdots \\ \text{cov}(\hat{\gamma}_P,\hat{\gamma}_1) & \text{cov}(\hat{\gamma}_P,\hat{\gamma}_2) & \cdots & \text{var}(\hat{\gamma}_P) \end{bmatrix}. \tag{5.14}$$

$\mathbf{V}$ estimates the true (asymptotic) covariance matrix, which appears in Chapter 6. Note that the variances and covariances in $\mathbf{V}$ are in the same positions as the corresponding second partial derivatives in $\mathbf{F}$. For the Class-B example, line 5 of Figure 2.1 shows the numerical value of this estimate of the covariance matrix. There the parameter is σ instead of C3. The added calculations for this substitution are not explained here; see Nelson (1982, p. 374).

Standard error. The standard error $\sigma(\hat{\gamma}_p)$ of $\hat{\gamma}_p$ is the standard deviation of its asymptotic normal distribution. The estimate of $\sigma(\hat{\gamma}_p)$ is

$$s(\hat{\gamma}_p) = [\text{var}(\hat{\gamma}_p)]^{1/2}. \tag{5.15}$$

This is used below for approximate confidence intervals for the true γ_p value. For the Class-B example, line 4 of Figure 2.1 shows these estimates of the standard errors.

Example. For the Arrhenius-lognormal example with complete data, the local estimate of the symmetric covariance matrix is

$$\mathbf{V} = \begin{bmatrix} \text{var}(\hat{\gamma}_1) & \text{cov}(\hat{\gamma}_1,\hat{\gamma}_2) & \text{cov}(\hat{\gamma}_1,\hat{\sigma}) \\ \text{cov}(\hat{\gamma}_2,\hat{\gamma}_1) & \text{var}(\hat{\gamma}_2) & \text{cov}(\hat{\gamma}_2,\hat{\sigma}) \\ \text{cov}(\hat{\sigma},\hat{\gamma}_1) & \text{cov}(\hat{\sigma},\hat{\gamma}_2) & \text{var}(\hat{\sigma}) \end{bmatrix}$$

$$= \mathbf{F}^{-1} = \begin{bmatrix} \hat{\sigma}^2 \left[\sum_{i=1}^{n} x_i^2/n\right]/S_{xx} & -\hat{\sigma}^2\bar{x}/S_{xx} & 0 \\ -\hat{\sigma}^2\bar{x}/S_{xx} & \hat{\sigma}^2/S_{xx} & 0 \\ 0 & 0 & \hat{\sigma}^2/(2n) \end{bmatrix};$$

here $S_{xx} = \sum\limits_{i=1}^{n} x_i^2 - n\bar{x}^2$. The estimates of the standard errors are

$$s(\hat{\gamma}_1) = \hat{\sigma}\left[\left(\sum_{i=1}^{n} x_i^2 / n\right) / S_{xx}\right]^{1/2}, \quad s(\hat{\gamma}_2) = \hat{\sigma}/S_{xx}^{1/2}, \quad s(\hat{\sigma}) = \hat{\sigma}/(2n)^{1/2}.$$

5.6. Estimate of a Function and Its Variance

Estimate. Often one wants to estimate the value of a given function $h = h(\gamma_1, \cdots, \gamma_P)$ of the model coefficients. Distribution parameters and percentiles are examples of such functions. The ML estimate $\hat{h}$ of h is the function evaluated at $\gamma_1 = \hat{\gamma}_1, \cdots, \gamma_P = \hat{\gamma}_P$; that is,

$$\hat{h} = h(\hat{\gamma}_1, \cdots, \hat{\gamma}_P). \tag{5.16}$$

For the Class-B example, lines 8 and 9 show such percentile estimates at two temperatures.

Variance. The estimate of the variance of $\hat{h}$ is calculated as follows. It is used later to obtain approximate confidence intervals for the true h value. Calculate the column vector of partial derivatives $\partial h / \partial \gamma_p$:

$$\hat{\mathbf{H}} = \begin{bmatrix} \partial\hat{h}/\partial\gamma_1 \\ \cdot \\ \cdot \\ \cdot \\ \partial\hat{h}/\partial\gamma_P \end{bmatrix}. \tag{5.17}$$

The caret $^\wedge$ indicates that the derivative is evaluated at $\gamma_1 = \hat{\gamma}_1, \cdots, \gamma_P = \hat{\gamma}_P$. The local estimate of the (asymptotic) variance of $\hat{h}$ is

$$\text{var}(\hat{h}) = \hat{\mathbf{H}}'\mathbf{V}\hat{\mathbf{H}} \tag{5.18}$$

$$= \sum_{p=1}^{P} (\partial\hat{h}/\partial\gamma_p)^2 \text{ var}(\hat{\gamma}_p) + 2\sum\sum_{p < p'}(\partial\hat{h}/\partial\gamma_p)(\partial\hat{h}/\partial\gamma_{p'})\text{cov}(\hat{\gamma}_p,\hat{\gamma}_{p'});$$

here $\mathbf{V}$ is the local estimate (5.14) of the covariance matrix, and $\hat{\mathbf{H}}'$ is the transpose of $\hat{\mathbf{H}}$ and is a row vector. The true theoretical (asymptotic) variance appears in Chapter 6. The estimate $s(\hat{h})$ of the standard error of $\hat{h}$ is

$$s(\hat{h}) = [\text{var}(\hat{h})]^{1/2}. \tag{5.19}$$

This is used below in approximate confidence intervals for the true h value. For the Class-B example, lines 8 and 9 of Figure 2.1 show such estimates of standard errors for percentile estimates.

Example. For the Arrhenius-lognormal model, one usually wishes to estimate the 100Pth percentile at a design temperature. Denote the reciprocal absolute design temperature by x_0. The ML estimate is

$$\hat{y}_P(x_0) = \hat{\gamma}_1 + \hat{\gamma}_2 x_0 + z_P \hat{\sigma};$$

here z_P is the standard normal $100P$th percentile. The partial derivatives are

$$\partial \hat{y}_P(x_0)/\partial \gamma_1 = 1, \quad \partial \hat{y}_P(x_0)/\partial \gamma_2 = x_0, \quad \partial \hat{y}_P(x_0)/\partial \sigma = z_P.$$

Then, for any types of data, the estimate of the variance is

$$\text{var}[\hat{y}_P(x_0)] = [1 \; x_0 \; z_P] \mathbf{V} \begin{bmatrix} 1 \\ x_0 \\ z_P \end{bmatrix}.$$

For complete data, this becomes

$$\text{var}[\hat{y}_P(x_0)] = \hat{\sigma}^2 \{(1/n) + [(x_0 - \bar{x})^2/S_{xx}] + [z_P^2/(2n)]\}.$$

The estimate $s[\hat{y}_P(x_0)]$ of the standard error is the square root of this.

5.7. Approximate Normal Confidence Intervals

Approximate normal $100\gamma\%$ confidence intervals for the true value of a model coefficient (or of a function of such coefficients) follow. The form of the interval depends on whether h is mathematically unbounded $(-\infty, \infty)$, positive, or a fraction in $(0,1)$; this assures that the limits are not outside the range of h. The approximation is poorer, the larger γ is and the smaller the number of failures is. Let $\hat{h}$ denote the ML estimate and let $s(\hat{h})$ denote the estimate of its standard error.

Unbounded limits. Suppose that the mathematically possible range of h is $(-\infty, \infty)$. Then a two-sided approximate normal $100\gamma\%$ confidence interval for the true h value has lower and upper limits

$$\underline{h} = \hat{h} - K_\gamma s(\hat{h}), \quad \tilde{h} = \hat{h} + K_\gamma s(\hat{h}); \tag{5.20}$$

here K_γ is the $100(1+\gamma)/2$th standard normal percentile. The limits require that the number of failures be large enough that the distribution of the estimate $\hat{h}$ is approximately normal. For the Class-B example, line 4 in Figure 2.1 shows the numerical values of such confidence limits for γ_1 and γ_2.

Example. For the Arrhenius-lognormal model, the $100P$th percentile of log life $y_P(x_0)$ is unbounded. Two-sided approximate $100\gamma\%$ confidence limits are

$$\underline{y}_P(x_0) = \hat{y}_P(x_0) - K_\gamma s[\hat{y}_P(x_0)], \quad \tilde{y}_P(x_0) = \hat{y}_P(x_0) + K_\gamma s[\hat{y}_P(x_0)].$$

Corresponding limits for the percentile are

$$\underline{t}_P(x_0) = \text{antilog}[\underline{y}_P(x_0)], \quad \tilde{t}_P(x_0) = \text{antilog}[\tilde{y}_P(x_0)].$$

For the Class-B example, lines 8 and 9 of Figure 2.1 show the numerical values of such confidence limits for percentiles at two temperatures.

Positive limits. If the range of h is $(0,\infty)$, appropriate two-sided approximate normal $100\gamma\%$ confidence limits are

$$\underline{h} = \hat{h} \exp[-K_\gamma s(\hat{h})/\hat{h}], \quad \tilde{h} = \hat{h} \exp[K_\gamma s(\hat{h})/\hat{h}]. \qquad (5.21)$$

These limits cannot take on negative values. They require that there be enough failures so that the distribution of $\ln(\hat{h})$ is approximately normal. A negative range $(-\infty, 0)$ requires the obvious modification of (5.21). For the Class-B example, line 4 of Figure 2.1 shows the values of such limits for σ.

Example. For the Arrhenius-lognormal model, σ must be positive. For complete data, two-sided approximate normal $100\gamma\%$ confidence limits are

$$\underline{\sigma} = \hat{\sigma} \exp\{-K_\gamma[\hat{\sigma}/(2n)^{1/2}]/\hat{\sigma}]\} = \hat{\sigma} \exp\{-K_\gamma/(2n)^{1/2}\},$$

$$\tilde{\sigma} = \hat{\sigma} \exp\{K_\gamma/(2n)^{1/2}\}.$$

Limits in (0,1). Suppose the mathematically possible range of h is $(0,1)$. Here h is usually the fraction failing or surviving. Then appropriate two-sided approximate normal $100\gamma\%$ confidence limits are

$$\underline{h} = \hat{h}/\{\hat{h}+(1-\hat{h}) \exp[K_\gamma s(\hat{h})/(\hat{h}(1-\hat{h}))] \},$$

$$\tilde{h} = \hat{h}/\{\hat{h}+(1-\hat{h}) \exp[-K_\gamma s(\hat{h})/(\hat{h}(1-\hat{h}))] \}. \qquad (5.22)$$

These limits cannot take on the values outside the range $(0,1)$. They require that there be enough failures so that the distribution of $\ln[\hat{h}/(1-\hat{h})]$ is approximately normal.

One-sided limits. A one-sided approximate $100\gamma\%$ confidence limit for h is the corresponding two-sided limit with K_γ replaced by z_γ, the 100γth standard normal percentile. For example, for the Arrhenius-lognormal model, a one-sided lower approximate $100\gamma\%$ confidence limit for $y_P(x_0)$ is $y_P(x_0) = \hat{y}_P(x_0) - z_\gamma s[\hat{y}_P(x_0)]$. The true confidence level of a normal two-sided interval is usually close to $100\gamma\%$. However, the true confidence level of such a one-sided interval can be far from $100\gamma\%$, since many sampling distributions are far from symmetric. For one-sided limits, the LR intervals (Section 5.8) are better. Vander Weil and Meeker (1988) and Doganaksoy (1989b) show this.

Improved limits. Authors have presented various improvements for the normal limits:

1. Transformations such as those above assure that such limits are in the natural range of the quantity estimated.
2. Other transformations make the sampling distribution of the ML estimator of the transformed parameter closer to normal. For example, one can find a transformation such that, at the maximum, the third derivative of the sample log likelihood with respect to the transformed parameter equals zero. Thus the log likelihood with the transformed parameter is

closer to quadratic, and the sampling distribution of the transformed ML estimate is closer to normal.

3. Corrections for the bias of ML estimators make one-sided limits more accurate. Most such corrected estimators are *mean* unbiased. *Median* unbiased corrections would be better, since the normal approximation then better matches percentiles.

4. Schneider and Weisfeld (1988) use selected functions of ML estimates to approximate asymmetric sampling distributions better.

The LR limits (Section 5.8) are generally better than normal limits, particularly one-sided limits.

Simultaneous limits. If one calculates L intervals each with $100\gamma\%$ confidence, then probability that they all *simultaneously* contain their true values is less than γ. L *simultaneous intervals* for $h_1, \cdots, h_L$ contain *all* L true values with a specified probability γ. Such intervals are described in detail by Miller (1966) and Nelson (1982, Chapter 10). They are usually used for simultaneous comparisons where the h_l's are differences or ratios. Suppose that the ML estimates for the L quantities $h_1, \cdots, h_L$ are $\hat{h}_1, \cdots, \hat{h}_L$, respectively, and their standard error estimates are $s(\hat{h}_1), \cdots, s(\hat{h}_L)$. Simultaneous approximate $100\gamma\%$ confidence intervals for $h_1, \cdots, h_L$ are

$$\underline{h}_1 = \hat{h}_1 - K_{\gamma'}s(\hat{h}_1), \quad \tilde{h}_1 = \hat{h}_1 + K_{\gamma'}s(\hat{h}_1),$$

$$\begin{matrix} \cdot & & \cdot \\ \cdot & & \cdot \\ \cdot & & \cdot \end{matrix} \qquad (5.23)$$

$$\underline{h}_L = \hat{h}_L - K_{\gamma'}s(\hat{h}_L), \quad \tilde{h}_L = \hat{h}_L + K_{\gamma'}s(\hat{h}_L),$$

where $\gamma' = 1 - [(1-\gamma)/(2L)]$. These intervals require the approximate normality of the sampling distributions of $\hat{h}_1, \cdots, \hat{h}_L$. They also depend on the Bonferroni inequality, described by Miller (1966). The intervals (5.21) and (5.22) can also be used as simultaneous intervals with $K_{\gamma'}$.

5.8. Approximate Likelihood Ratio Intervals

This section presents LR confidence limits and how to calculate them with ML packages of Section 1.

Properties. The preceding confidence intervals all employ a normal approximation to the sampling distribution of a ML estimator $\hat{h}$ (or to some transformation of it, such as $\ln(\hat{h})$). For such *one-sided* limits, especially for samples with few failures, the true confidence level may differ much from the intended one. A better interval employs the likelihood ratio as described by Lawless (1982, Sections 6.4 and 6.5), Vander Wiel and Meeker (1988), Doganaksoy (1989a,b), and Ostrouchov and Meeker (1988). This "LR interval" has other advantages besides a confidence level closer to the stated level in each tail. First, the limits are always in the natural range of the quantity being estimated. Thus an artificial transformation, such as $\ln(\hat{h})$, is not needed.

Second, the limits are invariant. That is, the same limits result no matter how h is parameterized. For example, the parameterizations $\sigma = \gamma_3$ and $\sigma = \exp(\gamma_3)$ yield the same LR limits, but they yield different limits based on the normal approximation. Third, the LR limits are asymmetric about the estimate $\hat{h}$ and roughly have the same probability in each tail, as do the exact limits where available. In contrast, the limits from the normal approximation are symmetric about $\hat{h}$. Disadvantages of LR limits are that they are more laborious to calculate, and standard packages do not yet calculate them. Vander Wiel and Meeker (1988) and Doganaksoy (1989a) present programs that calculate such limits. Doganaksoy (1989b) investigates improvements of LR intervals, including the corrected signed square root of the log LR statistic and Bartlett's correction to the log LR statistic. He concludes that in general the simpler LR intervals perform as well as the corrected ones.

LR method. Suppose the confidence interval is to be calculated for a coefficient or parameter γ_1, and suppose the remaining model coefficients are $\gamma_2, \cdots, \gamma_P$. Denote the sample log likelihood by $\mathcal{L}(\gamma_1, \gamma_2, \cdots, \gamma_P)$. Define the *constrained maximum log likelihood* for γ_1 as

$$\mathcal{L}_1(\gamma_1) \equiv \max_{\gamma_2, \cdots, \gamma_P} \mathcal{L}(\gamma_1, \gamma_2, \cdots, \gamma_P). \tag{5.24}$$

This is the log likelihood maximized with respect to all other parameters for a fixed value of γ_1. The ML estimate $\hat{\gamma}_1$ maximizes $\mathcal{L}_1(\gamma_1)$. $\mathcal{L}_1(\hat{\gamma}_1) = \mathcal{L}(\hat{\gamma}_1, \hat{\gamma}_2, \cdots, \hat{\gamma}_P)$ is the global maximum. If γ_1 is the true value, the statistic

$$T = 2[\mathcal{L}_1(\hat{\gamma}_1) - \mathcal{L}_1(\gamma_1)]$$

has a distribution that is approximately chi square with one degree of freedom. Thus, the γ_1 values that satisfy

$$T = 2[\mathcal{L}_1(\hat{\gamma}_1) - \mathcal{L}_1(\gamma_1)] \leq \chi^2(\varepsilon;1) \tag{5.25}$$

are an approximate $100\varepsilon\%$ confidence interval for γ_1. The lower and upper *LR confidence limits* $\underline{\gamma}_1$ and $\bar{\gamma}_1$ are the γ_1 values that satisfy

$$\mathcal{L}_1(\gamma_1) = \mathcal{L}_1(\hat{\gamma}_1) - (1/2)\chi^2(\varepsilon;1). \tag{5.26}$$

Figure 5.2 depicts $\mathcal{L}_1(\gamma_1)$ as a heavy curve and shows $\hat{\gamma}_1$, $\underline{\gamma}_1$, and $\bar{\gamma}_1$. These limits usually must be found by numerical search. Under the asymptotic normal theory the sample log likelihood is approximately quadratic near its maximum and given by

$$\mathcal{L}(\gamma_1, \hat{\gamma}_2, \cdots, \hat{\gamma}_P) \approx \mathcal{L}(\hat{\gamma}_1) + (1/2)[\partial^2 \mathcal{L}(\hat{\gamma}_1, \hat{\gamma}_2, \cdots, \hat{\gamma}_P)/\partial\gamma_1^2](\gamma_1 - \hat{\gamma}_1)^2 ;$$

here the derivative is evaluated at the ML estimates of all the coefficients. Figure 5.2 depicts this quadratic function and the corresponding limits $\underline{\gamma}_1$ and $\bar{\gamma}_1$ from the approximate normal theory. The quadratic approximation gets closer to the true $\mathcal{L}_1(\gamma_1)$ over the range of interest as the number of failures in the sample gets large.

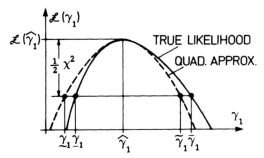

Figure 5.2.   LR and normal confidence limits.

Example. For the Arrhenius-lognormal model and complete data, explicit LR limits for σ can be obtained as follows.

$$\mathcal{L}(\gamma_1,\gamma_2,\sigma) = \sum_{i=1}^{n}\left[-(1/2)\ln(2\pi)-\ln(\sigma)-\frac{1}{2\sigma^2}(y_i-\gamma_1-\gamma_2 x_i)^2\right],$$

$$\mathcal{L}_\sigma(\sigma) = (-n/2)\ln(2\pi)-n\ln(\sigma)-\frac{1}{2\sigma^2}\sum_{i=1}^{n}(y_i-\hat{\gamma}_1-\hat{\gamma}_2 x_i)^2$$

$$= (-n/2)\ln(2\pi)-n\ln(\sigma)-\frac{1}{2\sigma^2}(n\hat{\sigma}^2),$$

$$\mathcal{L}_\sigma(\hat{\sigma}) = (-n/2)\ln(2\pi)-n\ln(\hat{\sigma})-(n/2)$$

where $\hat{\sigma}^2$ is given earlier. The $100\varepsilon\%$ LR limits are the two solutions of $\mathcal{L}_\sigma(\sigma) = \mathcal{L}_\sigma(\hat{\sigma})-(1/2)\chi^2(\varepsilon;1)$; this simplifies to calculating the two solutions $\rho_1 < \rho_2$ of

$$-\ln(\rho)-1 + \rho = \chi^2(\varepsilon;1)/n$$

where $\rho = (\hat{\sigma}/\sigma)^2$. Thus the limits are $\underline{\sigma} = \hat{\sigma}/\sqrt{\rho_2}$ and $\bar{\sigma} = \hat{\sigma}/\sqrt{\rho_1}$.

Calculation. Most computer programs (Section 1) do not yet calculate LR intervals. However, they can be used to solve (5.26) for a LR limit for a parameter. As follows, use a program to evaluate the constrained maximum log likelihood (5.24) for the parameter at selected parameter values near a LR limit. Then calculate the limit that satisfies (5.26) by interpolation. Figure 5.3 depicts such calculations. For example, suppose one wants a LR interval for a coefficient γ_1 in a linear relationship for the location parameter

$$\mu = \gamma_0 + \gamma_1 x_1 + \gamma_2 x_2 + \cdots + \gamma_P x_P.$$

Also, suppose that specimen i has *log* time y_i (observed, censored, or interval) and variable values $(x_{1i}, \cdots, x_{Pi})$, $i = 1, 2, \cdots, n$. For a trial limit value γ_1', calculate the transformed observations

$$y_i' = y_i - \gamma_1' x_{1i}.$$

For the intercept coefficient γ_0, calculate $y_i' = y_i - \gamma_0'$. Then use a computer package to fit the model to the transformed data but with the relationship

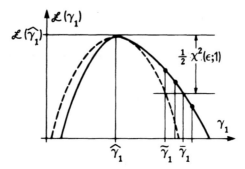

Figure 5.3.   LR limit from interpolation of constrained log likelihood.

$$\mu = \gamma_0 + \gamma_2 x_2 + \cdots + \gamma_P x_P,$$

this lacks the term with γ_1. The resulting maximum log likelihood is the constrained log likelihood $\mathcal{L}_1(\gamma_1)$ evaluated at $\gamma_1 = \gamma_1'$. Repeat this calculation for other γ_1' values near the limit, and interpolate to solve (5.26) for the limit. The LR limit is near the normal limit, which should be the first trial value.

Example.   For the insulating fluid data, use the power-Weibull model. The method above yields two-sided 95% LR limits for the power coefficient, as follows. Its ML estimate from SAS is $\hat{\gamma}_1 = -17.73$, and the SAS maximum log likelihood is $\mathcal{L}_1(\hat{\gamma}_1) = -137.748$. Its upper normal limit is $\bar{\gamma}_1 = -14.58$. The LR limits are the solutions of (2.6), namely, of

$$\mathcal{L}_1(\gamma_1) = -137.748 - (1/2)\chi^2(0.95;1) = -139.668.$$

Constrained maximum log likelihoods for nearby power values are $\mathcal{L}_1(-14.58) = -139.715$, $\mathcal{L}_1(-14.61) = -139.677$, and $\mathcal{L}_1(-14.64) = -139.641$. Linear interpolation yields the upper LR limit $\bar{\gamma}_1 = -14.62$. It is an upper one-sided approximate 97.5% confidence limit. Figure 5.3 depicts these calculations and interpolation, which can be done graphically. Similarly, the lower normal limit is $\gamma_1 = -20.88$. Nearby values are $\mathcal{L}_1(-20.88) = -139.532$, $\mathcal{L}_1(-20.95) = -139.609$, and $\mathcal{L}_1(-21.00) = -139.664$. Interpolation yields the lower LR limit $\gamma_1 = -21.00$. It is a lower one-sided approximate 97.5% confidence limit.

Other parameters.   The above calculation of LR limits extends to a linear function of such coefficients. An example is the location parameter at specified values of $x_1', \cdots, x_P'$, say, a design condition. To do this, reparameterize the relationship so

$$\mu = \gamma_0 + \gamma_1(x_1 - x_1') + \cdots + \gamma_P(x_P - x_P').$$

Then γ_0 is the desired value of the location parameter. Similarly calculate LR limits for a Weibull shape parameter β as follows. For specimen i, use the life t_i – not the log life. Use a trial value β' to calculate transformed lives

$$f_i' = t_i^{\beta'}.$$

Fit the same model but with an *exponential* distribution to the transformed t'_i data. The resulting maximum log likelihood is the constrained maximum log likelihood (5.24) evaluated at $\beta = \beta'$. As above, repeat this for other β' values and interpolate between them to get a LR limit.

Example. For the insulating fluid, the above method yields the following 95% LR limits for the characteristic life at 15 kV and for the Weibull shape parameter. The maximum likelihood estimates from SAS are $\hat{\gamma}_0 = 64.85$ and $\hat{\beta} = 0.7766$, and the SAS maximum log likelihood is $\hat{\mathcal{L}} = -137.748$. The normal 95% limits are $\underline{\beta} = 0.6535$ and $\bar{\beta} = 0.9228$. The LR limits for β are the solutions of (2.6), namely, of

$$\mathcal{L}_\beta(\beta) = -137.748 - (1/2)\chi^2(0.95;1) = -139.668.$$

Constrained log likelihood values near a lower limit for β are $\mathcal{L}_\beta(0.6535) = -139.506$, $\mathcal{L}_\beta(0.6450) = -139.769$, and $\mathcal{L}_\beta(0.6480) = -139.674$. Interpolation yields $\underline{\beta} = 0.6482$. Similarly, the upper LR limit is $\bar{\beta} = 0.9160$. For LR limits for the characteristic life at 15 kV, the reparameterized relationship is

$$\mu = \gamma_0 + \gamma_1[\ln(V) - \ln(15)].$$

Here γ_0 is the ln characteristic life at $V = 15$ kV. For LR limits for γ_0, solve

$$\mathcal{L}_0(\gamma_0) = -137.748 - (1/2)\chi^2(0.95;1) = -139.668.$$

Constrained log likelihood values near a lower limit for γ_0 are $\mathcal{L}_0(53.8) = -139.732$, $\mathcal{L}_0(54.0) = -139.661$, and $\mathcal{L}_0(54.2) = -139.591$. Interpolation yields $\underline{\gamma}_0 = 54.0$. Similarly, the upper LR limit is $\bar{\gamma}_0 = 76.3$.

A function. The LR limits are more difficult to calculate if the quantity h is not simply a coefficient or parameter of the model. For example, for the Arrhenius-lognormal model, one may seek such limits for a percentile at stress level x_0, namely, $\eta_P = \gamma_1 + \gamma_2 x_0 + z_P\sigma$, a function of the usual model coefficients (or parameters) γ_1, γ_2, and σ. One method of getting LR limits is to reparameterize the model, using η_P as a parameter in place of γ_1. That is, substitute $\gamma_1 = \eta_P - \gamma_2 x_0 - z_P\sigma$ into the likelihood (for complete data) to get

$$\mathcal{L}(\eta_P, \gamma_2, \sigma) = (-n/2)\ln(2\pi) - n\ln(\sigma) - \frac{1}{2\sigma^2}\sum_{i=1}^{n} [y_i - (\eta_P - \gamma_2 x_0 - z_P\sigma) - \gamma_2 x_i]^2.$$

Use this likelihood as described above to get LR limits for y_P. Another method of getting LR limits is to use constrained optimization as described by Lawless (1982, Section 5.1.3). In the future, statistical packages will automatically calculate such confidence limits, and users will not need to be concerned about the complex calculations.

Joint confidence region. One can use a LR approach to obtain a joint confidence region for two or more model coefficients or parameters as described by Lawless (1982, Section 5.1.3). Such regions are rarely needed in accelerated test analyses. Figure 3.4 of Chapter 9 depicts such a joint confidence region. The simultaneous limits (5.23) provide a "rectangular" joint confidence region.

PROBLEMS (* denotes difficult or laborious)

5.1. Capacitors. Meeker and Duke (1982) give simulated data from a voltage-accelerated life test. The data on 150 capacitors are singly censored at 300 days. The purpose is to estimate life in days at the design voltage of 20 V. At 20 V, 0 of 25 failed; at 26 V, 11 of 36 failed; at 29 V, 13 of 52 failed; at 32 V, 30 of 37 failed. Use the following CENSOR output on the power-Weibull model to answer the questions. Output is in terms of the extreme value distribution. The assumed relationship is

$$\xi(V) = \ln[\alpha(V)] = \gamma_0 + \gamma_1 \ln(V).$$

```
MAXIMUM VALUE OF THE LOGLIKELIHOOD IS          -122.1596

PARAMETER ESTIMATES
FOR THE SMALLEST EXTREME VALUE DISTRIBUTION

                                              95.00% CONFIDENCE LIMITS
   TERM        ESTIMATE     STANDARD ERROR      LOWER        UPPER
   SCALE        1.2103         0.1349           0.9728       1.5058
   B 0         67.9456         8.1513          51.9659      83.9253
   B 1        -18.5460         2.3956         -23.2423     -13.8497

VARIANCE-COVARIANCE MATRIX OF THE ESTIMATED PARAMETERS

               SCALE         B 0          B 1
   SCALE     0.182D-01    0.597D 00   -0.175D 00
   B 0       0.597D 00    0.664D 02   -0.195D 02
   B 1      -0.175D 00   -0.195D 02    0.574D 01

CORRELATION MATRIX OF THE ESTIMATED PARAMETERS

               SCALE         B 0          B 1
   SCALE      1.0000       0.5429      -0.5422
   B 0        0.5429       1.0000      -0.9998
   B 1       -0.5422      -0.9998       1.0000

PERCENTILE ESTIMATES FOR TEST CONDITIONS
             C I          C
   1        2.9957      20 V.
SMALLEST EXTREME VALUE DISTRIBUTION
LOCATION PARAMETER =     12.3867 AND SCALE PARAMETER =        1.2103

             WEIBULL DISTRIBUTION (T = EXP(Y))

                                              95.00% CONFIDENCE LIMITS
PERCENTILE       THAT          S(THAT)          LOWER         UPPER
  0.0100        3.4535         3.7662          0.4072        29.2915
  0.0500       24.2289        23.3553          3.6614       160.3338
  0.1000       56.0794        51.5864          9.2391       340.3924
  0.5000      394.2983       334.7625         74.6430      2082.8641
  1.0000      915.1363       763.4020        178.3453      4695.8033
  5.0000     6580.1697      5530.4572       1266.6820     34182.7178
 10.0000    15725.3249     13543.8629       2906.1499     85090.5336
 20.0000    38998.3391     34854.3601       6762.7965    224887.8047
 30.0000    68797.2696     63232.8222      11351.2676    416963.5023
 40.0000   106262.2483     99989.5935      16797.6193    672218.1995
 50.0000   153747.6239    147761.7507      23365.0614   1011695.6847
 60.0000   215529.6184    211347.4767      31523.8049   1473585.3284
 70.0000   299936.3631    300147.4416      42173.6575   2133128.2912
 80.0000   426184.1660    436028.6702      57351.1639   3167031.5125
 90.0000   657430.8036    691672.1697      83583.0013   5171090.4686
 95.0000   904008.7763    971146.2133     110040.9717   7426614.4242
 99.0000  1521207.3321   1691579.9219     171967.2135  13456470.5695
 99.9000  2484921.5866   2855484.3760     261192.7069  23640917.7132
```

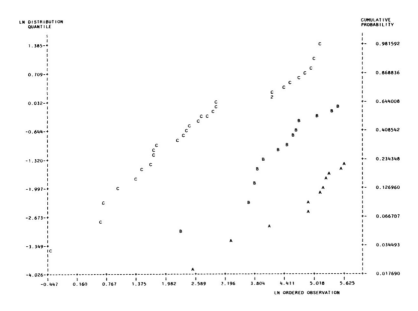

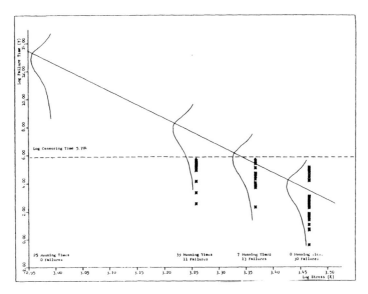

(a) Examine the Weibull plot of the data at the three test stresses. Is the Weibull distribution adequate? Why?

(b) Calculate the ML estimate and confidence limits for the Weibull shape parameter from those for the extreme value scale parameter.

(c) On Weibull paper, plot the estimate and confidence limits for the life distribution at 20 volts.

(d) Examine the log-log plot of the data. Use selected sample percentiles to judge whether the relationship is linear.

(e) Plot the fitted relationships for the 0.1, 10, and 90% lines on log-log paper.
(f) Examine the correlation matrix and note any high correlations. Explain the reason for each high correlation and how to reduce it.
(g) Suggest further analyses.

5.2. Insulating fluid. The power-Weibull model was fitted to the insulating fluid data of Table 3.1 of Chapter 3. Prof. William Meeker provided the following STAR output on the ML fit from Meeker (1984c). Note that output is in terms of the extreme value distribution for ln life.

(a) Convert all estimates and confidence limits to those for the Weibull distribution and inverse power law.
(b) Plot the selected data and the fitted distribution lines on Weibull probability paper.


```
                 Distribution is Weibull

The natural logs of these observations follow a[n] extreme-value
                                              distribution.
Intercept term included in model.

Maximum value of the log-likelihood is  -137.7476

Parameter estimates for the extreme-value distribution:

                       95.0% Confidence Limits

              Estimate    Std Error     Lower        Upper
Scale         1.287739   0.1133354    1.083667     1.530239
Intercept    64.84719    5.619756    53.83025     75.86414
VOLTAGE     -17.72958    1.606833   -20.87961    -14.57955

Variance-Covariance Matrix:
                 Scale        Intercept       VOLTAGE
Scale         0.01284491   -0.00888925     0.000924538
Intercept    -0.008889254  31.58166       -9.026538
VOLTAGE       0.000924538.  -9.026538       2.581914

Correlation Matrix:
                 Scale        Intercept       VOLTAGE
Scale         1.000000     -0.01395668     0.00507678
Intercept    -0.01395668    1.000000      -0.9996150
VOLTAGE       0.00507678   -0.9996150      1.000000
```

20 kV   Insulating Fluid ALT

WEIBULL QUANTILE ESTIMATES WITH 95% CONFIDENCE LIMITS

QUANTILE	MINUTES	STD ERROR	LOWER CL	UPPER CL
0.01	333.728	333.521	47.04691	2367.302
0.05	2722.430	2465.738	461.1517	16071.99
0.10	6879.012	6009.730	1240.865	38135.30
0.20	18080.41	15319.34	3434.279	95187.78
0.30	33075.44	27611.80	6438.064	169924.5
0.40	52528.29	43464.10	10373.33	265991.7
0.50	77819.17	64022.72	15510.97	390421.7

WEIBULL PROBABILITY PLOT

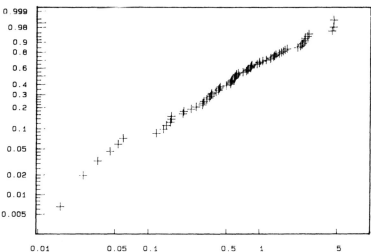

(c) Plot all the data and selected percentile lines on log-log paper.

(d) Calculate the ML estimate of the characteristic life at 20 kV.

(e*) Calculate approximate 95% confidence limits for (d).

(f) In view of the plot of residuals, comment on the adequacy of the assumed Weibull distribution.

(g) Comment on the high correlation -0.9996150. How can the relationship be reparameterized to reduce the correlation?

5.3. Bearings. The power-Weibull model was fitted by ML to the bearing data of Problem 3.5 without the outlier. STATPAC output follows.

(a) Compare the ML estimates and confidence limits for the power and Weibull shape parameters with the graphical and least-squares ones from Problems 3.5 and 4.6. In particular, are the estimates consistent with each other relative to the widths of the confidence limits? Are the ML confidence limits shorter as predicted by theory?

(b) Is the shape parameter estimate consistent with lab experience? That is, are the data consistent with a true value in the range 1.3 to 1.5? Is there convincing evidence that the failure rate is increasing?

(c) Calculate the ML estimate of the 10th percentile at a design load 0.75.

(d*) Calculate approximate 95% confidence limits for the 10th percentile at the design load 0.75.

(e) Examine the extreme value plot of the unitized log residuals, and comment on the adequacy of the Weibull distribution.

(f) Examine the crossplot of the residuals against LSTRES (ln(stress)), and comment on the linearity of the relationship. Comment on any other noteworthy features of the plot, e.g., which way each sample distribution is skewed, constant spread (β), etc.

MAXIMUM LOG LIKELIHOOD =   -54.138784

* FISHER MATRIX

COEFFICIENTS	C 1	C 2	C 3
C 1	60.30235		
C 2	-0.1596908E-03	0.6684998	
C 3	-22.05603	0.8500802	73.92661

* MAXIMUM LIKELIHOOD ESTIMATES FOR DIST. PARAMETERS
  WITH APPROXIMATE 95% CONFIDENCE LIMITS

PARAMETERS	ESTIMATE	LOWER LIMIT	UPPER LIMIT
C 1	0.4991979	0.2315432	0.7668526
C 2 β	-13.85280	-16.26991	-11.43569
SPREAD β	1.243396	0.9746493	1.586245

* COVARIANCE MATRIX

PARAMETERS	C 1	C 2	SPREAD
C 1	0.1864823E-01		
C 2 β	-0.7175376E-02	1.520832	
SPREAD β	0.7020492E-02	-0.2440648E-01	0.2386648E-01

* CORRELATION MATRIX

PARAMETERS	C 1	C 2	SPREAD
C 1	1.0000000		
C 2	-0.4260742E-01	1.000000	
SPREAD	0.3327780	-0.1281063	1.0000000

* MAXIMUM LIKELIHOOD ESTIMATE 'OF THE FITTED EQUATION

CENTER $\ln(\hat{\alpha})$
= 0.4991979

+ (LSTRES - 0.2404153E-01) * (-13.85280)

SPREAD $\hat{\beta}$
= 1.243396

PCTILES(ALL)

LSTRES

$-0.13926207 = \ln(0.87)$

* MAXIMUM LIKELIHOOD ESTIMATES FOR DIST. PCTILES
  WITH APPROXIMATE 95% CONFIDENCE LIMITS

PCT. P	ESTIMATE $\hat{y}_p$	LOWER LIMIT y_p	UPPER LIMIT y_p
0.1	0.6119165E-01	0.1281440E-01	0.2922040
0.5	0.2236344	0.6307659E-01	0.7928829
1	0.3913096	0.1251383	1.223632
5	1.451552	0.6150606	3.425682
10	2.589731	1.230454	5.450596
20	4.735488	2.506414	8.946984
50	11.78282	7.064818	19.65159
80	23.19973	14.52764	37.04851
90	30.94404	19.41020	49.33149
95	38.23777	23.85527	61.29158
99	54.03563	33.02691	88.40821

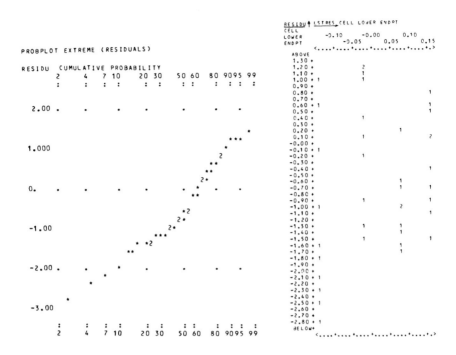

The following table summarizes various ML fits to the data. Use them to assess the model as follows.

Model	Max. ln(L) $\hat{\mathcal{L}}$	Weibull Shape Parameter $\hat{\beta}$	β	$\tilde{\beta}$	Million Revolutions $\hat{\tau}_{.10}$	$\tau_{.10}$	$\tilde{\tau}_{.10}$
1) Separ. α's, common β	−49.012	1.430	1.118	1.832	2.680	1.345	5.342
					0.8856	0.4743	1.654
					0.0795	0.0423	0.1496
					0.0512	0.0277	0.0946
2) Separ. dist's	−33.528	0.953	0.600	1.514	0.969	0.221	4.253
	−22.398	1.574	0.943	2.627	1.046	0.405	2.702
	2.900	1.950	1.232	3.085	0.1298	0.0623	0.2704
	7.282	1.963	1.197	3.221	0.0837	0.0397	0.1763

(g)   Make a separate Weibull probability plot of each of these two fitted models and of the fitted power-Weibull model. Include the data in each plot. Subjectively comment on the adequacy of the models.

(h)   Carry out the LR test for a constant Weibull shape parameter. Plot all shape parameter estimates and confidence limits from the various models on appropriate paper. State conclusions.

(i)   Use McCool's (1981,1974) exact test in eq. (3.2) for (h).

(j)   Plot the fitted relationship and the separate tabled estimates (Model 2) of the 10th percentiles on log-log paper. Include confidence limits. Subjectively comment on the fit of the relationship.

(k)   Calculate the likelihood ratio test for adequacy (linearity) of the power relationship, assuming a common β. State conclusions.

(l)   Carry out (k), assuming the β's differ. See Chapter 9.
(m) Suggest further analyses and other models.
(n)   Carry out (m).

   5.4. Transistor. A temperature-accelerated life test of 27 microwave power transistors yielded the following data. The main aim was to estimate reliability for 10 years at 100°C. Only "graceful" failures are treated as failures. Others are treated as censoring times (denoted +). Most others were caused by test equipment failures. Use the Arrhenius-Weibull model.

(a) Plot the data on Weibull and Arrhenius papers. Comment on the adequacy of the model.
(b) Graphically estimate the Weibull shape parameter and comment on its accuracy. Is the failure rate increasing or decreasing with age? Is the shape value convincingly different from 1?
(c) Graphically estimate the activation energy. Gauge its uncertainty.

215°C		190°C	
Hours	Reason	Hours	Reason
346+	TWT failure	2403	Graceful
346+	TWT failure	2668+	Removed - fixture failed
1416	Graceful	3669+	Removed - fixture failed
2197+	Removed - fixture failed	3863	Graceful
2533	Graceful	4400+	TWT failure
2630	Graceful	4400+	Source failure
2701+	Removed - power failed	4767	Graceful
3000	Graceful	4767	Graceful
3489	Graceful	5219	Graceful
6720+	Removed - test stopped	5276+	Removed - floor failed
		5276+	Removed - floor failed
		5276+	Removed - floor failed
		5276+	Removed - floor failed
		7517	Graceful
		7517	Graceful
		7840	Graceful
		8025	Graceful
		8025	Graceful
		8571+	Removed - test stopped

(d) Use a ML program to fit the model to the data. How well do the ML estimates and confidence limits compare with graphical ones?
(e) Obtain the ML estimate and confidence limits for reliability for 10 years at 100°C. Comment on the magnitude of the uncertainty.
(f) Plot estimates and confidence limits on Weibull and log-log paper.
(g) Plot the (log) residuals on extreme value paper, and comment on the adequacy of the Weibull distribution.

(h) Suggest further analyses.

(i) Carry out (h).

5.5. Turn failures. Do the following analyses of the Turn failure data in Table 5.1 of Chapter 3. Below is a summary of the ML fit of the Arrhenius-lognormal model to data from all four temperatures.

Parameter	Estimate	95% Conf. Limits	
γ_0'	3.526684	3.499721	3.553648
γ_1	3.478935	3.171346	3.786524
σ	0.07423483	0.05710498	0.09650315

$$\hat{\mu}(x) = 3.526684 + 3.478935(x - 2.044926)$$
Maximum log likelihood: 17.25

(a) Review the data plots in Section 5 of Chapter 3. Comment on the validity of the model and data.

(b) Calculate the log residuals about the fitted relationship.

(c) Plot the log residuals on normal hazard paper. Comment on the plot and the adequacy of the lognormal distribution.

(d) Crossplot the observed and censored residuals against temperature on square grid. Comment on the appearance of the plot.

Below is a summary of the ML fit of a model with a separate log mean for each temperature (no relationship) and a common σ.

Temp.	Log Mean Estimate	95%Conf. Limits	
190	3.9396	3.8964	3.9828
220	3.4220	3.3777	3.4664
240	3.2183	3.1746	3.2621
260	3.2246	3.1755	3.2737
σ:	0.06976	0.05272	0.08680

Maximum log likelihood: 36.42

(e) Calculate the log residuals about the log means.

(f) Plot the log residuals on normal hazard paper. Comment on this plot, the plot from (c), and the adequacy of the lognormal distribution.

(g) Calculate the likelihood ratio test statistic for linearity. Compare it with the appropriate χ^2 percentiles, and comment on the linearity. In view of all plots here and in Section 5 of Chapter 3, is a formal test of linearity necessary? Why?

(h) Is a formal test of homogeneity of σ necessary in view of such plots? What do you conclude about the homogeneity of σ?

(i) Suggest further analyses.

(j) Carry out (i).

5.6. Class-B without outliers. Re-analyze the Class-B data without the two low outliers. Redo all analyses applied in this chapter and in Chapter 3 to the Class-B data.

5.7. Later Class-B insulation data. Redo Problem 3.8 using ML analyses. Include analyses of residuals and tests for a common σ and for linearity. Repeat all analyses using a Weibull distribution; note where results are comparable and differ much. Suitably plot estimates and confidence limits.

5.8. Relay. Redo Problem 3.11 using ML analyses. Suitably plot estimates and confidence limits. Include analyses of residuals and tests for a common shape parameter and for linearity. Write a short report for engineers to summarize your findings, incorporating plots, output, etc.

5.9. Transformer turn data. Redo Problem 3.12 using ML analyses. Suitably plot estimates and confidence limits. Include analyses of residuals and tests for a common shape parameter and for linearity.

5.10. Lubricant. For the data of Table 2.1, do the following.
(a) Recalculate the equivalent times at 220°F, assuming that life increases by a factor of 1.5 for each 18°F decrease in temperature.
(b) Plot the equivalent times on 3-cycle Weibull paper.
(c) Plot the equivalent times (including nonfailures) on Arrhenius paper.
(d) Repeat (a) for a factor of 2.5.
(e) Plot the data from (d) on the same Weibull paper.
(f) Plot the data from (d), including nonfailures, on the Arrhenius paper.
(g) Plot the original (untransformed) data on the Weibull and Arrhenius papers. Also, plot the equivalent data based on a factor of 2 (Table 2.1) on both papers
(h) Compare the sets of equivalent data with respect to the percentage failing by 1,100 hours. Is the conclusion sensitive to the factor?
(i) Estimate the Weibull shape parameter, and comment on the nature of the failure rate.
(j) Fit a Weibull distribution to each transformed set of data and plot the ML estimate and confidence limits on Weibull papers with the data.
(k) Suggest further analyses.
(l) Carry out (k).

5.11. Quantal-response data on a turbine disk. Each of 356 turbine disks were inspected once for fatigue cracking, each at a different age. 100 were found cracked. The power-Weibull model was used to describe time to crack initiation as a function of the thickness of the web that was cracking. Use the adjoining STATPAC output on the fit of the model to the quantal-response data. Hopefully, thickness could be used to predict crack initiation better. Nelson (1979) presents and analyzes these data in more detail.
(a) Comment on the nature of the failure rate (increasing or decreasing

```
MAXIMUM LOG LIKELIHOOD =        -164.65188
```

• MAXIMUM LIKELIHOOD ESTIMATES FOR DIST. PARAMETERS
  WITH APPROXIMATE 95% CONFIDENCE LIMITS

PARAMETERS	ESTIMATE	LOWER LIMIT	UPPER LIMIT	STANDARD ERROR
C00001	8.402218	8.274504	8.529931	0.6515980E-01
C00002	-3.610256	-9.426808	2.206297	2.967629
SPREAD β	2.180689	1.687274	2.818396	0.2854109

• COVARIANCE MATRIX

PARAMETERS	C00001	C00002	β SPREAD
C00001	0.4245799E-02		
C00002	-0.7911342E-01	8.806821	
SPREAD	-0.1244511E-01	0.3106038	0.8145937E-01

• MAXIMUM LIKELIHOOD ESTIMATE OF THE FITTED EQUATION

CENTER $\ln(\hat{\alpha})$
```
    =      8.402218
```

```
    + (LWEB    -    -1.565025) •    (-3.610256)
```

SPREAD
β =   2.180689

PCTILES(-1.61)

```
             LWEB

         -1.6100000
```

• MAXIMUM LIKELIHOOD ESTIMATES FOR DIST. PCTILES
  WITH APPROXIMATE 95% CONFIDENCE LIMITS

PCT.	ESTIMATE	LOWER LIMIT	UPPER LIMIT	STANDARD ERROR
0.1	220.7706	110.6198	440.6053	77.83574
0.5	462.2401	273.8198	780.3157	123.4875
1	635.9341	402.7728	1004.070	148.1863
5	1342.842	967.5963	1863.614	224.5348
10	1868.023	1392.931	2505.156	279.6988
20	2635.319	1988.874	3491.880	378.4047
50	4431.600	3238.882	6063.536	708.9032
80	6521.208	4512.781	9423.491	1224.876
90	7685.192	5168.986	11426.26	1555.150
95	8670.887	5702.808	13183.73	1853.678
99	10560.83	6681.860	16691.64	2466.472

with age?). Is the evidence (confidence interval) convincing? Why?

(b)   Note the sign of the coefficient $-3.61\cdots$ for LWEB (ln web thickness). Is it consistent with the theory that thicker webs tend to take longer to crack? Comment on the fact that the confidence interval for the coefficient encloses zero.

(c*)   Use the coefficient estimates and their covariance matrix to calculate a (positive) percentile estimate and approximate confidence limits at LWEB = -1.61. Compare them with the output.

(d)   Describe how to test for adequacy of the model and how to calculate and analyze the residuals.

(e*) Carry out (d).

(f)   Suggest further analyses.

(g)   Carry out (f).

5.12. Left and right censored data. Analyze the data of Problem 3.10 using a ML program. Suitably plot estimates and confidence limits. Include analyses of residuals, and assess the model and data.

5.13. Thermal cycling. 18 cable specimens attached to an electronic module were thermally cycled over ranges of 190, 140, and 100°C. Specimens were inspected after 12, 50, 100, and 200 cycles. The resulting interval data appear below. The purpose is to estimate cable life for a 40°C cycle. Use the power-lognormal model (i.e., the Coffin-Manson relationship).

$\Delta T(C°)$	(0,12)	(12,50)	(50,100)	(100,200)	(200,∞)
190	1	1	2	1	1
140			2	1	3
100					6

(a)   Plot the data as intervals on lognormal probability paper and on log-log relationship paper. Locate the failures evenly over their interval.

(b)   Obtain ML estimates and 95% confidence limits for the model parameters from the interval data. Is the confidence interval for the power parameter consistent with a value of 2, the rule of thumb?

(c)   Obtain ML estimates and confidence intervals for percentiles of the life distribution at a temperature range of 40°C and at each test range.

(d)   Plot the estimates and confidence limits for the distribution lines on probability paper with the data. Comment on consistency of the fitted distributions and the data.

(e)   Calculate "interval" and censored (log) residuals. Locate failure residuals evenly over their intervals and treat them as observed. Pool (log) residuals and plot them on normal probability paper. Comment on the fit relative to the sample size and interval nature of the data.

(f*) Make a Peto plot of the interval residuals, using a computer package.

(g)   Calculate and plot the estimate of the life distribution for $\Delta T = 40°C$, assuming the power is 2. Plot confidence limits.

(h)   Suggest further analyses.

(i)   Carry out (h).

5.14. Fatigue limits. Steel specimens were fatigue tested in a four-point bending test with stress ratio -1. There were two steels, each in two forms – Standard and Induction Hardened. The purpose of the testing was to compare the fatigue curves of all four steels, particularly with respect to fatigue limit. In the table below, stress is in ksi and life is in 10^6 cycles. Specimens unfailed at 10×10^6 cycles were runouts (censored $+$).

Steel A (Std)		Steel A (I.H.)		Steel B (Std)		Steel B (I.H.)	
Stress	Cycles	Stress	Cycles	Stress	Cycles	Stress	Cycles
57.0	0.036	85.0	0.042	75.0	0.073	85.0	0.126
52.0	0.105	82.5	0.059	72.5	0.115	82.5	0.165
42.0	10.+	80.0	0.165	70.0	0.144	80.0	0.313
47.0	0.188	80.0	0.223	67.5	0.184	75.0	0.180
42.0	2.221	75.0	0.191	65.0	0.334	65.0	10.+
40.0	0.532	75.0	0.330	62.5	0.276	70.0	10.+
40.0	10.+	74.0	10.+	60.0	0.846	91.6	0.068
45.0	0.355	72.0	10.+	57.5	0.555	72.5	1.020
47.0	0.111	70.0	10.+	57.5	3.674	75.0	0.6009
45.0	0.241	65.0	10.+	57.5	10.+		
				55.0	0.398		
				55.0	10.+		
				55.0	10.+		

(a) Plot both data sets from Steel A on semi-log paper. Place the linear scale vertical as the stress scale. Use the horizontal log scale for life. Comment whether the fatigue curves of the Standard and Induction Hardened forms differ convincingly, in particular the fatigue limits.

(b) Repeat (a) for Steel B on another plot.

(c) Graphically compare Standard Steels A and B. Comment whether they differ convincingly.

(d) Graphically compare Induction Hardened Steels A and B. Comment whether they differ convincingly.

(e) Obtain graphical estimates of the four endurance limits. Plot them as horizontal lines on the plots. Are these estimates convincingly different from zero?

The following table presents key results from separate fits of the following endurance limit model to the four steels. Life is assumed to have a lognormal distribution. The fitted equation for median life is

$$\tau_{.50} = \begin{cases} 10^{\gamma_1}/(S-\gamma_3)^{\gamma_2} & , S > \gamma_3, \\ \infty & , S \leq \gamma_3; \end{cases}$$

here S is the stress in ksi and γ_3 is the fatigue limit.

Steel	$\hat{\ell}$	$\hat{\gamma}_1$	$\hat{\gamma}_2$	$\hat{\gamma}_3$	$\underline{\gamma}_3$	$\tilde{\gamma}_3$	$\hat{\sigma}$
A Std	−3.347	3.696	3.877	34.2	17.1	51.4	0.533
A IH	7.847	−0.542	0.563	74.0	do not exist		0.212
B Std	−7.775	3.440	3.237	48.6	25.3	71.9	0.529
B IH	6.380	0.258	0.998	70.5	61.9	79.2	0.174
pooled IH	2.067	0.153	1.034	72.0	71.8	72.2	

(f) On the data plots, plot the estimates $\hat{\gamma}_3$ and the 95% confidence limits $(\underline{\gamma}_3, \tilde{\gamma}_3)$. Comment why the upper limit is above a stress where failures were observed. That is, explain why the approximate theory for confidence limits fails. Are the fatigue limits convincingly different from zero? Why?

(g*)   Calculate LR limits for γ_3.

(h*)   Suggest why the Fisher matrix for Induction Hardened Steel A is singular, and thus approximate confidence limits cannot be calculated.

(i)   Calculate and plot the median curve for each of the four steels.

(j)   For each of the four steels, calculate the *log* residuals (log observation minus log median) and make a normal hazard plot. Is the lognormal distribution adequate? Why?

(k)   Pool the residuals from both Standard steels and make a normal hazard plot. Is the lognormal distribution adequate? Why? Do this for the residuals from both Induction Hardened steels.

(l)   Explain why it is not suitable to pool raw residuals from all four steels.

(m)   Crossplot all Standard residuals (including censored ones) against stress. Comment on the appearance of the plot.

(n)   Do (m) for the Induction Hardened residuals.

(o)   Divide the Standard residuals in to two equal size groups (low versus high stress). Make a hazard plot of each group, and comment on their appearance. For most metal fatigue, σ is greater at lower stress than at high. Can this be seen here?

(p)   Do (o) for the Induction Hardened residuals.

(q)   Calculate unitized residuals for the four steels. Pool them all and crossplot them against stress. Comment on the plot.

(r)   Make a normal plot of the residuals from (q). Comment on the plot.

(s)   Repeat (o) for the residuals from (q).

The following table present key results of separately fitting the power-lognormal model ($\gamma_3 = 0$) to the four steels. Here $\tau_{.50} = 10^{\gamma_1}/S^{\gamma_2}$.

Steel	$\hat{\ell}$	$\hat{\gamma_1}$	$\hat{\gamma_2}$	$\hat{\sigma}$
A Std	−3.717	−0.2524	−15.07	0.552
A IH	−0.437	0.1975	−37.93	0.598
B Std	−8.056	−0.0701	−14.88	0.536
B IH	−1.090	−0.1950	−17.71	0.397
pooled IH	−4.289	−0.0212	−24.91	0.578

(t)   For each of the four steels, use the likelihood ratio test for $\gamma_3 = 0$.

(u)   Suggest further analyses.

(v)   Comment on the width of the confidence limits for γ_3. How much do you think an improved choice of test stresses would reduce the width? Can you justify your opinion?

(w*)   Write the expression for the log likelihood for (i) a failure at a stress above the fatigue limit, (ii) a survivor at a stress above the fatigue limit, and (iii) a survivor at stress below the fatigue limit.

(x*)   Sketch the general appearance of sample log likelihoods as a function of γ_3, using a "typical" set of data, say, one of the four above.

5.15. **$1,000,000 experiment.** Redo Problem 3.9 using ML analyses.

Properly analyze the data as interval data; failures times as given in Problem 3.9 are the midpoints of intervals. Analyze the residuals; for simplicity treat each failure as if it is at the center of its (log) residual interval. Test for a common σ and for adequacy of any models you propose and fit.

5.16. Oil. Do ML analyses for Problem 4.10.

5.17.* Simple exponential model. This problem concerns ML theory and methods for fitting an exponential distribution and a log linear relationship to right censored data. The assumed relationship for the mean $\theta(x)$ as a function of the (possibly transformed) stress x is

$$\ln[\theta(x)] = \gamma_0 + \gamma_1 x.$$

(a)   Write the likelihood for a sample of n units with r failure times and $n - r$ right censoring times. t_i and x_i denote time and stress for unit i.
(b)   Write the sample log likelihood.
(c)   Derive the likelihood equations.
(d)   Derive explicit ML estimates of the coefficients for a test with just two test levels of stress. Express these estimates in terms of the separate ML estimates of the two means at those two stress levels. Show that the fitted equation passes through those estimates.
(e)   For a general test, calculate the matrix of negative second partial derivatives.
(f)   For a test with two stress levels, evaluate the matrix (e) using the ML estimates for the unknown coefficient values. Express this local information matrix in terms of the separate ML estimates of the two means.
(g)   Calculate the inverse of matrix (f) to get the local estimate of the covariance matrix.
(h*) For a general test, calculate the expectation of (e), as described in Chapter 6, to get the theoretical Fisher information matrix.
(i)   Calculate the inverse of (h) to get the theoretical covariance matrix.
(j)   For a test with two stress levels, evaluate the matrix (i) using the ML estimates for the coefficients. Express this matrix in terms of the separate ML estimates of the two means.
(k)   Give the ML estimates for the mean and failure rate at stress level x_0.
(l)   Calculate the theoretical variances of the estimates (k).
(m)   Calculate approximate normal confidence limits for (k) that are positive. Calculate LR limits.
(n*) Repeat (a)-(m) for interval data.
(o*) Repeat (a)-(m) for quantal-response data, a special case of (n).

5.18.* Simple Weibull model with known β. Repeat Problem 5.17 using a Weibull distribution in place of the exponential, assuming the shape parameter β has a known value. For (k)-(m), estimate the $100P$th percentile.

5.19.* Simple Weibull model with unknown β. Repeat Problem 5.18 assuming the shape parameter β is unknown and must be estimated.

5.20.* Endurance limit model. As follows, derive the ML theory for the fatigue limit model of Section 4.2. Use a lognormal life distribution.

(a) Write the log likelihood for a sample of n specimens with r failure times and $n - r$ right censoring times. Let t_i and S_i denote the time and stress for specimen i.

(b) Derive the likelihood equations.

(c) Calculate negative second partial derivatives for the local Fisher matrix.

(d) Assume you have the local estimate of the covariance matrix (the inverse of (c) evaluated at the ML estimates). Derive approximate confidence limits for (i) the fatigue limit γ_3 and (ii) the 100Pth percentile at stress S_0.

5.21.* σ depends on stress. Derive the ML theory for the model (4.2.2) and (4.2.3) of Section 4.2. Do (a)-(c) of Problem 5.20.

(d) Assume you have the local estimate of the covariance matrix. Derive approximate confidence limits for (i) γ_5 and (ii) the 100Pth percentile at stress S_0.

6

Test Plans

Purpose. This chapter presents accelerated test plans. It also gives the accuracy (standard errors) of estimates from such plans and guidance on how many specimens to test. Presented here are optimal, traditional, and good compromise plans. Optimal plans yield the most accurate estimates of life at the design stress. Traditional plans consist of equally spaced test stress levels, each with the same number of specimens. Test engineers traditionally use such plans, which yield less accurate estimates than optimum and good compromise plans. As this chapter shows, traditional plans generally require 25 to 50% more specimens for the same accuracy as good plans. So they rarely should be used. Good compromise plans run *more specimens at low stress* than at high stress – an important general principle. Good plans based on principles presented here will yield better results for given test cost and time. Poor plans waste time and money and may not even yield the desired information. Nelson (1990) presents the most basic and useful test plans.

Background. The simple models of Chapter 2 are essential background. Chapters 4 and 5 are desirable background, since the plans employ least squares and maximum likelihood estimates. However, this chapter can be read before Chapters 4 and 5. Indeed, in practice, one must decide on a test plan before analyzing the data. The test plans here are best suited to products with a single failure mode, but they can be used for products with any number of failure modes. Sections 4 and 6 of Chapter 1 describe many important considerations in planning a test which are not dealt with below.

Overview. Section 1 presents traditional, optimum, and good compromise plans for the simple model and complete data. Section 2 does so for singly censored data, the most important case in practice. Section 3 explains a general method for evaluating the performance of any test plan with any model and type of data. Section 4 surveys the literature on accelerated test plans. Section 5 provides the maximum likelihood (ML) theory used to evaluate test plans and to derive optimum ones.

1. PLANS FOR THE SIMPLE MODEL AND COMPLETE DATA

Purpose. This section deals with test plans for the simple linear model and complete data. It shows how to evaluate the accuracy of a plan and how to determine sample size. This section also compares optimum and traditional plans. Generally, tests yielding complete data run too long. However, theory for complete data appears here because it is simple and yields results that extend to more complex models and data. Chapter 4 is needed background. However, Section 1.1 here briefly presents what is needed. Readers may prefer to skip to Section 2 on plans for censored data.

Literature. There is much literature on optimum test plans for least squares fitting of linear relationships to complete data. For the simple linear relationship considered below, Gaylor and Sweeny (1965) summarize that literature and derive the optimum plan presented below. Daniel and Heerema (1950) investigate plans for optimal extrapolation and coefficient estimation. Stigler (1971) presents good compromise plans. Little and Jebe (1969) present optimal plans for extrapolation when specimens are tested individually one after another, subject to a fixed total test time, where the standard deviation of log life may not be constant. Hoel (1958) derives optimum plans for polynomial extrapolation, rarely useful in accelerated testing.

Overview. Section 1.1 presents needed background on the assumed model, estimates, and test constraints. Section 1.2 derives the optimum plan and its properties. Section 1.3 presents traditional plans, and Section 1.4 compares them with each other and with the optimum plan. Section 1.5 describes "good" test plans.

1.1. Background

Purpose. This section reviews needed background for Section 1. This includes the simple linear-lognormal model, least squares estimates, the variance of the least squares estimate (this variance is minimized by the optimum plan), and test constraints.

Model. The assumptions of the simple linear-lognormal model are:

1. Specimen life t at any stress level has a lognormal distribution.
2. The standard deviation σ of log life is constant.
3. The mean log life at a (possibly transformed) stress x is

$$\mu(x) = \gamma_0 + \gamma_1 x. \tag{1.1}$$

Here γ_0, γ_1, and σ are parameters to be estimated from data. It is also assumed that

4. The random variations in the specimen lives are statistically independent.

The following results extend to any simple linear model with obvious modifications (Chapter 4). Of course, the following results are valid to the extent that the model adequately represents product life.

Estimates. Suppose n specimens are tested at stress levels $x_1, x_2, \cdots, x_n$ (some may be equal), and the observed log lives are $y_1, y_2, \cdots, y_n$. From

Chapter 4, least squares estimates of γ_1 and γ_0 are

$$c_1 = [\sum y_i(x_i - \bar{x})]/[\sum(x_i - \bar{x})^2], \quad c_0 = \bar{y} - c_1\bar{x}; \tag{1.2}$$

here each sum runs over the n specimens, and $\bar{x} = (x_1 + x_2 + \cdots + x_n)/n$ and $\bar{y} = (y_1 + y_2 + \cdots + y_n)/n$. The least squares estimate of the mean log life $\mu(x_0)$ at a specified x_0 (say, design stress) is

$$m(x_0) = c_0 + c_1 x_0. \tag{1.3}$$

This is often the estimate of greatest interest. This estimate is statistically unbiased. That is, the mean of its sampling distribution equals the true value $\mu(x_0)$. The standard deviation of the sampling distribution is the *standard error* of the estimate in (1.5).

Accuracy. The variance of $m(x_0)$ is

$$\text{Var}[m(x_0)] = \{1 + (x_0 - \bar{x})^2[n/\sum(x_i - \bar{x})^2]\}\sigma^2/n. \tag{1.4}$$

The square root of this variance is the standard error of $m(x_0)$; namely,

$$\sigma[m(x_0)] = \{1 + (x_0 - \bar{x})^2[n/\sum(x_i - \bar{x})^2]\}^{1/2}\sigma/\sqrt{n}. \tag{1.5}$$

Statisticians often work with the variance, because it is proportional to $1/n$; thus it is more convenient for comparing sample sizes. However, the standard error is easier to interpret for most people. In particular, the estimate $m(x_0)$ is within $\pm K_\gamma \sigma[m(x_0)]$ of the true $\mu(x_0)$ with probability γ; here K_γ is the standard normal $100(1+\gamma)/2$th percentile. The accuracy of $m(x_0)$ for any test plan $x_1, \cdots, x_n$ is calculated from (1.4) (or equivalently (1.5)) as shown in the following example. Section 1.2 presents the optimum choice of $x_1, \cdots, x_n$ to minimize (1.4) or equivalently (1.5). Section 4 discusses other criteria used to optimize tests.

Heater example. For a test of sheathed tubular heaters, an engineer chose the four test temperatures (1520, 1620, 1660, and 1708°F) and allocated each six heaters – a traditional test. The Arrhenius-lognormal model is used. The variance of the estimate of mean log life at the design temperature of 1100°F is calculated as follows. The reciprocal absolute temperatures are

$$\begin{aligned}
x_1 &= \cdots = x_6 = 1{,}000/(1520+460) = 0.5051, \\
x_7 &= \cdots = x_{12} = 1{,}000/(1620+460) = 0.4808, \\
x_{13} &= \cdots = x_{18} = 1{,}000/(1660+460) = 0.4717, \\
x_{19} &= \cdots = x_{24} = 1{,}000/(1708+460) = 0.4613, \\
x_0 &= 1{,}000/(1100+460) = 0.6410.
\end{aligned}$$

Then $\bar{x} = [6(0.5051)+6(0.4808)+6(0.4717)+6(0.4613)]/24 = 0.47975$, and

$$\sum(x_i - \bar{x})^2/n = [6(0.5051 - 0.47975)^2 + 6(0.4808 - 0.47975)^2$$

$$+ 6(0.4717 - 0.47975)^2 + 6(0.4613 - 0.47975)^2]/24$$

$$= 0.00026183.$$

By (1.4), for the actual plan,

$$\text{Var}[m(0.6410)] = \{1 + [(0.6410 - 0.47975)^2 / 0.00026183]\} \, \sigma^2 / 24 = 4.18\sigma^2.$$

The optimum variance below is $2.16\sigma^2$, about half of this variance.

Test constraints. If the test stresses $x_1, \cdots, x_n$ are unconstrained, (1.4) and (1.5) are minimized by running all specimens at x_0; that is $x_1 = \cdots = x_n = x_0$. Then the test is not accelerated and is usually much too long. Thus it is necessary to choose a lowest allowed test stress level x_L, which produces long life. This is chosen as low as possible to minimize (1.4). But x_L must be high enough so the specimens all fail soon enough. The highest allowed test stress x_H, which produces short life, is chosen as high as possible to minimize (1.4). But x_H must not be so high that it causes other failure modes or that the linear relationship (1.1) is inadequate. Then one wants to minimize (1.4) subject to $x_1, \cdots, x_n$ all being in the test range x_L to x_H.

1.2. The Optimum Plan

Optimum stress levels. The following optimum test plan for a simple linear-lognormal model (such as the Arrhenius-lognormal model) and complete data appears in Daniel and Heerema (1950) and in Gaylor and Sweeny (1965). The optimum stress levels are the minimum and maximum in the allowed test range. Intermediate levels are not used. For comparisons with other plans, the optimum allocation of n test units to those two levels is derived here. The minimum variance of the least squares estimate (1.3) of the mean log life for the specified design stress level is also given. This optimum plan does not take into account the length or cost of the test.

Extrapolation factor. For a stress level x, define the **extrapolation factor**

$$\xi \equiv (x_H - x)/(x_H - x_L); \tag{1.6}$$

here x_L is the lowest allowed stress level and x_H is the highest one. Figure 1.1 depicts the meaning of ξ as the ratio of the length of the long arrow over the

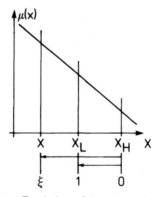

Figure 1.1.   Depiction of the extrapolation factor ξ.

short arrow. Thus ξ is how far x is from x_H as a multiple of the test range (x_H-x_L). A large ξ corresponds to long extrapolation, and a ξ close to 1 corresponds to short extrapolation. Then $\xi_H=0$ for $x=x_H$, and $\xi_L=1$ for $x=x_L$. For the design stress x_0, $\xi_0 \equiv (x_H-x_0)/(x_H-x_L)$ is greater than 1, since the design stress level is assumed to be below the test range.

Optimum allocation. A derivation of the optimum allocation of n test specimens follows. Let p denote the proportion tested at x_L ($\xi=1$); $1-p$ are tested at x_H ($\xi=0$). Thus the $\xi_i \equiv (x_H-x_i)/(x_H-x_L)$ values of the specimens are $\xi_1 = \cdots = \xi_{pn} = 1$ and $\xi_{pn+1} = \cdots = \xi_n = 0$, and

$$\text{Var}[m(x_0)] = \left[1+\frac{(\xi_0-p)^2}{p(1-p)}\right]\sigma^2/n, \tag{1.7}$$

since $\sum\xi_i=np$, $\bar{\xi}=p$, and $\sum\xi_i^2=np$. Such variance formulas contain the unknown true σ. The fraction p^* that minimizes (1.7) is

$$p^*=\xi_0/(2\xi_0-1). \tag{1.8}$$

This always exceeds 1/2 for extrapolation ($\xi_0>1$). Thus, test more specimens at the low stress than at the high one. The minimum variance is

$$\text{Var}_2^*[m(x_0)]=[1+4\xi_0(\xi_0-1)]\sigma^2/n. \tag{1.9}$$

The number of specimens allocated to the low test stress is the integer nearest to np^*. Such rounding results in a variance slightly larger than (1.9). The optimum plan may be impractical in many applications, as explained in Section 1.5. This allocation also minimizes the variance of the least squares estimate of the 100Pth percentile of log life $y_P(x_0) = m(x_0) + z_P s$. For the simple linear-lognormal model (1.1) and complete data, this optimum plan does not depend on the true, unknown values of the parameters γ_0, γ_1, and σ.

Extreme allocations. Two extreme cases of (1.9) are informative. First, if the lowest test stress equals the design stress ($x_L=x_0$), then $\xi_0=1$ and $p^*=1$. That is, all specimens should be run at the lowest test stress x_L. Second, if x_0 is much below x_L (that is, as $\xi_0\rightarrow\infty$), then $p^*=1/2$. That is, the specimens are allocated equally to x_L and x_H. The curve for $\text{Var}_2^*[m(x_0)]$ versus ξ_0 appears in Figure 1.2.

Class-H example. For the Class-H insulation discussed in Chapters 3 and 4, the optimum allocation and variance follow. The design temperature is $T_0=180°C$ ($x_0 = 1,000/453.2°K = 2.207$), and the extreme allowed test temperatures are $T_L=190°C$ ($x_L=1,000/463.2°K = 2.159$) and $T_H=260°C$ ($x_H=1,000/533.2°K = 1.875$). The extrapolation factor is $\xi_0=(1.875-2.207)/(1.875-2.159)=1.17$. Here ξ_0 is close to 1; that is, the design temperature is close to the test range. The optimum fraction of specimens allocated to 190°C is $p^*=1.17/(2(1.17)-1)=0.87$. That is, 87% of the specimens are tested at 190°C and 13% at 260°C, a very unequal allocation due to the short extrapolation. Of 40 specimens, $0.87(40) \approx 35$ would be tested at 190°C and $0.13(40) \approx 5$ at 260°C. The minimum variance from (1.9) is

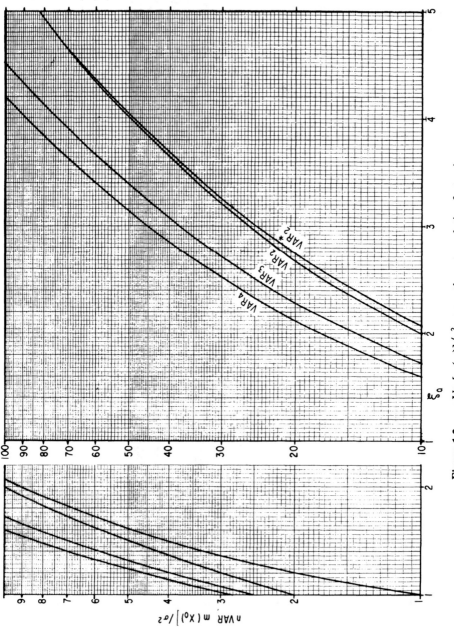

Figure 1.2. $n\,\mathrm{Var}[m(x_0)]/\sigma^2$ versus the extrapolation factor ξ_0.

$$\text{Var}_2^*[m(x_0)] - [1 + 4(1.17)(1.17 - 1)][\sigma^2/n] = 1.80\sigma^2/n.$$

The value 1.80 can be read from Figure 1.2. Enter the figure on the horizontal scale at $\xi_0 = 1.17$, go up to the curve labelled Var_2^* and then horizontally to the variance scale to read 1.80. For $n = 40$, $\text{Var}_2^*[m(2.207)] = 0.045\sigma^2$.

Heater example. An accelerated life test was to estimate the life of sheathed tubular heaters at a design temperature of $T_0 = 1100°F$ $(x_0 = 1000/(1100 + 460) = 0.6410)$. The extreme allowed test temperatures were $T_L = 1520°F$ $(x_L = 1000/(1520 + 460) = 0.5051)$ and $T_H = 1708°F$ $(x_H = 1000/(1708 + 460) = 0.4613)$. Then $\xi_0 = (0.4613 - 0.6410)/(0.4613 - 0.5051) = 4.10$. Here the design temperature is far from the test range. The optimum fraction of specimens at T_L is $p^* = 4.10/[2(4.10) - 1] = 0.57$. That is, 57% of the specimens are to be tested at 1520°F and 43% at 1708°F, almost equal allocation. Of 24 test units, $0.57(24) \approx 14$ would be tested at 1520°F and $0.43(24) \approx 10$ at 1708°F. The minimum variance from (1.9) is

$$\text{Var}_2^*[m(0.6410)] = [1 + 4(4.10)(4.10 - 1)] \sigma^2/n = 51.8 \sigma^2/n.$$

51.8 can be read from Figure 1.2. This variance is 29 times that for the previous example. This high multiple results from much greater extrapolation here. For $n = 24$, $\text{Var}_2^*[m(0.6410)] = 2.16\sigma^2$.

While optimum (minimum variance) allocation is seldom used in practice, it provides useful guidance in the choice of a test plan. A good test plan for estimating mean log life at the design stress level should use unequal allocation and be close to the optimum one within the practical constraints discussed in Section 1.5.

1.3. Traditional Plans

Definition. A commonly used test plan has equally spaced test stress levels and equal numbers of specimens at each. Such *traditional plans* with equal allocation of specimens to two, three, and four stress levels are described here for comparison with better plans. Traditional plans are **not** recommended. Better plans yield more accurate estimates of the life distribution at the design stress level. Optimum and good compromise plans with unequal allocation should be used instead.

Two stress levels. For two levels with equal allocation ($n/2$ specimens at each level), $\xi_1 = \cdots = \xi_{(n/2)} = 1$, and $\xi_{(n/2)+1} = \cdots = \xi_n = 0$. Then

$$\text{Var}_2[m(x_0)] = [1 + 4(\xi_0 - 0.5)^2]\sigma^2/n, \tag{1.10}$$

since $\sum \xi_i = n/2$, $\bar{\xi} = 0.5$, and $\sum \xi_i^2 = n/2$. $\text{Var}_2[m(x_0)]$ as a function of ξ_0 appears in Figure 1.2 as the curve Var_2.

Three stress levels. For three equally spaced levels with equal allocation ($n/3$ specimens at each level), $\xi_1 = \cdots = \xi_{n/3} = 1$, $\xi_{(n/3)+1} = \cdots =$

$\xi_{2n/3} \approx 1/2, \xi_{(2n/3)+1} = \cdots = \xi_n = 0$. For the theory here, spacing is equal on the transformed x scale – not on the scale of the original accelerating variable. This small difference can be ignored for simplicity. Then

$$\text{Var}_3[m(x_0)] = [1+6(\xi_0 - 0.5)^2]\sigma^2/n, \quad (1.11)$$

since $\sum \xi_i = n/2$, $\bar{\xi} = 0.5$, and $\sum \xi_i^2 = 5n/12$. $\text{Var}_3[m(x_0)]$ versus ξ_0 appears in Figure 1.2 as the curve Var_3.

Four stress levels For four equally spaced levels with equal allocation ($n/4$ specimens at each level), $\xi_1 = \cdots = \xi_{n/4} = 1$, $\xi_{(n/4)+1} = \cdots = \xi_{n/2} \approx 2/3$, $\xi_{(n/2)+1} = \cdots = \xi_{3n/4} \approx 1/3$, and $\xi_{(3n/4)+1} = \cdots = \xi_n = 0$. Then

$$\text{Var}_4[m(x_0)] = [1+(36/5)(\xi_0 - 0.5)^2]\sigma^2/n, \quad (1.12)$$

since $\sum \xi_i = n/2$, $\bar{\xi} = 1/2$, and $\sum \xi_i^2 = 7n/18$. $\text{Var}_4[m(x_0)]$ versus ξ_0 appears in Figure 1.2 as the curve Var_4.

Class-H example. The test plan for the Class-H insulation consisted of four roughly equally spaced test temperatures with equal allocation. For the design temperature 180°C, $\xi_0 = 1.17$, and

$$\text{Var}_4[m(x_0)] \approx [1+(36/5)(1.17-0.5)^2]\sigma^2/n = 4.23\,\sigma^2/n,$$

The value 4.23 can be read from Figure 1.2 from the Var_4 curve.

1.4. Comparison of the Test Plans

Comparisons. The test plan variances in Figure 1.2 can be compared by looking at their ratios for limiting values. For $\xi_0 = 1$, the design stress level is at the lower end of the test range. For $\xi_0 \to \infty$, the design stress level is far below the test range. These ratios appear in Table 1.1. These ratios and their reciprocals are graphed versus ξ_0 in Figure 1.3. The reciprocals give the fraction of the sample needed with the better plan to achieve the same accuracy as the poorer plan. For any extrapolation, ξ_0, Figure 1.3 shows $\text{Var}_2^* < \text{Var}_2 < \text{Var}_3 < \text{Var}_4$. Table 1.1 and Figure 1.3 show the following. For a given variance, equal allocation with two levels requires 0 to 100% more specimens than the optimum plan. Similarly, equal allocation with three lev-

Table 1.1. Extreme Variance Ratios for Optimum and Traditional Plans

Ratio	$\xi_0 = 1$	$\xi_0 \to \infty$
$\text{Var}_2/\text{Var}_2^*$	2.00	1.00
$\text{Var}_3/\text{Var}_2^*$	2.50	1.50
$\text{Var}_4/\text{Var}_2^*$	2.80	1.80
$\text{Var}_3/\text{Var}_2$	1.25	1.50
$\text{Var}_4/\text{Var}_2$	1.40	1.80
$\text{Var}_4/\text{Var}_3$	1.12	1.20

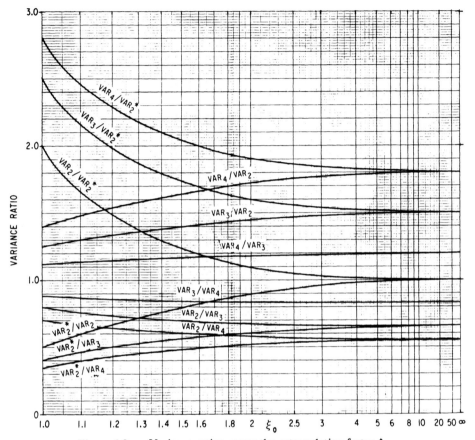

Figure 1.3.   Variance ratios versus the extrapolation factor ξ_0.

els requires 50 to 150% more specimens, and equal allocation with four levels requires 80 to 180% more. For a given standard error of an estimate, a traditional plan with three levels requires 25 to 50% more specimens than one with two levels. A traditional plan with four levels requires 40 to 80% more specimens than one with three levels.

Class-H example. The Class-H insulation test employed four roughly equally spaced test temperatures with equal allocation. The ratio of the variances of the optimum plan (1.9) and this plan (1.12) for $\xi_0 = 1.17$ is

$$\text{Var}_4[m(x_0)]/\text{Var}_2^*[m(x_0)] = 4.23/1.80 = 2.35.$$

2.35 can be read from Figure 1.3 for $\xi_0 = 1.17$. Thus, the actual plan requires 135% more specimens than the optimum one for the same accuracy. Equivalently, the accuracy achieved with 40 specimens with this traditional plan could have been achieved with the optimal plan with $40/2.35 \approx 17$ specimens. This statement ignores the rounding of the optimum allocation to an integer; namely $17(0.87) \approx 15$ specimens at 190°C and $17(0.13) \approx 2$ at 260°C.

1.5. Good Test Plans

Drawbacks. An optimum (minimum variance) test plan has two stress levels and yields the most accurate estimate of the mean log life, **provided** the model and data are valid. However, an optimum plan has drawbacks, and there are practical reasons to employ more stress levels as follows.

- The high stress may cause some failures from modes different from those at the design stress. Then those data are less informative. Also, they must be analyzed with the methods of Chapter 7. This suggests there should be at least three test levels.
- The assumed simple linear relationship may be inadequate. There must be at least three test levels to check linearity or to fit a nonlinear relationship.
- Specimens at a test level may have to be discarded due to some problem with them. For example, one of the test ovens may not properly hold the temperature, and those specimens must be omitted from analyses.
- It is possible that no specimens at the low stress fail by the time the data must be analyzed. Then the relationship cannot be estimated. Specimens at an intermediate stress are likely to fail in time.

Figure 1.3 shows that little additional accuracy is lost with four rather than three levels. Three or four test stress levels give a plan robustness against such difficulties. For extrapolation, the variance of the estimate is still greater with more than four levels. The main drawback of traditional plans is that they have poor accuracy.

Good plans. A good plan should be multi-purpose and robust and provide accurate estimates. Such a plan consists of three or four equally spaced test levels with *unequal* allocation. Such unequal allocation puts more specimens at the extremes of the test range and fewer in the middle. Also, allocate more specimens to the lowest level; this is especially effective if the design stress is close to the test range. The accuracy of such a plan is obtained by evaluating (1.4) or (1.5). Of course, more specimens at the lowest test level results in longer test time until all specimens fail. The time to complete the test can be controlled in part by the choice of the lowest test stress level. In summary, if one is sure that there will be no difficulties, then use the optimum plan. Otherwise, use more than two test stress levels and unequal allocation so the plan will be robust. Stigler (1971) describes some compromise plans.

Wire varnish example. A temperature-accelerated test of wire varnish (electrical insulation) for motors was planned to estimate median life of varnish at 180°C, the design temperature. The chosen test range was 220 to 260°C. The proposed compromise plan had 16 twisted-pair specimens at 220°C, 6 at 240°, and 8 at 260°. The variance of the estimate of mean log life at 180° is calculated from (1.4) as $0.469\sigma^2$. The variance of the optimum plan is $0.375\sigma^2$ by (1.9); then 65% of the specimens are at 220°C and 35% at 260°C. This same 2 to 1 ratio is used in the compromise plan, that is, 16 at 220°C to 8 at 260°C. The variance of the traditional plan with 10 specimens at each temperature is $0.595\sigma^2$ by (1.11). The ratio $0.595/0.469 = 1.27$ indicates that the traditional plan needs 27% more specimens than the compromise plan to achieve the same accuracy. Similarly, the compromise plan requires 25% more specimens than the optimum plan.

1.6. Sample Size

Purpose. Preceding discussions compare plans when the number n of specimens is given. As follows, one can determine the number n that achieves a desired accuracy of the estimate of the mean log life or median life at a design stress. Odeh and Fox (1975) give charts to determine sample size for linear and quadratic relationships.

Mean log life. To determine n, one can specify that the estimate $m(x_0)$ is to be within $\pm w$ of the true $\mu(x_0)$ with high probability γ. For any test plan, the n that achieves this is (1.5) rewritten as

$$n = \{1 + (x_0 - \bar{x})^2 [n / \sum (x_i - \bar{x})^2]\}(K_\gamma \sigma / w)^2. \tag{1.13}$$

The quantity in [ ] does not depend on n. As before, σ must be guessed or estimated from comparable data.

Class-H example. For Class-H insulation, suppose that $m(180°)$ is to be within $w = 0.10$ of $\mu(180°)$ with probability $\gamma = 0.95$. Also, suppose that equal allocation will be used with the same four test temperatures. The needed sample size is obtained from (1.12) rewritten as $n \approx [1 + 7.2(1.17 - 0.5)^2]$ $(1.96 \cdot 0.1053)^2 / (0.10)^2 \approx 19$; here $\xi_0 = 1.17$, and $s = 0.1053$ estimates σ. This would be rounded to 20 to get 5 specimens for each test temperature.

Other estimates. The preceding method for determining an appropriate n extends to any parameter estimate. Formulas for the standard errors of the various estimates appear in Chapter 4. They can be rewritten to express the desired n in terms of a specified multiple K_γ of the standard error.

Median life. An equivalent formula for determining sample size n uses the relationship $\tau_{.50}(x_0) = \text{antilog}[\mu(x_0)]$. One can specify that the estimate $t_{.50}(x_0) = \text{antilog}[m(x_0)]$ be within a factor r of $\tau_{.50}(x_0)$ with high probability γ. That is, $t_{.50}(x_0)$ is in the range $\tau_{.50}(x_0)/r$ to $r\tau_{.50}(x_0)$. For any test plan, the sample size that achieves this is (1.13) rewritten as

$$n = \{1 + (x_0 - \bar{x})^2 [n / \sum (x_i - \bar{x})^2]\}[K_\gamma \sigma / \log(r)]^2. \tag{1.14}$$

Class-H example. Suppose that the estimate of median life at 180°C is to be within a factor $r = 1.10$ (within about 10%) of the true median with 95% probability. Also, suppose that the test plan uses four equally spaced temperatures as before. The sample size that achieves this is (1.12) rewritten as

$$n = [1 + 7.2(1.17 - 0.5)^2][1.96 \cdot 0.1053 / \log(1.10)]^2 \approx 105; \tag{1.15}$$

here $\xi_0 = 1.17$ and 0.1053 estimates σ.

2. PLANS FOR THE SIMPLE MODEL AND SINGLY CENSORED DATA

2.1. Introduction

Purpose. This section presents optimum accelerated life test plans for ML estimates of median life with a simple model and singly censored data. Also, "traditional plans" (with equal numbers of specimens at equally spaced test stresses) and good compromise plans (including the Meeker-Hahn plans) are presented and compared with the optimum plans. The plans are illustrated with a temperature-accelerated life test of Class-B insulation, assumed to have a lognormal life distribution. Experience with such insulation supports this assumption. The plans indicate a general result. Namely, to get more accurate estimates of the life distribution at a low design stress, run more specimens at the lowest test stress than at the highest one.

Literature. Various authors have studied accelerated test plans for the simple linear model and singly censored data. References are listed below according to the distribution.

- *Exponential.* Chernoff (1962) presents optimum plans for estimating the mean or, equivalently, the failure rate, assuming a log-linear relationship and others.
- *Lognormal.* Kielpinski and Nelson (1975) present optimum and best traditional plans for estimating a simple linear relationship for mean log life. Nelson and Kielpinski (1976) derive the ML theory for such plans. Meeker (1984) and Meeker and Hahn (1985) present good compromise plans for estimating a percentile; they compare their plans with optimum and traditional plans. Barton (1987) adapts the optimum plan to minimize the maximum test stress while achieving a desired accuracy. Meeter and Meeker (1989) investigate plans for models with nonconstant σ.
- *Weibull.* Meeker and Nelson (1975) present optimum plans for estimating selected percentiles (1st, 10th, and 50th), assuming a log-linear relationship for the characteristic life α. Nelson and Meeker (1978) derive the ML theory for such plans. Meeker (1984) and Meeker and Hahn (1985) present good compromise plans for estimating a percentile; they compare their plans with optimum and traditional plans. Meeter and Meeker (1989) investigate plans for models with nonconstant β.
- *Logistic.* Meeker and Hahn (1977) present optimum plans for estimating a low failure probability from quantal-response data.

Computer program. Jensen (1985) provides a computer program that evaluates all tests plans in Section 2 and others listed below. The user specifies a simple (transformed) linear-Weibull or -lognormal model, its assumed parameter values, the common censoring time, the highest test stress level, the design stress level, and a percentile of interest at the design stress level. The program optimizes and calculates the asymptotic variance of the ML estimate of that percentile for the following plans:

1. Completely specified user plan (Section 5). The user specifies all test stress levels and the allocation of specimens to them. Here the program does not optimize the plan.
2. Optimum plan with two test stress levels (Section 2.4). The program optimizes the low stress level and the allocation.
3. Best traditional plan with three equally spaced test stress levels and equal allocation (Section 2.3). The program optimizes the low stress level.
4. Best compromise plan with three equally spaced test stress levels and 10% or 20% of the specimens allocated to the middle level (not in this book). The program optimizes the low test stress and the allocation of the remaining 80 or 90% to the low and high stress levels.
5. Best plan with three equally spaced test stress levels and (A) the same expected number failing at each (not in this book, as it compares poorly with other plans). The program optimizes the low test stress while satisfying (A).
6. Meeker-Hahn plans with three equally spaced test stress levels and 4:2:1 allocation (Section 2.6). The program optimizes the low test stress.
7. Adjusted Meeker-Hahn plans (Section 2.6), which use a lower low stress

The program runs on IBM PCs and compatibles.

Overview. Section 2.2 presents the accelerated test problem, including the test method and model; it also discusses ML estimation and other background material. Section 2.3 presents the best traditional plans with equally spaced test stresses, each with the same number of specimens. Section 2.4 presents the optimum plans. Section 2.5 compares the traditional and optimum plans and suggests good compromise plans. Section 2.6 presents good compromise plans of Meeker and Hahn (1985).

2.2. Background

Purpose. This section presents the test method, the assumed model, an example, estimation methods for censored data, the criterion for an optimum test, and other needed background.

Test method. It is assumed that

1. Each test unit runs a specified test time τ (the censoring time) if it does not fail sooner. That is, the data are time censored.
2. The highest test stress x_H is specified.
3. The specified design stress x_0 is below the test stresses.

The test time τ should be as long as practically and economically feasible to minimize the variance of estimates from the test. A test could continue beyond τ for a later analysis. $\eta \equiv \log(\tau)$ is the log of the censoring time. In practice, sometimes high-stress specimens are removed from test before time τ; then the estimates are less accurate. The highest test stress x_H should be as high as possible. This minimizes the standard error of the estimate of any percentile at the design stress. However, the highest stress should not cause failure modes different from those at the design stress (Chapter 7), and the model should be valid over the range of the design and test stresses.

Model. The assumptions of the simple linear-lognormal model are:

1. At any stress, life has a (base 10) lognormal distribution.
2. The standard deviation σ of log life is a constant. Meeter and Meeker (1989) investigate optimum plans for nonconstant σ.
3. The mean log life is a linear function of a (possibly transformed) stress x:

$$\mu(x) = \gamma_0 + \gamma_1 x. \tag{2.1}$$

The model parameters γ_0, γ_1, and σ are to be estimated from test data.
4. The lives of the test units are statistically independent.

Then the 100Pth percentile of life $\tau_P(x)$ or log life $\eta_P(x)$ at a stress x is

$$\eta_P(x) = \log[\tau_P(x)] = \mu(x) + z_P\sigma = \gamma_0 + \gamma_1 x + z_P\sigma; \tag{2.2}$$

here z_P is the standard normal 100Pth percentile. $\tau_{.50}(x) = \text{antilog}[\mu(x)]$ is the median and is commonly used as a typical life. The following results extend to other life distributions, as described in the literature above.

Estimates. ML estimates are used here, rather than linear estimates or the other estimates, because:

1. Optimum plans for ML estimates are easier to calculate than those for linear and other estimates.
2. ML estimates have a minimum standard error for large samples. For small samples, the ML estimates are generally as accurate as others.
3. Linear and other methods are more suited to data with failure censoring. They are not strictly correct for the time censoring considered here, whereas ML methods are correct.
4. The optimum design for ML estimates is close to optimum for any other estimates, even graphical ones.
5. In addition, available computer programs do the laborious ML calculations. There are no such programs for the laborious linear estimates.

Hahn and Nelson (1974) compare various estimation methods in detail.

Optimization criterion. Here an optimum test plan minimizes the variance (or standard error) of the ML estimate of the median life at a specified (design) stress x_0. Median life is of greatest interest in the example and many other applications. The estimate of another percentile (2.2) could be optimized; this would require a different plan. Meeker (1984) presents optimum plans for estimating the 1st and 10th lognormal percentiles. Meeker and Nelson (1975) present such optimum plans for estimating percentiles with the linear-Weibull model.

Class-B example. An insulation life test illustrates the test plans. To evaluate a new Class-B insulation for electric motors, a temperature-accelerated life test was run. The purpose was to estimate the median life of such insulation at the design temperature of 130°C using the Arrhenius-lognormal model, which is a special case of the model above. Here stress is $x = 1,000/T$, where T is the absolute temperature. Ten insulation specimens were run at each of four test temperatures (150°C, 170°C, 190°C, 220°C). Crawford (1970) gives the data, which are time censored at 8,064 hours. Following sections present test plans that are better than the actual one. For convenience, specimens at different temperatures started on test at different times and so have different censoring times. However, if started together, they would have yielded more information earlier. For illustrative purposes, we assume that all specimens start together.

Notation. Suppose that the specified highest test stress is x_H, and the mean log life is $\mu_H = \gamma_0 + \gamma_1 x_H$. Also, suppose that the specified design stress is x_0, and the mean log life is $\mu_0 = \gamma_0 + \gamma_1 x_0$. Figure 2.1 depicts these quantities and the model in a relationship plot. We use the standardized censoring time and slope

$$a \equiv (\eta - \mu_H)/\sigma = (\eta - \gamma_0 - \gamma_1 x_H)/\sigma, \quad b \equiv (\mu_0 - \mu_H)/\sigma = \gamma_1(x_0 - x_H)/\sigma. \quad (2.4)$$

a and b characterize a test plan and must be approximated. Optimum plans for censored data require one to specify values for the (usually unknown) true model parameters γ_0, γ_1, and σ to calculate a and b. In contrast, optimum plans for linear models and complete data do not depend on parameter values. So for censored tests one must approximate parameter values from experience, similar data, or a preliminary test. An optimum test plan using crude values is generally better than a plan devised by other means. Chernoff (1953,1962) calls such plans "locally optimum," since they are optimum only for the assumed parameter values.

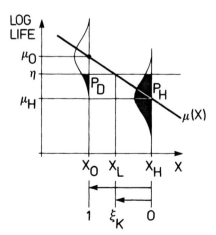

Figure 2.1.   Relationship plot of model and notation.

Class-B example. For the Class-B insulation of Chapter 5, $\eta =$ $\log(8,064) = 3.9066$, $\hat{\gamma}_0 = -6.0134$, $\hat{\gamma}_1 = 4.3083$, $\hat{\sigma} = 0.2592$, $x_H = 1,000$ $/(220 + 273.16) = 2.0277$, and $x_0 = 1,000/(130 + 273.16) = 2.4804$. Thus,

$$a = [3.9066 - (-6.0134) - 4.3083(2.0277)]/0.2592 = 4.57,$$

$$b = 4.3083(2.4804 - 2.0277)/0.2592 = 7.52.$$

2.3. Best Traditional Plans

Traditional plans. This section presents traditional plans, which are commonly used. Traditional plans have K equally spaced test stresses, each with the same number of test units. The highest test stress x_H must be specified. The "best" plan uses a lowest test stress x_L that minimizes the standard error of the ML estimate of the log mean at a specified design stress x_0. Best traditional plans are **not** recommended despite their wide use. It is better to use an optimum plan (Section 2.4) or a good compromise plan (Sections 2.5 and 2.6). Traditional plans are presented only to show they are generally inferior and to discourage their use.

Lowest test stress. The best lowest test stress x_L is given by Nelson and Kielpinski (1976) as

$$x_L = x_H + \xi_K (x_0 - x_H) ; \tag{2.5}$$

Figure 2.1 depicts ξ_K. Figures 2.2A and B give ξ_2 and ξ_4 as a function of K, a, and b. $\xi_3 \simeq (\xi_2 + \xi_4)/2$; Nelson and Kielpinski (1972) give such a graph of ξ_3. Their graphs for $K = 2$, 3, and 4 include b values down to 0.1. To use Figure 2.2, find the b value on the horizontal scale; go straight up to the curve for the a value (interpolate); and then go horizontally to the vertical axis to read the ξ_K value. Note that the definition of ξ_K here differs from that of ξ in Section 2.1. A transformed value x_L must be converted to the stress value.

Class-B example. For $K = 4$ test stresses and $a = 4.57$ and $b = 7.52$, $\xi_4 = 0.72$ from Figure 2.2B. For the highest test temperature of 220°C, $x_H = 2.0277$; for the design temperature of 130°C, $x_0 = 2.4804$. So the best lowest stress is $x_L = 2.0277 + 0.72(2.4804 - 2.0277) = 2.3536$, which is $T_L = (1,000/2.3536) - 273.16 = 152$°C. The experimenter had used 150°C, and four roughly equally spaced temperatures. But each temperature had a different censoring time; this is ignored for the purpose of an example.

Several censoring times. For some tests, the data are analyzed at various points in time. For example, $T_L = 152$°C is best only for the censoring time of 8,064 hours (a time when the data were analyzed). The test was terminated at 17,661 hours, and the data were reanalyzed. For this censoring time, $T_L = 140$°C is best. There may be several censoring times when the data are analyzed. Then choose the most important time or a compromise, and use its best plan.

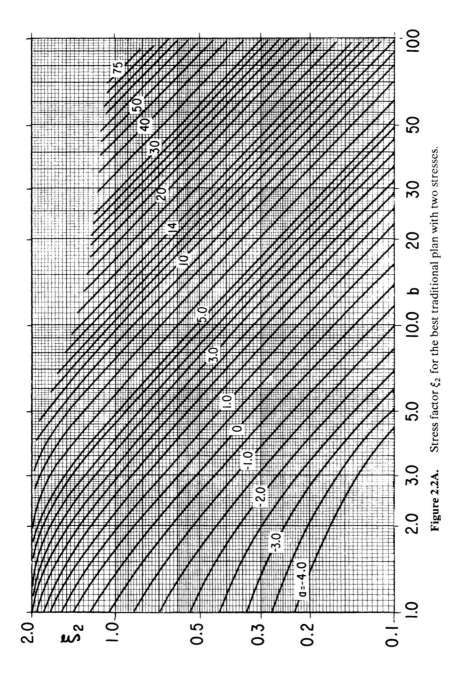

Figure 2.2A.   Stress factor ξ_2 for the best traditional plan with two stresses.

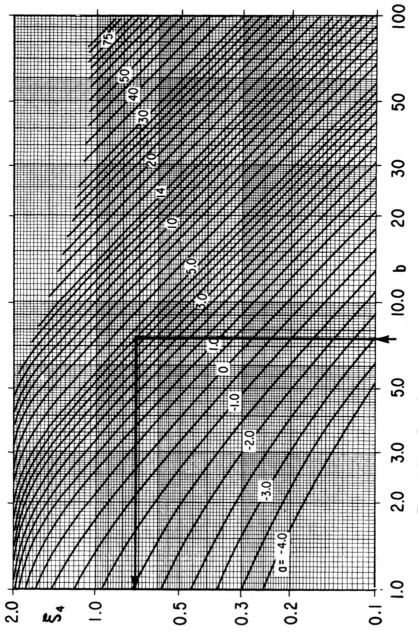

Figure 2.2B.   Stress factor ξ_4 for the best traditional plan with four stresses.

Equal spacing. The charts for best traditional plans assume that the transformed stresses are equally spaced. For the Arrhenius relationship, reciprocal absolute temperatures are equally spaced rather than temperatures. In practice, either can be equally spaced, and the charts are close enough.

Standard error. For such a best plan with n specimens, the large-sample standard error of the ML estimate $\hat{\mu}_0 = \hat{\gamma}_0 + \hat{\gamma}_1 x_0$ of $\mu_0 = \gamma_0 + \gamma_1 x_0$ is

$$\sigma(\hat{\mu}_0) = \sigma \cdot (V_K/n)^{1/2} ; \tag{2.6}$$

here V_K depends on K, a, and b. Figures 2.3 and 2.4 are graphs of V_2 and V_4. $V_3 \simeq (V_2 + V_4)/2$, and Kielpinski and Nelson (1975) give such a graph of V_3. For the example, $a = 4.57$ and $b = 7.52$, and Figure 2.4 gives $V_4 = 10.2$. V_K is a decreasing function of a and an increasing function of b. The minimum possible V_K value is 1. For practical values of a and b, the best traditional plan with two stresses is more accurate (smaller $\sigma(\hat{\mu}_0)$) than that with three

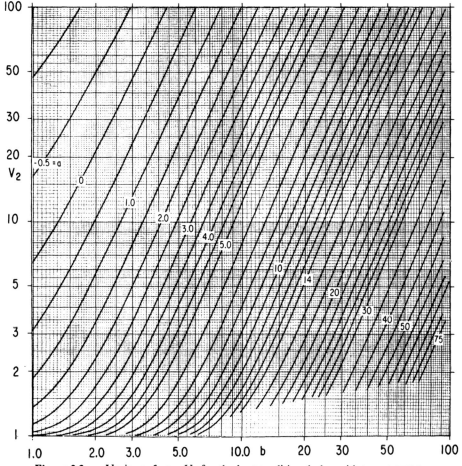

Figure 2.3.   Variance factor V_2 for the best traditional plan with two stresses.

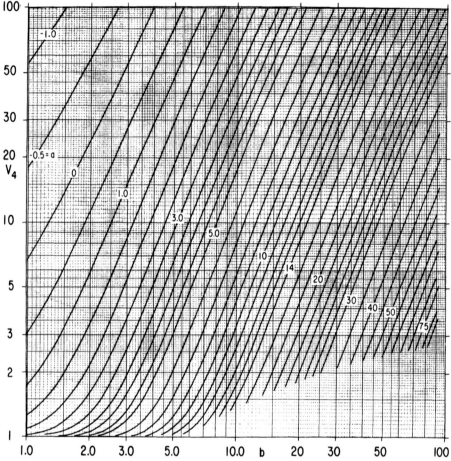

Figure 2.4.   Variance factor V_4 for the best traditional plan with four stresses.

stresses, which is more accurate than that with four stresses. That is, $V_2 < V_3 < V_4$ for practical a and b.

Sample size. Determine a sample size as follows. One requires that, with a desired high probability γ, $\hat{\mu}_0$ fall within $\pm w$ of the true μ_0. The sample size n_K that achieves this is approximately

$$n_K \simeq V_K(K_\gamma \sigma / w)^2 \tag{2.7}$$

To determine n_K, one must approximate the model parameters, since their true values are not known. After the data are collected, one obtains a confidence interval for μ_0 and ignores w.

Too big a sample. If the calculated n_K is impractically large, one must be content with fewer specimens and less accuracy. Otherwise, one must use a better plan which a) has fewer test stresses, b) has a higher highest test stress, c) has a higher censoring time, or d) is closer to an optimum plan.

Median. One can specify the accuracy of the ML estimate of the median $\hat{\tau}_{.50}(x_0) = \text{antilog}(\hat{\mu}_0)$ as follows. One requires that, with a high probability γ, the estimate falls between $\tau_{.50}(x_0)/r$ and $r \cdot \tau_{.50}(x_0)$. For r near 1, $100 \, | \, r - 1 \, |$ is the approximate percent error in the estimate. Then use $w = \log(r)$ in (2.7) to calculate n_K.

Class-B example. Suppose the median estimate for 130°C is to be within 20% of the true median with 90% probability. That is, $r = 1.20$ and $w = \log(1.20) = 0.0792$. Also, $V_4 = 10.2$ and $\sigma = 0.259$. Then $n_4 = 10.2 \, [1.645(0.259)/0.0792]^2 = 295$. The actual sample size was 40; this corresponds to a factor $r \simeq \text{antilog}[1.645(0.259)(10.2/40)^{1/2}] = 1.7$. Thus, with 90% probability, the estimate $\hat{\tau}_{.50}(2.4804)$ will be within a factor of 1.7 of the true median.

2.4. Optimum Test Plans

Purpose. This section describes optimum test plans. They use two test stresses with unequal numbers of specimens. It is assumed that the high test stress x_H is specified. The low test stress x_L and its proportion of the specimens are chosen to minimize the standard error of the ML estimate $\hat{\mu}_0$ of the median at a specified stress x_0. Such optimum plans are also good for comparing medians of a number of products at x_0; then an optimum plan yields the most precise estimate of each product median at the stress.

Optimum stress. The optimum low test stress is given by Nelson and Kielpinski (1976) as

$$x_L^* = x_H + \xi^*(x_0 - x_H) ; \tag{2.8}$$

here ξ^* is a function of a and b and is given in Figure 2.5. ξ^* corresponds to ξ_K in Figure 2.1.

Class-B example. For the insulation, $a = 4.57$, $b = 7.52$, $x_H = 2.0277$, and $x_0 = 2.4804$. Figure 2.5 yields $\xi^* = 0.63$. Then $x_L^* = 2.0277 + 0.63(2.4804 - 2.0277) = 2.3129$, and $T_L^* = (1,000/2.3129) - 273.16 = 159°C$. The actual lowest test temperature was 150°C. 159°C is optimum only for the censoring time of 8,064 hours (a time when the data were analyzed). The test was terminated at 17,661 hours; then the optimum low test temperature is 148°C.

Optimum allocation. The optimum proportion p^* of specimens at x_L^* depends on a and b as shown in Figure 2.6. Charts of Nelson and Kielpinski (1972) cover b values down to 0.1. For the example, Figure 2.6 yields $p^* = 0.735$. That is 73.5% of the specimens would be tested at 159°C. For a sample of 40, $(0.735)40 \simeq 29$ specimens would be tested at 159°C and 11 at 220°C. $p^* > 0.50$ for practical a and b values.

Standard error. For an optimum plan with n specimens, the large-sample standard error of $\hat{\mu}_0$ is

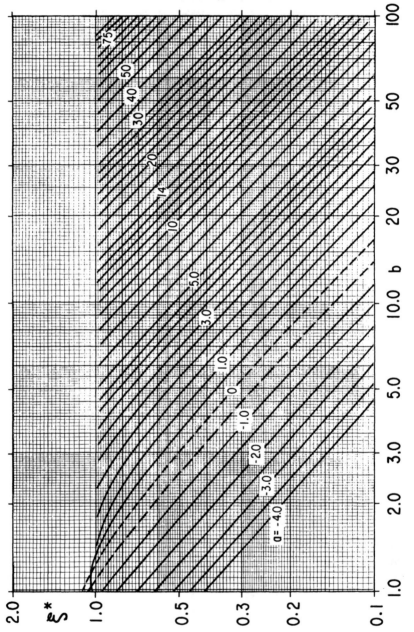

Figure 2.5.   Stress factor ξ^* for the optimum plan.

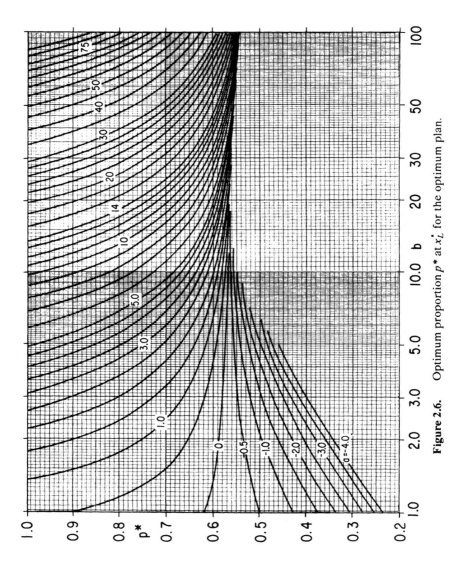

Figure 2.6.   Optimum proportion p^* at x'_L for the optimum plan.

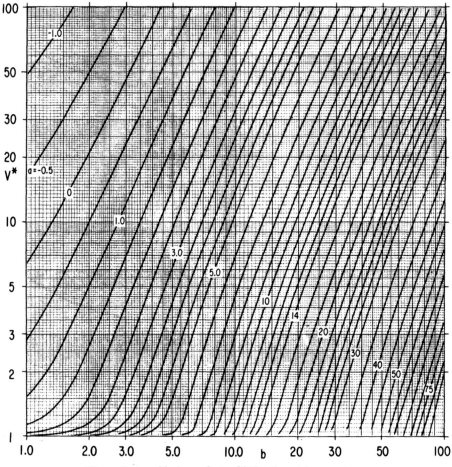

Figure 2.7.   Variance factor V^* for the optimum plan.

$$\sigma(\hat{\mu}_0) = \sigma \cdot (V^*/n)^{1/2} \; ; \qquad (2.9)$$

here V^* depends on a and b and is given in Figure 2.7. For the example, $V^* = 6.5$. For any a and b, the optimum plan has a smaller standard error than any plan with the same x_H and log censoring time η. Comparisons of the best traditional and optimum plans appear in Section 2.5.

Sample size. (2.7) gives the desired number n^* of specimens, but use V^* in place of V_K. For the example, suppose that the median estimate at 130° is to be within 20% of the true median with 90% probability. Then $r = 1.20$ and $V^* = 6.5$. So, for the optimum plan, $n^* = 6.5[1.645(0.259) / \log(1.20)]^2 = 188$. The corresponding sample size for the best traditional plan with four test temperatures is $n_4 = 295$. Here the best traditional plan requires 57% more specimens than the optimum plan for the same accuracy.

No failure. With the optimum plan, there is a possibility that none of the $n_L^* = p^* n$ specimens at low test stress fail. That is, each specimen's log life Y

exceeds the log censoring time η. Then γ_0 and γ_1 cannot be estimated. It is useful to calculate the probability of this to assure it is negligible; namely,

$$P\{none\} = [P(Y > \eta)]^{n\mathtt{L}} = \{\Phi[-(\eta - \mu_L)/\sigma]\}^{n\mathtt{L}}. \tag{2.10}$$

For the Class-B example, $\mu_L = \gamma_0 + \gamma_1 x_L^* = -6.0134 + 4.3083(2.3129) = 3.9513$ and $P\{none\} = \{\Phi[-(3.9066 - 3.9513)/0.2592]\}^{29} = 7.7 \times 10^{-8}$. This is small enough to ignore, provided the parameter values are accurate enough. It is useful to recalculate this probability with other parameter values.

2.5. Comparison of Optimum, Best Traditional, and Good Compromise Plans

Purpose. This section compares the optimum and best traditional plans with respect to:

1. their standard errors (2.6) and (2.9) and
2. their robustness to an incorrect model or data.

Compromise plans with both good accuracy and robustness are suggested.

Compare accuracy. Nelson and Kielpinski (1972) give graphs for $n_2/n^* = V_2/V^*$, $n_3/n^* = V_3/V^*$, and $n_4/n^* = V_4/V^*$; these are the ratios of the sample sizes of traditional and optimum plans with equal standard errors (2.6) and (2.9). For the example, we compare the optimum plan and the best traditional plan $(K = 4)$. Since $a = 4.57$ and $b = 4.52$, $n_4/n^* = V_4/V^* = 1.57$ from their graph. Thus the best traditional plan here needs 57% more specimens than the optimum plan.

Plans with two stresses. Figure 2.5 for ξ^* and Figure 2.2A for ξ_2 are similar. For most practical a and b values, the two plans have about the same low test stress. However, the optimum plan allocates more specimens there and has a slightly lower stress.

Drawbacks of optimum and traditional plans. Section 1.5 lists drawbacks of optimum plans for complete data. For censored data, there are additional drawbacks.

- If the x_L^* of an optimum plan is too low, there may be no failures by the censoring time. Then the relationship cannot be estimated.
- The assumption of a constant σ cannot be checked if there is no or one failure at x_L^*.

Both drawbacks can be avoided with a third test stress level between the two optimum levels. It must be high enough to insure enough failures. The optimum plan is suitable only if the preliminary parameter values and all data are valid. As discussed in Section 1.5, the main drawback of traditional plans is their poorer accuracy.

Compromise plans. The following discussion explains how to choose good compromise plans. As before, it is assume that the highest test stress

x_H has been chosen on the basis of criteria discussed earlier. Then one must chose the lowest test stress x_L , the intermediate stress or two, and the allocation (number) of specimens to each as follows.

A good compromise plan needs three or four test stresses and at least two stresses should yield failures by the censoring time. For more accurate estimates at the design stress, the lowest test stress should have more specimens than the highest one, as suggested by optimum plans. A reasonable lowest test stress for a compromise plan is between those for the optimum and the best traditional plan with three or four test stresses. The relative numbers of specimens for the two stresses for the optimum plan could be used for the extreme stresses for a compromise plan. The proportions of the sample at intermediate stresses should be small for accuracy or large for robustness and early failures. Meeker (1984) and Meeker and Hahn (1985) present compromise plans, which appear in Section 2.6.

Compromise standard errors. For a compromise plan, the standard error of the log mean estimate is generally between those for the optimum plan and the best traditional plan. The standard error can be evaluated with the program of Jensen (1985), simulation methods of Section 3, or the ML theory of Section 5.

Optimum anyhow. A compromise plan is usually preferable to the optimum plan, unless the model, assumed parameters values, and the data are likely all valid. An optimum plan was nevertheless used in a temperature-accelerated demonstration test of GaAs FETs, high reliability solid-state devices for a satellite application (Problem 6.19). Those responsible for the test felt that the accuracy, even with the optimum plan, was marginal. Specimens were very costly and were not to be used for fundamental research to verify the assumed Arrhenius relationship by allocating specimens to a third test temperature. They were content with a reasonable test demonstration that would provide a legal basis for evaluating the GaAs FET reliability.

Sensitivity analysis. The calculated accuracy or sample size for any plan – optimum, traditional, or compromise – depends on the assumed values of the model parameters. Of course, the assumed values differ from the true ones. Thus the calculated accuracy or sample size differs from the correct one. It is useful to re-evaluate a plan using other assumed values, changing one parameter at a time. Use at least a 20% change in a parameter to reflect the uncertainty in its true value. Hopefully, the plan and its accuracy are little affected. If the plan or accuracy is sensitive to a parameter value, then one must choose a compromise plan or a conservative one. Such an analysis can also be carried out on other characteristics of the plan, such as the probability of no failures at the lowest test stress level. Meeker (1984) illustrates such sensitivity analyses for a range of a and b values.

2.6. Meeker-Hahn Plans

Purpose. This section briefly presents the compromise plans of Meeker and Hahn (1985) for a censored test. Their plans apply to the simple linear-lognormal (Section 2.2) and linear-Weibull models. The following includes the plans, how to determine a plan, an example, standard error, sample size, adjusted plans, and remarks. Meeker and Hahn (1985) propose these plans merely as a starting point for developing plans – not as final plans. Their plans are more robust than optimum plans and usually more efficient than traditional plans. Jensen's (1985) program calculates their plans.

The plans. The plans contain three (transformed) stress levels – x_H, x_M, and x_L – High, Middle, and Low. x_H is specified from practical considerations above. $x_M = (x_H + x_L)/2$ is midway between the others. The best x_L' is chosen to minimize the asymptotic variance of ML estimate of the $100P$th percentile of log life at the (transformed) design stress x_D, denoted above by x_0. The allocation of specimens is $\pi_L = 4/7$, $\pi_M = 2/7$, and $\pi_H = 1/7$ for all plans. These are in the ratio 4:2:1. This unequal allocation is a compromise that extrapolates reasonably well. For a sample of n specimens, $n_L = 4n/7$, $n_M = 2n/7$, and $n_H = n/7$. Other notation is like that in previous sections.

Class-B example. For the Class-B example above, $x_H = 2.0277$ (220°C), $x_D = 2.4808$ (130°C), and $n = 40$. Thus the allocation is $n_L = 4(40)/7 = 23$, $n_M = 2(40)/7 = 11$, and $n_H = 6$, rounded to the nearest integer.

Best x_L. Determine the best low test stress x_L' as follows. Determine p_D and p_H, the probabilities of failure by the log censoring time η at x_D and x_H. Figure 2.1 depicts p_D and p_H. In terms of a and b of Section 2.2,

$$p_H = \Phi(a) = \Phi[(\eta - \gamma_0 - \gamma_1 x_H)/\sigma], \quad p_D = \Phi(a-b) = \Phi[(\eta - \gamma_0 - \gamma_1 x_D)/\sigma]. \quad (2.11)$$

As before, one assumes values for the unknown parameters. The best x_L is

$$x_L' = x_D + \varsigma'(x_H - x_D). \quad (2.12)$$

Here the factor ς' is a function of P, p_H, and p_D; it appears in Tables 2.1 and 2.2 for lognormal and Weibull distributions. Note that the previous formulas (2.5) and (2.8) and factor ξ differ from those in (2.12). ς' is in the range (0,1). For $\varsigma' = 0, x_L' = x_D$, the design stress; for $\varsigma' = 1, x_L' = x_H$, the high stress.

Class-B example. For the example, $\eta = \log(8,064) = 3.9066$, $a = 4.57$, and $b = 7.52$. Thus, $p_H = \Phi(4.57) = 1.0000$, and $p_D = \Phi(4.57 - 7.52) = 0.0016$. We wish to estimate the 50th percentile ($P = 0.50$) at 130°. However, Table 2.1 does not contain this percentile. Instead we find the best such plan for estimating the 10th percentile ($P = 0.10$). From Table 2.1, for $p_H = 1$ and $p_D = 0.0010$, $\varsigma' = 0.324$. That is, the low stress is 0.324 of the distance from x_D to x_H. The value 0.324 is close enough for the example. Interpolation is difficult, because the table is sparse. Then $x_L' = 2.4804 + 0.324$

Table 2.1.   Meeker-Hahn Plans for a (Log)Normal Distribution (Courtesy of Wm. Meeker and G.J. Hahn (1985) and with permission of the Amer. Soc. for Quality Control)

P	p_D	p_H	Optimum Plan					Best 4:2:1 Allocation						Adjusted 4:2:1 Allocation(0.80)					
			ξ^*	π^*_L	p_L	Er^*_L	V^*	ζ'_L	p_L	ζ_M	p_M	Er^*_L	R'	ζ''_L	p_L	ζ_M	p_M	Er^*_L	R''
.0001	.0001	0.250	507	.828	.015	12	61.42	448	.009	.724	.065	5	1.24	.359	.004	.679	.049	3	1.05
.0001	.0001	0.400	504	.817	.024	19	43.53	451	.016	.725	.114	8	1.27	.361	.007	.680	.087	3	1.06
.0001	.0001	0.600	493	.802	.039	31	31.06	447	.026	.724	.199	14	1.29	.358	.011	.679	.153	6	1.07
.0001	.0001	0.800	473	.783	.059	46	23.05	437	.042	.719	.329	24	1.31	.350	.017	.675	.261	9	1.08
.0001	.0001	0.900	456	.766	.075	57	19.54	427	.057	.714	.440	32	1.31	.342	.022	.671	.358	12	1.08
.0001	.0001	0.990	411	.722	.109	78	15.31	397	.094	.699	.693	53	1.27	.318	.036	.659	.604	20	1.10
.0001	.0001	1.000	356	.669	.143	95	13.18	351	.133	.675	.904	76	1.16	.280	.051	.640	.852	29	1.10
.0010	.0001	0.250	521	.807	.016	13	63.93	455	.010	.727	.066	5	1.23	.364	.005	.682	.050	2	1.05
.0010	.0001	0.400	514	.807	.026	21	44.00	455	.016	.728	.116	9	1.26	.364	.007	.682	.088	4	1.06
.0010	.0001	0.600	499	.805	.041	33	30.38	450	.027	.725	.200	15	1.30	.360	.011	.680	.154	6	1.07
.0010	.0001	0.800	476	.800	.061	48	21.68	439	.043	.719	.330	24	1.33	.351	.017	.675	.261	9	1.08
.0010	.0001	0.900	457	.794	.076	60	17.83	428	.057	.714	.441	32	1.35	.342	.022	.671	.358	12	1.08
.0010	.0001	0.990	410	.774	.107	82	13.07	398	.095	.699	.694	54	1.33	.319	.037	.659	.605	20	1.10
.0010	.0001	1.000	354	.740	.139	102	10.62	354	.139	.677	.906	79	1.24	.283	.054	.642	.854	30	1.11
.0010	.0010	0.250	379	.866	.015	12	35.33	317	.010	.659	.067	5	1.24	.254	.007	.627	.058	3	1.02
.0010	.0010	0.400	394	.852	.024	20	26.81	339	.017	.669	.117	9	1.26	.271	.010	.635	.099	5	1.02
.0010	.0010	0.600	397	.836	.039	32	20.25	350	.027	.675	.202	15	1.28	.280	.016	.640	.171	8	1.03
.0010	.0010	0.800	388	.818	.059	48	15.70	352	.044	.676	.333	25	1.30	.282	.024	.641	.284	13	1.04
.0010	.0010	0.900	377	.803	.074	59	13.59	349	.059	.674	.444	33	1.30	.279	.031	.640	.384	17	1.04
.0010	.0010	0.990	342	.765	.107	82	10.94	330	.097	.665	.696	55	1.27	.264	.049	.632	.631	27	1.06
.0010	.0010	1.000	296	.720	.142	102	9.532	294	.139	.647	.906	79	1.18	.235	.069	.618	.868	39	1.06
.0100	.0001	0.250	549	.735	.020	14	76.73	480	.012	.740	.071	6	1.23	.384	.005	.692	.053	3	1.05
.0100	.0001	0.400	540	.748	.032	24	50.70	474	.019	.737	.122	10	1.25	.379	.008	.690	.092	4	1.06
.0100	.0001	0.600	522	.760	.050	37	33.51	463	.030	.732	.208	17	1.27	.371	.012	.685	.159	7	1.07
.0100	.0001	0.800	497	.770	.073	56	22.76	448	.047	.724	.338	26	1.31	.358	.019	.679	.267	10	1.08
.0100	.0001	0.900	476	.776	.091	70	17.99	436	.062	.718	.449	35	1.34	.349	.024	.674	.364	13	1.08
.0100	.0001	0.990	425	.786	.125	98	11.94	405	.102	.702	.701	58	1.38	.324	.039	.662	.611	22	1.10
.0100	.0001	1.000	363	.791	.153	121	8.641	362	.153	.681	.911	87	1.36	.290	.059	.645	.860	33	1.12
.0100	.0010	0.250	425	.795	.019	15	41.23	340	.012	.670	.071	6	1.19	.272	.007	.636	.060	4	1.02
.0100	.0010	0.400	428	.804	.030	24	29.41	354	.018	.677	.121	10	1.22	.283	.011	.642	.102	6	1.03
.0100	.0010	0.600	422	.811	.046	37	20.91	360	.030	.679	.207	16	1.26	.288	.017	.644	.174	9	1.03
.0100	.0010	0.800	406	.816	.067	54	15.21	359	.046	.679	.337	26	1.30	.287	.025	.643	.288	14	1.04
.0100	.0010	0.900	390	.817	.083	67	12.57	354	.061	.677	.448	35	1.33	.283	.032	.642	.388	18	1.05
.0100	.0010	0.990	348	.816	.114	93	9.109	335	.101	.668	.701	57	1.36	.268	.051	.634	.635	29	1.06
.0100	.0010	1.000	297	.810	.143	115	7.189	302	.151	.651	.911	86	1.32	.242	.075	.621	.872	42	1.07
.0100	.0100	0.250	099	.961	.015	14	13.38	055	.013	.528	.073	7	1.29	.044	.012	.522	.072	6	1.00
.0100	.0100	0.400	173	.930	.025	22	11.98	128	.020	.564	.124	11	1.28	.103	.017	.551	.118	9	1.00
.0100	.0100	0.600	219	.905	.039	35	10.25	179	.031	.589	.210	17	1.29	.143	.025	.572	.197	14	1.01
.0100	.0100	0.800	240	.884	.059	51	8.684	210	.048	.605	.341	27	1.31	.168	.036	.584	.317	20	1.01
.0100	.0100	0.900	244	.870	.074	64	7.838	222	.063	.611	.451	36	1.31	.177	.046	.589	.420	26	1.01
.0100	.0100	0.990	233	.842	.107	89	6.630	228	.103	.614	.702	58	1.30	.183	.070	.591	.665	39	1.02
.0100	.0100	1.000	206	.810	.140	113	5.909	214	.150	.607	.910	85	1.23	.171	.098	.585	.887	56	1.03

Table 2.1. Continued

			Optimum Plan					Best 4:2:1 Allocation						Adjusted 4:2:1 Allocation(0.80)					
P	P_D	P_H	ξ	π_L	p_L	Er_L^*	V^*	ζ_L'	p_L	ζ_M'	p_M	Er_L^*	R'	ζ_L''	p_L	ζ_M''	p_M	Er_L^*	R''
0500	.0001	0.250	569	.669	.024	15	93.37	525	.017	.800	.100	9	1.27	420	.007	.800	.100	4	1.07
0500	.0001	0.400	559	.588	.038	25	60.16	501	.024	.751	.132	13	1.26	401	.010	.703	.100	5	1.06
0500	.0001	0.600	543	.706	.059	41	38.76	484	.036	.742	.220	20	1.27	387	.015	.694	.168	8	1.07
0500	.0001	0.800	519	.724	.088	63	25.61	464	.055	.732	.352	31	1.30	372	.021	.686	.277	12	1.07
0500	.0001	0.900	500	.734	.111	81	19.82	450	.071	.725	.463	40	1.32	360	.027	.680	.375	15	1.08
0500	.0001	0.990	451	.756	.160	121	12.40	416	.115	.708	.713	65	1.36	333	.044	.667	.622	25	1.10
0500	.0001	1.000	388	.778	.203	158	8.207	374	.175	.687	.918	99	1.39	299	.068	.650	.867	38	1.12
0500	.0010	0.250	462	.701	.024	16	53.55	409	.018	.749	.100	10	1.23	327	.011	.749	.100	6	1.03
0500	.0010	0.400	463	.724	.038	27	36.45	389	.023	.695	.131	13	1.22	311	.014	.654	.109	7	1.03
0500	.0010	0.600	456	.744	.058	43	24.77	385	.036	.693	.219	20	1.24	308	.020	.654	.183	11	1.03
0500	.0010	0.800	439	.761	.086	65	17.20	377	.054	.689	.351	30	1.26	302	.028	.651	.298	16	1.04
0500	.0010	0.900	423	.771	.108	82	13.74	369	.070	.685	.461	39	1.29	295	.036	.648	.398	20	1.05
0500	.0010	0.990	380	.791	.152	119	9.137	347	.113	.674	.712	64	1.35	278	.057	.639	.645	32	1.06
0500	.0010	1.000	325	.811	.189	153	6.441	315	.172	.658	.917	98	1.38	252	.085	.626	.880	48	1.07
0500	.0100	0.250	237	.788	.026	20	20.06	164	.020	.632	.100	11	1.16	131	.017	.632	.100	9	1.00
0500	.0100	0.400	272	.809	.039	31	15.60	168	.024	.584	.132	13	1.16	135	.020	.567	.125	11	1.00
0500	.0100	0.600	290	.825	.057	47	11.98	204	.036	.602	.220	20	1.21	164	.028	.582	.204	16	1.01
0500	.0100	0.800	293	.836	.081	67	9.260	227	.054	.614	.351	30	1.26	182	.040	.591	.325	22	1.01
0500	.0100	0.900	288	.842	.099	83	7.886	236	.070	.618	.461	40	1.29	189	.050	.594	.428	28	1.01
0500	.0100	0.990	263	.852	.135	115	5.907	240	.113	.620	.712	64	1.36	192	.076	.596	.673	43	1.02
0500	.0100	1.000	225	.862	.167	143	4.659	227	.171	.614	.917	97	1.38	182	.110	.591	.894	62	1.03
1000	.0001	0.250	577	.636	.025	15	103.7	560	.022	.945	.200	12	1.13	448	.009	.945	.200	5	1.10
1000	.0001	0.400	568	.658	.040	26	66.16	528	.029	.830	.200	16	1.28	422	.012	.830	.200	6	1.09
1000	.0001	0.600	552	.678	.063	43	42.22	498	.041	.749	.228	23	1.28	398	.016	.724	.200	9	1.08
1000	.0001	0.800	530	.698	.096	67	27.64	476	.061	.738	.362	34	1.29	381	.024	.690	.284	13	1.07
1000	.0001	0.900	512	.710	.123	87	21.26	460	.078	.730	.472	44	1.31	368	.030	.684	.382	17	1.08
1000	.0001	0.990	465	.734	.183	134	13.07	425	.125	.712	.722	71	1.35	340	.048	.670	.630	27	1.10
1000	.0001	1.000	405	.760	.239	181	8.396	382	.190	.691	.922	108	1.38	306	.074	.653	.872	42	1.12
1000	.0010	0.250	475	.657	.026	17	61.64	452	.023	.931	.200	13	1.13	361	.013	.931	.200	7	1.04
1000	.0010	0.400	477	.683	.041	28	41.23	429	.031	.793	.200	17	1.25	343	.017	.793	.200	9	1.04
1000	.0010	0.600	470	.707	.064	45	27.56	404	.041	.702	.228	23	1.24	323	.022	.673	.200	12	1.04
1000	.0010	0.800	455	.728	.097	70	18.85	391	.060	.696	.361	34	1.26	313	.031	.656	.305	17	1.04
1000	.0010	0.900	440	.741	.122	90	14.90	381	.077	.691	.472	44	1.28	305	.039	.653	.406	22	1.05
1000	.0010	0.990	400	.766	.178	136	9.648	357	.124	.679	.721	70	1.33	286	.061	.643	.652	35	1.06
1000	.0010	1.000	345	.791	.229	181	6.530	324	.189	.662	.922	107	1.37	259	.093	.630	.884	52	1.08
1000	.0100	0.250	267	.707	.030	20	25.41	223	.025	.899	.200	14	1.13	178	.021	.899	.200	12	1.01
1000	.0100	0.400	303	.741	.045	33	18.91	238	.033	.716	.200	19	1.19	191	.027	.716	.200	15	1.01
1000	.0100	0.600	321	.768	.067	51	13.96	231	.042	.616	.230	23	1.18	185	.032	.593	.212	18	1.01
1000	.0100	0.800	324	.789	.097	76	10.43	246	.061	.623	.362	34	1.22	197	.044	.598	.333	25	1.01
1000	.0100	0.900	319	.801	.120	96	8.688	251	.078	.626	.472	44	1.25	201	.055	.600	.436	31	1.02
1000	.0100	0.990	294	.822	.169	138	6.197	252	.124	.626	.721	70	1.32	201	.082	.601	.680	46	1.02
1000	.0100	1.000	253	.841	.213	178	4.598	238	.188	.619	.922	107	1.38	190	.120	.595	.898	68	1.03

Table 2.2.   Meeker-Hahn Plans for a Weibull (Extreme Value) Distribution (Courtesy of Wm. Meeker and G.J. Hahn (1985) and with permission of the Amer. Soc. for Quality Control)

			Optimum Plan					Best 4:2:1 Allocation						Adjusted 4:2:1 Allocation(0.80)					
P	P_D	P_H	ξ^*	π^*_L	P_L	Er^*_L	V^*_L	ζ'_L	P_L	ζ'_M	P_M	Er^*_L	R^*	ζ''_L	P_L	ζ''_M	P_M	Er^*_L	R''
.0001	.0001	0.250	.673	.837	.021	17	635.0	.606	.012	.803	.058	7	1.15	.485	.005	.742	.036	2	1.10
.0001	.0001	0.400	.690	.829	.036	29	426.1	.625	.021	.812	.098	11	1.16	.500	.007	.750	.059	4	1.12
.0001	.0001	0.600	.703	.818	.059	48	287.0	.638	.033	.819	.161	18	1.17	.510	.010	.755	.093	5	1.13
.0001	.0001	0.800	.709	.803	.091	73	201.3	.643	.050	.822	.249	28	1.19	.515	.015	.757	.142	8	1.14
.0001	.0001	0.900	.709	.789	.116	91	164.8	.643	.062	.822	.319	35	1.21	.515	.017	.757	.182	9	1.14
.0001	.0001	0.990	.700	.749	.168	125	121.8	.640	.092	.820	.487	52	1.26	.512	.024	.756	.285	13	1.12
.0001	.0001	1.000	.681	.698	.214	149	101.3	.645	.147	.823	.702	84	1.28	.516	.036	.758	.440	20	1.12
.0010	.0001	0.250	.691	.795	.024	19	703.5	.619	.014	.809	.061	7	1.15	.495	.005	.748	.038	2	1.10
.0010	.0001	0.400	.705	.795	.040	32	461.1	.636	.023	.818	.102	12	1.16	.509	.008	.754	.061	4	1.11
.0010	.0001	0.600	.715	.794	.066	52	302.0	.647	.036	.823	.167	20	1.17	.517	.011	.759	.096	6	1.13
.0010	.0001	0.800	.718	.790	.100	78	204.2	.650	.053	.825	.256	30	1.19	.519	.015	.760	.146	8	1.14
.0010	.0001	0.900	.717	.785	.125	98	161.9	.649	.065	.824	.326	37	1.21	.519	.018	.759	.186	10	1.14
.0010	.0001	0.990	.705	.771	.176	135	109.8	.638	.090	.819	.483	51	1.27	.510	.024	.755	.283	13	1.13
.0010	.0001	1.000	.683	.749	.217	162	82.68	.632	.129	.816	.676	73	1.33	.506	.032	.753	.421	18	1.11
.0010	.0010	0.250	.539	.865	.021	18	300.4	.459	.013	.730	.060	7	1.15	.367	.008	.684	.047	4	1.03
.0010	.0010	0.400	.576	.853	.036	30	214.5	.497	.022	.749	.101	12	1.16	.398	.012	.699	.075	6	1.05
.0010	.0010	0.600	.602	.840	.059	49	152.0	.524	.035	.762	.166	20	1.17	.419	.017	.710	.119	9	1.06
.0010	.0010	0.800	.618	.824	.091	75	111.0	.540	.052	.770	.255	29	1.18	.432	.024	.716	.179	13	1.06
.0010	.0010	0.900	.622	.810	.116	94	92.66	.544	.065	.772	.326	37	1.20	.435	.029	.718	.228	16	1.07
.0010	.0010	0.990	.618	.774	.167	129	70.32	.546	.095	.773	.493	54	1.24	.437	.039	.718	.348	22	1.07
.0010	.0010	1.000	.600	.729	.213	155	59.24	.555	.146	.777	.701	83	1.26	.444	.056	.722	.517	31	1.07
.0100	.0001	0.250	.716	.717	.029	21	879.8	.654	.018	.827	.070	10	1.18	.524	.006	.762	.042	3	1.10
.0100	.0001	0.400	.728	.725	.049	35	563.1	.666	.029	.833	.116	16	1.18	.533	.009	.766	.067	5	1.12
.0100	.0001	0.600	.736	.731	.079	57	359.8	.673	.045	.837	.187	25	1.19	.539	.014	.769	.106	7	1.13
.0100	.0001	0.800	.739	.734	.121	88	236.7	.674	.066	.837	.283	37	1.21	.540	.018	.770	.159	10	1.15
.0100	.0001	0.900	.738	.735	.153	112	183.4	.671	.081	.836	.357	46	1.22	.537	.022	.769	.202	12	1.15
.0100	.0001	0.990	.727	.734	.217	159	116.6	.657	.110	.829	.519	62	1.27	.526	.028	.763	.303	15	1.14
.0100	.0001	1.000	.705	.731	.270	197	79.76	.642	.142	.821	.695	81	1.35	.513	.035	.757	.435	19	1.12
.0100	.0010	0.250	.588	.780	.028	21	372.5	.496	.016	.748	.067	9	1.13	.396	.009	.698	.051	5	1.04
.0100	.0010	0.400	.614	.785	.045	35	253.7	.526	.026	.763	.110	14	1.14	.420	.014	.710	.080	7	1.05
.0100	.0010	0.600	.633	.788	.072	56	171.7	.547	.041	.774	.178	23	1.15	.438	.020	.719	.126	11	1.06
.0100	.0010	0.800	.643	.788	.109	85	118.9	.559	.060	.779	.271	34	1.18	.447	.027	.723	.188	15	1.07
.0100	.0010	0.900	.644	.787	.137	107	95.19	.560	.074	.780	.343	42	1.20	.448	.032	.724	.238	18	1.07
.0100	.0010	0.990	.635	.780	.191	148	64.76	.552	.100	.776	.502	57	1.25	.442	.041	.721	.354	23	1.07
.0100	.0010	1.000	.612	.770	.235	180	47.87	.545	.135	.772	.685	76	1.32	.436	.052	.718	.504	29.	1.06
.0100	.0100	0.250	.223	.939	.021	19	89.36	.148	.016	.574	.067	9	1.19	.118	.015	.559	.063	8	1.00
.0100	.0100	0.400	.326	.911	.036	32	74.62	.243	.026	.621	.109	14	1.17	.194	.021	.597	.100	12	1.01
.0100	.0100	0.600	.399	.888	.059	52	59.51	.311	.040	.655	.176	22	1.17	.249	.030	.624	.155	17	1.01
.0100	.0100	0.800	.444	.868	.091	79	47.26	.353	.059	.677	.268	33	1.18	.283	.041	.641	.229	23	1.02
.0100	.0100	0.900	.461	.853	.116	98	41.09	.370	.072	.685	.340	41	1.19	.296	.049	.648	.288	27	1.02
.0100	.0100	0.990	.473	.821	.167	136	32.86	.388	.102	.694	.506	58	1.22	.310	.065	.655	.427	37	1.02
.0100	.0100	1.000	.464	.785	.212	166	28.36	.406	.148	.703	.703	84	1.24	.325	.088	.662	.602	50	1.03

Table 2.2. Continued

P	P_D	P_H	Optimum Plan ξ*	π*_L	ρ_L	Er*_L	V*	Best 4:2:1 Allocation ζ'_L	ζ'_L	ζ_M	P_M	Er*_L	R'	Adjusted 4:2:1 Allocation(0.80) ζ''_L	ρ_L	ζ_M	P_M	Er*_L	R''
.0500	.0001	0.250	.729	.663	.033	21	1043.	.692	.024	.874	.100	13	1.24	.554	.008	.823	.100	4	1.12
.0500	.0001	0.400	.740	.674	.054	36	660.0	.693	.037	.847	.129	20	1.22	.555	.011	.815	.100	6	1.13
.0500	.0001	0.600	.749	.683	.088	60	417.0	.698	.057	.849	.206	32	1.22	.558	.016	.779	.115	9	1.11
.0500	.0001	0.800	.753	.690	.136	94	271.4	.698	.082	.849	.311	47	1.23	.558	.022	.779	.172	12	1.15
.0500	.0001	0.900	.752	.693	.174	120	208.7	.694	.101	.847	.391	57	1.24	.555	.026	.778	.219	14	1.16
.0500	.0001	0.990	.743	.697	.254	176	129.7	.680	.138	.840	.562	78	1.29	.544	.034	.772	.328	19	1.16
.0500	.0001	1.000	.724	.697	.325	226	85.47	.662	.175	.831	.736	100	1.35	.529	.042	.765	.465	23	1.14
.0500	.0010	0.250	.619	.697	.033	22	478.5	.569	.025	.823	.100	14	1.20	.455	.013	.823	.100	7	1.05
.0500	.0010	0.400	.643	.710	.054	38	316.7	.571	.035	.786	.126	19	1.17	.457	.017	.747	.100	9	1.06
.0500	.0010	0.600	.660	.721	.086	62	208.7	.586	.053	.793	.200	30	1.18	.469	.024	.734	.139	13	1.07
.0500	.0010	0.800	.671	.728	.131	95	140.9	.592	.076	.796	.301	43	1.19	.474	.033	.737	.206	18	1.07
.0500	.0010	0.900	.671	.732	.165	121	110.7	.592	.093	.796	.378	53	1.21	.474	.038	.737	.259	21	1.08
.0500	.0010	0.990	.664	.735	.237	174	71.73	.581	.126	.791	.545	72	1.25	.465	.049	.733	.383	28	1.08
.0500	.0010	1.000	.643	.735	.299	219	49.39	.566	.161	.783	.720	91	1.32	.453	.061	.726	.532	34	1.07
.0500	.0100	0.250	.361	.791	.033	26	130.6	.286	.026	.701	.100	14	1.14	.229	.021	.701	.100	12	1.01
.0500	.0100	0.400	.423	.802	.052	41	97.93	.300	.032	.650	.121	18	1.11	.240	.025	.620	.108	14	1.01
.0500	.0100	0.600	.468	.808	.080	64	71.95	.353	.048	.676	.192	27	1.13	.282	.035	.641	.166	20	1.01
.0500	.0100	0.800	.497	.811	.118	95	53.23	.387	.069	.693	.288	39	1.15	.309	.047	.655	.243	26	1.02
.0500	.0100	0.900	.507	.811	.146	118	44.14	.399	.084	.700	.362	48	1.17	.319	.055	.660	.304	31	1.02
.0500	.0100	0.990	.509	.807	.204	164	31.60	.406	.114	.703	.526	65	1.22	.325	.071	.663	.441	40	1.03
.0500	.0100	1.000	.492	.802	.251	201	24.04	.407	.149	.704	.705	85	1.28	.326	.089	.663	.603	50	1.03
.1000	.0001	0.250	.733	.641	.034	21	1124.	.715	.029	.968	.200	16	1.14	.572	.009	.968	.200	5	1.18
.1000	.0001	0.400	.745	.653	.056	36	707.9	.710	.042	.903	.200	24	1.25	.568	.013	.903	.200	7	1.16
.1000	.0001	0.600	.753	.663	.092	61	445.7	.709	.062	.855	.216	35	1.25	.567	.018	.845	.200	10	1.16
.1000	.0001	0.800	.758	.671	.143	95	289.1	.708	.091	.854	.324	51	1.25	.567	.024	.796	.200	13	1.16
.1000	.0001	0.900	.758	.675	.183	123	221.8	.705	.112	.852	.407	63	1.26	.564	.028	.782	.227	16	1.17
.1000	.0001	0.990	.750	.680	.270	183	137.1	.691	.153	.845	.583	87	1.30	.552	.037	.776	.341	21	1.17
.1000	.0001	1.000	.732	.681	.351	239	89.53	.672	.196	.836	.757	111	1.36	.538	.046	.769	.481	26	1.15
.1000	.0010	0.250	.629	.663	.035	22	534.0	.603	.030	.955	.200	17	1.13	.482	.015	.955	.200	8	1.07
.1000	.0010	0.400	.652	.679	.057	38	350.2	.605	.043	.867	.200	24	1.22	.484	.020	.867	.200	11	1.07
.1000	.0010	0.600	.669	.692	.092	63	228.8	.605	.060	.803	.212	34	1.20	.484	.027	.793	.200	15	1.09
.1000	.0010	0.800	.679	.701	.140	98	153.4	.610	.086	.805	.317	49	1.21	.488	.036	.744	.216	20	1.08
.1000	.0010	0.900	.682	.706	.178	125	120.0	.609	.106	.804	.398	60	1.22	.487	.043	.744	.271	24	1.09
.1000	.0010	0.990	.676	.712	.259	184	76.78	.598	.143	.799	.570	81	1.26	.478	.055	.739	.399	31	1.09
.1000	.0010	1.000	.657	.714	.332	236	51.90	.581	.182	.791	.744	104	1.32	.465	.067	.732	.551	38	1.08
.1000	.0100	0.250	.394	.742	.037	26	161.1	.346	.032	.924	.200	18	1.14	.276	.025	.924	.200	14	1.01
.1000	.0100	0.400	.454	.757	.058	43	116.4	.383	.044	.789	.200	25	1.18	.306	.033	.789	.200	18	1.02
.1000	.0100	0.600	.497	.767	.090	68	83.42	.390	.057	.695	.207	32	1.13	.312	.040	.687	.200	23	1.03
.1000	.0100	0.800	.524	.771	.134	102	60.26	.417	.080	.709	.307	45	1.15	.334	.053	.667	.257	30	1.02
.1000	.0100	0.900	.534	.775	.168	129	49.21	.427	.097	.714	.385	55	1.16	.342	.062	.671	.320	35	1.03
.1000	.0100	0.990	.537	.775	.236	183	34.04	.430	.131	.715	.553	74	1.21	.344	.080	.672	.461	45	1.03
.1000	.0100	1.000	.520	.775	.295	228	24.80	.426	.167	.713	.727	95	1.28	.340	.097	.670	.621	55	1.03

$(2.0277 - 2.4804) = 2.3337$, and $T_L' = (1000/2.3337) - 273.16 = 155°C$. Also, $x_M = (2.0277 + 2.3337)/2 = 2.1807$ (185°C).

Standard error. The asymptotic standard error of the ML estimate of the 100Pth percentile of log life at x_D is

$$\sigma[\hat{\eta}_P(x_D)] = \sigma \cdot (R'V^*/n)^{1/2}. \tag{2.13}$$

Here σ is the log standard deviation for the lognormal distribution, and $\sigma = 1/\beta$ for the Weibull distribution. The factors R' and V^* appear in Tables 2.1 and 2.2; they are functions of P, p_H, and p_D. V^* is the variance factor (2.9) for the optimum plan of Section 2.4, and R' the ratio of the compromise variance over the optimum variance.

Class-B example. From Table 2.1, for $p_H = 1$ and $p_D = 0.0010$, $R'V^* = 1.37(6.530) = 8.95$. Thus $\sigma[\hat{\eta}_{.10}(2.4804)] = \sigma(8.95/40)^{1/2} = 0.473\sigma$.

Sample size. Determine sample size n as follows. Suppose that with high probability γ, $\hat{\eta}_P(x_D)$ is to be within $\pm w$ of the true value. The approximate sample size n' that achieves this is

$$n' = R'V^*(K_\gamma \sigma/w)^2. \tag{2.14}$$

Here K_γ is the standard normal $100(1+\gamma)/2$ percentile. After the data are collected, obtain a confidence interval from the data and ignore w.

Class-B example. Suppose that the estimate is to be within ± 0.10 of the true 10th percentile of log life with 95% probability ($\gamma = 0.95$). The sample size that achieves this is $n' = 1.37 \cdot 6.530(1.96 \cdot 0.259/0.10)^{1/2} = 57$ specimens.

Tables. The meanings of other quantities in Tables 2.1 and 2.2 follow. For the optimum plan (Section 2.4),

ζ^*	is optimum the factor in (2.12).
π^*	is the optimum sample fraction at the low stress.
p_L	is the fraction failing at the low stress by the censoring time.
$E(r_L^*)$	$= 1000\pi^*p_L$ is the expected number of failures at x_L when $n = 1000$.
V^*	is the optimum variance factor.

For the Meeker-Hahn plan (optimized 4:2:1 allocation),

ζ'	is the best factor in (2.12).
p_L	is the fraction failing at the low stress by the censoring time.
p_M	is the fraction failing at the middle stress by the censoring time.
$E(r_L^*)$	$= 1000(4/7)p_L$ is the expected number of failures at the low stress when $n = 1000$.
R'	is the ratio of the Meeker-Hahn variance over the optimum variance.

Adjusted plans. Meeker and Hahn (1985) offer adjusted plans. These plans use a lower x_L' to reduce extrapolation in stress and possible model error. Their adjusted plans use a fraction of ζ'' of ζ' in (2.12). They table plans for fractions 0.90, 0.80, 0.70, 0.60. The adjusted plan for $\varsigma'' = 0.80\varsigma'$ appears in Tables 2.1 and 2.2. For the Class-B example, the adjusted plan uses $\varsigma'' = 0.80(0.324) = 0.292$. Then $x_L'' = 2.4804 + 0.292(2.0277 - 2.4804) = 2.3482$ or 153°C. The standard error of the percentile estimate is

$$\sigma[\hat{\eta}_P(x_D)] = \sigma \cdot (R''R'V^*/n)^{1/2}.$$

The factor R'' appears in Tables 2.1 and 2.2.

Remarks. Some useful remarks on the plans follow.

1. As discussed in previous sections, the plans depend on assumed values of model parameters. Thus it is useful to do a sensitivity analysis with other values to see if they affect the plan much.
2. The Meeker-Hahn (4:2:1) plans use very unequal allocation, which is close to optimum for short extrapolation. For long extrapolation, traditional plans (Section 2.3) with three test stress levels may be more accurate and as robust.
3. The Meeker-Hahn plans use a fixed (4:2:1) allocation and do not try to optimize allocation. A promising class of plans with three equally spaced stress levels follows. Allocate a sample proportion π each to x_H and x_M and $(1 - 2\pi)$ to x_L. Then choose π and x_L to minimize the asymptotic variance of the ML estimate of a percentile of log life at x_D. These plans should yield a smaller variance and be as robust. Moreover, if the specimens at x_L are lost, the equal allocation at x_M and x_H is good for the long extrapolation to x_D. Such plans merit investigation.

3. EVALUATION OF A TEST PLAN BY SIMULATION

Purpose. Many test plans are too complex to evaluate by means of the ML theory of Section 5. This section presents a simple alternative – simulation, as presented by Nelson (1983b) but previously unpublished. It applies to most models and forms of censoring. It provides standard errors of estimates of interest. These standard errors allow one to judge whether the estimates and test plan are accurate enough for the application. Here one analyzes the (simulated) data before running the test – a practice recommended in Chapter 1 to avoid unpleasant surprises.

Overview. Section 3.1 presents a proposed test plan for insulation life. Section 3.2 presents models used in the computer simulation of data from the plan. Such models would be fitted to the actual experimental data. Section 3.3 describes the computer simulation of test data. Section 3.4 presents computer output from fitting the models to such data. The output, for example, shows how accurately 1) insulation life is estimated at design stresses, 2) how accurately the effect of insulation thickness is estimated, 3) how accurately conductors are compared, and 4) how accurately another plan estimates such quantities.

3.1. The Proposed Plan

The plan. The proposed test plan involves 170 test specimens. The specimens have various combinations of three insulation thicknesses (0.163 cm, 0.266 cm, and 0.355 cm), four voltage stresses (200, 175, 150, and 120 volts/mm), and four types of conductor (S, SS, G, and SO). Table 3.1 shows the number of specimens at each combination of insulation thickness, voltage stress, and conductor. For example, there are five specimens with insulation thickness 0.163 cm, voltage stress 200 volts/mm, and conductor S. Figure 3.1 displays the number of specimens of S conductors at each combination of thickness and voltage stress. There are 12 test combinations of thickness (THICK) and voltage stress (VPM). The two X's in the plot show the extrapolated conditions of 65 and 80 volts/mm at a thickness of 0.266 cm.

Purpose. The investigators chose this test plan for various purposes. They sought to estimate insulation life at the extrapolated conditions, to estimate the effect of insulation thickness on life, and to compare the effects of the conductors on insulation life. The test range of thickness is greater than that in practice; thus this experiment would provide a more accurate estimate of the effect of thickness than would production specimens. The center thickness 0.266 cm is allocated more specimens, because it is a proposed thickness. Following good practice for more accurate extrapolation with respect to stress, this plan has more specimens at low test stress than at high, as suggested by optimum plans. Engineering judgment and statistical design considerations (Chapter 1) were used to choose the experimental combinations of stress, thickness, and conductor and the number of specimens for each.

Caution. Statistical theory for traditional experimental design is correct *only* for *complete* data. Do *not* assume that properties of standard experimental designs (2^{n-p}, Box central composite, etc.) hold for censored and interval data. They usually do not hold. For example, ML (and other) estimates of certain model coefficients ("effects") may not exist with censored data. Coefficient estimates that are orthogonal for complete data may be correlated for censored data. Also, aliasing of effects depends on the censoring. In addition, the variance of an estimate of a model coefficient depends

Table 3.1. Numbers of Specimens at the Test Conditions

Thickness:	.163cm		.266cm				.355cm	
Conductor:	S	SS	S	SS	G	SO	S	SS
V/mm								
200	5	1	10	1	1	3	5	1
175	8	2	14	2	2	4	8	2
150	8	2	14	2	2	4	8	2
120	11	3	18	3	3	7	11	3

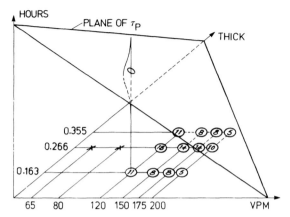

Figure 3.1.   Test conditions for S conductors and the τ_P percentile plane – all scales are logarithmic.

on the amount of censoring at all test conditions and on the true values of (possibly all) model coefficients. Thus the censoring times at each test condition are part of an experimental design and affect its statistical properties.

3.2. Models

Model. The assumed model has a Weibull distribution for the scatter in insulation life. The $100P$th percentile τ_P, as function of an insulation thickness (*THICK*) and voltage stress (*VPM*), is

$$\ln(\tau_P) = C1 + C2(LVPM - \overline{LVPM}) + C3(LTHICK - \overline{LTHICK})$$
$$+ (1/\beta) \ln[-\ln(1-P)]. \tag{3.1}$$

Here $LVPM = \ln(VPM)$, and $\overline{LVPM}$ is the average $LVPM$ over the test specimens; $LTHICK = \ln(THICK)$, and $\overline{LTHICK}$ is the average $LTHICK$ over the test specimens. (3.1) is an inverse power function of VPM and $THICK$. Experience suggests that coefficient $C3$ for thickness is negative. That is, thick insulation tends to have shorter life than thin insulation for the same voltage stress. Figure 3.1 depicts τ_P as a plane in three dimensions with log scales.

Lognormal. For a model with the lognormal distribution, the $100P$th percentile has the same equation (3.1), but log base 10 is used, and $z_P\sigma$ replaces $(1/\beta)\ln[-\ln(1-P)]$. That is,

$$\log(\tau_P) = C1 + C2(LVPM - \overline{LVPM}) + C3(LTHICK - \overline{LTHICK}) + z_P\sigma; \tag{3.2}$$

here σ is the standard deviation of log life, z_P is the standard normal $100P$th percentile. The lognormal percentile planes fitted to such test data are much closer together then Weibull ones. The Weibull distribution is used below.

Quadratic relationship. Previous data suggest that the relationship between life and voltage stress on a log-log plot is slightly curved, producing longer life at low stress than the inverse power law does. To model this curvature, a squared term $LVPM\,2$ in ln volts/mm is added to (3.1); namely,

$$\ln(\tau_P) = C1 + C2(LVPM - \overline{LVPM}) + C3(LVPM\,2 - \overline{LVPM\,2})$$
$$+ C4(LTHICK - \overline{LTHICK}) + (1/\beta)\,\ln[-\ln(1-P)]. \tag{3.3}$$

Here $LVPM\,2 = (LVPM - \overline{LVPM})^2$, and $\overline{LVPM\,2}$ is the average $LVPM\,2$ over the test specimens. $C1$ is the intercept coefficient. $C2$ no longer corresponds to a power in the relationship between log life and log stress. $C3$ is the coefficient for the quadratic term. $C4$ is the coefficient for the effect of thickness. The final term in (3.3) incorporates the Weibull distribution.

The models (3.1) and (3.3) are fitted by computer to simulated test data as described in Section 3.3. The program output provides standard errors of estimates of parameters, percentiles, and other quantities of interest for the test plan. The standard errors indicate the accuracy of the estimates, as described in Section 3.4.

3.3. Simulation

Simulated life. The STATPAC program was used as follows to generate a Monte Carlo life time of each specimen, according to its voltage stress and insulation thickness. The assumed numerical value of (3.1) is

$$\ln(\tau_P) = 7.416910 - 12.27645(LVPM - 5.090002)$$
$$- 1.296141[LTHICK - (-1.552109)] + 0.673422\,\ln[-\ln(1-P)]. \tag{3.1'}$$

The equation comes from a test of comparable insulation. Here the shape parameter is $\beta = 1/0.673422 = 1.484953$. For each specimen in the test plan, calculate the simulated random Weibull life t_i for specimen i as

$$\ln(t_i) = 7.416910 - 12.27645(LVPM_i - 5.090002) \tag{3.4}$$
$$- 1.296141[LTHICK_i - (-1.552109)] + 0.673422\,\ln[-\ln(u_i)].$$

Here u_i is a random observation from a uniform distribution on the interval

Table 3.2.   Test Conditions and Simulated Lives of 170 Specimens

1	V_1	T_1	t_1	15	V_1	T_2	t_{15}	165	V_4	T_3	t_{165}
2	V_1	T_1	t_2	16	V_1	T_2	t_{16}	166	V_4	T_3	t_{166}

1 V_1 T_1 t_1 15 V_1 T_2 t_{15} 165 V_4 T_3 t_{165}

2 V_1 T_1 t_2 16 V_1 T_2 t_{16} 166 V_4 T_3 t_{166}

.   .   .   .

.   .   .   .   etc.   .   .   .   .

.   .   .   .

14 V_1 T_1 t_{14} 45 V_1 T_2 t_{45} 170 V_4 T_3 t_{170}

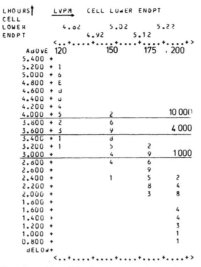

Figure 3.2.   Simulated log life data versus log voltage stress, LVPM.

[0,1]. Table 3.2 depicts such simulated Weibull lives for the 170 specimens. Various packages (Section 1 of Chapter 5) can generate such random observations. Ripley (1987) presents simulation theory and methods.

Example. Figure 3.2 depicts such simulated log life data at four log voltage stresses. This plot does not distinguish differences in specimen thickness. The lines drawn at 1,000, 4,000, and 10,000 hours show which failures are observed by those censoring time. This plot shows that, at 120 volts/mm, only two failures are observed by 4,000 hours and only seven by 10,000 hours. A computer package is used as explained below to fit the various models to such data. The package output gives the accuracy (standard errors) of ML estimates of various quantities.

Lognormal simulation. To simulate a lognormal life from (3.2), replace z_P by a random observation from a standard normal distribution (with a mean of zero and a standard deviation of one).

3.4. Accuracy of Estimates

Overview. This section presents the accuracy of estimates from two test plans for various censoring times and models, in particular for:
- The linear relationship (3.1) – the accuracy of the estimates of the Weibull β, the power $C2$ in the power law, the thickness coefficient $C3$, and percentage points of the Weibull life distribution at 65 and at 80 volts/mm.
- A test plan with the 120-vpm specimens reallocated to the higher test stresses. Few specimens fail at 120 vpm and thus yield little information. The preceding information on accuracy is desired.

- The quadratic relationship (3.3) − the preceding information plus the accuracy of the estimate of the quadratic coefficient is desired.
- A comparison of two types of conductors − the accuracy of the estimate of the difference between their intercept coefficients $C1$, their Weibull shape parameters β, and their powers $C2$ in the inverse power law, all under an assumed linear relationship (3.1).

Model fitting. A statistical package is used to fit a model such as (3.1) to the simulated data. Then the confidence intervals and standard errors indicate the accuracy of the estimates. For a simulated set of data on 170 specimens, the data are analyzed four times: 1) all specimens are run to failure, 2) the data are censored at 10,000 hours (output in Figure 3.3), 3) at 4,000 hours, and 4) at 1,000 hours.

Output. Figure 3.3 shows STATPAC output from fitting (3.1) to a simulated data set censored at 10,000 hours. In the figure, Line 1 points out that 114 of the 170 specimen lives are observed by 10,000 hours. Line 2 shows the ML estimates, approximate confidence limits, and local estimates of asymptotic standard errors. Line 3 shows percentile estimates with their confidence limits and standard errors. It is better to work with $\hat{y}_P = \ln(\hat{\tau}_P)$ and its standard error $s(\hat{y}_P) = s(\hat{\tau}_P)/\hat{\tau}_P$. For example, at 65 vpm and 0.266 cm, the estimate $\hat{y}_{.001} = \ln(\hat{\tau}_{.001})$ has $s(\hat{y}_{.001}) = 567,750.8/1,197,913 = 0.4739$; such standard errors are ordinarily rounded to two significant figures. The approximate *95% uncertainty* is 1.96 times a standard error; here it is $1.96 \times 0.4739 = 0.93$, which appears in Table 3.3 Such standard errors were obtained for the following.

Linear relationship. The linear relationship (3.1) was fitted to all 170 specimens, as if they are all from the same population. This ignores possible differences between conductors. The resulting standard errors are smaller than and thus a lower bound for those obtained from just one conductor. The 95% uncertainties (twice the standard errors) of various estimates appear in Table 3.3 for two such simulated data sets. The approximate 95% confidence limits are the estimate plus or minus the tabled uncertainty. A plot of such uncertainties can help one see subtler differences and patterns. As expected, Table 3.3 shows that the earlier the censoring time, the larger the uncertainty. These uncertainties can be examined to see if they are small enough for practical purposes. For example, the assumed $C3 = 1.30$ for *LTHICK* is an estimate from comparable data, but that estimate was not statistically significant (i.e., different from zero). Table 3.3 shows 95% uncertainties from 0.36 to 0.62. Thus, if the proposed plan yields the same estimate $\hat{C3} = 1.30$ with 95% uncertainty ± 0.62, it will be statistically significant. The added accuracy of $\hat{C3}$ is mostly due to the wide range of specimen thicknesses in the proposed test − much wider than in the comparable data.

Accuracy of standard errors. Table 3.3 displays *estimates* of the 95% uncertainties (2 times the standard errors). The accuracy of such estimates can

1) CASES
   114 WITH OBSERVED VALUES.
   56 WITH VALUES CENSORED ON THE RIGHT.
   ----
   170 IN ALL

MAXIMUM LOG LIKELIHOOD =   -867.28671

* MAXIMUM LIKELIHOOD ESTIMATES FOR DIST. PARAMETERS
  WITH APPROXIMATE 95% CONFIDENCE LIMITS

PARAMETERS		ESTIMATE	LOWER LIMIT	UPPER LIMIT	STANDARD ERROR
2) C	1	8.162629	8.012836	8.312422	0.7642508E-01
C	2	-11.98761	-12.86361	-11.11161	0.4469395
C	3	-1.360631	-1.798193	-0.9230682	0.2232462
SPREAD		1.643543	1.436191	1.880832	0.1130857

* MAXIMUM LIKELIHOOD ESTIMATE 'OF THE FITTED EQUATION

CENTER $= ln(\hat{\alpha})$
   $=$ 8.162629

   + (LVPM   -   5.016966   ) *   -11.98761

   + (LTHICK -   -1.371582) *   -1.360631

SPREAD $= \hat{\beta}$
   $=$ 1.643543

PCTILES(4.174387 -1.324259)

$$ln(65) \overset{LVPM}{=} \qquad ln(0.266) \overset{LTHICK}{=}$$
$$\quad 4.1743870 \qquad\qquad -1.3242590$$

* MAXIMUM LIKELIHOOD ESTIMATES FOR DIST. PCTILES
  WITH APPROXIMATE 95% CONFIDENCE LIMITS

PCT.	ESTIMATE	LOWER LIMIT	UPPER LIMIT	STANDARD ERROR
3) 0.1	1197913.	473141.4	3032911.	567750.8
0.5	3193283.	1328727.	7674307.	1428552.
1	4875974.	2064032.	0.1151877E 08	2138608.
5	0.1314513E 08	5712061.	0.3025082E 08	5589846.
10	0.2036922E 08	8894941.	0.4664507E 08	8610591.
20	0.3215655E 08	0.1405173E 08	0.7358833E 08	0.1358237E 08
50	0.6408691E 08	0.2779166E 08	0.1477829E 09	0.2731881E 08
80	0.1069958E 09	0.4583390E 08	0.2497739E 09	0.4627930E 08
90	0.1330469E 09	0.5660396E 08	0.3127251E 09	0.5801272E 08
95	0.1561500E 09	0.6605885E 08	0.3691075E 09	0.6853637E 08
99	0.2028448E 09	0.8493471E 08	0.4844430E 09	0.9009606E 08

PCTILES(4.382027 -1.324259)

$$ln(80) \overset{LVPM}{=} \qquad ln(0.266) \overset{LTHICK}{=}$$
$$\quad 4.3820270 \qquad\qquad -1.3242590$$

* MAXIMUM LIKELIHOOD ESTIMATES FOR DIST. PCTILES
  WITH APPROXIMATE 95% CONFIDENCE LIMITS

PCT.	ESTIMATE	LOWER LIMIT	UPPER LIMIT	STANDARD ERROR
0.1	99407.66	44845.15	220355.7	40372.38
0.5	264991.6	127944.9	548834.0	98438.71
1	404628.0	200039.9	818455.7	145428.9
5	1090836.	560914.0	2121401.	370178.8
10	1690320.	877366.1	3256544.	565523.0
20	2668480.	1390914.	5119500.	887062.3
50	5318191.	2759185.	0.1025055E 08	1780503.
80	8878947.	4552203.	0.1731814E 08	3026414.
90	0.1104077E 08	5620329.	0.2168887E 08	3803459.
95	0.1295796E 08	6556815.	0.2560826E 08	4503585.
99	0.1683289E 08	8423159.	0.3363892E 08	5946040.

Figure 3.3.   STATPAC fit of (3.1) to simulated data censored at 10,000 hours.

Table 3.3.   95% Uncertainties with All Data and the Linear Relationship

Cens. Time	No. Failed	$\hat{\beta}$	$2s(\hat{\beta})$	LVPM $2s(\hat{C2})$	LTHICK $2s(\hat{C3})$	65vpm,0.266cm $2s(\ln(\hat{t}_{.001}))$	80vpm,0.266cm $2s(\ln(\hat{t}_{.001}))$
∞	170	1.60	0.19	0.48	0.35	0.69	0.64
	170	1.53	0.18	0.52	0.37	0.71	0.65
10,000	114	1.64	0.22	0.88	0.44	0.93	0.80
	113	1.47	0.20	1.01	0.48	1.03	0.88
4,000	98	1.73	0.28	1.10	0.46	1.06	0.88
	104	1.49	0.22	1.18	0.51	1.20	1.01
1,000	66	1.66	0.34	2.11	0.63	1.90	1.50
	71	1.45	0.27	1.96	0.62	1.91	1.56

be gauged as follows. One could reasonably assume that the estimates come from a sampling distribution that is approximately normal with a mean equal to the true uncertainty. Then one uses the two (or more) such estimates from the two (or more) simulations to calculate a confidence interval for the true uncertainty. This interval is the usual one for the mean of a normal distribution and employs a t percentile. When there are two estimates, the estimates themselves are then the endpoints of a 50% confidence interval, and then an 80% confidence interval is three times as wide.

More accuracy. The estimates of certain standard errors can be made more accurate as follows. Suppose the model distribution has a scale parameter σ, and its assumed value in a simulation is σ'. Also, suppose that the estimate of σ' from the simulation is $\hat{\sigma}$. Suppose the estimate of a standard error (or 95% uncertainty) is s. The adjusted (more accurate) standard error is

$$s' = (\sigma'/\hat{\sigma})s.$$

For a Weibull distribution, the estimate and β' are inverted:

$$s' = (\hat{\beta}/\beta')s.$$

This is so because the shape parameter β is the reciprocal of the scale parameter of the corresponding extreme value distribution of ln life. For example, the estimates of the 95% uncertainties of $\hat{C2}$ at 10,000 hours are 0.88 and 1.00 in Table 3.3. The adjusted 95% uncertainties are $0.88(1.64/1.485) = 0.97$ and $1.01(1.47/1.485) = 1.00$. These adjusted uncertainties are closer to each other and likely to the true 95% uncertainty. These adjusted uncertainties can be used as described above to calculate confidence limits for the true 95% uncertainty. The adjustment is valid if the theoretical standard error of an estimate is proportional to the distribution scale parameter σ. This is roughly so for any coefficient in a linear relationship for the location parameter, for a constant scale parameter, and for any linear combination of such coefficients and σ (for example, a percentile).

Larger samples. The large sample of 170 specimens, most of which fail, assures that the simulated data yields estimates of asymptotic standard errors that are accurate enough for practical purposes. If a test involves fewer than, say, 50 failures, accuracy of standard errors can also be improved as follows. In the simulation, use four specimens for each specimen of the actual test. The standard errors from the bigger sample must be multiplied by $4^{1/2} = 2$ to get the correct standard errors for the actual test. Greater multiples like 9 and 16 can be used, if necessary, with multiples $9^{1/2} = 3$ and $16^{1/2} = 4$. Of course, for small samples, the (asymptotic) large sample theory tends to give standard errors and confidence intervals that are too small. Then the LR confidence limits are better approximations. However, few ML fitting packages provide such limits yet.

Sensitivity analyses. Such simulations employ assumed models and assumed values for the model parameters and coefficients. Possibly the standard errors of interest are sensitive to such assumptions. This can be checked with further simulations using different parameter values or models. Hopefully a proposed plan performs well with different parameter values and models. Otherwise, one must find a more robust plan or be content with the possibility that the actual standard errors are not know accurately enough.

Failures are informative. By 10,000 hours few specimens at 120 vpm fail. Thus they contribute little to the accuracy of estimates. In general, if there are few failures in a data set, the resulting estimates have poor accuracy. Many engineers prefer to have no failures in their test, since that indicates that a product is reliable. A statistician prefers many failures, since they provide more accurate estimates. It is important to recognize that many accelerated tests are intended to *measure* reliability, whether good or bad, and to do so accurately. The way the test is run and the number of observed failures do not determine how good the product is; they merely measure it. Of course, engineering design and manufacturing determine how good the product is. So it is generally best to have failures when measuring reliability.

Reallocated plan. In view of this, the 59 specimens at 120 vpm were reallocated to the lower stress levels, where there were originally 111 specimens. Thus at each remaining test condition the new number of specimens is 170/111 times the original number. Specimen lives were simulated for this reallocated plan, using the assumed linear-Weibull model (3.4). The resulting (unadjusted) standard errors of various estimates appear in Table 3.4, which includes the average (unadjusted) standard errors of the original plan. The tabled $\hat{\beta}$ is from the simulation with the reallocated plan. The following observations on Table 3.4 are generally true for such a reallocation.

- $s(\hat{\beta})$ for the reallocated plan is smaller if there is censoring. This is so because $s(\hat{\beta})$ decreases strongly with an increase in the sample fraction failing at a stress level, and the reallocated plan uses stresses where this fraction is larger than at 120 vpm.

**Table 3.4. Standard Errors with/without Reallocation of 120 vpm Speci-
mens – Linear Relationship**

Cens. Time	$\hat{\beta}$	$s(\hat{\beta})$	LVPM $s(\hat{C}2)$	LTHICK $s(\hat{C}3)$	65vpm,.266cm $s(\ln(\hat{\tau}_{.001}))$	80vpm,.266cm $s(\ln(\hat{\tau}_{.001}))$
∞	1.54	.093/.094	0.44/0.26	0.18/0.18	0.51/0.36	0.44/0.33
10,000	1.54	.095/.107	0.45/0.48	0.19/0.23	0.52/0.50	0.44/0.43
4,000	1.54	.102/.127	0.53/0.58	0.20/0.25	0.55/0.58	0.46/0.48
1,000	1.45	.117/.156	0.91/1.04	0.28/0.32	0.87/0.97	0.70/0.78

- $s(\hat{C}2)$ for *LVPM* for the reallocated plan is smaller for censoring at 10,000 hours or earlier. This is so because $s(\hat{C}2)$ decreases strongly with the width of the test range of *LVPM*, as well as with the number of failures. The test range for the original plan is effectively 150-200 vpm up to 10,000 hours since there are so few failures at 120 vpm. Of course, for the uncensored data (∞), $s(\hat{C}2)$ is much smaller for the original plan.
- $s(\hat{C}3)$ for *LTHICK* for the reallocated plan is slightly smaller with censoring. Without censoring (∞), $s(\hat{C}3)$ is the same for both plans. The dependence of life on *LTHICK* over the test range is much smaller than that on *LVPM*. Thus, the effective width of the test range of *LTHICK* is little affected by the censoring, and the reduced number of failures as a result of censoring has less effect than the test range.
- For both extrapolated conditions, $s(\ln(\hat{\tau}_{.001}))$ is slightly smaller for the reallocated plan for early censoring. Thus, for censoring before 10,000 hours, the few failures at 120 vpm with the original plan little affect the accuracy of extrapolation. Of course, for the original plan, $s(\ln(\hat{\tau}_{.001}))$ is smaller for no censoring or high censoring time.

In general, for the linear relationship (3.1), the reallocated plan performs better in the 10,000 hours allotted the test. However, this conclusion is quite different with the quadratic relationship, as shown below.

Quadratic relationship. The quadratic relationship was fitted to simulated data from the linear relationship. That is, the quadratic coefficient $C3=0$. This is reasonable, since in practice such relationships have only slight curvature and $C3$ is close to zero. Table 3.5 shows (unadjusted) standard errors of selected estimates for the test plans with and without reallocation of the specimens at 120 vpm. The same set of data was used for both test plans and all four censoring times. Thus comparisons of standards errors are within the same sample, and their ratios should be more accurate than those from different samples. The simulated lives of the 111 specimens below 120 vpm were used to obtain standard errors of the reallocated plan. But those standard errors must be multiplied by $(111/170)^{1/2}$ to get the standard errors for 170 specimens reallocated below 120 vpm in the same proportions as in the original plan. Comments on Table 3.5 follow.

Table 3.5.   Standard Errors for the Quadratic Relationship with/without Reallocation of 120 vpm Specimens

Cens. Time	$\hat{\beta}$	$s(\hat{\beta})$	LVPM^2 $s(\hat{C3})$	LTHICK $s(\hat{C4})$	65vpm,.266cm $s(\ln(\hat{\tau}_{.001}))$	80vpm,.266cm $s(\ln(\hat{\tau}_{.001}))$
∞	1.55/1.48	.093/.090	5.1/1.9	0.18/0.19	4.7/1.35	0.76/0.38
10,000	1.54/1.53	.094/.112	5.1/2.7	0.19/0.23	4.8/2.2	0.76/0.41
4,000	1.54/1.55	.104/.127	5.4/3.9	0.18/0.25	5.1/3.4	0.83/0.56
1,000	1.47/1.46	.123/.152	8.1/8.5	0.28/0.34	8.3/8.6	1.63/1.71

- Note that the $s(\hat{\beta})$ estimates are close to those in Table 3.4 for the linear relationship. This illustrates a general fact; namely, the chosen form of the relationship for the location parameter does not affect the accuracy of the estimate of a constant scale parameter. Only the amount of censoring affects that accuracy (the censoring is the same in both tables).
- The same is true for $s(\hat{C4})$, the thickness coefficient.
- $s(\hat{C3})$ for the quadratic term in $LVPM$ is smaller with the original plan (with 59 specimens at 120 vpm) with a censoring time of 4,000 hours or more. This is especially so without censoring (∞), since the effective range of $LVPM$ is greatest then (120-200 vpm). For early censoring (1,000 hours), $s(\hat{C3})$ is slightly smaller with the reallocated plan, as expected, since there are no failures at 120 vpm. Then the original plan effectively has 111 specimens and the reallocated plan has 170.
- $s(\ln(\hat{\tau}_{.001}))$ at the two extrapolated conditions is larger with the reallocated plan, except for the earliest censoring (1,000 hours). This is expected because the 120 vpm data are closer to the extrapolated conditions, and the extrapolation is shorter. These standard errors are much larger than those in Table 3.4 for the linear relationship, especially so at 65 vpm, a greater extrapolation. This exemplifies another general fact; namely, the accuracy of an estimate depends not just on the test plan but also on the model fitted to the data. Note that a good test plan for one model (or relationship) may not be a good plan for a different one.

The comments above show that the 120 vpm specimens contribute more to the accuracy of estimates with the quadratic relationship than with the linear one. Thus one must decide whether the linear or quadratic model is more appropriate before deciding whether to run specimens at 120 volts/mm.

Comparison of conductors. One test purpose is to compare the effects of the conductors on insulation life. Conductors S and SS are compared below with respect to their two intercept coefficients ($C1$ and $C1'$), two power parameters ($C2$ and $C2'$) in the power law, and two Weibull shape parameters (β and β'). Such comparisons use relationship (3.1) and assume certain parameters are equal or differ for the two conductors, as depicted in Figure 3.4. Four such assumed models were fitted to simulated data. Table 3.6 shows

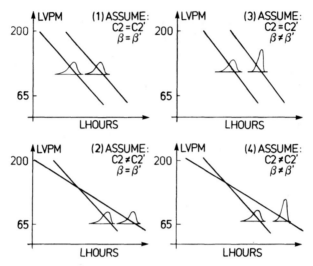

Figure 3.4.   Various assumed models to compare two conductors.

95% uncertainties (2 standard errors) for differences under the four models. Indicator variables were used in these models for such differences as described in Chapter 2. There are many specimens of S and SS conductors, and few of the other conductors. Thus comparisons of other pairs of conductors have much greater uncertainty. Table 3.6 shows that the uncertainties of estimates are the same under similar models as follows.

- The column for $2s(\hat{\beta})$ is the same under models 1) and 2), where the two conductors have a common β. Similarly the column for $s(\ln\hat{\beta}' - \ln\hat{\beta})$ is the same under models 3) and 4), where the two conductors have different β's.
- The column for $2s(\hat{C1}' - \hat{C1})$ is the same under models 1) and 3). Similarly

Table 3.6.   95% Uncertainties for Comparing S and SS Conductors

Cens. Time	Model: 1) Common C2 & β			3) Common C2		
	$\hat{C1}' - \hat{C1}$	$\hat{C2}$	$\hat{\beta}$	$\hat{C1}' - \hat{C1}$	$\hat{C2}$	$\ln\hat{\beta}' - \ln\hat{\beta}$
∞	.30	0.6	.20	.29	0.6	.35
10,000	.39	1.2	.22	.39	1.2	.40
4,000	.42	1.4	.25	.42	1.4	.43
1,000	.55	2.5	.27	.57	2.5	.52

Cens. Time	2) Common β			4) Different C2's & β's		
	$\hat{C1}' - \hat{C1}$	$\hat{C2}' - \hat{C2}$	$\hat{\beta}$	$\hat{C1}' - \hat{C1}$	$\hat{C2}' - \hat{C2}$	$\ln\hat{\beta}' - \ln\hat{\beta}$
∞	.30	1.7	.20	.30	1.6	.36
10,000	.59	4.0	.22	.57	4.0	.42
4,000	.76	4.9	.25	.76	5.0	.44
1,000	1.5	8.0	.28	1.5	8.2	.55

the column for $2s(\hat{C1}' - \hat{C1})$ is the same and larger under models 2) and 4). All four models have different $C1$'s for the two conductors.

- The columns for $2s(\hat{C2})$ are the same under models 1) and 3), where the two conductors have a common $C2$. Similarly, the columns for $2s(\hat{C2}' - \hat{C2})$ are the same under models 2) and 4), where the two conductors have different $C2$'s.

Lognormal standard errors. The standard errors above are based on an assumed Weibull distribution for insulation life. The analyses could be be redone assuming a lognormal distribution. The standard errors for the power in the power law and other coefficients in the relationships would be about the same. However, those for estimates of low percentiles are greater for the Weibull than for the lognormal distribution. Thus, a lognormal simulation and analysis is needed to obtain standard errors of estimates of low lognormal percentiles. In general, one may evaluate a test plan using any number of models (distributions and relationships).

Concluding remarks. Overall the test plan satisfied the responsible engineers. They chose to run the original test with the 120 vpm specimens. Their reasons were:

- The accuracy of the original plan is comparable to that of the reallocated plan for the linear relationship and censoring at 10,000 hours (the key time when decisions would be made).
- The quadratic relationship, if needed, would be estimated more accurately with the original plan.

4. SURVEY OF TEST PLANS

Purpose. This section surveys further literature on plans for accelerated testing, that is, for extrapolation. Sections 1, 2, and 3 are helpful background. Much further work on plans is needed.

General literature. There is much literature on design of experiments for regression models with complete data and least squares fitting to estimate the relationship for the mean. Most regression texts discuss test plans that are not suitable for extrapolation. References on general theory include Ford, Titterington, and Kitsos (1989), Elfving (1952), Chernoff (1953,1972), Silvey (1980), Karlin and Studden (1966), Herzberg and Cox (1972), and Fedorov (1972). Applications to metal fatigue include Little (1972) and Little and Jebe (1975).

Estimation method. Most references here assume that the model is fitted to the data by maximum likelihood or least squares. There are also optimum plans for best linear estimation using order statistics. Mann (1972) gives such theory for plans to estimate the log characteristic life α (of a Weibull distribution) which is a polynomial function of a stress. Escobar and Meeker (1986) do the same to estimate percentiles of the (log) normal,

Weibull, and other location-scale distributions; they assume the location parameter is multivariable linear function. Such theory is strictly correct only for failure censoring, which is mathematically tractable but rare in practice.

Simple model. Test plans for the simple model appear in Sections 1 and 2. Section 1 references literature for complete data and least squares estimation, and Section 2 references literature for singly censored data and ML estimation. Such simple models are widely used; thus such plans are important.

Other relationships. There are test plans for other relationships, more complex than the simple linear one. Test plans for multivariable relationships are studied by Escobar and Meeker (1986) and by Elfving (1952). The following authors give test plans for nonlinear relationships in a single variable. Hoel (1958) and Hoel and Levine (1964) derive optimum plans for a polynomial for the mean, fitted by least squares to complete data; similarly Stigler (1971) presents good compromise plans for a quadratic relationship. Mann (1972) derives optimum plans for fitting a polynomial for the log characteristic Weibull life α to failure censored data. Chernoff (1962) presents optimum plans for ML estimation of a quadratic relationship for the failure rate of an exponential distribution from singly censored data. Little and Jebe (1969) derive test plans for a model with a nonconstant standard deviation and a simple linear relationship for the mean, estimated from complete data by least squares. Meeter and Meeker (1989) investigate test plans for a simple linear model where the scale parameter is a loglinear function of stress.

Failure or time censoring. In this section most plans are for **time** censored data. Such data are common in practice and result from analyzing the data at a point in time when some specimens have not failed. **Failure** censoring is rare in practice. It requires that the test at each stress continue until a specified number of failures occurs. Failure censoring is merely mathematically more tractable. Only Mann (1972) and Escobar and Meeker (1986) give plans for failure censoring.

Inspection data. The references in this section contain plans for complete or singly censored data. Schatzoff and Lane (1987) for a Weibull distribution and Yum and Choi (1987) for an exponential distribution investigate optimum plans for inspection data (interval or quantal-response) and fitting a regression model. Meeker and Hahn (1977,1978) develop plans for quantal-response data and a linear-logistic model. Interval data arise in many tests of electrical insulation and solid state electronics. The following references for estimating a *single* distribution from interval data provide guidance on the choice of inspection times for an accelerated test. For an exponential distribution, Kulldorff (1961), Ehrenfeld (1962), and Nelson (1977) present optimum inspection times for ML estimation. For a (log) normal distribution, Kulldorff (1961) presents optimum inspection times for ML estimation of the mean and standard deviation. For a Weibull distribution, Meeker (1986) presents optimum inspection times for ML estimation of percentiles; he also discusses other inspection times and provides guidance on good inspection times. Equally spaced inspection times are convenient in practice, but the resulting estimates usually have poorer accuracy.

Simultaneous versus successive. In most tests, specimens are tested *simultaneously*. That is, all specimens start and run at the same time. For example, in the Class-B insulation test, specimens for the four test temperature were running simultaneously in four separate ovens. Most references here concern such testing. In contrast, for some tests, there is only one test apparatus, and it can test just one specimen at a time. So specimens must be tested *successively* one after the other. The insulating fluid data come from such a test. References on test plans for successive testing are Chernoff (1962) and Little and Jebe (1969). Sequential testing is successive testing but is concerned with when to stop testing; Bessler and others (1962) apply it to accelerated testing. Another situation is studied by Disch (1983). His work concerns a single large vacuum chamber which runs at elevated temperatures. The available test time for the chamber had to be divided into intervals. During each interval, a different group of specimens (batteries) was run at a different high temperature. His problem was to choose the duration, temperature, and number of specimens for each interval, subject to the constraint that the sum of the interval lengths equal the available time.

Optimization criteria. For complete data, most references here minimize the variance of the least squares estimate of the (log) mean at the design stress. For censored data, most references minimize the asymptotic variance of the ML estimate of a (log) percentile at the design stress. Other possible criteria are:
- Minimize the variance of an estimate over a range of stress. For the simple model, Gaylor and Sweeny (1965) minimize the maximum variance of the least squares estimate of the (log) mean over a specified range.
- Minimize the variance of the estimate of a particular coefficient. Daniel and Heerema (1950) do this for complete data and the least squares estimate of a multivariable linear function for the mean. For example, the plan for the simulation example of Section 3 is far from optimum for estimating the Thickness coefficient.
- Minimize the variance of the estimate of the scale parameter (lognormal σ or Weibull β).
- Be most sensitive for detecting nonlinearity of the relationship. For complete data, Stigler (1971) develops plans that minimize the variance of the least squares estimate of a quadratic coefficient, while achieving a specified variance for the estimate of the mean at the design stress.

Cost. Most references above do not take cost into account. Menzefricke (1988), Chernoff (1962), and Little and Jebe (1969) model the cost of specimens and running time. Subject to a fixed total cost, they minimize the variance of their estimates.

Stress loading. All references above concern constant stress testing.

Miller and Nelson (1983) present optimum plans for simple step-stress tests with two stress levels.

5. ML THEORY FOR TEST PLANS

Purpose. This advanced section presents maximum likelihood (ML) theory for evaluating good plans or optimizing test plans with data singly time censored on the right. In particular, this section shows how to calculate the theoretical asymptotic variance of the ML estimate for a quantity of interest, usually a percentile of life at the design stress. As shown here, one can minimize such a variance by optimally choosing the test stresses and the number of specimens allocated to each. The computer program of Jensen (1985), described in Section 2.1, does these calculations. Such theory extends to interval and other data. For example, Schatzoff and Lane (1987) present such theory for accelerated test plans with interval data. Nelson (1982, Chap. 9) presents general theory for interval and quantal-response data.

Background. This advanced section requires extensive background, particularly the ML theory of Section 5 of Chapter 5. Needed mathematical background includes partial differentiation, integral calculus, and simple matrix algebra. In addition, one needs to know the basic properties of statistical expectation. Also, one needs to read about test plans for censored data in Section 2 of this chapter. This section does not state the regularity conditions on the model that must be satisfied for the theory here to hold. For such conditions, consult Hoadley (1971), Rao (1987), and Rao (1973). Such conditions hold for most models used in practice.

Overview. This section covers the following main topics:
- a general model for accelerated testing,
- specimen and sample likelihoods for censored data,
- ML estimates and likelihood equations,
- the matrix of second partial derivatives,
- the theoretical Fisher information matrix,
- the theoretical covariance matrix of the ML estimates,
- variances and covariances of function estimates,
- optimum plans.

The order of presentation is much like that of Section 5 of Chapter 5.

Model. For specimen i, the random (log) time to failure is y_i, and the values of its K stress or other variables are $x_{1i}, x_{2i}, \cdots, x_{Ki}$ for $i = 1, 2, \cdots, n$. The cumulative distribution of (log) time y_i to failure is $F(y_i; \theta, x_i)$ where $x_i = (x_{1i}, x_{2i}, \cdots, x_{Ki})$ is the vector of variable values and $\theta = (\theta_1, \theta_2, \cdots, \theta_M)$ is the vector of M model parameters. Such parameters may be distribution parameters or coefficients in a relationship. The probability density $f(y_i; \theta, x_i) = dF(y_i; \theta, x_i)/dy_i$ is assumed to exist, and the reliability function is

$R(y_i;\theta,\mathbf{x}_i) = 1 - F(y_i;\theta,\mathbf{x}_i)$. This model is general enough to include all others in this book.

Example model. The simple linear-normal model has

$$F(y_i;\gamma_0,\gamma_1,\sigma,x) = \Phi[(y_i - \gamma_0 - \gamma_1 x)/\sigma],$$

$$f(y_i;\gamma_0,\gamma_1,\sigma,x) = (1/\sigma)\,\phi[(y_i - \gamma_0 - \gamma_1 x)/\sigma], \qquad (5.1)$$

$$R(y_i;\gamma_0,\gamma_1,\sigma,x) = 1 - \Phi[(y_i - \gamma_0 - \gamma_1 x)/\sigma].$$

Here $\phi[\ ]$ is the standard normal probability density, and the rest of the notation follows Sections 1 and 2. The model is expressed here in terms of log life y_i and the normal distribution. It could also be expressed in terms of life and the lognormal distribution, which is equivalent but more complex. Regardless, the results (estimates and their variances) are the same. Nelson and Kielpinski (1976) present the following theory for this model. Nelson and Meeker (1978) do so for the simple linear-Weibull model.

Specimen likelihood. If specimen i fails at age y_i, its log likelihood $\mathcal{L}_i$ is the natural log of the probability density at y_i, namely,

$$\mathcal{L}_i(y_i;\theta,\mathbf{x}_i) = \ln[f(y_i;\theta,\mathbf{x}_i)]. \qquad (5.2)$$

Suppose specimen i is censored at age y_i. That is, it is censored on the right, and its failure time is above y_i. Then its log likelihood is the natural log of the distribution probability above y_i (i.e., of the reliability at age y_i); namely,

$$\mathcal{L}_i = \mathcal{L}_i(y_i;\theta,\mathbf{x}_i) = \ln[R(y_i;\theta,\mathbf{x}_i)]. \qquad (5.3)$$

Each $\mathcal{L}_i$ is a function of the time y_i, the parameters $\theta_1, \cdots, \theta_M$, and variable values $x_{1i}, \cdots, x_{Ki}$. Suppose specimen i will be censored at (log) age η_i, if it does not fail sooner. Let $I_i = I_i(y_i)$ be an indicator function such that $I_i(y_i) = 1$ if $y_i < \eta_i$ (a failure is observed), and $I_i(y_i) = 0$ if $y_i \geq \eta_i$ (the observation is censored). In the following I_i and y_i are regarded as *random variables*, and the censoring time η_i may differ for each specimen. Then the **theoretical log likelihood** for specimen i is reexpressed as

$$\mathcal{L}_i = I_i(y_i)\ln[f(y_i;\theta,\mathbf{x}_i)] + [1 - I_i(y_i)]\ln[R(\eta_i;\theta,\mathbf{x}_i)]. \qquad (5.4)$$

Sample likelihood. For a sample of n specimens, the **theoretical sample log likelihood** is

$$\mathcal{L} = \mathcal{L}_1 + \mathcal{L}_2 + \cdots + \mathcal{L}_n, \qquad (5.5)$$

since the $y_1, \cdots, y_n$ are assumed statistically independent. It is a function of the random variables $y_1, \cdots, y_n$ and the constants $\theta_1, \cdots, \theta_M$ and $\mathbf{x}_1, \cdots, \mathbf{x}_n$. That is, in this section, the $y_1, \cdots, y_n$ are regarded as random quantities that will be observed. In Chapter 5, the same function was used but contained the actual observed data, regarded as given numbers. In Chapter 5, it would be better to call $\mathcal{L}$ the **observed sample log likelihood**.

Example likelihood. For the simple linear-normal model, define

$$z_i \equiv [y_i - \mu(x_i)]/\sigma = (y_i - \gamma_0 - \gamma_1 x_i)/\sigma,$$
$$\varsigma_i \equiv [\eta_i - \mu(x_i)]/\sigma = (\eta_i - \gamma_0 - \gamma_1 x_i)/\sigma, \qquad (5.6)$$
$$I_i = I(y_i), \quad \phi_i \equiv \phi(\varsigma_i), \quad \Phi_i \equiv \Phi(\varsigma_i).$$

Then, for singly right censored data, the theoretical specimen likelihood is

$$\mathcal{L}_i = I_i \left[-\ln(\sigma) - \frac{1}{2}\ln(2\pi) - \frac{1}{2}z_i^2 \right] + (1 - I_i)\ln(1 - \Phi_i). \qquad (5.7)$$

Note that z_i, ς_i, ϕ_i, and Φ_i are functions of the parameters γ_0, γ_1, and σ.

ML estimates. As stated in Chapter 5, the ML estimates $\hat{\theta}_1, \cdots, \hat{\theta}_M$ are the $\theta_1, \cdots, \theta_M$ values that maximize the observed sample log likelihood, $\mathcal{L}$, that is, (5.5) evaluated at the observed values of $y_1, \cdots, y_n$. Alternatively, one can calculate the M partial derivatives of the observed $\mathcal{L}$ with respect to $\theta_1, \cdots, \theta_M$ and set them equal to zero to get the **likelihood equations:**

$$\partial \mathcal{L}/\partial \theta_1 = 0, \cdots, \quad \partial \mathcal{L}/\partial \theta_M = 0. \qquad (5.8)$$

The solutions of these M simultaneous nonlinear equations in $\theta_1, \cdots, \theta_M$ are the ML estimates. These equations can usually be iteratively solved by numerical methods described in Chapter 5 and in more detail in Nelson (1982, Chap. 8, Sec. 6). However, usually one finds them numerically by maximimizing $\mathcal{L}$.

Matrix of derivatives. One must calculate the symmetric matrix of negative second partial derivatives; namely,

$$\mathbf{D} = \begin{bmatrix} -\partial^2 \mathcal{L}/\partial \theta_1^2 & -\partial^2 \mathcal{L}/\partial \theta_1 \partial \theta_2 & \cdots & -\partial^2 \mathcal{L}/\partial \theta_1 \partial \theta_M \\ -\partial^2 \mathcal{L}/\partial \theta_2 \partial \theta_1 & -\partial^2 \mathcal{L}/\partial \theta_2^2 & \cdots & -\partial^2 \mathcal{L}/\partial \theta_2 \partial \theta_M \\ \vdots & \vdots & & \vdots \\ -\partial^2 \mathcal{L}/\partial \theta_M \partial \theta_1 & -\partial^2 \mathcal{L}/\partial \theta_M \partial \theta_2 & \cdots & -\partial^2 \mathcal{L}/\partial \theta_M^2 \end{bmatrix}. \qquad (5.9)$$

Evaluated at $\theta_1 = \hat{\theta}_1, \cdots, \theta_M = \hat{\theta}_M$, (5.9) is the *local estimate of the Fisher information matrix.* In Chapter 5, this estimate was used to obtain approximate confidence limits.

Example derivatives. For specimen i, the first partial derivatives are

$$\partial \mathcal{L}_i/\partial \gamma_0 = (1/\sigma)\left\{ I_i z_i + (1 - I_i)\frac{\phi_i}{1 - \Phi_i} \right\},$$

$$\partial \mathcal{L}_i/\partial \gamma_1 = x_i \partial \mathcal{L}_i/\partial \gamma_0 = (x_i/\sigma)\left\{ I_i z_i + (1 - I_i)\frac{\phi_i}{1 - \Phi_i} \right\}, \qquad (5.10)$$

$$\partial \mathcal{L}_i/\partial \sigma = (1/\sigma)\left\{ I_i [z_i^2 - 1] + (1 - I_i)\frac{\varsigma_i \phi_i}{1 - \Phi_i} \right\}.$$

These set equal to zero are the likelihood equations. The six second partial derivatives are

$$\partial^2 \mathcal{L}_i / \partial \gamma_0^2 = (1/\sigma^2) \left\{ -I_i + (1-I_i) \left[\frac{\varsigma_i \phi_i}{1-\Phi_i} - \frac{\phi_i^2}{(1-\Phi_i)^2} \right] \right\},$$

$$\partial^2 \mathcal{L}_i / \partial \gamma_1^2 = x_i^2 (\partial \mathcal{L}_i / \partial \gamma_0^2), \quad \partial^2 \mathcal{L}_i / \partial \gamma_0 \partial \gamma_1 = x_i (\partial^2 \mathcal{L}_i / \partial \gamma_0^2), \tag{5.11}$$

$$\partial^2 \mathcal{L}_i / \partial \gamma_0 \partial \sigma = -(1/\sigma)(\partial \mathcal{L}_i / \partial \gamma_0) + (1/\sigma^2) \left\{ -I_i z_i + (1-I_i) \left[\frac{\varsigma_i^2 \phi_i}{1-\Phi_i} - \frac{\varsigma_i \phi_i^2}{(1-\Phi_i)^2} \right] \right\},$$

$$\partial^2 \mathcal{L}_i / \partial \gamma_1 \partial \sigma = -(1/\sigma)(\partial \mathcal{L}_i / \partial \gamma_1) + (1/\sigma^2) \left\{ -I_i z_i + (1-I_i) \left[\frac{\varsigma_i^2 \phi_i}{1-\Phi_i} - \frac{\varsigma_i \phi_i^2}{(1-\Phi_i)^2} \right] \right\},$$

$$\partial^2 \mathcal{L}_i / \partial \sigma^2 = -(1/\sigma)(\partial \mathcal{L}_i / \partial \sigma)$$

$$+ (1/\sigma^2) \left\{ -2I_i z_i^2 + (1-I_i) \left[-\frac{\varsigma_i \phi_i}{1-\Phi_i} + \frac{\varsigma_i^2 \phi_i}{1-\Phi_i} - \frac{\varsigma_i^2 \phi_i^2}{(1-\Phi_i)^2} \right] \right\}.$$

These are functions of the random quantities I_i and z_i and of the model parameters.

Fisher matrix. The **true** or **theoretical Fisher information matrix** $\mathbf{F}$ is the expectation of $\mathbf{D}$ in (5.9), where the observations y_i are treated as random variables. That is,

$$\mathbf{F} = \begin{bmatrix} E\{-\partial^2 \mathcal{L}/\partial \theta_1^2\} & E\{-\partial^2 \mathcal{L}/\partial \theta_1 \partial \theta_2\} & \cdots & E\{-\partial^2 \mathcal{L}/\partial \theta_1 \partial \theta_M\} \\ E\{-\partial^2 \mathcal{L}/\partial \theta_2 \partial \theta_1\} & E\{-\partial^2 \mathcal{L}/\partial \theta_2^2\} & \cdots & E\{-\partial^2 \mathcal{L}/\partial \theta_2 \partial \theta_M\} \\ \vdots & \vdots & \ddots & \vdots \\ E\{-\partial^2 \mathcal{L}/\partial \theta_M \partial \theta_1\} & E\{-\partial^2 \mathcal{L}/\partial \theta_M \partial \theta_2\} & \cdots & E\{-\partial^2 \mathcal{L}/\partial \theta_M^2\} \end{bmatrix}. \tag{5.12}$$

Here all derivatives are evaluated at the true parameter values.

Expectations. The expectations are calculated as follows. In general, such an expectation of any function $g(y_i; \theta, x_i)$ of an observation y_i is

$$E\{g(y_i; \theta, x_i)\} = \int_{-\infty}^{\infty} g(y_i; \theta, x_i) f(y_i; \theta, x_i) dy_i. \tag{5.13}$$

Here the integral runs over the range of the distribution if it is not $(-\infty, \infty)$. In particular, the expectations of second partial derivatives of a singly censored likelihood (5.4) for specimen i are

$$E\{-\partial^2 \mathcal{L}_i / \partial \theta_m^2\} = \int_{-\infty}^{\infty} \{-\partial^2 \mathcal{L}_i / \partial \theta_m^2\} f(y_i; \theta, x_i) dy_i$$

$$= \int_{-\infty}^{\eta_i} \{-\partial^2 \ln[f(y_i; \theta, x_i)]/\partial \theta_m^2\} f(y_i; \theta, x_i) dy_i$$

$$+ \{-\partial^2 \ln[R(\eta_i; \theta, x_i)]/\partial \theta_m^2\} R(\eta_i; \theta, x_i) dy_i,$$

$$E\{-\partial^2 \mathcal{L}_i / \partial \theta_m \partial \theta_{m'}\} = E\{-\partial^2 \mathcal{L}_i / \partial \theta_{m'} \partial \theta_m\} \tag{5.14}$$

$$= \int_{-\infty}^{\infty} \{-\partial^2 \mathcal{L}_i/\partial\theta_m\theta_m'\} f(y_i;\theta,x_i)dy_i,$$

$$+ \{-\partial^2 \ln[R(\eta_i;\theta,x_i)]/\partial\theta_m\theta_m'\} R(\eta_i;\theta,x_i)$$

where m and m' run over $1, 2, \cdots, M$. Since $\mathcal{L} = \sum_i \mathcal{L}_i$, the expectations for the derivatives of the sample log likelihoods are

$$E\{-\partial^2 \mathcal{L}/\partial\theta_m^2\} = \sum_i E\{-\partial^2 \mathcal{L}_i/\partial\theta_m^2\},$$

$$E\{-\partial^2 \mathcal{L}/\partial\theta_m\partial\theta_m'\} = \sum_i E\{-\partial^2 \mathcal{L}_i/\partial\theta_m\partial\theta_m'\}. \tag{5.15}$$

The expectations (5.14) can be equivalently calculated as

$$E\{-\partial^2 \mathcal{L}_i/\partial\theta_m^2\} = E\{(\partial \mathcal{L}_i/\partial\theta_m)^2\} = \int_{-\infty}^{\infty} (\partial \mathcal{L}_i/\partial\theta_m)^2 f(y_i;\theta,x_i)dy_i,$$

$$E\{-\partial^2 \mathcal{L}_i/\partial\theta_m\partial\theta_m'\} = E\{(\partial \mathcal{L}_i/\partial\theta_m)(\partial \mathcal{L}_i/\partial\theta_m')\} \tag{5.16}$$

$$= \int_{-\infty}^{\infty} (\partial \mathcal{L}_i/\partial\theta_m)(\partial \mathcal{L}_i/\partial\theta_m') f(y_i;\theta,x_i)dy_i.$$

These may be easier to evaluate than (5.14). The expectations of the M first derivatives always satisfy

$$E\{\partial \mathcal{L}_i/\partial\theta_m\} = 0, \tag{5.17}$$

which are similar to the likelihood equations (5.8). Relations (5.16) and (5.17) often simplify the calculation of theoretical results.

Example expectations. For the simple linear-normal model,

$$E\{-\partial^2 \mathcal{L}_i/\partial\gamma_0^2\} = (1/\sigma^2)\left\{\Phi_i - \left[\varsigma_i - \frac{\phi_i}{1-\Phi_i}\right]\phi_i\right\},$$

$$E\{-\partial^2 \mathcal{L}_i/\partial\gamma_0\partial\gamma_1\} = x_i E\{-\partial^2 \mathcal{L}_i/\partial\gamma_0^2\},$$

$$E\{-\partial^2 \mathcal{L}_i/\partial\gamma_1^2\} = x_i^2 E\{-\partial^2 \mathcal{L}_i/\partial\gamma_0^2\},$$

$$E\{-\partial^2 \mathcal{L}_i/\partial\gamma_0\partial\sigma\} = (1/\sigma^2)\left\{-\phi_i\left[1+\varsigma_i\left(\varsigma_i - \frac{\phi_i}{1-\Phi_i}\right)\right]\right\}, \tag{5.18}$$

$$E\{-\partial^2 \mathcal{L}_i/\partial\gamma_1\partial\sigma\} = x_i E\{-\partial^2 \mathcal{L}_i/\partial\gamma_0\partial\sigma\},$$

$$E\{-\partial^2 \mathcal{L}_i/\partial\sigma^2\} = (1/\sigma^2)\left\{2\Phi_i - \varsigma_i\phi_i\left[1+\varsigma_i^2 - \frac{\varsigma_i\phi_i}{1-\Phi_i}\right]\right\}.$$

These expectations are calculated from (5.14) with the aid of the expectations $E[I_i] = \Phi_i$ (which is a consequence of the definition of I_i) and $E\{\partial \mathcal{L}_i/\partial\gamma_0\} = E\{\partial \mathcal{L}_i/\partial\gamma_1\} = E\{\partial \mathcal{L}_i/\partial\sigma\} = 0$ from (5.17). Since ϕ_i and Φ_i are function of just ς_i, the formulas in braces $\{\ \}$ are functions only of ς_i. Denote them by $A(\varsigma_i)$, $B(\varsigma_i)$, and $C(\varsigma_i)$, respectively. Thus, for the example, the true Fisher information matrix F_{x_i} for an observation at x_i has the form

$$F_{x_i} = (1/\sigma^2) \begin{bmatrix} A(\varsigma_i) & x_i A(\varsigma_i) & B(\varsigma_i) \\ x_i A(\varsigma_i) & x_i^2 A(\varsigma_i) & x_i B(\varsigma_i) \\ B(\varsigma_i) & x_i B(\varsigma_i) & C(\varsigma_i) \end{bmatrix}. \tag{5.19}$$

Escobar and Meeker (1986, 1989) give a computer algorithm for calculating $A(\ )$, $B(\ )$, and $C(\ )$ for the extreme value (Weibull) distribution.

For a test with n specimens, suppose that the n test stress levels are $x_1, x_2, \cdots, x_n$. Then the true Fisher information matrix (5.19) of the sample is

$$\mathbf{F} = \mathbf{F}_{x_1} + \mathbf{F}_{x_2} + \cdots + \mathbf{F}_{x_n} = (1/\sigma^2) \begin{bmatrix} \Sigma A(\varsigma_i) & \Sigma x_i A(\varsigma_i) & \Sigma B(\varsigma_i) \\ \Sigma x_i A(\varsigma_i) & \Sigma x_i^2 A(\varsigma_i) & \Sigma x_i B(\varsigma_i) \\ \Sigma B(\varsigma_i) & \Sigma x_i B(\varsigma_i) & \Sigma C(\varsigma_i) \end{bmatrix}. \tag{5.20}$$

For example, consider a plan with single censoring at (log) time η and with two stresses — a low stress x_L and a high stress x_H. Also suppose that a fraction p of the n specimens are to run at x_L and the rest at x_H. Then the true Fisher information matrix for that test plan is

$$\mathbf{F}_2 = \frac{n}{\sigma^2} \begin{bmatrix} pA(\varsigma_L)+(1-p)A(\varsigma_H) & px_L A(\varsigma_L)+(1-p)x_H A(\varsigma_H) & pB(\varsigma_L)+(1-p)B(\varsigma_H) \\ & px_L^2 A(\varsigma_L)+(1-p)x_H^2 A(\varsigma_H) & px_L B(\varsigma_L)+(1-p)x_H B(\varsigma_H) \\ \text{symmetric} & & pC(\varsigma_L)+(1-p)C(\varsigma_H) \end{bmatrix}. \tag{5.21}$$

Here $\varsigma_L \equiv (\eta - \gamma_0 - \gamma_1 x_L)/\sigma$ and $\varsigma_H \equiv (\eta - \gamma_0 - \gamma_1 x_H)/\sigma$.

Covariance matrix. The true (asymptotic) covariance matrix $\boldsymbol{\Sigma}$ of the ML estimates $\hat{\theta}_1, \cdots, \hat{\theta}_M$ is the inverse of the true Fisher information matrix $\mathbf{F}$:

$$\boldsymbol{\Sigma} \equiv \begin{bmatrix} \mathrm{Var}(\hat{\theta}_1) & \mathrm{Cov}(\hat{\theta}_1,\hat{\theta}_2) & \cdots & \mathrm{Cov}(\hat{\theta}_1,\hat{\theta}_M) \\ \mathrm{Cov}(\hat{\theta}_2,\hat{\theta}_1) & \mathrm{Var}(\hat{\theta}_2) & \cdots & \mathrm{Cov}(\hat{\theta}_2,\hat{\theta}_M) \\ \vdots & \vdots & \ddots & \vdots \\ \mathrm{Cov}(\hat{\theta}_M,\hat{\theta}_1) & \mathrm{Cov}(\hat{\theta}_M,\hat{\theta}_2) & \cdots & \mathrm{Var}(\hat{\theta}_M) \end{bmatrix} = \mathbf{F}^{-1}. \tag{5.22}$$

The true variances and covariances of the parameter estimates are in the same positions in the covariance matrix as the corresponding partial derivatives in the Fisher matrix. For example, $\mathrm{Var}(\hat{\theta}_1)$ is in the upper left corner of $\boldsymbol{\Sigma}$, since $E\{-\partial^2 \mathcal{L}/\partial \theta_1^2\}$ is in the upper left corner of $\mathbf{F}$. The variances all appear on the main diagonal of $\boldsymbol{\Sigma}$, and $\boldsymbol{\Sigma}$ is symmetric about that diagonal.

The variances and covariance in this matrix are generally functions of the true values $\theta_1, \cdots, \theta_M$. For most models, the formulas for these variances and covariances are too complex to express analytically. So in practice one evaluates them numerically. In Chapter 5, the local estimate of this covariance matrix was used to obtain approximate confidence limits. A better estimate of $\boldsymbol{\Sigma}$ is the true matrix (5.22) evaluated at $\theta_1 = \hat{\theta}_1, \cdots, \theta_M = \hat{\theta}_M$. It is called the *ML estimate for* $\boldsymbol{\Sigma}$ and is denoted by $\hat{\boldsymbol{\Sigma}}$. In practice the local and ML estimates are usually numerically comparable, and either can be used to calculate approximate confidence limits.

Variance of a function estimate. Suppose that $h = h(\theta_1, \cdots, \theta_M)$ is a continuous function of $\theta_1, \cdots, \theta_M$. For the linear-normal model, such a function is the $100P$th percentile $\eta_P(x) = \gamma_0 + \gamma_1 x + z_P \sigma$ at stress x (z_P is the standard normal $100P$th percentile). So is the reliability $R(y;x) = 1 - \Phi[(y - \gamma_0 - \gamma_1 x)/\sigma]$. The ML estimate for the true value $h = h(\theta_1, \cdots, \theta_M)$ is $\hat{h} = h(\hat{\theta}_1, \cdots, \hat{\theta}_M)$, the function evaluated at $\theta_1 = \hat{\theta}_1, \cdots, \theta_M = \hat{\theta}_M$. For samples with many failures, the cumulative distribution function of $\hat{h}$ is close to a normal one, with mean h and a true (asymptotic) variance

$$\text{Var}(\hat{h}) = \sum_m (\partial h / \partial \theta_m)^2 \text{Var}(\hat{\theta}_m)$$

$$+ 2\sum_{m < m'}\sum (\partial h / \partial \theta_m)(\partial h / \partial \theta_{m'})\text{Cov}(\hat{\theta}_m, \hat{\theta}_{m'}) = \mathbf{H} \mathbf{\Sigma} \mathbf{H}' . \tag{5.23}$$

Here m and m' run over $1, 2, \cdots, M$, and $\mathbf{H} = (\partial h / \partial \theta_1, \partial h / \partial \theta_2, \cdots, \partial h / \partial \theta_M)$ is the row vector of partial derivatives evaluated at the true parameter values. To evaluate a test plan for estimating h, one calculates this theoretical variance of a ML estimate of interest. The complexity of this calculation is why Section 3 presents simulation as a simpler alternative. The partial derivatives must be continuous functions of $\theta_1, \cdots, \theta_M$ in the neighborhood of their true values. The ML and local estimates for $\text{Var}(\hat{h})$ are obtained by using respectively the ML and local estimate (Chapter 5) of the variances and covariances in (5.23) and using the estimates $\hat{\theta}_1, \cdots, \hat{\theta}_M$ for $\theta_1, \cdots, \theta_M$ in the partial derivatives. (5.23) and (5.24) are based on propagation of error (Taylor expansions); see Hahn and Shapiro (1967) and Rao (1973). The square root of (5.23) is the **true (asymptotic) standard error** of $\hat{h}$ and is denoted by $\sigma(\hat{h})$. Its local estimate (Chapter 5) or its ML estimate is used to obtain approximate confidence limits as described in Chapter 5.

Covariance of function estimates. Suppose that $g = g(\theta_1, \cdots, \theta_M)$ is another function of $\theta_1, \cdots, \theta_M$. Then the asymptotic covariance of $\hat{g} = g(\hat{\theta}_1, \cdots, \hat{\theta}_M)$ and $\hat{h}$ is

$$\text{Cov}(\hat{g}, \hat{h}) = \sum_m (\partial g / \partial \theta_m)(\partial h / \partial \theta_m)\text{Var}(\hat{\theta}_m) \tag{5.24}$$

$$+ \sum_{m < m'}\sum [(\partial g / \partial \theta_m)(\partial h / \partial \theta_{m'}) + (\partial g / \partial \theta_{m'})(\partial h / \partial \theta_m)]\text{Cov}(\hat{\theta}_m, \hat{\theta}_{m'})$$

$$= \mathbf{H} \mathbf{\Sigma} \mathbf{G}' .$$

Here m and m' run over $1, 2, \cdots, M$, and $\mathbf{G} = (\partial g / \partial \theta_1, \partial g / \partial \theta_2, \cdots, \partial g / \partial \theta_M)$ is the row vector of partial derivatives evaluated at the true parameter values. For large samples, the joint cumulative distribution of $\hat{g}$ and $\hat{h}$ is approximately a joint normal one with means equal to the true g and h and with variances from (5.23) and covariance (5.24). The preceding extends to the joint distribution of any number of such ML estimates.

Example. The ML estimate of the $100P$th (log) percentile at the design stress x_D is $\hat{\eta}_P(x_D) = \hat{\gamma}_0 + \hat{\gamma}_1 x_D + z_P \hat{\sigma}$; here z_P is the standard normal $100P$th percentile. The needed derivatives for its variance are

$$\partial\eta_P/\partial\gamma_0 = 1, \quad \partial\eta_P/\partial\gamma_1 = x_D, \quad \partial\eta_P/\partial\sigma = z_P. \tag{5.25}$$

The variance is

$$\mathrm{Var}[\hat{\eta}_P(x_D)] = [1 \, x_D \, z_P]\boldsymbol{\Sigma}[1 \, x_D \, z_P]' . \tag{5.26}$$

The plans in Section 2 minimize this variance.

Optimum plans. For an optimum plan, the n test stresses $x_1, \cdots, x_n$ (subject to the constraints) are chosen as follows to minimize the variance (5.26). For the example and the plan with two stresses x_L and x_H,

$$\mathrm{Var}[\hat{\eta}_P(x_D)] = [1 \, x_D \, z_P]\mathbf{F}_2^{-1}[1 \, x_D \, z_P]' , \tag{5.27}$$

where $\mathbf{F}_2$ appears in (5.21). For $P = 0.50$, the optimum plan in Section 2 minimizes this with respect to x_L and p. The optimum x_L^* and p^* must be found by numerical search, as described by Meeker and Nelson (1976).

Interval data. Theory above extends readily to interval and quantal-response data. The necessary theoretical log likelihoods and expectations of their second partial derivatives appear in Nelson (1982, Chap. 9) and Meeker (1986). All other aspects of the theory are the same as for singly censored data. Schatzoff and Lane (1987) present optimum plans for interval data.

Normal approximation. The ML theory above is asymptotic theory. That is, it is valid for samples with many **failures**. Many authors mislead by saying large samples, seeming to imply large n. However, for a large sample with few failures, the asymptotic normal theory does not provide a good approximation to the sampling distribution of a ML estimate. In practice, 20 failures often suffice. Nelson and Kielpinski (1976) present simulations of censored samples with $n = 40$; they show that the sampling distribution of the ML estimate of the median at design stress is well approximated by the asymptotic normal distribution for the situation they consider. Other authors examine the accuracy of the approximate normal ML confidence limits (which is another matter) and recommend larger numbers of failures than 20.

PROBLEMS (* denotes difficult or laborious)

6.1. Wire varnish - censored data. Do the following for the wire varnish example of Section 1.5. Assume that the test is to run 1,800 hours, and that $\mu(260°) = \log(300)$, $\mu(180°) = \log(20,000)$, $\sigma = 0.10$, and that the highest allowed test temperature is 260°C.
(a) Calculate the standardized a and b.
(b) For the optimum plan, determine the optimum low test temperature.

(c) Determine the optimum proportion of the sample to test at the low test temperature.

(d) Determine the optimum variance of the ML estimate of $\mu(180°C)$.

(e) Compare this variance with that for complete data. In particular, does the censoring affect the accuracy much?

(f) Repeat (a)-(e) for the Meeker-Hahn plan.

6.2. Wire varnish - complete data. Do the following for the wire varnish example of Section 1.5, assuming all specimens run to failure.

(a) For the given test plan, calculate the variance of the LS estimate of mean log life at 180°C.

(b) Calculate the allocation for the optimum plan for the given range of test temperatures.

(c) Calculate the variance for the optimum plan.

(d) For equal allocation of the specimens to the three test temperatures, calculate the variance of the traditional plan.

(e*) For the given plan and the traditional plan (d), calculate the variance, assuming i) the 220° specimens are lost, ii) the 240° specimens are lost, and iii) the 260° specimens are lost.

(f) Compare the variances from (a), (c), (d), and (e) with each other.

6.3. Insulating fluid. For the insulating fluid test (Section 3 of Chapter 3), do (a) through (f) of Problem 6.2, using the appropriate design and test voltage stresses.

(g*) Suppose specimens run successively (one at a time). Calculate the total expected test time for the optimum plan and the actual plan.

6.4. Lost specimens. For the Heater example of Section 1.1, calculate the resulting variance of the LS estimate of mean log life at 1100°F as follows.

(a) Assume that the data on the 1520° specimens are not valid (or lost).

(b) Do (a) but without the 1620° data.

(c) Do (a) but without the 1660° data.

(d) Do (a) but without the 1708° data.

(e) Compare the preceding variances with each other and with the variance for the complete sample.

6.5. Linear-Weibull model. For complete data, rewrite the LS results (key equations) of Section 1 in terms of the simple linear-Weibull model and its parameters. Note where results are approximate.

6.6. Linear-exponential model. Repeat 6.5 for the simple linear-exponential model and complete data.

6.7. Sensitivity analysis. For the Class-B insulation example,

(a) Recalculate the optimum plan, its variance, and the probability of no failures at the low test temperature using a σ that is 20% larger.

(b) Do (a) for the best traditional plan with four stress levels.

(c) Do (a) using a γ_1 that is 20% larger.

(d) Do (c) for the best traditional plan with four stress levels.
(e) Do (a) using a value of antilog(γ_0) that is 20% larger.
(f) Do (e) for the best traditional plan with four stress levels.
(g) Comment on your findings.
(h) Suggest further analyses.
(i) Carry out (h).

6.8. Adjusted 95% uncertainties. Use Table 3.3 for the following.
(a) Calculate the adjusted 95% uncertainties for the entire table. Are the adjusted uncertainties of the two simulations mostly closer to each other than are the unadjusted uncertainties?
(b) Provide the 50% confidence interval for each true uncertainty.
(c*) Calculate the 95% confidence interval for each true uncertainty.
(d) State your conclusions on viewing (a)-(d).
(e*) Calculate adjusted uncertainties (standard errors) for Tables 3.4 and 3.5.

6.9. Equal allocation. Redo the simulation of Section 3 to generate a new Table 3.3. However, use equal allocation of specimens to the four voltage stresses. Keep the same unequal allocations to conductors and thicknesses. Comment on the differences between your table and Table 3.3. Which allocation do you prefer and why?

6.10. Lognormal life. Redo the simulation of Section 3 to generate a new Table 3.3. However, use a lognormal life distribution that "matches" the Weibull distribution used in the simulation. Comment on differences between your table and Table 3.3. To help you do this, devise informative and revealing ways to plot the tables.

6.11. Linear-normal - censored data. For the example of Section 5, supply all missing steps in the derivation. Write out all formulas in complete detail, showing all parameters and data values.

6.12. Linear-normal - complete data. For the example of Section 5, derive all formulas there for a complete sample.

6.13. Exponential life - complete data. Derive an optimum test plan for the following model and complete data. Specimen i has an observed failure time t_i from an exponential distribution with mean $\theta_i = \exp(\gamma_0 + \gamma_1 x_i)$ where x_i is the (possibly transformed) stress, $i = 1, 2, \cdots, n$. γ_0 and γ_1 are parameters to be estimated from the data.
(a) Write the likelihood and log likelihood for specimen i.
(b) Write the sample log likelihood.
(c) Calculate the likelihood equations; and discuss how to solve them.
(d) Calculate second partial derivatives of the log likelihood of specimen i.
(e) Evaluate the negative expectations of (d) to obtain the theoretical Fisher information matrix for specimen i and for the sample.
(f) Calculate the theoretical covariance matrix for $\hat{\gamma}_0$ and $\hat{\gamma}_1$.

(g) For a design stress of x_D, give the ML estimate of the log mean $\mu_D = \ln(\theta_D) = \gamma_0 + \gamma_1 x_D$.

(h) Calculate the theoretical variance of $\hat{\mu}_D$.

(i) Assume that np specimens are tested at x_L, the lowest allowed test stress, and $n(1-p)$ at x_H, the highest allowed test stress. Calculate $\text{Var}(\hat{\mu}_0)$ from (h).

(j) Find the p value that minimizes (i).

(k) Calculate the minimum value of (i).

(l) Express (j) and (k) in terms of $\xi = (x_D - x_H)/(x_L - x_H)$.

(m) Compare the previous results with those in Section 1.

(n) Calculate the ML estimate of $\lambda(x_D) = 1/\theta(x_D)$ and its theoretical variance. Is that variance minimized?

6.14.* Exponential life - censored data. Repeat Problem 6.13 (a)-(n) but assume the data are censored at time τ.

(o) For a two-stress plan with specified stresses x_L and x_H and allocations p_L and $p_H = 1 - p_L$, derive the variance as a function of τ.

(p) For a particular ξ value, calculate and plot the variance versus τ.

6.15.* Quadratic relationship. For the following, use a normal distribution for log life y. Assume the mean is a quadratic function of a single (possibly transformed) stress x; that is, $\mu(x) = \gamma_0 + \gamma_1(x - \bar{x}) + \gamma_2(x - \bar{x})^2$. Assume that the standard deviation σ is a constant, and the data are complete. Thus the data on n specimens are $(y_1, x_1), \cdots, (y_n, x_n)$. Here $\bar{x} = (x_1 + \cdots + x_n)/n$.

(a) Write the likelihood and log likelihood for specimen i.

(b) Write the sample log likelihood.

(c) Calculate the likelihood equations and solve them for $\hat{\gamma}_0, \hat{\gamma}_1, \hat{\gamma}_2$, and $\hat{\sigma}$.

(d) Calculate second partial derivatives of the log likelihood of specimen i.

(e) Evaluate the negative expectations of (d) to obtain the theoretical information matrix for specimen i and for the sample.

(f) Calculate the theoretical (asymptotic) variance for $\hat{\gamma}_2$.

(g) Evaluate this variance for the wire varnish plan of Section 1.5.

(h) Evaluate (f) for traditional plans with (i) three and (ii) four stresses. Find the optimum test plan to minimize $\text{Var}(\hat{\gamma}_2)$ as follows. Assume that the test stresses must be in the range x_L to x_H. Assume that there are three test stresses x_L, x_M, and x_H with proportions p_L, p_M, and p_H of the n sample specimens $(p_L + p_M + p_H = 1)$.

(i) Evaluate $\text{Var}(\hat{\gamma}_2)$ for this test plan.

(j) Determine x_M, p_H, and p_M that minimize $\text{Var}(\hat{\gamma}_2)$.

6.16.* Linear-Weibull. For the simple linear-extreme value model, derive the theory of Section 5.

6.17.* Expectations. Do the following for the simple linear-normal model and singly censored data of Section 5.

(a) Verify $E\{\partial\mathcal{L}_i/\partial\gamma_0\} = 0$, $E\{\partial\mathcal{L}_i/\partial\gamma_1\} = 0$, and $E\{\partial\mathcal{L}_i/\partial\sigma\} = 0$, using properties of expectations or by evaluating the integrals.

(b)   Verify $E\{-\partial^2\mathcal{L}_i/\partial\gamma_0^2\} = E\{(\partial\mathcal{L}_i/\partial\gamma_0)^2\}$ by evaluating each side.

(c*)   Repeat (b) for each of the second partial derivatives.

(d*)   Prove (5.17) for a general model. Do not concern yourself with the validity of interchanging the order of integration and differentiation.

(e*)   Repeat (d) for (5.16).

6.18. Capacitor test. By simulation, evaluate the following plan for a capacitor life test. "Hours" is the censoring time at that test condition. Assume life has a Weibull distribution with a constant shape parameter β and

$$\ln[\alpha(T,V)] = \gamma_0 + \gamma_1\cdot\ln(V) + (\gamma_2/T) ;$$

here V is the voltage and T is the absolute temperature in °K. Use $\gamma_0 = -23.1$, $\gamma_1 = -12.8$, $\gamma_2 = 40,000$°K, and $\beta = 0.29$. The main purpose is to estimate early life at the design (and test) condition (290 V, 75°C).

Volts:	550	550	550	440	440	440	380	380	290
°C:	70	85	100	70	85	100	70	85	75
No.:	40	20	20	70	50	40	100	50	1342
Hours:	2300	2300	126	2320	2320	126	2140	2220	3119

(a)   On suitable paper, plot the test plan. That is, plot the number of specimens at each voltage-temperature condition. Explain whether the plan is reasonable.

(b)   On (a), plot contours of constant 1% life.

Run a simulation using 1) the actual censoring hours and 2) censoring at 10,000 hours (with the two conditions still terminated at 126 hours). Give 95% uncertainties for:

(c)   Each model parameter.

(d)   The 0.1%, 0.5%, and 1% points at the design condition.

(e)   Repeat (c) and (d) for a plan with all specimens at the design condition.

(f)   Repeat (c) and (d) for a test plan where the 1342 specimens at 290 V are proportionally distributed among the other 8 test conditions.

(g)   Discuss the preceding results and pros and cons of each plan.

(h)   Discuss pros and cons of running the test to 10,000 hours.

6.19. GaAs FET demonstration. A new GaAs FET (semiconductor device) was to undergo a temperature-accelerated reliability demonstration test. The device was to demonstrate a median life of 10^7 hours at a design temperature of 125°C with 95% confidence. The test was to run 24 devices and to terminate at 5000 hours. Use the Arrhenius-lognormal model, and assume that the median life is 10^7 hours at 125°C. Determine and evaluate optimum test plans under the following assumptions. That is, evaluate T_L, p_L^*, n_L^*, n_H^*, and V^*.

(a)   Assume $T_H = 295$°C, $\sigma = 0.421$, and $E = 1.8$ eV.

(b)   Assume $T_H = 275$°C, $\sigma = 0.482$, and $E = 1.5$ eV.

(c)   Assume $T_H = 250$°C, $\sigma = 0.482$, and $E = 1.5$ eV.

(d)   Plot the assumed models (a)-(c) on Arrhenius paper. Show design and test temperatures and the 5000-hour termination time.

(e)   For (a)-(c), calculate the value that the estimate of median life must exceed to pass a demonstration test with 95, 90, and 60% confidence.

(f)   For (a)-(c), calculate (1) the population fraction failing at T_L by 5000 hours and (2) the probability that none of the n_L^* specimens fail.

(g)   Compare the pros and cons of (a)-(c).

(h*)   In a satellite application, there are 22 such devices, all of which must survive a long mission. Discuss and derive a more suitable demonstration test in terms of device reliability for the mission.

(i)   Repeat (a)-(g) for the Meeker-Hahn plans.

7

Competing Failure Modes and Size Effect

Introduction. Many products have more than one cause of failure. Any such cause is called a *failure mode* or *failure mechanism*. Examples include:
- Fatigue specimens of a certain sintered superalloy can fail from a surface defect or an interior one.
- In ball bearing assemblies, a ball or the race can fail.
- In motors, the Turn, Phase, or Ground insulation can fail.
- A cylindrical fatigue specimen can fail in the cylindrical portion, in the fillet (or radius), or in the grip.
- Any solder joint on a circuit board can fail.
- A semiconductor device can fail at a junction or at a lead.
- Specimens fail from a cause of interest or from an extraneous cause such as test equipment failure that damages specimens.

Purpose. Accelerated tests that yield a mix of failure modes long troubled experimenters. They lacked a valid way to extrapolate such data to estimate the product life distribution at a design stress. It is clearly wrong to use data on one failure mode to estimate the life distribution of another failure mode. Yurkowski and others (1967) state this was a major unsolved problem at that time. This chapter presents valid models and graphical and maximum likelihood analyses for such data; namely,
- Series-system model.
- Analyses of data on individual failure modes.
- Estimation of product life when all failure modes act.
- Estimation of product life with some failure modes eliminated.
- Checks on the model and data.

These models and analyses are not yet widely known in engineering and other fields. Nelson (1990) briefly presents the basics of this topic.

Background. Needed background for this chapter is the basic models (Chapter 2) and graphical (Chapter 3) and maximum likelihood (Chapter 5) analyses of multiply censored data. The theory (Section 7) requires partial differentiation and basic matrix algebra.

Overview. Section 1 presents the series-system model for competing failure modes; this model is basic to the rest of this chapter. Section 2 extends the model to systems of identical parts. Section 3 extends the model to a product that comes in differing sizes. Section 4 extends the model to a product subjected to a nonuniform stress across it. Sections 2, 3, and 4 are specialized and may be skipped. Section 5 presents basic plots for analysis of such data. Section 6 explains maximum likelihood (ML) analyses of such data. Section 7 presents the advanced theory for such ML analyses.

1. SERIES-SYSTEM MODEL

This section presents the series-system model for products with a number of failure modes. Included are: series systems, the product rule for reliability, the addition law for failure rates, and the resulting distribution when some failure modes are eliminated. This presentation applies to product life at any stress level, say, the design stress. However, the stress level does not appear in the notation.

Series system. Suppose that a product has a potential time to failure from each of M causes (also called *competing risks* or *failure modes*). Such a product is called a **series system** if its life is the smallest of those M potential times to failure. Such a system fails when the first failure mode occurs. In other words, if $T_1, \cdots, T_M$ are the potential random times to failure for the M causes (or modes), then the system random time T to failure is

$$T = \min(T_1, \cdots, T_M). \tag{1.1}$$

In reliability, the term "series system" does not imply physical series connections of electrical or mechanical components. It refers only to how such product failure depends on component failure. Such a product is also called a *weakest link system*, a more descriptive term. Here a "failure mode" is defined in whatever way is useful. For example, it can be any failure of a particular subassembly or component, or it can be a failure cause within a component. Many consumer and industrial products are series systems of components or subsystems.

Product rule. For a particular stress level, let $R(t)$ denote the reliability function of the product at age t. Let $R_1(t), \cdots, R_M(t)$ denote the reliability functions of the M causes, each in the absence of all other causes. Suppose that the times to failure for the different causes (components or failure modes) are statistically **independent**. Such a product is said to have **independent failure modes** or to be a **series system with independent failure modes**. For such a system,

$$R(t) = P\{T > t\} = P\{T_1 > t \text{ and } T_2 > t \text{ and } \cdots \text{ and } T_M > t\}$$

$$= P\{T_1 > t\}P\{T_2 > t\} \cdots P\{T_M > t\},$$

since $T_1, \cdots, T_M$ are statistically independent. Thus

$$R(t) = R_1(t)R_2(t) \cdots R_M(t). \tag{1.2}$$

This key result is the **product rule** for reliability of series systems (with *independent* components). In contrast, for a mixture of distributions (Section 7 of Chapter 2), units come from different subpopulations, and the population reliability function is a weighted sum of their reliability functions. Cox (1959) and Hahn and Meeker (1982) make this distinction.

Class-H example. A Class-H insulation is a series system with three failure modes – Turn, Phase, and Ground failures. The Arrhenius-lognormal models for each mode are:

$$R_T(t;T) = \Phi\{-[\log(t) - (-3.587481) - (3478.935/T)]/0.07423483\},$$

$$R_P(t;T) = \Phi\{-[\log(t) - (-1.639364) - (2660.203/T)]/0.2034712\},$$

$$R_G(t;T) = \Phi\{-[\log(t) - (-5.660613) - (4624.660/T)]/0.2110285\}.$$

Evaluated at $t = 10,000$ hours and $T = 180 + 273.16 = 453.16°$K, the reliabilities are 0.886, 0.872, and 0.995 respectively. By the product rule, the system reliability is $R(10,000;453.16°K) = 0.886 \times 0.872 \times 0.995 = 0.769$. Repeat this calculation for a number of t values to obtain the reliability function at 180°C (or any other temperature), displayed in Figure 5.5A. In this example, the three failure modes all occur in the same test. For products subjected to several accelerated tests, failure modes may occur in different tests with different accelerating variables and models.

Literature. Many authors have used the series-system model for products with more than one cause of failure. Examples include
- McCool (1978) for bearings,
- Sidik and others (1980) for battery cells,
- Nelson (1983a) for metal fatigue,
- IEEE Standard 101 for electrical insulation.

Redundancy. Some products, particularly some military and aerospace products, are not series systems. They have redundant (duplicate) components. A redundant component is one whose failure does not cause system failure. For example, a car has two headlights and can still operate at night if one fails. Such redundancy is a means of improving system reliability. Shooman (1968) among others describes different types of redundancy and their statistical models. Redundancy is not treated in this book.

Addition law for failure rates. For series systems with statistically independent failure modes, the hazard function (failure rate) and cumulative hazard functions are simple. Consider the cumulative hazard functions $H(t)$ of the system and $H_1(t), \cdots, H_M(t)$ of each of the M causes at a stress level. The product rule (1.2) can be written as

$$\exp[-H(t)] = \exp[-H_1(t)]\exp[-H_2(t)] \cdots \exp[-H_M(t)]$$

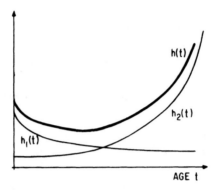

AGE t

Figure 1.1.   Hazard functions of a series system and of its two failure modes.

or

$$H(t) = H_1(t) + H_2(t) + \cdots + H_M(t). \tag{1.3}$$

Its derivative is the system hazard function (instantaneous failure rate)

$$h(t) = h_1(t) + h_2(t) + \cdots + h_M(t). \tag{1.4}$$

This is called the **addition law for failure rates** for *statistically independent* failure modes (or competing risks). So, for such series systems, failure rates add. Figure 1.1 depicts this addition law. It shows the hazard functions of a system and its two failure modes at a particular stress level.

Appearance on plotting paper. Suppose a probability or hazard plot of data is curved as in Figure 1.2. Such curvature may indicate that another failure cause with an increasing failure rate becomes dominant as the product ages. There the lower tail of the distribution spreads over a wide age range. The upper part of the distribution extends over a narrow age range; this reflects the higher failure rate there. The curve in Figure 1.1 is the derivative of that in Figure 1.2. Curvature in the other direction suggests that the distribution is a mixture or that the plotting paper is not suitable.

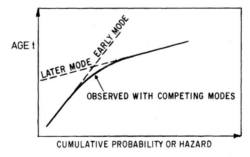

Figure 1.2.   Cumulative distribution with competing failure modes.

Exponential distributions. Suppose that a product is a series system with M statistically independent failure modes with exponential life distributions and constant failure rates $\lambda_1, \cdots, \lambda_M$. Then, by the addition law, the product has an exponential life distribution with a constant failure rate

$$\lambda = \lambda_1 + \cdots + \lambda_M. \tag{1.5}$$

This simple relationship is widely used to model many components and systems. Use of constant failure rates is correct only if all M distributions are exponential. Similarly, the mean time to failure for such series systems is

$$\theta = [(1/\theta_1) + \cdots + (1/\theta_M)]^{-1}; \tag{1.6}$$

here the exponential mean lives are $\theta_1 = 1/\lambda_1, \cdots, \theta_M = 1/\lambda_M$.

MIL-HDBK-217. On pages 5.1.2.4-1, this handbook expresses the failure rate λ of monolithic MOS, bipolar, and CCD memories as

$$\lambda = \lambda_1 + \lambda_2. \tag{1.7}$$

Here λ_1 represents electrical failure due to temperature and voltage, and λ_2 represents mechanical failure due to vibration. Many other devices in this handbook have failure rates expressed as such a sum.

Weibull distributions. Suppose that a product is a series system with M independent failure modes with Weibull life distributions with scale parameters $\alpha_1, \cdots, \alpha_M$ and the **same** shape parameter β. Then the product has a Weibull life distribution with a scale parameter

$$\alpha = 1/[(1/\alpha_1^\beta) + \cdots + (1/\alpha_M^\beta)]^{1/\beta} \tag{1.8}$$

and the same shape parameter β. This can be seen from the hazard function

$$h(t) = \beta(t^{\beta-1}/\alpha_1^\beta) + \cdots + \beta(t^{\beta-1}/\alpha_M^\beta) = \beta(t^{\beta-1}/\alpha^\beta), \tag{1.9}$$

which is the hazard function for a Weibull distribution.

Boltholes. Cracks initiate in boltholes in a flange in an engine. The distribution of time to crack initiation of individual boltholes was modeled with a Weibull distribution with $\alpha = 31{,}699$ hours and $\beta = 2.1914$ (which are estimates and uncertain). A flange is regarded as cracked when the first of its 30 boltholes cracks. The series-system model for the time t (in hours) to crack initiation of flanges is

$$R_{30}(t) = \exp\{-30(t/31{,}699)^{2.1914}\} .$$

Actually the boltholes in a flange do not have independent times to crack initiation. Thus this formula predicts lower reliability than is observed. Flange life is roughly that of series system of about 5 statistically independent boltholes. A later paragraph discusses dependent component lifetimes.

Other distributions. For a series system whose failure modes have a lognormal distribution, the system distribution is **not** lognormal. This is true of most other distributions. Similarly, (1.8) does not hold for Weibull distributions with different shape parameters.

Elimination of failure modes. Often it is important to know how design changes that would eliminate failure modes would affect a product life distribution. As before, suppose those causes are independent and that, say, cause 1 is eliminated. Cause 1 may be a group of causes. Then $R_1(t) = 1$, $h_1(t) = 0$, $H_1(t) = 0$, and $F_1(t) = 0$. The life distribution with the remaining causes has

$$R^*(t) = R_2(t) \times \cdots \times R_M(t), \quad F^*(t) = 1 - [1 - F_2(t)] \times \cdots \times [1 - F_M(t)],$$

$$h^*(t) = h_2(t) + \cdots + h_M(t), \quad H^*(y) = H_2(t) + \cdots + H_M(t). \quad (1.10)$$

These equations assume (1) that the distributions of the remaining failure modes are unaffected, (2) that no other failure modes are introduced by the design change, and (3) the eliminated failure mode is completely eliminated (over any time span of interest). Nelson (1982, pp. 182-185) shows analyses of data where a failure mode is not completely eliminated.

Class-H insulation. Suppose that the Class-H Turn failures could be eliminated. Then the resulting reliability for 10,000 hours at 180°C would be $0.872 \times 0.995 = 0.868$. This is the Phase reliability 0.872 times the Ground reliability 0.995. This calculation can be repeated for a number of survival times to calculate the reliability function at 180°C shown in Figure 5.6A.

Dependent failure modes. Some series systems have failure modes with statistically **dependent** lifetimes. For example, adjoining segments of a cable may have positively correlated lives, that is, have similar lives. Models for dependent failure modes are complicated. General theory is given by Birnbaum (1979), David and Moeschberger (1979), Harter (1977), and Moeschberger (1974). Galambos (1978) presents theory for asymptotically large systems with many failure modes that may be statistically dependent. Nadas (1969) uses the bivariate lognormal distribution to model two dependent failure modes. For a multivariate exponential distributions for such systems, Proschan and Sullo (1976) give some data analyses and references on a fatal shock model. Barlow and Proschan (1975, Chap. 5) present some work on multivariate life distributions. Block and Savits (1981) survey multivariate distributions. The multivariate lognormal distribution has satisfactory properties. However, multivariate exponential and Weibull distributions thus far invented (for example, shock models) have properties not suited to most applications. Harter (1977) comprehensively surveys models for the effect of size. There are simple upper and lower limits for the system life distribution when lives of failure modes are positively correlated. The lower limit is the life distribution for a series system of independent failure modes. This first approximation (using the product rule) is pessimistic. The upper limit is the lowest life distribution for a single failure mode. These crude limits may bracket the true distribution accurately enough for some practical purposes.

Other models. Other models have been used to describe competing failure modes. For example, Derringer (1982,1989) provides such a model that is similar to a mixture of distributions. His model gives just an equation for mean log life and does not employ an assumed distribution of life. MIL-HDBK-217 uses other models for the effect of size, say, of capacitors.

2. SERIES SYSTEMS OF IDENTICAL PARTS

Introduction. Some series systems consist of nominally identical parts, each from the same life distribution. Examples include:
- Tandem specimens are sometimes used in creep-rupture studies of an alloy. Pairs of specimens are linked end to end and stressed until one ruptures. Such pairs hasten the test and yield more information on the lower tail of the distribution of time to rupture.
- A power cable may be regarded as a series system of a large number of small segments of cable. The cable life is the life of its first segment to fail.
- The life of a battery is the life of its first cell to fail.
- An assembly of ball bearings fails when the first ball fails.
- Some integrated circuits (ICs) fail when the first gate fails. Such gate failures may not be statistically independent. More recent IC designs have redundant gates which can fail and not cause circuit failure; such circuits are not series systems.
- An amplifier in a satellite has a series system of 22 GaAs FETs (transistors).

Either the product or the test specimens may be series systems. Tests of series-system specimens are called "sudden death tests" by some authors. The following theory for such systems is a special case of the general theory in Section 1. As before, the theory applies to the distribution at any stress level. Many readers may prefer to skip this specialized section and go directly to Sections 5 and 6.

System life distribution. Suppose that a series system consists of M statistically **independent** components from the same life distribution. Suppose individual components have a reliability function $R_1(t)$, a cumulative distribution function $F_1(t)$, a hazard function $h_1(t)$, and a cumulative hazard function $H_1(t)$. For a series system of M such components, let $R_M(t)$ denote the system reliability function, $F_M(t)$ the system cumulative distribution function, $h_M(t)$ the system hazard function, and $H_M(t)$ the system cumulative hazard function. Then

$$R_M(t) = [R_1(t)]^M, \quad F_M(t) = 1 - R_M(t) = 1 - [1 - F_1(t)]^M,$$
$$h_M(t) = M \cdot h_1(t), \quad H_M(t) = M \cdot H_1(t). \tag{2.1}$$

These are special cases of (1.2), (1.3), and (1.4). The life of such a system is the smallest of a random sample of M (component) lives from the same component distribution.

Tandem specimens. A model for the life of creep-rupture specimens employs a lognormal life distribution for time t to failure. Specimen mean log life as a function of stress S is modeled with $\mu(S) = \gamma_0 + \gamma_1 \log(S)$, and σ is assumed constant. Pairs of such specimens are linked end to end (in tandem) and loaded with the test stress and run until one fails. This doubles the number of specimens on the test equipment, thereby utilizing it better. By (2.1) the reliability function for first specimen failure at stress level S is

$$R_2(t) = \{\Phi[-(\log(t) - \gamma_0 - \gamma_1 \log(S))/\sigma]\}^2.$$

Exponential components. Suppose components have **independent** life times from the same exponential life distribution with failure rate λ_1. Then series systems of M such components have an exponential distribution with

$$\lambda_M = M\lambda_1. \tag{2.2}$$

This is a special case of (1.5). Similarly, the mean system life θ_M is

$$\theta_M = \theta_1/M; \tag{2.3}$$

here $\theta_1 = 1/\lambda_1$ is the mean component life.

Weibull components. Suppose the identical components have a Weibull life distribution with shape parameter β and scale parameter α_1. Then series systems of M such independent components have a Weibull life distribution with the same β and a scale parameter

$$\alpha_M = \alpha_1/M^{1/\beta}. \tag{2.4}$$

This is a special case of (1.8).

Bearings. Morrison and others (1984) report a load-accelerated test of ball bearings. The distribution of time to first ball failure in test bearings is described with a Weibull distribution with $\beta = 1.40$, and $\alpha_1 = 2016$ million revolutions for a load of 4.45kN. In applications, such bearings have twice as many balls. Thus the bearing life distribution is Weibull with $\beta = 1.40$ and $\alpha_2 = 2016/2^{1/1.40} = 1229$ million revs.

Extreme value components. Suppose **statistically independent** components have a smallest extreme value distribution with a scale parameter δ and a location parameter ξ_1. Then series systems of M such components have an extreme value distribution with the same δ and a location parameter

$$\xi_M = \xi_1 - \delta \ln(M). \tag{2.5}$$

Residuals example. Nelson and Hendrickson (1972) give an example of $M = 360$ standardized residuals from a smallest extreme value distribution with $\xi_1 = 0$ and $\delta = 1$. The smallest residual is -8.90 and seems too small. This residual comes from a smallest extreme value distribution with $\xi_{360} = 0 - 1 \cdot \ln(360) = -5.886$ and $\delta = 1$. The probability that the smallest such residual is -8.90 or less is $F(-8.90)=1-\exp(-\exp\{[-8.90-(-5.886)] /1\})=0.049$. This probability is small enough to suspect the residual as a possible outlier worth investigating. An "outlier" is an observation that is not consistent with the model and the bulk of the data.

Other component distributions. For lognormal component life, the life distribution of series systems of M such independent components is not lognormal. However, for any distribution (satisfying mild restrictions), the distribution of system life is approximately Weibull or extreme value for large M, according to whether the original distribution is respectively bounded below or not. This is an important limit theorem for extreme values (Galambos, 1978 and Gumbel, 1958). The result suggests that life of some large series systems may be adequately described by the Weibull or extreme value distribution, for example, cables.

Dependent components. Little work has been done on life of series systems of identical *statistically dependent* components. Components in a system may tend to have similar lives because they are made at the same time and from the same batch of materials. However, systems may differ more, because they are made at different times and from different batches of components and materials. Also such dependence may exist because components in a particular system are under the same environment, but the environment differs from system to system. MIL-HDBK-217 uses an empirical relationship to model the failure rate λ_M of an integrated circuit containing M gates:

$$\lambda_M = \lambda_1 M^p; \tag{2.6}$$

here λ_1 and p are constants characteristic of the IC. Typically, $0.3 \le p \le 0.5$.

3. SIZE EFFECT

Some products come in various sizes. They may have failure rates that are proportional to product size. Examples include:
- The failure rate of a capacitor dielectric is often assumed to be proportional to the area of the dielectric.
- The failure rate of cable insulation is often assumed to be proportional to the cable length.
- The failure rate of conductor in microelectronics is assumed proportional to its length (and number of bends).
- The failure rate of electrical insulation is sometimes assumed proportional to its thickness.

For some products, the specimen and product sizes differ. Such size differences are erroneously ignored for some products. Examples of differences include:
- Test motorettes differ somewhat from motors with respect to size and geometry of insulation.
- Insulating oil is tested between parallel disk electrodes. Transformers con-

taining such oil are much bigger and have complicated geometries.
- Specimens of Class-F generator insulation come in different short lengths, whereas generators contain much greater length.
- Fatigue specimens of a superalloy have hourglass and cylindrical shapes, whereas actual parts are bigger and have other geometries.
- Wire varnish (insulation) is tested on twisted pairs of wires, whereas a great length of such wire is wound into a motor.
- Specimens of generator insulation come in different thicknesses, whereas a generator has a single thickness of insulation.

Harter (1977) surveys models for the effect of size on material strength. Many readers may prefer to skip this specialized section and go directly to Sections 5 and 6.

The model. In general, if $h_0(t)$ is the failure rate for an amount A_0 of a product, then the failure rate $h(t)$ of an amount A is

$$h(t) = (A/A_0)h_0(t). \tag{3.1}$$

That is, such products are regarded as **series systems** of $M = (A/A_0)$ nominally identical components from the same life distribution. A/A_0 need not be an integer and may be less than 1. (3.1) assumes that adjoining portions of the product have statistically **independent** lifetimes. Other formulas for the life distribution of an amount A of product at a stress level are

$$R(t) = [R_0(t)]^{A/A_0}, \quad H(t) = (A/A_0)H_0(t),$$
$$F(t) = 1 - R(t) = 1 - [1 - F_0(t)]^{A/A_0}; \tag{3.2}$$

here the zero subscript denotes the distribution for an amount A_0. These formulas are equivalent to (2.1) where A/A_0 replaces M.

Exponential life. Suppose an amount A_0 of product has an exponential life distribution with failure rate λ_0. Then (3.1) implies that an amount A has an exponential life distribution with failure rate

$$\lambda = (A/A_0)\lambda_0. \tag{3.3}$$

The product mean life is

$$\theta = \theta_0/(A/A_0); \tag{3.4}$$

here $\theta = 1/\lambda_0$ is the mean life of an amount A_0 of product.

Motor insulation. Test specimens of motor insulation were modeled with an exponential life distribution with $\lambda_0 = 3.0$ failures per million hours at design conditions. The insulation area of specimens is $A_0 = 6$ in.^{2}, and its area in motors is $A = 500$ in.^{2}. Then the life of motor insulation is exponential, with $\lambda = (500/6)3.0 = 250$ failures per million hours.

Weibull life. Suppose an amount A_0 of product has a Weibull life distribution with shape parameter β and scale parameter α_0. Then an amount A has a Weibull distribution with the same β and scale parameter

$$\alpha = \alpha_0/(A/A_0)^{1/\beta}. \qquad (3.5)$$

This follows from (3.1) and is like (2.4).

Capacitors. At design voltage, time to dielectric breakdown of a type of 100-pf capacitor is modeled with a Weibull distribution with $\beta=0.5$ and $\alpha_0 = 100,000$ hours. The 500 pf capacitor has the same design but $A/A_0 = 5$ times the dielectric area. Thus the dielectric life of 500 pf capacitors is Weibull with $\beta=0.5$ and $\alpha=100,000/5^{1/0.5}=4000$ hours. For $\beta<1$, the decrease in life is big.

Cryogenic cable. Accelerated tests of cryogenic cable insulation indicated that specimen life at design conditions approximately has a Weibull distribution with $\alpha_0 = 1.05\times10^{11}$ years and $\beta=0.95$. The volume of dielectric of such specimens is 0.12 in.^{3}, and the volume of dielectric of a particular cable is 1.984×10^7 in.^{3}. Suppose that the cable is a **series system** of **independent** specimens. Then its (approximate) life distribution is Weibull, with $\alpha=1.05\times10^{11}/(1.984\times10^7/\ 0.12)^{1/0.95}=234$ years and $\beta=0.95$. The cable engineers wanted to know the 1% point of this distribution. It is $t_{.01} = 234 \cdot [-\ln(1-0.01)]^{1/0.95}=1.8$ years. If adjoining "specimen lengths" of the cable have positively correlated lives, the cable life distribution is greater than predicted by the series system model. However, the cable life distribution cannot exceed that for the life of a single specimen. These two distributions differ appreciably because (A/A_0) is large, but even the pessimistic series-system distribution showed that the design was adequate. Brookes (1974) proposes a model for cable life involving length and conductor size.

Insulating fluid. A life test of insulating fluid was run as follows. A pair of parallel disk electrodes was immersed in the fluid, and the voltage across the disks was raised linearly with time until the fluid broke down (a spark passing through the fluid between the electrodes). Time to breakdown has a Weibull distribution. Disks with areas of 1 and 9 in.^{2} were used. Thus the characteristic lives with these disks satisfy $\alpha_9 = \alpha_1/9^{1/\beta}$. Nelson (1982, p. 190) and Problem 4.10 present such data.

Dependent lives. Adjoining portions of such a product may tend to have positively correlated lives. MIL-HDBK-217 uses an empirical relationship to model the failure rate λ_c of capacitors of capacitance C; namely,

$$\lambda_c = \lambda_1 C^p; \qquad (3.6)$$

here λ_1 and p are constants characteristic of the type of capacitor. Typically $0.3 \le p \le 0.5$.

4. NONUNIFORM STRESS

4.1. Introduction

Purpose. Some products (or specimens) are subjected to a stress that is not uniform across the product. The example below concerns insulation for an electrical machine. At one end of a machine circuit, the insulation is under high voltage, and at the other end it is under low voltage. For such products, engineers often (and incorrectly) use only the maximum stress. Also, they implicitly assume that failure occurs at the point of maximum stress. Thus they ignore the size of the product and possible failure at other points. Previous models assume that the product is under uniform stress or that the specimen and product have the exact same shape and nonuniform stress and differ only in size. Not previously published, this section presents a model for the life of products (or specimens) under nonuniform stress. It is an extension of the series-system model. The model also extends to products and specimens whose geometries differ. Many readers may prefer to skip this specialized section and go directly to Sections 5 and 6.

More complex modeling. The following theory is presented with the aid of a simple application. Namely, for such machine insulation, the stress is nonuniform along one dimension (length). The theory readily extends to more complicated products where geometry and stress may be nonuniform in three dimensions. For example, ceramic and metal fatigue specimens are usually simple cylinders. Moreover, they are usually subjected to simple torque, bending, or uniaxial load. Actual parts have complex geometries and experience complex tensor loads in use. The life of such specimens only crudely approximates the life of actual parts. Someday, using extensions of the theory here, ceramicists, mechanical engineers and metallurgists will do a finite-element type of analysis to predict the fatigue life of actual parts. Similarly, engineers and physicists will predict the reliability of microelectronic circuitry from conductor lengths and number of bends, from local temperatures and voltages, and from the numbers of diodes, connections, etc.

Overview. Section 4.2 summarizes the uniform-stress model and its assumptions. Section 4.3 presents the model for life of a *product* under nonuniform stress in terms of the life of specimens under uniform stress. The theory also extends to *specimens* under nonuniform stress.

4.2. Uniform-Stress Model

Overview. This section reviews the assumptions of the uniform-stress model. This is necessary background for the nonuniform-stress model. Machine insulation is used as an example. Topics include 1) the insulation life distribution, 2) the effect of the size (length here) of insulation on its life distribution, and 3) the relationship between insulation life and voltage stress.

Distribution. It is assumed that insulation specimens of length L_0 have a Weibull life distribution with characteristic life α_0 and shape parameter β at a particular uniform stress level. The characteristic life α_0 (and possibly β) depends on voltage stress, insulation thickness, environment, etc. The reliability function of such specimens at age t is

$$R(t;L_0) = \exp[(t/\alpha_0)^\beta], \quad t > 0. \tag{4.2.1}$$

Other distributions, such as the lognormal, can be used to model the life of insulation, dielectrics, metals, semiconductors, and other products. The Weibull distribution is used here for various reasons. 1) The distribution is widely used for the life of insulation and other materials; thus, it provides a common consistent basis for comparison. 2) The distribution adequately fits life data on many materials, including such insulation. 3) The Weibull model and its results here are simple. A lognormal life distribution leads to a more complicated model and results. For the insulation example, the lognormal distribution represents life more accurately, however.

Size effect. By (3.2) and (3.5), insulation of length L has the Weibull reliability function

$$R(t;L) = \exp[-(L/L_0)(t/\alpha_0)^\beta]. \tag{4.2.2}$$

This assumes that adjoining lengths of insulation have independent lives. This is a reasonable first approximation for most materials. Adjoining pieces tend to have similar (correlated) lives, and distant pieces tend to have unrelated lives. This is especially so of cable insulation and metal parts made as a continuous single piece. However, such a machine consists of many independently fabricated, insulated conductors. Thus, the assumption of independence appears satisfied here.

Size. The measure of "size" depends on the application. For the machine insulation, size is its length. Experience shows that the conductor cross-section, insulation thickness, and other characteristics of insulation negligibly affect life. For some products, material volume or exposed area is the size. For example, for fatigue of metals that fail from interior (surface) defects, specimen volume (area) is the size.

Life-stress relationship. For specimens of length L_0 under a stress V, the characteristic life is assumed given by the inverse power relationship

$$\alpha_0 = K/V^p; \tag{4.2.3}$$

here the factor K and power p are positive parameters characteristic of the material and test conditions. Other relationships between life and stress could be used, for example, Arrhcnius and multivariable relationships. The following results extend to other relationships.

Example. For specimens with $L_0 = 21$ cm, the parameter estimates are $K = 3.3770915\times10^{26}$, $p = 10.18498$, and $\beta = 1.140976$. Thus the reliability function (4.2.2) of insulation of length L at a voltage stress V is

$$R(t;L) = \exp\{-(L/21)[tV^{10.18498}/(3.3770915 \times 10^{26})]^{1.140976}\}.$$

Here t is in hours, L is in cm, and V is in volts per mm.

4.3. Nonuniform Stress Model

Overview. This section presents the nonuniform stress model, namely:

1. the reliability function (4.3.1) of machines represented by I discrete lengths L_i of insulation with differing voltage stresses V_i, $i = 1, 2, \cdots, I$,
2. the reliability function (4.3.5) of machines whose voltage stress is a function $V(l)$ of position l along the length of the circuit,
3. the probability (4.3.7) of machine survival during its design life,
4. the equivalent length L^{**} (4.3.4) of insulation at the maximum stress V^* that would have the same life distribution as a machine.

Differing stress levels. Suppose that a machine is conceptually divided into I pieces of insulation, and piece i has length L_i and is at a uniform (or nominal) stress V_i. Let α_i denote the characteristic life of a standard (or specimen) length L_0 of insulation at uniform stress V_i. Suppose that machines are series systems of such pieces. Then the reliability function of machines is the product of the reliabilities from (4.2.2); namely,

$$R^*(t) = \exp\left\{ - \sum_{i=1}^{I} (L_i/L_0)\, (t/\alpha_i)^\beta \right\}. \tag{4.3.1}$$

Accurately represent the stress pattern by using very small lengths in (4.3.1) at high voltage stress or an accurate equation like (4.3.5) below.

Example. It follows from (4.2.3) and (4.3.1) that the reliability function of such a machine is approximately

$$R^*(t) = \exp\left\{ - \frac{t^\beta}{L_0 K^\beta} \sum_{i=1}^{I} L_i V_i^{p\beta} \right\}. \tag{4.3.2}$$

This is a Weibull reliability function with a shape parameter of β and a scale parameter

$$\alpha^* = K / \left[\sum_{i=1}^{I} (L_i/L_0) V_i^{p\beta} \right]^{1/\beta}. \tag{4.3.3}$$

That is, $R^*(t) = \exp[-(t/\alpha^*)^\beta]$.

Bounds. In a machine, the voltage stress is a continuous function of position along a circuit. Thus the sum in (4.3.1) is a discrete approximation to a continuous stress. Lower and upper bounds on the exact α^* and distribution result from using, respectively, the highest and lowest voltage stresses on each length L_i in (4.3.3).

Stress pattern. One can also describe the voltage stress as a continuous function $V(l)$ of distance l from one end of the machine. For the insulation, voltage stress is zero at one end ($l=0$) of the circuit, and it increases linearly to its maximum value V^* at the other end ($l=L^*$). That is,

$$V(l) = V^*(l/L^*), \quad 0 \le l \le L^*. \tag{4.3.4}$$

Continuous voltage stress. Suppose the voltage stress $V(l)$ is a continuous function of position l along the length of the circuit. Then in (4.3.1), the differential dl replaces L_i, $V(l)$ replaces V_i, and the integral $\int$ replaces the sum $\sum$ in (4.3.2). That is,

$$R^*(t) = \exp\{-[t^\beta/(L_0 K^\beta)] \int_0^{L^*} [V(l)]^{p\beta}\, dl\} ; \qquad (4.3.5)$$

here the integral runs over the length 0 to L^* of the machine insulation. This equation shows that insulation life of a machine has a Weibull distribution with a shape parameter β and a scale parameter

$$\alpha^* = K[L_0 / \int_0^{L^*} [V(l)]^{p\beta} dl]^{1/\beta} . \qquad (4.3.6)$$

Example. For the linear stress pattern (4.3.4), (4.3.5), and (4.3.6) are evaluated to yield

$$R^*(t) = \exp\{-(t/\alpha^*)^\beta\} \qquad (4.3.7)$$

where

$$\alpha^* = K[(p\beta + 1)L_0/L^*]^{1/\beta}/(V^*)^p . \qquad (4.3.8)$$

Here t can be the design life, warranty period, or any other time. For a typical machine, the total length of insulation is $L^* = 30{,}240$ cm. For a maximum voltage stress of $V^* = 65$ volts per mm,

$$\alpha^* = 3.3770915 \times 10^{26}[(10.18498 \cdot 1.140976 + 1)21/30{,}240]^{1/1.140976}/(65)^{10.18498}$$

$$= 1{,}823{,}782.5 \text{ hours} = 208.0519 \text{ years.}$$

Machine reliability is $R^*(t) = \exp[-(t/208.0519)^{1.140976}]$, where t is in years. For $t=40$ years (design life), $R^*(40)=\exp[-(40/208.0519)^{1.140976}] = 0.859$ (85.9%). For $t = 5$ years (typical current age in service), $R^*(5) = \exp[-(5/208.0519)^{1.140976}] = 0.986$ (98.6%).

Equivalent length. The total length L^* of insulation is not at a single uniform voltage stress. For comparisons, it is useful to know what length of insulation L^{**} at the (uniform) maximum voltage stress V^* has the same life distribution (4.3.7) as a machine. Equate (4.3.7) and (4.2.3) to get

$$L^{**} = L^*/(P\beta+1). \qquad (4.3.9)$$

L^{**} does not depend on V^*. If K, p, and β are not constants but depend on V, then their values at V^* should be used in (4.3.9) and previous equations.

Example. For the typical machine, $L^* = 30{,}240$ cm. For $p = 10.18498$ and $\beta = 1.140976$, $L^{**} = 30{,}240/(10.18498 \cdot 1.140976 + 1) \simeq 2{,}396$ cm.

Extensions. The results above can be extended in a number of ways. (1) Another distribution could be used, for example, the lognormal distribution; the resulting equations are more complex. (2) In other applications, an appropriate measure of size may be area or volume; the theory readily extends to such applications. (3) The theory above extends to situations where specimens **and** product are under nonuniform stress. (4) The machine insulation can fail anywhere along the circuit length, but failure is more likely near the high voltage end. One can determine the probability that the failure occurs on the conductors with the highest voltage stress. This would indicate the potential value of using better insulation in the high stress locations, for example, insulation not made by the third production shift.

5. GRAPHICAL ANALYSIS

This section presents graphical methods to

1. Analyze data on a failure mode.
2. Estimate a product life distribution when all failure modes act.
3. Estimate a product life distribution with certain failure modes eliminated.
4. Check the model and data.

Section 6 presents analytic methods (maximum likelihood) for such purposes. It is most informative to use graphical methods first and then analytic methods as necessary. The validity of the data, model, and analytic methods should first be assessed with graphical methods. Both methods check the other and provide information not provided by the other. The methods require that the mode (cause) of each failure be identified. These methods allow higher test stresses that save time, since other failure modes are properly taken into account. Data from a temperature-accelerated life test of motor insulation illustrate the methods.

Literature. Graphical analyses of such data were developed by Nelson (1973,1975b) and Peck (1971). Nelson (1983a) presents an application to metal fatigue. Peck (1971) presents semiconductor applications.

5.1. Data and Model

Test purpose. Data in Table 5.1 illustrate the methods here. The data are times to failure of a Class-H insulation system in motorettes tested at high temperatures of 190, 220, 240, 260°C. A test purpose was to estimate the median life of such insulation at its design temperature of 180°C (Section 5.3). A median life of 20,000 hours was required. Other purposes were to determine the main cause of failure at the design temperature (Section 5.2) and to determine if redesign to eliminate that cause would improve the life distribution appreciably (Section 5.4).

Data. Ten motorettes were run at each temperature and periodically inspected for failure. The recorded time in Table 5.1 is midway between the inspection time when the failure was found and the previous inspection time. Times between inspections are short enough that the effect of rounding to the midpoint is small. Table 5.1 gives the failure time for each cause (Turn,

Table 5.1. Class-H Failure Mode Data (Hours)

190°C Motor	Turn	Phase	Ground	240°C Motor	Turn	Phase	Ground
1	7228	10511	10511+	21	1175	1175+	1175
2	7228	11855	11855+	22	1881+	1881+	1175
3	7228	11855	11855+	23	1521	1881+	1881+
4	8448	11855	11855+	24	1569	1761	1761+
5	9167	12191+	12191+	25	1617	1881+	1881+
6	9167	12191+	12191+	26	1665	1881+	1881+
7	9167	12191+	12191+	27	1665	1881+	1881+
8	9167	12191+	12191+	28	1713	1881+	1881+
9	10511	12191+	12191+	29	1761	1881+	1881+
10	10511	12191+	12191+	30	1953	1953+	1953+

220°C Motor	Turn	Phase	Ground	260°C Motor	Turn	Phase	Ground
11	1764	2436	2436	31	1632+	1632+	600
12	2436	2436	2490	32	1632+	1632+	744
13	2436	2436	2436	33	1632+	1632+	744
14	2436+	2772+	2772	34	1632+	1632+	744
15	2436	2436	2436	35	1632+	1632+	912
16	2436	4116+	4116+	36	1128	1128+	1128
17	3108	4116+	4116+	37	1512	1512+	1320
18	3108	4116+	4116+	38	1464	1632+	1632+
19	3108	3108	3108+	39	1608	1608+	1608
20	3108	4116+	4116+	40	1896	1896	1896

Phase, or Ground); each occurs on a separate part of the insulation system. Each failed part was isolated electrically and could not fail again, and the motorette was kept on test and run to a second or third failure. In actual use, the first failure from any cause ends the life of the motor. For most products, only one failure mode is observed on each specimen. However, the methods here apply to any observed number of failure modes on each specimen. Figure 5.1 depicts these data. The history of each motorette is represented by a line. Each failure appears on the line, and the length of the line is the length of the test on that motorette.

First failure. Here the data on time to first failure are *complete*; that is, each motorette has a failure time. Complete data are usually analyzed with standard least-squares regression analysis (Chapter 4). Such analysis may be misleading for data with a mix of failure modes. Use the analyses below.

Model. The model for these data with competing failure modes has:

1. A separate Arrhenius-lognormal model (Chapter 2) for each failure mode. The following data analyses apply to other models.
2. A series-system model (Section 1) for the relationship between the failure times for the different failure modes and the failure time for a specimen.

These models are briefly reviewed below. The following data analysis methods extend to other life-stress models in the obvious way.

Arrhenius-lognormal. The assumptions of the Arrhenius-lognormal model for failure mode m are:

1. For (absolute) temperature T, life has a lognormal distribution (base 10).

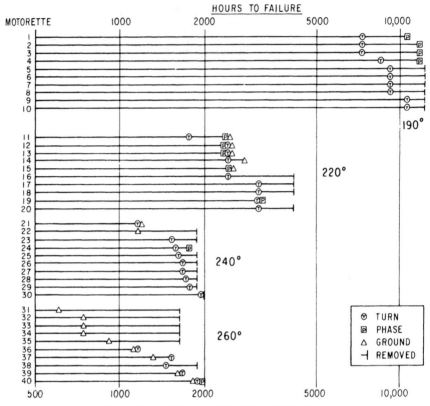

Figure 5.1.   Display of the Class-H life data.

2. The log standard deviation σ_m is a constant.
3. The mean log life as a function of $x = 1000/T$ is

$$\mu_m(x) = \alpha_m + \beta_m x. \qquad (5.1)$$

4. The random variations in specimen lifetimes are statistically independent.

α_m, β_m, and σ_m are parameters characteristic of the failure mode, product, and test method. The antilog of $\mu_m(x)$ is the median life and is regarded as a nominal life. Equation (5.1) is an Arrhenius relationship. This model is depicted in Figure 5.2A on Arrhenius plotting paper and in Figure 5.2B on lognormal probability paper. For units run at a stress x, the probability that failure mode m survives time t is the reliability function

$$R_m(t) = \Phi\{-[\log(t)-\mu_m(x)]/\sigma_m\}. \qquad (5.2)$$

See the Class-H example of Section 1. The following data analysis methods extend to other life-stress models in the obvious way. Such models may include different tests and accelerating variables for different failure modes and different distributions and relationships (including multivariable ones). Model parameters are usually estimated from data as shown below. Also, some parameters (e.g., activation energy and Weibull shape) may be assigned values from handbooks, the literature, similar products, or educated guesses.

Series system. The series-system model assumes:

1. Each unit has M potential times to failure, one from each mode.
2. The M times are statistically independent.
3. The time to failure of a system is the smallest of its M potential times.

The reliability $R_m(t)$ at a stress level is the probability that failure mode m does not occur by time t, if only mode m were acting. For the insulation, each reliability function comes from the Arrhenius-lognormal model (5.2) for that mode. The reliability function $R(t)$ of specimens by the *product rule* is

$$R(t) = R_1(t)R_2(t) \cdots R_M(t). \tag{5.3}$$

This is used in Sections 5.3 and 5.4 to estimate life distributions from data with competing failure modes. If a separate Arrhenius-lognormal model describes each failure mode, then an Arrhenius-lognormal model does not describe the life of specimens when all failure modes act. The correct model appears in Figures 5.5 and 5.6. Section 5.3 describes how to estimate it.

5.2. Analysis of a Failure Mode

This section describes graphical methods that, for each failure mode, estimate its life-stress model and life distribution at any stress. The methods require that the cause of each failure be known.

Point of view. The following point of view simplifies the data analyses on a particular failure mode. Each specimen has a time to failure with that mode or else a running time without that mode. Such a running time results when the specimen fails by another mode or when it is removed from the test. The running time is a censoring time for the failure mode, since the failure time for that mode is beyond the running time. Table 5.2A shows such data for Turn failures at the four test temperatures. The tabulations of Turn, Phase, or Ground failure data come from Table 5.1. Such data with intermixed failure and censoring times are multiply censored data.

Graphical analysis. Graphical analysis of data with a mix of failure modes involves the usual two plots (Chapter 3) for each failure mode: 1) a hazard plot of the multiply censored data and 2) a relationship plot. These plots provide the desired information. Also, they can be used to assess the validity of the model and data as described in Chapter 3. Table 5.2A shows the hazard calculations for the Turn data, and Figure 5.2A shows the hazard plots. Similar calculations for the Ground and Phase failures appear in Tables 5.2B and C. Plots appear in Figures 5.3A and 5.4A.

Interpretation. Such a hazard plot is used and interpreted as described in Chapter 3. For example, the estimate of the median time (50th percentile) to Turn failure at 220°C is 2,900 hours from Figure 5.2A; there median esti-

Table 5.2. Hazard Calculations for Class-H Failure Modes

A. Turn

190° Hrs.	Haz.	Cum. Haz.	220° Hrs.	Haz.	Cum. Haz.	240° Hrs.	Haz.	Cum. Haz.	260° Hrs.	Haz.	Cum. Haz.
7228	10.0	10.0	1764	10.0	10.0	1175	10.0	10.0	1128	10.0	10.0
7228	11.1	21.1	2436	11.1	21.1	1521	11.1	21.1	1464	11.1	21.1
7228	12.5	33.6	2436	12.5	33.6	1569	12.5	33.6	1512	12.5	33.6
8448	14.3	47.9	2436+			1617	14.3	47.9	1608	14.3	47.9
9167	16.7	64.6	2436	16.7	50.3	1665	16.7	64.6	1632+		
9167	20.0	84.6	2436	20.0	70.3	1665	20.0	84.6	1632+		
9167	25.0	109.6	3108	25.0	95.3	1713	25.0	109.6	1632+		
9167	33.3	142.9	3108	33.3	128.6	1761	33.3	142.9	1632+		
10511	50.0	192.9	3108	50.0	178.6	1881+			1632+		
10511	100.0	292.9	3108	100.0	278.6	1953	100.0	242.9	1896	100.0	147.9

B. Phase

190° Hrs.	Haz.	Cum. Haz.	220° Hrs.	Haz.	Cum. Haz.	240° Hrs.	Haz.	Cum. Haz.	260° Hrs.	Haz.	Cum. Haz.
10511	10.0	10.0	2436	10.0	10.0	1175+			1128+		
11855	11.1	21.1	2436	11.1	21.1	1761	11.1	11.1	1512+		
11855	12.5	33.6	2436	12.5	33.6	1881+			1608+		
11855	14.3	47.9	2436	14.3	47.9	1881+			1632+		
12191+			2772+			1881+			1632+		
12191+			3108	20.0	67.9	1881+			1632+		
12191+			4116+			1881+			1632+		
12191+			4116+			1881+			1632+		
12191+			4116+			1881+			1632+		
12191+			4116+			1953+			1896	100.0	100.0

C. Ground

190° Hrs.	Haz.	Cum. Haz.	220° Hrs.	Haz.	Cum. Haz.	240° Hrs.	Haz.	Cum. Haz.	260° Hrs.	Haz.	Cum. Haz.
10511+			2436	10.0	10.0	1175	10.0	10.0	600	10.0	10.0
11855+	None		2436	11.1	21.1	1175	11.1	21.1	744	11.1	21.1
11855+			2436	12.5	33.6	1761+			744	12.5	33.6
11855+			2490	14.3	47.9	1881+			744	14.3	47.9
12191+			2772	16.7	64.6	1881+			912	16.7	64.6
12191+			3108+			1881+			1128	20.0	84.6
12191+			4116+			1881+			1320	25.0	109.6
12191+			4116+			1881+			1608	33.3	142.9
12191+			4116+			1881+			1632+		
12191+			4116+			1953+			1896	100.0	242.9

mates are shown as crosses. Also, for example, the estimate of the percentage failing at 220°C by 3,000 hours is 55% from Figure 5.2A. The corresponding reliability estimate is 100% − 55% = 45%. The estimate of σ is the difference between the log of the 84th and 50th percentiles at a temperature. For 220°C, these estimates are 3,400 and 2,900 hours. The estimate of σ is $\log(3400) - \log(2900) = 0.069$.

A discovery. Figure 5.2A shows two noteworthy features. 1) The data plots at the four temperatures are parallel; this indicates that the log standard deviation for Turn failures has the same value at all test temperatures. In contrast, the Class-H data with all failure modes acting (Chapter 4) has a larger log standard deviation at 260°C. 2) The 260°C data coincide with the 240°C data but should be lower. There are two possible explanations for this. a) The motorettes tested at 260°C were made after the other motorettes, and

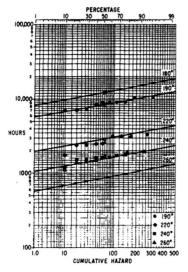

Figure 5.2A. Lognormal hazard plot of Turn failures.

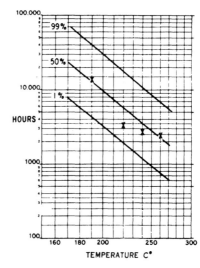

Figure 5.2B. Arrhenius plot of Turn failure medians ×.

failures. If cycled daily, the 260°C motorettes might have had a shorter life they may differ with respect to materials and handling. b) During the test, the motorettes are heated going into the oven and cooled coming out for inspection. The cycle is 7 days at 190°C, 4 days at 220°C, and 2 days at both 240°C and 260°C. One day at 260°C would be consistent with other cycle lengths and IEEE Std. 117. Frequent thermal cycling may accelerate Turn

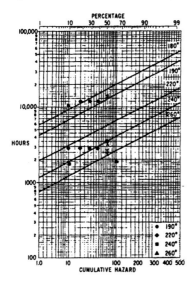
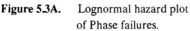

Figure 5.3A. Lognormal hazard plot of Phase failures.

Figure 5.3B. Arrhenius plot of Phase failure medians ×.

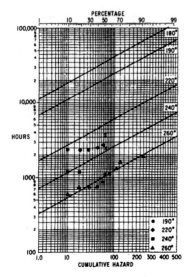

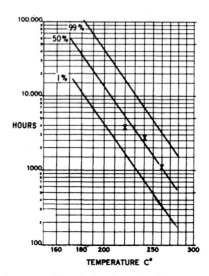

Figure 5.4A. Lognormal hazard plot of Ground failures.

Figure 5.4B. Arrhenius plot of Ground failure medians ×.

consistent with the data at the other test temperatures. A subsequent experiment (Problem 7.8) showed that insulation life depends on cycle length. The graphical analyses were valuable because they revealed problems in the data and experimental method. This discovery lead to a $1,000,000 yearly cost savings. Namely, motors that run continuously in service (not thermally cycled) can be built with a cheaper insulation.

Relationship plot. The relationship between (median) life for a failure mode and temperature is graphically estimated as described in Chapter 3. Figure 5.2B shows such an Arrhenius plot of the median times to Turn failure against test temperature. A straight line fitted to the medians estimates the Arrhenius relationship for Turn failures. For example, the estimate of the median Turn life at the design temperature of 180°C is 12,300 hours. The 260°C Turn failure data do not jibe with the model and were not used to estimate the line. A proper analysis (Problem 7.8) takes thermal cycling into account. Similar plots in Figures 5.3B and 5.4B yield median estimates of 17,000 hours for Phase failure and 19,000 hours for Ground failure at 180°C.

Eliminate Turn failures? Turn failures have the lowest median life at 180°C. Thus they are the main cause of failure there. If Turn failures could be eliminated through redesign, the Phase failures would have the lowest median (17,000 hours). This is still below the required 20,000 hour life. These results indicated that the insulation system could not be improved enough. So it was abandoned.

Life at any stress. An estimate of the entire Turn failure distribution at any temperature is obtained as in Chapter 3. For example, for 180°C, get the estimate of the median from the Arrhenius plot in Figure 5.2B. Mark this median as an × on the hazard plot (Figure 5.2A). Then draw a straight line through the median so its slope is the same as those for the test temperatures. This line in Figure 5.2A estimates the Turn life distribution at 180°C.

Maximum likelihood. Lines fitted to the data plots would ordinarily be fitted by eye. However, the models could also be fitted by the method of maximum likelihood as described in Section 6. The plotted lines are the maximum likelihood estimates. A benefit of maximum likelihood fitting is it provides confidence intervals that indicate the uncertainty in the estimates. Such intervals (say, for medians) are often quite wide.

5.3. The Life Distribution When All Failure Modes Act

Purpose. When a product is in actual use, all failure modes act. Thus, interest centers on the life distribution at a design stress when all modes act. This section presents methods for estimating such a distribution at any stress. The methods employ the estimates of the life distribution of each failure mode (Section 5.2).

Simple distribution estimate. The following is a simple estimate of the life distribution at a stress when all failure modes act. Examine the separate estimates of the life distributions (from Section 5.2) for each mode. The life distribution when all modes act is slightly below the shortest life distribution, especially if the other modes have much longer lives. For example, Turn failures have the lowest median (12,300 hours) at 180°C. So the median life when all failure modes act is slightly below 12,300 hours.

Exact distribution estimate. The following is an exact method to estimate the life distribution with all failure modes acting. First separately estimate the reliability at an age and temperature of interest for each failure mode as described in Section 5.2. Then the estimate of reliability at that age for specimens with all failure modes acting is the product of those reliabilities. This is an application of the product rule (5.3). For example, for the insulation, estimate the reliability at 10,000 hours and 180°C as follows. The reliability for Turn failure is estimated from Figure 5.2A as 0.886. Similarly, the reliability is 0.872 for Phase failure and 0.995 for Ground failure. Thus the estimate of reliability with all failure modes acting is $0.886 \times 0.872 \times 0.995 = 0.769$. The estimate of the fraction failing is $1 - 0.769 = 0.231$. This fraction is plotted as an × against 10,000 hours on lognormal paper in Figure 5.5A. There the 180°C curve estimates the life distribution with all failure modes acting. It and the other curves were obtained by such calculations for other ages and temperatures as shown in Table 5.3. These are *curves*, rather than straight lines, since the distributions are not lognormal. The plotted data on Figure 5.5A are explained below.

Life-stress relationship. The life-temperature relationship is depicted on Arrhenius paper in Figure 5.5B. This plot differs from the Arrhenius model

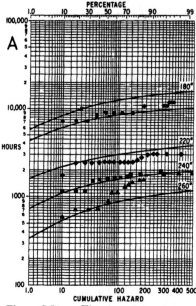
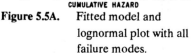
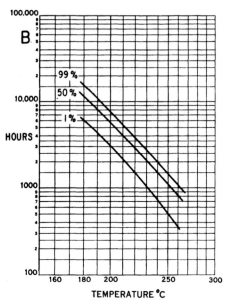

Figure 5.5A.   Fitted model and lognormal plot with all failure modes.

Figure 5.5B.   Arrhenius plot of the model with all failure modes.

for a single failure mode depicted in Figure 5.2A. In particular, the relationships in Figure 5.5B are curves. Also, the distribution is not symmetric; that is, the 1% and 99% lines are not equally far from the 50% line. Figure 5.5B shows an estimate of the median life at 180°C of 11,600 hours with all failure modes acting. The median curve is obtained by (1) estimating the medians for different temperatures from the lognormal plot (Figure 5.5A), (2) putting them on the Arrhenius plot, and (3) drawing a smooth curve through the medians. A relationship for any other percentile is estimated the same way. Such percentile curves are concave downwards for competing failure modes. If one fits a simple linear model and ignores the cause of failure, the straight-line estimate of a percentile at a low stress is generally above the correct estimate.

Hazard plot. Most data sets contain no more than *one* failure per specimen. For such data, make a hazard plot for each test stress level, using all failures and ignoring their failure modes. Assess the above estimate of the life distribution with all failure modes acting by examining a plot of it with the data. The Class-H data could be plotted this way by using just the first failure on each specimen, ignoring its failure mode and subsequent failures on a specimen. The following plot (Figure 5.5A) makes use of the information in the subsequent failures. The method sums the sample cumulative hazard functions of each failure mode (Section 5.2) to obtain a sample cumulative hazard function with all failure modes acting. To do this for a temperature, order *all* failure times in Table 5.2 from smallest to largest without regard to the failure mode. Table 5.4 shows this for each test temperature. Include with each failure its hazard value from Table 5.2. Then cumulate the hazard values as shown in Table 5.4. If desired, modify these cumulative hazard values by averaging each with the previous one. Plot each failure time against its (modified) cumulative hazard value on hazard paper. This plot is an estimate of the distribution with all failure modes acting. Figure 5.5A shows such plots for each test temperature. There the agreement between the plotted data and the calculated curves is satisfactory. This plot provides a check on the data, the model, and the calculated estimate. Poor agreement in the upper tail should be expected when the distributions of the competing failure modes do not overlap appreciably. The upper tail of the 260° data does not agree well with the calculated curve, because 260° Turn failures occur late due to less frequent thermal cycling at 260°. Regardless, points above $100(n - 0.5)/n\%$ should be ignored; here n is the number of specimens at that stress level.

5.4. The Life Distribution with Certain Failure Modes Eliminated

Purpose. To improve products, certain failure modes can be eliminated by design changes. This section shows how to graphically estimate the result-

Table 5.3. Reliability Calculations for All Modes Acting

Hours	Reliability Turn	Phase	Ground	All Acting	Hours	Reliability Turn	Phase	Ground	All Acting
180°					220°				
5,000	1.000 x	.996 x	1.000	= .996	1,300	1.000 x	.999 x	.998	= .997
6,000	1.000 x	.987 x	1.000	= .987	1,600	1.000 x	.997 x	.992	= .986
7,000	.999 x	.971 x	1.000	= .970	2,000	.987 x	.987 x	.976	= .951
8,000	.994 x	.946 x	.999	= .940	2,500	.823 x	.960 x	.935	= .739
9,000	.966 x	.913 x	.997	= .880	3,000	.445 x	.914 x	.872	= .355
10,000	.886 x	.872 x	.995	= .769	4,000	.034 x	.774 x	.707	= .019
11,000	.742 x	.824 x	.991	= .606	5,000	.0009 x	.608 x	.534	= .0003
12,000	.556 x	.772 x	.986	= .423					
13,000	.372 x	.717 x	.979	= .261					
14,000	.223 x	.662 x	.971	= .143	240°				
16,000	.061 x	.553 x	.947	= .032					
18,000	.013 x	.452 x	.915	= .005	500	1.000 x	1.000 x	.999	= .999
20,000	.002 x	.365 x	.876	= .001	600	1.000 x	1.000 x	.997	= .997
					700	1.000 x	1.000 x	.992	= .992
190°					800	1.000 x	.999 x	.983	= .982
3,000	1.000 x	.999 x	1.000	= .999	900	.999 x	.998 x	.970	= .967
4,000	1.000 x	.993 x	1.000	= .993	1,000	.995 x	.996 x	.952	= .944
5,000	.999 x	.977 x	.998	= .974	1,300	.853 x	.983 x	.870	= .729
6,000	.975 x	.945 x	.995	= .918	1,600	.435 x	.953 x	.758	= .314
7,000	.856 x	.899 x	.988	= .760	2,000	.071 x	.884 x	.595	= .037
8,000	.610 x	.839 x	.977	= .500	2,500	.003 x	.764 x	.413	= .001
9,000	.341 x	.769 x	.960	= .252					
10,000	.152 x	.696 x	.938	= .100	260°				
11,000	.057 x	.621 x	.910	= .032					
12,000	.018 x	.549 x	.877	= .009	250	1.000 x	1.000 x	.998	= .998
13,000	.005 x	.481 x	.841	= .002	300	1.000 x	1.000 x	.994	= .994
					400	1.000 x	1.000 x	.974	= .974
					500	.999 x	.999 x	.932	= .930
					600	.984 x	.998 x	.868	= .852
					700	.894 x	.993 x	.787	= .699
					800	.679 x	.986 x	.699	= .468
					900	.411 x	.974 x	.610	= .245
					1,000	.200 x	.957 x	.525	= .101
					1,300	.009 x	.877 x	.317	= .002

Table 5.4. Cumulative Hazard Calculations with All Failure Modes

Hours	Cause	Hazard	Cum. Hazard	Hours	Cause	Hazard	Cum. Hazard
190°				240°			
7,228	Turn	10.0	10.0	1,175	Ground	10.0	10.0
7,228	Turn	11.1	21.1	1,175	Turn	10.0	20.0
7,228	Turn	12.5	33.6	1,175	Ground	11.1	31.1
8,448	Turn	14.3	47.9	1,521	Turn	11.1	42.2
9,167	Turn	16.7	64.6	1,569	Turn	12.5	54.7
9,167	Turn	20.0	84.6	1,617	Turn	14.3	69.0
9,167	Turn	25.0	109.6	1,665	Turn	16.7	85.7
9,167	Turn	33.3	142.9	1,665	Turn	20.0	105.7
10,511	Turn	50.0	192.9	1,713	Turn	25.0	130.7
10,511	Turn	100.0	292.9	1,761	Phase	11.1	141.8
10,511	Phase	10.0	302.9	1,761	Turn	33.3	175.1
11,855	Phase	11.1	314.0	1,953	Turn	100.0	275.1
11,855	Phase	12.5	326.5				
11,855	Phase	14.3	340.8				
220°				260°			
1,764	Turn	10.0	10.0	600	Ground	10.0	10.0
2,436	Phase	10.0	20.0	744	Ground	11.1	21.1
2,436	Turn	11.1	31.1	744	Ground	12.5	33.6
2,436	Ground	10.0	41.1	744	Ground	14.3	47.9
2,436	Phase	11.1	52.2	912	Ground	16.7	64.6
2,436	Turn	12.5	64.7	1,128	Turn	10.0	74.6
2,436	Ground	11.1	75.8	1,128	Ground	20.0	94.6
2,436	Phase	12.5	88.3	1,320	Ground	25.0	119.6
2,436	Turn	16.7	105.0	1,464	Turn	11.1	130.7
2,436	Ground	12.5	117.5	1,512	Turn	12.5	143.2
2,436	Phase	14.3	131.8	1,608	Ground	33.3	176.5
2,436	Turn	20.0	151.8	1,608	Turn	14.3	190.8
2,490	Ground	14.3	166.1	1,896	Ground	100.0	290.8
2,772	Ground	16.7	182.8	1,896	Turn	100.0	390.8
3,108	Turn	25.0	207.8	1,896	Phase	100.0	490.8
3,108	Turn	33.3	241.1				
3,108	Phase	20.0	261.1				
3,108	Turn	50.0	311.1				
3,108	Turn	100.0	411.1				

ing life distribution at any stress level from the existing data. This estimate avoids the time and expense of making the redesign and testing it. This estimate employs the estimate of the life distribution of each remaining mode (Section 5.2). This estimate assumes such modes are completely eliminated for all practical purposes, and all remaining failure modes are unaffected.

Simple distribution estimate. The following is a simple estimate of the improved life distribution at a stress when certain failure modes are eliminated. Examine the separate fitted models (Section 5.2) for the modes left in the product. Determine the remaining mode that has the shortest life at the stress level. The improved life distribution is slightly below the distribution of that mode, especially if the other modes have much longer lives. For example, after elimination of Turn failures from the insulation, Phase failures have the shortest life at 180°C and a median of about 17,000 hours. The improved distribution is slightly below that.

Exact distribution estimate. The following exact method estimates the improved life distribution when certain failure modes are eliminated. Suppose that the reliability of the redesign at a given age and temperature is to be estimated. Obtain the estimate of the reliability at that age and temperature for each remaining failure mode as described in Section 5.2. Calculate the estimate of the reliability for the redesign as the product of those reliabilities. This is an application of the product rule (5.3) where the reliability of an eliminated failure mode is 1.

Class-H example. For the insulation, the earliest failure mode at 180°C is Turn failure. Turn failures could be eliminated through a redesign of the insulation system. Suppose that the reliability of the redesign at 10,000 hours at 180°C is to be estimated. Obtain the corresponding reliability estimates 0.872 for Phase failure and 0.995 for Ground failure from Figures 5.3A and 5.4A. Thus the reliability estimate for the redesign at 10,000 hours at 180°C is 0.872 × 0.995 = 0.868, and the fraction failing is 0.132. This is plotted as an × on the lognormal paper in Figure 5.6A. The 180°C curve estimates the life distribution of the redesign. It and the other curves result from repeated calculation of the reliability for different ages and temperatures as shown in Table 5.5. These are curves rather than straight lines, since they are not lognormal distributions. The plotted data on this figure are explained below. The 180° curve yields a median estimate of 16,400 hours. This is much below the desired 20,000 hours. Thus elimination of Turn failures would not provide a satisfactory insulation, and the insulation was abandoned.

Life-stress relationship. The Arrhenius plot in Figure 5.6B is obtained from Figure 5.6A as described in Section 5.3. The relationship is not an Arrhenius one; it is curved and always concave downward on Arrhenius paper.

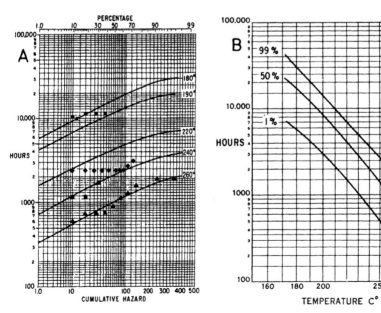

Figure 5.6A. Fitted model and lognormal plot with Turn failures eliminated.

Figure 5.6B. Arrhenius plot of the model with Turn failures eliminated.

Table 5.5. Reliability Calculations with Turn Failures Eliminated

Hours	Reliability Phase	Ground	Without Turn	Hours	Reliability Phase	Ground	Without Turn
180°				**220°**			
4,000	.999 x 1.000	=	.999	1,300	.999 x .998	=	.997
5,000	.996 x 1.000	=	.996	1,600	.997 x .992	=	.989
6,000	.987 x 1.000	=	.987	2,000	.987 x .976	=	.963
7,000	.971 x 1.000	=	.971	2,500	.960 x .935	=	.898
8,000	.946 x .999	=	.945	3,000	.914 x .872	=	.797
9,000	.913 x .997	=	.911	4,000	.774 x .707	=	.547
10,000	.872 x .995	=	.868	5,000	.608 x .534	=	.325
11,000	.824 x .991	=	.817	6,000	.454 x .386	=	.175
12,000	.772 x .986	=	.762	7,000	.329 x .272	=	.089
13,000	.717 x .979	=	.703	8,000	.233 x .189	=	.044
14,000	.662 x .971	=	.642	9,000	.163 x .130	=	.021
16,000	.553 x .947	=	.523	10,000	.114 x .090	=	.010
18,000	.452 x .915	=	.414	11,000	.080 x .062	=	.005
20,000	.365 x .876	=	.320	12,000	.055 x .043	=	.002
25,000	.206 x .757	=	.156	13,000	.039 x .030	=	.001
30,000	.113 x .626	=	.071				
40,000	.034 x .393	=	.013	**240°**			
50,000	.011 x .232	=	.002	500	1.000 x .999	=	.999
				600	1.000 x .997	=	.997
190°				700	1.000 x .992	=	.992
3,000	.999 x 1.000	=	.999	800	.999 x .983	=	.982
4,000	.993 x 1.000	=	.993	900	.998 x .970	=	.968
5,000	.977 x .998	=	.975	1,000	.996 x .952	=	.949
6,000	.945 x .995	=	.940	1,300	.983 x .870	=	.855
7,000	.899 x .988	=	.888	1,600	.953 x .758	=	.722
8,000	.839 x .977	=	.819	2,000	.884 x .595	=	.526
9,000	.769 x .960	=	.739	2,500	.764 x .413	=	.316
10,000	.696 x .938	=	.653	3,000	.630 x .276	=	.174
11,000	.621 x .910	=	.565	4,000	.389 x .118	=	.046
12,000	.549 x .877	=	.482	5,000	.224 x .050	=	.011
13,000	.481 x .841	=	.404	6,000	.126 x .022	=	.003
14,000	.418 x .801	=	.268	7,000	.070 x .010	=	.001
16,000	.312 x .716	=	.223				
18,000	.229 x .628	=	.144	**260°**			
20,000	.167 x .544	=	.091				
25,000	.074 x .364	=	.027	250	1.000 x .998	=	.998
30,000	.033 x .235	=	.008	300	1.000 x .994	=	.994
40,000	.007 x .094	=	.001	400	1.000 x .974	=	.974
				500	.999 x .932	=	.931
				600	.998 x .867	=	.865
				700	.993 x .787	=	.782
				800	.986 x .699	=	.690
				900	.974 x .610	=	.595
				1,000	.957 x .525	=	.503
				1,300	.877 x .317	=	.278
				1,600	.764 x .183	=	.140
				2,000	.595 x .086	=	.051
				2,500	.407 x .034	=	.014
				3,000	.266 x .014	=	.004

Hazard plot. For each test stress, one can make a hazard plot of the data with certain failure modes eliminated. The method provides a check on the validity of the data, the model, and the calculated estimate. Section 5.2 provides a separate estimate of the cumulative hazard function for each remaining failure mode. Sum them to get the sample cumulative hazard function as follows. For the example, assume that Turn failures will be eliminated. In Table 5.2 identify the failure times from Table 5.2 with the remaining failure modes. Order them from smallest to largest as shown in Table 5.6 for the Ground and Phase failures. Include with each failure its hazard value from Table 5.2. Then cumulate the hazard values as shown in Table 5.6. Plot each such failure time on hazard paper against its (modified) cumulative hazard value. This plot is an estimate of the life distribution with just the remaining failure modes. Such plots of the data appear in Figure 5.6A. The plotted data and the calculated curves can be compared. As before, poor agreement may be found in the upper tail.

Partial elimination. Sometimes a redesign does not completely eliminate a failure mode. Subsequent testing may reveal this. The estimate of the new distribution for that mode may be combined with the distributions of other remaining modes as described above. Nelson (1982, pp. 182-5) gives an example of this.

5.5. Checks on the Model and Data

Purpose. This section briefly describes other graphical analyses of data with competing failure modes. These include checks for independence of the competing failure modes and for the validity of the model and data.

Table 5.6. Cumulative Hazard Calculations with Turn Failures Eliminated

Hours	Cause	Hazard	Cum. Hazard	Hours	Cause	Hazard	Cum. Hazard
190°				240°			
10,511	Phase	10.0	10.0	1,175	Ground	10.0	10.0
11,855	Phase	11.1	21.1	1,175	Ground	11.1	21.1
11,855	Phase	12.5	33.6	1,761	Phase	11.1	32.2
11,855	Phase	14.3	47.9				
220°				260°			
2,436	Phase	10.0	10.0	600	Ground	10.0	10.0
2,436	Ground	10.0	20.0	744	Ground	11.1	21.1
2,436	Phase	11.1	31.1	744	Ground	12.5	33.6
2,436	Ground	11.1	42.2	744	Ground	14.3	47.9
2,436	Phase	12.5	54.7	912	Ground	16.7	64.6
2,436	Ground	12.5	67.2	1,128	Ground	20.0	84.6
2,436	Phase	14.3	81.5	1,320	Ground	25.0	109.6
2,490	Ground	14.3	95.8	1,608	Ground	33.3	142.9
2,772	Ground	16.7	112.5	1,896	Ground	100.0	242.9
3,108	Phase	20.0	132.5	1,896	Phase	100.0	342.9

Independence. The series-system model assumes that the failure times for the different modes in a specimen are statistically independent. The following briefly describes how to check this assumption and what to do if the assumption is not satisfied.

Life distribution with dependence. When there are correlations between the lifetimes for different failure modes, they are usually positive. Positive correlation means that long-lived specimens with respect to one failure mode tend to be long lived with respect to another mode, and short-lived specimens with respect to one failure mode tend to be short lived with respect to another mode. When a failure mode is positively correlated with other failure modes, the estimate of Section 5.2 for its life distribution is biased toward long life. Thus, that estimate is an upper bound for the distribution. To obtain a lower bound, treat the times for all other failure modes correlated with that mode as if they were from that mode and use the estimate of Section 5.2. A correct estimate would be between the two bounds. The two bounds may suffice if they are close for practical purposes.

Crossplot. Independence may be checked with a simple crossplot of the pairs of failure times for two modes at a test temperature. This works if there are enough specimens with both failure times. Figure 5.7 shows such crossplots for Turn and Ground failures for two of the test temperatures. An arrow points in the direction of a censored failure mode. If all failure times were observed, one would look for a positive trend in the plot, since that indicates a correlation. One must imagine where the censored times would fall. The three plots in Figure 5.7 do not have convincing trends. Thus Turn and Ground failure times appear statistically independent.

Another check of independence. Nadas (1969) gives a check for independence for normal and lognormal life distributions. His graphical method es-

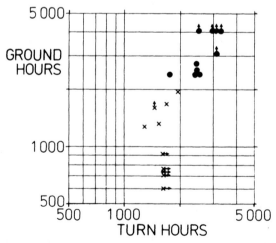

Figure 5.7. Crossplot of Turn and Ground failure times.

timates the correlation coefficient of a joint (log) normal distribution for two failure modes. A very positive or negative estimate indicates dependence. His method applies if there is only one failure mode per specimen.

Checks on each model and data. The graphical methods of Chapter 3 provide checks on the assumed model and the data for each failure mode.

6. ML ANALYSIS FOR COMPETING FAILURE MODES

This section describes maximum likelihood (ML) methods for analysis of such data with competing failure modes, namely,

- An estimate for a separate relationship between life and stress for each failure mode.
- An estimate of the life distribution at any stress with all modes acting.
- An estimate of the life distribution that would result if certain modes are eliminated.

The value of ML methods is that they provide objective estimates, confidence limits, and hypothesis tests (to check the model). They are best used along with graphical methods (Section 5). The Class-H data illustrate the methods. Chapter 5 and Sections 1 and 5 provide needed background.

Literature. ML analyses for data from a single population (test condition) with competing failure modes are given by Cox (1959), Moeschberger (1974), Moeschberger and David (1971), Herman and Patell (1971), Nelson (1982), and Birnbaum (1979). For fitting regression models to such data, Allen (1967) briefly suggests the ML method. Nelson (1971,1974) and Nelson and Hendrickson (1972) develop the details of the ML method, present applications, and provide the STATPAC computer package for the ML regression calculations. Sidik and others (1980) apply the method to battery cells. Glasser (1967) and Klein and Basu (1981) present ML theory for a model with exponential distributions. Sidik (1979) and Klein and Basu (1982) present ML theory for models with Weibull distributions. Section 1 of Chapter 5 describes ML programs that fit such models to data.

Identified modes. The following method requires that the mode of each failure be *identified*. If each failure mode is *not* identified, ML fitting of the model is more complex. Nelson (1982, Problem 8.15) gives an example of ML fitting to data with unidentified failure modes. Faraone (1986) gives such data from an accelerated test.

Overview. This section presents

1. The illustrative data and model.
2. The ML estimate of the model for each failure mode.
3. The ML estimate of the model when all failure modes act.
4. The ML estimate of product life that would result if certain failure modes were eliminated by product redesign.

5. Methods for checking the validity of the model and the data.

Section 7 presents ML theory for analysis of such data.

6.1. Data and Model

Data. Data in Table 5.1 illustrate the ML methods. The data are hours to failure of the Class-H insulation tested in motorettes at high temperatures of 190, 220, 240, and 260°C. Table 5.1 gives the cause of each failure, namely, Turn, Phase, and Ground failures – separate parts of the insulation system. The cause is determined by engineering examination. More than one cause of failure may occur in an inspection period. Each failed part of a specimen was isolated electrically and could not fail again by that cause. Such a specimen was put back on test and run to a second or third failure. Figure 5.1 depicts the data.

In use, the first failure from any cause ends the life of a motor. For most products, only one of the failure modes is observed on a specimen. The methods apply whether one or more failure modes occur on each specimen.

A purpose was to estimate the median life of such insulation at its design temperature of 180°C. A median life of 20,000 hours was desired. Another purpose was to determine which cause of failure is dominant (earliest) at the design temperature and to determine if eliminating that cause would improve the life distribution enough.

Model. The model for the data is the one in Section 5.1. Namely, there is a separate life-stress model for each failure mode. The Arrhenius-lognormal model is used for each failure mode in the example. Also, one uses the series-system model for the relationship between the life distribution of the product and those of the failure modes. An Arrhenius-lognormal model does not describe product life when more than one failure mode is acting. Figure 5.5 depicts the correct model. The following describes how to estimate the correct model by ML.

6.2. ML Estimates for Each Failure Mode

For each failure mode, the following ML methods estimate its Arrhenius-lognormal model and its life distribution at any temperature. The methods are easy carry out with the computer programs of Chapter 5.

Censoring. For an analysis of the data on a failure mode, each specimen has a time to failure with that mode or else a running time without that failure mode. Such a running time results when the specimen fails by another mode or is removed unfailed from test. Such a running time is a censoring time for the failure mode. Table 5.2 shows such multiply censored data for insulation Turn, Phase, and Ground failures. The data for each failure mode are separately analyzed with ML methods for multiply censored data as described in Chapter 5.

ML fit. An accelerated-test model, such as the Arrhenius-lognormal model, is fitted to such multiply censored data on a failure mode by a ML program. ML fitting of the model to the data for each failure mode was done by the STATPAC program of Nelson and Hendrickson (1972). The 260°C data on Turn failures were omitted from the fitting. Table 6.1 shows the estimates and confidence limits for the model parameters, their (estimated) covariance matrix, the estimates of the Arrhenius relationships, and estimates and confidence intervals for the median lives at the design temperature of 180°C. For example, the ML estimate of the mean log life for Turn failures at 180°C is $\hat{\mu}(2.2067261) = 3.526684 + (2.2067261 - 2.044926)$

Table 6.1. ML Fits to Failure Mode Data

TURN FAILURES	α'	β	σ	$180°$Median
Estimate:	3.526684	3.478935	0.07423483	12,290
Lower 95% Limit:	3.499721	3.171346	0.05710498	10,797
Upper 95% Limit:	3.553648	3.786524	0.09650315	13,990

Cov. Matrix:	$\hat{\alpha}'$	$\hat{\beta}$	$\hat{\sigma}$
$\hat{\alpha}'$	0.001892482		symmetric
$\hat{\beta}$	-0.00003268012	0.02462799	
$\hat{\sigma}$	0.00000460619	-0.00003613455	0.0000987278

Maximum Log Likelihood: -219.69875
$\mu(x) = 3.526684 + (x - 2.044926)3.478935$

PHASE FAILURES	α'	β	σ	$180°$Median
Estimate:	3.693313	2.660203	0.2034712	17,020
Lower 95% Limit:	3.574233	1.850073	0.1288536	11,650
Upper 95% Limit:	3.812393	3.470332	0.3212990	24,866

Cov. Matrix:	$\hat{\alpha}'$	$\hat{\beta}$	$\hat{\sigma}$
$\hat{\alpha}'$	0.00361172		symmetric
$\hat{\beta}$	-0.008941888	0.1708428	
$\hat{\sigma}$	0.2013194	-0.005933364	0.002249245

Maximum Log Likelihood: -32.534989
$\mu(x) = 3.693313 + (x - 2.004613)2.660203$

GROUND FAILURES	α'	β	σ	$180°$Median
Estimate:	3.610041	4.624660	0.2110285	35,054
Lower 95% Limit:	3.502661	3.667628	0.1442328	19,023
Upper 95% Limit:	3.717421	5.581691	0.3087580	64,595

Cov. Matrix:	$\hat{\alpha}$	$\hat{\beta}$	$\hat{\sigma}$
$\hat{\alpha}'$	0.0030001463		symmetric
$\hat{\beta}$	0.01385714	0.2384186	
$\hat{\sigma}$	0.001352738	0.008563337	0.001678907

Maximum Log Likelihood: -136.97398
$\mu(x) = 3.610041 + (x - 2.004613)4.624660$

3.478935 = 4.089576; here 2.2067261 = $1000/(273.16 + 180)$ is the reciprocal absolute temperature. The ML estimate of the median is anti-log(4.089576) = 12,290 hours. 95% confidence limits are 10,797 and 13,990 hours. The interval is well below the desired 20,000 hour-life. Also, the ML estimate of the fraction failing by 10,000 hours at 180°C, if only Turn failures are active, is $\hat{F}(10,000) = \Phi\{[\log(10,000) - 4.089576]/0.07423483\} = 0.114$. Confidence limits for estimated quantities are calculated as described in Chapter 5. Lognormal and Arrhenius plots of the fitted models appear in Figures 5.2, 5.3, and 5.4.

6.3. ML Estimate When All Failure Modes Act

When a product is in actual use, all failure modes act. Presented below are two methods for estimating the product life distribution at a stress level when all failure modes act. Both methods employ the ML estimate of the life distribution of each failure mode.

Simple estimate. Examine the plotted estimate of the distribution for each failure mode at that stress level. The distribution with all modes acting is generally slightly below the earliest such distribution. For the insulation, Turn failures are earliest at 180°C. Thus, the insulation median life is slightly below that for Turn failures (12,290 hours). The exact estimate (obtained below) is 11,600 hours. Approximate other percentiles this way.

ML estimate. Estimate the reliability at some age and stress level as follows. Calculate the ML estimate of the reliability at that age and stress level for each failure mode as described above. The product of those reliability estimates is the ML estimate of product reliability with all failure modes acting. For example, estimate insulation reliability for 10,000 hours at 180°C as follows. Calculate the ML estimate of that reliability as 0.886 for Turn failure, 0.872 for Phase failure, and 0.995 for Ground failures. Calculate the ML estimate of insulation reliability as 0.886×0.872×0.995 = 0.769. The ML estimate of the fraction failing is $1 - 0.769 = 0.231$. This is plotted as an × on lognormal paper in Figure 5.5A. There the 180° curve is the ML estimate of the insulation life distribution with all modes acting. Such calculations (Table 5.3) for various ages yield the distribution estimate. For any temperature, such calculations yield a distribution curve in Figure 5.5A, not a straight line; such a distribution is not lognormal.

Life-stress relationship. The relationship between insulation life and temperature appears on Arrhenius paper in Figure 5.5B as percentile curves. Figure 5.5A yields an ML estimate of 11,600 hours for the median life at 180° with all modes acting. Similarly, obtain such median estimates for other temperatures and plot them on Arrhenius paper as in Figure 5.5B. Then draw a smooth curve through the median estimates. This curve could also be calculated. Estimate the relationship for any other percentile the same way. Not an Arrhenius relationship, such a curve for competing modes is always concave down on Arrhenius paper. Thus a fitted Arrhenius relationship ignoring failure modes (as in Problem 7.6(a)-(f)) usually yields estimates of percentiles at low (design) temperatures that are biased high. Confidence limits for such relationships are calculated as described in Section 7.

6.4. ML Estimate with Some Failure Modes Eliminated

Some failure modes may be eliminated by design changes. Presented below are two estimates of the resulting improved life distribution with the remaining failure modes. These methods avoid the time and expense of making the redesign and testing it. Both methods employ the ML estimate of the life distribution of each remaining failure mode.

Simple estimate. For a stress level, determine which remaining failure mode has the earliest distribution. The improved distribution when all remaining modes act is slightly below the earliest one. For example, after elimination of Turn failures, Phase failures have the earliest distribution at 180°C; the median estimates in Table 6.1 show this. Then insulation median life would be slightly below that of Phase failures – about 17,020 hours. The exact value (obtained below) is 16,400 hours.

ML estimate. Estimate the improved reliability at some age and stress as follows. Calculate the ML estimate of reliability at that age and stress level for each remaining failure mode as described above. The product of those reliability estimates is the ML estimate of the reliability of the redesign. This method assumes that reliability for an eliminated failure mode is 1. For example, the dominant insulation failure mode at 180°C is Turn failure. Turn failures could be eliminated through redesign. (The method applies to elimination of any number of failure modes.) The ML estimate of the improved reliability for 10,000 hours at 180°C follows. The ML estimate for reliability for 10,000 hours at 180°C is 0.872 for Phase failures and 0.995 for Ground failures. Thus the ML estimate of the reliability of the redesign is 0.872 × 0.995 = 0.868. The estimate of the fraction failing is 1 − 0.868 = 0.132. This estimate is plotted as an × on the 180° curve on lognormal paper in Figure 5.6A. There the 180°C curve is the ML estimate of the life distribution of the redesign. The ML curves there come from such calculations (Table 5.5) for various ages and temperatures. These are curves, since they are not lognormal distributions. In Figure 5.6A, the estimate of the median at 180°C is 16,400 hours. This is well below the desired 20,000 hours. Thus elimination of Turn failures would not improve the insulation enough. The Arrhenius plot in Figure 5.6B is obtained graphically from Figure 5.6A or numerically. Confidence limits for such curves are described in Section 7.

6.5. ML Checks on the Model and Data

Other ML analyses of such data include checking the life-stress relationship, the distribution, and independence of the failure modes. Section 5.5 presents graphical checks, which are effective combined with ML checks.

Linearity. The likelihood ratio test for linearity (Chapter 5) applies to the data on a failure mode. The following example is a test of fit of the Arrhenius relationship to the Turn failure data (Table 5.1). The fitted model has three parameters and a maximum log likelihood of 17.25. Fitting a model with a different log mean for each of the four test temperatures and a common σ to the same data yields a maximum log likelihood of 36.42. The test statistic is $T = 2(36.42 - 17.25) = 38.34$. Its number of degrees of freedom is $5 - 3 = 2$. Since $T = 38.34 > 13.82 = \chi^2(0.999;2)$, there is very highly significant lack of fit of the Arrhenius relationship at the 0.1% level. Figure 5.2B shows that the 260° data are not consistent with the rest of the Turn data. The 260° data were omitted, and the Arrhenius relationship was fitted to the remaining Turn failure data and provided a satisfactory fit. Of course, this ignores the effect of cycling rate (Problem 7.8).

Distribution fit. Adequacy of the Weibull and lognormal distributions can be assessed by fitting a model with a log gamma distribution. Farewell and Prentice (1977) describe ML theory and a computer program for such fitting. SAS, STAR, and SURVCALC (Chap. 5, Sec. 1) do such fitting.

Residuals. It is useful to calculate residuals for each failure mode from the fitted model and to graphically analyze them as described in Chapter 5. Such plots help one assess the distribution, relationship, and data. For crossplots of residuals against independent variables, it may be useful to replace censored residuals by their (conditional) expected values, as described by Lawless (1982, pp. 281-282).

Independence. The following provides a check on the independence assumption of the series-system model. Namely, the failure times for the different modes in a specimen are statistically independent. Maximum likelihood may be used to estimate the joint distribution of correlated failure modes. For example, a joint lognormal distribution could be used, and the estimates of the correlations examined to determine if they are significantly different from zero. General theory given by Moeschberger and David (1971) extends to the regression models here. Section 5.5 describes graphical checks of independence.

Correlated modes. Correlations between the lifetimes for different modes are usually positive. This means that long (short) lives of one failure mode tend to go with long (short) lives of another mode. That is, a specimen is generally strong or generally weak or sees severe or mild usage. With such correlation, the ML estimate in Section 6.2 for the life distribution of a particular failure mode is usually biased toward long life. Thus, that ML estimate is an upper bound for the distribution of that mode. For a lower bound, treat the failure times for the correlated failure modes as if they were from that mode and estimate the distribution. If close, the two bounds may serve practical purposes.

7. ML THEORY FOR COMPETING MODES

This section presents ML theory for fitting life-stress models to data with competing failure modes. The theory provides estimates, confidence intervals, and tests of hypotheses. Results appear without proof. David and Moeschberger (1979), Rao (1973), and Lawless (1982) derive such theory. This section covers the general model, log likelihood, ML estimates, Fisher and covariance matrices, and estimate and confidence limits for a function. ML theory in Section 5 of Chapter 5 is needed background.

7.1. General Model

A general model for life data with independent competing failure modes follows. Suppose that each of M failure modes has a separate distribution of time to failure. For mode m, denote the reliability function by $R_m(y; \mu_m, \sigma_m)$ at (log) age y, where μ_m and σ_m are the distribution parameters, $m = 1, \cdots, M$. Denote the probability density by $f_m(y; \mu_m, \sigma_m)$. For the insulation, each of the $M = 3$ failure modes has a lognormal distribution of time to failure. For concreteness, two parameters appear here, but one may use a distribution with any number.

Denote the J independent variables by $x_1, \cdots, x_J$. For the insulation, there is $J = 1$, namely, temperature. For mode m, the parameters μ_m and σ_m are given functions of the independent variables; namely,

$$\mu_m = \mu_m(x_1, \cdots, x_J; \gamma_{m1}, \cdots, \gamma_{mP_m}),$$
$$\sigma_m = \sigma_m(x_1, \cdots, x_J; \gamma_{m1}, \cdots, \gamma_{mP_m}). \tag{7.1}$$

Here $\gamma_{m1}, \cdots, \gamma_{mP_m}$ are P_m unknown coefficients to be estimated from the data. Some coefficients may appear in both functions. Also, one function may have some coefficients, and the other has the rest. However, the coefficients in $\mu_m(\ )$ and $\sigma_m(\ )$ are distinct from those in $\mu_{m'}(\ )$ and $\sigma_{m'}(\ )$. Let $P = P_1 + \cdots + P_M$ denote the total number of coefficients γ_{mp}. For the insulation, $P_1 = P_2 = P_3 = 3$ coefficients for each model, and $P = 9$. Independent variables may be indicator (zero-one) variables for category variables in analysis-of-variance relationships.

7.2. Log Likelihood

The sample log likelihood follows. Suppose there are n specimens in the sample, and they are statistically independent of each other. First consider specimens that fail once from some cause and do not run further. Let $x_{1i}, \cdots, x_{Ji}$ denote values of the independent variables for specimen i. Similarly, for $i = 1, \cdots, n$ and $m = 1, \cdots, M$, let

$$\mu_{mi} = \mu_m(x_{1i}, \cdots, x_{Ji}; \gamma_{m1}, \cdots, \gamma_{mP_m}), \tag{7.2}$$

$$\sigma_{mi} = \sigma_m(x_{1i}, \cdots, x_{Ji}; \gamma_{m1}, \cdots, \gamma_{mP_m}).$$

Throughout, μ_{mi} and σ_{mi} depend on $\gamma_{m1}, \cdots, \gamma_{mP_m}$. The following likelihood for failed specimen i takes into account its cause of failure and that other failure modes have not occurred. Suppose that specimen i fails at (log) time y_i by mode m. Then its likelihood is

$$L_i = R_{1i} \cdots R_{m-1,i} \, f_{mi} \, R_{m+1,i} \cdots R_{Mi};$$

here $R_{m'i} = R_{m'}(y_i; \mu_{m'i}, \sigma_{m'i})$ and $f_{mi} = f_m(y_i; \mu_{mi}, \sigma_{mi})$. If specimen i does not fail from any cause by censoring age y_i, its likelihood is

$$L_i = R_{1i} R_{2i} \cdots R_{Mi}.$$

These likelihoods are products of factors for the failure modes, because the modes are statistically independent. These L_i are correct for both Type I and Type II censored data. The ML theory of Moeschberger and David (1971) for dependent modes can be extended to the regression models here.

Next consider a specimen that may fail from more than one cause. Its likelihood takes into account the times of failure modes that occurred and the running times of those that did not occur. Suppose that specimen i fails by M_i modes in the set $\mathbf{M}_i = \{m_{1i}, m_{2i}, \cdots, m_{M_i i}\}$ at (log) times $y_{m_1 i}, y_{m_2 i}, \cdots, y_{M_i i}$. Also, suppose, it does not fail by the other modes by a (log) time y_i'. Then its likelihood is

$$L_i = \prod_{m \in \mathbf{M}_i} f_{mi}(y_{mi}) \prod_{m \notin \mathbf{M}_i} R_{mi}(y_i'); \tag{7.4}$$

here the notation has the obvious meaning. Some insulation motorettes failed by more than one mode. The log likelihood for specimen i is

$$\mathcal{L}_i = \ln(L_i). \tag{7.5}$$

The log likelihood $\mathcal{L}$ for the sample of I statistically independent units is

$$\mathcal{L} = \mathcal{L}_1 + \cdots + \mathcal{L}_I. \tag{7.6}$$

7.3. ML Estimates

Estimates. The ML estimates $\hat{\gamma}_{mp}$ of the P model coefficients γ_{mp} are the coefficient values that maximize $\mathcal{L}$. Under *regularity conditions* on the model and data (that usually hold), the ML estimates are unique. Also, for asymptotic sample sizes (many failures), the ML estimates are usually approximately jointly normally distributed with a mean vector equal to the true coefficient values and an asymptotic covariance matrix given later. Moreover, their asymptotic variances are smaller than those for any other asymptotically normally distributed estimates.

Calculation. The estimates must usually be obtained from iterative numerical optimization of $\mathcal{L}$ with respect to the P coefficients γ_{mp}. If P is large, the calculations are time consuming and convergence may be uncertain. The following method simplifies the calculations. It splits the problem into simpler ones, which can be solved with existing computer programs – one failure mode at a time.

Likelihood equations. For some models, $\mathcal{L}$ can be maximized by the usual calculus method. Namely, the P partial derivatives of $\mathcal{L}$ with respect to each γ_{mp} are set equal to zero:

$$\partial\mathcal{L}/\partial\gamma_{mp} = 0, \quad m = 1, \cdots, M \text{ and } p = 1, \cdots, P_m. \tag{7.7}$$

These are called the *likelihood equations*. The γ_{mp} values that solve this set of P simultaneous nonlinear equations are the ML estimates $\hat{\gamma}_{mp}$. A simple approach to solving (7.7) follows.

Separate modes. Suppose that the coefficients γ_{mp} for mode m differ from those $\gamma_{m'p}$ for mode m'. That is, each mode has a *distinct* model with no common parameter values. Then (7.7) can be rewritten and solved as separate small sets of P_m equations for the coefficients of the model for a single failure mode. The following argument applies to specimens with one observed failure mode. However, it extends to specimens subject to more than one observed failure mode. The P_m likelihood equations for failure mode m are obtained from (7.7), (7.6), and (7.3) or (7.4) as

$$0 = \partial\mathcal{L}/\partial\gamma_{m1} = \sum_i \frac{\partial}{\partial\gamma_{m1}}\ln(f_{mi}) + \sum_i' \frac{\partial}{\partial\gamma_{mP_m}}\ln(R_{mi}),$$

$$\begin{array}{c} . \\ . \\ . \end{array} \tag{7.8}$$

$$0 = \partial\mathcal{L}/\partial\gamma_{mP_m} = \sum_i \frac{\partial}{\partial\gamma_{mP_m}}\ln(f_{mi}) + \sum_i' \frac{\partial}{\partial\gamma_{mP_m}}\ln(R_{mi}).$$

Here the first (second) sum runs over specimens that do (not) have failure mode m. These P_m equations are solved for $\hat{\gamma}_{m1}, \hat{\gamma}_{m2}, \cdots, \hat{\gamma}_{mP_m}$. The probability densities, reliability functions, and model coefficients for other failure modes do not appear in these equations for mode m. This is so because their log likelihoods do not contain coefficients for mode m, and thus their partial derivatives are all zero. Thus one can solve equations for one failure mode at a time. Moreover, these equations for a failure mode are equations for a multiply censored sample, and they may be solved with a computer program for such data.

Optimize likelihood. The preceding result also comes from an argument involving maximizing the log likelihood function. Then the likelihood terms are regrouped so all coefficients for only one failure mode are in a group, and the likelihoods for such groups are separately maximized.

7.4. Fisher and Covariance Matrices

Purpose. The Fisher information matrix and the asymptotic covariance matrix for the $\hat{\gamma}_{mp}$ yield approximate confidence limits for the γ_{mp} and functions of them. Such functions include, for example, a percentile of the life distribution at the design stress level.

Derivatives. First calculate the P by P matrix of the negative second partial derivatives

$$-\partial^2 \mathcal{L}/\partial \gamma_{mp} \partial \gamma_{m'p'} \text{ for } m,m' = 1, \cdots ,M,$$
$$p = 1, \cdots ,P_m, \ p' = 1, \cdots ,P_{m'}, \tag{7.9}$$

For $m \neq m'$, these derivatives are zero, since $\partial \mathcal{L}/\partial \gamma_{m'p'}$ does not contain coefficients γ_{mp} for failure mode m. Suppose the matrix is arranged so the derivatives for mode 1 are in the first P_1 rows and columns, the derivatives for mode 2 are in the next P_2 rows and columns, etc. Then the matrix is block diagonal, with zeros off the block diagonal.

Fisher matrix. The expectation of the matrix (7.9) with elements

$$E\left\{ - \frac{\partial^2 \mathcal{L}}{\partial \gamma_{mp} \ \partial \gamma_{m'p'}} \right\}, \tag{7.10}$$

evaluated at the true coefficient values, is the *true Fisher information matrix* and is denoted by **F**. An expected value depends on the type of censoring (Type I, Type II, or other) and on all of the models for all failure modes. General expressions for such expectations for a single population are given by Moeschberger and David (1971). Those expressions extend to the regression models here but are so complex, they are not given here.

Covariance matrix. The inverse $\mathbf{\Sigma} = \mathbf{F}^{-1}$ of Fisher matrix is the *true (asymptotic) covariance matrix* of the $\hat{\gamma}_{mp}$. A covariance $\text{Cov}(\hat{\gamma}_{mp},\hat{\gamma}_{m'p'})$ in $\mathbf{\Sigma}$ is in the same position as corresponding term (7.10) in **F**. Similarly, a variance $\text{Var}(\hat{\gamma}_{mp})$ in $\mathbf{\Sigma}$ is in the same position as the corresponding term $E\{-\partial^2 \mathcal{L}/\partial \gamma_{mp}^2\}$. The *ML estimate for the covariance matrix* is obtained by substituting the $\hat{\gamma}_{mp}$ for the unknown true γ_{mp}. Often one estimates $\mathbf{\Sigma}$ with the simpler local estimate below.

Uncorrelated. The Fisher information matrix is block diagonal. Thus its inverse, the covariance matrix, is block diagonal and zero elsewhere. Thus the ML estimates of coefficients for different failure modes are asymptotically statistically uncorrelated. Moreover, when the ML estimates are asymptotically jointly normally distributed, the estimates of coefficients for *different* modes are asymptotically statistically independent. Estimates of coefficients for the *same* failure mode generally are correlated.

Local estimate. The matrix (7.9) of negative second partial derivatives evaluated for $\gamma_{mp} = \hat{\gamma}_{mp}$ is the *local information matrix*. Its inverse is the *local estimate of* $\mathbf{\Sigma}$. It is easier to calculate than the ML estimate, since it does not require the difficult expectations (7.10). An estimate of $\mathbf{\Sigma}$ yields as follows approximate confidence intervals.

7.5. Estimate and Confidence Limits for a Function

Estimate. Hereafter the model coefficients are relabelled γ_1, γ_2, $\cdots$,γ_P where P is the total number of coefficients for all failure modes. Suppose that an estimate is desired for the value of a function $h = h(\gamma_1, \cdots, \gamma_P)$ of the model coefficients $\gamma_1, \cdots, \gamma_P$. Model coefficients and distribution parameters and percentiles are such functions. For the insulation, such a function is the median time to Turn failure at 180°C, namely, $h = $ antilog$\{\gamma_1 + \gamma_2[1000/(273.16+180)]\}$. The ML estimate $\hat{h}$ is the function evaluated at $\gamma_1 = \hat{\gamma}_1, \cdots, \gamma_P = \hat{\gamma}_P$; that is, $\hat{h} = h(\hat{\gamma}_1, \cdots, \hat{\gamma}_P)$.

Variance. Calculate the estimate of the variance of $\hat{h}$ as follows. First calculate the column vector of first partial derivatives of h with respect to each coefficient γ_p, namely,

$$\mathbf{H} = [\partial\hat{h}/\partial\gamma_1 \cdots \partial\hat{h}/\partial\gamma_P]'. \tag{7.11}$$

Here the prime ´ denotes the transpose and the caret indicates that each derivative is evaluated at $\gamma_1 = \hat{\gamma}_1, \cdots, \gamma_P = \hat{\gamma}_P$. By propagation of error (Rao (1973)), the variance estimate is

$$\text{Var}(\hat{h}) = \mathbf{H}'\hat{\boldsymbol{\Sigma}}\mathbf{H} = \sum_{p=1}^{P}(\partial\hat{h}/\partial\gamma_p)^2\,\text{Var}(\hat{\gamma}_p) \tag{7.12}$$

$$+ 2\sum\sum_{p<p'}(\partial\hat{h}/\partial\gamma_p)(\partial\hat{h}/\partial\gamma_{p'})\,\text{Cov}(\hat{\gamma}_p,\hat{\gamma}_{p'});$$

here $\hat{\boldsymbol{\Sigma}}$ is an estimate of the covariance matrix of the $\hat{\gamma}_p$. The estimate $s(\hat{h})$ of the standard error of $\hat{h}$ is

$$s(\hat{h}) = [\text{Var}(\hat{h})]^{1/2}. \tag{7.13}$$

This standard error is used to calculate normal confidence limits as described in Section 5.7 of Chapter 5.

PROBLEMS (* denotes difficult or laborious)

7.1. Class-H insulation. Verify selected calculations for the life distributions with all failure modes acting in Table 5.3.

7.2. Turn failure eliminated. Verify selected calculations for the life distributions at the design and test temperatures for the Class-H insulation, assuming that Turn failures are eliminated.

7.3. Tandem specimens. Nelson and Hahn (1973) give the following data on hours to first failure of five pairs of tandem specimens in a creep-rupture test of a metal. When the first specimen of a pair fails, the other specimen is removed from test. These data are failure censored – rare in practice.

Stress (ksi):	29	32	34	37	44
Hours:	11,495	8322	5578	2435	1350

(a) Plot the data (including runouts) on log-log paper. Comment on the adequacy of the inverse power relationship.

(b) Appropriately plot the data on lognormal *probability* paper.

(c) By maximum likelihood, fit the power-lognormal model for individual specimens to the 10 data points.

(d) Estimate the life distribution of individual specimens at a design load of 25 ksi. In particular, calculate the ML estimate and 95% confidence limits for the 5th percentile.

(e) Plot the fitted model (life distributions) for individual specimens for the test and design loads on the lognormal paper.

(f) Plot selected percentile lines for individual specimens on (a).

(g) Calculate the (log) residuals and plot them on normal paper. Comment on the adequacy of the (log) normal distribution.

(h) Crossplot the residuals versus stress, comment on the plot.

(i) Suggest further analyses.

(j) Carry out (h).

(k) Repeat (a)-(i) using a power-Weibull model. Which distribution fits better and how much better?

7.4. Class-B insulation. The temperature-accelerated life test of the Class-B motor insulation yielded the following data in hours on its three failure modes. The main purpose was to estimate insulation median life at the design temperature of 130°C. Ignore the cycle length.

(a) Plot the data as in Figure 5.1. Use different colors for the three failure modes. Comment on what you see.

(b) Graphically analyze the data on each failure mode, and estimate each

	150°C (28 days)				190°C (4 days)		
Motor	Turn	Phase	Ground	Motor	Turn	Phase	Ground
1	12453	12453+	11781	21	552	552+	408
2	14637	14637	12453	22	600	408	600+
3	14637	14637+	13897	23	1764	1764+	1440
4	15309	14637	15309+	24	2112+	2112	1344
5	15645+	15645+	15645+	25	2208	1344	2208+
6	15645+	15645+	15645+	26	1920	2232+	2232+
7	15645+	15645+	15645+	27	2304	2304+	2208
8	15645+	15645+	15645+	28	2496	2400+	2400
9	15645+	15645+	15645+	29	2592	3264+	3264
10	15645+	15645+	15645+	30	3360	3360	3360+

	170°C (7 days)				220°C (2 days)		
Motor	Turn	Phase	Ground	Motor	Turn	Phase	Ground
11	1932	1932+	1764	31	504	504+	408
12	2970+	2970	2722	32	600	600+	504
13	3612	3612+	3444	33	648	648+	648
14	3780	3780+	3780	34	504	696+	696

15	3948	3948+	2612	35	504	696+	696
16	4680	5196	5196+	36	696+	696	600
17	5796	5796+	5796	37	696	696+	648
18	6204	6204+	6204	38	408	768+	768+
19	7716	9648	9818	39	600	768+	768+
20	9648	9900+	7884	40	696	768+	768+

life distribution at 130°C. Use modified hazard plotting positions.

(c) Viewing (b), comment on the validity of the model and data for each failure mode.

(d) By ML fit a model to the data on each failure mode. Plot each model and confidence limits on separate lognormal and Arrhenius papers.

(e) Calculate and plot the log residuals from (d) for each failure mode on normal paper. Comment on each plot.

(f) Assess (1) the Arrhenius relationship and (2) constant σ for each failure mode, using a LR or other test.

(g) For all failure modes acting, calculate the estimate of the insulation life distribution at the test and design temperatures. Plot these distributions on lognormal paper. Make hazard plots of the data with all modes acting for each test temperature, (1) using only the first failure on each specimen and (2) using all subsequent failures. Comment on the agreement between the data and fitted model.

(h) Determine the dominant (earliest) failure mode at 130°C. Calculate the ML estimate of the life distribution at 130° that would result if that failure mode were eliminated. Plot it on (g).

7.5. Ball bearings. Morrison and others (1984) give the following data on a load-accelerated life test of ceramic ball bearings. They analyze the data with the power-Weibull model. Each of the 59 test bearings contains 7 balls, and 1 or more balls in a bearing may fail during the test.

Bearing Life in Millions of Revolutions (No. of failed balls)

4.45 kN	5.00 kN		6.45 kN		9.56 kN
88(1)	217(1)	499+	47.1(1)	240.0+	14.5(1)
144(2)	236+	561+	68.1(1)	240.0(1)	25.6(3)
492(1)	281(1)	574+	68.1(1)	278.0+	26.2(1)
492+	346(1)	574+	90.8(1)	278.0+	52.4+
582+	346+	699+	103.6(1)	289.0+	66.3(1)
631(1)	411+	699+	106.0(1)	289.0(1)	69.3+
631(1)	414+	998+	115.0(1)	367.0(1)	69.3+
638+	414+	998+	126.0(2)	385.9+	69.8+
769+	423(1)	1041+	146.6(1)	392.0+	76.2(1)
769+	423+	1041+	229.0+	505.0+	

(a) Graphically analyze the data on the life of individual balls, using all ball failures. Use modified hazard plotting positions. In particular, estimate the life distribution of such balls at the test loads and the power relationship for the 10th percentile.

(b)   Using (a) comment on the validity of the model and data.
(c)   By ML, fit the model to the data on individual balls. Plot the fitted life distributions and confidence limits for the test loads on Weibull paper. Plot the estimate and confidence limits for the 10th percentile line for balls on log-log paper.
(d)   Graphically analyze the residuals from (c), and comment on the data and Weibull distribution.
(e)   Assess whether the shape parameter β is constant using (1) the LR test and (2) a ln-linear relationship for β as a function of load.
(f)   Assess the fit of the power relationship using the LR test for linearity.
(g)   Use the results of (c) to estimate the model for the life of actual *bearings* with 14 balls. Plot the estimates of the actual bearing life distributions for the test loads. Plot the estimate of the 10th percentile line for actual bearings on log-log paper.
(h)   A bearing test "fails" when its vibration reaches a specified level. Then its balls are examined to determine how many were damaged ("failed"). Make a case for ignoring the number of failed balls (pros and cons).
(i)   Repeat (a)-(f) using data on test bearing failure (ignoring the number of failed balls). Redo (g) and comment on differences.
(j)   Suggest further analyses.
(k)   Do (j).

7.6. Heaters. The following data are from a temperature-accelerated life test of industrial heaters, which have two failure modes – open and short. Use the Arrhenius-lognormal model.

Heater	Hours	Cause		Heater	Hours	Cause
1820°F 116	72.7	Short	1675°F	108	1532.0	Open
119	343.9	Short		105	2125.0	Open
117	347.6	Short		107	2212.0+	Censored
				106	2242.0+	Censored
1750°F 111	320.0	Short				
110	1035.0	Short	1600°F	103	1547.0	Open
112	1154.4	Short		104	1726.5	Open
113	1979.0	Short		101	1729.3	Short
				102	2539.0+	Censored

(a)   Ignoring failure modes, make Arrhenius and lognormal plots of all data. Use modified hazard plotting positions.
(b)   Assess the Arrhenius-lognormal model and data using (a).
(c)   Using (a), graphically estimate the life distribution at the design temperature of 1150°F.
(d)   By ML, fit the Arrhenius-lognormal model to all data, ignoring failure modes. Estimate the life distributions for the design and test temperatures. Plot the estimates of these distributions and 95% confidence limits on lognormal paper.

(e)   Calculate (log) residuals from (d). Plot them on normal paper, and comment on the adequacy of the lognormal distribution.

(f)   Crossplot residuals versus heater number. Comment on the effect of no randomization in the assignment of heaters to test temperatures.

(g)   Repeat (a)-(f) for Short failures, breakdown of the electrical insulation.

(h)   Repeat (a)-(f) for Open failures, breaking of the heating element wire.

(i)   Assess the fit of the Arrhenius relationship for each failure mode, using the LR test for a "quadratic" relationship.

(j)   Analytically assess whether σ is constant for each failure mode.

(k)   At the design and test temperatures, estimate the heater life distribution with both failure modes acting, using the product rule (1.2).

(l)   Plot the distributions from (k) on lognormal paper. How do these distributions compare with those from (d).

(m)   Plot selected percentile lines of the fitted model with both failure modes on Arrhenius paper. Comment on how this plot differs from that from (a).

(n)   The responsible engineers concluded that Open failures "occurred due to differential [thermal] expansion combined with low wire strength at the test temperatures. We further decided that at normal operating conditions, this would not be a problem." What are the practical implications of this respect to heater life?

(o)   Specify a better test plan.

(p*)   Assume that failure modes were not identified. Fit a model with two competing failure modes to the data. Plot the fitted model and compare it with (m).

7.7. Other heaters. The following data on hours to failure are from a temperature-accelerated life test of industrial heaters like those of Problem 7.6. (S) denotes Short, and (O) denotes Open.

1750°F:	108.2(S)	181.8(S)	232.2(S)	476.0(S)
1675°F:	450.9(O)	501.0(S)	515.2(O)	608.1(O)
1600°F:	487.1(S)	575.5(O)	600.6(O)	702.7(O)
1525°F:	557.0(O)	1171.2(S)		
1750°F*:	24.1(S)	24.6(S)	35.7(S)	83.6(S)

The four heaters marked * are from a different production lot. Assume that their log mean is not consistent with those of the other heaters, but their log standard deviation is. Repeat the analyses of Problem 7.6 for these data.

7.8. $1,000,000 experiment. Problem 3.9 gives data on hours to Turn failure on test of a Class-H insulation. Ground and Phase failure data from that test follow. This experiment showed the effect of oven cycle length on life. Moreover, it showed that cheaper insulation can be used in applications with no cycling, thus saving $1,000,000 yearly.

Ground failure hours:

200°C/7 days: 9 survived 7392+

215°C/28 days:	8400, 7 survived 11424+
215°C/2 days:	6 survived 2784+
230°C/7 days:	2451, 2955, 3444, 3780, 3948, 1 survived 6216+
245°C/28 days:	3 at 2352, 3696, 4368, 5712
245°C/2 days:	892, 2 at 988, 1 survived 3072+
260°C/7 days:	1088, 1256, 1592, 1764

Phase failure hours:

200°C/7 days:	9 survived 7392+
215°C/28 days:	10416, 7 survived 11424+
215°C/2 days:	6 survived 2784+
230°C/7 days:	2 at 4620, 4 survived 5040+
245°C/28 days:	6384, 5040, 4 survived 6048+
245°C/2 days:	4 survived 1824+
260°C/7 days:	1424, 1592, 2 survived 2018+

(a)   Graphically analyze the data for each failure mode, and estimate the life distribution for each at the design temperature 180°C. Comment on the validity of the data and model.

(b)   How does cycle length effect each failure mode?

(c)   Choose a relationship to model the combined effect on life of temperature and cycle length. Separately fit it by ML to the data on each failure mode.

(d)   Analyze the residuals for each failure mode.

(e)   Assess the fit of your relationship for each failure mode, using the LR test or a more general relationship.

(f)   Analytically assess whether σ is constant for each failure mode.

(g)   Write a short report for engineers on your findings.

(h)   Suggest further analyses.

(i)   Carry out (h).

(j)   Do the preceding analyses for the Class-H data of Section 5.

7.9. Power lognormal. Suppose that a standard size A_0 of a product has a lognormal life distribution with log mean μ and log standard deviation σ. That is, the reliability function is $R(t;\mu,\sigma,A_0) = \Phi\{-[\log(t)-\mu]/\sigma\}$. A size $A = \rho A_0$ of the product then has reliability function $R(t;\mu,\sigma,\rho A_0) = (\Phi\{-[\log(t)-\mu]/\sigma\})^\rho$. Derive the properties of this *power lognormal* distribution and make suitable plots of its density, reliability, and hazard functions.

7.10. Nonuniform stress. Develop the theory for Section 4 using the power lognormal distribution in place of the Weibull distribution. The resulting integrals can be evaluated only numerically.

7.11.* Specimen size. Suppose that product life is described by a linear-exponential model. Also, suppose that life depends on size A according to the model of Section 3. Suppose that the 100Pth percentile of ln life $\eta_P(x_0)$ for product of size A_0 is to be estimated at (transformed) stress level x_0. Also, suppose that specimen i may have any length A_i ($i = 1, \cdots, n$), that the test is censored at time τ, and that the highest allowed test stress is x_H.

(a) Find the optimum test plan with two stresses that minimizes the asymptotic variance assuming the A_i are specified.

(b) Derive the optimum choice of the A_i.

(c) Repeat (a) and (b) for the linear-Weibull model with β unknown.

7.12. Size compensation. A voltage-accelerated life test of electrical insulation involved specimens of two lengths L and L'. The power-Weibull model was employed. To combine and analyze data on both specimen sizes, an engineer used the following. For specimen i of length L', t_i' (the failure or censoring time) was converted to an equivalent time $t_i = t_i' (L'/L)^{1/\beta^*}$ where β^* is an estimate of the Weibull shape parameter. These equivalent times were combined with the times t_i on specimens of length L, and the model was fitted to the combined data.

(a) Comment on the pros and cons of this analysis. This analysis gets around the need for a special computer program that fits a size-effect model.

(b) Suggest other analyses, including those possible with a standard computer package.

7.13.* Turn failures eliminated. For the Class-H insulation with Turn failures eliminated, Phase and Ground failures would remain.

(a) Following the theory of Section 7, develop all the theoretical equations there for the model with two remaining failure modes.

(b) Develop the estimate and confidence limits for median life at the design temperature of 180°C.

7.14.* Unidentified modes. Assume failure modes are independent.

(a) Develop all equations for ML fitting of a model with two failure modes to data where each specimen can fail only once and the failure mode is not identified. Use an Arrhenius-lognormal model for each mode.

(b) For the Class-H data, assume that Turn failures will be eliminated. Fit the model to the first time to Phase or Ground failure for each specimen, assuming the cause is unknown.

(c) Plot the fitted model for each failure mode and the model with both modes acting. Compare with previous fits.

(d) Calculate confidence limits for quantities of interest.

(e) Comment on the relative size of confidence intervals from analyses with failure modes identified and not identified.

(f) Suggest how to calculate and plot suitable residuals for such a model.

(g) Carry out (f) and comment on the plots.

7.15. Two modes. For two competing failure modes with Arrhenius-exponential models,

(a) Calculate and plot the series-system model on Arrhenius and Weibull papers, assuming $E_1 = 0.3$ eV and $E_2 = 0.8$ eV, and $\theta_1(150°C) = \theta_2(150°C) = 5,000$ hours.

For two Arrhenius-Weibull models,

(b) Repeat (a) with $\beta_1 = 0.5$ and $\beta_2 = 2.0$.

7.16. Hazard function. Reexpress the theory of Section 4.3 in terms of cumulative hazard functions rather than reliability functions.

8

Least-Squares Comparisons
for Complete Data

Purpose. This chapter presents graphical and least squares (LS) comparisons for complete data. They are used to compare product designs, materials, suppliers, production periods, test labs, and other such populations. Also, such comparisons are used in demonstration tests to assess whether a product meets reliability specifications. A combination of graphical and analytic methods is always most informative.

Background. Needed background for this chapter includes the graphical methods of Chapter 3 and the LS methods of Chapter 4. Also, previous acquaintance with statistical hypothesis tests is helpful, as their introduction in Section 1 is brief.

Why least squares? The maximum likelihood comparisons of Chapter 9 can be used for complete data. Indeed least squares methods are a special case of maximum likelihood methods. Nevertheless, LS methods merit separate presentation because:

1. The methods are well known, and many would prefer to use these familiar methods, when they suffice. This is true for those who do such analyses and for their clients.
2. These familiar methods serve to introduce the maximum likelihood comparisons of Chapter 9, which are less familiar and more complex.
3. The methods are exact (if the assumed linear-(log)normal model is correct). That is, estimates are unbiased, and confidence limits and hypothesis tests are exact. ML methods are approximate.
4. Computer programs for the calculations are widely available.
5. Certain LS comparisons often yield good approximations when the distribution is not (log) normal. That is, such comparisons are robust.
6. The methods apply to aging-degradation data (Chapter 11).

Overview. Section 1 briefly introduces comparisons with confidence intervals and hypothesis tests. Section 2 presents graphical comparisons. Sections 3, 4, and 5 present least squares comparisons of (log) standard deviations, means, and relationships for the simple linear-(log)normal model. Section 6 extends these comparisons to multivariable relationships. This chapter treats in detail the simple linear-(log)normal model with one accelerating variable, because it is widely used. Moreover, this model avoids the complexity of least squares theory for multivariable relationships. The least squares notation follows that of Chapter 4.

1. HYPOTHESIS TESTS AND CONFIDENCE INTERVALS

The following paragraphs briefly review basics of comparisons with hypothesis tests and confidence intervals. The basics include: reasons for comparisons, models, hypotheses, actions, tests, confidence limits, significance (statistical and practical), test performance, and sample size. Introductory statistical texts discuss hypothesis testing in detail. Draper and Smith (1981) and Neter, Wasserman, and Kutner (1985) give intermediate discussions. Without such basic background, readers may find this and the next chapter difficult. Lehman (1986) presents advanced theory of hypothesis testing. Many readers can skip this section. Others may wish to return to it after reading later sections, as it is relatively general and abstract.

Reasons for comparisons. The following are some reasons for comparisons. (1) In reliability demonstration, a product must demonstrate that its reliability, mean life, failure rate, or other parameter is better than a specified value. (2) To verify engineering theory, one may check estimates of model parameters are consistent with theoretical values; for example, a Weibull shape parameter equals 1. (3) In development work, one may compare two or more designs to select one. (4) In analyzing sets of data collected over time, one wants to confirm that model parameters are not changing; this is often done before pooling data to get a more precise pooled estimate of a parameter. Such comparisons have two basic objectives. One is to **demonstrate** that a product parameter **surpasses** a requirement or that a product surpasses others. The other is to assess (a) whether an estimate of a product parameter is **consistent** with a specified value or (b) whether corresponding estimates of a number of products are **comparable** (equal). Here "parameter" means any model value, including coefficients, percentiles, and reliabilities.

Models. Below, data are assumed to be random observations described by a parametric model, which is assumed correct. In practice, one usually does not know this. Such a model must first be assessed through data plots or formal tests of fit (Chapters 4 and 5). Below, the model is assumed adequate for the intended purposes. Engineering experience and theory indicate that certain models adequately describe certain products.

Hypotheses. A **hypothesis** is a proposed statement about the value(s) of one or more model (population) parameters. Some examples are:

1. The mean life (of an exponential distribution) at the design stress level exceeds a specified value. This is common in reliability demonstration.
2. Product reliability at a specified age and stress level exceeds a given value.

3. A Weibull shape parameter equals 1. That is, product life has an exponential distribution.

4. The median of a (log) normal distribution for product 1 exceeds that for product 2 at a particular stress level. This is common in comparing designs, materials, methods of manufacture, manufacturing periods, etc.

5. The activation energies for a number of designs are equal.

6. The (log) standard deviations of a number of (log) normal distributions are equal.

7. The shape parameters of a number of Weibull distributions are equal.

8. The 10th percentiles of a number of Weibull distributions are equal (a common hypothesis in ball bearing life tests).

The **alternative** (hypothesis) is the statement that the hypothesis is not true. In contrast, the "hypothesis" above is also called the **null hypothesis**. Examples of alternatives are:

1. The mean life is below the specified value.

3. The Weibull shape parameter differs from 1 (greater or smaller).

4. The median of product 1 is below that of product 2.

5. Two or more of the true activation energies differ.

7. Two or more of the Weibull shape parameters differ.

Such a hypothesis (or its alternative) about a parameter may be **one sided**. That is, the parameter is above (below) a specified value. (1) and (2) are examples of this. (4) is a one-sided example concerning two parameters. Also, a hypothesis (or alternative) may be **two sided**. That is, (a) a parameter has a specified value, or (b) parameters of different populations are equal. (3) and (5) through (8) are examples of two-sided (or **equality**) hypotheses (and alternatives). In practice, one must decide whether a one- or two-sided hypothesis is appropriate to the application. The choice is determined by the practical consequences of the true parameter value(s).

Actions. If the hypothesis is true, the engineer wants to take one course of action. If the alternative is true, the engineer wants to take another, depending on the parameter values. Some examples are:

1. In reliability demonstration, a product with a mean life that "exceeds a specified mean" (the hypothesis) is accepted by the customer. Otherwise, the customer rejects the product (the alternative action). Then the product must be redesigned, abandoned, or the contract renegotiated.

4. If the median of product 1 exceeds that of the standard product 2, then product 1 replaces 2 (the hypothesis action); otherwise, product 2 is retained (the alternative action).

5. If the activation energies of designs are equal (the hypothesis), their data may be pooled to estimate the common activation energy. Otherwise, estimate their activation energies separately (the alternative action).

7. If the Weibull shape parameters are equal (the hypothesis), then the data can be pooled to estimate the common value. Otherwise, use a separate estimate for each product (the alternative action). Such pooling is often

considered for data collected on the same product under different conditions or from different production periods.

Hypothesis test. Of course, model parameter values are not known, and one must take actions based on data. One wants convincing evidence in the data that an action is appropriate. For example, an observed difference between a sample estimate for a parameter and a specified value should be greater than the normal random variation in the estimate; then it is convincing that the observed difference is due to a real difference between the true and specified parameter values. A **statistical test** involves a **test statistic,** which is a suitable function of the data. Such statistics include sample means, medians, and *t* statistics. If the hypothesized parameter values are the true ones, the statistic has a known ("null") sampling distribution. For true parameter values under the alternative, the statistic tends to have larger (or smaller) values. An observed value of the statistic, if unusual (in an extreme tail of its null distribution), is evidence that the hypothesis is false, and an alternative action is appropriate. If the observed statistic is beyond the upper (or lower) 5% point, it is said to be **statistically significant,** that is, convincing evidence. If beyond the (0.1%) 1% point, it is said to be **(very) highly statistically significant,** that is, more convincing evidence. The percentage of the null distribution beyond the observed statistic is called the **significance level** or *p* **value** of the statistic. The smaller the *p* value, the more convincing the evidence that the alternative is true.

Confidence intervals. Most comparisons are best made with confidence intervals. Most such intervals are equivalent to but more informative than a corresponding hypothesis test. Such intervals can (1) indicate that the data are consistent with specified parameter values or (2) demonstrate that the data surpass (or fall short of) specified parameter values as follows:

1. A $100\gamma\%$ confidence interval for a parameter is **consistent** with a specified parameter value if the interval encloses that value. If such an interval does not enclose that value, then the corresponding hypothesis test shows a statistically significant difference at the $100(1-\gamma)\%$ level. For example, if a confidence interval for a Weibull shape parameter encloses 1, the data are consistent with an assumed exponential life distribution (for this way of assessing adequacy of the exponential distribution).

2. A confidence interval for a parameter **demonstrates** a specified parameter value (or **better**) if the interval encloses only "better" parameter values. For example, a specified mean θ^* of an exponential life distribution is "demonstrated with $100\gamma\%$ confidence" if the $100\gamma\%$ lower confidence limit $\underline{\theta}$ for the true θ is above θ^*.

3. A confidence interval for the difference (or ratio) of corresponding parameters of two products is consistent with their equality if the interval encloses zero (one). Similarly, such an interval that does not enclose zero (one) **"demonstrates"** that one product parameter exceeds the other. For example, one wants convincing evidence that the mean life of a new design surpasses that of the standard design before adopting the new one.

4. Simultaneous confidence intervals for differences (or ratios) of corresponding parameters of K products are consistent with their equality, if all such intervals enclose zero (one).

Significance. It is important to distinguish between practical and statistical significance. Observed sample differences are **statistically significant** if they are greater than would normally be observed by chance. That is, such observed differences are large and convincing compared to the normal random variation in the data. Hence, they are presumed due to a real difference in the products. Observed differences are **practically significant** if they are large enough to be important in real life. Observed differences can be statistically significant but not practically significant; that is, they are convincing but so small that they have no practical value. This can happen for large samples that reveal even small differences. Then the corresponding true product parameters, although convincingly different, are equal for practical purposes. Observed differences can be practically significant (important) but not statistically significant (convincing). This can happen when sample sizes are small. Then a larger sample is needed to resolve whether the observed important differences are real and not due just to normal random variation in the data. In practice, one needs observed differences that are **both** statistically and practically significant, that is, convincing and important. Confidence intervals are most informative for judging both practical and statistical significance. Hypothesis tests merely indicate statistical significance. A confidence interval for a difference should ideally be smaller than an important practical difference. If it is not, one needs more data or a better test plan to discriminate adequately. Mace (1974) shows how to choose sample size to obtain confidence intervals of desired length.

Performance and sample size. The performance of a confidence interval is usually judged by its "typical" length. That of a hypothesis test is judged by its *operating characteristic (OC) function*, defined in most statistics texts. Such performance, of course, depends on the assumed model(s), the parameter(s) compared, the sample statistic(s) used, the sample size(s), and test plan(s). Mace (1974) gives methods for choosing sample sizes for confidence intervals. Cohen (1988), IDEA WORKS (1988), Odeh and Fox (1975), Brush (1988), Kraemer and Thieman (1987), Bowker and Lieberman (1972), and Lehmann (1986) give methods for choosing sample sizes for hypothesis tests. This chapter assumes that the sample size has been determined. In practice, it is often determined by nonstatistical considerations such as limited budget, time, and number of specimens or test fixtures.

2. GRAPHICAL COMPARISONS

This section presents simple graphical comparisons for distribution percentiles at a stress level, relationships over a range of stress, the slope coefficients of the relationships, and (log) standard deviations. Chapter 3 provides needed background on graphical methods.

Table 2.1. Life Data (Hours) on Three Insulations

Insulation System 1

Hours to Failure				Plotting Positions	
200°C	225°C	250°C	Rank i	$100(i\text{-}0.5)/n$	$100i/(n+1)$
1176	624	204	1	10	16.7
1512	624	228	2	30	33.3
1512	624	252	3	50	50.0
1512	816	300	4	70	66.7
3528	1296	324	5	90	83.3

Insulation System 2

Hours to Failure				Plotting Positions	
200°C	225°C	250°C	Rank i	$100(i\text{-}0.5)/n$	$100i/(n+1)$
2520	816	300	1	10	16.7
2856	912	324	2	30	33.3
3192	1296	372	3	50	50.0
3192	1392	372	4	70	66.7
3258	1488	444	5	90	83.3

Insulation System 3

Hours to Failure				Plotting Positions	
200°C	225°C	250°C	Rank i	$100(i\text{-}0.5)/n$	$100i/(n+1)$
3528	720	252	1	16.7	25
3528	1296	300	2	50.0	50
3528	1488	324	3	83.3	75

Graphical methods are subjective. Sometimes there is a question whether such subjectively observed differences are convincing. Then use the analytical methods in following sections.

2.1. Data and Model

Data. Three sets of data in Table 2.1 from three types of motor insulation illustrate the graphical methods. The data are the times (hours) to failure of specimens at test temperatures of 200, 225, and 250°C. A failure time is the midpoint of the inspection period in which the failure occurred. Such rounding of the data is negligible here and is ignored. The test purpose was to compare the median lives (a) at the design temperature of 200°C and (b) over the range of test temperatures, since the aerospace application sometimes involves temperatures up to 250°C.

Model. The Arrhenius-lognormal model is used. However, the graphical methods apply to other relationships and distributions.

2.2. Compare Percentiles at a Stress Level

The following graphically compares a chosen percentile of the products at a stress level. For each product, make a separate relationship plot of the data as described in Chapter 3. In each plot fit a line for the percentile. Then compare those lines at the stress level of interest. For example, Figure 2.1 shows the data on the three insulations on Arrhenius paper. Comparison of the three median lines at 200°C indicates that insulations 2 and 3 have comparable medians, and insulation 1 has a slightly lower one. Usually such plots are best on separate papers. Then stack up the papers and hold them up to the light to compare them. Transparencies of plots (especially in different colors) make this comparison easier and clearer.

2.3. Compare Relationships

The following method compares the life-stress relationships of products. In particular, it compares the intercept and slope coefficients.

To compare the entire relationships, make a relationship plot of the data for each product as described in Chapter 3. Fit a line to each plot. Stack them up and hold them up to the light. If the fitted lines roughly coincide, then the relationships are the same for practical purposes. Figure 2.1 shows that the three insulations have roughly the same temperature dependence (slope) from 200 to 250°C, but insulation 1 has slightly lower life. The analytic comparison in Section 5 indicates that three intercepts differ statistically significantly (convincingly).

2.4. Compare Slope Coefficients

For many products, the slope coefficient of the life-stress relationship has physical meaning. For example, the slope of the Arrhenius relationship is proportional to the activation energy of the failure process. The following graphical method compares slope coefficients. Equal slopes may result when products contain the same materials but have different geometry or usage. Seldom used, comparisons of intercept coefficients are not given here.

Make relationship plots on separate papers. Stack them up and hold them up to the light to compare the slopes. Figure 2.1 suggests that the three insulations have slightly different slopes (activation energies). Insulation 1 has a low slope. The analytic method of Section 5 is needed to assess whether they differ statistically significantly. They do not.

2.5. Compare Distribution Spreads

Compare the spreads of the life distributions of products as follows. Do this before using analytic methods, since they assume that all spreads are equal. The following method assesses equality of log standard deviations for the example. This method also applies to Weibull shape parameters.

Probability plots. For each product, make a probability plot of the data at each stress level as described in Chapter 3. Fit parallel lines to the data for a product. The spread of (log) life at a stress level corresponds to the slope

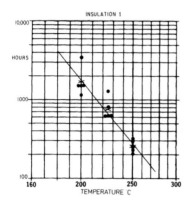

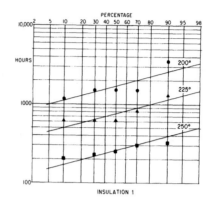

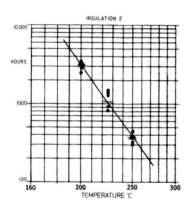

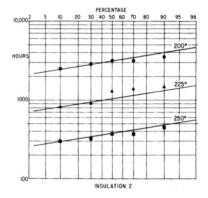

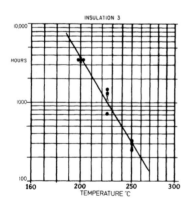

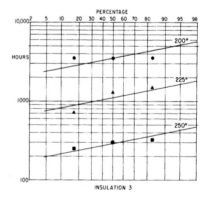

Figure 2.1.   Arrhenius plots for three insulations.

Figure 2.2.   Lognormal plots for three insulations.

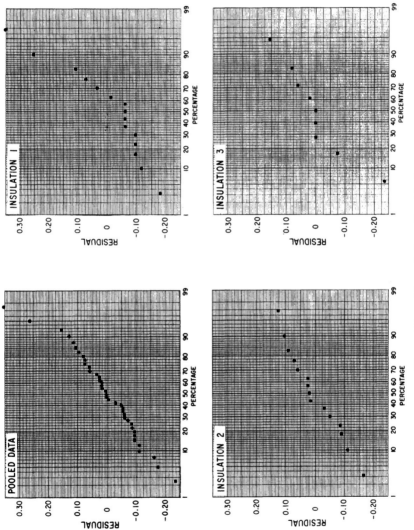

Figure 2.3.   Normal plots of residuals.

of the plotted data. Then stack up the plots and hold them up to the light. Nearly parallel plots indicate that the products have comparable spread of (log) life. Such samples usually have few specimens. So the slopes of the fitted lines for the different test stresses and products usually differ much, even when the true product spreads are all equal. Thus, only pronounced differences in the slopes are evidence of real differences among products.

Figure 2.2 shows lognormal plots for the three insulations. The samples are small, and the slopes are roughly equal. This is consistent with equal true log standard deviations of the insulations.

Residual plots. Probability plots of the (log) residuals are a more sensitive means of comparing (log) spreads. For a product, calculate and pool such residuals and plot them as described in Chapter 4. Compare the slopes of such plots for each product, since the slopes correspond to the spread in log life. Figure 2.3 shows such plots for the three insulations on normal probability papers. The slopes are roughly comparable, consistent with equal log standard deviations.

3. COMPARE LOG STANDARD DEVIATIONS

Purpose. Compare the (log) standard deviations (σ's) of different products, because:

1. The σ's are a measure of the spread in log life of the products.
2. The σ's indicate the behavior of the product failure rates.
3. Other statistical methods that compare products assume that the true product σ's are all equal.

The following comparisons assume that the σ for each product does not depend on stress. Methods for checking this assumption appear in Chapter 4. The following methods compare the σ's of one, two, and K products. Nelson (1982, Chap. 10) gives further methods and examples for such comparisons, particularly for simultaneous comparisons.

Robustness. All comparisons of standard deviations below are valid **only** for (log) normal distributions. If the true distribution is far from (log) normal, the comparisons are crude no matter how large the samples or degrees of freedom. Methods for checking (log)normality appear in Chapter 4. Also, it is important to note that the estimate s' for σ can be biased too large and be misleading if the assumed relationship does not adequately fit the data.

Example. The least squares comparisons are illustrated with the data on three insulations in Section 2. The Arrhenius-lognormal model is used for each insulation. The linear relationship for the mean log life $\mu_k(x)$ of insulation k is written as

$$\mu_k(x) = \alpha_k + \beta_k x;$$

Table 3.1. Summary of the LS Calculations for the Three Insulations

	INSULATION			Pooled
	1	2	3	Data
n	15	15	9	39
$\bar{y}_j$ (2.2.1)				
200°	3.23134	3.48258	3.54750	3.40093
225°	2.88198	3.06010	3.04750	2.98868
250°	2.41130	2.55520	2.46300	2.95606
s_j (2.2.2)				
200°	0.1832	0.0558	0	–
225°	0.1384	0.1166	0.1674	–
250°	0.0829	0.0653	0.0559	–
$\bar{x}$ (2.2.4)	2.01133	2.01133	2.01133	2.01133
$\bar{y}$ (2.2.5)	2.85154	3.03263	3.01933	2.95606
S_{yy} (2.2.6)	1.93144	2.23969	1.83009	6.32192
S_{xx} (2.2.7)	0.102093	0.102093	0.061256	0.265443
S_{xy} (2.2.8)	0.413109	0.467640	0.328181	1.20893
b (2.2.9)	4.04639	4.58052	5.35753	4.55439
a (2.2.10)	−5.29709	−6.18032	−7.75645	−6.20434
s (2.2.11)	0.1408	0.0836	0.1019	–
$\nu = n - 3$	12	12	6	–
s' (2.2.12)	0.1414	0.0867	0.1013	0.1485
$\nu' = n - 2$	13	13	7	37
$m(x_o)$ (2.2.13)				
200°	3.25697	3.50289	3.56937	3.42365
225°	2.82805	3.01736	3.00147	2.94088
250°	2.43960	2.57763	2.48715	2.50366
Antilog[$m(x_o)$]				
200°	1,807	3,183	3,710	2,652
225°	673	1,041	1,003	873
250°	275	378	307	319

here $x = 1000/T$ where T is the absolute temperature in °K. The corresponding log standard deviation is σ_k. Table 3.1 summarizes the least squares calculations for the log life data for each insulation and for the pooled data. Equations numbers in the table are for equations in Chapter 4. In practice, such results are calculated with a standard least squares program.

3.1. Compare One σ

Interval. To compare a sample (log) standard deviation s (or s') with a specified value σ_0, use the confidence interval (2.3.6) for σ from Chapter 4. If a $100\gamma\%$ confidence interval encloses σ_0, then s is consistent with σ_0. Otherwise, s differs convincingly (statistically significantly) from σ_0 at the $100(1-\gamma)\%$ level. Such an interval may be one- or two-sided. Such a comparison is rare in practice.

3.2. Compare Two σ's

Interval. Suppose that the estimates of σ_1 and σ_2 for two products with (log) normal life distributions are s_1 and s_2, and they have ν_1 and ν_2 degrees of freedom. A $100\gamma\%$ confidence interval for the ratio σ_1/σ_2 has lower and upper limits

$$(s_1/s_2)/\{F[(1+\gamma)/2;\nu_1,\nu_2]\}^{1/2}, \quad (s_1/s_2)\times\{F[(1+\gamma)/2;\nu_2,\nu_1]\}^{1/2}; \quad (3.1)$$

here $F[(1+\gamma)/2;\nu_1,\nu_2]$ is the $(1+\gamma)/2$ point of the F distribution with ν_1 degrees of freedom in the numerator and ν_2 in the denominator. Such F values are tabled in Appendix A6. ν_1 and ν_2 are reversed in the upper and lower limits. If this interval encloses 1, then the sample (log) standard deviations do not differ convincingly (statistically significantly) at the $100(1-\gamma)\%$ level. Otherwise, they differ convincingly at the $100(1-\gamma)\%$ level. One must determine whether a convincing difference is large enough to be important in practice. A convincing difference indicates that the statistical methods in following sections may be inaccurate.

Example. For insulations 2 and 3, the observed ratio (Table 3.1) is $(0.0836/0.1019) = 0.820$; here the numerator has 12 degrees of freedom and the denominator has 6. The 95% confidence interval for the true ratio has limits $0.820/(5.37)^{1/2} = 0.35$ and $0.820(3.73)^{1/2} = 1.58$; here 5.37 (3.73) is the 97.5 F-percentile with 12 (6) degrees of freedom in the numerator and 6 (12) in the denominator. This interval does enclose 1. So the two log standard deviations do not convincingly differ.

The following table shows the estimates and 95% confidence limits for such ratios for each pair of insulations. All of these intervals enclose 1. Thus no two log standard deviations differ convincingly.

Ratio	Estimate	95% Conf. Interval	
σ_1/σ_2	1.68	0.93	3.04
σ_2/σ_3	0.82	0.35	1.58
σ_3/σ_1	0.72	0.38	1.68

3.3. Compare K σ's

Bartlett's test. The following hypothesis test compares the σ_k of K products with (log) normal life distributions. Suppose that, for product k, the estimate for σ_k is s_k, which has ν_k degrees of freedom. Either estimate s_k or s_k' could be used. Calculate the *pooled sample standard deviation*

$$s^* = [(\nu_1 s_1^2 + \cdots + \nu_K s_K^2)/\nu^*]^{1/2}; \quad (3.2)$$

here $\nu^* = \nu_1 + \cdots + \nu_K$ is its pooled number of degrees of freedom. Calculate *Bartlett's test statistic*:

$$Q = C\{\nu^*\log(s^*) - [\nu_1\log(s_1) + \cdots + \nu_K\log(s_K)]\}; \quad (3.3)$$

here base 10 logs are used, and

$$C = 4.605/\left\{1 + \frac{1}{3(K-1)}\left[\frac{1}{\nu_1} + \cdots + \frac{1}{\nu_K} - \frac{1}{\nu^*}\right]\right\}.$$

The approximate level α test for equality of the σ_k is

1) If $Q \leq \chi^2(1-\alpha;K-1)$, the s_k do *not* differ convincingly (statistically significantly) at the $100\alpha\%$ level.

2) If $Q > \chi^2(1-\alpha;K-1)$, they differ statistically significantly at the $100\alpha\%$ level.

Here $\chi^2(1-\alpha;K-1)$ is the $100(1-\alpha)$th chi-square percentile with $(K-1)$ degrees of freedom. This is *Bartlett's test* for homogeneity of variance. The same test appears in Chapter 4, where its purpose is different.

If the s_k differ convincingly, examine them and their confidence limits to determine how they differ. Take those differences into account to interpret the data. For example, then one might not act on marginally significant results from methods in later sections.

Example. For the three insulations (Table 3.1), the calculations are:

$$s^* = \{[12(0.1408)^2 + 12(0.0836)^2 + 6(0.1019)^2]/30\}^{1/2} = 0.1132,$$

$$C = 4.605/\left\{1 + \frac{1}{3(3-1)}\left[\frac{1}{12} + \frac{1}{12} + \frac{1}{6} - \frac{1}{30}\right]\right\} = 4.386, \text{ and}$$

$$Q = 4.386\{30\log(0.1132) - [12\log(0.1408)$$
$$+ 12\log(0.0836) + 6\log(0.1019)]\} = 3.13.$$

The chi-square distribution for Q has $K - 1 = 2$ degrees of freedom. Since $Q = 3.13 < 5.99 = \chi^2(0.95;2)$, the three s_k do not differ convincingly at the 5% level. So comparisons in following sections appear accurate enough.

Maximum F ratio. Sometimes the estimates s_k have nearly the same number ν_k of degrees of freedom. If so, they can be simultaneously compared graphically with the maximum F ratio, as described by Nelson (1982, Chap. 10). If the ν_k differ much, the s_k can be compared by means of the Bonferroni inequality, as described by Nelson (1982, Chap. 10). Such simultaneous pairwise comparisons are described in general by Miller (1966) and Hochberg and Tamhane (1987).

4. COMPARE (LOG) MEANS

The following analytic methods compare mean (log) lives of one, two, and K products at a stress level. Such comparisons often help decide which product is best. Nelson (1982, Chap. 10) gives further methods and examples for such comparisons, particularly for simultaneous comparisons.

Robustness. The following methods often are good approximations if the life distribution is not (log) normal. This is so provided the sample size is large enough so the sample (log) means have sampling distributions that are near normal. Such normality of sample (log) means is a consequence of the central limit theorem. Bootstrapping of sample estimates can be used to assess normality of such a sampling distribution. Also, note that the assumed relationship, if not adequate, may make the comparisons crude. The following comparisons assume that the true product σ's are all equal. However, the comparisons are not sensitive to modest differences in the σ's. In particular, if the sample log standard deviations do not differ statistically significantly, then one may assume that the methods are satisfactory. Moreover, one can apply the ML methods of Chapter 9 when the true σ's differ.

Percentiles. Below it is assumed that the products σ's are equal. Under that assumption, if product means are equal, then corresponding percentiles are equal. Thus, the following comparisons of means are also comparisons of percentiles. If the product σ's differ, the percentiles must be compared with the maximum likelihood methods of Chapter 9.

4.1. Compare One (Log) Mean

Interval. Suppose $m(x_0)$ is the LS estimate of the log mean at a specified (transformed) stress level x_0. To compare it with a specified value μ_0, use the confidence interval (2.3.3) of Chapter 4. If a $100\gamma\%$ confidence interval encloses μ_0, then $m(x_0)$ is consistent with μ_0. Otherwise, $m(x_0)$ differs convincingly (statistically significantly) from μ_0 at the $100(1-\gamma)\%$ level. Such an interval may be one- or two-sided. In a *demonstration test*, use a one-sided lower limit. The limit must be above μ_0 for the product to "demonstrate a life μ_0 with $100\gamma\%$ confidence."

4.2. Compare Two (Log) Means

Interval. Denote the LS estimates of two (log) means $\mu_1(x_0)$ and $\mu_2(x_0)$ at the stress level x_0 by m_1 and m_2. Denote the sample (log) standard deviations by s_1 and s_2, which have ν_1 and ν_2 degrees of freedom. A two-sided $100\gamma\%$ confidence interval for the difference $\mu_1(x_0)-\mu_2(x_0)$ is

$$(m_1-m_2) \pm t[(1+\gamma)/2;\nu^*]\{(1/n_1)+(x_0-\bar{x}_1)^2 S_{xx1}^{-1} \tag{4.1}$$
$$+(1/n_2)+(x_0-\bar{x}_2)^2 S_{xx2}^{-1}\}^{1/2}s^*.$$

Here $t[(1+\gamma)/2;\nu^*]$ is the $100(1+\gamma)/2$th t-percentile with $\nu^* = \nu_1+\nu_2$ degrees of freedom. The sums of squares S_{xx1} and S_{xx2} are each defined by (2.2.7) of Chapter 4; $\bar{x}_1$ and $\bar{x}_2$ are the two sample averages; n_1 and n_2 are the two sample sizes. Also,

$$s^* = \{(\nu_1 s_1^2 + \nu_2 s_2^2)/\nu^*\}^{1/2} \tag{4.2}$$

is the *pooled estimate of* σ and has ν^* degrees of freedom.

If this interval encloses zero, then m_1 and m_2 do not differ convincingly at the $100(1-\gamma)\%$ significance level. If it does not enclose zero, then m_1 and m_2 differ convincingly at the $100(1-\gamma)\%$ significance level.

Example. For insulations 2 and 3 in Table 3.1, $s^* = \{[12(0.0836)^2 + 6(0.1019)^2]/18\}^{1/2} = 0.0912$, which has $\nu^* = 12+6 = 18$ degrees of freedom. The 95% confidence interval for the difference in true means at 200°C is

$$(3.503-3.569) \pm 2.101[(1/15)+(2.114-2.011)^2 0.1021^{-1}$$

$$+ (1/9)+(2.114-2.011)^2 0.0613^{-1}]^{1/2} 0.0912 = -0.066 \pm 0.129$$

or -0.195 to 0.063. Here $t[(1+0.95)/2;18] = 2.101$ is the 97.5th t-percentile with 18 degrees of freedom. This 95% confidence interval encloses 0. So there is no convincing evidence at a 5% significance level that the two true means at 200°C differ. Thus the median lives of insulations 2 and 3 appear comparable at 200°C.

Such 95% intervals for each pair of insulations means appear below.

Insulations	Estimate and 95% Conf. Interval
1 − 2	−0.246 ± 0.136
2 − 3	−0.066 ± 0.129
3 − 1	0.312 ± 0.175

The first and third intervals do not enclose 0. This means that insulations 2 and 3 are convincingly better than insulation 1. The middle interval encloses 0. This means that insulations 2 and 3 do not convincingly differ. Examine such convincing differences for engineering importance.

The example is unusual in that there are data at the temperature where the comparison is made. Hahn and Schmee (1980) discuss some advantages of using just the data at that temperature. For example, one need not specify an assumed (and possibly inaccurate) relationship.

4.3. Compare K Log Means

The following method simultaneously compares the mean (log) lives of K products at a stress level x_0. Suppose, for product k, the LS estimate of the mean at x_0 is m_k. Also, suppose the estimate of σ is s_k and has ν_k degrees of freedom. Calculate the *pooled sample mean* at x_0:

$$m^* = (N_1 m_1 + \cdots + N_K m_K)/(N_1 + \cdots + N_K). \qquad (4.3)$$

Here

$$N_k = 1/\{(1/n_k) + [(x_0 - \bar{x}_k)^2/S_{xxk}]\}$$

is the "*equivalent sample size*" at x_0. It is not an integer. The notation follows Chapter 4. Calculate the *pooled sample log standard deviation*

$$s* = \{[\nu_1 s_1^2 + \cdots + \nu_K s_K^2]/\nu^*\}^{1/2}. \tag{4.4}$$

This has $\nu^* = \nu_1 + \cdots + \nu_K$ degrees of freedom. Calculate the *sum of squares for the means*:

$$M = N_1(m_1 - m^*)^2 + \cdots + N_K(m_K - m^*)^2, \tag{4.5}$$

which has $(K-1)$ degrees of freedom.

F test. Calculate the *F statistic for the means*

$$F = [M/(K-1)]/s^{*2}. \tag{4.6}$$

The level α test for equality of the true (log) means is:

1) If $F \leq F(1-\alpha; K-1, \nu^*)$, the (log) means do *not* differ convincingly (statistically significantly) at the $100\alpha\%$ level.

2) If $F > F(1-\alpha; K-1, \nu^*)$, the means differ convincingly at the $100\alpha\%$ level.

Here $F(1-\alpha; K-1, \nu^*)$ is the $1-\alpha$ point of the F distribution with $(K-1)$ degrees of freedom in the numerator and ν^* in the denominator. F values are tabled in Appendix A6. This test is like that for one-way analysis of variance with unequal sample sizes N_k.

If the means differ convincingly, determine how they differ and whether the differences are important in practice. The confidence interval (4.1) for pairs of means often helps determine which differences are convincing and important. Also, the confidence intervals (2.3.3) of Chapter 4 for individual means may yield insight. Plot such estimates and intervals for insights.

Example. For the three insulations (Table 3.1), the calculations to compare the log means at 200°C are

$N_1 = N_2 = 1/\{(1/15) + [(2.114 - 2.011)^2/0.1021]\} = 5.863$,

$N_3 = 1/\{(1/9) + [(2.114 - 2.011)^2/0.0613]\} = 3.519$,

$m^* = [5.863(3.257) + 5.863(3.503) + 3.519(3.569)]/[5.863 + 5.863 + 3.519]$

$\quad = 3.424$,

$\nu^* = 12 + 12 + 6 = 30$,

$s^{*2} = [12(0.1408)^2 + 12(0.0836)^2 + 6(0.1019)^2]/30 = 0.01281$,

$M = 5.863(3.257 - 3.424)^2 + 5.863(3.503 - 3.424)^2 + 3.519(3.569 - 3.424)^2$

$\quad = 0.275$.

M has $K - 1 = 3 - 1 = 2$ degrees of freedom. Then $F = [0.275/(3-1)]/0.01281 = 10.7$. Since $F = 10.7 > 8.77 = F(0.999; 2, 30)$, the sample means differ very highly significantly (0.1% level). The previous comparison of pairs of means indicates that the log means of insulations 2 and 3 at 200°C are comparable and convincingly higher than that of insulation 1. One must then decide whether such convincing differences are large enough to be important.

Tukey's comparison. Sometimes the estimates of m_k of the means have nearly the same true standard errors, and the s_k have nearly the same number ν_k of degrees of freedom. Then the m_k can be simultaneously compared with Tukey's method, as described by Nelson (1982, Chap. 10). Sometimes the true standard errors or the ν_k differ much. Then the m_k can be simultaneously compared using the Bonferroni inequality, as described by Nelson (1982, Chap. 10). Such simultaneous pairwise comparisons are described in general by Miller (1966) and Hochberg and Tamhane (1987).

5. COMPARE SIMPLE RELATIONSHIPS

Presented below are product comparisons for simple linear relationships (one accelerating variable) and for slope and intercept coefficients. The discussion of robustness in Section 4 applies here.

5.1. Compare Slope Coefficients

Purpose. For many products, the slope coefficient β has physical meaning. For example, in the Arrhenius relationship, β is proportional to the activation energy of a failure mechanism. Products made of the same materials, but differing with respect to geometry or usage, may have the same β value. If so, one can pool their data to estimate β more accurately. The following methods compare one, two, and K slope coefficients.

One slope. Suppose b is the LS estimate for a slope coefficient. To compare it with a specified value β_0, use the confidence interval (2.3.12) of Chapter 4 for β. If a $100\gamma\%$ confidence interval encloses β_0, then b is consistent with β_0. Otherwise, b differs convincingly (statistically significantly) from β_0 at the $100(1-\gamma)\%$ level. Such an interval may be one- or two-sided.

Two slopes. Suppose that the LS estimates of β_1 and β_2 are b_1 and b_2. Also, suppose the corresponding estimates of σ are s_1 and s_2 with ν_1 and ν_2 degrees of freedom. A $100\gamma\%$ confidence interval for $(\beta_1 - \beta_2)$ is

$$[b_1 - b_2] \pm t[(1+\gamma)/2; \nu^*][(1/S_{xx1}) + (1/S_{xx2})]^{1/2} s^*. \tag{5.1}$$

Here $t[(1+\gamma)/2; \nu^*]$ is the $100(1+\gamma)/2$th t-percentile with $\nu^* = \nu_1 + \nu_2$ degrees of freedom. S_{xx1} and S_{xx2} for the two samples are defined in (2.2.7) of Chapter 4.

$$s^* = [(\nu_1 s_1^2 + \nu_2 s_2^2)/\nu^*]^{1/2} \tag{5.2}$$

is the *pooled estimate of* σ and has ν^* degrees of freedom.

If this interval encloses 0, b_1 and b_2 do *not* differ convincingly (statistically significantly) at a $100(1-\gamma)\%$ level. Otherwise, they differ convincingly at that level. Determine whether a convincing difference is large enough to be important in practice. Also, the interval should be narrow enough to detect a difference $\beta_1 - \beta_2$ that is important in practice. If not, use Chapter 6 to choose a larger sample or better test plan before the test is run.

Example. For insulations 2 and 3 (Table 3.1), $s^* = 0.0912$ and $\nu^* = 12+6 = 18$. The 95% confidence interval is

$$(4.581-5.358)\pm2.101[(1/0.1021)+(1/0.0613)]^{1/2}0.0912 = -0.777\pm0.978;$$

here $t[(1+0.95)/2;18] = 2.101$. This interval encloses 0; thus the two coefficients do not differ convincingly at the 5% level. One must decide whether the uncertainty ±0.978 is small enough for practical purposes. The 95% confidence intervals for the slopes of each pair of insulations follow.

	95% Conf. Interval	
$\beta_1 - \beta_2$	-0.535	± 1.074
$\beta_2 - \beta_3$	-0.777	± 0.978
$\beta_3 - \beta_1$	1.312	± 1.385

Each interval encloses 0. Thus the coefficients of each pair of insulations do not differ convincingly. It is useful to plot the differences and their confidence limits.

K slopes. The following method simultaneously compares slope coefficients of K products. Suppose that, for product k, b_k is the LS estimate of β_k. Also, suppose the estimate of σ is s_k and has ν_k degrees of freedom. Calculate the *pooled slope coefficient*

$$b^* = [b_1 S_{xx1} + \cdots + b_K S_{xxK})]/[S_{xx1} + \cdots + S_{xxK}]; \qquad (5.3)$$

here the notation follows Chapter 4. Calculate the *pooled sample (log) standard deviation* as

$$s^* = [(\nu_1 s_1^2 + \cdots + \nu_K s_K^2)/\nu^*]^{1/2}; \qquad (5.4)$$

$\nu^* = \nu_1 + \cdots + \nu_K$ is its number of degrees of freedom. Calculate the *sum of squares for the slope coefficients* as

$$B = S_{xx1}[b_1 - b^*]^2 + \cdots + S_{xxK}[b_K - b^*]^2. \qquad (5.5)$$

F test. Calculate the F *statistic for the slopes*

$$F = [B/(K-1)]/s^{*2}. \qquad (5.6)$$

The level α test for equality of the slopes is:

1) If $F \leq F(1-\alpha;K-1, \nu^*)$, the sample slope coefficients do *not* differ convincingly (statistically significantly) at the $100\alpha\%$ level.

2) If $F > F(1-\alpha;K-1, \nu^*)$, they differ convincingly at that level.

Here $F(1-\alpha;K-1, \nu^*)$ is the $1-\alpha$ point of the F distribution with $(K-1)$ degrees of freedom in the numerator and ν^* in the denominator. This test is like that for a one-way analysis of variance with unequal sample sizes S_{xxk}.

If the slopes b_k differ convincingly, examine them to determine how they differ and whether the differences are important. Confidence intervals (2.3.12) of Chapter 4 for the β_k help determine this. Also, the confidence intervals (5.1) for comparing pairs of slopes help.

Example. For the three insulations (Table 3.1), the calculations to compare the slopes are

$b* = [4.046(0.1021) + 4.581(0.1021) + 5.358(0.0613)]/[0.1021 + 0.1021 + 0.0613]$

$\quad = 4.554,$

$\nu* = 12 + 12 + 6 = 30,$

$s* = \{[12(0.1408)^2 + 12(0.0836)^2 + 6(0.1019)^2]/30\}^{1/2} = 0.1132,$

$B = 0.1021(4.046 - 4.554)^2 + 0.1021(4.581 - 4.554)^2 + 0.0613(5.358 - 4.554)^2$

$\quad = 0.0659,$

$F = [0.0659/(3 - 1)]/(0.1132)^2 = 2.57.$

Since $F = 2.57 < 3.32 = F(0.95; 2, 30)$, the three slopes do not differ convincingly at the 5% level.

5.2. Compare Intercept Coefficients

Methods for comparing the intercept coefficients α_k of products are not given here. Such comparisons are seldom used for accelerated tests since the intercept rarely have a useful physical meaning. The comparisons are the same as those in Section 4 for (log) means at $x_0 = 0$.

5.3. Simultaneously Compare Slopes and Intercepts

The following test compares the simple relationships of $K \geq 2$ products. It tests for equality of all intercepts ($\alpha_1 = \alpha_2 = \cdots = \alpha_K$) *and simultaneously* for equality of all slopes ($\beta_1 = \beta_2 = \cdots = \beta_K$). The discussion of robustness in Section 4 applies here. The notation follows Chapter 4.

Product calculations. For product k, suppose the LS coefficient estimates are a_k and b_k, based on n_k specimens. Calculate the *sum of squares about the relationship* for product k as

$$A_k = S_{yyk} - b_k S_{xyk} = (n_k - 2)s_k'^2. \tag{5.7}$$

Here $s_k'^2$ is the standard deviation about the fitted relationship for product k. Most regression programs give s_k'. Calculate the *total sum of squares about the relationships*

$$A = A_1 + \cdots + A_K = (n_1 - 2)s_1'^2 + \cdots + (n_K - 2)s_K'^2. \tag{5.8}$$

Pooled calculations. Pool the data on all K products, and treat the data as a single sample of $n = n_1 + \cdots + n_K$ specimens. For the pooled data, calculate the grand sample averages $\bar{x}$ and $\bar{y}$ and the grand sums of squared deviations S_{yy}, S_{xx}, and S_{xy}, as in Table 3.1. Calculate the *pooled coefficient estimates*

$$b^* = S_{xy}/S_{xx}, \quad a^* = \bar{y} - b^*\bar{x}. \tag{5.9}$$

Calculate the *pooled sum of squares about the relationship*

$$A^* = S_{yy} - b^*S_{xy} = (n-2)s'^2. \tag{5.10}$$

Here s' is the standard deviation about the fitted relationship for the pooled data. Most regression programs give s'. Table 3.1 shows all such calculations for the pooled data on the three insulations.

F test. Calculate the *F statistic for equality of the relationships*

$$F = [(A^* - A)/(2K - 2)]/[A/(n - 2K)]. \tag{5.11}$$

Carry extra significant figures (say, six) in all preceding calculations. This helps assure that the F value is accurate to three figures. The level α test for equality of the relationships is

1) If $F \leq F(1-\alpha;2K-2, n-2K)$, the relationships do not differ convincingly (statistically significantly) at the $100\alpha\%$ level.

2) If $F > F(1-\alpha;2K-2, n-2K)$, they differ convincingly at the $100\alpha\%$ level.

Here $F(1-\alpha;2K-2, n-2K)$ is the $1-\alpha$ point of the F distribution with $(2K-2)$ degrees of freedom in the numerator and $(n-2K)$ in the denominator. If the relationships differ significantly, examine plots of the a_k and b_k and their confidence intervals. Determine how the coefficients differ and whether the differences are important in practice.

Example. For the three insulations, compare their relationships as follows. This assesses whether the insulations have comparable mean log life over operating temperatures from 200 to 250°C. The calculations are

$A_1 = 1.93144 - 4.04639(0.413109) = 0.259840,$

$A_2 = 2.23969 - 4.58052(0.467640) = 0.097656,$

$A_3 = 1.83009 - 5.35753(0.328181) = 0.071850,$

$A = 0.259840 + 0.097656 + 0.071850 = 0.429346,$

$n = 15 + 15 + 9 = 39,$

$S_{yy} = 6.32192, \quad S_{xx} = 0.265443, \quad S_{xy} = 1.20893,$

$b^* = 1.20893/0.265443 = 4.55439,$

$A^* = 6.32192 - 4.55439(1.20893) = 0.815981,$

$F = \{(0.815981 - 0.429346)/[2(3) - 2]\}/\{0.429346/[39 - 2(3)]\} = 7.43.$

Since $F = 7.43 > 5.99 = F(0.999;4,33)$, the Arrhenius relationships differ very highly convincingly (0.1% level). Such differences may be among the slope coefficients, the intercept coefficients, or both. To determine which, examine the estimates and their confidence intervals. Also, use confidence limits for the difference of each pair of coefficients. Such comparisons above indicate that the slopes are comparable, that insulations 2 and 3 have comparable intercepts, and insulation 1 has a convincingly lower intercept.

6. COMPARE MULTIVARIABLE RELATIONSHIPS

Purpose. This section extends the comparisons of preceding sections to multivariable relationships. Such comparisons include 1) log standard deviations, 2) log means, 3) coefficients, and 4) relationships. Where robust in preceding sections, such comparisons are robust here. Section 6 of Chapter 4 provides needed background. Theory appears in Draper and Smith (1981) and Neter, Wasserman, and Kutner (1985). Only analytic comparisons appear here, as graphical ones are generally cumbersome.

Model. The assumed model follows. Life has a lognormal distribution. The standard deviation σ of log life is a constant. The mean log life is a linear function of P engineering variables $x_1, x_2, \cdots, x_P$; namely,

$$\mu(x_1, \cdots, x_P) = \gamma_0 + \gamma_1 x_1 + \cdots + \gamma_P x_P . \qquad (6.1)$$

Such an x_p may be an indicator variable $(0-1)$ for a categorical variable. The coefficients γ_p are estimated from the data. If there are two or more populations, the same model is separately fitted to the data from each population. The following comparisons assume that the relationship (6.1) is correct for each population. If not correct, the significance level of comparisons may be in error. If σ is a function of $x_1, \cdots, x_P$, use ML fitting (Chapters 5 and 9).

6.1. Compare Log Standard Deviations

Section 3 presents comparisons for one, two, and K standard deviations. These comparisons apply to multivariable relationships. Compare estimates s_k (based on pure error) or s'_k (based on lack of fit) for population k. Least squares regression programs always calculate such an s'_k. However, for (6.1), the number of degrees of freedom for s'_k is $\nu'_k = n_k - Q$; here n_k is the number of specimens from population k, and $Q = P + 1$ is the number of coefficients in the relationship, including the intercept. Such comparisons can also be made graphically, using residuals as described in Section 2.5. If the standard deviations differ significantly, then the fitted models need to reflect that. Also, then it is better to make comparisons with maximum likelihood methods (Chapter 9). However, the following comparisons are robust to moderate differences among the σ_k.

6.2. Compare (Log) Means

This section extends the comparisons of means of Section 4 to multivariable relationships. Below m_k denotes the LS estimate of the log mean of population k at a specified condition (values of the P variables). Many LS programs calculate m_k. Its variance is $V_k\sigma^2$. Some LS programs calculate $V_k s'^2$. Denote a pooled estimate of the log standard deviation by s, and denote its degrees of freedom by ν.

One log mean. Calculate $100\gamma\%$ confidence limits for such an estimate m; namely,

$$\underset{\sim}{\mu} = m - t[(1+\gamma)/2;\nu]V^{1/2}s, \quad \tilde{\mu} = m + t[(1+\gamma)/2;\nu]V^{1/2}s. \quad (6.2)$$

Here $t[(1+\gamma)/2;\nu]$ is the $100(1+\gamma)/2$ t-percentile with ν degrees of freedom. Compare this interval with a specified value μ_0 as follows. If the interval encloses μ_0, then m is consistent with μ_0. Otherwise, m differs convincingly (statistically significantly) from μ_0. For a one-sided $100\gamma\%$ interval, replace $(1+\gamma)/2$ by γ in the desired limit (6.2). In a *demonstration test*, use a one-sided lower limit. The limit $\underset{\sim}{\mu}$ must be above μ_0 for the product to "demonstrate a life (better than) μ_0 with $100\gamma\%$ confidence."

Two log means. Suppose data from two populations yield estimates m_1 and m_2 for the mean log lives μ_1 and μ_2 at a particular condition. Two-sided $100\gamma\%$ confidence limits for the true difference $\Delta = \mu_1 - \mu_2$ are

$$\underset{\sim}{\Delta} = (m_1 - m_2) - t[(1+\gamma)/2;\nu][V_1 + V_2]^{1/2}s,$$
$$\tilde{\Delta} = (m_1 - m_2) + t[(1+\gamma)/2;\nu][V_1 + V_2]^{1/2}s. \quad (6.3)$$

If this interval encloses zero, m_1 and m_2 do not convincingly differ at the $100(1-\gamma)\%$ level. Otherwise, they differ convincingly. Here $\nu = \nu_1 + \nu_2$, and $s = [(\nu_1 s_1^2 + \nu_2 s_2^2)/\nu]^{1/2}$.

K log means. To compare K log means, use the F test of Section 4.3. Replace N_k there by $1/V_k$. The pooled estimate of a common mean is

$$m^* = [(m_1/V_1) + \cdots + (m_K/V_K)]/[(1/V_1) + \cdots + (1/V_K)]. \quad (6.4)$$

6.3. Compare Coefficients

This section extends the coefficient comparisons of Section 5.1 to multivariable relationships. For population k, c_k denotes the LS estimate of a particular coefficient (say, γ_1) in (6.1). The variance of c_k equals $D_k\sigma^2$. Most LS regression programs calculate $D_k s'^2$, which appears as diagonal term of the covariance matrix of coefficient estimates.

One coefficient. Sometimes there is a specified or theoretical value γ' for a particular coefficient, say, γ_1. To compare the LS estimate c_1 with γ', use a $100\varepsilon\%$ confidence interval for γ_1, namely,

$$\gamma_1 = c_1 - t[(1+\gamma)/2;\nu]D_1^{1/2}s, \quad \tilde{\gamma}_1 = c_1 + t[(1+\gamma)/2;\nu]D_1^{1/2}s. \quad (6.5)$$

Here $t[(1+\varepsilon)/2;\nu]$ is the $100(1+\varepsilon)/2$ t-percentile with ν degrees of freedom. Most LS regression programs calculate such an interval. If this interval contains γ', c_1 is consistent with γ'. Otherwise, c_1 differs convincingly (statistically significantly) from γ' at the $100(1-\varepsilon)\%$ level.

Two coefficients. Suppose that data from two populations yield LS estimates c_1 and c_2 of the same coefficient. Calculate the $100\gamma\%$ confidence limits for the true difference $\Delta = \gamma_1 - \gamma_2$ of the two population coefficients as

$$\underset{\sim}{\Delta} = (c_1-c_2) - t[(1+\gamma)/2;\nu][D_1+D_2]^{1/2}s,$$
$$\tilde{\Delta} = (c_1-c_2) + t[(1+\gamma)/2;\nu][D_1+D_2]^{1/2}s. \tag{6.6}$$

$\nu = \nu_1 + \nu_2$, and $s = [(\nu_1 s_1^2 + \nu_2 s_2^2)/\nu]^{1/2}$. If this interval encloses zero, c_1 and c_2 do not differ convincingly at the $100(1-\gamma)\%$ level. Otherwise, they do.

K coefficients. Suppose that data from K populations yield LS estimates $c_1, c_2, \cdots, c_K$ for the same coefficient in (6.1), say, γ_1. The following F test for their equality extends the F test of Section 5.1. Calculate the *pooled estimate of the coefficient*

$$c^* = [(c_1/D_1)+ \cdots +(c_K/D_K)] / [(1/D_1)+ \cdots (1/D_K)]. \tag{6.7}$$

Calculate the *sum of squares for the K coefficients*

$$D = (1/D_1)(c_1-c^*)^2 + \cdots + (1/D_K)(c_K-c^*)^2. \tag{6.8}$$

Use (3.2) to calculate the pooled estimate s^* of the sample K sample standard deviations with ν^* degrees of freedom. Calculate the *F statistic for the coefficients*

$$F = [D/(K-1)] / s^{*2}. \tag{6.9}$$

The level α test for equality of the K population coefficients is

1) If $F \le F(1-\alpha;K-1,\nu^*)$, the K coefficient estimates do *not* differ convincingly (statistically significantly) at the $100\alpha\%$ level.

2) If $F > F(1-\alpha;K-1,\nu^*)$, they differ convincingly.

Here $F(1-\alpha;K-1,\nu^*)$ is the $1-\alpha$ point of the F distribution with $K-1$ degrees of freedom in the numerator and ν^* in the denominator. To understand how such c_k differ, examine them as described in Section 5.1.

6.4. Compare Relationships

The following comparison of K relationships extends that in Section 5.3 to multivariable relationships. For each of the K data sets, separately fit the relationship to the data. The assumed relationship has $Q = P+1$ coefficients, including the intercept. Its fit to sample k yields an estimate s'_k of the standard deviation based on lack of fit. Suppose sample k has n_k specimens, and $n = n_1 + \cdots + n_K$. Calculate the pooled estimate

$$s'^2 = [(n_1 - Q)s_1'^2 + \cdots + (n_K - Q)s_K'^2]/(n - KQ). \qquad (6.10)$$

Next pool the K data sets, and fit the relationship to the pooled data. Suppose that s_0' is the resulting estimate of the standard deviation based on lack of fit. Calculate the F *statistic for equality of the K relationships*

$$F = \{[(n - Q)s_0'^2 - (n - KQ)s'^2]/[(K-1)Q]\} / s'^2. \qquad (6.11)$$

Carry at least six significant figures in this and preceding calculations. This helps assure that this F value is accurate to three figures. The level α test for equality of the K relationships is

1) If $F \leq F[1-\alpha;(K-1)Q,n-KQ]$, the K relationships do not differ convincingly (statistically significantly) at the $100\alpha\%$ level.
2) If $F > F[1-\alpha;(K-1)Q,n-KQ]$, they differ convincingly.

Here $F[1-\alpha;(K-1)Q,n-KQ]$ is the $1-\alpha$ point of the F distribution with $(K-1)Q$ degrees of freedom in the numerator and $n-KQ$ in the denominator. If the relationships differ convincingly, examine their coefficients, aided by confidence limits, to determine which differences are important.

PROBLEMS (* denotes difficult or laborious)

8.1. Insulations revisited. Further analyze the data on the three insulations (Tables 2.1 and 3.1).
(a) Calculate 95% confidence limits for the difference of each pair of log means at 250°C. Suitably plot the estimates of the differences and the confidence limits. Comment on how the insulations compare at 250°C.
(b) Simultaneously compare the three log means at 250°C with the F test. Interpret the results.
(c) Repeat (a) and (b) for 225°C.

8.2. Insulations with s'. Reanalyze the example data in this chapter.
(a) Throughout use the estimate s_k' in place of s_k. Comment whether confidence limits or hypothesis tests based on s_k' give appreciably different results. Do you prefer s_k or s_k'? Why?
(b) In comparisons of two insulations, use the pooled standard deviation from all three and its degrees of freedom. Comment whether results differ appreciably from those in the chapter. Comment on the validity of using this pooled standard deviation for such comparisons.
(c) Suggest further analyses.
(d) Carry out (c).

8.3. Au-Al bonds. A temperature-accelerated life test of gold-aluminum bonds in three encapsulating resins for integrated circuits yielded the following failure data in hours, kindly provided by Dr. Muhib Khan of AMD. The purpose of the test was to compare the effects of the resins on bond life at 120°C. Use the Arrhenius relationship and a suitable distribution.

Resin	Temp.	Life in Hours							
1)	175°C	110.0	82.2	99.1	82.9	71.3	91.7	76.0	79.2
	194	45.8	51.3	26.5	58.0	45.3	40.8	35.8	45.6
	213	33.8	34.8	24.2	20.5	22.5	18.8	18.2	24.2
	231	14.2	16.7	14.8	14.6	16.2	18.9	14.8	
	250	18.0	6.7	12.0	10.5	12.2	11.4		
2)	175	12.3	17.7	12.3	4.3	6.8	8.5	5.7	
	194	4.0	4.0	3.1	4.9	6.3	6.2	3.5	3.7
	250	3.6	1.7	3.8	4.1	1.3	3.4	1.2	2.8
3)	200	240.1	238.3	140.4	142.1	223.4	173.1		
	225	63.0	102.0	125.0	67.8	81.0	101.0	76.0	
	250	40.3	33.0	34.3	38.3	29.5	35.3	39.5	

(a) Graphically compare the three resins. State conclusions.

(b) Compare the resins using least squares methods. State conclusions. In view of (a), are least squares comparisons necessary? Why?

(c) Write a brief report on your findings for the materials engineer.

(d) Suggest further analyses.

(e) Carry out (d).

8.4. Transformer oil. Data Set 2 below came from a repeat of the experiment of Problem 4.10 (Data Set 1). The following analyses assess consistency of the two data sets. Where possible use a computer.

```
1 Sq. Inch Electrodes
10 volts/sec Rate of Rise
40 45 44 40 51 36 47 41 44 45
41 42 47 44 35 46 40 51 47 38
41 35 39 46 48 46 39 38 49 41
42 48 38 41 46 48 46 46 48 52
45 38 45 50 36 49 36 47 41 43
46 41 50 50 50 49 46 45 48 52
```

```
9 Sq. Inch Electrodes
10 volts/sec Rate of Rise
42 38 40 42 37 33 41 37 39 35
43 42 41 34 40 40 38 38 25 36
36 39 41 29 37 37 40 38 39 37
40 38 40 38 40 37 42 42 44 39
40 41 44 39 38 43 46 40 38 40
42 44 50 40 38 42 45 37 43 48
```

```
1 Sq. Inch Electrodes
100 volts/sec Rate of Rise
40 51 52 51 47 54 55 52 50 31
44 50 39 53 50 51 47 46 57·55
53 51 50 42 43 45 42 43 47 58
51 53 55 52 48 40 41 54 57 53
57 48 52 55 56 49 43 52 52 49
43 44 49 54 52 50 56 48 52 50
```

```
9 Sq. Inch Electrodes
100 volts/sec Rate of Rise
39 45 46 46 47 45 32 46 47 44
45 44 47 47 47 45 48 50 45 44
48 50 44 44 34 43 34 50 46 52
47 51 51 44 43 50 50 46 43 54
46 49 46 51 49 45 51 51 49 45
54 53 50 49 52 52 53 53 44 48
```

```
1 Sq. Inch Electrodes
1000 volts/sec Rate of Rise
57 59 56 56 58 64 58 55 58 54
65 61 64 65 65 52 53 60 58 63
60 62 54 63 60 52 62 50 60 57
63 57 57 58 52 67 52 62 56 59
55 65 63 57 67 64 62 58 66 60
57 64 66 52 65 57 58 62 60 59
```

```
9 Sq. Inch Electrodes
1000 volts/sec Rate of Rise
57 49 49 41 52 40 48 48 43 45
57 54 49 49 52 53 51 46 55 54
49 41 50 49 51 49 47 55 49 51
51 50 50 55 46 55 57 53 54 54
54 41 60 50 55 54 53 54 53 46
55 50 59 58 60 55 55 56 59 51
```

(a)   Graphically compare the two data sets for consistency. State and explain your findings.

(b*)   For Data Set 2, repeat the analyses of Problem 4.1.

(c)   Fit relationship (1) of Problem 4.10 to Data Set 2, and calculate separate confidence limits for the differences between the three pairs of corresponding coefficients γ_0, γ_1, and γ_2, for the two data sets. State and interpret your findings.

(d)   Test for equality of the two fitted relationships (1). State and interpret your findings.

(e)   Repeat (c) for relationship (2) of Problem 4.10.

(f)   Repeat (d) for relationship (2).

(g)   Pool the two data sets, and fit relationships (1) and (2). Comment on the validity of pooling the data and resulting estimates.

(h)   Compare the Weibull shape parameters for the two data sets, using plots of the ln residuals.

(i)   Write a short report summarizing your findings on consistency of the two data sets.

(j)   Suggest further analyses.

(k)   Carry out (j).

8.5*. Two means. Develop a method for comparing the LS estimates of the means of two products at a design stress level. Assume that each is described by a simple linear-(log)normal model, where the two models have differing (log) standard deviations. This is the Behrens-Fisher problem in a regression setting. Use your method to compare two insulations of Table 2.1.

9

Maximum Likelihood Comparisons for Censored and Other Data

1. INTRODUCTION

Purpose. This advanced chapter presents maximum likelihood (ML) methods for comparing samples. These methods include hypothesis tests and confidence intervals. Presented are comparisons for one sample (Section 2), two samples (Section 3), and K samples (Section 4). Section 5 explains the general theory for likelihood ratio (LR) tests. But one can use the ML methods without reading Section 5. Needed background includes knowledge of ML estimation (Chapter 5) and basics of hypothesis testing (Chapter 8).

ML comparisons. ML comparisons have good properties. Most important, they are versatile. They apply to most models and types of data, including interval and multiply (time or failure) censored data. Such models include the constant stress models used in the examples in this chapter and cumulative exposure models for step-stress tests (Chapter 10). ML comparisons have good statistical properties. Asymptotically they provide the shortest confidence intervals and the (locally) most powerful hypothesis tests. For small samples, they generally perform well. In particular, for complete data, ML methods yield more accurate estimates and more sensitive comparisons than least squares methods for models with Weibull and exponential distributions. Computer programs that do the laborious ML calculations are described in Chapter 5. The ML methods are approximate, since tables for exact confidence limits and hypothesis tests have not been developed for censored and interval data.

Graphical analysis. A combination of ML and graphical comparisons is most effective. ML analyses without graphical analyses may be misleading. This happens when the model does not adequately describe the data, or when the data contain outliers or other peculiarities. Graphical comparisons may be made as described in Section 2 of Chapter 8. Of course, then the probability plots must take into account the censored and interval data, and the relationship plot is more difficult to interpret. Also, check the model and data as described in Section 3 of Chapter 5. This chapter assumes that such checks have been made.

2. ONE-SAMPLE COMPARISONS

Purpose. This section presents ML comparisons for a model fitted to a single sample, namely,

1. Compare a ML estimate of a "quantity" for equality with a specified value. Examples include a slope or intercept coefficient, a log standard deviation, and the median life at a design stress condition. Section 2.1 presents such comparisons.
2. Compare a one-sided confidence limit for a quantity with a specified value to demonstrate that the product surpasses that value – a demonstration test (Section 2.1).
3. Simultaneously compare a number of ML estimates each with a (possibly different) specified value. For example, compare separate estimates of the Weibull shape from each test condition with the value 1. Section 2.2 presents such comparisons.
4. Compare a number of ML estimates for equality to each other. Examples include separate estimates of the log standard deviations or Weibull shape parameters from each test condition. Section 2.3 presents such comparisons.

Many such comparisons appear in Chapter 5, mostly as checks on the model.

Example. The insulating oil data (Table 3.1 of Chapter 3) and the power-Weibull model illustrate the comparisons. The following comparisons assess 1) whether the exponential distribution (suggested by engineering theory) adequately fits the data and 2) whether the Weibull shape parameter is the same at all seven voltages. Table 2.1 summarizes ML fits of various models to the data. This example has complete data and a simple linear model. However, the methods apply to censored and interval data and complex models with multiple variables.

2.1. Compare an Estimate with a Specified Value

This section presents comparisons of a single ML estimate with a specified value, using a confidence interval or LR test.

Confidence interval. Chapter 5 presents approximate normal and LR confidence intervals for a quantity. If such an interval contains the specified value, then the estimate is consistent with that value. Otherwise, it is not consistent with that value. To test for consistency, two-sided intervals are usually used. In contrast, a product passes a *demonstration test* if the interval con-

Table 2.1. ML Fits to Insulating Oil Data

1. Separate Weibull Distributions

Voltage	n_j	$\hat{\mathcal{L}}_j$	$\hat{\beta}_j$	99% limits		$\hat{\alpha}_j$
26	3	−23.717	0.545	0.16	1.84	955.7
28	5	−34.376	0.979	0.45	2.15	352.5
30	11	−58.578	1.059	0.61	1.84	77.59
32	15	−65.737	0.561	0.35	0.91	25.93
34	19	−68.386	0.771	0.51	1.16	12.22
36	15	−37.691	0.889	0.58	1.36	4.292
38	8	−6.765	1.363	0.71	2.60	1.001
		−295.250	Total			

Voltage	n_j	2. Sep. Expo. Dists. $\hat{\mathcal{L}}_j$	$\hat{\theta}_j$	3. Weib. Dists., Common β $\hat{\alpha}_j$	
26	3	−24.517	1303.	1174.	$0.799 = \hat{\beta}$
28	5	−34.378	356.2	349.9	
30	11	−58.606	75.79	104.7	$0.670 = \underline{\beta}$
32	15	−70.763	41.16	68.81	
34	19	−69.623	14.36	38.88	$0.953 = \tilde{\beta}$
36	15	−37.910	4.606	12.41	
38	8	−7.300	0.916	1.851	$-299.648 = \hat{\mathcal{L}}$
		−303.097	Total		

4. Power-Weibull Model

$\hat{\mathcal{L}} = -137.748$ $\hat{\beta} = 0.777\ (0.653, 0.923)$
$\hat{\gamma}_0 = 64.84719$ $\hat{\gamma}_1 = -17.72958$

5. Single Weibull Distribution (Pooled Data)

$\hat{\mathcal{L}} = -339.654$ $\hat{\alpha} = 26.61$ $\hat{\beta} = 0.4375$

6. Single Exponential Distribution (Pooled Data)

$\hat{\mathcal{L}} = -424.889$ $\hat{\theta} = 98.56$ $(\beta = 1)$

tains only values that surpass the specified value. Usually a one-sided interval is used for demonstration tests.

Example. Seven Weibull distributions with a common shape parameter were fitted to the oil data at the seven test voltages (Table 2.1), model 3. The ML estimate of the common shape parameter is $\hat{\beta} = 0.799$, and the normal approximate 95% confidence limits are 0.670 and 0.953. 99% limits are 0.558 and 1.144. The 95% interval does not enclose 1. Thus there is a convincing evidence (5% level) that the life distribution is not exponential. As noted before, this may result from varying test conditions that increase the scatter in the life data, thereby reducing the shape estimate. The samples at some voltages are small (3 and 5 failures). Thus the normal approximate interval may be much too short. The LR interval would be much better here.

Another such comparison consists of fitting the power-Weibull model to the data (Problem 5.2), model 4 in Table 2.1. Then $\hat{\beta} = 0.777$, and normal 95% limits are 0.653 and 0.923. 99% limits are 0.632 and 0.953. Thus there is very convincing evidence (1% level) against an exponential distribution. If the assumed relationship is inadequate, this β estimate is biased low, and the confidence interval is inaccurate. Thus the previous comparison, which does not employ a relationship, is preferred in most applications, as that β estimate is not biased small by possible lack of fit of the relationship.

LR statistic. The following is the LR test that a model parameter θ equals a specified value θ'. Calculate the ML estimate $\hat{\theta}$ and the corresponding maximum log likelihood $\hat{\mathcal{L}}$. Also, calculate the constrained maximum log likelihood $\hat{\mathcal{L}}'$ with $\theta = \theta'$. Both likelihoods are also maximized with respect to all other parameters. The LR test statistic is $T = 2(\hat{\mathcal{L}} - \hat{\mathcal{L}}')$.

LR test. When $\theta = \theta'$, the large-sample distribution of T is approximately chi square with one degree of freedom. If $\theta \neq \theta'$, then T tends to have larger values. Thus the approximate LR test is

- If $T \leq \chi^2(1-\alpha;1)$, then the ML estimate $\hat{\theta}$ is consistent with θ'.
- If $T > \chi^2(1-\alpha;1)$, then the estimate $\hat{\theta}$ differs convincingly from θ' at the $100\alpha\%$ significance level.

Here $\chi^2(1-\alpha;1)$ is the $100(1-\alpha)$th chi-square percentile with one degree of freedom. Figure 2.1 depicts T and the sample likelihood $\mathcal{L}(\theta)$ as a function of just θ, maximized with respect to all other parameters. $T/2$ is the difference between the log likelihood at $\theta = \hat{\theta}$ and at $\theta = \theta'$. This two-sided test is equivalent to using the LR interval for θ with $100(1-\alpha)\%$ confidence. This test, suitably modified to be one sided, can be used as a demonstration test to determine whether θ convincingly *surpasses* a specified value θ'

2.2. Compare Q Estimates with Specified Values

This section presents simultaneous comparisons of Q ML estimates each with a (possibly different) specified value. In particular, suppose that $\hat{\theta}_q$ is

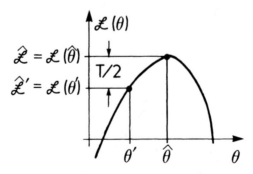

Figure 2.1.   Likelihood $\mathcal{L}(\theta)$ and test statistic T for $\theta = \theta'$.

the ML estimate for a quantity θ_q, and θ_q' is the specified value, $q = 1, \cdots, Q$. θ_q may be a model parameter or coefficient or a function of them. Such comparisons employ simultaneous confidence intervals or a LR test as follows.

Simultaneous intervals. Q parameter estimates can be compared with specified values with *simultaneous* $100\gamma\%$ confidence limits. For each θ_q, calculate a separate $100\gamma'\%$ confidence interval (exact, normal, or LR) where $\gamma' = 1 - [(1-\gamma)/Q]$. Such intervals may be one- or two-sided, as appropriate. Then the probability that *all Q* intervals *simultaneously* enclose their true parameter values is at least γ. This is a result of the Bonferroni inequality. If all such intervals enclose their specified values, then they are all consistent with those values. If an interval does not enclose its specified value θ_q', then that estimate convincingly differs from its specified value.

Example. Table 2.1, model 1, shows the separate ML estimates and 99% confidence limits (normal approximation) for the $Q = 7$ Weibull shape parameters at the test voltages. Here $\gamma' = 0.99$, and working backwards, $\gamma = 1 - 7(1 - \gamma') = 0.93$. Thus the intervals are simultaneous 93% confidence intervals. Only the interval $(0.35, 0.91)$ for 32 kV does not enclose 1. Thus that estimate 0.561 differs convincingly from 1 at the 7% significance level. A Weibull plot of the 32 kV data (Figure 3.1 of Chapter 3) merely shows that those data have a lower slope (shape estimate). So the low estimate is not explained by an outlier or other problem data. The numbers of specimens are small (3 and 5) at some test voltages. So these intervals are too short here, and the true simultaneous confidence is below 93%. LR intervals would be better, and exact intervals of McCool (1981) would be correct.

LR statistic. What follows is the LR test for the null hypothesis that the Q parameters simultaneously equal their specified values: $\theta_1 = \theta_1', \cdots, \theta_Q = \theta_Q'$. Fit the model with those parameters equal to their specified values. This yields the constrained maximum log likelihood $\hat{\mathcal{L}}'$, which is maximized with respect to all other model parameters. Also, fit the unconstrained model

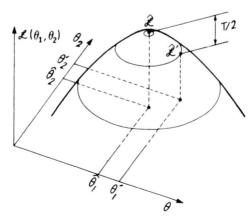

Figure 2.2.   Likelihood $\mathcal{L}(\theta_1, \theta_2)$ and test statistic T for $\theta_1 = \theta_1'$ and $\theta_2 = \theta_2'$.

with ML estimates of the Q parameters and all other model parameters. This yields a maximum log likelihood $\hat{\mathcal{L}}$. Often either model consists of Q identical models fitted to Q samples. Then $\hat{\mathcal{L}} = \hat{\mathcal{L}}_1 + \cdots + \hat{\mathcal{L}}_Q$, where $\hat{\mathcal{L}}_q$ is the maximum log likelihood for sample q. This equation may apply to $\mathcal{L}'$ with the specified parameter values. The *LR test statistic* is $T = 2(\hat{\mathcal{L}} - \mathcal{L}')$. Figure 2.2 depicts such a sample likelihood and LR statistic. That likelihood is a function of $Q = 2$ parameters θ_1 and θ_2 with specified values θ_1' and θ_2'. $T/2$ is the difference between the maximum log likelihood at $(\hat{\theta}_1, \hat{\theta}_2)$ and at (θ_1', θ_2').

LR test. If the null hypothesis is true, T asymptotically has a chi-square distribution with Q degrees of freedom. The alternative is that one or more parameters differ from their specified values. Under the alternative, T tends to have larger values. Thus the approximate LR test is
- If $T \leq \chi^2(1-\alpha; Q)$, then all Q estimates are consistent with their specified values.
- If $T > \chi^2(1-\alpha; Q)$, then some estimates differ convincingly from their specified values at the $100\alpha\%$ significance level.

Here $\chi^2(1-\alpha; Q)$ is the $100(1-\alpha)$th chi-square percentile with Q degrees of freedom. If T is significant, examine individual confidence intervals for the θ_q to determine which are responsible.

Example. For the oil data, we test the hypothesis that the Weibull shape parameter at each test voltage equals 1; that is, the life distribution is exponential. A separate Weibull distribution was fitted to the subset of data from each of the $Q = 7$ test voltages. From Table 2.1, model 1, the sum of the 7 log likelihoods is $\hat{\mathcal{L}} = -23.717 - 34.376 - 58.578 - 65.737 - 68.386 - 37.691 - 6.765 = -295.250$. A separate exponential distribution ($\beta_q' = 1$) was fitted to the data from each voltage (Table 2.1, model 2), and $\hat{\mathcal{L}}' = -24.517 - 34.378 - 58.606 - 70.763 \ -69.623 - 37.910 - 7.300 = -303.097$. Then $T = 2[-295.250 - (-303.097)] = 15.69$. This has $Q = 7$ degrees of freedom. Since $T = 15.69 > 14.07 = \chi^2(0.95; 7)$, some estimates convincingly differ from 1 at the 5% significance level. Because there are so few failures (3 and 5) at some voltages, the chi-square approximation is crude, and the result should be viewed as marginal. The individual confidence intervals for the β_q (Table 2.1, model 1) merely show that the data at 32 kV are not consistent with $\beta_{32}' = 1$. There is no trend of the β_q estimate versus voltage or any other pattern. This can best be seen from a plot of the estimates and confidence intervals.

2.3. Compare Estimates for Equality

This section presents the LR test for equality of J model parameters $\theta_1, \cdots, \theta_J$. That is, the null hypothesis is $\theta_1 = \cdots = \theta_J$. For example, one can test for equality of the Weibull shape parameters (or lognormal log standard deviations) at all test conditions.

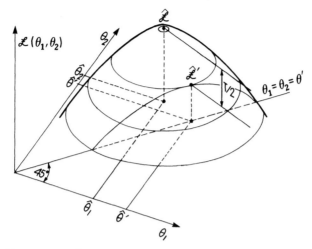

Figure 2.3.   Likelihood $\mathcal{L}(\theta_1,\theta_2)$ and test statistic T for $\theta_1=\theta_2$.

LR statistic. Fit the model with separate ML estimates of the J parameters; this yields a maximum log likelihood $\hat{\mathcal{L}}$. Also, fit the model with a common estimate for $\theta_1 = \cdots = \theta_J = \theta'$; this yields a constrained maximum log likelihood $\hat{\mathcal{L}}'$. Often the model consists of separate models, which are fitted to J distinct and exhaustive subsets of the sample. If $\hat{\mathcal{L}}_j$ is the maximum log likelihood for subset j, then $\hat{\mathcal{L}} = \hat{\mathcal{L}}_1 + \cdots + \hat{\mathcal{L}}_J$. The *LR test statistic* is $T = 2(\hat{\mathcal{L}} - \hat{\mathcal{L}}')$. Figure 2.3 depicts such a LR statistic and sample likelihood, which is a function of $J=2$ parameters θ_1 and θ_2. $T/2$ is the difference between the global maximum $\hat{\mathcal{L}}$ and the constrained maximum $\hat{\mathcal{L}}'$ along the line $\theta_1 = \theta_2$.

LR test. Under the null hypothesis, T asymptotically has a chi-square distribution with $J-1$ degrees of freedom. Under the alternative of inequality, T tends to have larger values. Thus the approximate LR test is
- If $T \leq \chi^2(1-\alpha;J-1)$, then the estimates are consistent with equality.
- If $T > \chi^2(1-\alpha;J-1)$, then some estimates differ convincingly from others at the $100\alpha\%$ significance level.

Here $\chi^2(1-\alpha;J-1)$ is the $100(1-\alpha)$th chi-square percentile with $J-1$ degrees of freedom.

Example. For the oil data, we test the equality hypothesis that the Weibull shape parameter has the same value at all $J=7$ test voltages. Fitting a separate Weibull distribution for each voltage (Table 2.1, model 1) yields $\hat{\mathcal{L}} = -23.717 - 34.376 - 58.578 - 65.737 - 68.386 - 37.691 - 6.765 = -295.250$. Fitting Weibull distributions with a common shape parameter and differing scale parameters (Table 2.1, model 3) yields $\hat{\mathcal{L}}' = -299.648$ and $\hat{\beta}' = 0.799$. Then $T = 2[-295.250 - (-299.648)] = 8.80$. This has $J-1 = 7-1 = 6$ degrees of freedom. Since $T = 8.80 < 10.64 = \chi^2(0.90;6)$, the 7 shape estimates do not convincingly differ even at the 10% significance level. Because there are so few failures (3 and 5) at some voltages, the chi-square approximation is crude. So the result should be viewed as even less significant.

3. TWO-SAMPLE COMPARISONS

Purpose. This section presents ML methods for comparing two samples. Here the same model is fitted to each sample, and pairs of estimates of corresponding quantities are compared for equality. Here "quantity" means any model coefficient, parameter, or function of them. Examples include slope coefficients and median lives at a design stress condition. Section 3.1 shows how to compare one pair of corresponding quantities, one from each sample. Section 3.2 shows how to simultaneously compare two or more pairs. Meeker (1980) presents such comparisons for accelerated test models. Nelson (1982) and Lawless (1982) do so for life distributions.

Example. The comparisons are illustrated with data on two types of insulation for an electrical machine. The life of insulation k is modeled with a lognormal distribution with a constant log standard deviation σ_k. The mean log life μ_k is described with a linear function of log voltage stress (LVPM) in volts per mm and insulation thickness (THICK) in cm. That is,

$$\mu_k = \gamma'_{1k} + \gamma_{2k}(LVPM - LVPM'_k) + \gamma_{3k}(THICK - THICK'_k). \qquad (3.1)$$

$\gamma'_{1k}, \gamma_{2k}, \gamma_{3k}$, and σ_k are estimated from the data. $LVPM'_k$ and $THICK'_k$ are specified values of the variables; here each equals its sample average. Such "centering" of the independent variables aids convergence of the iterative fitting and a more accurate fit. (3.1) is an inverse power relationship between life and voltage stress. Figure 3.1 shows output on ML fits of the model to the data. These results are used in comparisons below.

3.1. Compare One Pair of Quantities

Each data set yields an estimate of a model quantity. The two estimates are compared below by 1) normal approximate confidence limits, 2) LR confidence limits, and 3) the LR test. These comparisons are roughly equivalent, especially for samples with many failures. So in practice, one usually uses just one of them.

Normal Limits

Difference. Suppose θ_k is the true value of a particular quantity for population k. Separately fit the model to each sample (no common parameter values). Also, suppose that $\hat{\theta}_k$ is its ML estimate, and V_k is the sample variance of that estimate. The normal approximation for two-sided $100\%\gamma$ confidence limits for the difference $\Delta = \theta_1 - \theta_2$ is

$$\underline{\Delta} = (\hat{\theta}_1 - \hat{\theta}_2) - K(V_1 + V_2)^{1/2}, \quad \tilde{\Delta} = (\hat{\theta}_1 - \hat{\theta}_2) + K(V_1 + V_2)^{1/2}. \qquad (3.2)$$

Here K is the $100(1+\gamma)/2$th standard normal percentile. If the interval encloses zero, the data are consistent with equality ($\theta_1 = \theta_2$). Otherwise, the estimates differ convincingly at the $100(1-\gamma)\%$ level. Such normal intervals tend to be too narrow. That is, the true confidence level is less than γ, but it

DISTRIBUTION LOGNORMAL(HOURS)

RELATION CENTER 1STORDER(LOGVPM THICK)

COEFFICIENTS(3.3 -8.0 -3.0 -.8;.01 .01 .01 .01)

FIT (POWELL 20 -1)

CASES
92 WITH OBSERVED VALUES.
7 WITH VALUES CENSORED ON THE RIGHT.
---
99 IN ALL

MAXIMUM LOG LIKELIHOOD = -853.65955 = $\hat{\mathfrak{L}}_2$

* MAXIMUM LIKELIHOOD ESTIMATES FOR DIST. PARAMETERS
  WITH APPROXIMATE 95% CONFIDENCE LIMITS

PARAMETERS	ESTIMATE	LOWER LIMIT	UPPER LIMIT	STANDARD ERROR
C 1	3.491614	3.414492	3.568736	0.3948016E-01
C 2	-10.60784	-13.67565	-7.540039	1.565206
C 3	-2.868524	-7.312165	1.575118	2.267164
SPREAD	0.3862163	0.3338898	0.4467433	0.2868772E-01

* COVARIANCE MATRIX

PARAMETERS	C 1	C 2	C 3	SPREAD
C 1	0.1548266E-02			
C 2	-0.1154081E-02	2.449870		
C 3	-0.5854316E-03	1.824424	5.140033	
SPREAD	0.3936731E-04	-0.1284458E-02	-0.2652347E-05	0.8229850E-03

* MAXIMUM LIKELIHOOD ESTIMATE OF THE FITTED EQUATION

CENTER μ_2

= 3.491614

+ (LOGVPM - 2.249733) * (-10.60784)

+ (THICK - 0.2123939) * (-2.868524)

SPREAD σ_2

= 0.3862163

DISTRIBUTION LOGNORMAL(HOURS)

RELATION CENTER 1STORDER(LOGVPM THICK)

COEFFICIENTS(3.3 -8.0 -3.0 -.8;.01 .01 .01 .01)

FIT (POWELL 20 -1)

CASES
106 WITH OBSERVED VALUES.
---
106 IN ALL

MAXIMUM LOG LIKELIHOOD = -905.45452 = $\hat{\mathfrak{L}}_1$

* MAXIMUM LIKELIHOOD ESTIMATES FOR DIST. PARAMETERS
  WITH APPROXIMATE 95% CONFIDENCE LIMITS

PARAMETERS	ESTIMATE	LOWER LIMIT	UPPER LIMIT	STANDARD ERROR
C 1	3.323656	3.275010	3.372302	0.2481937E-01
C 2	-9.692341	-11.42652	-7.958166	0.8847833
C 3	-2.085675	-4.063851	-0.1074986	1.009274
SPREAD	0.2555302	0.2233645	0.2923280	0.1753968E-01

* COVARIANCE MATRIX

PARAMETERS	C 1	C 2	C 3	SPREAD
C 1	0.6160013E-03			
C 2	0.1469285E-07	0.7828414		
C 3	-0.1520959E-07	0.4576534	1.018633	
SPREAD	-0.1010996E-08	-0.3414217E-05	-0.2192421E-05	0.3076440E-03

* MAXIMUM LIKELIHOOD ESTIMATE OF THE FITTED EQUATION

CENTER μ_1

= 3.323656

+ (LOGVPM - 2.248131) * (-9.692341)

+ (THICK - 0.2133585) * (-2.085675)

SPREAD σ_1

= 0.2555302

Figure 3.1.   ML fits to data on two insulations.

is closer to γ for larger numbers of failures. Also, sometimes the approximation improves with a transformed parameter whose estimate has a sampling distribution closer to normal (Chapter 5).

Example. For the two insulations, such 95% confidence limits for $\Delta_2 = \gamma_{21} - \gamma_{22}$ (the LVPM coefficients) are

$$\underset{\sim}{\Delta}_2 = [-9.692341 - (-10.60784)] - 1.960(0.7828414 + 2.449870)^{1/2} = -2.609,$$

$$\tilde{\Delta}_2 = 0.915 + 3.524 = 4.439.$$

Equivalently, the limits are 0.915 ± 3.524. These limits enclose zero. Thus these coefficient estimates do not differ convincingly. That is, both insulations appear to have the same power for the inverse power relationship. This is physically reasonable, as the insulations contain similar materials. Similarly, such 95% confidence limits for $\Delta_3 = \gamma_{31} - \gamma_{32}$ (*THICK* coefficients) are -4.08 and 5.66 (0.79 ± 4.87). These limits enclose zero. Thus the thickness coefficients do not differ convincingly. The intercept coefficients γ_{11}' and γ_{12}' cannot be readily compared this way, since $LVPM_1' \neq LVPM_2'$ and $THICK_1' \neq THICK_2'$ for the two samples.

Ratio. To compare some parameters, it is better to use their ratio $\rho = \theta_1/\theta_2$. This is so if θ_k must be **mathematically** positive, for example, a log standard deviation or Weibull shape parameter. Then $\ln(\hat{\theta}_k)$ is treated as approximately normally distributed. Suppose that the model is separately fitted to each sample, and the sample variance of $\hat{\theta}_k$ is V_k. Normal approximate two-sided $100\gamma\%$ confidence limits for ρ are

$$\underset{\sim}{\rho} = (\hat{\theta}_1/\hat{\theta}_2)/\exp\{K[(V_1/\hat{\theta}_1^2) + (V_2/\hat{\theta}_2^2)]^{1/2}\}, \tag{3.3}$$

$$\tilde{\rho} = (\hat{\theta}_1/\hat{\theta}_2) \cdot \exp\{K[(V_1/\hat{\theta}_1^2) + (V_2/\hat{\theta}_2^2)]^{1/2}\}.$$

Here K is the $100(1+\gamma)/2$th standard normal percentile. If the interval encloses 1, the data are consistent with $\theta_1 = \theta_2$. Otherwise, the estimates differ convincingly at the $100(1-\gamma)\%$ level.

Example. For the insulations, 95% confidence limits for $\rho = \sigma_1/\sigma_2$ are

$$\underset{\sim}{\rho} = (0.2555/0.3862)/\exp\{1.960[(0.003076/0.2555^2)$$

$$+ (0.0008230/0.3862^2)]^{1/2}\} = 0.5426,$$

$$\tilde{\rho} = (0.6616) \cdot 1.219 = 0.8067.$$

The limits can be written as $0.6616 \times /1.219$. Similarly, 99% limits are 0.5098 and 0.8585 ($0.6616 \times /1.298$). Neither interval encloses 1. So the log standard deviations differ very convincingly at (at least) the 1% level. Insulation 1 has a smaller log standard deviation. This needs to be taken into account in modelling and comparing the insulations.

Common parameter values. The simple limits above employ separate fits of the model to each data set. This assures that the two estimates of a quantity are statistically independent, and the above formulas for the limits are correct. In contrast, one could fit a combined model where the two populations have the same value for some other pair(s) of parameters. One does this, if physically and statistically plausible, to more accurately estimate the common value or other parameter values. Also, most ML programs insist on fitting a single common value of the lognormal σ (or Weibull shape parameter) in a combined model for two populations, whether physically plausible or not. For most combined models, estimates of pairs of parameters are statistically *dependent*, and the formulas above are not valid. The following LR methods are always valid. In practice, one often analyzes the data with both separate and combined models. Usually they lead to the same conclusion. If not, they yield further insight into the data and product.

Combined model. The limits above employ separate fits of the model to the two data sets and a hand calculation. Such normal limits for a difference of a pair of coefficients from the two fits can be directly calculated with some ML packages. To do this, specify a combined model with an indicator variable (0-1) for the two populations; see Section 4.3 of Chapter 5 for this. Such a combined model may have common values for some parameters. Specify the model so an *interaction* coefficient equals the desired difference of the two coefficients. Fit the model. Then the package's approximate normal limits for that coefficient are the correct limits for the difference, even if some other pairs of parameters have a common value. Most packages fit only models with a common log standard deviation (or Weibull shape parameter) for all populations. Thus they could not fit a proper combined model for the two insulations, which have different σ_k's.

LR Limits

As discussed in Chapter 5, the LR interval is often a better approximation than the normal one. Most packages for ML fitting do not yet automatically calculate such LR intervals. Section 5.8 of Chapter 5 explains how to manipulate such packages to obtain LR limits for a coefficient. To obtain LR limits for a difference of coefficients, use a combined model where an interaction coefficient is that difference, as described above. Such a combined model may have common values for some pairs of other parameters.

LR Test

LR statistic. The LR test for equality of corresponding parameters θ_1 and θ_2 of two populations follows. While a confidence interval for their difference (or ratio) is more informative, this test is usually easier to calculate and often suffices. The model for the two populations may be a combined one, and some of the other pairs of model coefficients may have common values. Suppose that the maximum log likelihood for the model with $\hat{\theta}' = \hat{\theta}_1 = \hat{\theta}_2$ is $\hat{\mathcal{L}}'$ and is $\hat{\mathcal{L}}$ with $\hat{\theta}_1 \neq \hat{\theta}_2$. Then the LR test statistic for the equality hypothesis is

$$T = 2(\hat{\mathcal{L}} - \hat{\mathcal{L}}') . \tag{3.4}$$

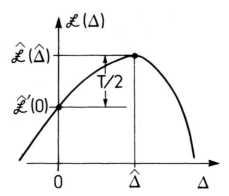

Figure 3.2.   Likelihood and test statistic T for $\Delta = \theta_1 - \theta_2 = 0$.

Sometimes none of the other pairs of coefficients in the model are assumed equal. Then the model is separately fitted to each data set, and sample k yields a maximum log likelihood $\hat{\mathcal{L}}_k$. The LR test statistic then is

$$T = 2(\hat{\mathcal{L}}_1 + \hat{\mathcal{L}}_2 - \hat{\mathcal{L}}').$$

Figure 3.2 depicts this statistic and the log likelihood as a function of $\Delta = \theta_1 - \theta_2$. Other model parameters are not depicted.

LR test. When $\theta_1 = \theta_2$, the large sample distribution of T is approximately chi square with one degree of freedom. Otherwise, T tends to have larger values. Thus the LR test for equality is

- If $T \leq \chi^2(1-\alpha;1)$, the two parameters do not differ convincingly (statistically significantly) at the $100\alpha\%$ level.
- If $T > \chi^2(1-\alpha;1)$, they differ convincingly at the $100\alpha\%$ level.

Here $\chi^2(1-\alpha;1)$ is the $100(1-\alpha)$th chi square percentile with 1 degree of freedom. If the parameters differ convincingly, examine their estimates to see why. Transforming a parameter to improve normality of its ML estimate improves the approximate normal limits. Such a transformation has no effect on the LR limits and LR test.

Example. For the two insulations, the intercept coefficients are compared as follows. A separate model was fitted to each data set; see (4.3.1)–(4.3.4) of Chapter 5. This yielded $\hat{\mathcal{L}}_1 = -905.455$ and $\mathcal{L}_2 = -853.660$ (Figure 3.1). Another model was fitted with a common value of the intercept coefficient; while all other coefficients (and σ) differed for the two populations. This model yielded $\hat{\mathcal{L}}' = -1759.269$. Then $T = 2[-905.455 - 853.660 - (-1759.269)] = 0.31$. Since $T = 0.31 < 3.841 = \chi^2(0.95;1)$, the intercept estimates do not differ convincingly at the 5% level.

3.2. Simultaneously Compare Q Pairs of Coefficients

Hypothesis and model. As follows, one simultaneously compares Q pairs of coefficients of the same model fitted to two samples. The equality (null) hypothesis for Q pairs is $\gamma_{11} = \gamma_{12},\ \cdots,\ \gamma_{Q1} = \gamma_{Q2}$. Other pairs of model coefficients may have common or different values, whatever is assumed. The following methods for such simultaneous comparisons are the LR test, simultaneous confidence intervals, and the LR confidence region. The Wald and Rao tests (Section 5) also apply.

LR statistic. Suppose that the constrained maximum log likelihood for the model with equality of those Q pairs of coefficients is $\hat{\mathcal{L}}'$. Also, suppose that the unconstrained maximum log likelihood for the model with those pairs unequal is $\hat{\mathcal{L}}$. Then the (log) LR statistic for the equality hypothesis is

$$T = 2(\hat{\mathcal{L}} - \hat{\mathcal{L}}').\qquad(3.5)$$

Sometimes none of the other pairs of coefficients in the model are assumed equal. Then the model is separately fitted to each data set, and sample k yields a maximum log likelihood $\hat{\mathcal{L}}_k$. Then the statistic is

$$T = 2(\hat{\mathcal{L}}_1 + \hat{\mathcal{L}}_2 - \hat{\mathcal{L}}').\qquad(3.6)$$

LR test. When the Q pairs are equal, the large sample distribution of T is approximately chi square with Q degrees of freedom. Otherwise, T tends to have larger values. Thus the LR test for equality is
- If $T \leq \chi^2(1-\alpha;Q)$, the Q coefficient pairs do not differ convincingly (statistically significantly) at the $100\alpha\%$ level.
- If $T > \chi^2(1-\alpha;Q)$, some pairs differ convincingly at the $100\alpha\%$ level.

Here $\chi^2(1-\alpha;Q)$ is the $100(1-\alpha)$th chi square percentile with Q degrees of freedom. If the parameters differ convincingly, examine their estimates and confidence limits to see why. Plotting confidence limits for individual pairs (Section 3.1) will help. For $Q = 1$, this LR test is one in Section 3.1.

Example. As follows, simultaneously compare the two insulations with respect to *all* $Q = 4$ pairs of coefficients (including σ_k). The maximum log likelihood for the model (with common parameter values) fitted to the pooled insulation data is $\hat{\mathcal{L}}' = -1774.986$. Separate fits of the model (3.1) yield $\hat{\mathcal{L}}_1 = -905.455$ and $\hat{\mathcal{L}}_2 = -853.660$. Then $T = 2[-905.455 - 853.660 - (-1774.986)] = 31.74$. Since $T = 31.74 > 16.27 = \chi^2(0.999;4)$, the coefficients differ very highly significantly (0.1% level). Confidence limits in Figure 4.9 of Chapter 5 show which pairs of coefficients differ convincingly, namely, just the two log standard deviations.

The $Q = 3$ pairs of coefficients in the relationship could be simultaneously compared as follows, assuming $\sigma_1 \neq \sigma_2$. Fit the model with common values of the three pairs of coefficients and $\sigma_1 \neq \sigma_2$ to get $\hat{\mathcal{L}}'$. Then fit the model with differing coefficients and $\sigma_1 \neq \sigma_2$ (separate fits) to get $\hat{\mathcal{L}}_1 = -907.544$ for Insulation 1 and $\hat{\mathcal{L}}_2 = -854.451$ for Insulation 2. Then use the LR test above.

Simultaneous intervals. Q pairs of parameters can be compared for equality by using *simultaneous* $100\gamma\%$ confidence intervals. For each of the Q differences Δ_q (or ratios), calculate a $100\gamma'\%$ confidence interval (normal

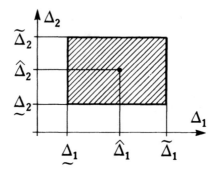

Figure 3.3.   Rectangular simultaneous confidence region (Bonferroni inequality).

or LR), where $\gamma' = 1-[(1-\gamma)/Q]$. Such intervals may be one- or two-sided, as appropriate. Then the probability that *all* Q intervals *simultaneously* enclose the Q true differences is at least γ. This is a result of the Bonferroni inequality. If the Q intervals all enclose zero for a difference (or one for a ratio), then all Q pairs do not differ statistically significantly at the $100(1-\gamma)\%$ level approximately. Otherwise, intervals that do not enclose zero correspond to convincing differences. This test is roughly equivalent to the LR test above. These intervals can also be viewed as a confidence region in a Q-dimensional parameter space of $(\Delta_1, \Delta_2, \cdots, \Delta_Q)$. The Q confidence intervals specify a "rectangular" $100\gamma\%$ confidence region consisting of all $(\Delta_1, \Delta_2, \cdots, \Delta_Q)$ such that $\underset{\sim}{\Delta}_1 \leq \Delta_1 \leq \tilde{\Delta}_1, \cdots, \underset{\sim}{\Delta}_Q \leq \Delta_Q \leq \tilde{\Delta}_Q$. The "center" of this region is the estimate $(\hat{\Delta}_1, \hat{\Delta}_2, \cdots, \hat{\Delta}_Q)$. If this region encloses the origin of the Q-dimensional space, then no pairs differ statistically significantly. Figure 3.3 shows such a region for $Q = 2$ differences.

Example. For the two insulations, the $Q = 2$ approximate 95% intervals above for $\Delta_2 = (\gamma_{21} - \gamma_{22})$ and $\Delta_3 = (\gamma_{31} - \gamma_{32})$ are 0.915 ± 3.524 and 0.79 ± 4.87. They are simultaneous 90% confidence intervals for both differences.

LR confidence region. As follows, Q pairs of parameters can be simultaneously be compared with a LR confidence region in the Q-dimensional space of the Q differences $(\Delta_1, \Delta_2, \cdots, \Delta_Q)$ or ratios. The combined log likelihood $\mathcal{L}(\Delta_1, \cdots, \Delta_Q)$ of the two samples can be written as a function of the Q differences (or ratios). The other model parameters in $\mathcal{L}(\ )$ are not shown in this notation, but are implicit. Such other pairs of corresponding parameters for the two populations may have common values. An approximate $100\gamma\%$ *simultaneous confidence region* for Q true differences are the points $(\Delta_1, \Delta_2, \cdots, \Delta_Q)$ satisfying.

$$2[\hat{\mathcal{L}} - \hat{\mathcal{L}}(\Delta_1, \cdots, \Delta_Q)] \leq \chi^2(\gamma; Q). \tag{3.7}$$

Here $\hat{\mathcal{L}}$ is the maximum log likelihood evaluated at $(\hat{\Delta}_1, \cdots, \hat{\Delta}_Q)$ and at the ML estimates of all other parameters. Also, $\hat{\mathcal{L}}(\Delta_1, \cdots, \Delta_Q)$ is the log likelihood at $\Delta_1, \cdots, \Delta_Q$ but maximized with respect to all other parameter

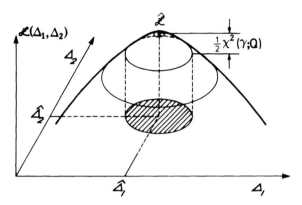

Figure 3.4.   Likelihood and simultaneous LR confidence region.

values. For $Q = 2$, some ML programs calculate and plot the perimeter of this region. Figure 3.4 depicts such region for $Q = 2$. There a plane intersects the log likelihood function $\mathcal{L}(\Delta_1,\Delta_2)$ a distance $(1/2)\chi^2(\gamma;2)$ below the maximum $\hat{\mathcal{L}} = \mathcal{L}(\hat{\Delta}_1,\hat{\Delta}_2)$. The projection of that intersection on the (Δ_1,Δ_2)-plane is the perimeter of the confidence region. If the confidence region encloses the origin of the space of Q differences, the Q differences are not convincingly (statistically significantly) different from zero at the $100(1-\gamma)\%$ level. This test for simultaneous equality of the Q pairs is equivalent to the LR test above. For large samples with many failures and normally distributed ML estimates, such a confidence region has a shape close to an ellipsoid in Q dimensions. In contrast, the simultaneous (Bonferroni) limits above yield a rectangular region. These two types of regions usually roughly coincide in practice.

4. *K*-SAMPLE COMPARISONS

Purpose. This section presents ML comparisons for K samples using the *same model* for each sample. Section 4.1 shows how to compare K estimates of a quantity for equality, one from each population. Section 4.2 shows how to simultaneously compare Q sets of K estimates for equality. Nelson (1982) and Lawless (1982) present such comparisons for life distributions.

Example. Complete data on three motor insulations (Table 2.1 of Chapter 8) illustrate the comparisons. The comparisons apply also to censored and inspection data and other models with more variables. The model for life of insulation k is a lognormal distribution with a log standard deviation σ_k. The assumed Arrhenius relationship for mean log life is

$$\mu_k(x) = \alpha_k + \beta_k x ; \qquad (4.1)$$

here $x = 1000/T$ and T is the absolute temperature. α_k, β_k, and σ_k are estimated from the data. Table 4.1 shows results of ML fits of three models

Table 4.1. ML Fits to Data on Three Insulations

k	1. Separate Fits			2. Common α, β; σ_1, σ_2, σ_3	3. Pooled Data Common α, β, σ
	1	2	3		
n_k	15	15	9	39	39
$\hat{\ell}_k$	9.128	16.463	8.960	$\hat{\ell}' = 24.710$	20.079
$\hat{\alpha}_k$	-5.302	-6.177	-7.758	$\tilde{\alpha} = -6.502$	-6.204
$\tilde{\alpha}_k$	-3.70	-5.20	-6.37		
$\underline{\alpha}_k$	-6.91	-7.15	-9.15		
$\hat{\beta}_k$	4.050	4.580	5.360	$\hat{\beta}' = 4.731$	4.554
$\bar{\beta}_k$	4.85	5.07	6.05		
$\underline{\beta}_k$	3.25	4.09	4.67		
$\hat{\sigma}_k$	0.1317	0.08075	0.08941	$\hat{\sigma}'_1 = 0.2222$	
$\bar{\sigma}_k$	0.1787	0.1096	0.1307	$\hat{\sigma}'_2 = 0.0844$	0.1446
$\underline{\sigma}_k$	0.0846	0.0519	0.0482	$\hat{\sigma}'_3 = 0.1036$	

to the data. These results are used in comparisons below. The exact comparisons of Chapter 8 apply to this model and are preferable. However, the following comparisons apply to models for which there are no exact methods.

4.1. Compare K Estimates of the Same Quantity

Suppose that sample k yields a ML estimate $\hat{\theta}_k$ of a quantity θ_k. The K estimates are compared below with 1) a quadratic statistic, 2) a LR ratio test, and 3) simultaneous confidence intervals for pairs of estimates. These comparisons are roughly equivalent, especially for samples with many failures. The *equality (or homogeneity) hypothesis* is $\theta_1 = \theta_2 = \cdots = \theta_K$. The *alternative* is some $\theta_k \neq \theta_{k'}$.

Quadratic Statistic

Statistic. Suppose that the model is separately fitted to each data set. Denote the sample variance of $\hat{\theta}_k$ by V_k. ML programs calculate such V_k. Calculate the **linearly pooled estimate** of a common θ:

$$\theta^* = V[(\hat{\theta}_1/V_1) + \cdots + (\hat{\theta}_K/V_K)]. \qquad (4.2)$$

$$V = 1/[(1/V_1) + \cdots + (1/V_K)] \qquad (4.3)$$

is an estimate of the variance of θ^*. The *quadratic statistic* is

$$Q = [(\hat{\theta}_1 - \theta^*)^2/V_1] + \cdots + [(\hat{\theta}_K - \theta^*)^2/V_K]. \qquad (4.4)$$

Test. If the equality hypothesis is true, the distribution of Q is approximately chi square with $K - 1$ degrees of freedom. If the alternative is true, Q tends to have larger values. Thus the approximate test is

- If $Q \leq \chi^2(1-\alpha;K-1)$, the K estimates do not differ convincingly (statistically significantly) at the $100\alpha\%$ level.
- If $Q > \chi^2(1-\alpha;K-1)$, they differ convincingly at the $100\alpha\%$ level.

Here $\chi^2(1-\alpha;K-1)$ is the $100(1-\alpha)$th chi square percentile with $K-1$ degrees of freedom. If the estimates differ convincingly, determine why. Examine plots of confidence limits for the estimates (Chapter 5) and for pairs of estimates (Section 3.1).

Example. For the three insulations, the ML estimates of the slope coefficients are $\hat{\beta}_1 = 4.050$, $\hat{\beta}_2 = 4.580$, and $\hat{\beta}_3 = 5.360$. Their sample variances are $V_1 = 0.1659$, $V_2 = 0.0614$, and $V_3 = 0.1239$. $V = 1/[(1/0.1659) + (1/0.0614) + (1/0.1239)] = 0.03291$. The pooled common slope estimate is $\beta^* = 0.03291[(4.050/0.1659) + (4.580/0.0614) + (5.360/0.1239)] = 4.682$. $Q = [(4.050-4.682)^2/0.1659] + [(4.580-4.682)^2/0.0614] + [(5.360-4.682)^2/0.1239] = 6.29$. Since $Q = 6.29 > 5.991 = \chi^2(0.95;2)$, the three estimates differ convincingly at the 5% level. Note that the exact test (Chapter 8) was not significant at the 5% level. This illustrates that the approximate test usually shows higher significance than the corresponding exact one.

Other parameters equal. The test above employs a *separate* fit of the model to each data set. This assures that the K estimates are statistically independent, and that the test is correct. In contrast, one could fit a combined model where the K populations have a common value for some other parameter(s). One does this, if physically and statistically plausible, to more accurately estimate the common value or other parameter values. For such combined models, most estimates of corresponding parameters are statistically dependent, and the formulas above are *not* valid. The LR test below is valid for a combined model with some common parameter values. In practice, one often analyzes the data with both separate and combined models. Usually they lead to the same conclusion. If not, they yield further insight into the data and product.

LR Test

LR statistic. Suppose that the corresponding model parameters of the K populations are $\theta_1, \cdots, \theta_K$. For example, these may be the K log standard deviations, the K intercept coefficients, or K slope coefficients. The LR test for their equality (a common value) follows. The model for the K populations may be a combined one; that is, some of the other model coefficients may have common values for the K populations. Suppose that the maximum log likelihood for the model with equality is $\hat{\mathcal{L}}'$ and is $\hat{\mathcal{L}}$ without equality. Then the LR test statistic for the equality hypothesis is

$$T = 2(\hat{\mathcal{L}} - \hat{\mathcal{L}}') . \tag{4.5}$$

Sometimes no other coefficient is assumed to have a common value. Then the model is separately fitted to each data set, and sample k yields a maximum log likelihood $\hat{\mathcal{L}}_k$. Then the statistic is

$$T = 2(\hat{\mathcal{L}}_1 + \cdots + \hat{\mathcal{L}}_K - \hat{\mathcal{L}}') . \tag{4.6}$$

LR test. Under equality, the large sample distribution of T is approximately chi square with $K - 1$ degrees of freedom. Otherwise, T tends to have larger values. Thus the LR test for equality is
- If $T \le \chi^2(1-\alpha; K-1)$, the K estimates do not differ convincingly (statistically significantly) at the $100\alpha\%$ level.
- If $T > \chi^2(1-\alpha; K-1)$, some estimates differ convincingly at the $100\alpha\%$ level.

Here $\chi^2(1-\alpha; K-1)$ is the $100(1-\alpha)$th chi square percentile with $K-1$ degrees of freedom. If the estimates differ convincingly, examine a plot of them and their confidence limits to see why.

Example. For the three insulations, compare the σ_k as follows. From Table 4.1, model 2 with a common slope, a common intercept, and different σ_k has $\hat{\mathcal{L}} = 24.710$. Model 3 with a common slope, a common intercept, and common σ has $\hat{\mathcal{L}}' = 20.079$. $T = 2(24.710 - 20.079) = 9.26 > 9.21 = \chi^2(0.99;2)$. Thus the three σ_k **estimates** convincingly differ at the 1% significance level. This is misleading. Keep in mind that such a test really tests equality of the estimates – not of the true parameter values. Using a common slope and intercept, when those coefficients do not have a common value, results in biased estimates of the σ_k. This test is really detecting differences among the relationships. This shows the need for care in using models with common parameter values and in interpreting hypothesis tests employing them.

Simultaneous Pairwise Confidence Intervals

Intervals. All pairs of the K estimates can be compared for equality with *simultaneous* $100\gamma\%$ confidence intervals. For each of the $D = K(K-1)/2$ such pairs, calculate a $100\gamma'\%$ confidence interval (Section 3.1) for the difference (or ratio) of the pair, where $\gamma' = 1 - [(1-\gamma)/D]$. Such limits may be one- or two-sided as appropriate. The Bonferroni inequality guarantees that *all* D intervals *simultaneously* enclose the true differences (or ratios) with probability at least γ. If all D intervals for differences (or ratios) enclose zero (one), then the K estimates do not differ convincingly at the $100(1-\gamma)\%$ significance level. Otherwise, some estimates differ convincingly, and the intervals not enclosing zero (one) show which pairs. A plot of such intervals and estimates for differences (or ratios) is revealing.

Example. For the three insulations, compare the σ_k as follows. There are $D = 3(3-1)/2 = 3$ pairs. For $100\gamma = 90\%$ simultaneous confidence, each interval has $100\gamma' = 1 - [(1-0.90)/3] = 96.6667\%$ confidence. The normal approximate intervals (3.3) for the ratios are: (1.02 , 2.62) for σ_1/σ_2, (0.39 , 1.56) for σ_2/σ_3, and (0.39 , 1.17) for σ_3/σ_1. Only the first interval barely does not enclose one. Discount that one as not significant, since normal intervals tend to be too short. Thus the estimates do not differ convincingly at even the 10% level, according to this method. Exact intervals (Chap. 8) are better.

4.2. Simultaneously Compare Q Sets of K Coefficients

Hypothesis and model. As follows, one simultaneously compares Q sets of K corresponding coefficients of the same model fitted to the K sets of data. Then there are Q simultaneous equality (null) hypotheses: $\gamma_{11} = \gamma_{12} = \cdots = \gamma_{1K}, \cdots, \gamma_{Q1} = \gamma_{Q2} = \cdots = \gamma_{QK}$. That is, there is a common value γ_q for each such coefficient. The alternative is some $\gamma_{qk} \neq \gamma_{qk'}$. Other sets of K corresponding model coefficients may have common or different values, whatever is assumed. The LR test for this null hypothesis follows. If $Q = P$, the number of model parameters, the test compares the K models for equality. The Wald and Rao tests (Section 5.3) also apply.

LR statistic. Suppose that the maximum log likelihood for the model with equality is $\hat{\mathcal{L}}'$. Also, suppose that the maximum log likelihood for the model without equality is $\hat{\mathcal{L}}$. Then the LR test statistic for equality is

$$T = 2(\hat{\mathcal{L}} - \hat{\mathcal{L}}'). \tag{4.7}$$

Suppose any other set of K corresponding coefficients in the model have distinct values. Then the model is separately fitted to each data set, and sample k yields a maximum log likelihood $\hat{\mathcal{L}}_k$. Then the LR statistic is

$$T = 2(\hat{\mathcal{L}}_1 + \cdots + \hat{\mathcal{L}}_K - \hat{\mathcal{L}}'). \tag{4.8}$$

LR test. When the Q sets of coefficients are equal, the large sample distribution of T is approximately chi square with $Q(K-1)$ degrees of freedom. Otherwise, T tends to have larger values. Thus the LR test for equality is
- If $T \leq \chi^2(1-\alpha; Q(K-1))$, the Q coefficient sets do not differ convincingly (statistically significantly) at the $100\alpha\%$ level.
- If $T > \chi^2(1-\alpha; Q(K-1))$, some differ convincingly at the $100\alpha\%$ level.
Here $\chi^2(1-\alpha; Q(K-1))$ is the $100(1-\alpha)$th chi square percentile with $Q(K-1)$ degrees of freedom. If the parameters differ convincingly, examine their estimates and confidence limits to see why. LR tests for $Q = 1$ set of coefficients (Section 4.1) will help.

Example. For the three insulations, test the hypothesis of a common relationship. That is, test $\alpha_1 = \alpha_2 = \alpha_3$ and $\beta_1 = \beta_2 = \beta_3$. This consists of $Q = 2$ simultaneous equality hypotheses of $K = 3$ parameters each. The σ_k are assumed to differ. From model 1 of Table 4.1, a separate fit to each data set yields $\hat{\mathcal{L}}_1 = 9.128$, $\hat{\mathcal{L}}_2 = 16.463$, and $\hat{\mathcal{L}}_3 = 8.960$. The fit of a common relationship with different σ_k (model 2) yields $\hat{\mathcal{L}}' = 24.710$. $T = 2(9.128 + 16.463 + 8.960 - 24.710) = 19.68$. It has $Q(K-1) = 2(3-1) = 4$ degrees of freedom. Since $T = 19.68 > 18.47 = \chi^2(0.999; 4)$, the relationships differ very highly significantly (0.1% level). Further data analysis indicates that the intercept coefficients differ convincingly.

The three models are compared for equality (common α, β, and σ) as follows. Fitting the model to the pooled data yields $\hat{\mathcal{L}}' = 20.079$ (Table 4.1, model 3). Then $T = 2(9.128 + 16.463 + 8.960 - 22.079) = 28.94$. It has $Q(K - 1) = 3(3-1) = 6$ degrees of freedom. Since, $T = 28.94 > 22.46 = \chi^2(0.999;6)$, the models differ very highly convincingly (0.1% level). Other analyses indicate that this is due to differences among the intercept coefficients.

5. THEORY FOR LR AND RELATED TESTS

Purpose. This advanced section informally presents theory for likelihood ratio (LR) tests. It is for those who seek a deeper understanding or wish to develop their own models and hypothesis tests. Rao (1973), Wilks (1962), and Lehmann (1986) rigorously present the formal theory. Nelson (1982) gives examples of LR tests for life data, and Lawless (1982) gives applications to accelerated testing. LR tests have good asymptotic properties; namely, they are consistent and uniformly or locally most powerful. Also, they generally perform well for small sample sizes. Chapter 8 is helpful background. Section 5 of Chapter 5 is essential background.

LR test. In principle, a LR test is simple. It is a means of assessing whether a general model fits a data set better than a constrained model. One fits to the data a general model with P separately estimated parameters. This yields a global maximum log likelihood $\hat{\mathcal{L}}$. Also, one fits a constrained model (a special case of the general model) with $P' < P$ separately estimated parameters. This yields a constrained maximum log likelihood $\hat{\mathcal{L}}'$. The constrained model fits the data nearly as well as the general model if the statistic $T = 2(\hat{\mathcal{L}} - \hat{\mathcal{L}}')$ is small. If the constrained model is the true one, then the sampling distribution of T is approximately chi square with $(P - P')$ degrees of freedom for samples with many failures. Otherwise, T tends to have larger values. Thus the approximate LR test is

- If $T \leq \chi^2(1-\alpha;P-P')$, accept (use) the constrained model.
- If $T > \chi^2(1-\alpha;P-P')$, reject the constrained model and use the general one.

Here $\chi^2(1-\alpha;P-P')$ is the $100(1-\alpha)$th chi square percentile with $P - P'$ degrees of freedom. This chapter presents many such LR comparisons of general and constrained models. This section presents underlying theory.

Overview. Section 5.1 states the hypothesis testing problem, namely, the general model, parameter space, null hypothesis, and alternative. Section 5.2 presents the LR test, exact and approximate critical values for it, and its OC function. Section 5.3 describes Rao's and Wald's tests, which are asymptotically equivalent to the LR test.

5.1. Problem Statement

Overview. This section describes the general hypothesis testing problem. It first presents the general model for a sample. Then it describes the parameter space and the subspaces of a null hypothesis and its alternative. It also shows how to specify constant, equality, and constraint hypotheses. Several examples illustrate these ideas.

General model. The following formulation of a general model includes all models considered in this book. Suppose that a sample of n specimens yields data values $y_1, \cdots, y_n$. Such a data value may be a failure or a censoring time or an interval. For simplicity at the moment, regard them as failure times. Censoring times and interval data are treated later. For specimen i, $x_i = (x_{1i}, x_{2i}, \cdots, x_{Ji})$ denotes its values of the J independent variables in the model, $i = 1, 2, \cdots, n$. Denote parametric joint probability density of $y_1, \cdots,$ y_n by $f_0(y_1, \cdots, y_n; \theta, x_1, \cdots, x_n)$. Here $\theta = (\theta_1, \cdots, \theta_P)$ denotes the P numerical parameters of the model. Such a density may be continuous, discrete, or a mix. Such a density may include independent samples from two or more populations. Then usually the same submodel is fitted to each sample, and the collection of submodels is the general model. Usually $y_1, \cdots, y_n$ are assumed *statistically independent*. Then their joint density is the product

$$f_0(y_1, \cdots, y_n; \theta, x_1, \cdots, x_n) = f_1(y_1; \theta, x_1) \times \cdots \times f_n(y_n; \theta, x_n).$$

Here $f_i(y_i; \theta, x_i)$ is the probability density for y_i. Often the $f_i()$ are the same for all y_i. However, $f_i()$ may be a different model for each specimen. That is, each $f_i()$ may involve a different distribution or different functions of the independent variables $(x_1, \cdots, x_J)$. This is so if specimens come from different populations or if they have a mix of failure modes each with its own distinct model. Also, a parameter θ_p may be a distribution parameter (e.g., lognormal σ or log-gamma shape parameter), a coefficient in a relationship, or whatever. The log-gamma shape parameter lets one compare how well the Weibull and lognormal distributions fit a sample. The model also can include AOV relationships with categorical (indicator) variables. The model may include a cumulative exposure model for step-stress testing (Chapter 10). Thus the model here is quite general.

Parameter space. The value of the parameters $\theta = (\theta_1, \cdots, \theta_P)$ is regarded as a point in a P-dimensional *parameter space* Ω which is the set of all theoretically possible parameter values. Ω is a subset of Euclidean P-space. Usually Ω is an open subset, excluding parameter values on its boundary, since such boundary points create theoretical complications. Both Ω and $f_i(y_i; \theta, x_i)$ are assumed specified. Examples follow.

Example A. Suppose that $t_1, \cdots, t_n$ are independent observations from a single exponential distribution with mean θ. The probability density for t_i is

$$f_i(t_i; \theta) = (1/\theta)\exp(-t_i/\theta).$$

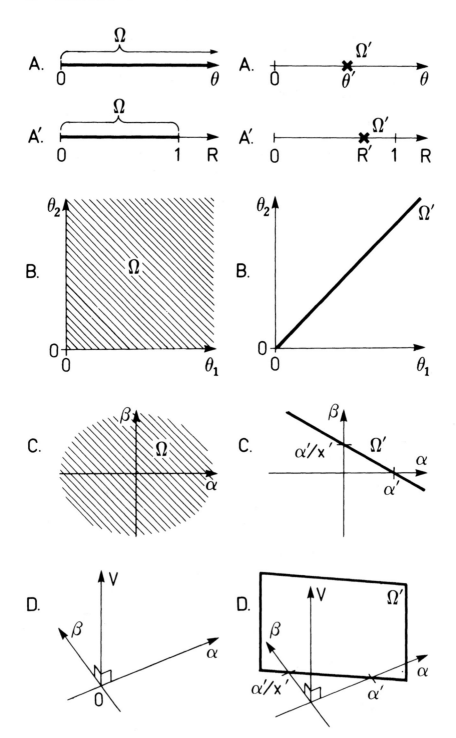

Figure 5.1.   Parameter spaces of examples.   **Figure 5.2.**   Null hypothesis spaces.

The parameter space Ω for θ is $(0,\infty)$, the positive half of the real line. Figure 5.1A depicts this parameter space.

Example B. Suppose that $t_1, \cdots, t_{n_1}$ are n_1 independent observations from a single exponential distribution with mean θ_1, and $t_1', \cdots, t_{n_2}'$ are n_2 independent observations from an exponential distribution with mean θ_2. The probability density for an observation t_i from population k is

$$f_i(t_i;\theta_k) = (1/\theta_k)\exp(-t_i/\theta_k), \quad k = 1,2.$$

The parameter space Ω of points (θ_1,θ_2) consists of the positive quadrant ($\theta_1 > 0$ and $\theta_2 > 0$) of the plane. Figure 5.1B depicts Ω.

Example C. Suppose that t_i is from an exponential distribution with mean $\theta(x_i) = \exp(\alpha + \beta x_i)$, $i = 1,2, \cdots, n$. x_i is the (possibly transformed) value of an independent variable. The probability density for t_i is

$$f_i(t_i;\alpha,\beta,x_i) = [1/\exp(\alpha + \beta x_i)] \exp[-t_i/\exp(\alpha + \beta x_i)].$$

Since $-\infty < \alpha < \infty$ and $-\infty < \beta < \infty$, the parameter space Ω of points (α, β) is the plane. Figure 5.1C depicts Ω.

Example D. Suppose that y_i is a (log) observation from a (log) normal distribution with variance v and mean $\mu(x_i) = \alpha + \beta x_i$. Here x_i is the (possibly transformed) value of an independent variable, $i = 1,2, \cdots, n$. The probability density for y_i is

$$f_i(y_i;\alpha,\beta,v,x_i) = (2\pi v)^{-1/2}\exp[-(y_i - \alpha - \beta x_i)^2/(2v)].$$

The parameter space of points (α, β, v) is half of 3-dimensional space where $v > 0$, $-\infty < \alpha < \infty$, and $-\infty < \beta < \infty$. Figure 5.1D depicts Ω.

Null hypothesis. To better understand the following view of a null hypothesis, reread this paragraph after reading the examples below. In terms of the general model and parameter space, a *null hypothesis* can be viewed as a statement that the model parameters $\theta = (\theta_1, \cdots, \theta_P)$ are a point in a P'-dimensional subspace Ω' of Ω. Ω' is called the *subspace of the null hypothesis*. If the true θ is in Ω', the null hypothesis is true. For LR tests, $P' < P$. That is, Ω' is a *lower* dimensional subspace of Ω. Ω' is usually a (hyper)plane, line, or point in Ω. In other words, the *constrained model* under the null hypothesis is a special case of the (unconstrained or full) general model. For example, the following theory for a likelihood ratio test cannot compare a normal distribution and an exponential distribution, since neither distribution is a special case of the other. It can compare a Weibull distribution and an exponential distribution, since the exponential distribution is a special case of the Weibull distribution with a shape parameter of 1.

Alternative. If the true θ is not in Ω', the alternative is true. Thus the *subspace of the alternative* consists of the points of Ω that are not in Ω'. This subspace is $\Omega - \Omega'$ in set notation. The natural alternative for a LR test is usually two-sided, as in the examples below. Generally, such a two-sided test can be suitably modified for a one-sided alternative as shown later.

Example A continued. Consider a null hypothesis that the exponential mean θ equals a specified value θ', that is, $\theta = \theta'$. Here the subspace of the null hypothesis Ω' consists of a single point θ'. Figure 5.2A depicts this subspace as a point at θ'. Here Ω has $P = 1$ dimension (a line), and Ω' has $P' = 0$ dimensions (a single point). The subspace of the alternative is the positive half line minus the point θ', that is, all positive points $\theta \neq \theta'$. However, in most applications, a one-sided alternative $(\theta > \theta')$ is required.

Example A'. Consider Example A but with a null hypothesis that the reliability $R(t')$ at age t' has a specified value R'. Then the parameter space Ω for $R(t')$ is the unit interval $(0,1)$, and the subspace of the null hypothesis consists of the single point R'. Figures 5.1A' and 5.2A' depict Ω and Ω'. Here Ω has $P = 1$ dimension (the unit interval), and Ω' has $P' = 0$ dimensions (a single point). This null hypothesis is equivalent to that in Example A where $\theta = \theta' = -t'\ln(R')$. The alternative is $R(t') \neq R'$ or equivalently $\theta \neq \theta'$. However, most applications have a one-sided alternative $(R(t') > R')$.

Example B continued. Consider a null hypothesis that the two exponential means are equal; that is, $\theta_1 = \theta_2$. Here the subspace of the null hypothesis Ω' consists of all points (θ_1, θ_2) such that $\theta_1 = \theta_2$. Figure 5.2B depicts this subspace as a 45° line through the origin in the positive quadrant of the plane. In this example, Ω has $P = 2$ dimensions, and Ω' has $P' = 1$ dimension. The subspace of the alternative is the positive quadrant minus the line, that is, the positive points (θ_1, θ_2) satisfying $\theta_1 \neq \theta_2$. Most applications use such a two-sided alternative.

Example C continued. For the simple linear-exponential model, consider a null hypothesis that the mean life at stress level x' has a specified value θ'; that is, $\theta' = \exp(\alpha + \beta x')$. Here the subspace of the null hypothesis Ω' consists of the (α, β) values that satisfy $\alpha + \beta x' = \ln(\theta') \equiv \alpha'$. That is, they are the points on a straight line as depicted in Figure 5.2C. In this example, Ω has $P = 2$ dimensions (the plane), and Ω' has $P' = 1$ dimension (a line). The subspace of the alternative consists of the points (α, β) satisfying $\alpha + \beta x' \neq \ln(\theta') \equiv \alpha'$. However, in most applications, a one-sided alternative is required, namely, $\alpha + \beta x' < \ln(\theta') \equiv \alpha'$.

Example D continued. Fot the simple linear-normal model, consider a null hypothesis that the mean (log) life at stress level x' has a specified value α'; that is, $\mu(x') = \alpha + \beta x' = \alpha'$. Here the subspace of the null hypothesis Ω' consists of the (α, β, v) values that satisfy $\alpha + \beta x' = \alpha'$. These values are the points on the half plane depicted in Figure 5.2D. In this example, Ω has $P = 3$ dimensions (half a 3-space), and Ω' has $P' = 2$ dimensions (half a plane). The subspace of the alternative consists of the points (α, β, v) satisfying $v > 0$ and $\alpha + \beta x' \neq \alpha'$. However, in most applications, a one-sided alternative is required, namely, $\alpha + \beta x' < \alpha'$.

Ways to specify Ω'. There are various ways to specify a null hypothesis, that is, its subspace. Three such ways follow. For many problems, they are equivalent, if the model is suitably reparameterized.

Constant hypothesis. Some null hypotheses are specified by setting certain parameters equal to specified *constants*. For example,

$$\theta_1 = \theta_1', \; \theta_2 = \theta_2', \; \cdots, \; \theta_Q = \theta_Q';$$

here the primed values are specified. Examples A and A' have such a null hypothesis. Such an Ω' is a hyperplane in Ω and has $P' = P - Q$ dimensions. P' is the number of unconstrained parameters $\theta_{Q+1}, \theta_{Q+2}, \cdots, \theta_P$.

Equality hypothesis. A null hypothesis for *equality* (a common value) of K parameters is

$$\theta_1 = \theta_2 = \cdots = \theta_K = \theta', \text{ the common value.}$$

Example B has such a null hypothesis. Also, when the same submodel is separately fitted to K samples, a common hypothesis is that the K values of the same parameter are equal, for example, equal intercept coefficients. Then Ω' is a hyperplane in Ω and has $P' = P - K + 1$ dimensions. P' is the number of unconstrained parameters; namely, $\theta_{K+1}, \cdots, \theta_P$ plus the common value $\theta' = \theta_1 = \cdots = \theta_K$. This equality hypothesis can be expressed as a constant hypothesis. Reparameterize the model with the $K - 1$ parameters $\Delta_1 = \theta_1 - \theta_K$, $\Delta_2 = \theta_2 - \theta_K, \cdots, \Delta_{K-1} = \theta_{K-1} - \theta_K$. Then the constant hypothesis

$$\Delta_1 = 0, \quad \Delta_2 = 0, \quad \cdots, \quad \Delta_{K-1} = 0$$

is equivalent to the equality hypothesis above.

Constraint hypothesis. Some null hypotheses are specified by setting Q *constraints* on parameters (functions of them) equal to zero. For example,

$$h_1(\theta_1, \cdots, \theta_P) = 0, \quad \cdots, \quad h_Q(\theta_1, \cdots, \theta_P) = 0.$$

Examples C and D have one such constraint. In general, then Ω' is a hypersurface (line or point) in Ω and has $P' = P - Q$ dimensions. The theory requires that each function have continuous first partial derivatives with respect to all parameters.

Later we maximize the sample log likelihood subject to such constraints. The method of Lagrange multipliers may help solve such constrained optimization. The theory here applies to hypotheses with equality constraints. Robertson and others (1988) treat in detail the subject of hypotheses with inequality constraints.

5.2. Likelihood Ratio Test

This section presents the LR test for the model and null hypotheses of Section 5.1. This section presents

- the sample likelihood,
- the ML estimates of parameters under the general model and under the constrained (null hypothesis) model,
- the LR test,
- its critical values (exact and approximate), and
- its OC function.

Examples from Section 5.1 illustrate these concepts.

Sample likelihood. Section 5 of Chapter 5 presents sample likelihoods for observed, censored, interval, and quantal-response data. Denote the *general sample likelihood for the model* and form of data by

$$L(\theta_1, \cdots, \theta_P) = L_0(y_1, \cdots, y_n; \theta_1, \cdots, \theta_P, x_1, \cdots, x_n).$$

Here the notation is the same as in Section 5.1. This likelihood includes the form of the data – observed, censored, interval, and quantal-response. The *sample log likelihood* is

$$\mathcal{L}(\theta_1, \cdots, \theta_P) = \ln[L_0(y_1, \cdots, y_n; \theta, x_1, \cdots, x_n)].$$

If $y_1, \cdots, y_n$ are *statistically independent*,

$$L(\theta_1, \cdots, \theta_P) = L_1(y_1; \theta, x_1) \times \cdots \times L_n(y_n; \theta, x_n),$$

and

$$\mathcal{L}(\theta_1, \cdots, \theta_P) = \mathcal{L}_1(y_1; \theta, x_1) + \cdots + \mathcal{L}_n(y_n; \theta, x_n),$$

where $\mathcal{L}_i(y_i, \theta, x_i) = \ln[L_i(y_i; \theta, x_i)]$ is the log likelihood for specimen i. The notation for the (log) likelihood explicitly shows that it is a function of the parameters $\theta = (\theta_1, \cdots, \theta_P)$, but it is also a function of the data $y_1, \cdots, y_n$, $x_1, \cdots, x_n$. Similarly, denote the *constrained log likelihood* under the null hypothesis as $\mathcal{L}'(\theta_1, \cdots, \theta_P)$ where the point $(\theta_1, \cdots, \theta_P)$ is constrained to Ω'. As shown below, the constraints of the null hypothesis are incorporated into this likelihood, which then is a function of the remaining unconstrained parameters. Examples follow.

Example A continued. To make this exponential example more general, assume that the data are multiply censored (on the right), that $t_1, \cdots, t_r$ are observed failure times, and $t_{r+1}, \cdots, t_n$ are censoring times. The general sample likelihood then is

$$L(\theta) = \prod_{i=1}^{r}(1/\theta)\exp(-t_i/\theta) \prod_{i=r+1}^{n} \exp(-t_i/\theta).$$

The general sample log likelihood is

$$\mathcal{L}(\theta) = -r\ln(\theta) - (T/\theta);$$

here $T = t_1 + \cdots + t_n$ is the total time for the sample. The constrained log likelihood under the null hypothesis ($\theta = \theta'$) is

$$\mathcal{L}'(\theta') = -r\ln(\theta') - (T/\theta').$$

Example A′ continued. As in Example A, assume that the data are multiply censored. Express the exponential likelihood in terms of the parameter $R = \exp(-t'/\theta)$. Then the general sample log likelihood is

$$\mathcal{L}(R) = -r\ln[-t'/\ln(R)] + (T/t')\ln(R);$$

here $T = t_1 + \cdots + t_n$ is the total of all n times. The constrained log likelihood under the null hypothesis $(R = R')$ is

$$\mathcal{L}'(R') = -r\ln[-t'/\ln(R')] + (T/t')\ln(R').$$

Example B continued. For the two exponential distributions, assume that the data are multiply censored (on the right). Then $t_1, \cdots, t_{r_1}$ are r_1 observed failures times, and $t_{r_1+1}, \cdots, t_{n_1}$ are $n_1 - r_1$ censoring times from sample 1. Similarly, $t'_1, \cdots, t'_{r_2}$ are r_2 observed failure times and $t'_{r_2+1}, \cdots, t'_{n_2}$ are $n_2 - r_2$ censoring times from sample 2. Since the samples are statistically independent, the sample log likelihood is the sum of two log likelihoods like that in Example A. That is,

$$\mathcal{L}(\theta_1, \theta_2) = -r_1\ln(\theta_1) - (T_1/\theta_1) - r_2\ln(\theta_2) - (T_2/\theta_2);$$

here T_k is the total time for sample k. The constrained log likelihood under the null hypothesis $(\theta_1 = \theta_2 = \theta')$, the common value) is

$$\mathcal{L}'(\theta') = \mathcal{L}(\theta', \theta') = -(r_1 + r_2)\ln(\theta') - [(T_1 + T_2)/\theta'].$$

Example C continued. For the simple linear-exponential model, assume that the sample is multiply censored (on the right). Then $t_1, \cdots, t_r$ are r observed failure times, and $t_{r+1}, \cdots, t_n$ are $n - r$ censoring times. The general sample log likelihood is

$$\mathcal{L}(\alpha, \beta) = -\sum_{i=1}^{r}(\alpha + \beta x_i) - \sum_{i=1}^{n} t_i \exp(-\alpha - \beta x_i).$$

The constraint for the null hypothesis is $\theta(x') = \exp(\alpha + \beta x') = \theta'$. Equivalently, $\alpha + \beta x' = \ln(\theta') \equiv \alpha'$ or $\alpha = \alpha' - \beta x'$. Thus the constrained log likelihood as a function of β' is

$$\mathcal{L}'(\beta') = \mathcal{L}(\alpha' - \beta'x', \beta') = -\sum_{i=1}^{r}[\alpha' + \beta'(x_i - x')] - \sum_{i=1}^{n} t_i \exp[-\alpha' - \beta'(x_i - x')].$$

Example D continued. For the simple linear-(log)normal model and complete data, the sample log likelihood is

$$\mathcal{L}(\alpha, \beta, v) = -(n/2)\ln(2\pi) - (n/2)\ln(v) - (0.5/v)\sum_{i=1}^{n}(y_i - \alpha - \beta x_i)^2.$$

The constraint for the null hypothesis is $\mu(x') = \alpha + \beta x' = \alpha'$ or $\alpha = \alpha' - \beta x'$. Thus the constrained log likelihood as a function of β' and v' is

$$\mathcal{L}'(\beta', v') = \mathcal{L}(\alpha' - \beta'x', \beta', v')$$

$$= -(n/2)\ln(2\pi) - (n/2)\ln(v') - (0.5/v')\sum_{i=1}^{n}[y_i - \alpha' - \beta'(x_i - x')]^2.$$

ML Estimate. The *ML estimate under the general model* for the parameters $\boldsymbol{\theta} = (\theta_1, \cdots, \theta_P)$ is the point $\hat{\boldsymbol{\theta}} = (\hat{\theta}_1, \cdots, \hat{\theta}_P)$ in the parameter space Ω that maximizes the sample likelihood $L(\theta_1, \cdots, \theta_P)$ or (equivalently) the log likelihood $\mathcal{L}(\theta_1, \cdots, \theta_P)$. The value $\hat{\boldsymbol{\theta}}' = (\hat{\theta}_1', \cdots, \hat{\theta}_P')$ in Ω' that maximizes the constrained likelihood $L'(\theta_1, \cdots, \theta_P)$ or $\mathcal{L}'(\theta_1, \cdots, \theta_P)$ is the constrained *ML estimate under the null hypothesis.* Each parameter estimate $\hat{\theta}_p$ or $\hat{\theta}_p'$ is a function of the data $y_1, \cdots, y_n, x_1, \cdots, x_n$. Usually the constrained ML estimate $\hat{\theta}_p'$ differs from the unconstrained estimate $\hat{\theta}_p$. Denote the maximum value of these (log) likelihoods by L' and L' ($\hat{\mathcal{L}}$ and $\hat{\mathcal{L}}'$). In theory, such ML estimates are found by the usual calculus method. Namely, set equal to zero the first partial derivative of the (log) likelihood with respect to each parameter and solve the resulting *likelihood equations;* this is described in Chapter 5. In practice, such equations usually cannot be solved explicitly. Then a computer program yields such estimates by an iterative numerical calculation. Nelson (1982, Chap. 8, Sec. 6) describes such calculations. Examples follow.

Example A continued. The general (unconstrained) ML estimate $\hat{\theta}$ of the exponential mean is the solution of the likelihood equation

$$0 = \partial \mathcal{L}/\partial \theta = -(r/\theta) + (T/\theta^2).$$

Thus $\hat{\theta} = T/r$, and $\hat{\mathcal{L}} = -r\ln(T/r) - r$. Under the null hypothesis, $\theta = \theta'$ and $\hat{\mathcal{L}}' = -r\ln(\theta') - (T/\theta')$.

Example B continued. For the two exponential distributions, the general (unconstrained) ML estimates $\hat{\theta}_1, \hat{\theta}_2$ are the solution of the likelihood equations

$$0 = \partial \mathcal{L}/\partial \theta_1 = -(r_1/\theta_1) + (T_1/\theta_1^2), \quad 0 = \partial \mathcal{L}/\partial \theta_2 = -(r_2/\theta_2) + (T_2/\theta_2^2).$$

Thus $\hat{\theta}_1 = T_1/r_1, \hat{\theta}_2 = T_2/r_2$, and $\hat{\mathcal{L}} = -r_1\ln(T_1/r_1) - r_1 - r_2\ln(T_2/r_2) - r_2$. The constrained ML estimate $\hat{\theta}'$ is the solution of the likelihood equation

$$0 = \partial \mathcal{L}'/\partial \theta' = -[(r_1+r_2)/\theta'] + [(T_1+T_2)/\theta'^2].$$

Thus $\hat{\theta}' = \hat{\theta}_1' = \hat{\theta}_2' = (T_1+T_2)/(r_1+r_2)$, and $\hat{\mathcal{L}}' = -(r_1+r_2)\ln[(T_1+T_2)/(r_1+r_2)] - (r_1+r_2)$.

Example C continued. For the simple linear-exponential model, the general (unconstrained) ML estimate $(\hat{\alpha}, \hat{\beta})$ is the solution of the likelihood equations

$$0 = \partial \mathcal{L}/\partial \alpha = -r + \sum_{i=1}^{n} t_i \exp(-\alpha - \beta x_i),$$

$$0 = \partial \mathcal{L}/\partial \beta = -\sum_{i=1}^{v} x_i + \sum_{i=1}^{n} t_i x_i \exp(-\alpha - \beta x_i).$$

In general, these equations cannot be solved explicitly for $\hat{\alpha}$ and $\hat{\beta}$. Suppose there are just two test stress levels x_1 and x_2. Also, suppose that n_1 and n_2 are the numbers of specimens, r_1 and r_2 the numbers of failures, and T_1 and T_2 the totals of the n_1 and n_2 (failure and censoring) times. Then

$$0 = \partial\mathcal{L}/\partial\alpha = -(r_1 + r_2) + T_1\exp(-\alpha-\beta x_1) + T_2\exp(-\alpha-\beta x_2),$$

$$0 = \partial\mathcal{L}/\partial\beta = -r_1 x_1 - r_2 x_2 + T_1 x_1 \exp(-\alpha-\beta x_1) + T_2 x_2 \exp(-\alpha-\beta x_2),$$

Their solution is $\hat{\beta} = [\ln(\hat{\theta}_2)-\ln(\hat{\theta}_1)]/(x_2-x_1)$ and $\hat{\alpha} = \ln\{[T_1 \exp(-\hat{\beta}x_1) + T_2 \exp(-\hat{\beta}x_2)]/(r_1+r_2)\}$ where $\hat{\theta}_k = T_k/r_k$. Then $\hat{\mathcal{L}} = -r_1\ln(T_1/r_1)-r_1 - r_2\ln(T_2/r_2)-r_2$. Under the null hypothesis ($\alpha = \alpha'-\beta x'$), the likelihood equation is

$$0 = \partial\mathcal{L}'/\partial\beta' = -\sum_{i=1}^{v}(x_i-x') + \sum_{i=1}^{n}t_i(x_i - x')\exp[-\alpha'-\beta'(x_i-x')]$$

$$= -r_1 x_1 - r_2 x_2 + (r_1+r_2)x' + T_1(x_1-x')\exp[-\alpha'-\beta'(x_1-x')]$$

$$+ T_2(x_2-x')\exp[-\alpha'-\beta'(x_2-x')].$$

This equation in β' cannot be solved explicitly for $\hat{\beta}'$ nor can $\hat{\mathcal{L}}'$ be evaluated explicitly. This requires numerical solution in practice.

Example D continued. For the simple linear (log)normal model, the general (unconstrained) ML estimates $\hat{\alpha}$, $\hat{\beta}$, $\hat{v}$ are the solution of the likelihood equations (in notation of Chapter 4)

$$0 = \partial\mathcal{L}/\partial\alpha = (1/v)\sum(y_i-\alpha-\beta x_i) = (n/v)(\bar{y}-\alpha-\beta\bar{x}),$$

$$0 = \partial\mathcal{L}/\partial\beta = (1/v)\sum x_i(y_i-\alpha-\beta x_i) = (1/v)(\sum x_i y_i - n\bar{x}\alpha+\beta\sum x_i^2),$$

$$0 = \partial\mathcal{L}/\partial v = -(n/2)(1/v)+(0.5/v^2)\sum(y_i-\alpha-\beta x_i)^2.$$

The solutions are $\hat{\beta} = S_{xy}/S_{xx}$ and $\hat{\alpha} = \bar{y}-\hat{\beta}\bar{x}$ (which are also LS estimates) and $\hat{v}=(1/n)\sum(y_i-\hat{\alpha}-\hat{\beta}x_i)^2$. Then $\hat{\mathcal{L}} = -(n/2)\ln(\hat{v})-(n/2)$. Under the null hypothesis constraint ($\alpha = \alpha'-\beta x'$), the likelihood equations are

$$0 = \partial\mathcal{L}'/\partial\beta' = (1/v')\sum(x_i-x')[y_i-\alpha'-\beta'(x_i-x')],$$

$$0 = \partial\mathcal{L}'/\partial v' = -(n/2)(1/v')+(0.5/v'^2)\sum[y_i-\alpha'-\beta'(x_i-x')]^2.$$

Their solution is $\hat{\beta}' = [\sum(x_i-x')(y_i-\alpha')]/[\sum(x_i-x')^2]$ and $\hat{v}' = (1/n)\sum[y_i-\alpha'-\hat{\beta}'(x_i-x')]^2$. Then $\hat{\mathcal{L}}' = -(n/2)\ln(\hat{v}')-(n/2)$.

Likelihood ratio. Consider the sample data $y_1, \cdots, y_n, x_1, \cdots, x_n$ and the assumed general model Ω. The **likelihood ratio** (LR) Λ for testing the null hypothesis Ω' is the ratio of the constrained and unconstrained *maximum* likelihoods; namely,

$$\Lambda = \Lambda(y_1, \cdots, y_n; x_1, \cdots, x_n) = \hat{L}'/\hat{L}$$

$$= [\max_{\theta\,in\,\Omega'} L'(\theta_1, \cdots, \theta_P)]/[\max_{\theta\,in\,\Omega} L(\theta_1, \cdots, \theta_P)].$$

The LR is a function of just the data, a sample statistic that does not depend on the unknown true parameter values $(\theta_1, \cdots, \theta_P)$. Usually one uses the maximum *log* likelihoods $\hat{\mathcal{L}}$ and $\hat{\mathcal{L}}'$ and the equivalent (*log*) LR test statistic

$$T \equiv -2\ln(\Lambda) = 2(\hat{\mathcal{L}} - \hat{\mathcal{L}}') = 2[\max_{\theta \text{ in } \Omega} \mathcal{L}(\theta_1, \cdots, \theta_P) - \max_{\theta \text{ in } \Omega'} \mathcal{L}'(\theta_1, \cdots, \theta_P)]$$

T is also loosely called the likelihood ratio. Examples follow.

Example A. For the exponential distribution and a specified constant $\theta = \theta'$, the (log) LR statistic is

$$U = 2\{[-r\ln(T/r) - r] - [-r\ln(\theta') - (T/\theta')]\} = 2r[-\ln(\hat{\theta}/\theta') - 1 + (\hat{\theta}/\theta')].$$

Example B. For the two exponential distributions, the (log) LR statistic is

$$T = 2\{[-r_1\ln(T_1/r_1) - r_1 - r_2\ln(T_2/r_2) - r_2]$$
$$-[-(r_1 + r_2)\ln[(T_1 + T_2)/(r_1 + r_2)] - (r_1 + r_2)]\}$$
$$= 2\{r_1\ln[r_1 + r_2(\hat{\theta}_2/\hat{\theta}_1)] + r_2\ln[r_1(\hat{\theta}_1/\hat{\theta}_2) + r_2] - (r_1 + r_2)\ln(r_1 + r_2)\}.$$

Example C. For the simple linear-exponential model, the sample (log) LR statistic $T = 2(\hat{\mathcal{L}} - \hat{\mathcal{L}}')$ cannot be written as an explicit formula. In practice, it must be numerically evaluated from $\hat{\mathcal{L}}$ and $\hat{\mathcal{L}}'$.

Example D. For the simple linear-(log)normal model with complete data, the (log) LR statistic is

$$T = 2\{-(n/2)\ln(\hat{v}) - (n/2) - [(n/2)\ln(\hat{v}') - (n/2)]\} = n\ln(\hat{v}'/\hat{v}).$$

Likelihood ratio test. For different random samples, the likelihood ratio Λ would take on different values between 0 and 1. That is, Λ is a function of random variables $y_1, \cdots, y_n$, and it is a random variable. An observed Λ value near 1 indicates that the corresponding data values $y_1, \cdots, y_n$, are likely under the constrained model Ω', and a Λ value near 0 indicates that the data $y_1, \cdots, y_n$ are not likely under Ω' compared to under the general model $\Omega - \Omega'$. This suggest the *likelihood ratio test*:

1. If $\Lambda > \lambda_\alpha$, accept Ω' (the constrained model). That is, the general model does not fit convincingly better than the constrained model.
2. If $\Lambda \leq \lambda_\alpha$, reject the null hypothesis Ω' (the constrained model). That is, the general model fits convincingly better than the constrained model.

Here λ_α is a chosen constant called the *critical value*. The (log) LR statistic $T = 2(\hat{\mathcal{L}} - \hat{\mathcal{L}}') = -2\ln(\Lambda)$ yields the equivalent test:

1. If $T \leq t_\alpha$, accept the null hypothesis Ω' (the constrained model).
2. If $T > t_\alpha$, reject the null hypothesis Ω'. That is, the general model fits the data convincingly better than the constrained one.

Here $t_\alpha = -2\ln(\lambda_\alpha)$ is the equivalent critical value. How to choose it is described below.

LR test depicted. Figure 5.3 adds insight to the LR test. There the sample log likelihood $\mathcal{L}(\alpha,\beta)$ on the vertical axis is a function of parameters (α,β). Ω is the horizontal plane and $P = 2$. The subspace Ω' of the null hypothesis $\beta=\beta'$, a specified constant, is the labeled line in the figure; $P'=1$. Figure 5.3 shows the (unconstrained) ML estimates $(\hat{\alpha},\hat{\beta})$ under Ω and the maximum log likelihood $\hat{\mathcal{L}}$. Figure 5.3 also shows the constrained ML estimate $\hat{\alpha}',\beta'$ under Ω' and $\hat{\mathcal{L}}'$. If $\hat{\mathcal{L}}$ is much above $\hat{\mathcal{L}}'$, the general model fits the data much better than the (constrained) null hypothesis model. Then $T = 2(\hat{\mathcal{L}} - \hat{\mathcal{L}}')$ is large. For different samples, the $\mathcal{L}(\alpha,\beta)$ differs because it is a function of the random data. Thus $\hat{\alpha}$, $\hat{\beta}$, $\hat{\alpha}'$, $\hat{\mathcal{L}}$, $\hat{\mathcal{L}}'$, and T differ from sample to sample and have sampling distributions.

Critical value. The critical value (λ_α or t_α) is chosen so the probability of rejecting the null hypothesis Ω' is small when Ω' is true. One chooses t_α so that the maximum rejection probability equals α, called the *level of the test*. That is,

$$\alpha = \max_{\theta \text{ in } \Omega'} Pr\{\text{reject } \Omega' ; \theta\} = \max_{\theta \text{ in } \Omega'} Pr\{T > t_\alpha ; \theta\}.$$

Thus the LR test is a level α test. Roughly speaking, t_α is the upper α point of the sampling distribution of T under a parameter value θ in the null hypothesis Ω'. Two methods provide this distribution and t_α:

1. Show that T is a function of a statistic U with a known sampling distribution. Then the LR test employs the critical value of the hypothesis test based on U.
2. Approximate the distribution of T under the null hypothesis. This yields an approximate critical value. For samples with many failures, the approximate large-sample distribution of T under the null hypothesis Ω' is chi square with $(P-P')$ degrees of freedom. That is, $t_\alpha \approx \chi^2(1-\alpha;P-P')$. Wilks (1962) and Rao (1973) state regularity conditions on the model and null hypothesis for the validity of this approximation. Even when such

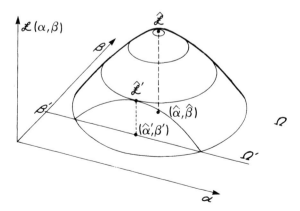

Figure 5.3.   Likelihood function and maxima for $\beta=\beta'$, specified.

mathematical conditions are satisfied, there is no simple rule of thumb for the number of failures that provide a satisfactory approximation. Thus the true level of the test is usually bigger than α. In practice, one often has only the approximation and must use it, since it is better than no test. Then marginally significant results are not convincing. Also, one can simulate the sampling distribution of T to assess the chi-square approximation.

Examples follow.

Example A. For the exponential distribution, the (log) LR statistic $U = 2r[-\ln(\hat{\theta}/\theta') - 1 + (\hat{\theta}/\theta')]$ is a function of the ML estimate $\hat{\theta} = T/r$. For $\hat{\theta} = \theta'$, $U = 0$, the minimum. Also, U increases as $\hat{\theta}$ goes away from θ'. Thus $\hat{\theta}$ is an equivalent statistic. For a failure censored sample, $2r\hat{\theta}/\theta'$ has a chi-square distribution with $2r$ degrees of freedom under the null hypothesis $\theta = \theta'$. For a time censored sample, the chi square distribution approximates the sampling distribution of $2r\hat{\theta}/\theta'$. Thus the *equivalent test* has the form:

1. If $\chi^2(\alpha'; 2r) \le 2r\hat{\theta}/\theta' \le \chi^2(1-\alpha''; 2r)$, accept the null hypothesis $\theta = \theta'$.
2. Otherwise, reject the equality hypothesis.

Here $\alpha = \alpha' + \alpha''$ is the level of the test. Usually $\alpha' = \alpha'' = \alpha/2$. This equivalent test suggests a one-sided *demonstration test* of the form

1'. If $2r\hat{\theta}/\theta' > \chi^2(1-\alpha; 2r)$, reject the null hypothesis $\theta = \theta'$. That is, accept the product as demonstrating that it surpasses the specified mean life θ'.
2'. If $2r\hat{\theta}/\theta' \le \chi^2(1-\alpha; 2r)$, accept the null hypothesis. That is, reject the product as not proving it surpasses the specified mean θ'.

The approximate *LR test* is

1''. If $U = 2r[-\ln(\hat{\theta}/\theta') - 1 + (\hat{\theta}/\theta')] \le \chi^2(1-\alpha; 1)$, accept the null hypothesis $\theta = \theta'$.
2''. Otherwise reject $\theta = \theta'$.

Here the number of degrees of freedom is $P - P' = 1 - 0 = 1$. This test is equivalent to

1'''. If $\rho' \le \hat{\theta}/\theta' \le \rho''$, accept the null hypothesis $\theta = \theta'$.
2'''. Otherwise, reject $\theta = \theta'$.

Here $\rho' < 1 < \rho''$ are the two solutions of $-\ln(\rho) + \rho = 1 + [\chi^2(1-\alpha; 1)/(2r)]$. This test is similar to the first one above.

Example B. For the two exponential distributions, the LR test statistic is $T = 2\{r_1 \ln[r_1 + r_2(\hat{\theta}_2/\hat{\theta}_1)] + r_2 \ln[r_1(\hat{\theta}_1/\hat{\theta}_2) + r_2] - (r_1 + r_2)\ln(r_1 + r_2)\}$. For $F \equiv \hat{\theta}_2/\hat{\theta}_1 = 1$, $T = 0$, the minimum. Also, T increases as F goes away from 1. Thus F is an equivalent statistic. For failure censored samples, F has an F distribution with $2r_2$ degrees of freedom in the numerator and $2r_1$ in the denominator, under the null hypothesis $\theta_1 = \theta_2$. For time censored data, the F

distribution is an approximation to the sampling distribution of $F = \hat{\theta}_2/\hat{\theta}_1$. Thus the *equivalent test* has the form:

1. If $F(\alpha';2r_2,2r_1) \leq F \leq F(1-\alpha'';2r_2,2r_1)$, accept the null hypothesis $\theta_1 = \theta_2$.
2. Otherwise, reject the equality hypothesis.

Here $\alpha = \alpha' + \alpha''$ is the level of the test, and $F(\alpha';2r_2,2r_1)$ is the $100\alpha'$th F percentile with $2r_2$ ($2r_1$) degrees of freedom in the numerator (denominator). This two-sided test suggests an obvious one-sided test. The approximate *LR test* is

1'. If $T = 2\{r_1\ln[r_1 + r_2 F] + r_2\ln[(r_1/F) + r_2] - (r_1 + r_2)\ln(r_1 + r_2)\} \leq \chi^2(1-\alpha;1)$, accept the null hypothesis $\theta_1 = \theta_2$.
2'. Otherwise, reject the equality hypothesis.

Here the number of degrees of freedom is $P - P' = 2 - 1 = 1$. This test is equivalent to

1''. If $F' \leq F \leq F''$, accept $\theta_1 = \theta_2$.
2''. Otherwise, reject $\theta_1 = \theta_2$.

Here $F' < 1 < F''$ are the two solutions of $2\{r_1\ln[r_1 + r_2 F] + r_2\ln[(r_1/F) + r_2] - (r_1 + r_2)\ln(r_1 + r_2)\} = \chi^2(1-\alpha;1)$.

Example C. For the simple linear-exponential model, the sampling distribution of the statistic T cannot be expressed in terms of another statistic whose distribution is known. The approximate LR test is

1. If $T \leq \chi^2(1-\alpha;1)$, accept the null hypothesis $\alpha + \beta x' = \alpha'$.
2. If $T > \chi^2(1-\alpha;1)$, reject $\alpha + \beta x' = \alpha'$.

Here the number of degrees of freedom is $P - P' = 2 - 1 = 1$.

Example D. For the simple linear-(log)normal model, after some heroic algebra (Problem 9.11), the LR statistic becomes $T = n \cdot \ln\{1 + [t^2/(n-2)]\}$. Here $t = (m' - \alpha')/\{\hat{v}[(1/n) + (x' - \bar{x})^2 / S_{xx}]\}^{1/2}$ is the t statistic ($n - 2$ degrees of freedom) for the difference $m' - \alpha'$, and m' is the LS estimate of $\mu(x')$. $T = 0$ at $t = 0$, that is, at $m' = \alpha'$. Also, T is an increasing function of $|t|$. Thus $|t|$ is an equivalent statistic. Thus the (exact) equivalent test is:

1. If $|t| \leq t[1 - (\alpha/2); n - 2]$, accept the null hypothesis $\mu(x') = \alpha'$.
2. Otherwise, reject equality.

Here $t[1 - (\alpha/2); n - 2]$ is the $100[1 - (\alpha/2)]$th t-percentile with $n - 2$ degrees of freedom, and α is the level of the test. This two-sided test suggests the following one-sided *demonstration test*

1'. If $t \leq t[1 - \alpha; n - 2]$, reject the product. It does not convincingly surpass $\mu(x') = \alpha'$.
2'. If $t > t[1 - \alpha; n - 2]$, accept the product at the $100(1-\alpha)\%$ confidence level. That is, the product convincingly surpasses $\mu(x') = \alpha'$.

The approximate two-sided LR test (level α) is

$1''$. If $n \cdot \ln\{1+[t^2/(n-2)]\} \leq \chi^2(1-\alpha;1)$, accept the null hypothesis $\mu(x')=\alpha'$.

$2''$. Otherwise, reject $\mu(x')=\alpha'$.

Here the number of degrees of freedom is $P-P'=3-2=1$. This could also yield an approximate one-sided test.

Test that K parameters are equal. The general theory yields the LR test (Section 4) for equality of K corresponding parameter values. Suppose that there are K independent samples, and sample k comes from the same model with J parameters $\gamma_{1k},\gamma_{2k}, \cdots ,\gamma_{Jk}$. The parameter space Ω consists of all allowed points $(\gamma_{11}, \cdots ,\gamma_{J1}; \cdots ; \gamma_{1K}, \cdots ,\gamma_{JK})$. Ω has $P=J \cdot K$ dimensions. The subspace Ω' for the null (equality) hypothesis $\gamma_{11}=\gamma_{12}= \cdots =\gamma_{1K}=\gamma_1$ (the common value) consists of all allowed points $(\gamma_1,\gamma_{21}, \cdots ,\gamma_{J1}; \cdots ; \gamma_1,\gamma_{2K}, \cdots ,\gamma_{JK})$. Ω' has $P'=JK-(K-1)$ dimensions.

Suppose that sample k has log likelihood $\mathcal{L}_k(\gamma_{1k}, \cdots ,\gamma_{Jk})$. Then the log likelihood for the general model is $\mathcal{L}(\gamma_{11}, \cdots ,\gamma_{J1}; \cdots ;\gamma_{1K}, \cdots ,\gamma_{JK}) = \mathcal{L}_1(\gamma_{11}, \cdots ,\gamma_{J1})+ \cdots +\mathcal{L}_K(\gamma_{1K}, \cdots ,\gamma_{JK})$. Also, the log likelihood for the null hypothesis model is

$$\mathcal{L}'(\gamma_1,\gamma_{21}, \cdots ,\gamma_{J1}; \cdots ;\gamma_1,\gamma_{2K}, \cdots ,\gamma_{JK}) = \mathcal{L}_1(\gamma_1,\gamma_{21}, \cdots ,\gamma_{J1})+ \cdots$$
$$+\mathcal{L}_K(\gamma_1,\gamma_{2K}, \cdots ,\gamma_{JK}).$$

Under Ω, the ML estimates $\hat{\gamma}_{1k}, \cdots ,\hat{\gamma}_{Jk}$ maximize the log likelihood $\mathcal{L}_k(\gamma_{1k}, \cdots ,\gamma_{Jk})$. That is, separately fit the model to sample k to obtain these ML estimates and the maximum value $\hat{\mathcal{L}}_k$. Then the unconstrained maximum log likelihood for the combined samples is $\hat{\mathcal{L}}=\hat{\mathcal{L}}_1+ \cdots +\hat{\mathcal{L}}_K$. Under Ω', the ML estimates $(\hat{\gamma}_1;\hat{\gamma}_{21}, \cdots ,\hat{\gamma}_{J1}; \cdots ;\hat{\gamma}_{2K}, \cdots ,\hat{\gamma}_{JK})$ maximize $\mathcal{L}'(\ )$, which has a maximum value of $\hat{\mathcal{L}}'$. The LR test statistic is $T=2(\hat{\mathcal{L}}-\hat{\mathcal{L}}')$. Under the null hypothesis, its distribution is approximately chi square with $P-P'=J \cdot K - [JK-(K-1)] = K-1$ degrees of freedom. Under the general model Ω, the parameter estimates from different samples are statistically independent, since the samples are statistically independent and the models for the samples have no common parameter values. Under the constrained model Ω', $\hat{\gamma}_1$ is a common estimate, and the other parameter estimates are generally correlated (not statistically independent).

Test that K models are identical. The general theory yields the LR test (Section 4) that K models are identical. Suppose that there are K independent samples, and sample k comes from the same model with J parameters $\gamma_{1k}, \cdots , \gamma_{Jk}$. The parameter space Ω consists of all allowed points $(\gamma_{11}, \cdots ,\gamma_{J1}; \cdots ;\gamma_{1K}, \cdots ,\gamma_{JK})$. Ω has $P=J \cdot K$ dimensions. The null (equality) hypothesis is $\gamma_{11} = \cdots = \gamma_{1K} = \gamma_1$ (the common value), $\cdots ,\gamma_{J1} = $

$\cdots = \gamma_{JK} = \gamma_J$ (the common value). Its subspace Ω' consists of all allowed points $(\gamma_1, \cdots, \gamma_J; \cdots; \gamma_1, \cdots, \gamma_J)$. Ω' has $P' = J$ dimensions.

Suppose that sample k has log likelihood $\ell_k(\gamma_{1k}, \cdots, \gamma_{Jk})$. Then the log likelihood for the general model is $\ell = \ell(\gamma_{11}, \cdots, \gamma_{J1}; \cdots; \gamma_{1K}, \cdots, \gamma_{JK}) = \ell_1(\gamma_{11}, \cdots, \gamma_{J1}) + \cdots + \ell_K(\gamma_{1K}, \cdots, \gamma_{JK})$. The log likelihood for the null hypothesis model is $\ell'(\gamma_1, \cdots, \gamma_J) = \ell_1(\gamma_1, \cdots, \gamma_J) + \cdots + \ell_K(\gamma_1, \cdots, \gamma_J)$. Under Ω, the ML estimates $\hat{\gamma}_{1k}, \cdots, \hat{\gamma}_{Jk}$ maximize $\ell_k(\gamma_{1k}, \cdots, \gamma_{Jk})$. That is, the model is separately fitted to sample k to obtain the ML estimates and the maximum log likelihood $\hat{\ell}_k$. Then $\hat{\ell} = \hat{\ell}_1 + \cdots + \hat{\ell}_K$. Under Ω', the ML estimates, $\hat{\gamma}'_1, \cdots, \hat{\gamma}'_J$ maximize $\ell'(\gamma_1, \cdots, \gamma_J)$, and its maximum value is $\hat{\ell}'$. That is, a single model is fitted to the pooled data from all K samples to get $\hat{\gamma}'_1, \cdots, \hat{\gamma}'_J$, since ℓ' is the likelihood for a single model and the pooled data. Under Ω', the test statistic $T = 2(\hat{\ell} - \hat{\ell}')$ has a distribution that is approximately chi square with $P - P' = JK - J = J(K-1)$ degrees of freedom.

OC function. The *Operating Characteristic (OC) function* of a hypothesis test is the probability $P\{T > t_\alpha; \theta\}$ that the test rejects the null hypothesis Ω' as a function of the parameter values $\theta = (\theta_1, \cdots, \theta_P)$. The OC function gives the performance of a LR test as a function of θ in Ω. Sometimes the distribution of T is known or the test is equivalent to a known one, as in Examples A, B, and D. Then one can get the OC function from various sources such as Kraemer and Thiemann (1987). If, as in Example C, the sampling distribution of T is not known, one cannot find the exact OC function. The following result assures that the approximate LR test is "good" (consistent) for a large sample with many failures.

Consistency. A level α test of a null hypothesis Ω' against the alternative $\Omega - \Omega'$ is called a **consistent test** if $P\{\text{reject } \Omega'; \theta\} \to 1$ as the sample size $n \to \infty$ for any θ in $\Omega - \Omega'$. This just says that the test is almost sure to reject the null hypothesis when it is false (i.e., θ is in $\Omega - \Omega'$) if the sample size is large enough. For example, consider $y_1, \cdots, y_n$ from a normal distribution with unknown mean μ and known standard deviation σ_0. Consider testing Ω'; $\mu = \mu_0$ against $\mu \neq \mu_0$. The level α test is: if $|\bar{y} - \mu_0| \leq K\sigma_0/n^{1/2}$, accept Ω', and if $|\bar{y} - \mu_0| > K\sigma_0/n^{1/2}$, reject Ω'. Here K is the $100[1 - (\alpha/2)]$th standard normal percentile. Then $P\{\text{reject } \Omega'; \mu\} = P\{|\bar{y} - \mu_0| > K\sigma_0/n^{1/2}; \mu\}$ $= \Phi\{[\mu_0 - K(\sigma_0/n^{1/2}) - \mu](n^{1/2}/\sigma_0)\} + 1 - \Phi\{[\mu_0 + K(\sigma_0/n^{1/2}) - \mu](n^{1/2}/\sigma_0)\}$. For $\mu \neq \mu_0$, $P\{\text{reject } \Omega'; \mu\} \to 1$ as $n \to \infty$. So the test is consistent. In general, the LR test is consistent under some mild conditions on the model $F(y_1, \cdots, y_n; \theta)$. That is, if T_n is the log LR test statistic based on n failures, then $P\{T_n > \chi^2(1 - \alpha, P - P'); \theta)\} \to 1$ as $n \to \infty$ for any θ in $\Omega - \Omega'$. Moreover, no other test achieves this limit faster. Consequently, the likelihood ratio test is said to be *"asymptotically" uniformly most powerful*. The previous sentence is not intended to be precise. Rao (1973) and Wilks (1962) state such regularity conditions and prove this consistency result. Also, they and Nelson (1977) show how to calculate an asymptotic OC function.

5.3. Related Tests

There are other tests that are asymptotically equivalent to the LR test. This section presents two such tests – Rao's and Wald's. For some applications, these tests are easier to calculate in practice than the LR test.

Rao's test. Rao (1973, p. 418) gives the following test statistic. Suppose that the $P\times1$ column vector of *scores* is (prime ´ denotes vector transpose)

$$S(\theta) = (\partial\mathcal{L}/\partial\theta_1, \cdots, \partial\mathcal{L}/\partial\theta_P)'.$$

Here $\mathcal{L}$ is the sample log likelihood, and $\theta = (\theta_1, \cdots, \theta_P)$ in Ω is the vector of the P parameters under the general model. Under the null hypothesis Ω', the expectation of the score vector is the zero vector. Otherwise, its expectation differs from the zero vector. Denote the $P\times P$ theoretical Fisher matrix under Ω by

$$F(\theta) = \{-E[\partial^2\mathcal{L}/\partial\theta_p\partial\theta_{p'}]\} = \{-E[(\partial\mathcal{L}/\partial\theta_p)(\partial\mathcal{L}/\partial\theta_{p'})]\}, \quad p,p' = 1, \cdots, P.$$

Suppose that $\hat{\theta}' = (\hat{\theta}_1', \cdots, \hat{\theta}_{P}')'$ is the $P\times1$ column vector of ML estimates under the null hypothesis Ω' with P' dimensions. For example, some $\hat{\theta}_p'$ will be (a) constants, (b) equal to other $\hat{\theta}_{p'}'$, or (c) functions of other $\hat{\theta}_p$. *Rao's statistic* for testing the null hypothesis that θ is in Ω' is the quadratic form

$$R = S'(\hat{\theta}')[F(\hat{\theta}')]^{-1}S(\hat{\theta}').$$

This statistic measures how far the observed score vector is from zero. Under the null hypothesis, R is asymptotically equal to the log LR statistic and to Wald's test statistic below. That is, R then has an asymptotic chi-square distribution with $P - P'$ degrees of freedom. R is not convenient for multiply censored data, since the expectations for $F(\theta)$ are difficult to calculate. Then one can use the local estimate of the Fisher information matrix provided by most ML programs. The SURVREG program of Preston and Clarkson (1980) does this. R employs only the ML estimates under the null hypothesis. Thus R avoids the sometimes greater labor of calculating the ML estimates for the general model.

Equality of Poisson λ_k. The following is an example of Rao's test. Suppose that $Y_1, \cdots, Y_K$ are independent Poisson counts where λ_k is the occurrence rate and t_k is the length of observation, $k = 1, \cdots, K$. The null (equality) hypothesis is $\lambda_1 = \cdots = \lambda_K$. The sample log likelihood under Ω is

$$\mathcal{L}(\lambda_1, \cdots, \lambda_K) = \sum_{k=1}^{K} [-\lambda_k t_k + Y_k\ln(\lambda_k t_k) - \ln(Y_k!)].$$

The kth score is $\partial\mathcal{L}/\partial\lambda_k = -t_k + (Y_k/\lambda_k)$, $k = 1, \cdots, K$. Under the null hypothesis, $\hat{\lambda}_1' = \cdots = \hat{\lambda}_K' = \hat{\lambda}' = (Y_1 + \cdots + Y_K)/(t_1 + \cdots + t_K)$, and the vector of scores is

$$S(\hat{\lambda}_1', \cdots, \hat{\lambda}_K') = [-t_1 + (Y_1/\hat{\lambda}'), \cdots, -t_K + (Y_K/\hat{\lambda}')]'.$$

The terms in the Fisher matrix are

$$E\{-\partial^2 \mathcal{L}/\partial\lambda_k^2\} = E\{Y_k/\lambda_k^2\} = \lambda_k t_k/\lambda_k^2 = t_k/\lambda_k,$$

$$E\{-\partial^2 \mathcal{L}/\partial\lambda_k\partial\lambda_{k'}\} = E\{0\} = 0 \quad \text{for } k \neq k'.$$

The Fisher matrix is diagonal. Its estimate under the null hypothesis has $\hat{\lambda}_k = \hat{\lambda}'$ in place of λ_k. Then Rao's test statistic is

$$R = [(Y_1/\hat{\lambda}') - t_1, \cdots, (Y_K/\hat{\lambda}') - t_K] \begin{bmatrix} t_1/\hat{\lambda}' & & 0 \\ & \cdot & \\ & & \cdot \\ 0 & & t_K/\hat{\lambda}' \end{bmatrix}^{-1} \begin{bmatrix} (Y_1/\hat{\lambda}') - t_1 \\ \cdot \\ \cdot \\ \cdot \\ (Y_K/\hat{\lambda}') - t_K \end{bmatrix}$$

$$= \sum_{k=1}^{K} (Y_k - \hat{\lambda}' t_k)^2 / (\hat{\lambda}' t_k).$$

This is the chi-square test statistic (quadratic form) for equality of Poisson occurrence rates (Nelson (1982, Chap. 10)). Under the null hypothesis, R has an asymptotic chi square distribution with $K-1$ degrees of freedom.

Wald's test. Rao (1973, p. 419) gives Wald's test statistic. It is asymptotically equivalent to the LR statistic. Suppose that the subspace Ω' of the null hypothesis is specified by Q constraints

$$h_1(\theta_1, \cdots, \theta_P) = 0, \quad \cdots, \quad h_Q(\theta_1, \cdots, \theta_P) = 0.$$

The test uses the $P \times Q$ matrix of partial derivatives

$$\mathbf{H}(\theta) = \{\partial h_q/\partial\theta_p\}, \quad q = 1, \cdots, Q, \quad p = 1, \cdots, P.$$

They depend on θ. Suppose that $\hat{\theta} = (\hat{\theta}_1, \cdots, \hat{\theta}_P)'$ are the ML estimates under Ω, the general model. Denote their asymptotic covariance matrix by $\sum_{\hat{\theta}}(\theta)$; it depends on θ. $\hat{\mathbf{h}} = \mathbf{h}(\hat{\theta})$ denotes the $Q \times 1$ vector of constraints evaluated at $\hat{\theta}$. The asymptotic covariance matrix of $\mathbf{h}(\hat{\theta})$ is the $Q \times Q$ matrix

$$\sum_{\hat{\mathbf{h}}}(\theta) = \mathbf{H}'(\theta) \sum_{\hat{\theta}}(\theta) \mathbf{H}(\theta).$$

This is estimated by using the derivatives at $\theta = \hat{\theta}$ and the ML or local estimate of $\sum_{\hat{\theta}}$. *Wald's statistic* for testing the null hypothesis (the Q constraints) is

$$W = \mathbf{h}'(\hat{\theta}) [\sum_{\hat{\mathbf{h}}}(\hat{\theta})]^{-1} \mathbf{h}(\hat{\theta}).$$

This is a quadratic form in the observed values of the constraints; it is a measure of how close they are to zero. Under the (constrained) null hypothesis, W is asymptotically equal to the log LR statistic and to Rao's test statistic. That is, W has a chi square distribution with Q degrees of freedom. W is convenient to use with multiply censored data. Then one can use the local estimate of $\sum_{\hat{\theta}}(\theta)$ in place of its ML estimate. Most ML programs give this local estimate. W employs only the ML estimates under the general model. This is convenient, since calculating estimates for the constrained null hypothesis model often requires special features not in some ML programs.

Snubber example. Life test data on an Old and a New type of snubber for a toaster were assumed to come from normal life distributions. One can compare the two types with a test of the hypothesis Ω': $\mu_O = \mu_N$, assuming $\sigma_O = \sigma_N = \sigma$. The alternative is: $\mu_O \neq \mu_N$, assuming $\sigma_O = \sigma_N = \sigma$. Expressed as $Q = 1$ constraint, the null hypothesis is $h_1(\mu_O, \mu_N, \sigma) = \mu_O - \mu_N = 0$. The partial derivatives are $\partial h_1 / \partial \mu_O = 1$, $\partial h_1 / \partial \mu_N = -1$, $\partial h_1 / \partial \sigma = 0$. The matrix of partial derivatives is (the column vector) $\mathbf{H} = (1 \quad -1 \quad 0)'$. The ML estimates under Ω are $\hat{\mu}_O = 974.3$, $\hat{\mu}_N = 1061.3$, and $\hat{\sigma} = 458.4$. The local estimate of the covariance matrix of the ML estimates under Ω is

$$
\begin{array}{ccc}
\hat{\mu}_O & \hat{\mu}_N & \hat{\sigma}
\end{array}
$$
$$
\sum_{\hat{\theta}}^* = \begin{bmatrix} 7930.61 & 1705.42 & 2325.45 \\ 1705.42 & 8515.93 & 2435.20 \\ 2325.45 & 2435.20 & 3320.58 \end{bmatrix}.
$$

The estimate of the 1×1 covariance matrix of $\hat{h}_1$ is $\sum_{\hat{h}}^* = (1 -1\ 0)\sum_{\hat{\theta}}^*$ $(1 -1\ 0)' = 13035.7$. The Wald statistic is $W = (974.3 - 1061.3)'(13035.7)^{-1}$ $(974.3 - 1061.3) = 0.58$. Under $\mu_O = \mu_N$, the distribution of W is approximately chi square with one degree of freedom. Since $W = 0.58 < 2.706 = \chi^2(0.90;1)$, the two means do not differ convincingly at even the 10% level.

PROBLEMS (* denotes difficult or laborious)

9.1. Insulating oil. Analyze the data of Section 2, omitting the 32 kV data, which have a much lower shape estimate.
(a) Repeat all analyses of Section 2. What conclusions change?
(b) Suitably plot all estimates and confidence limits from (a).

9.2. Power-lognormal. For the insulating oil data, repeat the analyses of Section 2, using the power-lognormal model. Note any differing conclusions.

9.3. Arrhenius-Weibull. Repeat the analyses of the data on the three motor insulations in Section 4, but use the Arrhenius-Weibull model. Note any markedly different results. Does the model fit adequately?

9.4.* Transformer oil. Use the transformer oil data of Problem 8.4. Use the first model of Problem 4.10.
(a) For the two samples, compare each coefficient in the relationship, using the LR test and confidence intervals. Plot individual and pairwise confidence limits.
(b) Do (a) for the Weibull shape parameter.
(c) Simultaneously compare all model parameters.
(d) Suggest further analyses, for example, using the other proposed relationship in Problem 4.10.
(e) Carry out (d).
(f) Make probability and relationship plots of the data.

9.5. $1,000,000 experiment. Do ML comparisons of the data of Problem 3.9. Treat the data from each cycling rate as a separate sample. Fit a separate Arrhenius-lognormal model to each sample. Treat the data as interval data if you wish.

(a) Compare the activation energies, using the LR test and confidence intervals. Plot individual and pairwise confidence limits.

(b) Do (a) for the intercept coefficients.

(c) Do (a) for the log standard deviations.

(d) Simultaneously compare all model parameters.

(e) Suggest further analyses.

(f) Carry out (e).

(g) Make probability and relationship plots of the data.

(h) For the insulation engineer, write a short report on your analyses and conclusions.

9.6. Au-Al bonds. Use the data of Problem 8.3.

(a) Carry out all the ML comparisons of Section 4.

(b) Suitably plot the exact confidence limits and test statistics of Chapter 8 and those of Section 4. Comment on how they compare. In particular, do the approximate ML methods yield different conclusions?

9.7. Exact limits. Calculate exact 99% confidence limits of McCool (1974,1981)for the seven Weibull shape parameters for the insulating oil (Section 2). Suitably plot these exact limits, normal approximate limits, and corresponding estimates. What do you conclude from these simultaneous 93% confidence intervals.

9.8. Relays. Make all ML comparisons of the data on two relays in Problem 3.11. Suitably plot estimates and confidence limits where possible. Also, make hazard and relationship plots of the data.

9.9.* Fatigue limits. Make ML comparisons of the fatigue data of Problem 5.14. Also make relationship plots of the data. Write a short report for a fatigue expert, summarizing your findings and incorporating output, plots, etc., as appropriate.

9.10.* Linear-exponential model. For Example C of Section 5, assume that there are two test stress levels and the data are multiply time censored.

(a) Derive formulas for the LR confidence interval for mean life at a specified stress level x_0.

(b) Do (a) for the slope coefficient.

(c) Do (a) for the intercept coefficient.

9.11.* Linear-lognormal model. For Example D of Section 5, supply all steps in the derivations of all equations. In particular, show that the LR statistic is a function of the Student t statistic.

9.12.* Behrens-Fisher problem. Suppose samples of two products are compared in a life test with one accelerating variable. Also, suppose that log life of each product is described with a simple linear-normal model, where the true model parameters differ for the two products. Also, suppose that the data for the two samples are complete.

(a) Derive the likelihood ratio test for equality of the two sigmas.

(b) Derive the normal and LR confidence intervals for the ratio of the two log standard deviations.

(c) Compare the widths of (b) with that of the exact interval based on the F statistic (Chapter 8).

(d) Derive the LR test for equality of the two log means at a design stress level, assuming the other parameters differ.

(e) Derive the normal and LR confidence intervals for the difference of the two mean log lives at a design stress level.

(f) Compare the widths of (e) with that of the usual Student's t interval, which assumes that the sigmas are equal.

(g) Derive the LR test for equality of the two slope coefficients, assuming the other parameters differ.

(h) Derive the normal and LR confidence intervals for the difference of the two slope coefficients.

(i) Compare the widths of (h) with that of the exact interval, which assumes that the sigmas are equal.

(g) Use these results to calculate such LR comparisons of data on two of the three motor insulations. Compare mean log lives at 200°C.

9.13.* Bartlett's test. Use a lognormal distribution, and assume that the data at the J test stress levels are complete. Derive the LR test for homogeneity of the J log standard deviations at those stress levels as follows.

(a) Write the sample log likelihood, assuming the σ_j and μ_j all differ.

(b) Derive the likelihood equations, and solve them for $\hat{\sigma}_j$ and $\hat{\mu}_j$.

(c) Evaluate the maximum log likelihood $\hat{\mathcal{L}}$ at the solutions (b).

(d) Write the sample log likelihood for the data, assuming a common σ' value and differing μ'_j.

(e) Derive the likelihood equations, and solve them for the common $\hat{\sigma}'$ and the $\hat{\mu}'_j$.

(f) Evaluate the maximum log likelihood $\hat{\mathcal{L}}'$ at the solutions (e).

(g) Evaluate the LR test statistic. It differs from Bartlett's test statistic (Chapter 8) by the factor C there, which improves the approximation.

(h) State the LR test and the number of degrees of freedom of the approximate chi square distribution.

(i*) Repeat (a)-(h) for a singly time censored sample at each stress level.

9.14. 9.13 revisited. Derive Wald's test for Problem 9.13 as follows. Use the $J-1$ constraints $h_q = \sigma_q - \sigma_{q+1} = 0$, $q = 1, 2, \cdots, J-1$.

(a) Evaluate the matrix of first partial derivatives of the h_q with respect to each of the model parameters σ_j and μ_j.

(b) Derive the theoretical covariance matrix of the ML estimates $\hat{\sigma}_j$ and $\hat{\mu}_j$ and invert it.

(c) Derive the Wald statistic, state the number of degrees of freedom of its chi square approximation, and state the Wald test.

(d) Apply this test to the Class-H data, and state your conclusions.

(e) Repeat (a)-(d), using the transformed parameter $\theta_j = \ln(\sigma_j)$.

(f) Repeat (a)-(d), using $\theta_j = \sigma_j^{1/3}$, a normalizing transformation to improve the approximation.

(g*) Repeat (a)-(f) for a singly time censored sample at each stress level.

9.15. Test for linearity. Use the LR theory of Section 5 to derive the test of linearity for the simple linear-lognormal relationship appearing in Section 3.3 of Chapter 5. State all models and assumptions.

9.16. GaAs FET demonstration. For a demonstration test of Problem 6.19, derive its asymptotic OC curve. Numerically evaluate and plot it as a function of median life at 125°C.

10

Models and Data Analyses for Step and Varying Stress

Purpose. This advanced chapter is an introduction to simple cumulative exposure models and data analyses for life tests with step and varying stress, described in Section 3 of Chapter 1. Such models are used here to estimate product reliability. However, they are useful background for those who run such "elephant" tests only to identify and fix failure modes (Chapter 1). These models also apply to the life of products under varying stress in service. Needed background includes the description of such tests in Chapter 1, constant-stress models of Chapter 2, ML fitting of Chapter 5, and possibly ML comparisons of Chapter 9.

Limitations. Such testing and models have limitations, including:
- As noted in Chapter 1, step-stress and ramp tests are used to assure failures quickly. However, the accuracy of estimates from such a test is inversely proportional to its length. Such a test yields no greater accuracy than a constant-stress test of the same length. However, the asymptotic theory is a better approximation when there are many failures.
- A model like that in Sections 2.2 and 3.2 applies to a *single failure mode*. If the product has a number of failure modes, each must be described with a separate model. Then such models must be combined as described in Chapter 7. The basic cumulative exposure model of Sections 2.2 and 3.2 is a simple, plausible one that has been used in some applications. It serves as a simple introduction to such models. In applications, one may need a more elaborate model.
- In practice, it is easier to hold a stress constant than to vary it exactly in a prescribed manner. Thus varying-stress tests have an added source of experimental error.

Overview. Section 1 briefly surveys theory and applications of cumulative exposure models. For step-stress tests, Section 2 presents example data, develops a cumulative exposure model, and presents ML fitting of the model. Section 3 does the same for a test with varying stress of any form. The general cumulative exposure model appears in equations (3.3) and (3.4). The rest of this chapter presents special cases and other formulations of it.

1. SURVEY OF THEORY FOR TESTS WITH VARYING STRESS

Purpose. This section briefly surveys a few references from the vast literature on cumulative damage models for accelerated tests with varying stress. Many authors note that such models have not been adequately verified by experiment. Thus all models need critical evaluation. The survey covers early work, metal fatigue, electronics, other products, and statistical developments.

Early work. Yurkowski and others (1967) survey early work on applications of such testing and physical and statistical theory for it. Since then there have been few major advances in physical models and their application. Also, since then statistical methods for fitting such models to data have been developed and computerized so that virtually any model can be properly fitted to such data.

Metal fatigue. Metal fatigue under varying load in *service* is a major research area. Fatigue researchers have developed and evaluated many cumulative damage models. The large ongoing effort in this area indicates that such fatigue is not well enough understood. Saunders (1970,1974) briefly surveys such models, and Murthy and Swartz (1972,1973) provide a bibliography. The simplest such model is Miner's rule, which Palmgren (1924) proposed before Miner (1945) popularized it. It is a deterministic model based on linear damage theory. Its inadequacies for metal fatigue are well known, but it is widely used. The basic cumulative damage model of this chapter is a probabilistic extension of Miner's rule. Fatigue books that treat such models include Bolotin (1969) and Bogdanoff and Kozin (1984). Prot (1948) was first to propose ramp stress for fatigue *testing*. Also, much fatigue testing involves a distribution (spectrum) of loads applied in a random order. Jaros and Zaludova (1972) and Holm and de Mare´ (1988) model fatigue life under such random loading.

Electronics. Step- and progressive-stress testing are widely used in electronics applications to reveal failure modes (elephant testing), so they can be designed out of the product. The following references treat another problem – that of estimating reliability of electronics. In early applications, Endicott and others (1961a,b,1965) and Starr and Endicott (1961) tested capacitors with a linearly increasing voltage (*"ramp stress"*). Hatch, Endicott, and others (1962) report that the cumulative exposure model they use (Section 3.2 here) and their ramp data were consistent with constant-stress data. Yurkowski and others (1967) reference a number of electronics applications.

Other products. Goba's (1969) bibliography on thermal aging of electrical insulation lists references on progressive-stress tests and models. Rosenberg and others (1986) and Yoshioka and others (1987) present models for stability of pharmaceuticals. Rabinowicz and others (1970) report on life tests of light bulbs, electric hand drills, electric motors, and bearing balls.

They repeatedly alternated the stress between a design level (a fraction of the time) and an overstress level (the other fraction). They conclude that Miner's rule describes the life of these products. Problem 10.17 provides a statistical version of Rabinowicz's deterministic model.

Statistical developments. Developed by experts in engineering and physical sciences, most cumulative damage models are deterministic. Statisticians have helped extend such models to probabilistic ones. For example, Birnbaum and Saunders (1968) develop a probabilistic version of Miner's rule, and Shaked and Singpurwalla (1983) generalize the basic model of Section 2. Also, statisticians have developed methods for ML fitting of such models to data and for evaluating their adequacy. Nowadays such methods can fit most any model to such data. Yurkowski and others (1967) survey early statistical methods. Allen (1959) presents a cumulative damage model and describes data analyses for it. Nelson (1980) was first to implement ML fitting of a cumulative damage model (Section 2) to step-stress data. Schatzoff and Lane (1987) extend that model to multiple accelerating stresses and interval (read-out) data. Also, they developed a computer program that calculates optimum test plans. Miller and Nelson (1983) present optimum test plans for simple step-stress tests. The textbook by Tobias and Trindade (1986) devotes a section to step-stress testing. This chapter is the first devoted to the topic from a statistical viewpoint.

2. STEP-STRESS MODEL AND DATA ANALYSES

Purpose. This section presents an example of and a basic model for step-stress data. This section also presents maximum likelihood (ML) fitting of the model to the data. This expands on work of Nelson (1980).

2.1. Step-Stress Data

Purpose. Data in Table 2.1 illustrate the basic model and analyses for step-stress data. A step-stress test of cryogenic cable insulation was run to estimate insulation life at a constant design stress of 400 volts/mil. Also, this insulation was to be compared with another insulation that was tested.

Data. Each specimen was first stressed for 10 minutes each at steps of 5kV, 10kV, 15kV, and 20kV before it went into step 5 at 26kV. Thereafter one group of specimens was stressed 15 minutes at each step (5 through 11 below), and three other groups were held 60, 240, and 960 minutes at each step. Thus there were four step-stress patterns.

Step:	5	6	7	8	9	10	11
Kilovolts:	26.0	28.5	31.0	33.4	36.0	38.5	41.0

Figure 3.2 of Chapter 1 depicts such step-stress patterns and data. Table 2.1

Table 2.1. Step-Stress Data on Insulation 1

Hold (min)	Final Step	Total Time to Failure (min)	Thickness (mils)
15	9	102	27
15	9	113	27
15	9	113	27
60	10	370+	29.5
60	10	345+	29.5
60	10	345	28
240	10	1249	29
240	10	1333	29
240	10	1333+	29
240	9	1106.4	29
240	10	1250.8	30
240	9	1097.9	29
960	7	2460.9	30
960	7	2460.9+	30
960	7	2703.4	30
960	8	2923.9	30
960	6	1160.0	30
960	7	1962.9	30
960	5	363.9+	30
960	5	898.4+	30
960	9	4142.1	30

+ denotes a running time without failure.

shows the step number and the total time on test when a specimen failed. All failures were due to the same failure mode. The stress on a specimen is the voltage divided by its insulation thickness. Specimens removed from test before failure are noted by +. Thus the data are censored.

2.2. Step-Stress Model

Purpose. This section presents a step-stress model, which consists of

1. the model for the distribution of life as a function of constant stress,
2. the model for the effect on life of the 'size' of a unit, and
3. the basic model for the cumulative effect of exposure in a step-stress test.

In applications, one needs to verify each part of the model. Such a model describes a *single failure mode*. Section 3.2 (equations (3.3) and (3.4)) presents an equivalent, simpler version of this model.

Constant stress model. The following power-Weibull model describes specimen life as a function of constant stress. Its assumptions are:

1. For any constant stress V (which must be positive), the life distribution is Weibull.
2. The Weibull shape parameter β is constant.
3. The Weibull scale parameter α is

$$\alpha(V) = (V_0/V)^p. \tag{2.1}$$

Here β, V_0, and p are positive parameters characteristic of the product and the test method. (2.1) is the inverse power law. Other constant-stress models could be used. Each product failure mode would be described with such a model. Then such models are combined as described in Chapter 7.

The assumptions imply that the population fraction $F(t;V)$ of specimens failing by time t under constant stress V is

$$F(t;V) = 1 - \exp[-\{t(V/V_0)^p\}^\beta], \quad t > 0. \tag{2.2}$$

The F fractile of the life distribution for a stress V is

$$\tau_F(V) = \exp[p\ln(V_0/V) + (1/\beta)\,u(F)]; \tag{2.3}$$

$u(F) \equiv \ln\{-\ln(1-F)\}$ is the standard extreme value F fractile.

Size effect. Here the specimens are smaller than actual cable. A^* is the size of a cable and A is the size of a specimen. A cable is modeled as a series system (Chapter 7) of A^*/A specimens with statistically independent life times. That is, the cable fails when the first "specimen" fails. Some insulation engineers use dielectric volume as size, and others use "exposed area." The series-system assumption and (2.2) imply that the fraction of cables of size A^* failing by time t under constant stress V is

$$F(t;V,A^*) = 1 - \exp[-(A^*/A)\{t(V/V_0)^p\}^\beta]. \tag{2.4}$$

This reduces to (2.2) if $A^* = A$.

Dependence. Eq. (2.4) may underestimate life of cables. Adjoining specimen-size pieces of cable can have dependent (positively correlated) life-times, instead of statistically independent ones. That is, if a piece has a short (long) life-time, pieces adjoining it tend to have short (long) lifetimes. Positive correlation implies longer cable life than predicted by (2.4). As an extreme, if pieces in a cable were perfectly correlated, all pieces would have the exact same lifetime, and the cable would have the lifetime of a specimen. Thus, for positive correlation, the life distribution (2.2) of specimens is an *upper bound* on the life distribution of cables. For some applications, such lower (2.4) and upper (2.2) bounds for the distribution suffice.

Cumulative exposure. For a step-stress pattern, there is a distribution $F_0(t)$ of time t to failure on test. Data from this distribution are observed in the test. But one usually wants the life distribution under constant stress, which units see in use. Thus one needs a cumulative exposure (or damage) model for a failure mode that relates the distribution (or cumulative exposure) under step-stressing to the distribution (or exposure) under constant stress. The following describes one such model.

Model inadequacy. The basic cumulative exposure model below does not display the following observed behavior of some products. Some products frequently fail during the brief time while the stress is being raised from one step to the next. According to the cumulative exposure model here, such rise times are too brief to produce so many failures on a rise. Such products are better described with a degradation model (Chapter 11). For example, suppose that the breakdown voltage of an insulation degrades with time during a voltage step-stress test. The breakdown voltage of a specimen at the time of a rise may be between the applied voltage on the previous step and the voltage on the next step. Such a specimen fails during the rise. The cumulative exposure model below is inadequate for such products.

Assumptions. The model for the failure mode assumes that the remaining life of specimens depends only on the current cumulative fraction failed and current stress – *regardless* how the fraction accumulated – a Markov property. Moreover, if held at the current stress, survivors will fail according to the cumulative distribution for that stress but starting at the previously accumulated fraction failed. Also, the change in stress has no effect on life – only the level of stress does. Thus, this model does not describe thermal cycling that produces failure. Nachlas (1986) proposes a cumulative damage model that includes the effect of cycling damage and of exposure at different levels of constant stress.

Depiction. Figure 2.1 depicts the *basic cumulative exposure model* for a failure mode. Part A depicts a step-stress pattern with four steps and failure and censoring times of specimens. Part B depicts the four cumulative distributions for the constant stresses (V_1, V_2, V_3, V_4). The arrows show that the specimens first follow the cumulative distribution for V_1 up to the first hold time t_1. When the stress increases from V_1 to V_2, the unfailed specimens follow the cumulative distribution for V_2, starting at the accumulated fraction failed. Similarly, when the stress increases from V_2 to V_3, from V_3 to V_4, etc., the unfailed specimens follow the next cumulative distribution, starting at the accumulated fraction failed. The cumulative distribution for life under the step-stress pattern appears in part C. It consists of the segments of the cumulative distributions for the constant stresses. In this *simple* way, this basic model takes into account the previous exposure history of a specimen. Note that this and other cumulative exposure models have not been adequately verified by experience.

Mathematical formulation. The basic cumulative exposure model for a failure mode is mathematically formulated as follows. This yields the cumulative distribution $F_0(t)$ of time to specimen failure under a particular step-stress pattern. A simpler equivalent formulation of $F_0(t)$ appears in (2.12). Those not interested in the mathematics can skip to Section 2.3. Suppose

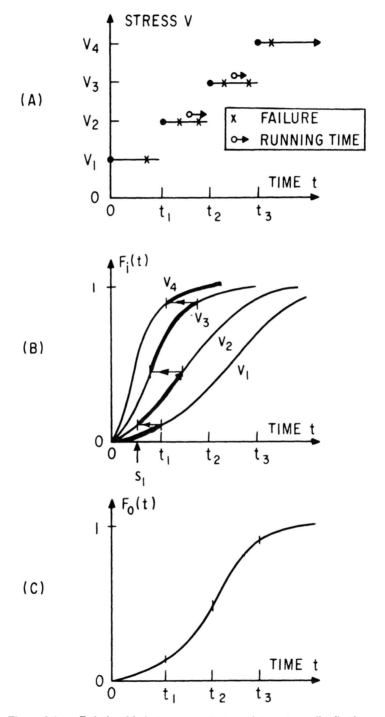

Figure 2.1.   Relationship between constant- and step-stress distributions.

that, for a particular pattern, step i runs at stress V_i, starts at time t_{i-1}, and runs to time t_i ($t_0 = 0$). The cumulative distribution for specimens at a constant stress V_i is $F_i(t)$. For the example with the power-Weibull model,

$$F_i(t) = 1 - \exp[-\{t(V_i/V_0)^P\}^\beta].$$

Step 1. The population cumulative fraction of specimens failing in step 1 is

$$F_0(t) = F_1(t), \quad 0 \le t \le t_1, \tag{2.5}$$

For the example for step 1,

$$F_0(t) = 1 - \exp[-\{t(V_1/V_0)^P\}^\beta], \quad 0 \le t \le t_1.$$

Step 2. Step 2 has an equivalent start time s_1 which would have produced the same population cumulative fraction failing (see Figure 2.1B). Thus, s_1 is the solution of

$$F_2(s_1) = F_1(t_1). \tag{2.6}$$

For the example, the equivalent time s_1 at V_2 is given by (2.6) as

$$s_1 = t_1(V_1/V_2)^P. \tag{2.6$'$}$$

The population cumulative fraction of specimens failing in step 2 by total time t is

$$F_0(t) = F_2[(t-t_1)+s_1], \quad t_1 \le t \le t_2. \tag{2.7}$$

For the example,

$$F_0(t) = 1 - \exp[-\{(t-t_1+s_1)(V_2/V_0)^P\}^\beta], \quad t_1 \le t \le t_2. \tag{2.7$'$}$$

Step 3. Similarly, step 3 has the equivalent start time s_2 given by

$$F_3(s_2) = F_2(t_2-t_1+s_1). \tag{2.8}$$

Then

$$F_0(t) = F_3[(t-t_2)+s_2], \quad t_2 \le t \le t_3. \tag{2.9}$$

For the example for step 3,

$$s_2 = (t_2-t_1+s_1)(V_2/V_3)^P, \tag{2.8$'$}$$

$$F_0(t) = 1 - \exp[-\{(t-t_2+s_2)(V_3/V_0)^P\}^\beta], \quad t_2 \le t \le t_3. \tag{2.9$'$}$$

Step i. In general, step i has the equivalent start time s_{i-1} given by

$$F_i(s_{i-1}) = F_{i-1}(t_{i-1}-t_{i-2}+s_{i-2}). \tag{2.10}$$

Then

$$F_0(t) = F_i[(t-t_{i-1})+s_{i-1}], \quad t_{i-1} \le t \le t_i. \tag{2.11}$$

Thus, $F_0(t)$ for a step-stress pattern consists of segments of the cumulative distributions $F_1(\ )$, $F_2(\ )$, etc., as shown in Figure 3.1C. A different step-stress pattern has a different $F_0(t)$ distribution for a failure mode. For the example for step i,

$$s_{i-1} = (t_{i-1}-t_{i-2}+s_{i-2})(V_{i-1}/V_i)^p, \tag{2.10'}$$

$$F_0(t) = 1 - \exp[-\{((t-t_{i-1}+s_{i-1})(V_i/V_0)^p\}^\beta], \quad t_{i-1} \le t \le t_i. \tag{2.11'}$$

Thus, $F_0(t)$ for a failure mode consists of segments of Weibull distributions.

Cumulative exposure. The cumulative exposure model above and the power-Weibull model can be expressed in a simpler, equivalent form to yield $F_0(t)$ for a failure mode. For the model, the fraction failed after any step-stress pattern is independent of the order of the steps as follows. Suppose that step i is at stress level V_i with corresponding characteristic life $\alpha_i = K/V_i^p$ for a time $\Delta_i = t_i - t_{i-1}$ $(t_0 = 0)$. Then it can be shown that the population fraction failed by time $t_I = \Delta_1 + \Delta_2 + \cdots + \Delta_I$ after I steps is

$$F_0(t_I) = 1 - \exp(-\varepsilon^\beta); \tag{2.12}$$

here the *"cumulative exposure"* ε for the failure mode is

$$\varepsilon = (\Delta_1/\alpha_1) + (\Delta_2/\alpha_2) + \cdots + (\Delta_I/\alpha_I). \tag{2.13}$$

Δ_I may be only a fraction of the planned time at step I. Moreover, the values of $F_0(t_I)$ and ε are the same *regardless* of the order of the I steps, each with its corresponding time Δ_i at the stress level V_i. (2.12) is an equivalent, simpler form of the basic cumulative exposure model. However, for some failure modes, products, and materials, the failure behavior depends on the order of the steps, called the *sequence effect*. (2.13) has no sequence effect. It is a probabilistic analog of Miner's rule, usually stated deterministically.

General model. The results (2.12) and (2.13) extend to any model where the life distribution $F(t;V)$ for a failure mode depends on constant stress V only through a scale parameter $\theta(V)$, namely,

$$F(t;V) = G[t/\theta(V)];$$

here $G[\ ]$ is the assumed cumulative distribution with the scale parameter set equal to 1. The simple linear-lognormal, linear-Weibull, and linear-exponential models have this property. The lognormal scale parameter is the median. The exponential scale parameter is the mean. Other distribution parameters are constants and are not explicitly shown here. Then $F_0(t_I) = G(\varepsilon)$ where $\varepsilon = [\Delta_1/\theta(V_1)] + [\Delta_2/\theta(V_2)] + \cdots + [\Delta_I/\theta(V_I)]$. For example, for the power-Weibull model above and $t_{I-1} < t \le t_I$,

$$\varepsilon(t) = \frac{t_1-0}{(V_0/V_1)^p} + \frac{t_2-t_1}{(V_0/V_2)^p} + \cdots + \frac{t-t_{I-1}}{(V_0/V_I)^p}.$$

These results hold for multivariable life-stress relationships and step-stressing with more than one accelerating variable and other engineering variables; see Schatzoff and Lane (1987).

Test plans. Little work has been done on optimum plans for step-stress tests. Miller and Nelson (1983) present optimum plans for a simple step-stress test with two stress levels where all specimens run to failure. They use the simple linear-exponential model and the basic cumulative exposure model here. Their plans minimize the (asymptotic) variance of the ML estimate of the mean life at a design stress. They consider two types of such a simple step-stress test: 1) a *time-step* test which runs a specified time at the first stress, and 2) a failure-step test which runs until a specified proportion of specimens fail at the first stress. Their results include: 1) the optimum time at the first stress for a time-step test, 2) the optimum proportion failing at the low stress for a failure-step test, and 3) the asymptotic variance of these optimum tests. Both the optimum time-step and failure-step tests have the same asymptotic variance as the corresponding optimum *constant*-stress test. Thus step-stress tests yield the same accuracy of estimates as constant-stress tests for their model with an exponential distribution. Schatzoff and Lane (1987) use the cumulative exposure model here, a Weibull life distribution, and a multistress relationship. For inspection (read-out) data, they optimize the ML estimate of a percentile at a constant design stress level. This requires their special computer program.

2.3. Maximum Likelihood Analyses

Purpose. For the cable insulation data, this section presents:
- A description of ML fitting.
- The ML fitted model.
- ML estimates and confidence limits for the model parameters (β, V_0, p) and fractiles $t_F(V)$.
- Comparisons with data on another type of cable insulation.

ML fitting. Estimates and confidence limits for the model parameters and functions of them are calculated with the ML methods of Chapter 5. The estimates are the parameter values that maximize the sample likelihood with the segmented distribution $F_0(t)$ in Figure 2.1C. From the data such as in Figure 3.2 of Chapter 1, one can calculate the sample cumulative distribution function for $F_0(t)$ and plot it on Figure 2.1C. Then ML fitting in some sense fits the $F_0(t)$ so it is close to the sample cumulative distribution. A specimen life may be 1) observed, 2) censored on the right (or left), or 3) in an interval (t', t''). The corresponding specimen likelihoods are 1) $f_0(t) = dF_0(t)/dt$, 2) $1 - F_0(t)$ (or $F_0(t)$), and 3) $F_0(t'') - F_0(t')$. If there is a number of step-stress patterns, the corresponding distribution is used for each, and the sample likelihood contains them. Such a likelihood must be programmed and added to a ML package. The ML theory also provides comparisons (hypothesis tests and confidence intervals), as described in Chapter 9. This section illustrates such fitting and comparisons. Of course, a complete analysis of such data also includes suitable graphical display and analysis.

Fitted model. Fitted by ML to the data of Table 2.1, the model for the fraction of cables that fail by age t in minutes is

$$F(t;V,A^*) = 1 - \exp[-(A^*/9.425)\{t(V/1619.4)^{19.937}\}^{0.75597}].$$

Here A^* is the area of the dielectric on a cable; test specimen area is 9.425 sq. in. $\hat{\beta} = 0.75597$ is the Weibull shape parameter, $\hat{p} = 19.937$ is the power in the power law for the failure mode. V is the constant stress in volts per mil. The parameters are not known as accurately as the number of significant figures might imply. The estimate of the $100P$th percentile of cable life in minutes at voltage stress V is

$$\hat{\tau}_P(V,A^*) = (1619.4/V)^{19.937} \{(9.425/A^*)[-\ln(1-P)]\}^{1/0.75597}.$$

Estimates. Tables 2.2 and 2.3 present ML estimates and approximate normal 95% confidence limits for the model parameters and the 1% point of specimen life at the design stress, 400 V/mil. For such a small sample, the intervals tend to be too narrow. Figure 2.2 depicts the first percentile of the failure mode versus stress on log-log paper. Other information in the table is explained later. ML fitting was done with a user-written likelihood function in STATPAC of Nelson and others (1972,1983).

Residuals. The ML fitting assumes that the model and data are valid. So they should be checked as described in Section 3 of Chapter 5. The residuals come from the distribution $F_0(t)$, which is not a standard distribution. For a

Table 2.2. ML Results for Cable Insulation 1

Parameter	Estimate	95% Conf. Limits Lower	95% Conf. Limits Upper
V_0	1616.4	1291.0	1941.8
p	19.937	6.2	33.7
β	0.75597	0.18	1.33
1% point (min) at 400 V/mil	2.81×10^9	2.65×10^4	2.98×10^{14}

Asymptotic Covariance Matrix

	$\hat{V}_0$	$\hat{p}$	$\hat{\beta}$
$\hat{V}_0$	27566.		symmetric
$\hat{p}$	-1145.7	49.004	
$\hat{\beta}$	41.572	-1.7561	0.086575

Maximum Log Likelihood = -103.53

Table 2.3. ML Results for Cable Insulation 2

Parameter	Estimate	95% Conf. Limits Lower	95% Conf. Limits Upper
V_0	3056.3	2177.6	3934.9
p	9.6015	5.6	13.6
β	0.96910	0.54	1.40
1% point (min) at 400 V/mil	2.62×10^6	2.96×10^4	2.32×10^8

Asymptotic Covariance Matrix

	$\hat{V}_0$	$\hat{p}$	$\hat{\beta}$
$\hat{V}_0$	200,957		symmetric
$\hat{p}$	−901.11	4.1599	
$\hat{\beta}$	61.017	−0.27439	0.047608

Maximum Log Likelihood = −141.66

failure or censoring time t_i, one can then use $u_i = \hat{F}_0(t_i)$ as a transformed residual. Such residuals come from a uniform distribution on the unit interval (0,1). Also, one can use $e_i = \exp(u_i)$, which come from a standard exponential distribution ($\theta = 1$). Such transformed residuals can be observed, censored, or in an interval. Plot such residuals on suitable distribution paper.

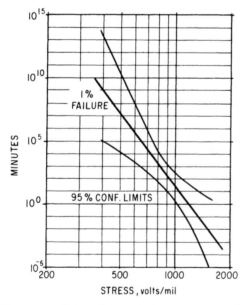

Figure 2.2.   1% line and 95% confidence limits vs. stress − cable insulation 1.

For example, use Weibull paper for the e_i; this suitably displays the lower tail of the sample. Also, crossplot the u_i (on a linear scale) or e_i (on a log scale) against other variables (on a suitable scale) to examine their effect. Such residuals are transformed to an artificial range. (3.13) below defines a better equivalent residual for any varying stress. That residual has the natural range of the time data.

Comparisons. The step-stress test was run on another type of cable insulation – insulation 2 – with a single failure mode. ML results appear in Tables 2.2 and 2.3. A test purpose was to compare the two insulations. Methods of Chapter 9 assess whether corresponding parameter estimates differ convincingly relative to their uncertainties. The two insulations are compared below with respect to their

1. shape parameters, β,
2. power parameters, p,
3. entire models (β, p, and V_0 simultaneously),
4. 1 percent points (in minutes) of life at 400 volts per mil.

Shapes. Each confidence interval for a β in Tables 2.2 and 2.3 overlaps the other estimate. Thus the two β estimates do not differ convincingly. The following formal comparison for the power parameters could be used for β.

Powers. The observed difference of the two power parameter estimates is $19.937 - 9.601 = 10.336$. These estimates are statistically independent. So the variance of their difference is the sum of their variances, $49.004 + 4.1599 = 53.1639$. Approximate 95% confidence limits for the true difference are $10.336 \pm 1.960(53.1639)^{1/2}$ or -3.955 and 24.627. This interval encloses zero. So the observed difference is not convincingly different from zero.

Models. One compares entire models for equality with the LR test, which simultaneously compares corresponding parameter estimates. For the two insulations, the sum of their maximum log likelihoods is $\hat{\mathcal{L}} = (-103.53) + (-141.66) = -245.19$. For the same model fitted to the pooled data, the maximum log likelihood is $\hat{\mathcal{L}}' = -265.15$. The LR test statistic is $T = 2\{-245.19 - (-265.15)\} = 39.92$. If the two models are the same, this statistic is approximately chi-square distributed with three degrees of freedom. Since $T = 39.93 > 16.27 = \chi^2(0.999;3)$, the parameter estimates of the two models differ very highly significantly (0.1% level). This is due largely to the difference between the V_0 estimates; they can be compared as the power parameters were.

1% points. The confidence interval for each 1% point of specimens in Tables 2.2 and 2.3 overlaps the other estimate. This indicates that the two estimates do not convincingly differ. The estimates could be formally compared with the method used for the power parameters.

3. VARYING-STRESS MODEL AND DATA ANALYSES

Purpose. Section 3.1 presents an example of varying-stress data, Section 3.2 extends the basic cumulative exposure model to such data, and Section 3.3 presents results of fitting such a model to the example data.

3.1. Varying-Stress Data

Purpose. The insulating oil data of Problem 4.10 illustrate the model and data analyses for a varying-stress test. The main purpose of the test and analyses was to estimate a model for time to oil breakdown under constant voltage. A second purpose was to assess whether the model and an exponential distribution adequately describe time to breakdown at constant voltage, as suggested by engineering theory.

Test method. The accelerated test employed a pair of parallel disk electrodes immersed in the oil. The voltage V across the electrodes was increased linearly with time t. That is, $V = Rt$ is a ramp voltage where R is the rate of rise in volts/sec. The voltage at oil breakdown was recorded (Problem 4.10). Equivalently time to breakdown could have been recorded. 60 breakdown voltages were observed at each of six combinations of three rates of rise R (10, 100, 1000 V/sec) and two electrode areas A (1 and 9 sq. in.). So the data consist of 360 breakdown voltages. Figure 3.3 of Chapter 1 depicts ramp stress and data.

Test plans. A test plan consists of selected stress patterns and the number of specimens to be subjected to each pattern. The oil breakdown example involves rate of voltage rise and electrode area. The plan involves two areas and three equally spaced rates of rise (on a log scale). Each of the six combinations of rate of rise and area has the same number of specimens. In statistical parlance, this is a 2×3 design with 60 replicates. Engineering test plans traditionally (and inefficiently) use equal allocation of specimens and equally spaced stress levels. There appears to be no work on optimum or efficient test plans for ramp tests or other tests with varying stress. A better plan for the oil breakdown test no doubt has unequally spaced rates of rise and unequal specimen allocation with respect to both rate of rise and area.

3.2. Varying-Stress Model

Overview. This section presents a general model for varying-stress testing. It extends the basic cumulative-damage model for step-stressing of Section 3.2 to varying stress. While plausible, this basic cumulative-damage model lacks adequate experimental verification. The varying-stress model consists of a constant-stress model and the basic cumulative-damage model. The model applies to a single failure mode.

Constant-stress model. For the theory here, the assumed constant-stress model follows. For simplicity, assume that life is a function of one accelerating stress V and possibly other (constant) variables x. Also, the model has a scale parameter $\alpha(V, x)$ that is a function of V, x, and coefficients. Those coefficients and all other model parameters are constants to be estimated from data. The power-Weibull and Arrhenius-lognormal models are such models. Also, the following theory extends to such constant-stress models with more than one accelerating variable.

Oil example. For the oil example, the constant-stress model consists of a Weibull life distribution with a constant shape parameter β. The characteristic life α_0 is an inverse power function of voltage stress V; namely, $\alpha_0(V) = (V_0/V)^p$ where V_0 and p are parameters characteristic of the oil and test method. The effect of the area A of the test electrodes is modeled as in Chapter 7. Thus the constant-stress model for time t breakdown is

$$F(t; V, A) = 1 - \exp\{-A[t(V/V_0)^p]^\beta\}. \tag{3.1}$$

Its characteristic life is a function of V and A, namely,

$$\alpha(V, A) = (V_0/V)^p / A^{1/\beta}. \tag{3.2}$$

Thus this model has one accelerating variable, voltage V, and one other variable, area A.

Types of varying stress. For purposes of this section, the applied accelerating stress $V(t)$ is any (integrable) function of time t. In practice, $V(t)$ usually is one of the following.

1. A step stress, as described in Section 2.
2. A *ramp* stress. That is, stress increases linearly (from zero) with time.
3. A cyclical stress. Examples include sinusoidal and square waves.
4. A randomly varying stress over time (stochastic loading) with a given load distribution and autocorrelation.
5. A nonrepeating pattern.

Also, the following theory readily extends to situations with more than one accelerating stress and other engineering variables.

Cumulative exposure. When the stress $V(t)$ is a function of time, the distribution scale parameter $\alpha(V, x)$ is a function of time; namely, $\alpha(t) = \alpha[V(t), x]$. The corresponding *cumulative exposure* $\varepsilon(t)$ (or damage), which appears as a sum in (2.13), becomes the integral

$$\varepsilon(t) = \int_0^t dt / \alpha[V(t), x]. \tag{3.3}$$

This is the limit of a step-stress approximation to $V(t)$ which approaches $V(t)$ as all intervals $\Delta_i \to 0$. $\varepsilon(t)$ is a function of $V(t), x$, and the model parameters. Some authors assume (3.3) as the *basic cumulative exposure model*.

This limiting argument shows that the step-stress model (with a segmented, cumulative distribution) of Section 2.2 is *equivalent* to (3.3). Most authors assume (3.3) and regard the step-stress model as a special case. Then the population fraction failed by time t under varying stress $V(t)$ is

$$F_0[t;"V(t)",x] = G[\varepsilon(t)]; \qquad (3.4)$$

here $G[\ ]$ is the assumed cumulative distribution with the scale parameter set equal to 1. "$V(t)$" in quotes emphasizes that $V(t)$ does *not* merely replace V in the constant-stress model. Other distribution parameters do not depend on V but may depend on x. All other results in this chapter are special cases of this model.

Oil example. The oil is subjected to a ramp voltage $V(t) = Rt$ where R is the *rate of rise* of voltage stress in volts per minute. The cumulative exposure at time t is

$$\varepsilon(t) = \int_0^t dt/\alpha[V(t),A] = \int_0^t A^{1/\beta}(Rt/V_0)^p dt = A^{1/\beta}(R/V_0)^p t^{p+1}/(p+1). \qquad (3.5)$$

The distribution of time t to oil breakdown is

$$\begin{aligned} F_0(t;"Rt",A) &= 1 - \exp\{-[\varepsilon(t)]^\beta\} \\ &= 1 - \exp\{-[t^{p+1}A^{1/\beta}(R/V_0)^p/(p+1)]^\beta\} \\ &= 1 - \exp\{-(t/\alpha')^{\beta'}\}. \end{aligned} \qquad (3.6)$$

This is a Weibull distribution with shape and scale parameters

$$\beta' = \beta(p+1), \quad \alpha' = [(p+1)(V_0/R)^p/A^{1/\beta}]^{1/(p+1)}. \qquad (3.7)$$

Yurkowski and others (1967) present this result. Equivalently, the distribution of breakdown voltage $V = Rt$ is

$$F_0(V;"Rt",A) = 1 - \exp[-(V/\alpha'')^{\beta''}]. \qquad (3.8)$$

This is a Weibull distribution with shape and scale parameters

$$\beta'' = \beta(p+1), \quad \alpha'' = [V_0^p(p+1)R/A^{1/\beta}]^{1/(p+1)}. \qquad (3.9)$$

Moreover, the distribution of $\ln(V)$ is extreme value with location parameter

$$\ln[\alpha''(R,A)] = \gamma_0 + \gamma_1\ln(R) + \gamma_2\ln(A). \qquad (3.10)$$

This is a linear relationship where

$$\gamma_2 = -1/[\beta(p+1)], \quad \gamma_1 = 1/(p+1), \quad \gamma_0 = [1/(p+1)]\ln[V_0^p(p+1)]. \qquad (3.11)$$

The extreme value scale parameter is

$$\delta = 1/\beta'' = 1/[\beta(p+1)] = -\gamma_2. \qquad (3.12)$$

For the linear relationship (3.10), the four parameters γ_0, γ_1, γ_2, and $\delta = -\gamma_2$ are fitted to the oil data in the next section, using standard features of a ML program. Directly fitting the model with three parameters V_0, p, and β yields more accurate estimates of those parameters but requires a user written likelihood for a ML package.

Varying use stress. Preceding material concerns varying *test* stress. Some products in *actual use* undergo a varying stress $V^*(t)$. The model and method above also yield the product life distribution $F_0[t; "V^*(t)"]$ under the stress pattern $V^*(t)$ in actual use. To estimate this distribution, first estimate the parameters of the constant-stress model from test data (constant or varying test stress). Then use those estimates in $F_0[t; "V^*(t)"]$ from (3.4) to estimate this distribution.

3.3. Maximum Likelihood Analyses

Overview. This section describes ML fitting of the basic varying-stress model (3.4) to data. This section also presents the results of the ML fit of the model (3.8) to the oil data and checks on the model and data. Of course, a complete analysis of such data includes graphical display and analysis.

ML fit. According to the model, the cumulative distribution of time t to failure of specimen i is $F_0[t_i; "V_i(t)", x_i] = G(\varepsilon_i)$ in the notation of Section 3.2. Here the exposure ε_i of specimen i is a function of its stress pattern $V_i(t)$, its other variable values x_i, and the model parameters. Usually $F_0(\ )$ is written as a function of the parameters of the constant-stress model. For the oil example, p, V_0, and β are those parameters. Also, $F_0(\ )$ can be written as a function of parameters more suited to the varying-stress model. For the oil example, γ_0, γ_1, and γ_2 in (3.10) are such parameters. The ML estimates of either set of parameters are the parameter values that maximize the sample likelihood with $F_0(\ )$. Of course, for an observed failure time t_i, the likelihood is $f_0[t_i; "V_i(t)", x_i]$, the probability density. For a right censored time t_i, the likelihood is $1 - F_0[t_i; "V_i(t_i)", x_i]$. Left censored and interval data have their corresponding likelihoods. The ML theory and methods of Chapter 5 apply to this likelihood and such varying-stress data. Such a likelihood must be programmed and added to a ML package.

Fitted models. For the oil breakdown data, the fitted model (3.6) for time t to breakdown under constant stress is

$$F(t; V, A) = 1 - \exp\{-A[t(V/42.298)^{16.40}]^{0.8204}\}.$$

The fitted model for (3.8) for voltage V at breakdown under ramp stress Rt is

$$F_0(V; "Rt", A) = 1 - \exp[-\exp\{[\ln(V) - 3.69370 - 0.05747\ln(R)$$
$$+ 0.07005\ln(A)]/0.07005\}].$$

Table 3.1 displays the ML estimates and (normal approximate) 95% confidence limits for both sets of model parameters. The confidence interval for β is $(0.7396, 0.9101)$. This does not enclose 1, which suggests that the life distribution at constant voltage is not exponential. On the other hand, varia-

Table 3.1. ML Fit for Oil Breakdown under Ramp Voltage

Progressive-Stress Model (3.11) and (3.12)

	ML Est.	95% Conf. Limits	
γ_0	3.69370	3.67176	3.71564
γ_1	0.05747	0.06180	0.05314
$\gamma_2 = -\delta$	−0.07005	−0.07475	−0.06535

Covariance Matrix (multiply each by 10^{-5})

	$\hat{\gamma}_0$	$\hat{\gamma}_1$	$\hat{\gamma}_2$
$\hat{\gamma}_0$	12.5354	−2.2702	−0.5331
$\hat{\gamma}_1$	−2.2702	0.4885	0.0298
$\hat{\gamma}_2$	−0.5331	0.0298	0.5761

Maximum Log Likelihood = 1050.4205

Constant-Stress Model

	ML Est.	95% Conf. Limits	
β	0.8204	0.7396	0.9101
p	16.40	15.09	17.71
V_0	42.298	−	−

bility of test conditions could increase the scatter in the data and thereby lower the β estimate below 1.

Assess model and data. Assess the model and data with the ML methods of Chapter 5. The following paragraphs include a test for equal shape parameters at all test conditions, a test of fit of the assumed relationship, and examination of residuals.

Equal shape parameters. The varying-stress model for breakdown voltage has the same shape parameter value $\beta'' = \beta(p+1)$ at all six test conditions. The following LR test (Chapter 5) assesses this assumption. Table 3.2 displays results of fitting 1) six separate Weibull distributions to the six data sets and 2) a model with six Weibull distributions with a common shape parameter and six separate scale parameters for the six data sets. The test statistic is $T = 2[-1023.92508 - (-1029.0448)] = 10.24$. The two models have 12 and 7 parameters. So the statistic has $\nu = 12-7 = 5$ degrees of freedom. Since $T = 10.24 < 11.07 = \chi^2(0.95;5)$, the shape parameter estimates do not differ significantly at the 5% level. However, $T = 10.24 > 9.236 = \chi^2(0.90;5)$, the shape parameter estimates differ significantly at the 10% level − very slight evidence. This suggests that further examination of such estimates may yield further insight. A plot of the six estimates and confidence limits suggests that β'' may depend on rate of rise.

Assess $1/\beta'' = -\gamma_2$. In the progressive-stress model for oil breakdown voltage (3.12), $1/\beta'' = -\gamma_2$. A LR test for this equality follows. The ramp-stress model has three parameters γ_0, γ_1, and $\gamma_2 = -1/\beta''$ and $\hat{\mathcal{L}} =$

Table 3.2. ML Fits to Oil Breakdown Data

1) SEPARATE WEIBULL DISTRIBUTION FOR EACH TEST CONDITION

Rate	Area	$\hat{\mathcal{L}}$	$\widehat{\alpha}$	$\hat{\beta}''$	var($\widehat{\alpha}$)	var($\hat{\beta}''$)	cov($\widehat{\alpha}'',\hat{\beta}''$)
10	1	-174.69513	44.56709	10.83883	0.3150676	1.133066	0.1941754
100	1	-176.59302	50.67680	12.52561	0.3013819	1.642074	0.2168922
1000	1	-181.65427	60.05503	13.31633	0.3764391	1.716795	0.2539894
10	9	-165.58370	39.69478	12.21972	0.1917803	1.678554	0.1638259
100	9	-155.44341	46.24801	16.45265	0.1455555	1.870928	0.1993759
1000	9	-169.95555	50.85969	14.68103	0.2182360	2.432529	0.2107394

$$-1023.92508 = \text{total}$$

2) COMMON $\hat{\beta}''$ AND SEPARATE $\widehat{\alpha}_j''$ FOR EACH TEST CONDITION

-1029.0448	12.99680	0.2932445

3) RELATIONSHIP (3.10) AND $1/\beta'' \neq -\gamma_2$ (4 PARAMETERS)

Covariance Matrix (multiply each term by 10^{-5})

	ML Est.	γ_0	γ_1	γ_2	$1/\beta''$
γ_0	3.673202	15.332557		symmetric	
γ_1	0.05843506	-2.401166	0.494523		
γ_1	-0.058626	-2.094359	0.103338	1.445912	
$1/\beta''$	0.07856677	-0.631478	0.024987	0.072873	1.059068

Maximum Log Likelihood $= -1035.4269$

-1050.4205 (from Table 3.1). The model (3.11) where $1/\beta'' \neq -\gamma_2$ has four parameters and $\hat{\mathcal{L}} = -1035.4269$ (from model 3 of Table 3.2). The LR statistic for the equality is $T = 2[-1035.4269 - (-1050.4205)] = 29.99$. This statistic has $4 - 3 = 1$ degree of freedom. Since $T = 29.99 > 10.83 = \chi^2(0.999;1)$, $1/\beta''$ differs very highly significantly (0.1% level) from $-\gamma_2$. This suggests that the ramp-stress model does not adequately fit the data. The separate estimates are $-\widehat{\gamma}_2 = 0.058626$ and $1/\widehat{\beta}'' = 0.07856677$ (Table 3.2). $1/\widehat{\beta}''$ is the ML estimate of the scale parameter of the extreme value distribution of ln voltage. It estimates the scatter of the data about the fitted relationship, whereas $\widehat{\gamma}_2$ estimates the slope of the relationship. Thus there is more scatter in the breakdown data than predicted by the ramp-stress model. The greater scatter may be due to varying test conditions. The estimate (model 2 of Table 3.2) of a common β'' (with a separate estimate for the scale parameter α_j'' for each test condition) is $1/\widehat{\beta}'' = 1/12.99680 = 0.076942$. It is another estimate of $1/\beta''$ which is not inflated by possible lack of fit of the relationship (3.10). The previous estimate (0.07856677) could be inflated by lack of fit. Both estimates of $1/\beta''$ reflect greater scatter in the ln voltage data than predicted by the progressive-stress model (3.8). The equality hypothesis above could equivalently be tested with Wald's test (Chapter 9). In conclusion, the four-parameter model represents the data better.

Assess the relationship. The progressive-stress model for breakdown voltage employs the relationship (3.10). The following LR test (Chapter 5) assesses adequacy of that relationship. Constrained model 3 with the relationship (3.10) and a *separate* shape parameter was fitted to the data. Thus here $1/\beta'' \neq -\gamma_2$, and this model has four parameters, whereas the model (3.6) has three. The resulting maximum log likelihood is $\hat{\mathcal{L}} = -1035.4269$. Model 2 with six separate scale parameters (for the six test conditions) and a common $\hat{\beta''}$ has $\hat{\mathcal{L}} = -1029.0448$ (model 2 of Table 3.2). The test statistic is $T = 2[-1029.0448 - (-1035.4269)] = 12.76$. The models have 4 and 7 parameters. So T has $\nu = 7-4 = 3$ degrees of freedom. Since $T = 12.76 > 11.34 = \chi^2(0.99;3)$, there is highly statistically significant evidence (1% level) that the relationship does not adequately fit the data. A possible explanation of this is that the 1 and 9 sq. in. electrodes may not have the exact same separation. An analysis (omitted here) shows that a 3% difference in this tiny separation (of a few mils) would produce such a lack of fit.

Residuals. For a test with varying stress, the definition of a residual is not obvious. Suppose, as above, that the constant-stress model has a cumulative distribution $G(t/\alpha;\beta)$ with a constant shape parameter β and scale parameter $\alpha(V, x)$ that is a function of the accelerating stress(es) V and other variable(s) x. For concreteness, regard $G(\ )$ as a Weibull distribution. Suppose that specimen i has varying stress $V_i(t)$, other values x_i, and observed life t_i. Its *"cumulative exposure" residual* is

$$e_i \equiv \int_0^{t_i} dt / \hat{\alpha}[V_i(t), x_i] . \qquad (3.13)$$

Here $\hat{\alpha}(V, x)$ is the ML estimate of the constant-stress relationship. These residuals are (approximately) a random sample from the constant-stress distribution $G(e;\beta)$ where $\alpha = 1$. To check the assumed distribution $G(\ )$, plot these residuals on paper for the distribution. For example, for the oil data, plot such residuals on Weibull paper. The definition (3.13) applies also to censored and interval residuals. Also, one can crossplot such residuals against any variables of interest to examine them. The residuals used in Chapters 4 and 5 are the logs of these residuals. Thus one could equivalently plot the ln residuals. For example, if $G(\ )$ is a Weibull distribution, then the ln residuals come from an extreme value distribution. Invented by the author, cumulative exposure residuals have not been previously published.

Oil example. For the oil data, (3.13) and (3.6) give the constant-stress residual for a specimen i at R_i, A_i as

$$e_i = A_i^{1/\hat{\beta}}(R_i/\hat{V}_0)^{\hat{p}} t_i^{\hat{p}+1} / (\hat{p}+1).$$

These residuals come from (approximately) a Weibull distribution with scale parameter 1 and shape parameter β. Equivalently expressed in terms of voltage $V_i = R_i t_i$,

$$e_i = A_i^{1/\hat{\beta}}(V_i/V_0)^{\hat{p}}(V_i/R_i) / (\hat{p}+1).$$

For the oil data model, one could also use the natural residuals $e_i'' = V_i/\hat{\alpha}''(R_i, A_i)$, which come from a Weibull distribution with scale parameter 1 and shape parameter $\beta'' = \beta(p+1)$. For a general $V(t)$, such simple natural residuals e_i'' often do not exist, and the e_i must be used.

Comparisons. Preceding paragraphs present ML comparisons to assess the model. ML comparisons of Chapter 9 can also be used to compare two or more data sets from varying-stress tests. For example, Problem 8.4 concerns least-squares comparisons of two sets of oil breakdown data. Problem 10.6 concerns ML comparisons of those two sets. Problem 10.8 concerns ML comparisons of two sets of oil breakdown data – one under ramp stress and the other under constant stress.

PROBLEMS (* denotes difficult or laborious)

10.1. Cable plots. Devise and make suitable probability, relationship, and other plots of the data on cable insulation 1 of Section 2. State your findings and conclusions from your plots.

10.2. Oil breakdown plots. Devise and make suitable probability, relationship, and other plots for the oil breakdown data of Section 3. State your findings and conclusions from your plots.

10.3. Cable fit. Use the data on cable insulation 1.
(a*) In a ML program, install an appropriate likelihood function for the three-parameter model of Section 2.2 and fit the model. Calculate and plot Figure 2.2.
(b) Carry out the analytic checks on the model.
(c) Calculate and plot residuals.
(d) Suggest further analyses.
(e) Carry out (d).
(f*) Do (a)-(e) using a lognormal distribution (see Problem 10.14). Comment on results different from the Weibull ones.

10.4. Oil fit. Use the oil breakdown data (Set 1) of Section 3 (Problem 4.10). Do (a)-(f) of Problem 10.3. The breakdown voltages are recorded to the nearest volt. Taking into account the interval data, do (a) and (b) of Problem 10.3. Comment on different results from the interval data analyses.

10.5. Oil set 2. Use the oil breakdown data (Set 2) of Problem 8.4. Suitably plot the data. Do (a)-(f) of Problem 10.3. The breakdown voltages are recorded to the nearest volt. Taking into account the interval data, do (a) and (b) of Problem 10.3. Comment on differences in results from the interval data analyses.

10.6. Oil comparisons. Compare Sets 1 and 2 of the oil breakdown data as follows.

(a) Compare probability and relationship plots of the data sets, and state your conclusions.
(b) Use ML methods to compare parameters for equality – separately and simultaneously. State your conclusions.
(c) Plot residuals and state your conclusions.
(d*) Do (a)-(c) using a lognormal distribution. Comment on results difference from those with the Weibull distribution.
(e) Suggest further analyses.
(f) Carry out (e).

10.7. Drift and correlation. The oil breakdown data of Section 3 were collected with same oil sample and the same two pairs of electrodes. The dielectric experts wish to find out whether repeated breakdown of the oil improved or degraded its breakdown strength, as the breakdowns could be removing defects or creating them. The data (Problem 4.10) were collected in the following order. The six test conditions were applied in a cycle with the order:

Order:	1	2	3	4	5	6
Rate of Rise (V/sec):	1000	1000	100	100	10	10
Electrode Area (in.^{2}):	9	1	9	1	9	1

Each cycle yielded six breakdown voltages, and the cycle was repeated 60 times to yield the 360 breakdown voltages. In Problem 4.10, for the time order of the data at a test condition, read across row 1, then row 2, etc.

(a) Graphically analyze the original data to assess whether there is a trend to the breakdown voltage at each test condition. Also, make a plot of cumulative voltage versus order.
(b) Do (a) for residuals.
(c) Use a relationship and analytic (ML) methods to assess for a trend.

All data analysis methods in this book (and most others) assume that the observations are *statistically independent*. The breakdown voltages may be positively serially correlated. That is, low voltage may tend to follow low voltage, and high voltage may tend to follow high voltage. Such serial correlation could arise from drifting test conditions, for example, drifting temperature.

(d) For a test condition, crossplot each observation against the one preceding it. Does the plot show evidence of trend (correlation)? Note that the marginal distributions are Weibull rather than (log) normal.
(e) Use the following nonparametric test for correlation. Divide the plot into four quadrants at the median observation on each scale. Equal numbers of observations should fall into each quadrant when there is no correlation. Use an appropriate contingency table test with fixed marginal numbers.

(f) Repeat (d) and (e) for the autocorrelation plot of each observation plotted against the second one preceding it (a lag of 2). In view of the two plots, is looking at the plot for lag 3 worthwhile? Why?

10.8. Assess oil model. To assess the cumulative-exposure model of Section 3, a ramp test and constant-stress test were run with the same oil sample and same two pairs of electrodes. The resulting data appear below. The constant-voltage data are censored on the right and the left. At high voltage, some breakdowns occurred before the applied voltage reached its constant value. These are noted with a failure time of $1-$, that is, less than one second. It may be best to treat all early failures as censored on the left, say, below 3, 4, or 5 seconds.

Constant-Stress Times (Seconds) to Breakdown

9 sq. in. electrode					1 sq. in. electrode				
45kV	40kV	35kV	30kV	25kV	50kV	45kV	40kV	35kV	30kV
1–	1	30	50	521	1–	1–	49	287	908
1–	1	33	134	2,517	1–	1–	60	301	908
1–	2	41	187	4,056	1–	5	133	531	2,458
2	3	87	882	12,553	2	15	211	582	3,245
2	12	93	1,448	40,290	3	23	245	966	3,263
3	25	98	1,468	50,560+	6	35	259	1,184	9,910
9	46	116	2,290	52,900+	21	50	274	1,208	38,990
13	56	258	2,932	67,270+	83	61	440	1,585	41,310
47	68	461	4,138	83,990+	112	93	619	2,036	44,170
50	109	1,182	15,750	85,500+	113	142	704	3,150	74,520+
55	323	1,350	29,180+	85,700+	154	143	776	4,962	78,750+
71	417	1,495	86,100+	86,420+	303	229	920	68,730+	86,620+

(a) Graphically display the data from both tests. Graphically estimate model parameters, and assess each model and the data. Are the estimates from the two data sets consistent with each other?

(b) By ML separately fit the appropriate model to each data set. Are the data consistent with a shape parameter of 1 in the constant-stress model? Why?

(c) Examine the residuals from (b), and comment on your findings.

(d) Make ML comparisons of corresponding parameter estimates from the two tests – separately and simultaneously. State conclusions.

(e) Describe how to simultaneously fit the two models to the two data sets to get pooled estimates of parameters.

(f) The ramp data appear in the order collected. Make plots to assess for trend and serial correlation, as described in Problem 10.7.

(g) Suggest further analyses.

(h) Carry out (g).

Ramp Test Breakdown Voltages

1 V/s		10 V/s		100 V/s		1000 V/s	
1	9	1	9	1	9	1	9
28	18	37	35	37	40	58	47
32	35	44	40	42	41	46	37
36	37	43	37	47	40	41	47
32	19	35	36	46	41	52	44
30	22	37	33	50	43	56	50
37	38	42	40	50	46	57	48
39	37	42	32	50	42	58	46
35	37	39	38	42	42	62	47
38	34	41	38	51	41	52	37
34	24	43	26	47	44	52	46
36	37	40	32	52	48	53	41
37	45	46	36	51	49	60	50
35	32	33	34	57	49	57	51
38	32	38	40	51	44	56	43
34	35	45	35	50	43	57	46
38	32	42	36	53	46	53	44
38	38	47	33	59	42	63	49
40	40	44	37	45	43	62	44
39	37	39	39	52	50	57	45
38	39	50	39	49	50	53	46
39	35	43	42	48	50	67	52
42	32	44	41	57	46	61	48
41	33	43	44	53	48	57	50
43	38	41	45	56	43	61	41
38	30	45	37	46	43	62	50

10.9.* Cable insulations. A voltage step-stress test of three types of cryogenic cable insulation yielded the following data. Each pair of lines shows the time (in minutes) at each voltage for a specimen.
(a) Separately fit the model of Section 2.2 to each type of cable.
(b) In view of the small number of specimens, do you think that assessing the model and residuals is worthwhile? Why?
(c) Compare corresponding parameter estimates – separately and simultaneously.
(d) Suggest further analyses.
(e) Carry out (d).

Insulation A

Volt.	250	500	768	845	929	1022
Min.	15	15	15	15	15	12.2

Volt.	285	500	768	845	929	
Min.	15	15	15	15	12.3	

Volt.	500	665	732	805	886	929	974	1022	1072	1123	1179
Min.	20	20	20	20	20	20	20	10	10	10	1.1

Volt.	665	732	805	886	929	974	1022	1072
Min.	25	10	15	15	15	15	15	0.75

Volt.	500	550	605	665	732	805	886	974	1072
Min.	15	15	15	15	15	15	15	15	13.55

Volt.	500	698	768	805	845	886	929	974	1022	1072
Min.	15	15	10	10	10	10	10	10	10	8.4

Insulation B

Volt.	250	500	768	805
Min.	15	15	15	3.0

Volt.	250	500	550	605	665	698	732	768	805
Min.	15	10	15	15	15	15	15	15	0.05

Volt.	500	605	665	732	768	805	845	886
Min.	15	15	15	15	15	15	15	0.01

Volt.	500	605	665	732	768	805
Min.	15	15	20	25	25	5.6

Insulation C

Volt.	250	500	768	845	929
Min.	15	15	15	15	0.4

Volt.	250	500	768	805	845	886
Min.	15	15	15	15	15	3.3

Volt.	500	605	698	768	805	845	886	929	974	1022	1072	1124
Min.	15	15	15	15	15	15	15	15	15	15	15	5.0

Volt.	500	605	698	805	886	929	974	1022	1072
Min.	15	15	15	15	15	15	15	15	5.4

Volt.	500	698	768	845	886	929	974	1022
Min.	15	15	15	15	15	15	15	7.5

Volt.	605	665	719	768	805	845	886	929	974	1022	1072	1124
Min.	1015	270	945	15	15	15	15	15	15	15	15	1.9

Volt.	605	665	732	805	886	929	974	1022
Min.	1020	15	15	15	10	20	15	14.4

10.10.* Derive model. Derive all formulas for the model and example of Section 3.2, showing all intermediate steps. Make a numbered list of all assumptions used in the derivation and model.

10.11.* Constant stress. Verify that the model of Section 3.2 reduces to the constant-stress model when the test stress is constant.

10.12.* Step stress. Verify that the model of Section 3.2 reduces to the step-stress model of Section 2.2 when the varying stress is a step stress.

10.13.* No sequential effect. Derive (2.12) and (2.13) from the model of Section 2.2, and show that (2.13) holds for any order of the I steps.

10.14.* Lognormal. Repeat the derivation of the model of Section 3.2, but use the lognormal distribution in place of the Weibull.

10.15. Lognormal fit. Use the oil breakdown data of Section 2.
(a*) With a ML program, fit the lognormal model of Problem 10.14.
(b)   Carry out the analytic checks on the model.
(c)   Calculate and plot residuals.
(d)   Suggest further analyses.
(e)   Carry out (d).
(f)   Point out and discuss differences between these results and those using a Weibull distribution.

10.16.* Read-out data. Use the step-stress model of Section 2.2 with the power-Weibull model. Assume that the n specimens are inspected for failure at the end of each step i at stress level V_i, $i = 1, 2, \cdots, I$.
(a)   Write the sample log likelihood.
(b)   Calculate the likelihood equations.
(c)   Calculate the true theoretical Fisher information matrix. The expectations are easy to evaluate, since the number of failures in an interval has a binomial distribution.
(d)   Calculate approximate normal confidence limits for a percentile of the life distribution at constant design stress V_0.
(e)   Do (a)-(c) for a test with removal of specimens. That is, n_1 specimens enter interval 1, n_2 specimens enter interval 2 (after removals at inspection 1), $\cdots$, n_I specimens enter interval I (after removals at inspection $I - 1$).
(f*) Do (a)-(e) for a general distribution where the ln of the scale parameter is a linear function of a (possibly transformed) stress x.

10.17.* Square-wave stress. Use the general model of Section 3.2 where the life distribution is $G[t/\theta(x)]$ and $\theta(x)$ is the scale parameter at stress level x. Other distribution parameters do not depend on x. Suppose that specimens on test run under a cyclic square-wave stress which has length τ and alternates between two levels, x' and x'' with scale parameters θ' and θ''. In each cycle, a specimen is at stress x' a time $f\tau$ and at x'' a time $(1-f)\tau$. Assume that the time t to specimen failure is much greater than the cycle length τ. The following is a probabilistic version of the deterministic model of Rabinowicz and others (1970). Note that this model makes no assumption about the life-stress relationship.
(a)   Calculate the approximate cumulative exposure to time t.
(b)   Give the distribution of time to failure on test. Give the expression for the scale parameter of the distribution as a function of f.
(c)   Suppose x' is the design stress and x'' is an elevated stress. Suppose there are two groups of specimens where n_i specimens run at x' a fraction of time f_i, $i = 1, 2$. Show how to use such data (complete or censored) to estimate θ' and the life distribution at x', which is constant, in actual use.
(d)   Write the sample log likelihood for (c) for complete data and a lognormal distribution with a constant σ. This likelihood is a function of the

two medians. Derive the likelihood equations, ML estimates, and the local estimate of asymptotic covariance matrix. Derive approximate confidence limits for a percentile at design stress x'.

(e*) Repeat (d) for a singly time censored sample (times t' and t'') for each group.

(f) Repeat (d) for an exponential distribution and time censored data. Formulas are simpler with failure rates rather than means.

(g*) Repeat (d) for a Weibull distribution and time censored data.

(h) Describe a test plan and hypothesis test for the validity of the relationship from (b) between the scale parameter θ and f.

(i*) Develop optimum test plans for such a model.

(j*) Assume that $\ln[\theta(x)]$ is a linear function of x. Repeat (a)-(h).

10.18.* Sinusoidal stress. Use the general model of Section 3.2 where the ln scale parameter is a simple linear function of a (possibly transformed) stress x. Suppose that specimens run under sinusoidal stress $x(t) = A\sin(2\pi t/\tau)$ with period τ. Assume that the time t to specimen failure is much greater than τ.

(a) Calculate the (approximate) cumulative exposure to time t.

(b) Give the general distribution of time to failure. Give the expression for the distribution scale parameter as a function of the amplitude A and period τ. The life of some products is a function of frequency $1/\tau$. Does this model have such behavior?

(c) Consider a test plan where specimen i runs at amplitude A_i, $i = 1, 2, \cdots, n$. Write the sample log likelihood for (b) for a lognormal distribution and complete data. Derive the likelihood equations, the ML estimates, and the asymptotic covariance matrix. Derive approximate confidence limits for a percentile at a design stress amplitude A'.

(d) Do (c) for an exponential distribution, specimens at two stress amplitudes, and multiply right censored data.

(e) Do (c) for a Weibull distribution with an assumed value of the shape parameter, specimens at two stress amplitudes, and multiply right censored data.

(f*) Do (c) for a Weibull distribution and multiply right censored data.

10.19.* Random stress. Suppose that stress on a specimen is random. That is, the amount of time a specimen is under stress level x is given by a known probability density $f(x)$ or "load spectrum." Use the general model of Section 3 where the ln scale parameter is a simple linear function of a (possibly transformed) stress x. Assume that a specimen undergoes the spectrum many times before failure. Do (a) to (f) of Problem 10.18.

10.20.* Optimum simple plan. A "simple" step-stress plan consists of two stress levels. Use the cumulative exposure model of Section 2.2, an exponential life distribution, and a simple ln linear relationship between mean life θ and a (possibly transformed) stress x, namely, $\ln\theta(x) = \alpha + \beta x$. Assume that the first step at level x_1 lasts a time τ, and the second step at level x_2 lasts until all specimens fail.

(a)   Write the sample log likelihood for n specimens.

(b)   Derive the likelihood equations and the ML estimates of α and β.

(c)   Derive their true asymptotic covariance matrix.

(d)   Derive the true asymptotic variance of the ML estimate of the mean at a design stress level x_0.

(e)   Derive the time τ^* that minimizes the variance (d) and its minimum value. Calculate and plot τ^* and the minimum variance V^* as a function of the standardized extrapolation $\varsigma = (x_0 - x_1)/(x_1 - x_2)$.

(f*)   Compare this minimum variance with that from an optimum constant-stress plan with the same two stress levels and n specimens. Comment on the results.

(g*)   Do (a)-(f) for a test where the second step is censored at time $\tau' > \tau$.

(h*)   Do (a)-(g) for a Weibull distribution.

(i)   Investigate such optimum plans to get further results.

11

Accelerated Degradation

Purpose. *Accelerated degradation* is concerned with models and data analyses for degradation of product performance over time at overstress and design conditions. This chapter briefly introduces basics of accelerated (aging) degradation. This vast topic merits entire books for each area of application.

Advantages. Accelerated degradation tests have some advantages over accelerated life tests. Performance degradation data can be analyzed earlier, for example, before any specimens "fail." This further accelerates the test. This is done by extrapolating performance degradation to estimate a time when performance reaches a failure level. Such extrapolation allows one to examine the effect on life of different design choices or assumptions about the performance level resulting in failure. Such performance degradation data may even yield more accurate life estimates than life test data with few failures. Also, performance degradation can yield better insight into the degradation process and how to improve it. However, most such advantages can be achieved *only if* one has a suitable model for extrapolation of performance degradation and an appropriate definition of failure in terms of performance. Because of insufficient knowledge of such models for adhesive degradation, Ballado-Perez (1986,1987) suggests treating such data as life data to simplify modeling and data analysis.

Overview. Section 1 briefly surveys applications of accelerated degradation. Section 2 presents basic accelerated degradation models. Section 3 shows how to analyze degradation data, using a specific application. Readers preferring simplicity and concreteness should first read Sections 2.1 and 3. Other sections, which are more abstract, can later be read more easily.

1. SURVEY OF APPLICATIONS

This section briefly surveys applications of accelerated (or aging) degradation. The survey includes bibliographies, various products and materials, and statistical methods. The survey in Section 1 of Chapter 1 is useful background and includes relevant technical societies, journals, and meetings.

Bibliographies. Devoted to degradation applications are bibliographies by Carey (1985) and Kulshreshtha (1976). General bibliographies that contain degradation among other topics include those by Meeker (1980), Losickij and Chernishov (1970), Goba (1969), and Yurkowski, Schafer, and Finkelstein (1969). Books and articles referenced in Section 1 of Chapter 1 contain such bibliographies and references on specific degradation applications, particularly Section 1.4, which surveys wear, corrosion, creep, etc.

Metals. Degradation of metal properties includes creep, crack initiation and propagation, wear, corrosion, oxidation, and rusting. References include ASTM STP 738 (1981) and 748 (1981), Bogdanoff and Kozin (1984), Yokobori and Ichikawa (1974), Zaludova (1981), and Zaludova and Zalud (1985). Many references in Section 1 of Chapter 1 pertain to such degradation.

Semiconductors and microelectronics. The degradation of electronic performance of many such devices is observed in accelerated tests. Then a device is often defined to "fail" when its performance degrades below a specified value. Many references of Section 1 of Chapter 1 are pertinent, for example, Howes and Morgan (1981). Recently Carey and Tortorella (1987) and LuValle and others (1986,1988a,b) developed probabilistic degradation models based on physical mechanisms.

Dielectrics and insulations. Measured properties include breakdown voltage, elongation, and tensile and flexural strength. Goba's (1969) bibliography includes degradation references. Selected references include Simoni (1974,1983), Vincent (1987), Bernstein (1981), Whitman and Doigan (1954), Vlkova and Rychtera (1974), and Veluzat and Goddet (1987).

Food and drugs. Virtually all accelerated food and drug testing is degradation testing for stability and shelf life. Performance includes the amount of active ingredient and bacteria level. All references of Section 1 of Chapter 1 are pertinent, particularly Young (1988), Labuza (1982), FDA (1987), Beal and Sheiner (1985), and pharmacokinetic references of Lu and Meeker (1989).

Plastics and polymers. Most accelerated testing of plastics and polymers is degradation testing of mechanical and other properties. Most references of Section 1 of Chapter 1 are pertinent.

Statistical methods. Standard regression methods apply to most aging degradation data, as such data are usually complete. Such methods are referenced in Chapter 4. Nonlinear regression methods are often needed; books include Seber and Wild (1989), Gallant (1987), Borowiak (1989), Ratkowsky (1983), and Bates and Watts (1988). Especially useful and versatile are maximum likelihood methods, which Chapters 5 and 9 cover. Lancaster (1990) presents statistical models that may extend to degradation processes.

2. DEGRADATION MODELS

Purpose. This section briefly surveys some basic degradation models for *constant* stress. For the large and rapidly developing literature on such models, see the references of Section 1 and Chapter 1. Here Section 2.1 presents the widely used Arrhenius rate relationship in detail. Section 2.2 presents a general simple relationship for constant degradation rate. Sections 2.3 and 2.4 extend that relationship to models with random coefficients and with random increments. Section 2.5 introduces mathematical rate functions. Section 2 draws from and expands on the expositions of Zaludova (1981) and Zaludova and Zalud (1985), which draw from Gertsbakh and Kordonskiy (1969). Lu and Meeker (1989) present such models more abstractly and include pharmacokinetic references and other data analysis methods. The following paragraphs provide general background.

Assumptions. Assumptions of all models below include:

1. Degradation is *not reversible*. That is, performance always gets monotonically worse. For example, cracks in metal continue to get longer, and breakdown voltage of insulation always decreases. Such models do not apply to products that can improve with exposure. For example, annealing can increase the life of fatigued metal or plastic.
2. Usually a model applies to a *single* degradation process (mechanism or failure mode). If there are simultaneous degradation processes and failure modes, each requires its own model. LuValle and others (1986,1988a,b) model competing processes.
3. Degradation of specimen performance before the test starts is negligible.

Another assumption that is separate from the degradation model is:

4. Performance is measured with *negligible* random error. Measurement error could be, but is not, included here in such models. Such error is important ("large") when performance changes little compared to it. Young (1988) notes that pharmaceutical measurement errors are large.

Statistical models. Much engineering work uses only a relationship between performance, age, and the accelerating variable(s). Thus such a relationship describes the "typical" performance. The statistical models below include a statistical distribution of performance around the typical value. Such a distribution is important in high reliability applications concerned with early failure in the lower tail.

Advances. For many applications, adequate engineering relationships for degradation have not been developed. Engineers and scientists with knowledge of degradation physics will advance such theory. Also, statisticians will contribute. Abdel-Hameed and others (1984) present some recent developments. Various models in the engineering literature have not been well validated by comparison with data. On the other hand, once a model is specified, most statistical planning of degradation tests and data analyses are routine matters in principle. That is, they are standard or easy to develop but may be computationally laborious.

Non-constant stress. Models for degradation testing with non-constant stress require a cumulative damage model (Chapter 10). Most such models assume that the product has a "Markov property." That is, the rate of damage (degradation) depends only on the current stress and cumulative damage and not on other features of the previous stress history. This is not so for metal fatigue and other phenomena. Bogdanoff and Kozin (1984) present such cumulative damage models for structural materials. Iuculano and Zanini (1986) apply such a model to metallic layer resistors. Rosenberg and others (1986) and Yoshioka and others (1987) present such a model for stability of pharmaceuticals. The bibliography of Goba (1969) contains references on such models for electrical insulations.

2.1. The Arrhenius Rate Relationship

Applications. The Arrhenius rate relationship below is widely used for temperature-accelerated degradation. Applications and references include:
- *Pharmaceuticals* – Bently (1970), Carstensen (1972), Connors and others (1979), Grimm (1987), FDA (1987), and Young (1988).
- *Insulations and dielectrics* – Whitman and Doigan (1954), Veluzat and Goddet (1987), and Goba (1969).
- *Plastics and polymers* – Hawkins (1971,1984).
- *Adhesives* – Beckwith (1979,1980) and Ballado-Perez (1986,1987).
- *Batteries and cells* – Linden (1984) and Gabano (1983).

The presentation below follows Nelson (1981).

Assumptions. The assumptions of the *Arrhenius rate model* are:

1. For any temperature and exposure time, the distribution of (positive) performance u is lognormal (base 10). Thus, the distribution of log performance $y = \log(u)$ is normal. Other distributions could be used.
2. The standard deviation σ of log performance is a constant. That is, σ does not depend on temperature or exposure time.
3. The relationship between the mean log performance μ (the log of median performance $u_{.50}$) and absolute temperature T and exposure time t is

$$\mu(t,T) = \log(u_{.50}) = \alpha - t \cdot \beta \cdot \exp(-\gamma/T). \qquad (2.1)$$

Log( ) is the base 10 logarithm here; the base e logarithm is an alternative. The parameters α, β, γ, and σ are characteristic of the product, degradation process, and test method. They are estimated from data (Section 3). The relationship is linear in parameters α and β and nonlinear in γ. Figure 2.1 depicts the relationship (2.1) on semilog paper. There the median performance $u_{.50}$ versus t is a straight line for each temperature. This figure also

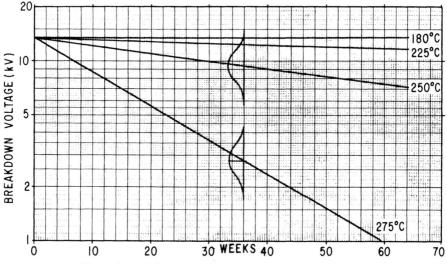

Figure 2.1.   Arrhenius rate model on semilog paper.

depicts assumptions 1 and 2 with normal distributions with the same spread. In practice, the linear dependence of (2.1) on time t may not adequately describe product behavior near time zero nor for great times; Section 2.2 discusses this further.

Interpretation. α is the log of median performance u_0 at time zero. In Figure 2.1, the median lines for all temperatures have the value u_0 at time zero. Then (2.1) becomes

$$\log(u_{.50}/u_0) = -t \cdot \beta \cdot \exp(-\gamma/T). \qquad (2.1')$$

An interpretation of this equation is that degradation is 1) governed by a simple first order chemical reaction where 2) the rate $\beta \exp(-\gamma/T)$ has an *Arrhenius dependence* on temperature. Whitman and Doigan (1954) present this equation. Here $\gamma = E/k$ where k is Boltzmann's constant and E is the activation energy for the reaction. Also, (2.1') is used to express the percentage $100(u_{.50}/u_0)$ of the initial performance u_0 remaining at age t.

Percentiles. The $100P$th population *percentile* of log performance $y_P(t,T) = \log[u_P(t,T)]$ at absolute temperature T and age t is

$$y_P(t,T) = \mu(t,T) + z_P \sigma = \alpha - t \cdot \beta \cdot \exp(-\gamma/T) + z_P \sigma; \qquad (2.2)$$

here z_P is the $100P$th standard normal percentile. Such percentile lines for a temperature are parallel to the median line in Figure 2.1.

Design temperature. In some applications, a *design temperature* T^* is determined as follows. The design life t^* is specified, and it is required that at most a specified percentage $100P^*$ be below a specified log performance y^*. The T^* that achieves this comes from (2.2) rewritten as

$$T^* = \gamma/\ln[\beta t^*/(\alpha+z_{P*}\sigma-y^*)]. \tag{2.3}$$

Median time. (2.1) yields the median time $t_{.50}$ for log performance to degrade to a specified value y^* (corresponding to "failure"); namely,

$$t_{.50} = [(\alpha-y^*)/\beta]\exp(\gamma/T).$$

This is the Arrhenius relationship for (median) *time* to failure (Chapter 2). It applies if the log performance must be above a log design value y^*. For example, y^* would be the log of the design voltage applied to insulation. For pharmaceuticals, y^* is the guaranteed (median) amount of drug on the package label, and $t_{.50}$ is the shelf life that assures that amount.

Life distribution. The following gives the product life distribution at design temperature T'. It is assumed that a specimen "fails" when its log performance degrades below $y' = \log(V')$, the design value. Figure 2.2 depicts the distributions of specimen life and of performance V. The line shows median performance as a function of time t at the design temperature T', which determines the slope of the line. The figure shows how the distribution of performance descends as the population ages, and how the population fraction of "failed" specimens increases with time. The population fraction $F(t)$ failed (with log performance below y') at time t is the shaded fraction of the performance distribution in Figure 2.2; namely,

$$F(t;T') = \Phi\{[y'-\alpha+t\beta\exp(-\gamma/T')]/\sigma\}; \tag{2.4}$$

here $\Phi\{\ \}$ is the standard normal cumulative distribution function. Thus time t to failure has a normal distribution with mean and standard deviation

$$\mu_t = [(\alpha-y')/\beta]\exp(\gamma/T'), \quad \sigma_t = (\sigma/\beta)\exp(\gamma/T'). \tag{2.5}$$

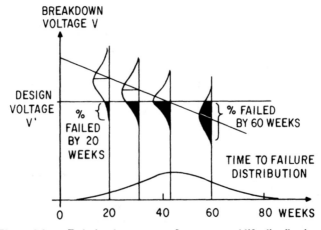

Figure 2.2.   Relation between performance and life distributions.

The 100Pth percentile of this normal life distribution is $t_P = \mu_t + z_P \sigma_t$, where z_P is the 100Pth standard normal percentile. Different performance levels y' at failure can be tried to see how sensitive the life distribution is to the assumed value. A more realistic model would have a distribution of performance y' at failure. For example, insulation with a design voltage y' sees slightly different voltage from one unit to the next. Moreover, the applied voltage may vary over time. Thus the actual σ_t tends to be larger.

Some readers may wish to skip to Section 3 where the model is fitted to insulation breakdown data.

2.2. Simple Constant Rate Models

In a general framework, this section presents simple constant rate models for degradation under constant stress. They include the Arrhenius rate relationship of Section 2.1.

Motivation. Figure 2.3 shows a "typical" plot of specimen performance (or its log) of some products versus age. Initially the rate (slope) of performance degradation is high – a wear-in period. After a short time, the rate of degradation becomes relatively constant and remains constant for a considerable time, corresponding to the straight portion of the plot. Finally, the rate (slope) of degradation may increase at great age as shown – a wear-out period. For some products, degradation rate is constant over all time. This motivates the following model, which applies to the straight portion of Figure 2.3. To fit a straight line to such data with an initial wear-in, one usually omits the performance data from time zero from the fitting.

General relationship. The simplest degradation relationship for "typical" *log* performance $\mu(t)$ is a simple *linear* function of product age t; namely,

$$\mu(t) = \alpha - \beta' t. \tag{2.6}$$

For concreteness, regard $\mu(t)$ as the mean or median of the population distribution of log performance at age t. Figure 2.1 depicts such relationships for

Figure 2.3.   "Typical" performance degradation versus specimen age.

several β' values. The intercept coefficient α is the "typical" log performance at age zero. The *degradation rate* β' is assumed constant over time. Moreover, β' depends on the constant accelerating stresses, as described below. Use of log performance assures that this degradation relationship does not yield negative performance values. Exponentiate (2.6) to get typical performance as $\exp(\alpha - \beta't)$, which decreases exponentially with age t. For some applications, $\mu(t)$ is typical performance rather than log performance, for example, for mechanical wear applications. If the amount of degradation is small, performance and its log work equally well. If "performance" increases with age (for example, crack length), use $+$ in place of $-$ in (2.6).

Failure. If the product fails when typical log performance degrades to a value μ^*, the typical failure time is (2.6) rewritten as

$$t = (\alpha - \mu^*)/\beta'. \tag{2.7}$$

Arrhenius dependence. For degradation accelerated by (absolute) temperature T, the *Arrhenius degradation rate* is

$$\beta' = \beta \cdot \exp(-\gamma/T). \tag{2.8}$$

Here β and γ are constants characteristic of the product and degradation process. With this rate parameter, (2.6) becomes the Arrhenius rate relationship (2.1). Then typical log performance is

$$\mu(t,T) = \alpha - t \cdot \beta \cdot \exp(-\gamma/T).$$

The typical failure time, when log performance reaches a value μ^*, is

$$t = [(\alpha - \mu^*)/\beta] \cdot \exp(\gamma/T).$$

This is the Arrhenius relationship for life (Chapter 2).

Power dependence. For some degradation processes, the degradation rate is represented by a *power function* of a positive stress V, namely,

$$\beta' = \beta V^\gamma . \tag{2.9}$$

Here β and γ are constants characteristic of the product and degradation process. This is often used for electronics and dielectrics where V is voltage. It is used in Taylor's model for wear of machine tools as a function of cutting velocity V; see Boothroyd (1975). Then the typical log performance is

$$\mu(t,V) = \alpha - t \cdot \beta V^\gamma .$$

The typical failure time, when log performance reaches a value μ^*, is

$$t = [(\alpha - \mu^*)/\beta]/V^\gamma .$$

This is the inverse power relationship for life (Chapter 2).

Exponential dependence. Some degradation rates are represented by an *exponential function* of a stress V; namely,

$$\beta' = \beta \cdot \exp(\gamma V) \ . \tag{2.10}$$

Here β and γ are constants characteristic of the product and degradation process. For example, this exponential function is used for weathering variables such as humidity. Then the typical log performance is

$$\mu(t,V) = \alpha - t \cdot \beta \cdot \exp(\gamma V) \ .$$

The typical failure time, when log performance reaches a value μ^*, is

$$t = [(\alpha - \mu^*)/\beta] \cdot \exp(-\gamma V) \ .$$

This is the exponential relationship for life (Chapter 2).

Eyring dependence. Some rate parameters are a function of absolute temperature T and a second (possibly transformed) stress V. The *generalized Eyring degradation rate* is

$$\beta' = \beta \cdot \exp[-(\gamma/T) - \delta V - \varepsilon(V/T)] \ . \tag{2.11}$$

Here β, γ, δ, and ε are constants characteristic of the product and degradation process. This is used, for example, for electronics and dielectrics, where V is voltage or ln voltage. Then the typical log performance is

$$\mu(t,T,V) = \alpha - t \cdot \beta \cdot \exp[-(\gamma/T) - \delta V - \varepsilon(V/T)] \ .$$

The typical failure time, when the log performance reaches a value μ^* is

$$t = [(\alpha - \mu^*)/\beta] \exp[(\gamma/T) + \delta V + \varepsilon(V/T)] \ .$$

This is the general Eyring relationship for life (Chapter 2).

Distribution. The preceding relationships model "typical" performance. At any population age, performance has a distribution. In Section 2.1, the distribution is lognormal and has a *constant* σ. (The models of Sections 2.3 and 2.4 below have non-constant spread.) Other distributions with a constant spread can be used with such relationships. For an assumed distribution of performance, the distribution percentiles can be written in terms of model parameters, analogous to (2.2). Also, a design level of stress can be determined from an equation like (2.3).

Other models. This paragraph briefly surveys some recent degradation models based on physical mechanisms. Carey and Tortorella (1987) present a model for degradation of MOS oxides. It employs a birth and death process for charge carriers. The model fits their data better than a simple birth process. LuValle and others (1986,1988a,b) present a variety of degradation models based on chemical kinetics and probabilistic considerations. By fitting a number of such plausible models to a data set and assessing which fit well, they gain physical insight. Also, in microelectronic applications, relatively small numbers of atoms are involved. So such probabilistic models appear suitable and yield insights. For example, they present a model with competing reactions where some specimens never degrade enough to fail.

That is, some specimens survive forever (Chapter 2), an observed phenomenon in some electronic devices.

Geometry. Note that most models based on kinetic theory do not include product geometry in the model. That is, such models implicitly *assume* that the reacting materials are homogeneously mixed at the start. This is often so for the Arrhenius rate relationship. In contrast, some degradation in microelectronics results from atomic diffusion of adjoining (different) materials, and material geometry affects the degradation process. Thus, where important, geometry should be modeled. However, it is often overlooked.

2.3. Random Coefficients Model

Model. Another simple degradation model for log performance y_i of specimen i as a function of age t is

$$y_i(t) = a_i - b_i t. \tag{2.12}$$

Here a_i and b_i are the constant intercept and constant degradation rate for specimen i. Thus each specimen exactly follows its own linear relationship as it ages, as depicted in Figure 2.4. The intercept a_i and rate b_i vary from specimen to specimen and have a joint distribution. Thus the model (2.12) is said to have *random coefficients* or, better said, a joint distribution of coefficients, since the coefficients a_i and b_i of specimen i are constant.

Mean and variance. Denote the population mean and standard deviation of the intercepts a_i by α and σ_a. Here α and σ_a are the mean and standard deviation of log performance at time zero. σ_a usually reflects variability in manufacturing tolerances and process variables. Denote the mean and standard deviation of the rates b_i by β' and σ_b. σ_b usually reflects variability in material composition, which determines degradation rate. Within a homogenous lot σ_b is usually small. But σ_b can be large when specimens come from a number of lots and there is lot-to-lot variability in composition. Also, one assumes that the constant mean rate β' depends on the accelerating stress(es), as described in Section 2.2. Assume that the coefficient pairs (a_i, b_i) are *uncorrelated*. Then the distribution of log performance $y_i(t)$ at time t has a mean and variance

$$\mu(t) = \alpha - \beta' t, \quad \sigma^2(t) = \sigma_a^2 + \sigma_b^2 t^2. \tag{2.13}$$

The relationship for the mean is the simple linear one (2.6), where β' depends on stress as described in Section 2.2. $\sigma(t)$ reduces to that for the previous model (constant σ) if $\sigma_b^2 = 0$, that is, if all specimens have the same rate $b_i = \beta'$. For example, specimens with the same composition may all degrade at the same rate. Other assumptions could lead to this mean and variance. Note that this model is not well verified. Beal and Scheiner (1988) investigate methods for fitting heteroscedastic models to data. Lu and Meeker (1989) survey references on random coefficients models and data analyses.

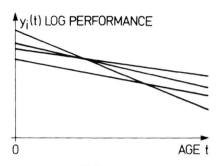

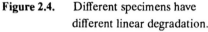

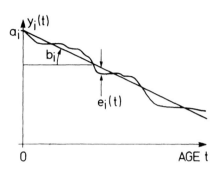

Figure 2.4.   Different specimens have different linear degradation.

Figure 2.5.   Random variation $e_i(t)$ around a linear degradation relationship.

Life distribution. If the a_i and b_i are further assumed to have *normal* distributions, the distribution of time to failure (when log performance reaches a value y^*) is

$$F(t) = \Phi[(y^* - \alpha + \beta't)/(\sigma_a^2 + \sigma_b^2 t^2)^{1/2}]. \tag{2.14}$$

This *Bernstein distribution* appears in Gertsbakh and Kordonskiy (1969), Levitanus (1973), and Peshes and Stepanova (1972). Ahmad and Shiekh (1981) describe how to estimate its parameters from censored data.

Random variation. In the preceding model, specimen i exactly follows its own linear relationship (2.12) as it ages. A more plausible model includes random variation $e_i(t)$ in log performance as the specimen performance wanders about its linear relationship, as depicted in Figure 2.5. That is, at age t, log performance $y_i(t)$ of specimen i differs from (2.12) by an amount $e_i(t)$. Assume, for any time t, the mean of $e_i(t)$ is zero and its variance is a constant σ_e^2. The model then is

$$y_i(t) = a_i - b_i t + e_i(t). \tag{2.12'}$$

If a_i, b_i, and $e_i(t)$ are assumed statistically independent, then the distribution of $y_i(t)$ has mean and variance

$$\mu(t) = \alpha - \beta't, \quad \sigma^2(t) = (\sigma_a^2 + \sigma_e^2) + \sigma_b^2 t^2. \tag{2.13'}$$

The relationship for the mean is the simple linear one (2.6). Also, $\sigma(t)$ is the same as before, but $(\sigma_a^2 + \sigma_e^2)$ is in place of σ_a^2. (2.13') with $(\sigma_a^2 + \sigma_e^2)$ in place of σ_a^2 gives a Bernstein distribution (2.14) for time to failure for this model.

Batches. The preceding model is appropriate for a single homogenous population. If there is batch-to-batch variability, the model can be extended to represent such variability.

2.4. Random Increments Model

Model. The following model involves random increments in the performance degradation. It is another type of model with random coefficients.

For specimen i, denote the (log) performance at time zero by a_i; that is, $y_i(0) = a_i$. To model (log) performance $y_i(t)$ at age t, divide the interval $(0,t)$ into J short intervals of equal length $\tau = t/J$. Suppose that the (log) performance of specimen i at time $j\tau$ $(j = 1, 2, \cdots, J)$ is

$$y_i(j\tau) = y_i[(j-1)\tau] - b_{ij}\tau.$$

Here $b_{ij}\tau$ is the random non-negative increment in (log) performance between time $(j-1)\tau$ and $j\tau$, and b_{ij} is the random rate of degradation of specimen i over interval j. Then, at time t,

$$y_i(t) = y_i(J\tau) = a_i - (b_{i1} + b_{i2} + \cdots + b_{iJ})\tau.$$

Zaludova and Zalud (1985) apply this model to wear of diesel engines, as measured by the metal debris content of their lubricating oil.

Mean and variance. Denote the population mean and standard deviation of the a_i by α and σ_a. Assume that the mean of any b_{ij} is the constant degradation rate β'. β' depends on stress as described in Section 2.2. Thus the mean of the population distribution of (log) performance $y_i(t)$ at time t is

$$\mu(t) = \alpha - (J\beta')\tau = \alpha - \beta't. \tag{2.15}$$

This relationship is independent of the number J of intervals. This is the simple linear relationship for the mean (2.6). To get a variance $\sigma^2(t)$ for the performance distribution at time t, assume that the $a_i, b_{i1}, b_{i2}, \cdots, b_{iJ}$ are *uncorrelated* and that the variance of any b_{ij} is a *constant* σ_b^2. Note that the meaning of σ_b^2 here differs from that in Section 2.3. Here σ_b^2 could be a function of stress. (The assumption that the $a_i, b_{i1}, b_{i2}, \cdots, b_{iJ}$ are uncorrelated also simplifies the fitting of this model to data.) Then the variance of the population distribution of $y_i(t)$ is

$$\sigma^2(t) = \sigma_a^2 + (J\sigma_b^2)\tau = \sigma_a^2 + \sigma_b^2 t. \tag{2.16}$$

This relationship is independent of the number J of intervals. This variance is a linear function of time t. The previous variance (2.13') of the random coefficients model is function of t^2. Other assumptions could yield (2.15) and (2.16). Thus, if these relationships fit data well, it does not necessarily follow that the model above is a suitable description. Tomsky (1982) describes how to estimate parameters of this model from repeat measurements on specimens over time.

Life distribution. Suppose a specimen "fails" when its (log) performance degrades below a specified value y^*. By (2.15) the typical age at failure is

$$t = (\alpha - y^*)/\beta' \ .$$

If $a_i, b_{i1}, b_{i2}, \cdots, b_{iJ}$ have *normal* distributions (an assumption inconsistent with non-negative increments), then $y_i(t)$ has a normal distribution, and the population fraction failed by age t is

$$F(t) = Pr\{y_i(t) \le y^*\} = \Phi[(y^* - \alpha + \beta't)/(\sigma_a^2 + \sigma_b^2 t)^{1/2}] \ . \tag{2.17}$$

For this distribution, the above typical age at failure is the median. Suppose that the measurement scale of performance is rescaled so $\alpha = 0$. Also, suppose that all specimens have the *same* initial performance a_i; that is, $\sigma_a^2 = 0$. Then this life distribution becomes the *Birnbaum-Saunders* (1969) *distribution*, which has been proposed for metal fatigue.

2.5. Mathematical Rate Functions

Many rate relationships fitted to degradation data are merely reasonable mathematical functions or curves that are not based on physical theory. Such curves should be mathematically sensible, as in the Weibull example below. They merely smooth the data and usually interpolate adequately. However, they may provide no physical insight and may extrapolate badly. This section gives examples of such curves.

Weibull relationship. Crow and Slater (1969) used the following "Weibull relationship" to model the degradation of mechanical properties of plastic building materials under accelerated weathering. The "typical" performance as a function of aging time t is

$$\mu(t) = \alpha \cdot \exp[-(t/\beta)^\gamma] + \delta . \tag{2.18}$$

Here the positive parameters α, β, γ, and δ are characteristic of the property (for example, percent elongation), the plastic, and the weathering variables (for example, humidity, ultraviolet, temperature). Here $\exp[-(t/\beta)^\gamma]$ looks like the Weibull reliability function. However, it is a relationship here – not a distribution. Crow and Slater do not use a distribution to describe the scatter in performance.

Interpretation. There is no physical basis for (2.18). It merely has suitable mathematical properties. For example, it is positive and monotone decreasing. The parameters can be interpreted as follows. $(\alpha + \delta)$ is the initial typical value at $t = 0$. δ is the ultimate typical value at great age. Usually δ is zero; then β is the age when the typical value reaches 36.8% of its initial value. Use of such a relationship is merely curve fitting, since it is not based on engineering theory. Crow and Slater also expressed the parameters as mathematical functions of the weathering variables with further curve fitting.

Larsen-Miller. Dieter (1961) gives various generalizations of the Larsen-Miller relationship for creep-rupture. Some merely provide an improved fit to data and have no physical basis or interpretation. They are merely mathematical functions.

Polynomials. Much data smoothing is done by fitting polynomials. Polynomials are often adequate for interpolation, but they are notoriously bad for extrapolation and sometimes for interpolation. For example, designers use materials properties curves for metals and ceramics to design components. Many such curves are polynomial fits. Also, for example, Zaludova and Zalud (1985) fit a cubic polynomial to wear data on diesel engines, and

Underwriters Lab (1975) fits a cubic polynomial to degradation data on polymer properties. Such a polynomial can be a function of time and any number of other variables.

3. ARRHENIUS ANALYSIS

Purpose. This section describes analyses for degradation data with the Arrhenius rate model (Section 2.1). The analyses provide estimates of 1) the model parameters, 2) the distribution of time for product performance to degrade to failure, and 3) the design temperature that achieves a specified small percentage failed over design life. The analyses are applied here to a specific degradation model and data. However, they apply to other products and other constant rate models. Here performance of each specimen is measured just *once* at some age, as breakdown is destructive. Analyses for data with repeat measurements on specimens are complicated; so they are not described here. For example, Tomsky (1982) analyzes repeat measurements.

Overview. Section 3.1 describes the data and model that illustrate the analyses. Section 3.2 presents graphical analyses and model fitting. Section 3.3 presents analytic model fitting, which requires special computer programs. Section 3.4 provides checks on the model and data. The presentation below follows Nelson (1981).

3.1. Insulation Breakdown Data and Model

Data. The following analyses are illustrated with data on the dielectric breakdown strength of insulation specimens. Each insulation specimen was held at a high test temperature for a specified time in weeks. Then its breakdown voltage was measured – a destructive test. Such performance data are typically complete (i.e., not censored). The data (Table 3.1) consist of the breakdown voltage (in kV) of four specimens (equal allocation) for each combination of four test temperatures (180, 225, 250, 275°C) and eight aging times (1, 2, 4, 8, 16, 32, 48, 64 weeks). For example, the first specimen in Table 3.1 ran 1 week at 180°C, and its breakdown voltage was then measured as 15.0 kV. Note that no specimens were measured at age 0 weeks. Beckwith (1979,1980) notes that estimates are more accurate if some specimens are measured at age 0. Of course, this is true only if the product degrades linearly and has no wear-in. Better said, there should be more specimens at early (and late) measurement ages, that is, unequal allocation. In all, there were $4{\times}4{\times}8 = 128$ specimens. The test purpose was to estimate the insulation life distribution. The insulation fails when its breakdown strength degrades below the design voltage 2.0 kV at a design temperature of 150°C.

Test plans. The test plan above is statistically inefficient for the Arrhenius rate model. Reduce the statistical uncertainty in estimates as follows:

Table 3.1. Insulation Breakdown Data

Wk	Temp	kV	Wk	Temp	kV	Wk	Temp	kV	Wk	Temp	kV
1.	180.	15.0	4.	180.	13.5	16.	180.	18.5	48.	180.	13.0
1.	180.	17.0	4.	180.	17.5	16.	180.	17.0	48.	180.	13.5
1.	180.	15.5	4.	180.	17.5	16.	180.	15.3	48.	180.	16.5
1.	180.	16.5	4.	180.	13.5	16.	180.	16.0	48.	180.	13.6
1.	225.	15.5	4.	225.	12.5	16.	225.	13.0	48.	225.	11.5
1.	225.	15.0	4.	225.	12.5	16.	225.	14.0	48.	225.	10.5
1.	225.	16.0	4.	225.	15.0	16.	225.	12.5	48.	225.	13.5
1.	225.	14.5	4.	225.	13.0	16.	225.	11.0	48.	225.	12.0
1.	250.	15.0	4.	250.	12.0	16.	250.	12.0	48.	250.	7.0
1.	250.	14.5	4.	250.	13.0	16.	250.	12.0	48.	250.	6.9
1.	250.	12.5	4.	250.	12.0	16.	250.	11.5	48.	250.	8.8
1.	250.	11.0	4.	250.	13.5	16.	250.	12.0	48.	250.	7.9
1.	275.	14.0	4.	275.	10.0	16.	275.	6.0	48.	275.	1.2
1.	275.	13.0	4.	275.	11.5	16.	275.	6.0	48.	275.	1.5
1.	275.	14.0	4.	275.	11.0	16.	275.	5.0	48.	275.	1.0
1.	275.	11.5	4.	275.	9.5	16.	275.	5.5	48.	275.	1.5
2.	180.	14.0	8.	180.	15.0	32.	180.	12.5	64.	180.	13.0
2.	180.	16.0	8.	180.	15.0	32.	180.	13.0	64.	180.	12.5
2.	180.	13.0	8.	180.	15.5	32.	180.	16.0	64.	180.	16.5
2.	180.	13.5	8.	180.	16.0	32.	180.	12.0	64.	180.	16.0
2.	225.	13.0	8.	225.	13.0	32.	225.	11.0	64.	225.	11.0
2.	225.	13.5	8.	225.	10.5	32.	225.	9.5	64.	225.	11.5
2.	225.	12.5	8.	225.	13.5	32.	225.	11.0	64.	225.	10.5
2.	225.	12.5	8.	225.	14.0	32.	225.	11.0	64.	225.	10.0
2.	250.	12.5	8.	250.	12.5	32.	250.	11.0	64.	250.	7.2667
2.	250.	12.0	8.	250.	12.0	32.	250.	10.0	64.	250.	7.5
2.	250.	11.5	8.	250.	11.5	32.	250.	10.5	64.	250.	6.7
2.	250.	12.0	8.	250.	11.5	32.	250.	10.5	64.	250.	7.6
2.	275.	13.0	8.	275.	6.5	32.	275.	2.7	64.	275.	1.5
2.	275.	11.5	8.	275.	5.5	32.	275.	2.7	64.	275.	1.0
2.	275.	13.0	8.	275.	6.0	32.	275.	2.5	64.	275.	1.2
2.	275.	12.5	8.	275.	6.0	32.	275.	2.4	64.	275.	1.2

1. Test specimens at or near time zero. This yields a better estimate of α, the median (log) performance at time zero. Also, it yields a better estimate of the slope β, since a wider range of time is used. Of course, this assumes there is no wear-in near time zero.

2. For each temperature, measure more specimens at or near time zero and at the longest test time, and fewer at intermediate times. This yields a more accurate estimate of β, since a wider range of time is used, provided that the model is suitable. Intermediate times allow for a test of linearity and provide some early information.

3. Test more specimens at the lowest test temperatures than at the highest. The lowest test temperature is nearest the design temperature. To extrapolate most accurately, use more specimens near the design condition.

4. Make the highest test temperature as high as possible. This yields a more accurate estimate of γ for the effect of temperature. However, the temperature should not introduce new failure modes or degradation inconsistent with that at design temperature.

Beckwith (1979,1980) studies improved test plans. Haynes and others (1984) describe sequential plans. Ford, Titterington, and Kitsos (1989) survey experimental design for nonlinear models for chemical reactions. Accuracy of estimates from proposed test plans can be studied by simulation (Chapter 6).

Model. The Arrhenius rate model of Section 2.1 (in brief) assumes

1. At any absolute temperature T and after exposure time t, the distribution of specimen breakdown voltage V is lognormal (base 10). Some authors use base e.
2. The log standard deviation σ is a constant for all temperatures and exposure times.
3. The mean log (base 10) breakdown voltage $y = \log(V)$ is

$$\mu(t,T) = \log(V_{.50}) = \alpha - t \cdot \beta \cdot \exp(-\gamma/T). \qquad (3.1)$$

Depiction. Figure 3.1 depicts the 32 test conditions (temperature-age combinations) as dots and the relationship (3.1) as a surface.

Independence. All of the data analyses in this book *assume* that the random variations in the observations of performance are statistically *independent*. This assumption simplifies data analyses. Indeed, most statistical theory makes this assumption. Moreover, the assumption is usually appropriate in practice. It is usually appropriate for performance degradation where each specimen is measured *once*. This is so when the measurement is destructive. For example, the insulation breakdown measurement is destructive. If specimen performance is repeatedly measured (nondestructively), successive measurements on a specimen may be statistically autocorrelated. Analysis of such autocorrelated data requires multivariate methods, which are more complex and less familiar. Methods developed for growth/wear curves apply to such correlated (dependent) data. Recent work by Timm (1980) and O'Rear and Leeper (1983) references the Kleinbaum (1973) model, the

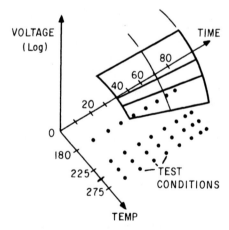

Figure 3.1.   Depiction of the relationship surface and test conditions .

Pothoff-Roy (1964) model, and early work by Box (1950) for such data. Tomsky (1982) treats a reliability application. Treating such successive repeat measurements as independent yields less accurate estimates and misleading confidence limits.

3.2.   Graphical Analyses

The following graphical analyses provide estimates of model parameters and the relationship between performance, exposure time, and stress. Also, they assess the validity of the model and data.

Relationship plot. Use plotting paper on which the theoretical relationship for product performance plots as a straight line against stress or exposure time. For the insulation, the assumed relationship (3.1) between breakdown voltage and exposure time is a straight line on semilog paper, as shown in Figure 2.1. Plot each (average log) performance value against its stress (or exposure time). Then fit a straight line to the data by eye. This is shown in Figure 3.2, which has a straight line relationship between (log) breakdown voltage and exposure time for each temperature. According to (3.1), the straight lines all pass through the same voltage at time zero. So fit the lines to pass through a common point.

Coefficient estimates. Graphically estimate the coefficients α, β, and γ as follows. The estimate α^* of α is the log of the estimate V_0^* of the common breakdown voltage V_0 at time zero. From Figure 3.2, $V_0^* = 13.5$ kV and $\alpha^* = \log(13.5) = 1.130$. To estimate β and γ, choose two (absolute) test temperatures T_1 and T_2, which are widely separated, and a long exposure time t.

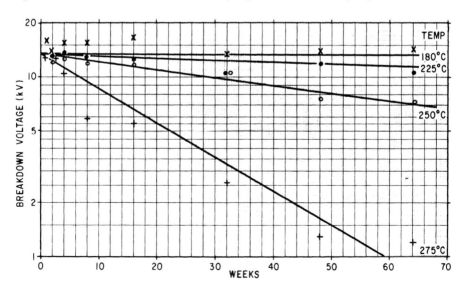

Figure 3.2.   Average log breakdown voltage versus age.

From the fitted lines for T_1 and T_2, obtain estimates V_1^* and V_2^* of the typical breakdown voltages V_1^* and V_2^* at time t. The true voltages V_1 and V_2 satisfy (Section 2.1):

$$\log(V_1) = \log(V_0) - \beta t \exp(-\gamma/T_1),$$
$$\log(V_2) = \log(V_0) - \beta t \exp(-\gamma/T_2). \qquad (3.2)$$

Rewritten these yield

$$\gamma = [T_1 T_2/(T_1 - T_2)] \ln[\log(V_0/V_1)/\log(V_0/V_2)],$$
$$\beta = (1/t) \exp(\gamma/T_1) \log(V_0/V_1). \qquad (3.3)$$

These yield estimates of γ and β from the estimates of V_0, V_1, V_2. For the insulation, use $t = 32$ weeks, $T_1 = 250°C = 523.16°K$ and $T_2 = 275°C = 548.16°K$. By Figure 3.1, $V_0^* = 13.5$ kV, $V_1^* = 9.8$ kV, and $V_2^* = 3.27$ kV. So

$$\gamma^* = [523.16(548.16)/(-25)]\times\ln[\log(13.5/9.8)/\log(13.5/3.27)] = 17065°K,$$

$$\beta^* = (1/32)\exp(17065/523.16)\log(13.5/9.8) = 6.375 \times 10^{11}/\text{week.}$$

σ estimate. The log standard deviation σ of log breakdown voltage is graphically estimated in Section 3.4 or analytically in Section 3.3.

Life distribution and design temperature. Estimate the mean and standard deviation of the normal life distribution, using the graphical estimates of the model coefficients in (2.5). Analytic estimates appear below. Similarly, estimate distribution percentiles (2.2) and a design temperature (2.3).

3.3. Analytic Methods

This section describes 1) fitting a degradation model to data by maximum likelihood and 2) the estimate of the distribution of time to specimen failure. Such fitting requires special computer programs.

Fitting. The following model fitting applies to other models. Here one estimates the unknown model parameters α, β, γ, and σ from the data. The relationship (3.1) is *not* a linear function of γ. Standard computer programs for linear least squares regression fitting cannot fit (3.1). To fit (3.1), use maximum likelihood or nonlinear least squares fitting. The STATPAC program of Nelson, Morgan, and Caporal (1983) was used to fit the model to the breakdown data. Young (1988) presents a computer program that calculates maximum likelihood estimates and confidence limits. Beckwith (1980) presents traditional engineering estimates for such data. Such estimates are less accurate than the ML ones for the Arrhenius rate model. Also, nonlinear least squares fitting applies to the model here. Bates and Watts (1988), Seber and Wild (1989), and Gallant (1987) present such fitting.

Example. STATPAC output from fitting (3.1) to the breakdown data appears in Figure 3.3. The estimates for α, β, γ, and σ are $a = 1.123568$, $b =$

```
* MAXIMUM LIKELIHOOD ESTIMATES FOR MODEL COEFFICIENTS
  WITH APPROXIMATE 95% CONFIDENCE LIMITS

COEFFICIENTS  ESTIMATE       LOWER LIMIT      UPPER LIMIT

C00001 a     1.123568        1.107296         1.139841
C00002 b     0.2961467E 12  -0.1047282E 13    0.1639575E 13
C00003 C     16652.63        14166.86         19138.41
C00004 S     0.7495586E-01   .0.6577759E-01   0.8413413E-01
```

```
* COVARIANCE MATRIX

COEFFICIENTS  C00001          C00002          C00003         C00004
              a               b               C              S
C00001   0.6892625E-04
C00002  -0.2595812E 10   0.4698043E 24
C00003      -4.846288    0.8692406E 15    1608462.
C00004  -0.3546367E-06   0.6395259E 08    0.1183286    0.2192853E-04
```

PCTILES(ALL)

```
     WEEKS          TEMP

       1.           180.
```

```
* MAXIMUM LIKELIHOOD ESTIMATES FOR DIST. PCTILES
  WITH APPROXIMATE 95% CONFIDENCE LIMITS

PCT.          ESTIMATE       LOWER LIMIT      UPPER LIMIT

0.1           7.796200       7.228703         8.408250
0.5           8.519846       7.972861         9.104358
1             8.894664       8.359331         9.464279
5             10.00501       9.504473         10.53191
10            10.65277       10.16968         11.15881
20            11.49380       11.02556         11.98193
50            13.29033       12.80197         13.79732
80            15.36766       14.74589         16.01565
90            16.58093       15.83532         17.36164
95            17.65444       16.77899         18.57557
99            19.85829       18.67322         21.11858
```

Figure 3.3.   STATPAC fit of the Arrhenius rate model to the breakdown data.

0.2961467×10^{12}, $c = 16652.63$, and $s = 0.07495586$. The output shows seven significant figures – more than justified by the accuracy of the data and estimates. However, all figures are retained in calculations, and final results are rounded appropriately. This minimizes round-off error. The fitted (3.1) is

$$m(t,T) = \log(V_{.50}) = 1.123568 - 2.961467 \times 10^{11} \cdot t \cdot \exp(-16{,}652.63/T); \qquad (3.4)$$

here t is in weeks, V in kV, and T in degrees Kelvin. This relationship is in Figure 3.1. It appears to fit the data reasonably well. Figure 3.3 includes confidence intervals for estimated quantities.

Percentiles. The fitted relationship (3.4) estimates the distribution median. The estimate V_P^* of the $100P$th percentile of the distribution of breakdown voltage at absolute temperature T and exposure time t is

$$V_P^* = \text{antilog}[a - b \cdot t \cdot \exp(-c/T) + z_P s];$$

here z_P is the $100P$th standard normal percentile.

Life distribution. Suppose "failure" occurs when product performance degrades to a specified value V'. For the insulation $V' = 2.0$ kV, its design voltage at a design temperature of $T' = 423.16°$K. (150°C). Section 2.1 shows that the distribution of time to failure is normal with the mean and standard deviation in (2.5). Substitute the parameter estimates into (2.5) to get

$$m_t = (1/2.961467 \times 10^{11}) [1.123568 - \log(V')]\cdot\exp(16,652.63/T'),$$

$$s_t = (0.07495586/2.961467 \times 10^{11})\cdot\exp(16,652.63/T').$$

Evaluated at $V' = 2.0$ kV ànd $T' = 423.16$, $m_t = 342,000$ weeks $\simeq 6560$ years and $s_t = 31,200$ weeks $\simeq 600$ years. Approximate confidence limits for μ_t, σ_t, and percentiles of this life distribution can be calculated with STATPAC or by hand from the covariance matrix in Figure 3.3.

3.4. Checks on the Model and Data

Purpose. This section describes checks on the model (assumptions 1, 2, and 3) and data. These include whether the degradation rate is constant, whether the assumed dependence of rate on stress is adequate, whether there is uncontrolled variation in performance, whether there are outliers, and whether σ is constant. These checks follow. They employ the relationship plot and the residuals. Such checks are essential for a complete data analysis.

Constant rate. If the constant rate relationship (2.6) describes degradation, the relationship plot (Figures 3.2 and 3.4) should be straight for each stress level. Systematic departure from a straight line at a number of stress levels indicates that the rate is not constant. For the insulation breakdown data, the plot is relatively straight for each temperature.

Rate dependence. In some applications, one does not know how the deg-

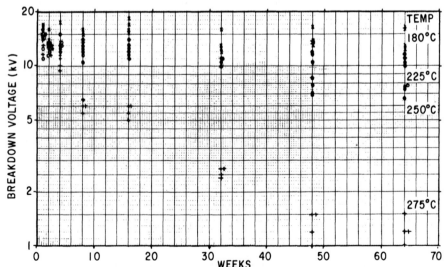

Figure 3.4.   Individual breakdown voltages versus age.

radation rate depends on the accelerating stress. Common practice is to sep-arately estimate the degradation rate β' in (2.6), using data from each test stress level. For example, for the breakdown data, use the relationship plot to estimate the slope (rate) of the straight line through the (log) data (by eye or least squares) for each test temperature. Figure 3.2 shows such lines for the four test temperatures. Then plot these rate estimates against their stress levels as in Figure 3.5 – a *rate plot*. Use linear, log-log, semilog, Arrhenius, or other paper to straighten the rate plot, and thereby determine which best describes the dependence of *rate* on stress. The straight lines in the relation-ship plot may be separately fitted by least squares to the data from each stress level; then each line has a different intercept at age 0. Alternatively, the lines in the relationship plot can be fitted through a common intercept at age 0; this usually yields more accurate rate estimates. For the breakdown data and separate intercepts, such rate estimates are .000669, .00174, .00408, and .0173 for 180, 225, 250, and 275°C, respectively. These LS estimates are plotted on Arrhenius paper in Figure 3.5.

The rate plot is relatively straight. This suggests that the Arrhenius rate dependence (2.8) adequately describes the data. Judging straightness is aid-ed by confidence limits for the rate parameter estimates at each stress level. Such 95% limits appear in Figure 3.5. The slope of the straight line in the figure is the ML estimate $c = 16652.63$. Engineers traditionally use such a plot (without confidence limits) to graphically estimate this slope; the result-

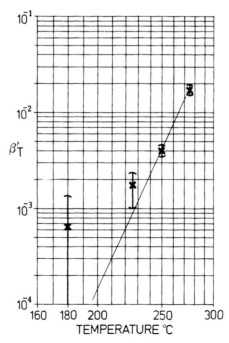

Figure 3.5. Arrhenius plot of degradation rate estimates versus temperature.

ing estimate has poor accuracy. A formal LS hypothesis test of linearity can be complex. It depends on the number of specimens at each temperature, the exposure ages, and whether there are separates estimates of the intercept coefficient or a common estimate. A LR test (Chapter 9) is simpler to do.

Uncontrolled variation. The fitted straight lines in Figure 3.2 do not adequately fit the data, according to an F test of linearity (Chapters 4 and 8). Moreover, the data do not have systematic curvature (different rate dependence). Thus the random variation of the average log performance about a straight line for a temperature is great compared to the variation between specimens within a test age. This suggests that uncontrolled factors in the test affect the breakdown voltages, and their cause should be sought. Better control of the test and measurement methods may reduce such variability and yield a better fit. Also, outliers may cause this.

Outliers. Questionable data points are appreciably out of line with the others. For example, in Figure 3.2, the average at 275°C and 64 weeks seems high. Review such "outliers" to determine their cause. Also, decide to include or exclude them from analyses. In practice, it is best to analyze both with and without suspect data to see if they appreciably affect results. Such outliers may also appear in residual plots below.

Constant σ. σ is assumed constant for all combinations of temperature and exposure time. The model fitting employs that assumption. Moreover, the confidence limits are inaccurate if the assumption is not satisfied. Check this assumption with the methods of Chapter 4 and with a relationship plot of individual observations (Figure 3.4). In that plot, the data scatter for each of 32 test conditions should be about the same. Pronounced trends in the scatter as a function of exposure time or temperature indicate that σ is not constant. No such trends appear in Figure 3.4. Thus σ appears constant, except Figure 3.4 slightly suggests that the 275° data at 48 and 64 weeks have greater scatter than the rest of the data. The greater scatter in those low voltage measurements may be due to the coarseness of the measurements (nearest 0.5 volts) at low voltage. A chi-square probability plot of the 32 variances would be informative.

Residuals. As defined in Chapter 4, a log residual is an observed log breakdown voltage minus the fitted log relationship at its test condition. STATPAC calculates these residuals. Another residual is the observed log voltage minus the average log voltage at its test condition. If the relationship fits the data badly, these residuals are better for assessing the assumed distribution and whether σ is constant. Plots of residuals are multipurpose and often more informative than formal tests of the fit of a relationship, of homogeneity of variance, of distributional assumption, etc. Moreover, such formal tests require special computer routines.

Distribution check. The log residuals should look like a sample from a normal distribution, if the lognormal distribution describes voltage break-

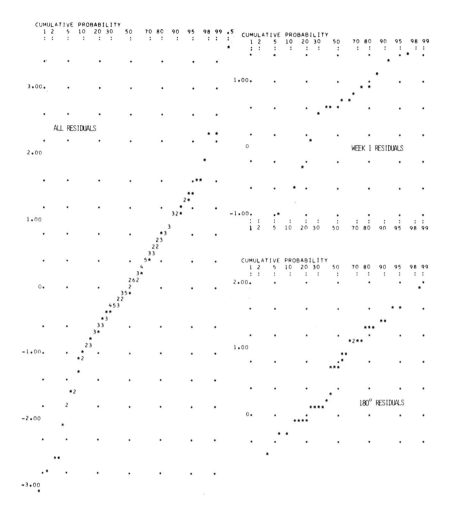

Figure 3.6.   Normal plots of log residuals.

down. Alternatively, they should look like a sample from an extreme value distribution, if the Weibull distribution adequately describes voltage breakdown. Selected normal probability plots of log residuals (about the relationship) appear in Figure 3.6. The plot of all residuals has tails that are too long. This indicates that a lognormal distribution does not adequately describe breakdown voltage. This may be due to outliers noted below. Such a probability plot also yields a graphical estimate of σ (Chapter 4).

Crossplots. Crossplots of the residuals against variables in or outside the model often reveal information. Figure 3.7 shows a crossplot of the residuals against their temperature values. If the relationship (3.1) adequately fits the data, the median of the residuals at each temperature should be close to zero. The figure suggests slight systematic variation in the medians (shown as −)

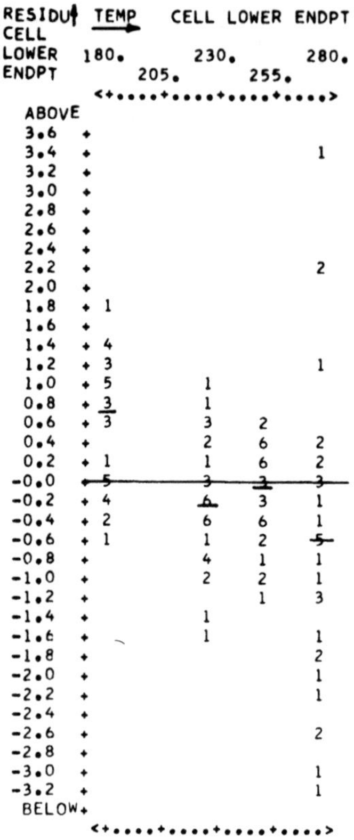

```
RESIDUAL TEMP    CELL LOWER ENDPT        RESIDUAL WEEKS   CELL LOWER ENDPT
CELL                                     CELL
LOWER  180.    230.      280.            LOWER  0.     20.      40.     60.
ENDPT      205.    255.                  ENDPT     10.     30.     50.
        <+....+....+....+....>                    <+....+....+....+....+....+....+..>
  ABOVE +                                  ABOVE +
  3.6 +                                    3.6 +
  3.4 +                1                    3.4 +                              1
  3.2 +                                    3.2 +
  3.0 +                                    3.0 +
  2.8 +                                    2.8 +
  2.6 +                                    2.6 +
  2.4 +                                    2.4 +
  2.2 +                2                    2.2 +                              2
  2.0 +                                    2.0 +
  1.8 + 1                                  1.8 +            1
  1.6 +                                    1.6 +
  1.4 + 4                                  1.4 + 1 2       1
  1.2 + 3              1                    1.2 + 1                     1       2
  1.0 + 5    1                             1.0 + 11  1     1     1              1
  0.8 + 3    1                             0.8 + 2   1     1
  0.6 + 3    3   2                         0.6 + 3 1 2           1     1
  0.4 +      2   6   2                     0.4 + 4         1     2     1       2
  0.2 + 1    1   6   2                     0.2 + 31 1      3     1             1
 -0.0 + 5    3   3   3                    -0.0 + 134 2     2                   2
 -0.2 + 4    4   3   1                    -0.2 + 22       1     1     3        3
 -0.4 + 2    6   6   1                    -0.4 + 144 2          1     2        2
 -0.6 + 1    1   2   5                    -0.6 + 12        2     1     2       1
 -0.8 +      4   1   1                    -0.8 + 11             3     1
 -1.0 +      2   2   1                    -1.0 + 1         1          2       1
 -1.2 +          1   3                    -1.2 + 1         1     2
 -1.4 +      1                            -1.4 + 1
 -1.6 +      1       1                    -1.6 +                2
 -1.8 +              2                    -1.8 +          1           1
 -2.0 +              1                    -2.0 +          1
 -2.2 +              1                    -2.2 + 1
 -2.4 +                                   -2.4 +
 -2.6 +              2                    -2.6 + 2
 -2.8 +                                   -2.8 +
 -3.0 +              1                    -3.0 +                      1
 -3.2 +              1                    -3.2 + 1
 BELOW+                                   BELOW+
        <+....+....+....+....>                    <+....+....+....+....+....+....+..>
```

Figure 3.7. Residuals versus temperature.

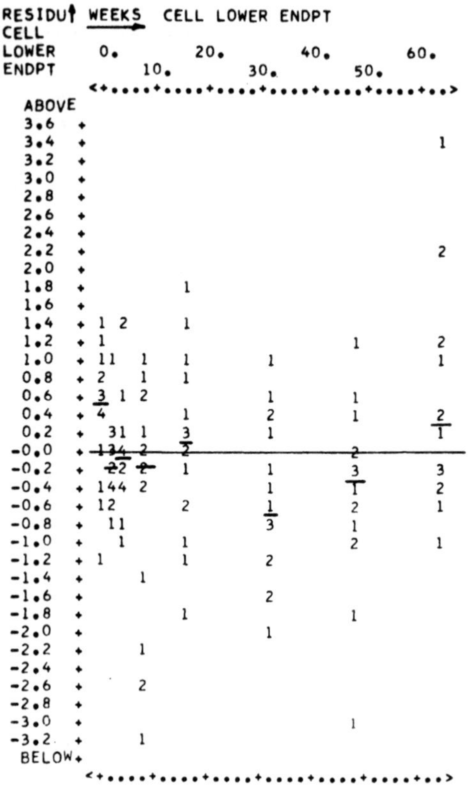

Figure 3.8. Residuals versus week.

over temperature. One possible explanation is that the actual test temperatures differ from the intended ones. The plot shows that the scatter at 275° is much larger than at other temperatures. This may be due to the previously noted high data at 275°C and 64 weeks. Figure 3.8 shows a crossplot of the residuals against their weeks on test. This plots shows no peculiarities. Thus, the relationship seems to fit the data with respect to exposure time.

Concluding remarks. The preceding analyses suggest reanalysis of the data, omitting data at 275°C and 64 weeks and possibly at 275°C and 32 weeks. Such reanalysis may yield better estimates and further insight.

PROBLEMS (* denotes difficult or laborious)

11.1. Literature. Do a literature search and develop a bibliography on an applied area of aging degradation.

11.2. Weibull. In Section 2.1, replace the lognormal distribution by a Weibull one and suitably modify all equations.

11.3. Random coefficients. For the random coefficients model, assume that the rate parameter has a power dependence on stress V. Express all results of Section 2.3 as functions of stress.

11.4. Random increments. For the random increments model, assume that the rate parameter has an Arrhenius dependence on temperature. Express all results of Section 2.4 as functions of temperature.

11.5. Birnbaum-Saunders. Assume that the rate parameter has a power dependence on stress V, and derive the Birnbaum-Saunders distribution of Section 2.2. Show how the distribution parameters depend on stress.

11.6.* Measurement error. Include random measurement error in a model. Young (1988) discusses the importance of such error in pharmaceutical measurements.
(a) Derive all results for the random coefficients model of Section 2.3.
(b) Derive all results for the constant rate model from (a), using the fact that the rate parameter is a constant.
(c) Repeat (a) for the random increments model of Section 2.4.

11.7.* Batches. Extend the random coefficients model to data from sample batches from a population of (infinitely) many differing batches.

11.8.* Breakdown data. Use the insulation breakdown data of Section 3.
(a) Repeat the analyses of Section 3, omitting data from 275°C, 64 weeks.
(b) Do (a) omitting data from any test conditions you think suitable.
(c) Use the methods of Chapter 9 to assess the fit of the relationship with a hypothesis test.
(d) Suggest further data plots and analyses.
(e) Carry out (d).
(f) Repeat all analyses using a Weibull distribution of breakdown voltage. Which distribution do you prefer and why?

11.9. Rate plot. Use the breakdown data of Section 3.1. For each test temperature, calculate the least squares estimate and 95% confidence limits for the rate coefficient β'.
(a) Assuming a separate intercept α for each temperature.
(b) Assuming a common intercept α for all temperatures.
(c) Plot these estimates and confidence limits on Arrhenius paper and comment on the plot.
(d*) Develop a hypothesis test of linearity of the Arrhenius rate dependence of such estimates.
(e) Apply (d) to the data.

11.10. Adhesive data. The following data on shear strength (in pounds) for adhesive specimens were provided courtesy of Prof. J. Phil Beckwith

Exposure	Shear Strength, Lb.		
	Group A	Group B	Group C
150°C A = 1 Day B = 4 Day C = 5 Day	8140 4550 8780 10060 5800	6080 5430 5060 2620 4700	4684 4713 4547 2222 4275
130°C A = 1 Day B = 5 Day C = 10 Day	7970 6650 7880 5000 8050	7280 9740 2390 5800 7190	4838 5916 6149 5669 6398
110°C A = 1 Day B = 10 Day C = 20 Day	7050 7190 11190 8810	7125 7418 7340 7930 9250	7760 7000 7960 7570 7590
100°C A = 1 Day B = 20 Day C = 40 Day	7740 7760 8670 6050 6450	7770 8380 6120 6020 6960	11010 7550 8350 9170 9380
90°C A = 1 Day B = 50 Day C = .100 Day	6840 11260 10920 11040 9150	10960 9400 11120 8520 4970	9840 5720 7320 7920 10568
80°C A = 1 Day B = 100 Day C = 200 Day	8300 7550 10380 8840 10580	9480 9080 9650 7940 11900	8110 11590 9150 8230 7480

(1980). Note that the three measurement ages (A, B, and C) differ from temperature to temperature.

(a)   Plot the data on semilog paper. Comment on the linearity of the plot for each temperature.

(b)   Use the constant rate relationship (2.6). For each temperature, separately estimate the degradation rate β' by least squares. Repeat this with a common intercept.

(c)   On Arrhenius paper, plot the rate estimates and their confidence limits. Does the Arrhenius rate relationship adequately describe the dependence of rate on temperature. That is, is the plot straight?

(d)   Fit the Arrhenius rate model of Section 2.1.

(e*)   Do all other analyses of Section 3.

(f) Calculate and plot trade-off curves of constant percent degradation on suitable time-temperature paper.

(g*) Delete all initial data (Group A), and redo (a)-(f). This is appropriate if the data show initial "wear-in" (Figure 2.3). Note results differing from those with all the data.

(h) Suggest and carry out further analyses.

(i) Write a brief report stating your findings for the adhesives experts.

11.11.* **Test plans.** Standard test plans for temperature degradation involves I (absolute) test temperatures T_i $(i = 1,2, \cdots ,I)$ and J_i ages t_{ij} $(j = 1,2, \cdots ,J_i)$ when specimen performance is measured once. Suppose that N_{ij} specimens are measured at test exposure (T_i,t_{ij}). Use the Arrhenius rate model of Section 2.1 with a normal distribution for log performance. Denote the measured log performance of specimen n at exposure (T_i,t_{ij}) by y_{ijn}.

(a) Write the sample log likelihood for all measurements.

(b) Derive the likelihood equations for the model parameters.

(c) Describe how to solve the likelihood equations for the ML estimates.

(d) Derive the theoretical Fisher information matrix. Evaluate it for the breakdown data of Section 3.

(e) Derive the theoretical covariance matrix of the ML estimates. Evaluate it for the breakdown data.

(f) Suppose that the product fails when its log performance degrades below a value y' at (absolute) design temperature T'. Give the ML estimates of the mean and standard deviation of the normal life distribution and derive their theoretical covariance matrix. Evaluate them for the breakdown data.

(g) Do (f) for the $100P$th percentile of the life distribution. Estimate the first percentile for the breakdown data.

(h) Assume that an optimum test plan use three test exposures $(T_0,0)$, (T_1,t_1), and (T_2,t_2) with numbers of specimens N_0, N_1, and N_2. The total number $N = N_0 + N_1 + N_2$ is fixed. Test constraints are $0 \leq t_1 \leq t_2 \leq t^*$ (a specified maximum test time) and $T_1 \leq T_2 \leq T^*$ (a specified maximum test temperature).

(i) For each parameter estimate, determine the optimum plan that minimizes its theoretical variance. That is, determine optimum T_0, T_1, t_1, T_2, t_2, N_0, N_1, and N_2. Evaluate them for an optimum breakdown test, and compare the variances.

(j) Determine the optimum plan that minimizes the variance for (g). Evaluate the optimum plan for the breakdown test.

(k) Compare the optimum plans with a standard plan with the same number N' of specimens (equal allocation) at each of $I \times J$ combinations of test temperatures $T_1, \cdots ,T_I$ and measurement ages $t_1, \cdots ,t_J$. The total number of specimens is $N = I \times J \times N'$. Apply these results to an optimum breakdown test.

(l) Assume that the degradation rate is a quadratic function of $1/T$. Find an optimum plan that minimizes the theoretical variance of the ML estimate of the coefficient for $(1/T)^2$. Such a plan would be sensitive for testing adequacy of the assumed Arrhenius dependence (2.8). Evaluate such an optimum breakdown test.

(m) Explain how to use the optimum plans above for other rate dependence (power, exponential, etc.).

(n) Suggest and investigate good compromise plans.

(o) Repeat the preceding, replacing $(T_0,0)$ with (T_0,t_0) where $0 < t_0^* \le t_0 \le t^*$, and t_0^* is a minimum measurement age.

11.12.* Consistent models. The lognormal-Arrhenius model of Chapter 2 has a lognormal life distribution. The life distribution for the model of Section 2.1 is normal. Show that these distributions are close, that is, that the models are quite similar.

11.13. Reaction order. According to chemical kinetic theory, the remaining amount $\mu(t)$ at time t of a chemical being consumed in a chemical reaction satisfies the differential equation $d\mu/dt = \beta' \mu^p$. Here β' and p are parameters characteristic of the reaction. β' is the rate parameter and depends on the accelerating stress, and p is a constant called the *order* of the reaction. Usually p equals 0, 1, or 2 and is the number of molecules in the rate determining step of the reaction. When there is more than one reaction taking place, p may not be integer. Young (1988) gives a computer program that fits the model for $p = 0,1,2$.

(a) Solve the differential equation for arbitrary p. Explicitly give the relationship $\mu(t)$ for 0th, 1st, and 2nd order reactions.

(b) Write the relationship $\mu(t)$ with a random term e_t to represent the distribution of the amount of chemical $u(t)$ about the typical amount $\mu(t)$.

(c) Describe how to fit 0th and 2nd order relationships to the voltage breakdown data.

(d*) Fit those relationships to the data. Which relation fits best and why?

11.14. Permalloy corrosion. In view of Chapter 11, do Problems 3.13 and 4.9.

11.15.* Distribution of design stress. The life distribution (2.4) assumes that every unit has the same (log) design stress y'. Extend the theory for the life distribution to a situation where the design stress applied to a unit comes from a distribution with mean μ' and standard deviation σ'. Compare your results with (2.4) and (2.5).

Appendix A
Statistical Tables

A1.   Standard Normal Cumulative Distribution Function $\Phi(u)$, 550

A2.   Standard Normal Percentiles z_P, 552

A3.   Standard Normal Two-Sided Factors K_P, 552

A4. t-Distribution Percentiles $t(P;v)$, 553

A5.   Chi-Square Percentiles $\chi^2(P;v)$, 554

A6a. F-Distribution 95% Points $F(0.95;v_1,v_2)$, 556

A6b. F-Distribution 99% Points $F(0.99;v_1,v_2)$, 558

A7.   Probability Plotting Positions $F_i = 100(i-0.5)/n$, 560

Appendix A1.   Standard Normal Cumulative Distribution Function $\Phi(u)$

u	·00	·01	·02	·03	·04	·05	·06	·07	·08	·09
− ·0	·5000	·4960	·4920	·4880	·4840	·4801	·4761	·4721	·4681	·4641
− ·1	·4602	·4562	·4522	·4483	·4443	·4404	·4364	·4325	·4286	·4247
− ·2	·4207	·4168	·4129	·4090	·4052	·4013	·3974	·3936	·3897	·3859
− ·3	·3821	·3783	·3745	·3707	·3669	·3632	·3594	·3557	·3520	·3483
− ·4	·3446	·3409	·3372	·3336	·3300	·3264	·3228	·3192	·3156	·3121
− ·5	·3085	·3050	·3015	·2981	·2946	·2912	·2877	·2843	·2810	·2776
− ·6	·2743	·2709	·2676	·2643	·2611	·2578	·2546	·2514	·2483	·2451
− ·7	·2420	·2389	·2358	·2327	·2297	·2266	·2236	·2206	·2177	·2148
− ·8	·2119	·2090	·2061	·2033	·2005	·1977	·1949	·1922	·1894	·1867
− ·9	·1841	·1814	·1788	·1762	·1736	·1711	·1685	·1660	·1635	·1611
−1·0	·1587	·1562	·1539	·1515	·1492	·1469	·1446	·1423	·1401	·1379
−1·1	·1357	·1335	·1314	·1292	·1271	·1251	·1230	·1210	·1190	·1170
−1·2	·1151	·1131	·1112	·1093	·1075	·1056	·1038	·1020	·1003	·09853
−1·3	·09680	·09510	·09342	·09176	·09012	·08851	·08691	·08534	·08379	·08226
−1·4	·08076	·07927	·07780	·07636	·07493	·07353	·07215	·07078	·06944	·06811
−1·5	·06681	·06552	·06426	·06301	·06178	·06057	·05938	·05821	·05705	·05592
−1·6	·05480	·05370	·05262	·05155	·05050	·04947	·04846	·04746	·04648	·04551
−1·7	·04457	·04363	·04272	·04182	·04093	·04006	·03920	·03836	·03754	·03673
−1·8	·03593	·03515	·03438	·03362	·03288	·03216	·03144	·03074	·03005	·02938
−1·9	·02872	·02807	·02743	·02680	·02619	·02559	·02500	·02442	·02385	·02330
−2·0	·02275	·02222	·02169	·02118	·02068	·02018	·01970	·01923	·01876	·01831
−2·1	·01786	·01743	·01700	·01659	·01618	·01578	·01539	·01500	·01463	·01426
−2·2	·01390	·01355	·01321	·01287	·01255	·01222	·01191	·01160	·01130	·01101
−2·3	·01072	·01044	·01017	$\cdot 0^2 9903$	$\cdot 0^2 9642$	$\cdot 0^2 9387$	$\cdot 0^2 9137$	$\cdot 0^2 8894$	$\cdot 0^2 8656$	$\cdot 0^2 8424$
−2·4	$\cdot 0^2 8198$	$\cdot 0^2 7976$	$\cdot 0^2 7760$	$\cdot 0^2 7549$	$\cdot 0^2 7344$	$\cdot 0^2 7143$	$\cdot 0^2 6947$	$\cdot 0^2 6756$	$\cdot 0^2 6569$	$\cdot 0^2 6387$
−2·5	$\cdot 0^2 6210$	$\cdot 0^2 6037$	$\cdot 0^2 5868$	$\cdot 0^2 5703$	$\cdot 0^2 5543$	$\cdot 0^2 5386$	$\cdot 0^2 5234$	$\cdot 0^2 5085$	$\cdot 0^2 4940$	$\cdot 0^2 4799$
−2·6	$\cdot 0^2 4661$	$\cdot 0^2 4527$	$\cdot 0^2 4396$	$\cdot 0^2 4269$	$\cdot 0^2 4145$	$\cdot 0^2 4025$	$\cdot 0^2 3907$	$\cdot 0^2 3793$	$\cdot 0^2 3681$	$\cdot 0^2 3573$
−2·7	$\cdot 0^2 3467$	$\cdot 0^2 3364$	$\cdot 0^2 3264$	$\cdot 0^2 3167$	$\cdot 0^2 3072$	$\cdot 0^2 2980$	$\cdot 0^2 2890$	$\cdot 0^2 2803$	$\cdot 0^2 2718$	$\cdot 0^2 2635$
−2·8	$\cdot 0^2 2555$	$\cdot 0^2 2477$	$\cdot 0^2 2401$	$\cdot 0^2 2327$	$\cdot 0^2 2256$	$\cdot 0^2 2186$	$\cdot 0^2 2118$	$\cdot 0^2 2052$	$\cdot 0^2 1988$	$\cdot 0^2 1926$
−2·9	$\cdot 0^2 1866$	$\cdot 0^2 1807$	$\cdot 0^2 1750$	$\cdot 0^2 1695$	$\cdot 0^2 1641$	$\cdot 0^2 1589$	$\cdot 0^2 1538$	$\cdot 0^2 1489$	$\cdot 0^2 1441$	$\cdot 0^2 1395$
−3·0	$\cdot 0^2 1350$	$\cdot 0^2 1306$	$\cdot 0^2 1264$	$\cdot 0^2 1223$	$\cdot 0^2 1183$	$\cdot 0^2 1144$	$\cdot 0^2 1107$	$\cdot 0^2 1070$	$\cdot 0^2 1035$	$\cdot 0^2 1001$
−3·1	$\cdot 0^3 9676$	$\cdot 0^3 9354$	$\cdot 0^3 9043$	$\cdot 0^3 8740$	$\cdot 0^3 8447$	$\cdot 0^3 8164$	$\cdot 0^3 7888$	$\cdot 0^3 7622$	$\cdot 0^3 7364$	$\cdot 0^3 7114$
−3·2	$\cdot 0^3 6871$	$\cdot 0^3 6637$	$\cdot 0^3 6410$	$\cdot 0^3 6190$	$\cdot 0^3 5976$	$\cdot 0^3 5770$	$\cdot 0^3 5571$	$\cdot 0^3 5377$	$\cdot 0^3 5190$	$\cdot 0^3 5009$
−3·3	$\cdot 0^3 4834$	$\cdot 0^3 4665$	$\cdot 0^3 4501$	$\cdot 0^3 4342$	$\cdot 0^3 4189$	$\cdot 0^3 4041$	$\cdot 0^3 3897$	$\cdot 0^3 3758$	$\cdot 0^3 3624$	$\cdot 0^3 3495$
−3·4	$\cdot 0^3 3369$	$\cdot 0^3 3248$	$\cdot 0^3 3131$	$\cdot 0^3 3018$	$\cdot 0^3 2909$	$\cdot 0^3 2803$	$\cdot 0^3 2701$	$\cdot 0^3 2602$	$\cdot 0^3 2507$	$\cdot 0^3 2415$
−3·5	$\cdot 0^3 2326$	$\cdot 0^3 2241$	$\cdot 0^3 2158$	$\cdot 0^3 2078$	$\cdot 0^3 2001$	$\cdot 0^3 1926$	$\cdot 0^3 1854$	$\cdot 0^3 1785$	$\cdot 0^3 1718$	$\cdot 0^3 1653$
−3·6	$\cdot 0^3 1591$	$\cdot 0^3 1531$	$\cdot 0^3 1473$	$\cdot 0^3 1417$	$\cdot 0^3 1363$	$\cdot 0^3 1311$	$\cdot 0^3 1261$	$\cdot 0^3 1213$	$\cdot 0^3 1166$	$\cdot 0^3 1121$
−3·7	$\cdot 0^3 1078$	$\cdot 0^3 1036$	$\cdot 0^4 9961$	$\cdot 0^4 9574$	$\cdot 0^4 9201$	$\cdot 0^4 8842$	$\cdot 0^4 8496$	$\cdot 0^4 8162$	$\cdot 0^4 7841$	$\cdot 0^4 7532$
−3·8	$\cdot 0^4 7235$	$\cdot 0^4 6948$	$\cdot 0^4 6673$	$\cdot 0^4 6407$	$\cdot 0^4 6152$	$\cdot 0^4 5906$	$\cdot 0^4 5669$	$\cdot 0^4 5442$	$\cdot 0^4 5223$	$\cdot 0^4 5012$
−3·9	$\cdot 0^4 4810$	$\cdot 0^4 4615$	$\cdot 0^4 4427$	$\cdot 0^4 4247$	$\cdot 0^4 4074$	$\cdot 0^4 3908$	$\cdot 0^4 3747$	$\cdot 0^4 3594$	$\cdot 0^4 3446$	$\cdot 0^4 3304$
−4·0	$\cdot 0^4 3167$	$\cdot 0^4 3036$	$\cdot 0^4 2910$	$\cdot 0^4 2789$	$\cdot 0^4 2673$	$\cdot 0^4 2561$	$\cdot 0^4 2454$	$\cdot 0^4 2351$	$\cdot 0^4 2252$	$\cdot 0^4 2157$
−4·1	$\cdot 0^4 2066$	$\cdot 0^4 1978$	$\cdot 0^4 1894$	$\cdot 0^4 1814$	$\cdot 0^4 1737$	$\cdot 0^4 1662$	$\cdot 0^4 1591$	$\cdot 0^4 1523$	$\cdot 0^4 1458$	$\cdot 0^4 1395$
−4·2	$\cdot 0^4 1335$	$\cdot 0^4 1277$	$\cdot 0^4 1222$	$\cdot 0^4 1168$	$\cdot 0^4 1118$	$\cdot 0^4 1069$	$\cdot 0^4 1022$	$\cdot 0^5 9774$	$\cdot 0^5 9345$	$\cdot 0^5 8934$
−4·3	$\cdot 0^5 8540$	$\cdot 0^5 8163$	$\cdot 0^5 7801$	$\cdot 0^5 7455$	$\cdot 0^5 7124$	$\cdot 0^5 6807$	$\cdot 0^5 6503$	$\cdot 0^5 6212$	$\cdot 0^5 5934$	$\cdot 0^5 5668$
−4·4	$\cdot 0^5 5413$	$\cdot 0^5 5169$	$\cdot 0^5 4935$	$\cdot 0^5 4712$	$\cdot 0^5 4498$	$\cdot 0^5 4294$	$\cdot 0^5 4098$	$\cdot 0^5 3911$	$\cdot 0^5 3732$	$\cdot 0^5 3561$
−4·5	$\cdot 0^5 3398$	$\cdot 0^5 3241$	$\cdot 0^5 3092$	$\cdot 0^5 2949$	$\cdot 0^5 2813$	$\cdot 0^5 2682$	$\cdot 0^5 2558$	$\cdot 0^5 2439$	$\cdot 0^5 2325$	$\cdot 0^5 2216$
−4·6	$\cdot 0^5 2112$	$\cdot 0^5 2013$	$\cdot 0^5 1919$	$\cdot 0^5 1828$	$\cdot 0^5 1742$	$\cdot 0^5 1660$	$\cdot 0^5 1581$	$\cdot 0^5 1506$	$\cdot 0^5 1434$	$\cdot 0^5 1366$
−4·7	$\cdot 0^5 1301$	$\cdot 0^5 1239$	$\cdot 0^5 1179$	$\cdot 0^5 1123$	$\cdot 0^5 1069$	$\cdot 0^5 1017$	$\cdot 0^6 9680$	$\cdot 0^6 9211$	$\cdot 0^6 8765$	$\cdot 0^6 8339$
−4·8	$\cdot 0^6 7933$	$\cdot 0^6 7547$	$\cdot 0^6 7178$	$\cdot 0^6 6827$	$\cdot 0^6 6492$	$\cdot 0^6 6173$	$\cdot 0^6 5869$	$\cdot 0^6 5580$	$\cdot 0^6 5304$	$\cdot 0^6 5042$
−4·9	$\cdot 0^6 4792$	$\cdot 0^6 4554$	$\cdot 0^6 4327$	$\cdot 0^6 4111$	$\cdot 0^6 3906$	$\cdot 0^6 3711$	$\cdot 0^6 3525$	$\cdot 0^6 3348$	$\cdot 0^6 3179$	$\cdot 0^6 3019$

Appendix A1. $\Phi(u)$ (Continued)

u	·00	·01	·02	·03	·04	·05	·06	·07	·08	·09
·0	·5000	·5040	·5080	·5120	·5160	·5199	·5239	·5279	·5319	·5359
·1	·5398	·5438	·5478	·5517	·5557	·5596	·5636	·5675	·5714	·5753
·2	·5793	·5832	·5871	·5910	·5948	·5987	·6026	·6064	·6103	·6141
·3	·6179	·6217	·6255	·6293	·6331	·6368	·6406	·6443	·6480	·6517
·4	·6554	·6591	·6628	·6664	·6700	·6736	·6772	·6808	·6844	·6879
·5	·6915	·6950	·6985	·7019	·7054	·7088	·7123	·7157	·7190	·7224
·6	·7257	·7291	·7324	·7357	·7389	·7422	·7454	·7486	·7517	·7549
·7	·7580	·7611	·7642	·7673	·7703	·7734	·7764	·7794	·7823	·7852
·8	·7881	·7910	·7939	·7967	·7995	·8023	·8051	·8078	·8106	·8133
·9	·8159	·8186	·8212	·8238	·8264	·8289	·8315	·8340	·8365	·8389
1·0	·8413	·8438	·8461	·8485	·8508	·8531	·8554	·8577	·8599	·8621
1·1	·8643	·8665	·8686	·8708	·8729	·8749	·8770	·8790	·8810	·8830
1·2	·8849	·8869	·8888	·8907	·8925	·8944	·8962	·8980	·8997	·90147
1·3	·90320	·90490	·90658	·90824	·90988	·91149	·91309	·91466	·91621	·91774
1·4	·91924	·92073	·92220	·92364	·92507	·92647	·92785	·92922	·93056	·93189
1·5	·93319	·93448	·93574	·93699	·93822	·93943	·94062	·94179	·94295	·94408
1·6	·94520	·94630	·94738	·94845	·94950	·95053	·95154	·95254	·95352	·95449
1·7	·95543	·95637	·95728	·95818	·95907	·95994	·96080	·96164	·96246	·96327
1·8	·96407	·96485	·96562	·96638	·96712	·96784	·96856	·96926	·96995	·97062
1·9	·97128	·97193	·97257	·97320	·97381	·97441	·97500	·97558	·97615	·97670
2·0	·97725	·97778	·97831	·97882	·97932	·97982	·98030	·98077	·98124	·98169
2·1	·98214	·98257	·98300	·98341	·98382	·98422	·98461	·98500	·98537	·98574
2·2	·98610	·98645	·98679	·98713	·98745	·98778	·98809	·98840	·98870	·98899
2·3	·98928	·98956	·98983	$\cdot9^{2}0097$	$\cdot9^{2}0358$	$\cdot9^{2}0613$	$\cdot9^{2}0863$	$\cdot9^{2}1106$	$\cdot9^{2}1344$	$\cdot9^{2}1576$
2·4	$\cdot9^{2}1802$	$\cdot9^{2}2024$	$\cdot9^{2}2240$	$\cdot9^{2}2451$	$\cdot9^{2}2656$	$\cdot9^{2}2857$	$\cdot9^{2}3053$	$\cdot9^{2}3244$	$\cdot9^{2}3431$	$\cdot9^{2}3613$
2·5	$\cdot9^{2}3790$	$\cdot9^{2}3963$	$\cdot9^{2}4132$	$\cdot9^{2}4297$	$\cdot9^{2}4457$	$\cdot9^{2}4614$	$\cdot9^{2}4766$	$\cdot9^{2}4915$	$\cdot9^{2}5060$	$\cdot9^{2}5201$
2·6	$\cdot9^{2}5339$	$\cdot9^{2}5473$	$\cdot9^{2}5604$	$\cdot9^{2}5731$	$\cdot9^{2}5855$	$\cdot9^{2}5975$	$\cdot9^{2}6093$	$\cdot9^{2}6207$	$\cdot9^{2}6319$	$\cdot9^{2}6427$
2·7	$\cdot9^{2}6533$	$\cdot9^{2}6636$	$\cdot9^{2}6736$	$\cdot9^{2}6833$	$\cdot9^{2}6928$	$\cdot9^{2}7020$	$\cdot9^{2}7110$	$\cdot9^{2}7197$	$\cdot9^{2}7282$	$\cdot9^{2}7365$
2·8	$\cdot9^{2}7445$	$\cdot9^{2}7523$	$\cdot9^{2}7599$	$\cdot9^{2}7673$	$\cdot9^{2}7744$	$\cdot9^{2}7814$	$\cdot9^{2}7882$	$\cdot9^{2}7948$	$\cdot9^{2}8012$	$\cdot9^{2}8074$
2·9	$\cdot9^{2}8134$	$\cdot9^{2}8193$	$\cdot9^{2}8250$	$\cdot9^{2}8305$	$\cdot9^{2}8359$	$\cdot9^{2}8411$	$\cdot9^{2}8462$	$\cdot9^{2}8511$	$\cdot9^{2}8559$	$\cdot9^{2}8605$
3·0	$\cdot9^{2}8650$	$\cdot9^{2}8694$	$\cdot9^{2}8736$	$\cdot9^{2}8777$	$\cdot9^{2}8817$	$\cdot9^{2}8856$	$\cdot9^{2}8893$	$\cdot9^{2}8930$	$\cdot9^{2}8965$	$\cdot9^{2}8999$
3·1	$\cdot9^{3}0324$	$\cdot9^{3}0646$	$\cdot9^{3}0957$	$\cdot9^{3}1260$	$\cdot9^{3}1553$	$\cdot9^{3}1836$	$\cdot9^{3}2112$	$\cdot9^{3}2378$	$\cdot9^{3}2636$	$\cdot9^{3}2886$
3·2	$\cdot9^{3}3129$	$\cdot9^{3}3363$	$\cdot9^{3}3590$	$\cdot9^{3}3810$	$\cdot9^{3}4024$	$\cdot9^{3}4230$	$\cdot9^{3}4429$	$\cdot9^{3}4623$	$\cdot9^{3}4810$	$\cdot9^{3}4991$
3·3	$\cdot9^{3}5166$	$\cdot9^{3}5335$	$\cdot9^{3}5499$	$\cdot9^{3}5658$	$\cdot9^{3}5811$	$\cdot9^{3}5959$	$\cdot9^{3}6103$	$\cdot9^{3}6242$	$\cdot9^{3}6376$	$\cdot9^{3}6505$
3·4	$\cdot9^{3}6631$	$\cdot9^{3}6752$	$\cdot9^{3}6869$	$\cdot9^{3}6982$	$\cdot9^{3}7091$	$\cdot9^{3}7197$	$\cdot9^{3}7299$	$\cdot9^{3}7398$	$\cdot9^{3}7493$	$\cdot9^{3}7585$
3·5	$\cdot9^{3}7674$	$\cdot9^{3}7759$	$\cdot9^{3}7842$	$\cdot9^{3}7922$	$\cdot9^{3}7999$	$\cdot9^{3}8074$	$\cdot9^{3}8146$	$\cdot9^{3}8215$	$\cdot9^{3}8282$	$\cdot9^{3}8347$
3·6	$\cdot9^{3}8409$	$\cdot9^{3}8469$	$\cdot9^{3}8527$	$\cdot9^{3}8583$	$\cdot9^{3}8637$	$\cdot9^{3}8689$	$\cdot9^{3}8739$	$\cdot9^{3}8787$	$\cdot9^{3}8834$	$\cdot9^{3}8879$
3·7	$\cdot9^{3}8922$	$\cdot9^{3}8964$	$\cdot9^{4}0039$	$\cdot9^{4}0426$	$\cdot9^{4}0799$	$\cdot9^{4}1158$	$\cdot9^{4}1504$	$\cdot9^{4}1838$	$\cdot9^{4}2159$	$\cdot9^{4}2468$
3·8	$\cdot9^{4}2765$	$\cdot9^{4}3052$	$\cdot9^{4}3327$	$\cdot9^{4}3593$	$\cdot9^{4}3848$	$\cdot9^{4}4094$	$\cdot9^{4}4331$	$\cdot9^{4}4558$	$\cdot9^{4}4777$	$\cdot9^{4}4988$
3·9	$\cdot9^{4}5190$	$\cdot9^{4}5385$	$\cdot9^{4}5573$	$\cdot9^{4}5753$	$\cdot9^{4}5926$	$\cdot9^{4}6092$	$\cdot9^{4}6253$	$\cdot9^{4}6406$	$\cdot9^{4}6554$	$\cdot9^{4}6696$
4·0	$\cdot9^{4}6833$	$\cdot9^{4}6964$	$\cdot9^{4}7090$	$\cdot9^{4}7211$	$\cdot9^{4}7327$	$\cdot9^{4}7439$	$\cdot9^{4}7546$	$\cdot9^{4}7649$	$\cdot9^{4}7748$	$\cdot9^{4}7843$
4·1	$\cdot9^{4}7934$	$\cdot9^{4}8022$	$\cdot9^{4}8106$	$\cdot9^{4}8186$	$\cdot9^{4}8263$	$\cdot9^{4}8338$	$\cdot9^{4}8409$	$\cdot9^{4}8477$	$\cdot9^{4}8542$	$\cdot9^{4}8605$
4·2	$\cdot9^{4}8665$	$\cdot9^{4}8723$	$\cdot9^{4}8778$	$\cdot9^{4}8832$	$\cdot9^{4}8882$	$\cdot9^{4}8931$	$\cdot9^{4}8978$	$\cdot9^{5}0226$	$\cdot9^{5}0655$	$\cdot9^{5}1066$
4·3	$\cdot9^{5}1460$	$\cdot9^{5}1837$	$\cdot9^{5}2199$	$\cdot9^{5}2545$	$\cdot9^{5}2876$	$\cdot9^{5}3193$	$\cdot9^{5}3497$	$\cdot9^{5}3788$	$\cdot9^{5}4066$	$\cdot9^{5}4332$
4·4	$\cdot9^{5}4587$	$\cdot9^{5}4831$	$\cdot9^{5}5065$	$\cdot9^{5}5288$	$\cdot9^{5}5502$	$\cdot9^{5}5706$	$\cdot9^{5}5902$	$\cdot9^{5}6089$	$\cdot9^{5}6268$	$\cdot9^{5}6439$
4·5	$\cdot9^{5}6602$	$\cdot9^{5}6759$	$\cdot9^{5}6908$	$\cdot9^{5}7051$	$\cdot9^{5}7187$	$\cdot9^{5}7318$	$\cdot9^{5}7442$	$\cdot9^{5}7561$	$\cdot9^{5}7675$	$\cdot9^{5}7784$
4·6	$\cdot9^{5}7888$	$\cdot9^{5}7987$	$\cdot9^{5}8081$	$\cdot9^{5}8172$	$\cdot9^{5}8258$	$\cdot9^{5}8340$	$\cdot9^{5}8419$	$\cdot9^{5}8494$	$\cdot9^{5}8566$	$\cdot9^{5}8634$
4·7	$\cdot9^{5}8699$	$\cdot9^{5}8761$	$\cdot9^{5}8821$	$\cdot9^{5}8877$	$\cdot9^{5}8931$	$\cdot9^{5}8983$	$\cdot9^{6}0320$	$\cdot9^{6}0789$	$\cdot9^{6}1235$	$\cdot9^{6}1661$
4·8	$\cdot9^{6}2067$	$\cdot9^{6}2453$	$\cdot9^{6}2822$	$\cdot9^{6}3173$	$\cdot9^{6}3508$	$\cdot9^{6}3827$	$\cdot9^{6}4131$	$\cdot9^{6}4420$	$\cdot9^{6}4696$	$\cdot9^{6}4958$
4·9	$\cdot9^{6}5208$	$\cdot9^{6}5446$	$\cdot9^{6}5673$	$\cdot9^{6}5889$	$\cdot9^{6}6094$	$\cdot9^{6}6289$	$\cdot9^{6}6475$	$\cdot9^{6}6652$	$\cdot9^{6}6821$	$\cdot9^{6}6981$

From A. Hald, *Statistical Tables and Formulas*, Wiley, New York, 1952, Table II. Reproduced by permission.

Appendix A2.   Standard Normal Percentiles z_P

$100P\%$	z_P	$100P\%$	z_P
10^{-4}	-4.753	50	0.
10^{-3}	-4.265	60	0.253
0.01	-3.719	70	0.524
0.02	-3.540	75	0.675
0.05	-3.291	80	0.842
0.1	-3.090	90	1.282
0.2	-2.878	95	1.645
0.5	-2.576	97.5	1.960
1.0	-2.326	98	2.054
2.0	-2.054	99	2.326
2.5	-1.960	99.5	2.576
5	-1.645	99.9	3.090
10	-1.282	99.95	3.291
20	-0.842	99.99	3.719
25	-0.675		
30	-0.524		
40	-0.253		

Appendix A3.   Standard Normal Two-Sided Factors K_P

$100P\%$	K_P
50	0.675
60	0.842
70	1.036
80	1.282
90	1.645
95	1.960
99	2.576
99.9	3.291
99.99	3.890

Appendix A4. t-Distribution Percentiles $t(P;v)$

v \ P	0.750	0.900	0.950	0.975	0.990	0.995	0.999	0.9995
1	1.000	3.078	6.314	12.706	31.821	63.657	318.31	636.62
2	0.816	1.886	2.920	4.303	6.965	9.925	22.326	31.598
3	0.765	1.638	2.353	3.182	4.541	5.841	10.213	12.924
4	0.741	1.533	2.132	2.776	3.747	4.604	7.173	8.610
5	0.727	1.476	2.015	2.571	3.365	4.032	5.893	6.869
6	0.718	1.440	1.943	2.447	3.143	3.707	5.208	5.959
7	0.711	1.415	1.895	2.365	2.998	3.499	4.785	5.408
8	0.706	1.397	1.860	2.306	2.896	3.355	4.501	5.041
9	0.703	1.383	1.833	2.262	2.821	3.250	4.297	4.781
10	0.700	1.372	1.812	2.228	2.764	3.169	4.144	4.587
11	0.697	1.363	1.796	2.201	2.718	3.106	4.025	4.437
12	0.695	1.356	1.782	2.179	2.681	3.055	3.930	4.318
13	0.694	1.350	1.771	2.160	2.650	3.012	3.852	4.221
14	0.692	1.345	1.761	2.145	2.624	2.977	3.787	4.140
15	0.691	1.341	1.753	2.131	2.602	2.947	3.733	4.073
16	0.690	1.337	1.746	2.120	2.583	2.921	3.686	4.015
17	0.689	1.333	1.740	2.110	2.567	2.898	3.646	3.965
18	0.688	1.330	1.734	2.101	2.552	2.878	3.610	3.922
19	0.688	1.328	1.729	2.093	2.539	2.861	3.579	3.883
20	0.687	1.325	1.725	2.086	2.528	2.845	3.552	3.850
21	0.686	1.323	1.721	2.080	2.518	2.831	3.527	3.819
22	0.686	1.321	1.717	2.074	2.508	2.819	3.505	3.792
23	0.685	1.319	1.714	2.069	2.500	2.807	3.485	3.767
24	0.685	1.318	1.711	2.064	2.492	2.797	3.467	3.745
25	0.684	1.316	1.708	2.060	2.485	2.787	3.450	3.725
26	0.684	1.315	1.706	2.056	2.479	2.779	3.435	3.707
27	0.684	1.314	1.703	2.052	2.473	2.771	3.421	3.690
28	0.683	1.313	1.701	2.048	2.467	2.763	3.408	3.674
29	0.683	1.311	1.699	2.045	2.462	2.756	3.396	3.659
30	0.683	1.310	1.697	2.042	2.457	2.750	3.385	3.646
40	0.681	1.303	1.684	2.021	2.423	2.704	3.307	3.551
60	0.679	1.296	1.671	2.000	2.390	2.660	3.232	3.460
120	0.677	1.289	1.658	1.980	2.358	2.617	3.160	3.373
∞	0.674	1.282	1.645	1.960	2.326	2.576	3.090	3.291

From N. L. Johnson and F. C. Leone, *Statistics and Experimental Design in Engineering and the Physical Sciences,* 2nd ed., Wiley, New York, 1977, Vol. 1, p. 466. Reproduced by permission of the publisher and the Biometrika Trustees. Use $t(P;v) = -t(1-P;v)$ for $P < 0.50$.

Appendix A5.   Chi-Square Percentiles $\chi^2(P;v)$

v \ P	0.005	0.010	0.025	0.050	0.100	0.250	0.500
1	0.00004	0.00016	0.00098	0.00393	0.01579	0.1015	0.4549
2	0.0100	0.0201	0.0506	0.1026	0.2107	0.5754	1.386
3	0.0717	0.1148	0.2158	0.3518	0.5844	1.213	2.366
4	0.2070	0.2971	0.4844	0.7107	1.064	1.923	3.357
5	0.4177	0.5543	0.8312	1.145	1.610	2.675	4.351
6	0.6757	0.8721	1.2373	1.635	2.204	3.455	5.348
7	0.9893	1.239	1.690	2.167	2.833	4.255	6.346
8	1.344	1.646	2.180	2.733	3.490	5.071	7.344
9	1.735	2.088	2.700	3.325	4.168	5.899	8.343
10	2.156	2.558	3.247	3.940	4.865	6.737	9.342
11	2.603	3.053	3.816	4.575	5.578	7.584	10.34
12	3.074	3.571	4.404	5.226	6.304	8.438	11.34
13	3.565	4.107	5.009	5.892	7.041	9.299	12.34
14	4.075	4.660	5.629	6.571	7.790	10.17	13.34
15	4.601	5.229	6.262	7.261	8.547	11.04	14.34
16	5.142	5.812	6.908	7.962	9.312	11.91	15.34
17	5.697	6.408	7.564	8.672	10.09	12.79	16.34
18	6.265	7.015	8.231	9.390	10.86	13.68	17.34
19	6.844	7.633	8.907	10.12	11.65	14.56	18.34
20	7.434	8.260	9.591	10.85	12.44	15.45	19.34
21	8.034	8.897	10.28	11.59	13.24	16.34	20.34
22	8.643	9.542	10.98	12.34	14.04	17.24	21.34
23	9.260	10.20	11.69	13.09	14.85	18.14	22.34
24	9.886	10.86	12.40	13.85	15.66	19.04	23.34
25	10.52	11.52	13.12	14.61	16.47	19.94	24.34
26	11.16	12.20	13.84	15.38	17.29	20.84	25.34
27	11.81	12.88	14.57	16.15	18.11	21.75	26.34
28	12.46	13.56	15.31	16.93	18.94	22.66	27.34
29	13.12	14.26	16.05	17.71	19.77	23.57	28.34
30	13.79	14.95	16.79	18.49	20.60	24.48	29.34
40	20.71	22.16	24.43	26.51	29.05	33.66	39.34
50	27.99	29.71	32.36	34.76	37.69	42.94	49.33
60	35.53	37.48	40.48	43.19	46.46	52.29	59.33
70	43.28	45.44	48.76	51.74	55.33	61.70	69.33
80	51.17	53.54	57.15	60.39	64.28	71.14	79.33
90	59.20	61.75	65.65	69.13	73.29	80.62	89.33
100	67.33	70.06	74.22	77.93	82.36	90.13	99.33

Appendix A5.   Chi-Square Percentiles $\chi^2(P;v)$ (*Continued*)

v / P	0.750	0.900	0.950	0.975	0.990	0.995	0.999
1	1.323	2.706	3.841	5.024	6.635	7.879	10.83
2	2.773	4.605	5.991	7.378	9.210	10.60	13.82
3	4.108	6.251	7.815	9.348	11.34	12.84	16.27
4	5.385	7.779	9.488	11.14	13.28	14.86	18.47
5	6.626	9.236	11.07	12.83	15.09	16.75	20.52
6	7.841	10.64	12.59	14.45	16.81	18.55	22.46
7	9.037	12.02	14.07	16.01	18.48	20.28	24.32
8	10.22	13.36	15.51	17.53	20.09	21.96	26.12
9	11.39	14.68	16.92	19.02	21.67	23.59	27.88
10	12.55	15.99	18.31	20.48	23.21	25.19	29.59
11	13.70	17.28	19.68	21.92	24.72	26.76	31.26
12	14.85	18.55	21.03	23.34	26.22	28.30	32.91
13	15.98	19.81	22.36	24.74	27.69	29.82	34.53
14	17.12	21.06	23.68	26.12	29.14	31.32	36.12
15	18.25	22.31	25.00	27.49	30.58	32.80	37.70
16	19.37	23.54	26.30	28.85	32.00	34.27	39.25
17	20.49	24.77	27.59	30.19	33.41	35.72	40.79
18	21.60	25.99	28.87	31.53	34.81	37.16	42.31
19	22.72	27.20	30.14	32.85	36.19	38.58	43.82
20	23.83	28.41	31.41	34.17	37.57	40.00	45.32
21	24.93	29.62	32.67	35.48	38.93	41.40	46.80
22	26.04	30.81	33.92	36.78	40.29	42.80	48.27
23	27.14	32.01	35.17	38.08	41.64	44.18	49.73
24	28.24	33.20	36.42	39.36	42.98	45.56	51.18
25	29.34	34.38	37.65	40.65	44.31	46.93	52.62
26	30.43	35.56	38.89	41.92	45.64	48.29	54.05
27	31.53	36.74	40.11	43.19	46.96	49.64	55.48
28	32.62	37.92	41.34	44.46	48.28	50.99	56.89
29	33.71	39.09	42.56	45.72	49.59	52.34	58.30
30	34.80	40.26	43.77	46.98	50.89	53.67	59.70
40	45.62	51.80	55.76	59.34	63.69	66.77	73.40
50	56.33	63.17	67.50	71.42	76.15	79.49	86.66
60	66.98	74.40	79.08	83.30	88.38	91.95	99.61
70	77.58	85.53	90.53	95.02	100.4	104.2	112.3
80	88.13	96.58	101.9	106.6	112.3	116.3	124.8
90	98.65	107.6	113.1	118.1	124.1	128.3	137.2
100	109.1	118.5	124.3	129.6	135.8	140.2	149.4

From N. L. Johnson and F. C. Leone, *Statistics and Experimental Design in Engineering and the Physical Sciences,* 2nd ed., Wiley, New York, 1977, Vol. 1, pp. 511–512. Reproduced by permission of the publisher and the Biometrika Trustees.

Appendix A6a.   *F*-Distribution 95% Points $F(0.95;v_1,v_2)$

v_2 \ v_1	1	2	3	4	5	6	7	8	9
1	161.45	199.50	215.71	224.58	230.16	233.99	236.77	238.88	240.54
2	18.513	19.000	19.164	19.247	19.296	19.330	19.353	19.371	19.385
3	10.128	9.5521	9.2766	9.1172	9.0135	8.9406	8.8868	8.8452	8.8123
4	7.7086	6.9443	6.5914	6.3883	6.2560	6.1631	6.0942	6.0410	5.9988
5	6.6079	5.7861	5.4095	5.1922	5.0503	4.9503	4.8759	4.8183	4.7725
6	5.9874	5.1433	4.7571	4.5337	4.3874	4.2839	4.2066	4.1468	4.0990
7	5.5914	4.7374	4.3468	4.1203	3.9715	3.8660	3.7870	3.7257	3.6767
8	5.3177	4.4590	4.0662	3.8378	3.6875	3.5806	3.5005	3.4381	3.3881
9	5.1174	4.2565	3.8626	3.6331	3.4817	3.3738	3.2927	3.2296	3.1789
10	4.9646	4.1028	3.7083	3.4780	3.3258	3.2172	3.1355	3.0717	3.0204
11	4.8443	3.9823	3.5874	3.3567	3.2039	3.0946	3.0123	2.9480	2.8962
12	4.7472	3.8853	3.4903	3.2592	3.1059	2.9961	2.9134	2.8486	2.7964
13	4.6672	3.8056	3.4105	3.1791	3.0254	2.9153	2.8321	2.7669	2.7144
14	4.6001	3.7389	3.3439	3.1122	2.9582	2.8477	2.7642	2.6987	2.6458
15	4.5431	3.6823	3.2874	3.0556	2.9013	2.7905	2.7066	2.6408	2.5876
16	4.4940	3.6337	3.2389	3.0069	2.8524	2.7413	2.6572	2.5911	2.5377
17	4.4513	3.5915	3.1968	2.9647	2.8100	2.6987	2.6143	2.5480	2.4943
18	4.4139	3.5546	3.1599	2.9277	2.7729	2.6613	2.5767	2.5102	2.4563
19	4.3808	3.5219	3.1274	2.8951	2.7401	2.6283	2.5435	2.4768	2.4227
20	4.3513	3.4928	3.0984	2.8661	2.7109	2.5990	2.5140	2.4471	2.3928
21	4.3248	3.4668	3.0725	2.8401	2.6848	2.5727	2.4876	2.4205	2.3661
22	4.3009	3.4434	3.0491	2.8167	2.6613	2.5491	2.4638	2.3965	2.3419
23	4.2793	3.4221	3.0280	2.7955	2.6400	2.5277	2.4422	2.3748	2.3201
24	4.2597	3.4028	3.0088	2.7763	2.6207	2.5082	2.4226	2.3551	2.3002
25	4.2417	3.3852	2.9912	2.7587	2.6030	2.4904	2.4047	2.3371	2.2821
26	4.2252	3.3690	2.9751	2.7426	2.5868	2.4741	2.3883	2.3205	2.2655
27	4.2100	3.3541	2.9604	2.7278	2.5719	2.4591	2.3732	2.3053	2.2501
28	4.1960	3.3404	2.9467	2.7141	2.5581	2.4453	2.3593	2.2913	2.2360
29	4.1830	3.3277	2.9340	2.7014	2.5454	2.4324	2.3463	2.2782	2.2229
30	4.1709	3.3158	2.9223	2.6896	2.5336	2.4205	2.3343	2.2662	2.2107
40	4.0848	3.2317	2.8387	2.6060	2.4495	2.3359	2.2490	2.1802	2.1240
60	4.0012	3.1504	2.7581	2.5252	2.3683	2.2540	2.1665	2.0970	2.0401
120	3.9201	3.0718	2.6802	2.4472	2.2900	2.1750	2.0867	2.0164	1.9588
∞	3.8415	2.9957	2.6049	2.3719	2.2141	2.0986	2.0096	1.9384	1.8799

Appendix A6a.   F-Distribution 95% Points $F(0.95;v_1,v_2)$ (Continued)

v_2 \ v_1	10	12	15	20	24	30	40	60	120	∞
1	241.88	243.91	245.95	248.01	249.05	250.09	251.14	252.20	253.25	254.32
2	19.396	19.413	19.429	19.446	19.454	19.462	19.471	19.479	19.487	19.496
3	8.7855	8.7446	8.7029	8.6602	8.6385	8.6166	8.5944	8.5720	8.5494	8.5265
4	5.9644	5.9117	5.8578	5.8025	5.7744	5.7459	5.7170	5.6878	5.6581	5.6281
5	4.7351	4.6777	4.6188	4.5581	4.5272	4.4957	4.4638	4.4314	4.3984	4.3650
6	4.0600	3.9999	3.9381	3.8742	3.8415	3.8082	3.7743	3.7398	3.7047	3.6688
7	3.6365	3.5747	3.5108	3.4445	3.4105	3.3758	3.3404	3.3043	3.2674	3.2298
8	3.3472	3.2840	3.2184	3.1503	3.1152	3.0794	3.0428	3.0053	2.9669	2.9276
9	3.1373	3.0729	3.0061	2.9365	2.9005	2.8637	2.8259	2.7872	2.7475	2.7067
10	2.9782	2.9130	2.8450	2.7740	2.7372	2.6996	2.6609	2.6211	2.5801	2.5379
11	2.8536	2.7876	2.7186	2.6464	2.6090	2.5705	2.5309	2.4901	2.4480	2.4045
12	2.7534	2.6866	2.6169	2.5436	2.5055	2.4663	2.4259	2.3842	2.3410	2.2962
13	2.6710	2.6037	2.5331	2.4589	2.4202	2.3803	2.3392	2.2966	2.2524	2.2064
14	2.6021	2.5342	2.4630	2.3879	2.3487	2.3082	2.2664	2.2230	2.1778	2.1307
15	2.5437	2.4753	2.4035	2.3275	2.2878	2.2468	2.2043	2.1601	2.1141	2.0658
16	2.4935	2.4247	2.3522	2.2756	2.2354	2.1938	2.1507	2.1058	2.0589	2.0096
17	2.4499	2.3807	2.3077	2.2304	2.1898	2.1477	2.1040	2.0584	2.0107	1.9604
18	2.4117	2.3421	2.2686	2.1906	2.1497	2.1071	2.0629	2.0166	1.9681	1.9168
19	2.3779	2.3080	2.2341	2.1555	2.1141	2.0712	2.0264	1.9796	1.9302	1.8780
20	2.3479	2.2776	2.2033	2.1242	2.0825	2.0391	1.9938	1.9464	1.8963	1.8432
21	2.3210	2.2504	2.1757	2.0960	2.0540	2.0102	1.9645	1.9165	1.8657	1.8117
22	2.2967	2.2258	2.1508	2.0707	2.0283	1.9842	1.9380	1.8895	1.8380	1.7831
23	2.2747	2.2036	2.1282	2.0476	2.0050	1.9605	1.9139	1.8649	1.8128	1.7570
24	2.2547	2.1834	2.1077	2.0267	1.9838	1.9390	1.8920	1.8424	1.7897	1.7331
25	2.2365	2.1649	2.0889	2.0075	1.9643	1.9192	1.8718	1.8217	1.7684	1.7110
26	2.2197	2.1479	2.0716	1.9898	1.9464	1.9010	1.8533	1.8027	1.7488	1.6906
27	2.2043	2.1323	2.0558	1.9736	1.9299	1.8842	1.8361	1.7851	1.7307	1.6717
28	2.1900	2.1179	2.0411	1.9586	1.9147	1.8687	1.8203	1.7689	1.7138	1.6541
29	2.1768	2.1045	2.0275	1.9446	1.9005	1.8543	1.8055	1.7537	1.6981	1.6377
30	2.1646	2.0921	2.0148	1.9317	1.8874	1.8409	1.7918	1.7396	1.6835	1.6223
40	2.0772	2.0035	1.9245	1.8389	1.7929	1.7444	1.6928	1.6373	1.5766	1.5089
60	1.9926	1.9174	1.8364	1.7480	1.7001	1.6491	1.5943	1.5343	1.4673	1.3893
120	1.9105	1.8337	1.7505	1.6587	1.6084	1.5543	1.4952	1.4290	1.3519	1.2539
∞	1.8307	1.7522	1.6664	1.5705	1.5173	1.4591	1.3940	1.3180	1.2214	1.0000

Appendix A6b.   F-Distribution 99% Points $F(0.99;v_1,v_2)$

v_1 / v_2	1	2	3	4	5	6	7	8	9
1	4052.2	4999.5	5403.3	5624.6	5763.7	5859.0	5928.3	5981.6	6022.5
2	98.503	99.000	99.166	99.249	99.299	99.332	99.356	99.374	99.388
3	34.116	30.817	29.457	28.710	28.237	27.911	27.672	27.489	27.345
4	21.198	18.000	16.694	15.977	15.522	15.207	14.976	14.799	14.659
5	16.258	13.274	12.060	11.392	10.967	10.672	10.456	10.289	10.158
6	13.745	10.925	9.7795	9.1483	8.7459	8.4661	8.2600	8.1016	7.9761
7	12.246	9.5466	8.4513	7.8467	7.4604	7.1914	6.9928	6.8401	6.7188
8	11.259	8.6491	7.5910	7.0060	6.6318	6.3707	6.1776	6.0289	5.9106
9	10.561	8.0215	6.9919	6.4221	6.0569	5.8018	5.6129	5.4671	5.3511
10	10.044	7.5594	6.5523	5.9943	5.6363	5.3858	5.2001	5.0567	4.9424
11	9.6460	7.2057	6.2167	5.6683	5.3160	5.0692	4.8861	4.7445	4.6315
12	9.3302	6.9266	5.9526	5.4119	5.0643	4.8206	4.6395	4.4994	4.3875
13	9.0738	6.7010	5.7394	5.2053	4.8616	4.6204	4.4410	4.3021	4.1911
14	8.8616	6.5149	5.5639	5.0354	4.6950	4.4558	4.2779	4.1399	4.0297
15	8.6831	6.3589	5.4170	4.8932	4.5556	4.3183	4.1415	4.0045	3.8948
16	8.5310	6.2262	5.2922	4.7726	4.4374	4.2016	4.0259	3.8896	3.7804
17	8.3997	6.1121	5.1850	4.6690	4.3359	4.1015	3.9267	3.7910	3.6822
18	8.2854	6.0129	5.0919	4.5790	4.2479	4.0146	3.8406	3.7054	3.5971
19	8.1850	5.9259	5.0103	4.5003	4.1708	3.9386	3.7653	3.6305	3.5225
20	8.0960	5.8489	4.9382	4.4307	4.1027	3.8714	3.6987	3.5644	3.4567
21	8.0166	5.7804	4.8740	4.3688	4.0421	3.8117	3.6396	3.5056	3.3981
22	7.9454	5.7190	4.8166	4.3134	3.9880	3.7583	3.5867	3.4530	3.3458
23	7.8811	5.6637	4.7649	4.2635	3.9392	3.7102	3.5390	3.4057	3.2986
24	7.8229	5.6136	4.7181	4.2184	3.8951	3.6667	3.4959	3.3629	3.2560
25	7.7698	5.5680	4.6755	4.1774	3.8550	3.6272	3.4568	3.3239	3.2172
26	7.7213	5.5263	4.6366	4.1400	3.8183	3.5911	3.4210	3.2884	3.1818
27	7.6767	5.4881	4.6009	4.1056	3.7848	3.5580	3.3882	3.2558	3.1494
28	7.6356	5.4529	4.5681	4.0740	3.7539	3.5276	3.3581	3.2259	3.1195
29	7.5976	5.4205	4.5378	4.0449	3.7254	3.4995	3.3302	3.1982	3.0920
30	7.5625	5.3904	4.5097	4.0179	3.6990	3.4735	3.3045	3.1726	3.0665
40	7.3141	5.1785	4.3126	3.8283	3.5138	3.2910	3.1238	2.9930	2.8876
60	7.0771	4.9774	4.1259	3.6491	3.3389	3.1187	2.9530	2.8233	2.7185
120	6.8510	4.7865	3.9493	3.4796	3.1735	2.9559	2.7918	2.6629	2.5586
∞	6.6349	4.6052	3.7816	3.3192	3.0173	2.8020	2.6393	2.5113	2.4073

Appendix A6b.   F-Distribution 99% Points $F(0.99;v_1,v_2)$ (Continued)

v_2 \ v_1	10	12	15	20	24	30	40	60	120	∞
1	6055.8	6106.3	6157.3	6208.7	6234.6	6260.7	6286.8	6313.0	6339.4	6366.0
2	99.399	99.416	99.432	99.449	99.458	99.466	99.474	99.483	99.491	99.501
3	27.229	27.052	26.872	26.690	26.598	26.505	26.411	26.316	26.221	26.125
4	14.546	14.374	14.198	14.020	13.929	13.838	13.745	13.652	13.558	13.463
5	10.051	9.8883	9.7222	9.5527	9.4665	9.3793	9.2912	9.2020	9.1118	9.0204
6	7.8741	7.7183	7.5590	7.3958	7.3127	7.2285	7.1432	7.0568	6.9690	6.8801
7	6.6201	6.4691	6.3143	6.1554	6.0743	5.9921	5.9084	5.8236	5.7372	5.6495
8	5.8143	5.6668	5.5151	5.3591	5.2793	5.1981	5.1156	5.0316	4.9460	4.8588
9	5.2565	5.1114	4.9621	4.8080	4.7290	4.6486	4.5667	4.4831	4.3978	4.3105
10	4.8492	4.7059	4.5582	4.4054	4.3269	4.2469	4.1653	4.0819	3.9965	3.9090
11	4.5393	4.3974	4.2509	4.0990	4.0209	3.9411	3.8596	3.7761	3.6904	3.6025
12	4.2961	4.1553	4.0096	3.8584	3.7805	3.7008	3.6192	3.5355	.3.4494	3.3608
13	4.1003	3.9603	3.8154	3.6646	3.5868	3.5070	3.4253	3.3413	3.2548	3.1654
14	3.9394	3.8001	3.6557	3.5052	3.4274	3.3476	3.2656	3.1813	3.0942	3.0040
15	3.8049	3.6662	3.5222	3.3719	3.2940	3.2141	3.1319	3.0471	2.9595	2.8684
16	3.6909	3.5527	3.4089	3.2588	3.1808	3.1007	3.0182	2.9330	2.8447	2.7528
17	3.5931	3.4552	3.3117	3.1615	3.0835	3.0032	2.9205	2.8348	2.7459	2.6530
18	3.5082	3.3706	3.2273	3.0771	2.9990	2.9185	2.8354	2.7493	2.6597	2.5660
19	3.4338	3.2965	3.1533	3.0031	2.9249	2.8442	2.7608	2.6742	2.5839	2.4893
20	3.3682	3.2311	3.0880	2.9377	2.8594	2.7785	2.6947	2.6077	2.5168	2.4212
21	3.3098	3.1729	3.0299	2.8796	2.8011	2.7200	2.6359	2.5484	2.4568	2.3603
22	3.2576	3.1209	2.9780	2.8274	2.7488	2.6675	2.5831	2.4951	2.4029	2.3055
23	3.2106	3.0740	2.9311	2.7805	2.7017	2.6202	2.5355	2.4471	2.3542	2.2559
24	3.1681	3.0316	2.8887	2.7380	2.6591	2.5773	2.4923	2.4035	2.3099	2.2107
25	3.1294	2.9931	2.8502	2.6993	2.6203	2.5383	2.4530	2.3637	2.2695	2.1694
26	3.0941	2.9579	2.8150	2.6640	2.5848	2.5026	2.4170	2.3273	2.2325	2.1315
27	3.0618	2.9256	2.7827	2.6316	2.5522	2.4699	2.3840	2.2938	2.1984	2.0965
28	3.0320	2.8959	2.7530	2.6017	2.5223	2.4397	2.3535	2.2629	2.1670	2.0642
29	3.0045	2.8685	2.7256	2.5742	2.4946	2.4118	2.3253	2.2344	2.1378	2.0342
30	2.9791	2.8431	2.7002	2.5487	2.4689	2.3860	2.2992	2.2079	2.1107	2.0062
40	2.8005	2.6648	2.5216	2.3689	2.2880	2.2034	2.1142	2.0194	1.9172	1.8047
60	2.6318	2.4961	2.3523	2.1978	2.1154	2.0285	1.9360	1.8363	1.7263	1.6006
120	2.4721	2.3363	2.1915	2.0346	1.9500	1.8600	1.7628	1.6557	1.5330	1.3805
∞	2.3209	2.1848	2.0385	1.8783	1.7908	1.6964	1.5923	1.4730	1.3246	1.0000

From C. A. Bennett and N. L. Franklin, *Statistical Analysis in Chemistry and the Chemical Industry*, Wiley, New York, 1954, pp. 702–705. Reproduced by permission of the publisher and the Biometrika Trustees.

Appendix A7.   Probability Plotting Positions $F_i = 100(i - 0.5)/n$

i	6	7	8	9	10	11	12	13	14	15	16	17	18	19	20
1	8.3	7.1	6.2	5.6	5.0	4.5	4.2	3.8	3.6	3.3	3.1	2.9	2.8	2.6	2.5
2	25.0	21.4	18.7	16.7	15.0	13.6	12.5	11.5	10.7	10.0	9.4	8.8	8.3	7.9	7.5
3	41.7	35.7	31.2	27.8	25.0	22.7	20.8	19.2	17.9	16.7	15.6	14.7	13.9	13.2	12.5
4	58.3	50.0	43.7	38.9	35.0	31.8	29.2	26.9	25.0	23.3	21.9	20.6	19.4	18.4	17.5
5	75.0	64.3	56.2	50.0	45.0	40.9	37.5	34.6	32.1	30.0	28.1	26.5	25.0	23.7	22.5
6	91.7	78.6	68.7	61.1	55.0	50.0	45.8	42.3	39.3	36.7	34.4	32.4	30.6	28.9	27.5
7		92.9	81.2	72.2	65.0	59.1	54.2	50.0	46.4	43.3	40.6	38.2	36.1	34.2	32.5
8			93.7	83.3	75.0	68.2	62.5	57.7	53.6	50.0	46.9	44.1	41.7	39.5	37.5
9				94.4	85.0	77.3	70.8	65.4	60.7	56.7	53.1	50.0	47.2	44.7	42.5
10					95.0	86.4	79.2	73.1	67.9	63.3	59.4	55.9	52.8	50.0	47.5
11						95.5	87.5	80.8	75.0	70.0	65.6	61.8	58.3	55.3	52.5
12							95.8	88.5	82.1	76.7	71.9	67.6	63.9	60.5	57.5
13								96.2	89.3	83.3	78.1	73.5	69.4	65.8	62.5
14									96.4	90.0	84.4	79.4	75.0	71.1	67.5
15										96.7	90.6	85.3	80.6	76.3	72.5
16											96.9	91.2	86.1	81.6	77.5
17												97.1	91.7	86.8	82.5
18													97.2	92.1	87.5
19														97.4	92.5
20															97.5

References

Abdel-Hameed, M. S., Cinlar, E., and Quinn, J., Eds. (1984), *Reliability Theory and Models*, Academic Press, New York.

Ahmad, M., and Sheikh, A. K. (1981), "Estimation of the Parameters of the Bernstein Distribution from Complete and Censored Samples," *Proc. 43rd Session of the International Statist. Inst.*, 225-228.

Ahmad, M., and Sheikh, A. K. (1983), "Accelerated Life Testing," presented at the 1983 Joint Statistical Meetings, Toronto.

Aitchison, J., and Brown, J. A. C. (1957), *The Lognormal Distribution*, Cambridge Univ. Press, New York and London.

Aitkin, M. (1981), "A Note on the Regression Analysis of Censored Data," *Technometrics* 23, 161-164.

Aitkin, M. and Clayton, D. (1980), "The Fitting of Exponential, Weibull, and Extreme Value Distributions to Complex Censored Survival Data Using GLIM," *Applied Statistics* 29, 155-163.

Allen, W. R. (1959), "Inference from Tests with Continuously Increasing Stress,"*Operations Research* 7, 303-312.

American Society for Testing and Materials (1962), *Manual on Fitting Straight Lines*, Special Technical Publication 313, 1916 Race St., Philadelphia, PA 19103, (215)299-5400 or -5428.

American Society for Testing and Materials (1963), *A Guide for Fatigue Testing and the Statistical Analysis of Fatigue Data*, Special Technical Publication No. 91-A (2nd ed.), 1916 Race St., Philadelphia, PA 19103, (215)299-5400 or -5428.

American Society for Testing and Materials (1970), "Standard Method of Testing Paper by Direct-Voltage Life Testing of Capacitors," Standard D2631-68, *Annual Book of American Society for Testing and Materials Standards, Part 29: Electrical Insulating Materials*, 1916 Race St., Philadelphia, PA 19103, (215)299-5400 or -5428.

American Society for Testing and Materials (1979), *Statistical Analysis of Fatigue Data*, by R. E. Little and J. C. Ekvall, Eds., Special Technical Publication 744, 1916 Race St., Philadelphia, PA 19103, (215)299-5400 or -5428.

American Society for Testing and Materials (1981a), "Tables for Estimating Median Fatigue Limits," Special Technical Publication 731, R. E. Little, Ed., 1916 Race St., Philadelphia, PA 19103, (215)299-5400 or -5428.

American Society for Testing and Materials (1981b), "Statistical Analysis of Linear or Linearized Stress-Life (S-N) and Strain-Life (ε-N) Fatigue Data," Special Technical Publication No. 744, 1916 Race St., Philadelphia, PA 19103, (215)299-5400 or -5428.

American Statistical Association (1987), *Current Index to Statistics: Applications, Methods and Theory*, 1429 Duke St., Alexandria, VA 22314, (703)684-1221. Yearly since 1975.

Ascher, H., and Feingold, H. (1984), *Repairable Systems Reliability – Modeling, Inference, Misconceptions and Their Causes*, Marcel Dekker, Inc., New York.

ASTM STP 648, Hoeppner, D. W., Ed. (1978), "Fatigue Testing of Weldments," Amer. Soc. for Testing and Materials, 1916 Race St., Philadelphia, PA 19103, (215)299-5585. Price $28.50.

ASTM STP 738 (1981), "Fatigue Crack Growth Measurement and Data Analysis," Amer. Soc. for Testing and Materials, 1916 Race St., Philadelphia, PA 19103, (215)299-5585, 371 pp. Price $39.00.

ASTM STP 748 (1981), "Methods and Models for Predicting Fatigue Crack Growth under Random Loading," Amer. Soc. for Testing and Materials, 1916 Race St., Philadelphia, PA 19103, (215)299-5585, 140 pp. Price $16.50.

Ballado-Perez, D. A. (1986), "Statistical Modeling of Accelerated Life Tests for Adhesive-Bonded Wood Composites," Dept. of Wood and Paper Science, North Carolina State Univ., Raleigh, NC.

Ballado-Perez, D. A. (1987), "Statistical Model for Accelerated Life Testing of Wood Composites – A Preliminary Evaluation," Dept. of Wood and Paper Science, North Carolina State Univ., Raleigh, NC.

Barnett, V., and Lewis, T. (1984), *Outliers in Statistical Data*, 2nd ed., Wiley, New York.

Bartnikas, R., Ed. (1987), *Engineering Dielectrics – Vol. II*, American Society for Testing and Materials Publication 04-78300-21, 1916 Race St., Philadelphia, PA 19103.

Bartnikas, R., and McMahon, E. J., Eds. (1979), *Corona Measurement and Interpretation*, American Society for Testing and Materials Special Technical Publication 669.

Barton, R. R. (1987), "Optimal Accelerated Lifetest Plans Which Minimize the Maximum Test Stress," author at School of OR and Indus. Engineering, Cornell Univ., Ithaca, NY 14853-7501. To appear in *IEEE Trans. on Reliability*.

Basu, A. P., and Ebrahimi, N. (1982), "Nonparametric Accelerated Life Testing," *IEEE Trans. on Reliability* R-31, 432-435.

Bates, D. M. and Watts, D. G. (1988), *Nonlinear Regression and Its Applications*, Wiley, New York, 384 pp.

Beal, S. L., and Sheiner, L. B. (1985), "Methodology of Population Pharmacokinetics," in *Drug Fate and Metabolism: Methods and Techniques*, Vol. 5, 135-183, E. R. Garrett and J. L. Hirtz, Eds., Marcel Dekker, New York.

Beal, S. L., and Sheiner, L. B. (1988), "Heteroscedastic Nonlinear Regression," *Technometrics* 30, 327-338.

Beckman, R. J., and Cook, R. D. (1983), "Outlier.........s," *Technometrics* 25, 119-163 includes discussions.

Beckwith, J. P. (1979), "Estimation of the Strength Remaining of a Material Which Decays with Time," private communication, Dept. of Math and Computer Science, Michigan Technological Univ. Houghton, MI 49931.

Beckwith, J. P. (1980), "An Estimator and Design Technique for the Estimation of the Rate Parameter in Accelerated Testing," paper at the Joint Statistical Meeting, author at Michigan Technological Univ., Houghton, MI 49931.

Bently, D. L. (1970), "Statistical Techniques in Predicting Thermal Stability," *J. of Pharmaceutical Sciences* 59, 464-468.

Bernstein, B. S. project manager (1981), "Experimental Techniques for Investigating the Degradation of Electrical Insulation," EL-1854, Technical Planning Study TPS 79-723 prepared for EPRI, 3412 Hillview Ave., Palo Alto, CA 94304.

Bessler, S., Chernoff, H., and Marshall, A. W. (1962), "An Optimal Sequential Accelerated Life Test," *Technometrics* 4, 367-379.

Beyer, O., Pieper, V., and Tiedge, J. (1980), "On Modeling Problems of Reliability and Maintenance of Components Subject to Wear," in German, *Proc. of Conf. STAQUAREL*, Prague, 31-37.

Birks, J. B., and Schulman, J. H. (1959), *Progress in Dielectrics*, Wiley, New York.

Birnbaum, Z. W. (1979), "On the Mathematics of Competing Risks," DHEW Publ. No. (PHS) 79-1351. For sale by the Superintendent of Documents, U.S. Gov't. Printing Office, Washington, DC 20402.

Birnbaum, Z. W., and Saunders, S. C. (1968), "A Probabilistic Interpretation of Miner's Rule," *SIAM J. of Applied Math.* 16, 637-652.

Birnbaum, Z. W., and Saunders, S. C. (1969), "A New Family of Life Distributions," *J. of Applied Probability* **6**, 319-327.

Black, J. R. (1969a), "Electromigration Failure Modes in Aluminum Metallization for Semiconductor Devices," *Proc. IEEE* **57**, 1587-1593.

Black, J. R. (1969b), "Electromigration – A Brief Survey and Some Recent Results," *IEEE Trans. on Electronic Devices* **ED-16**, 338-.

Block, H. W., and Savits, T. H. (1981), "Multivariate Distributions in Reliability Theory and Life Testing," Technical Report No. 81-13, Inst. for Statistics and Applications, Dept. of Math. and Statistics, Univ. of Pittsburgh, Pittsburgh, PA 15260.

Bogdanoff, J. L., and Kozin, F. (1984), *Probabalistic Models of Cumulative Damage*, Wiley, New York.

Bolotin, V. V. (1969), *Statistical Methods in Structural Mechanics*, Holden-Day, San Francisco.

Boothroyd, G. (1975), *Fundamentals of Metal Machining and Machine Tools*, McGraw-Hill, New York.

Borowiak, D. S. (1989), *Model Discrimination for Nonlinear Regression Models*, Dekker, New York, 200 pp.

Bowker, A. H. and Lieberman, G. J. (1972), *Engineering Statistics*, 2nd ed., Prentice-Hall, Englewood Cliffs, NJ.

Bowman, K. O., and Shenton, L. R. (1987), *Properties of Estimators for the Gamma Distribution*, Marcel Dekker, New York.

Box, G. E. P. (1950), "Problems in the Analysis of Growth and Wear Curves," *Biometrics* **6**, 362-389.

Box, G. E. P. and Draper, N. R. (1987), *Empirical Model Building and Response Surfaces*, Wiley, New York.

Box, G. E. P., Hunter, W. G., and Hunter, J. S. (1978), *Statistics for Experimenters*, Wiley, New York.

Brancato, E. L., Johnson, L. M., Campbell, F. G., and Walker, H. P. (1977), "Reliability Prediction Studies on Electrical Insulation: Navy Summary Report," Naval Research Laboratory Report 8095, available from the National Technical Information Service, U.S. Dept. of Commerce, Springfield, VA 22161.

Breslow, N. E., and Day, N. E. (1980), *Statistical Methods in Cancer Research, Volume 1 - The Analysis of Case-Control Studies*, International Agency for Research on Cancer, World Health Organization, Lyon, France.

Brookes, A. S. (1974), "The Weibull Distribution: Effect of Length and Conductor Size of Test Cables," *Electra* **33**, 49-61.

Brostow, W., and Corneliussen, J., Eds. (1986), *Failure of Plastics*, Hanser, Munich.

Brush, G. G. (1988), *How to Choose the Proper Sample Size*, Publication T3512, 115 pages, Amer. Soc. for Quality Control, 310 W. Wisconsin Ave., Milwaukee, WI 53203, (800)952-6587.

Bugaighis, M. M. (1988), "Efficiencies of MLE and BLUE for Parameters of an Accelerated Life-Test Model," *IEEE Trans. on Reliability* **37**, 230-233.

Buswell, G. D., Meeker, W. Q., and Myers, D. H. (1984), "STAR – Statistical Reliability Analysis," internal AT&T Bell Labs document; contact Dr. J.H. Hooper, Room 2K537, AT&T Bell Labs, Holmdell, NJ 07733, (201)949-1996.

Buswell, G. D., Meeker, W. Q., Myers, D. H., and Gibson, C. L. (1985), "STAR – Software for the Analysis and Presentation of Reliability Data," *Amer. Statist. Assoc. 1985 Proceedings of the Statistical Computing Section*.

Carey, M. (1985), "Bibliography on Aging Degradation," private communication, Rm. 2K-503, AT&T Bell Labs, Crawfords Corner Rd., Holmdel, NJ 07733-1988.

Carey, M. and Tortorella, M. (1987), "Analysis of Degradation Data in Reliability," paper at the Annual Joint Statistical Meeting, San Francisco. Authors at AT&T Bell Labs, Crawfords Corner Rd., Holmdel, NJ 07733-1988.

Carstensen, J. T. (1972) *Theory of Pharmaceutical Systems*, Academic Press, New York.

Carter, A. D. S. (1985), *Mechanical Reliability*, 2nd ed., Macmillian, London.

Chambers, J. M. (1977), *Computational Methods for Data Analysis*, Wiley, New York.

Chambers, J. M.; Cleveland, W. S.; Kleiner, B.; Tukey, P. A. (1983), *Graphical Methods for Data Analysis*, (hard cover) Wadsworth, Monterey, CA; (paperback) Duxbury Press, Boston.

Chartpak (1988), *Catalog*, One River Road, Leeds, MA 01053, $3.95, (800)628-1910.

Chernoff, H. (1953), "Locally Optimum Designs for Estimating Parameters," *Ann. Math. Statist.* **24**, 586-602.

Chernoff, H. (1962), "Optimal Accelerated Life Designs for Estimation," *Technometrics* **4**, 381-408.

Chernoff, H. (1972), *Sequential Analysis and Optimal Design*, Conference Board of the Mathematical Sciences Regional Conference Series in Applied Mathematics, No. 8, Soc. for Industrial and Applied Mathematics, 1916 Race St., Philadelphia, PA.

Clark, J. E., and Slater, J. A. (1969), "Outdoor Performance of Plastics: III. Statistical Model for Predicting Weatherability," National Bureau of Standards Report 10 116, Washington, DC 20234.

Cleveland, W. S. (1985), *The Elements of Graphing Data*, Wadsworth & Brooks/Cole, Pacific Grove, CA 93950, (408)373-0728.

CODEX (1988), "CODEX Chart and Graph Papers," Catalog DG, Codex Book Co., 74 Broadway, Norwood, MA 02062, (617)769-1050.

Coffin, Jr., L. F. (1954), "A Study of the Effects of Cyclic Thermal Stresses on a Ductile Metal," *Trans. ASME* **76**, 923-950.

Coffin, Jr., L. F. (1974), "Fatigue at High Temperature – – Prediction and Interpretation," James Clayton Memorial Lecture, *Proc. Inst. Mech. Eng. (London)* **188**, 109-127.

Cohen, A. C. and Whitten, B. J. (1988), *Parameter Estimation in Reliability and Life Span Models*, Marcel Dekker, New York.

Cohen, J. (1988), *Statistical Power Analysis*, 2nd ed., Lawrence Assocs., 365 Broadway, Hillsdale, NJ 07642, (201)666-4110. The publisher also offers a companion "MicroComputer program for Power Analysis," for IBM compatible PCs, by J. Cohen and M. Borenstein.

Collins, J. A. (1981), *Failure of Materials in Mechanical Design*, Wiley-Interscience, New York.

Connors, K. A., Amidon, G. L., and Kennon, L. (1979), *Chemical Stability of Pharmaceuticals*, Wiley, New York.

Cox, D. R. (1958), *Planning of Experiments*, Wiley, New York.

Cox, D. R. (1959), "The Analysis of Exponentially Distributed Life-Times with Two Types of Failures," *J. Royal Statist. Soc. B* **21**, 411-421.

Cox, D. R., and Oakes, D. (1984), *Analysis of Survival Data*, Methuen (Chapman and Hall), New York.

Cox, D. R., and Snell, E. J. (1968), "A General Definition of Residuals," *J. of the Royal Statist. Soc., Series B* **30**, 248-275.

Cramp, M. G. (1959), "A Statistical Basis for Transformer Oil Breakdown," Masters thesis, Dept. of Electrical Engineering, Rensselaer Polytechnic Inst., Troy, NY 12181.

Craver, J. S. (1980), *Graph Paper from Your Copier*, HP Books, P.O. Box 5367, Tucson, AZ 85703, (602)888-2150. Price $12.95.

Crawford, D. E. (1970), "Analysis of Incomplete Life Test Data on Motorettes," *Insulation/Circuits* **16**, 43-48.

Crow, E. L., and Shimizu, K., Eds. (1988), *Lognormal Distributions*, Marcel Dekker, New York.

D'Agostino, R. B., and Stephens, M. A. (1986), *Goodness-of-Fit Techniques*, Marcel Dekker, New York. Price $79.95.

Dakin, T. W. (1948), "Electrical Insulation Deterioration Treated As a Chemical Reaction Rate Phenomenon," *AIEE Trans.* **67**, 113-122.

Dallal, G. E. (1988), "Statistical Microcomputing – Like It Is," *The Amer. Statistician* **42**, 212-216.

Daniel, C. (1976), *Applications of Statistics to Industrial Experimentation*, Wiley, New York.

Daniel, C., and Heerema, N. (1950), "Design of Experiments for the Most Precise Slope Estimation or Linear Extrapolation," *J. Amer. Statist. Assoc.* **45**, 546-556.

Daniel, C., and Wood, F. S. (1980), *Fitting Equations to Data: Computer Analysis of Multifactor Data*, Wiley, New York.

David, H. A., and Moeschberger, M. L. (1979), *The Theory of Competing Risks*, Griffin's Statistical Monograph No. 39, Methuen, London.

Department of Defense (1981), "Information Analysis Centers Profiles for Specialized Technical Information," Defense Technical Information Center, Defense Logistics Agency, Cameron Station, Alexandria, VA 22314. Data Centers for Concrete, Metals and Ceramics, Mech. Props, Plastics, Relia, & others.

Department of Defense (1985), "Information Analysis Centers Directory," Defense Technical Information Center, Defense Logistics Agency, Cameron Station, Alexandria, VA 22304-6145. General information: (202)274-6434.

DePaul, D. J., Ed. (1957), *Corrosion and Wear Handbook*, Sponsored by Naval Reactors Branch Division of Reactor Development, U. S. Atomic Energy Comm., McGraw-Hill, New York.

Derringer, G. C. (1982), "A Proposed Model for Accelerated-Stress Life-Test Data Exhibiting Two Failure Modes," *Polymer Engineering and Science* 22, 354-357.

Derringer, G. C. (1989), "A Model for Service Life of Polyethylene Pipe Exhibiting Ductile-Brittle Transition in Fracture Mode," *J. Applied Polymer Science* 37, 215-224.

d'Heurle, F. M. and Ho, P. S. (1978), "Electromigration in Thin Films," pp. 243-303 of *Thin Films – Interdiffusion and Reactions*, edited by J. M. Poate, K. N. Tu, and J. W. Mayer, Wiley, New York.

Diamond, W. (1981), *Practical Experimental Designs for Engineers and Scientists*, Lifetime Learning Publications, Belmont, CA.

Dieter, Jr., G. E. (1961), *Mechanical Metallurgy*, McGraw-Hill, New York.

Dietzgen (1988), "Quality Graph Papers," Dietzgen Corp., 250 Wille Rd., Des Plaines, IL 60018, (315)635-5200.

Disch, D. (1983), "Optimum Accelerated Sequential Life Tests When Total Testing Time is Limited," private communication, Dept. of Mathematics, Rose-Hulman Inst. of Technology, 5500 Wabash Ave., Terre Haute, IN 47803.

Dixon, W. J., Ed. (1985), *BMDP Statistical Software*, Univ. of Calif. Press, Los Angeles, CA. For information contact: BMDP Statistical Software, Inc., 1440 Sepulveda Blvd., Los Angeles, CA 90025, (213)479-7799.

Doganaksoy, N. (1989a), "A Computer Program to Fit Weibull Accelerated Life Test Models to Multiply Right Censored Data," author at Graduate Mgt. Inst., Union College, Schenectady, NY 12308.

Doganaksoy, N. (1989b), "Approximate Confidence Intervals for the Weibull and Smallest Extreme Value Distribution Parameters and Quantiles in Single Stress Models and Simple Accelerated-Stress Regression Models with Censored Data," Ph.D. thesis, Graduate Mgt. Inst., Union College, Schenectady, NY 12308.

Draper, N. R., and Smith, H. (1981), *Applied Regression Analysis*, 2nd ed., Wiley, New York.

Efron, B. (1986), "Why Isn't Everyone a Bayesian?," *The Amer. Statistician* 40, 1-11.

EG&G Electro-Optics (1984), "Flashlamps – 1984 Catalog," available from Mr. Clayton Van Buren, Robert F. Lamb Co. Inc., 4515 Culver Rd., Rochester, NY 14622.

Ehrenfeld, S. (1962), "Some Experimental Design Problems in Attribute Life Testing," *J. Amer. Statist. Assoc.* 57, 668-679.

Elfving, G. (1952), "Optimum Allocation in Linear Regression Theory," *Ann. Math. Statist.* 23, 255-262.

Endicott, H. S., and Starr, W. T. (1961), "Progressive Stress – A New Accelerated Approach to Voltage Endurance," *Trans. of AIEE (Power and Apparatus Systems)* 80, 515-522.

Endicott, H. S., and Zoellner, J. A. (1961), "A Preliminary Investigation of the Steady and Progressive Stress Testing of Mica Capacitors," *Proc. of the 7th National Symposium on Reliability and Quality Control*, 229-235.

Endicott, H. S., Hatch, B. D., and Schmer, R. G. (1965), "Application of the Eyring Model to Capacitor Aging Data," *IEEE Trans. on Component Parts*, CP-12, 34-41.

Escobar, L. A., and Meeker, Jr., W. Q. (1986a), "Algorithm AS 218 Elements of the Fisher Information Matrix for the Smallest Extreme Value Distribution and Censored Data," *Applied Statistics* 35, 80-86.

Escobar, L. A., and Meeker, Jr., W. Q. (1986b), "Optimum Accelerated Life Tests with Type II Censored Data," *J. of Statistical Computation and Simulation* 23, 273-297.

Escobar, L. A., and Meeker, Jr., W. Q. (1988), "Assessing Local Influence in Regression Analysis with Censored Data," presented at the 1988 Joint Statistical Meeting, New Orleans.

Prof. Escobar is at Dept. of Exp'l Statistics, Louisiana State Univ., Baton Rouge, LA 70803.

Escobar, L. A., and Meeker, W. Q. (1989), "Elements of the Fisher Information Matrix for Location-Scale Distributions," Dept. of Experimental Statistics, Louisiana St. Univ., Baton Rouge, LA 70803-5606.

Evans, R. A. (1969), "The Analysis of Accelerated-Temperature-Tests," *Proceedings of the 1969 Annual Symposium on Reliability*, 294-302.

Everitt, B. S., and Hand, D. J. (1981), *Finite Mixture Distributions*, Chapman and Hall, New York.

Faraone, L. (1986), "Endurance of 9.3 mm EEPROM Tunnel Oxide," in *Insulating Films on Semiconductors*, edited by J. J. Simonne and J. Buxo, Elsevier Science Publ.

Farewell, V. T., and Prentice, R. L. (1977), "A Study of Distributional Shape in Life Testing," *Technometrics* 19, 69-75.

FDA Center for Drugs and Biologics (1987), "Guidelines for Submitting Documentation for the Stability of Human Drugs and Biologics," Office of Drug Research and Review, 5600 Fishers Lane, Rockville, MD 20857, (301)443-4330.

Fedorov, V. V. (1972), *Theory of Optimal Experiments*, Academic Press, New York.

Fiegl, P., and Zelen, M. (1965), "Estimation of Exponential Survival Probabilities with Concomitant Information," *Biometrics* 21, 826-838.

Finney, D. J. (1968), *Probit Analysis*, 3rd ed., Cambridge Univ. Press.

Flack, V. F., and Flores, R. A. (1989), "Using Simulated Envelopes in the Evaluation of Normal Probability Plots of Regression Residuals," *Technometrics* 31, 219-225.

Ford, I., Titterington, D. M., and Kitsos, C. P. (1989), "Recent Advances in Nonlinear Experimental Design," *Technometrics* 31, 49-60.

Freeman, D. A. (1981), "Bootstrapping Regression Models," *Annals of Statistics* 9, 1218-1228.

Frieman, S. W. (1980), "Fracture Mechanics of Glass," in Glass Science and Technology Vol. 5, *Elasticity and Strength in Glasses*, D. R. Ullman and N. J. Kreidl, Eds., pp. 21-78, Academic Press, New York.

Fuller, W. A. (1987), *Measurement Error Models*, Wiley, New York.

Gabano, J. P. (1983), *Lithium Batteries*, Academic Press, New York.

Galambos, J. (1978), *The Asymptotic Theory of Extreme Order Statistics*, Wiley-Interscience, New York.

Gallant, A. R. (1987), *Nonlinear Statistical Models*, Wiley, New York.

Gaylor, D. W., and Sweeny, H. C. (1965), "Design for Optimal Prediction in Simple Linear Regression," *J. Amer. Statist. Assoc.* 60, 205-216.

Gertsbakh, L. B., and Kordonskiy, K. B. (1969), *Models of Failure*, Springer Verlag, Berlin.

Ghate, P. B. (1982), "Electromigration-Induced Failures in VLSI Interconnects," *Proc. International Reliability Physics Symp.* 20, 292-299.

Gillespie, R. H. (1965), "Accelerated Aging of Adhesives in Plywood-Type Joints," *Forest Products J.*, 369-378.

Glaser, R. E. (1984), "Estimation for a Weibull Accelerated Life Testing Model," *Naval Research Logistics Quarterly* 31, No. 4, 559-570.

Glasser, M. (1965), "Regression Analysis with Dependent Variable Censored," *Biometrics* 21, 300-307.

Glasser, M. (1967), "Exponential Survival with Covariance," *J. Amer. Statist. Assoc.* 62, 561-568.

Glasstone, S., Laidler, K. J., and Eyring, H. E. (1941), *The Theory of Rate Processes*, McGraw-Hill, New York.

Goba, F. A. (1969), "Bibliography on Thermal Aging of Electrical Insulation," *IEEE Trans. on Electrical Insulation* EI-4, 31-58.

Goldhoff, R. M., and Hahn, G. J. (1968), "Correlation and Extrapolation of Creep-Rupture Data of Several Steels and Superalloys Using Time-Temperature Parameters," ASM Publ. No. D8-100, Amer. Soc. for Metals, Metals Park, OH 44073.

Goldhoff, R. M., and others (1979), "Development of a Standard Methodology for the Correlation and Extrapolation of Elevated Temperature Creep and Rupture Data," EPRI FP-1062, Project 638-1 Final Report, Electric Power Research Inst., Palo Alto, CA.

Graham, J. A., Ed. (1968), *Fatigue Design Handbook*, Soc. of Automotive Engineers, Inc., 400 Commonwealth Dr., Warrendale, PA 15096, (412)776-4970.

Grange, J. M. (1971), "Study on the Validity of Electronic Parts Stress Models," *IEEE Trans. on Reliability* **R-20**, 136-142.

Greene, W. H. (1986), "Analysis of Survival and Failure Time Data with LIMDEP," *The American Statistician* **40**, 228-229.

Greenwood, M. (1926), "The Natural Duration of Cancer," *Reports of Public Health and Medical Subjects* **33**, Her Majesty's Stationery Office, London.

Grimm, W., Ed. (1987), *Stability Testing of Drug Products*, Wissenschaftliche Verlagsgesellschaft mbH, Stuttgart.

Gumbel, E. J. (1958), *Statistics of Extremes*, Columbia Univ. Press, New York.

Hahn, G. J. (1979), "Statistical Methods for Creep, Fatigue and Fracture Data Analysis," *J. of Engineering Materials and Technology* **101**, 344-348.

Hahn, G. J., and Meeker, W. Q. (1982), "Pitfalls and Practical Considerations in Product Life Analysis," "Part I: Basic Concepts and Dangers of Extrapolation," "Part II: Mixtures of Product Populations and More General Models," *J. of Quality Technology* **14**, 144-152 and 177-185.

Hahn, G. J., and Meeker, W. Q. (1990), *Statistical Intervals: A Guide for Practitioners*, Wiley, New York.

Hahn, G. J., and Miller, J. M. (1968a), "Methods and Computer Program for Estimating Parameters in a Regression Model from Censored Data," General Electric Research & Development Center TIS Report 68-C-277.

Hahn, G. J., and Miller, J. M. (1968b), "Time-Sharing Computer Programs for Estimating Parameters of Several Normal Populations and for Regression Estimation from Censored Data," General Electric Research & Development Center TIS Report 68-C-366.

Hahn, G. J., and Nelson, W. (1974), "A Comparison of Methods for Analyzing Censored Life Data to Estimate Relationships between Stress and Product Life," *IEEE Trans. on Reliability* **R-23**, 2-10.

Hahn, G. J., and Schmee, J. (1980), "Regression Estimates versus Separate Estimation at Individual Test Conditions," *J. of Quality Technology* **12**, 25-35.

Hahn, G. J., and Shapiro, S. S. (1967), *Statistical Models in Engineering*, Wiley, New York.

Hahn, G. J., Morgan, C., and Nelson, W. (1985), "More Accurate Estimates of the Lower Tail of a Fatigue Life Distribution," ASM Metals/Material Technology Series 8515-002 from ASM's Materials Week '85, Amer. Soc. for Metals, Metals Park, OH 44073. Based on General Electric Corporate Research & Development TIS Report 85CRD004.

Hamada, M. (1988), "The Costs of Using Incomplete Response Data for the Exponential Regression Model," Tech. Report STAT-88-05, Dept. of Statist. and Acturial Sci., Univ. of Waterloo, Waterloo, Ontario N2L 3G1.

Harrell, Jr., F. E. (1987), "A Survey of Microcomputer Survival Analysis Software," from author, Div. of Biometry, Duke Univ. Medical Center, Box 3363, Durham, NC 27710.

Harris, T. A. (1984), *Rolling Bearing Analysis*, 2nd ed., Wiley Interscience, New York.

Harter, H. L. (1977), "A Survey of the Literature on the Size Effect on Material Strength," Report No. AFFDL-TR-77-11, Air Force Flight Dynamics Lab. AFSC, Wright-Patterson AFB, OH 45433.

Hasselblad, V., and Stead, A. G. (1982), "DISFIT: A Distribution Fitting System, 2. Continuous Distributions," Biometry Div., Health Effects Res. Lab., U.S. Environmental Protection Agency, Research Triangle Park, NC 27711.

Hatch, B. D., Endicott, H. S. et al. (1962), "Long Life Satellite Reliability Program," General Electric Spacecraft Dept. Docu. No. 62SD4299, Philadelphia, PA.

Hawkins, W. L. (1971), *Polymer Stabilization*, Wiley, New York.

Hawkins, W. L. (1984), *Polymer Degradation and Stabilization*, Springer-Verlag, Heidelberg.

Haynes, J., Simpson, J., Krueger, J., and Callahan, J. (1984), "Optimization of Experimental Designs for Two Cases in Elevated Temperature Stability Studies," *Drug Devel. and Indus. Pharmacy* **10**, 1505-1526.

Herzberg, A. M. and Cox, D. R. (1972), "Some Optimal Designs for Interpolation and Extrapolation," *Biometrika* **59**, 551-561.

Heymen, J. S. (1988), *Electronics Reliability and Measurement Technology*, Noyes Publs., Park Ridge, NJ, 128 pp.

Hitz, M., Hudec, M., and Müllner, W. (1985), "PROSA: a Software Package for the Analysis of Censored Survival Data," *Statistical Software Newsletter* **11**, No. 2, 43-54.

Hoadley, B. (1971), "Asymptotic Properties of Maximum Likelihood Estimators for the Independent Not Identically Distributed Case," *Ann. Math. Statist.* **42**, 1977-1991.

Hochberg, Y., and Tamhane, A. C. (1987), *Multiple Comparison Procedures*, Wiley, New York.

Hoel, P. G. (1958), "Efficiency Problems in Polynomial Estimation," *Ann. Math. Statist.* **29**, 1134-1145.

Hoel, P. G., and Levine, A. (1964), "Optimal Spacing and Weighting in Polynomial Prediction," *Ann. Math. Statist.* **35**, 1553-1560.

Holm, S., and de Mare´, J. (1988) "A Simple Model for Fatigue Life," *IEEE Trans. on Reliability* **R-37**, 314-322.

Howes, M. J., and Morgan, D.V., Eds. (1981), *Reliability and Degradation – Semiconductor Devices and Circuits*, The Wiley Series in Solid State Devices and Circuits Vol. 6, New York.

IDEA WORKS (1988), "Ex-Sample," ad for computer package in *Amstat News*, No. 150, p. 2, 1(800)537-4866.

IEC Publ. 64 (1974), "Tungsten Filament Lamps for General Service," International Electrotechnical Commission, 1 rue de Varembe, Geneva, Switzerland.

IEC Publ. 82 (1980), "Ballasts for Tubular Fluorescent Lamps," International Electrotechnical Commission, 1 rue de Varembe, Geneva, Switzerland.

IEEE Index (1988), *The 1988 Index to IEEE Publications*, IEEE Service Center, P.O. Box 1331, Piscataway, NJ 08855-1331, (201)981-1393 and -9535.

IEEE Standard 101 (1988), "Guide for the Statistical Analysis of Thermal Life Test Data," by H. Rosen (Chr.), W. Nelson, and others of the Statistics Tech. Comm. of the IEEE Dielectrics and Electrical Insulation Soc., IEEE Service Center, P.O. Box 1331, Piscataway, NJ 08854-1331, (201)981-0060.

IEEE Standard 117 (1974), "Standard Test Procedure for Evaluation of Systems of Insulating Materials for Random-Wound AC Machinery," IEEE Service Center, P.O. Box 1331, Piscataway, NJ 08855-1331, (201)981-1393 and -9535.

IEEE Standard 930 (1987), "IEEE Guide for the Statistical Analysis of Electrical Insulation Voltage Endurance Data," by G. C. Stone (Chr.), W. Nelson, and others of the Statistical Technical Comm. of the IEEE Dielectrics and Electrical Insulation Soc. Purchase from IEEE Service Center, P.O. Box 1331, Piscataway, NJ 08855-1331, (201)981-0060.

Intel Corp. (1988), *Quality and Reliability Handbook*, Intel Literature Sales, P.O. Box 58130, Santa Clara, CA 95052-8130, (800)548-4725.

Ireson, W. G., and Coombs, Jr., C. F., Eds. (1988), *Handbook of Reliability Engineering and Management*, McGraw-Hill, New York.

Iuculano, G. and Zanini, A. (1986), "Evaluation of Failure Models through Step-Stress Tests," *IEEE Trans. on Reliability* **R-35**, 409-413.

Jaros, F. and Zaludova, A. H. (1972), "The Estimation of Guaranteed Fatigue Life under Random Loading," *Statistica Neerlandica* **3**, 171-181.

Jensen, F., and Peterson, N. E. (1982), *Burn-in: An Engineering Approach to the Design and Analysis of Burn-in Procedures*, Wiley, New York.

Jensen, K. L. (1985), "ALTPLAN – Microcomputer Software for Developing and Evaluating Accelerated Life Test Plans," Dept. of Statistics, Iowa State Univ., Ames, Iowa 50011. Contact Prof. Wm. Meeker.

Johnson, L. G. (1964), *The Statistical Treatment of Fatigue Experiments*, Elsevier Publ., New York.

Johnson, N. L., and Kotz, S. (1970), *Distributions in Statistics: Continuous Univariate Distributions*, Vols. 1 and 2, Houghlin-Mifflin, Boston.

Johnston, D. R., LaForte, J. T., Podhorez, P. E., and Galpern, H. N. (1979), "Frequency Acceleration of Voltage Endurance," *IEEE Trans. on Electrical Insulation* **EI-14**, 121-126.

Kalbfleisch, J. D., and Prentice, R. L. (1980), *The Statistical Analysis of Failure Time Data*, Wiley, New York.

Karlin, S., and Studden, W. J. (1966), "Optimum Experimental Designs," *Ann. Math. Statist.* **37**, 783-815.

Kaufman, R. B., and Meador, J. R. (1968), "Dielectric Tests for EHV Transformers," *IEEE Trans. Power Apparatus and Systems*, PAS-87, No. 1, 1895-1896.

Kennedy, Jr., W. J. and Gentle, J. E. (1980), *Statistical Computing*, Marcel Dekker, New York.

Keuffel & Esser (1988), "Graphic Charting and Digital Plotter Media," Catalog 4. Also "Graph Sheets Selection Guide," Keuffel & Esser Co., 20 Whippany Rd., Morristown, NJ 07960, (800)538-3355.

Khan, M., Fatemi, H., Romero, J., and Delenia, J. (1988), "Effect of High Thermal Stability Mold Material on the Gold-Aluminum Bond Reliability in Epoxy Encapsulated VLSI Devices," *Proceedings of the 1988 International Reliability Physics Symposium*, 40-49.

Kielpinski, T. J., and Nelson, W. (1975), "Optimum Censored Accelerated Life Tests for Normal and Lognormal Life Distributions," *IEEE Trans. on Reliability* R-24, 310-320.

King, J.R. (1971), *Probability Charts for Decision Making*, Industrial Press, New York.

Klein, J. P., and Basu, A. P. (1981), "Weibull Accelerated Life Tests When There are Competing Causes of Failure," *Communications in Statistical Methods and Theory* A10, 2073-2100.

Klein, J. P., and Basu, A. P. (1982), "Accelerated Life Testing under Competing Exponential Failure Distributions," *IAPQR Trans.* 7, 1-20.

Kleinbaum, D. G. (1973), "A Generalization of the Growth Curve Model Which Allows Missing Data," *J. of Multivariate Anal.* 3, 117-124.

Kraemer, H. C., and Thiemann, S. (1987), *HOW MANY SUBJECTS? Statistical Power Analysis in Research*, Sage Publications, Newbury Park, CA (805)499-0721.

Krause, B. (1974), "Modern Techniques for Resistor Reliability Testing," *Electronic Components* 16, 28-29.

Kulldorff, G. (1961), *Estimation from Grouped and Partially Grouped Samples*, Wiley, New York.

Kulshreshtha, H. K. (1976), "Use of Kinetic Methods in Storage Stability Studies on Drugs and Pharmaceuticals," *Defence Science Journal* 26, (No. 4), 189-204, India.

Labuza, T. P. (1982), *Shelf-Life Dating of Food*, Food & Nutrition Press, Westport, CT 06880.

Lachenbruch, P. A. (1985), "SURVCALC User's Manual," Wiley Professional Software, 605 Third Ave., New York, NY 10158, (212)850-6788.

Lancaster, T. (1990), *The Economic Analysis of Transition Data*, in preparation.

Lawless, J. F. (1976), "Confidence Interval Estimation in the Inverse Power Law Model," *Applied Statistics* 25, 128-138.

Lawless, J. F. (1982), *Statistical Models and Methods for Lifetime Data*, Wiley, New York.

Lee, E. T. (1980), *Statistical Methods for Survival Data Analysis*, Lifetime Learning (Wadsworth), Belmont, CA.

Lefkowitz, J. M. (1985), *Introduction to Statistical Computer Packages*, an outstanding primer on the basics of SAS, SPSS*, Minitab, and BMDP, 159 pp., Duxbury Press/PWS Publications, Statler Office Bldg., 20 Park Plaza, Boston, MA 02116.

Lehmann, E. L. (1986), *Testing Statistical Hypotheses*, Wiley, New York.

Letraset (1986), *Graphic Arts Reference Manual*, 40 Eisenhower Dr., Paramus, NJ 07653, $3.95, (800)526-9073.

Levitanus, A. D. (1973), "Accelerated Tests for Tractors and Their Components," in Russian, *Mashinostrojenie*, Moscow, 208 pp.

Lieblein, J., and Zelen, M. (1956), "Statistical Investigation of the Fatigue Life of Deep-Groove Ball Bearings," *J. of Research of the Nat'l. Bur. of Standards* 57, 273-316.

Linden, D. (1984), *Handbook of Batteries and Fuel Cells*, McGraw-Hill, New York, 1088 pp.

Little, R. E. (1972), *Manual on Statistical Planning and Analysis of Fatigue Experiments*, American Society for Testing and Materials Special Technical Publication 588, 1916 Race St., Philadelphia, PA 19103, (215)299-5400 or -5428.

Little, R. E. (1981), *Tables for Estimating Median Fatigue Limits*, American Society for Testing and Materials Special Technical Publication 731, 1916 Race St., Philadelphia, PA 19103, (215)299-5400 or -5428.

Little, R. E., and Jebe, E. H. (1969), "A Note on the Gain in Precision for Optimal Allocation in Regression As Applied to Extrapolation in S-N Fatigue Testing," *Technometrics* 11, 389-392.

Little, R. E., and Jebe, E. H. (1975), *Statistical Design of Fatigue Experiments*, Halstead Press (Wiley), New York.

Little, R. J. A., and Rubin, D. B. (1987), *Statistical Analysis with Missing Data*, Wiley, New York.

Lipson, C., and Sheth, N. C. (1973), *Statistical Design and Analysis of Engineering Experiments*, McGraw-Hill, New York.

Losickij, O. G. and Chernishov, A. S. (1970), "Bibliography on Problems of Accelerated Testing," in Russian, *Nadezhnost i Kontrol Kachestva 6* and 7, (Supplement to Soviet Journal *Standarty i Kachestva - Standards and Quality* - publication of the Soviet State Committee for Standardization) Moscow, USSR.

Lu, C.J., and Meeker, Jr., W. Q. (1989), "Using Degradation Measures to Assess Reliability," Dept. of Statistics, Iowa State Univ., Ames, Iowa 50011. Also, presented at the 1989 Joint Statistical Meeting, Washington, DC.

LuValle, M. J. and Welsher, T. L. (1988a), "An Example of Analyzing an Accelerated Life Test Using a Kinetic Model," AT&T Tech. Memorandum 52415-880929-O1TM, Rm 4C347, AT&T Bell Labs, Whippany, NJ 07981-0903, (201)386-2244.

LuValle, M. J., Welsher, T. L., and Mitchell, J. P. (1986), "A New Approach to the Extrapolation of Accelerated Life Test Data," *Proceedings of the 5th International Conference on Reliability and Maintainability*, 630-635.

LuValle, M. J., Welsher, T. L., and Svoboda, K. (1988b), "Acceleration Transforms and Statistical Kinetic Models," *J. of Statistical Physics* **52**, 311-330.

Mace, A. E. (1974), *Sample Size Determination*, Reinhold, New York.

McCallum, J., Thomas, R. E., Waite, J. H. (1973), "Accelerated Testing of Space Batteries," NASA report SP-323, National Technical Information Service, Springfield, VA 22151.

McCool, J. I. (1974), "Inferential Techniques for Weibull Populations," Aerospace Research Laboratories Report ARL TR 74-0180, National Technical Information Services Clearinghouse, Springfield, VA 22151, publication AD A 009 645.

McCool, J. I. (1978), "Competing Risk and Multiple Comparison Analysis for Bearing Fatigue Tests," *ASLE Trans.* **21**, 271-284.

McCool, J. I. (1980), "Confidence Limits for Weibull Regression with Censored Data," *IEEE Trans. on Reliability* **R-29**, 145-150.

McCool, J. I. (1981), *Life Test and Weibull Analysis*, unpublished course material, Penn State Great Valley, 30 E. Swedesford Rd., Malvern, PA 19355.

McCool, J. I. (1986), "Using Weibull Regression to Estimate the Load-Life Relationship for Rolling Bearings," *The Amer. Soc. of Lubr. Eng'rs. Trans.* **29**, 91-101.

McCoun, K. L., Davenport, J. M., and Kolarik, W. J. (1987), "Conditional Confidence Interval Estimation for the Arrhenius and Eyring Models," presented at the Joint Statistical Meetings, San Francisco. Authors at Dept. of Math., Texas Tech., Lubbock, TX 79409.

McLachlan, G. J., and Basford, K. E. (1987), *Mixture Models*, Marcel Dekker, New York.

Maindonald, J. H. (1984), *Statistical Computation*, Wiley, New York.

Mann, N. R. (1972), "Design of Over-Stress Life-Test Experiments When Failure Times Have a Two-Parameter Weibull Distribution," *Technometrics* **14**, 437-451.

Mann, N. R., Schafer, R. E., and Singpurwalla, N. D. (1974), *Methods for Statistical Analysis of Reliability and Life Data*, Wiley, New York.

Manson, S. S. (1953), "Behavior of Materials under Conditions of Thermal Stress," NACA-TN-2933 from NASA, Lewis Research Center, Cleveland, OH 44135.

Manson, S. S. (1966), *Thermal Stress and Low Cycle Fatigue*, McGraw-Hill, New York.

Mark, H., Ed. (1985), *Encyclopedia of Polymer Science and Engineering* 19 volumes, Wiley, New York.

Martz, H. F. and Waller, R. A. (1982), *Bayesian Reliability Analysis*, Wiley, New York.

Meeker, Jr., W. Q. (1980a), "Bibliography on Accelerated Testing," Statistics Dept., Iowa State Univ., Ames, IA 50011, (515)294-5336.

Meeker, Jr., W.Q. (1980b), "Large-Sample Accelerated Life Test Procedures for Comparing Two Products," private communication, Dept. of Statistics, Iowa State Univ., Ames, IA 50011.

Meeker, Jr., W. Q. (1984a), "A Comparison of Accelerated Life Test Plans for Weibull and Lognormal Distributions and Type I Censoring," *Technometrics* 26, 157-172.

Meeker, Jr., W. Q. (1984b), "GENMAX – A Computer Program for Maximum Likelihood Estimation," Dept of Statistics, Iowa State Univ., Ames, IA 50011.

Meeker, Jr., W. Q. (1984c), "A Review of the Statistical Aspects of Accelerated Life Testing," Proc. of the 1984 Statistical Symposium on National Energy Issues, Seattle, WA.

Meeker, Jr., W. Q. (1985), "Limited Failure Population Life Tests: Application to Integrated Circuit Reliability," private communication from author at the Dept. of Statistics, Iowa State Univ., Ames, IA 50011.

Meeker, Jr., W. Q. (1986), "Planning Life Tests in Which Units Are Inspected for Failure," *IEEE Trans. on Reliability* R-35, 571-578.

Meeker, Jr., W.Q. (1987), "Limited Failure Population Life Tests: Application to Integrated Circuit Reliability," *Technometrics* 29, 51-65.

Meeker, Jr., W. Q., and Duke, S. D. (1981), "CENSOR – A User-Oriented Program for Life Data Analysis," *The Amer. Statistician* 35, 112.

Meeker, Jr., W. Q., and Duke, S. D. (1982), "User's Manual for CENSOR – A User-Oriented Computer Program for Life Data Analysis," Statistical Laboratory, Iowa State Univ., Ames, IA 50011, (515)294-5336 or -1076. Price $5.00.

Meeker, Jr., W. Q., and Hahn, G. J. (1977), "Asymptotically Optimum Over-Stress Tests to Estimate the Survival Probability at a Condition with a Low Expected Failure Probability," *Technometrics* 19, 381-399.

Meeker, Jr., W. Q., and Hahn, G. J. (1978), "A Comparison of Accelerated Life Test Plans to Estimate the Survival Probability at a Design Stress," *Technometrics* 20, 245-247.

Meeker, Jr., W. Q., and Hahn, G. J. (1985), *How to Plan an Accelerated Life Test – Some Practical Guidelines*, Volume 10 of the ASQC Basic References in Quality Control: Statistical Techniques. Available from the Amer. Soc. for Quality Control, 310 W. Wisconsin Ave., Milwaukee, WI 53203, (800)952-6587.

Meeker, Jr., W. Q., and Nelson, W. (1975), "Optimum Accelerated Life Tests for Weibull and Extreme Value Distributions and Censored Data," *IEEE Trans. on Reliability* R-24, 321-332.

Meeter, C. A., and Meeker, W. Q. (1989), "Optimum Accelerated Life Tests with Nonconstant σ," Dept. of Statistics, Iowa State Univ., Ames, Iowa 50011.

Menon, M. V. (1963), "Estimation of the Shape and Scale Parameters of the Weibull Distribution," *Technometrics* 5, 175-182.

Menzefricke, U., (1988), "On Sample Size Determination for Accelerated Life Tests under a Normal Model with Type II Censoring," at Faculty of Mgt., Univ. of Toronto, 246 Bloor St. West, Toronto, Ont. M5S 1V4.

Metals and Ceramics Information Center (1984), "User's Guide and Materials Information Publication List," P.O. Box 8128, Columbus, OH 43201-9988.

MIL-HDBK-217E (27 Oct. 1986), "Reliability Prediction of Electronic Equipment," available from Naval Publications and Forms Center, 5801 Tabor Ave., Philadelphia, PA 19120, (215)697-3321.

MIL-STD-883 (29 Nov. 1985), "Test Methods and Procedures for Microelectronics," available from Naval Publications and Forms Center, 5801 Tabor Ave., Philadelphia, PA 19120, (215)697-3321.

Miller, R. (1966,1981), *Simultaneous Statistical Inference*, McGraw-Hill, New York.

Miller, R. (1981), *Survival Analysis,* Wiley, New York.

Miller, Robert, and Nelson, Wayne (1983), "Optimum Simple Step-Stress Plans for Accelerated Life Testing," *IEEE Trans. on Reliability* R-32, 59-65.

Miller, Jr., R. G., Efron, B., Brown, Jr., B. W., and Moses, L.E. (1980), *Biostatistics Casebook*, Wiley, New York.

Millet, M. A. (1975), "Precision of Rate-Process Method for Predicting Life Expectancy," *Proc. 1975 Symp. on Adhesives for Products from Wood*, 113-141, Forest Products Lab., USDA, 1 Pinchot Dr., Madison, WI.

Miner, M. A. (1945), "Cumulative Damage in Fatigue," *J. of Applied Mechanics* 12, A159-A164.

Moeschberger, M. L. (1974), "Life Tests under Dependent Competing Causes of Failure," *Technometrics* **16**, 39-47.

Montanari, G. C., and Cacciari, M. (1984), "Application of the Weibull Probability Function to Life Prediction of Insulating Materials Subjected to Combined Thermal-Electrical Stress," private communication, Instituto di Elettrotecnica, Industriale, Univ. of Bologna, Italy.

Morgan, C. B. (1982), "Analysis of Censored Data from an Extreme Value Distribution Using an Iterative Least Squares Technique," Ph.D thesis, Inst. of Administration and Management, Union College, Schenectady, NY 12308.

Morrison, F. R., McCool, J. I., Yonushonis, T. M., and Weinberg, P. (1984), "The Load-Life Relationship for M50 Steel Bearings with Silicon Nitride Ceramic Balls," *J. Amer. Soc. of Lubrication Engineers* **40**, 153-159.

Morton, M., Ed. (1987), *Rubber Technology*, Van Nostrand Reinhold, New York.

Murthy, V. K., and Swartz, G. B. (1972), "Annotated Bibliography on Cumulative Fatigue Damage and Structural Reliability Models," Aerospace Research Laboratories Report ARL 72-0161, sold by National Technical Information Services Clearinghouse, Springfield, VA 22151.

Murthy, V. K., and Swartz, G. B. (1973), "Cumulative Fatigue Damage Theory and Models," Aerospace Research Laboratories Report ARL 73-0170, available from National Technical Info. Services Clearinghouse, Springfield, VA 22151.

Nachlas, J. A. (1986), "A General Model for Age Acceleration During Thermal Cycling," *Quality and Reliability Engineering Internat'l* **2**, 3-6.

Nadas, A. (1969), "A Graphical Procedure for Estimating All Parameters of a Life Distribution in the Presence of Two Dependent Death Mechanisms, Each Having a Lognormally Distributed Killing Time," private communication from the author at IBM Corp., East Fishkill Facility, Hopewell Junction, NY.

NAG (1984), "GLIM-3 Users' Manual," Numerical Algorithms Group, Inc., The GLIM Coordinator, NAG Central Office, Mayfield House, 256 Banbury Rd., Oxford OX2 7DE, England.

Nelson, W. B. (1970), "Statistical Methods for Accelerated Life Test Data – The Inverse Power Law Model," General Electric Co. Corp. Research & Development TIS Report 71-C-001. Graphical methods appear in Nelson (1972a); least-squares methods appear in Nelson (1975a).

Nelson, W. B. (1971), "Analysis of Residuals from Censored Data – with Applications to Life and Accelerated Test Data," General Electric Co. Corp. Research & Development TIS Report 71-C-120. Part published in *Technometrics* **15**, (Nov. 1973) 697-715.

Nelson, Wayne (1972a), "Graphical Analysis of Accelerated Life Test Data with the Inverse Power Law," *IEEE Trans. on Reliability* **R-21**, 2-11; correction (Aug. 1972), 195.

Nelson, Wayne (1972b), "Theory and Application of Hazard Plotting for Censored Failure Data," *Technometrics* **14**, 945-966.

Nelson, Wayne (1972c), "A Short Life Test for Comparing a Sample with Previous Accelerated Test Results," *Technometrics* **14**, 175-185.

Nelson, W. B. (1973a), "Graphical Analysis of Accelerated Life Test Data with Different Failure Modes," General Electric Co. Corp. Research & Development TIS Report 73-CRD-001.

Nelson, Wayne (1973b), "Analysis of Residuals from Censored Data," *Technometrics* **15**, 697-715.

Nelson, W. B. (1974), "Analysis of Accelerated Life Test Data with a Mix of Failure Modes by Maximum Likelihood," General Electric Co. Corp. Research & Development TIS Report 74-CRD-160.

Nelson, Wayne (1975a). "Analysis of Accelerated Life Test Data – Least Squares Methods for the Inverse Power Law Model," *IEEE Trans. on Reliability* **R-24**, 103-107.

Nelson, Wayne (1975b), "Graphical Analysis of Accelerated Life Test Data with a Mix of Failure Modes," *IEEE Trans. on Reliability* **R-24**, 230-237.

Nelson, Wayne (1977), "Optimum Demonstration Tests with Grouped Inspection Data," *IEEE Trans. on Reliability* **R-26**, 226-231.

Nelson, W. B. (1979), "Analysis of Life Data as a Function of Other Variabiles When Each Unit Is Inspected Once," General Electric Co. Corp. Research & Development TIS Report 79CRD216.

Nelson, Wayne (1980), "Accelerated Life Testing – Step-Stress Model and Data Analyses," *IEEE Trans. on Reliability* **R-29**, 103-108.

Nelson, Wayne (1981), "Analysis of Performance Degradation Data from Accelerated Tests," *IEEE Trans. on Reliability* **R-30**, 149-155.

Nelson, Wayne (1982), *Applied Life Data Analysis*, Wiley, New York, (877)762-2974.

Nelson, W. B. (1983a), "Prediction of Fatigue Life that Would Result If Defects Are Eliminated," General Electric Co. Corp. Research & Development TIS Report 83CRD187.

Nelson, Wayne (1983b), "Monte Carlo Evaluation of Accelerated Life Test Plans," presented at the Joint Statistical Meeting, Toronto.

Nelson, Wayne (1983c), *How to Analyze Reliability Data*, Volume 6 of the ASQC Basic References in Quality Control: Statistical Techniques, Order Entry Dept., Amer. Soc. for Quality Control, 611 E. Wisconsin Ave., Milwaukee, WI 53201, (800)248-1946.

Nelson, Wayne (1984), "Fitting of Fatigue Curves with Nonconstant Standard Deviation to Data with Runouts," *J. of Testing and Evaluation* **12**, 69-77.

Nelson, Wayne (1985), "Weibull Analysis of Reliability Data with Few or No Failures," *J. of Quality Technology* **17**, 140-146.

Nelson, Wayne (1988), "Graphical Analysis of System Repair Data," *J. of Quality Technology* **20**, 24-35.

Nelson, Wayne (1990), *How to Plan and Analyze Accelerated Tests*, ASQC Basic References in Quality Control: Statistical Techniques, Order Entry Dept., Amer. Soc. for Quality Control, 611 E. Wisconsin Ave., Milwaukee, WI 53201, (800)248-1946.

Nelson, Wayne, and Hahn, G. J. (1972), "Linear Estimation of a Regression Relationship from Censored Data – Part I. Simple Methods and Their Application," *Technometrics* **14**, 247-269.

Nelson, Wayne, and Hahn, G. J. (1973), "Linear Estimation of a Regression Relationship from Censored Data – Part II. Best Linear Unbiased Estimation and Theory," *Technometrics* **15**, 133-150.

Nelson, W. B., and Hendrickson, R. (1972), "1972 User Manual for STATPAC – A General Purpose Program for Data Analysis and for Fitting Statistical Models to Data," General Electric Co. Corp. Research & Development TIS Reports 72GEN009, 73GEN012, and 77GEN032.

Nelson, W. B., and Kielpinski, T.J. (1972), "Optimum Accelerated Life Tests for Normal and Lognormal Life Distributions," General Electric Co. Corp. Research & Development TIS Report 72CRD215. Also, published as Kielpinski and Nelson (1975).

Nelson, Wayne, and Kielpinski, T. J. (1976), "Theory for Optimum Censored Accelerated Tests for Normal and Lognormal Life Distributions," *Technometrics* **18**, 105-114.

Nelson, Wayne, and Meeker, W. Q. (1978), "Theory for Optimum Censored Accelerated Life Tests for Weibull and Extreme Value Distributions," *Technometrics* **20**, 171-177.

Nelson, W. B., Morgan, C. B., and Caporal, P. (1983), "1983 STATPAC Simplified – A Short Introduction to How to Run STATPAC, a General Statistical Package for Data Analysis," General Electric Co. Corp. Research & Development TIS Report 83CRD146.

Neter, J., Wasserman, W., and Kutner, M. H. (1983), *Applied Linear Regression Models*, Richard D. Irwin Co., Homewood, IL.

Neter, J., Wasserman, W., and Kutner, M. H. (1985), *Applied Linear Statistical Models*, 2nd ed., Richard D. Irwin Co., Homewood, IL. Price $43.95.

Nishimura, A., Tatemichi, A., Miura, H., and Sakamoto, T. (1987), "Life Estimation for IC Plastic Packages under Temperature Cycling Based on Fracture Mechanics," *IEEE Trans. on Compo., Hybrids, and Mfg. Tech.* **CHMT-12**, 637-642.

O'Connor, P. D. T. (1985), *Practical Reliability Engineering*, 2nd ed., Wiley, New York.

Odeh, R. E. and Fox, M. (1975), *Sample Size Choice*, Marcel Dekker, New York.

O'Rear, M. R., and Leeper, J. D. (1983), "Analysis of Incomplete Growth/Wear Curve Data," presented at the 1983 Joint Statistical Meetings, Toronto. Authors at College of Community Health Sciences, Univ. of Alabama, P. O. Box 6291, University, Alabama 35486.

Ostrouchov, G., and Meeker, Jr., W. Q. (1988), "Accuracy of Approximate Confidence Bounds Computed from Interval Censored Weibull and Lognormal Data," *J. of Statist. Computation and Simulation* **29**, 43-76.

Owen, D. B. (1968), "A Survey of Properties and Applications of the Noncentral t-Distribution," *Technometrics* 10, 445-478.

Palmgren, A. (1924), "Die Lebensdauer von Kugellagern," *Z. Verein. Deutschland Ingeniur* 58, 339-341. In German.

Paris, P. C. and Erdogan, F. (1963), "A Critical Analysis of Crack Propagation," *Trans. ASME, Series D,* 85, 528-534.

Peck, D. S. (1971), "The Analysis of Data from Accelerated Stress Tests," *9th Annual Proceedings – Reliability Physics 1971,* IEEE Catalog No. 71-C-9-Phy, 69-78.

Peck, D. S. (1986), "Comprehensive Model for Humidity Testing Correlation," *Proc. International Reliability Physics Symp.* 24, 44-50.

Peck, D. S., and Trapp, O. D. (1978), *Accelerated Testing Handbook,* Technology Assoc's., 51 Hillbrook Dr., Portola Valley, CA 94025, (415)941-8272. Revised 1987.

Peck, D. S. and Zierdt, Jr., C. H. (1974), "The Reliability of Semiconductor Devices in the Bell System," *Proceedings of the IEEE* 62, 185-211.

Peduzzi, P. N., Holford, T. R., and Hardy, R. J. (1980), "A Stepwise Variable Selection Procedure for Survival Models," *Biometrics* 36, 511-516.

Peiper, V. and Thum, H. (1979), "Zuverlässigkeitsuntersuchungen unter Anwendung der Mathematischen Modellierung des Verschliessprozesses," *Schmierungstechnik* 10, 134-136.

Peshes, L. J. and Stepanova, M. D. (1972), *Models for Accelerated Testing,* in Russian, Nauka i Technika, Minsk, USSR, 165 p.

Peterson, M. and Winer, W., Eds. (1980), *Wear Control Handbook,* ASME Order Dept., 22 Law Dr., P.O. Box 2300, Fairfield, NJ 07007, (201)882-1167, Cat. No. G00169, $85.

Peto, R. (1973), "Experimental Survival Curves for Interval-Censored Data," *Applied Statistics* 22, 86-91.

Potthoff, R. F. and Roy, S.N. (1964), "A Generalized Multivariate Analysis of a Variance Model Useful Especially for Growth Curve Problems," *Biometrika* 51, 313-326.

Preston, D. L., and Clarkson, D. B. (1980), "A User's Guide to SURVREG: Survival Analysis with Regression," contact Dr. Douglas B. Clarkson, IMSL Inc., 7500 Bellaire Blvd., Houston, TX 77036, (800)222-IMSL.

Preston, D. L., and Clarkson, D. B. (1983), "SURVREG: A Program for Interactive Analysis of Survival Regression Models," *The Amer. Statistician* 37, 174.

Proschan, F. (1963), "Theoretical Explanation of Observed Decreasing Failure Rate," *Technometrics* 5, 375-383.

Proschan, F., and Singpurwalla, N. D. (1979), "Accelerating Life Testing – A Pragmatic Bayesian Approach," in *Optimization in Statistics,* J. S. Rustagi, ed., Academic Press, New York.

Proschan, F., and Sullo, P. (1976), "Estimating the Parameters of a Multivariate Exponential Distribution," *J. Amer. Statist. Assoc.* 71, 465-472.

Prot, E. M. (1948), "Fatigue Testing Under Progressive Loading; A New Technique for Testing Materials," *Revue de Metallurgie* XLV, No. 12, 481-489 (in French). Translation in WADC TR 52-148 (Sept. 1952).

Quality Progress (1988), "Equipment Overview: Environmental Test Chambers," *Quality Progress* (Aug 1988) 84-87.

Rabinowicz, E. (1988), *Friction and Wear of Metals,* 2nd ed., Wiley, New York.

Rabinowitz, E., McEntire, R. H., and Shiralkar, B. (1970), "A Technique for Accelerated Life Testing," *Trans. ASME J. of Engineering for Industry,* 706-710.

Rao, B. L. S. P. (1987), *Asymptotic Theory of Statistical Inference,* Wiley, New York.

Rao, C. R. (1973), *Linear Statistical Inference and Its Applications,* 2nd ed., Wiley, New York.

Ratkowsky, D. A. (1983), *Nonlinear Regression Modeling, A Unified Practical Approach,* Dekker, New York, 288 pp.

Reynolds, F. H. (1977), "Accelerated-Test Procedures for Semiconductor Components, (Invited Review)," *15th Annual Proceedings: Reliability Physics 1977,* Las Vegas, 168-178. IEEE Catalog No. 80CH1531-3, IEEE Service Center, P.O. Box 1331, Piscataway, NJ 08854-1331, (201)981-0060.

Ripley, B.D. (1987), *Stochastic Simulation,* Wiley, New York.

Rivers, B. H., Gillespie, R. H., and Baker, A. J. (1981), "Accelerated Aging of Phenolic-Bonded Hardboards and Fiberboards," Forest Products Lab., USDA, Research Paper FPL 400. 1 Pinchot Dr., Madison, WI.

Robertson, T., Wright, F. T., and Dykstra, R. L. (1988), *Statistical Inference under Inequality Constraints*, Wiley, New York.

Robinson, J.A. (1983), "Bootstrap Confidence Intervals in Location-Scale Models with Progressive Censoring," *Technometrics* 25, 179-188.

Rosenberg, L. S., Pelland, D. W., Black, G. D., Aunet, C. K., Hostetler, C. K., and Wagenknecht, D. M. (1986) "Nonisothermal Methods for Stability Prediction," *J. of Parenteral Science and Technology* 40, 164-168.

Ross, G. J. S. (1990), *Nonlinear Estimation*, Springer-Verlag, New York, (800)777-4643.

Rychtera, M. (1985), *Atmospheric Deterioration of Technological Materials, a Technoclimatic Atlas, Part A: Africa*, Academia (Prague) and co-published by Elsevier Science Publ. Co. (Amsterdam), 225 pp.

SAE Handbook AE-4 (1968), *Fatigue Design Handbook*, Soc. of Automotive Engineers, 400 Commonwealth Dr., Warrendale, PA 15096.

SAS Institute, Inc. (1985), *SAS User's Guide: Statistics*, Box 8000, Cary, NC 27511-8000, (919)467-8000.

Saunders, S.C. (1970), "A Review of Miner's Rule and Subsequent Generalizations for Calculating Expected Fatigue Life," Boeing Aircraft Co. Document D1-82-1019.

Saunders, S.C. (1974), "The Theory Relating the Wöhler Equation to Cumulative Damage in the Distribution of Fatigue Life," Aerospace Research Laboratories Report ARL 74-0016. Wright-Patterson AFB, OH 45433.

Schatzoff, M. (1985), "Regression Analysis in GRAFSTAT," available from Publ's Dept., T.J. Watson Research Center, IBM, Yorktown Heights, NY 10596.

Schatzoff, M., and Lane, T. P. (1986), "Reliability Analysis in GRAFSTAT," IBM Research Report RC 11655, IBM Distribution, 73F04 Stormytown Rd., Ossining, NY 10562, (914)241-4273.

Schatzoff, M., and Lane, T. P. (1987), "A General Step Stress Model for Accelerated Life Testing," private communication from Dr. Schatzoff, IBM Cambridge Scientific Center, 101 Main St., Cambridge, MA 02142. Also, presented at the Joint Statistical Meeting, Aug. 1987, San Francisco.

Schmee, J., and Hahn, G. J. (1979), "A Simple Method for Regression Analysis with Censored Data," *Technometrics* 21, 417-432.

Schmee, J., and Hahn, G. J. (1981), "A Computer Program for Simple Regression with Censored Data," *J. of Quality Technology* 13, 264-269.

Schneider, H. (1986), *Truncated and Censored Samples for Normal Populations*, Marcel Dekker, New York.

Schneider, H. and Weissfeld, L. (1987), "Interval Estimation for Accelerated Life Tests Based on the Lognormal Model," authors at Louisiana State Univ., Baton Rouge, LA 70803. Also in *Technometrics* 21 (Jan. 1989) 24-31.

Seber, G. A. F. and Wild, C. J., (1989), *Nonlinear Regression*, Wiley, New York.

Serensen, S. V., Garf, M. E., and Kuz´menko, V. A. (1967), *Dinamika Mashin Dlia Ispytanii Na Ustalost´*, in Russian, Mashinostaroenie, Moscow, 459 pp.

Shaked, M., and Singpurwalla, N. D. (1983), "Inference for Step-Stress Accelerated Life Tests," *J. of Statistical Planning and Inference* 7, 295-306.

Shaked, M., Zimmer, W. J., and Ball, C. A. (1979), "A Nonparametric Approach to Accelerated Life Testing," *J. of the Amer. Statistical Assoc.* 79, 694-699.

Shatzkes, M. and Lloyd, J. R. (1986), "A Model for Conductor Failure Considering Diffusion Concurrently with Electromigration Resulting in Current Exponent of 2," *J. Applied Physics* 59, 3890-3893.

Shorack, G. R. (1982), "Bootstrapping Robust Regression," *Communications in Statistics, Part A − Theory and Methods* 11, 961-972.

Sidik, S. M. (1979), "Maximum Likelihood Estimation for Life Distributions with Competing Failure Modes," NASA Technical Memorandum TM-79126, National Aeronautics and Space Administration, Lewis Research Center, Cleveland, OH 44135.

Sidik, S. M., Leibecki, H. F., and Bozek, J. M. (1980), "Cycles to Failure of Silver-Zinc Cells with Competing Failure Modes – Preliminary Data Analysis," NASA Technical Memorandum 81556, Lewis Research Center, Cleveland, OH 44135. Also presented at the 1980 Joint Statistical Meetings, Houston.

Sillars, R. W. (1973), *Electrical Insulating Materials and Their Application*, Peter Peregrinus Ltd., Southgate House, Stevenage, Herts. SG1 1HQ, England, published for the Institution of Electrical Engineers.

Silvey, S. D. (1980), *Optimal Design*, Chapman and Hall, New York.

Simoni, L. (1974), *Voltage Endurance of Electrical Insulation*, Tecnoprint, Bologna.

Simoni, L. (1983), *Fundamentals of Endurance of Electrical Insulating Materials*, CLUEB, Bologna.

Singpurwalla, N. D. (1971), "Inference from Accelerated Life Tests When Observations Are Obtained from Censored Samples," *Technometrics* 13, 161-170.

Singpurwalla, N. D. (1975), "Annotated Bibliography on Some Physical Models in Accelerated Life Testing and Models for Fatigue Failure," George Washington Univ. Technical Memorandum TM-64901. Also, published as Aerospace Research Laboratories Report ARL 75-0158, Wright-Patterson Air Force Base, OH 45433.

Singpurwalla, N. D., and Al-Khayyal, F. A. (1977), "Accelerated Life Tests Using the Power Law Model for the Weibull Distribution," *The Theory and Applications of Reliability with Emphasis on Bayesian and Nonparametric Methods*, C. P. Tsokos and I. N. Shimi, Eds., Academic Press.

Skelton, R. P., Ed. (1982), *Fatigue at High Temperature*, Elsevier Science Publ. Co., New York.

SKF (1981), *General Catalogue*, Number 3200 E, 1100 First Ave., King of Prussia, PA 19406-1352, (215)265-1900.

SPSS, Inc. (1986), *SPPS^x User's Guide*, 2nd ed., 444 N. Michigan Ave., Suite 3300, Chicago, IL 60611, (312)329-2400.

Starr, W. T., and Endicott, H. S. (1961), "Progressive Stress – A New Accelerated Approach to Voltage Endurance," *Trans. of the AIEE* 80, Part 3, 515-522.

Steinberg, D. and Colla, P. (1988), "SURVIVAL: A Supplementary Module for SYSTAT," SYSTAT, Inc. 1800 Sherman Ave., Evanston, IL 60201, (312)864-5670.

Stigler, S. M., (1971), "Optimal Experimental Design for Polynomial Regression," *J. Amer. Statist. Assoc.* 66, 311-320.

Strauss, S. H. (1980), "STATPAC: A General Purpose Package for Data Analysis and Fitting Statistical Models to Data," *The Amer. Statistician* 34, 59-60.

Taguchi, G. (1987), *System of Experimental Design*, Amer. Supplier Inst., Six Parklane Blvd., Suite 411, Dearborn, MI 48126, (313)271-4200.

TEAM (1988), "1988 Catalog and Price List," Technical and Engineering Aids for Management, Box 25, Tamworth, NH 03886, (603)323-8843.

Thisted, R. A. (1987), *Elements of Statistical Computing – Numerical Computation*, Chapman and Hall, New York.

Thomas, D. R. and Grunkemeier, G. L. (1975), "Confidence Interval Estimation of Survival Probabilities for Censored Data," *J. Amer. Statist. Assoc.* 70, 865-871.

Timm, N. H. (1980), "Multivariate Analysis of Variance of Repeated Measurements," in *Handbook of Statistics, Volume I: Analysis of Variance*, P. R. Krishnaiah, Ed., 41-87, North-Holland, New York.

Titterington, D. M., Smith, A. F. M., and Makov, U. E. (1986), *Statistical Analysis of Finite Mixture Distributions*, Wiley, New York.

Tobias, P. A., and Trindade, D. (1986), *Applied Reliability*, Van Nostrand Reinhold Co., New York.

Tomsky, J. (1982), "Regression Models for Detecting Reliability Degradation," *Proc. 1982 Reliability and Maintainability Symp.*, 238-245.

Tufte, E. R. (1983), *The Visual Display of Quantitative Information*, Graphics Press, Box 430, Cheshire, CT 06410.

Turnbull, B. W. (1976), "The Empirical Distribution Function with Arbitrarily Grouped, Censored, and Truncated Data," *J. Royal Statist. Soc. B* 38, 290-295.

Tustin, W. (1986), "Recipe for Reliability: Shake and Bake," *IEEE Spectrum* **23**, no. 12, 37-42.

Uhlig, H. H., and Revie, R. W. (1985), *Corrosion and Corrosion Control*, Wiley-Interscience, 458 pp. Price $56.55.

Underwriters Laboratories, Inc. (1975), "Polymeric Materials – Long Term Property Evaluations," Standard for Safety UL 746B, 1285 Walt Whitman Rd., Melville, NY 11747.

Vander Wiel, S. A. and Meeker, W. Q. (1988), "Accuracy of Approximate Confidence Bounds Using Censored Weibull Regression Data from Accelerated Life Tests," authors at Dept. of Statistics, Iowa State Univ., Ames, Iowa 50011. To appear in *IEEE Trans on Reliability*.

Vaupel, J. W., and Yashin, A. I. (1985), "Heterogeneity's Ruses: Some Surprising Effects of Selection on Population Dynamics," *The Amer. Statistician* **39**, 176-185.

Veluzat, P. and Goddet, T. (1987), "New Trends in Rigid Insulation of Turbine Generators," *IEEE Electrical Insulation Magazine* **3**, 24-26.

Viertl, R. (1987), "Bayesian Inference in Accelerated Life Testing," invited paper, 46th Session of the ISI, Tokyo. From the author, Technische Univ. Wien, A-1040 Wien, Austria.

Viertl, R. (1988), *Statistical Methods in Accelerated Life Testing*, Vandenhoeck & Ruprecht, Göttingen.

Vincent, G. A. (1987) "A Guide to Testing Liquid Dielectrics in Simple Combination with Solid Dielectrics," a guide to tests and standards, *IEEE Electrical Insulation* **3**, 10-20.

Vlkova, M. and Rychtera, M. (1978), "Control of Thermo-oxidative Aging of High-Voltage Insulations," *Elektrotechn. Obzor* **67**, 225-229.

Wagner, A. E., and Meeker, Jr., W. Q. (1985), "A Survey of Statistical Software for Life Data Analysis," private communication from Prof. Meeker, Dept. of Statistics, Snedecor Hall, Iowa State Univ., Ames, IA 50011.

Weibull, W. (1961), *Fatigue Testing and Analysis of Results*, Pergamon Press, New York.

Weisberg, S. (1985), *Applied Linear Regression*, 2nd ed., Wiley, New York.

Whitman, L. C., and Doigan, P. (1954), "Calculation of Life Characteristics of Insulation," *AIEE Transactions* **73**, 193-198.

Wilks, S. S. (1962), *Mathematical Statistics*, Wiley, New York.

Winspear, G., Ed. (1968), *Vanderbilt Rubber Handbook*, 230 Park Ave., New York, NY.

Yokobori, T. and Ichikawa, M. (1974), "Non-linear Cumulative Damage Law for Time-Dependent Fracture Based on Stochastic Approach," *Strength and Fracture of Materials* **10**, 1-14.

Yoshioka, S., Aso, Y., and Uchiyama, M. (1987), "Statistical Evaluation of Nonisothermal Prediction of Drug Stability, *J. of Pharm. Sci.* **76**, 794-798.

Young, W. R. (1988), "Accelerated Temperature Pharmaceutical Product Stability Determinations," private communication, author at Dept. 916, Lederle Labs, Pearl River, NY 10965. To appear in *Drug Development and Industrial Pharmacy*.

Yum, B.-J. and Choi, S. C. (1987), "Optimal Design of Accelerated Life Tests under Periodic Inspection," at Dept. of Indus. Engineering, Korea Advanced Inst. of Science and Technology, P.O. Box 150, Chongryang, Seoul, Korea.

Yurkowski, W., Schafer, R. E., and Finkelstein, J. M. (1967), "Accelerated Testing Technology," Rome Air Development Center Tech. Rep. RADC-TR-67-420, Griffiss AFB, NY.

Zalud, F. H. (1971), "Accelerated Testing in Automobile Development," *Proc. EOQC Seminar on Quality Control in the Automobile Industry*, Torino.

Zaludova, A. H. (1981), "Designing for Reliability Using Failure Mechanism Model," *Proc. EOQC Conference*, Paris, 17-26.

Zaludova, A. H., and Zalud, F. H. (1985), "New Developments in Accelerated Testing," *Proc. EOQC Conference*, Estoril, 10-24.

Zelen, M. (1959), "Factorial Experiments in Life Testing," *Technometrics* **1**, 269-288.

Zhurkov, S. N. (1965), "Kinetic Concept of Strength of Solids," *Internat'l J. of Fracture Mechanics* **I**, 311-.

Index

Many phrases are listed according to the main noun, rather than the first word, for example: Distribution, normal; Data, multiply censored; Comparison, binomial. For distribution properties, consult: Distribution, name. Many data analysis methods are under the name of the method, for example: Least squares and maximum likelihood.

A priori distribution, 241
  non-informative, 241
A-ratio, 21
Absolute temperature, 76, 78
Accelerated testing, 3*ff*
Accelerated tests, 37*ff*
Accelerating stresses, 6, 7, 29
Acceleration, by censoring, 17
  high usage rate, 15
  types of, 15*ff*
Acceleration factor, 77, 151, 251, 253, 283
  Arrhenius, 77, 152
  assumptions, 42
  definition, 40
  estimated, 41, 253
  generalized Eyring, 111
  inverse power, 88
  known, 41
  multiple, 154
  separate factors, 254
  traditional value, 253
  uncertainty, 154
  value, 253
Acceptance sampling, 23
Accuracy, computational, 240
  ML estimate, 266
  numerical, 240
Actions, 427
Activation energy, 76, 122, 525
  assumed value, 152
Addition law, exponential failure rates, 381
  for failure rates, 379
  Weibull failure rates, 381
Adhesives, 7, 102, 524
Allocation, 36
  equal, 321, 373

optimum, 321
  unequal, 321, 326, 350
Alternative, hypothesis, 427, 473
  subspace, 473
Analyses, variety of, 114
Analysis, least squares, 167*ff*
  maximum likelihood, 233*ff*
  sensitivity, 252
Analysis of variance, one-way, 440
Analytic methods, 113, 168
Area effect, 231, 386, 507
Arrhenius, acceleration factor, 77, 152
  activation energy assumed, 154
  degradation, 524
  degradation data analysis, 534
  degradation rate, 528
  life relationship, 76
  paper, 78, 120, 142
  papers, 122
  rate law, 76
  relationship, 75
  relationship plot, 120, 138, 142
Arrhenius-exponential model, 83*ff*
  failure rate, 84
  fraction failed, 84
  percentiles, 85
Arrhenius-lognormal model, assess, 138, 144
  assumptions, 80
  data checks, 189
  fraction failed, 80
  graphical analysis, 114
  graphical estimates, 138, 143
  ML fit, 243
  model, 79, 114, 124, 285*ff*, 393
  percentiles, 81
  plots, 114

Arrhenius-Weibull model, assumptions, 82
  design temperature, 83
  fraction failed, 83
  model, 82
  percentile, 83
Artificial censoring, 135
Assess, Arrhenius-lognormal, 124, 144
  constant lognormal σ, 125, 184, 257
  constant scale parameter, 257, 268
  constant standard deviation, 125, 184, 257
  constant Weibull shape β, 204, 257
  data, 124, 223
  data with failure modes, 405, 411
  degradation data, 540
  degradation model, 540
  distribution, 117, 211, 261, 268, 412
  distribution of degradation, 542
  independence of failure modes, 406, 412
  independent observations, 514
  linearity, 183, 204, 260, 412
  lognormal distribution, 125, 185
  model, 124
  model for failure modes, 405, 411
  power-Weibull model, 134
  relationship, 223, 280, 512
  right censored data, 255
  scale parameter is constant, 280
  simple model, 255
  varying stress data, 510
  varying stress model, 510
  Weibull distribution, 205
Assumed coefficient value, 250, 255
Assumptions, degradation, 523
  model, see each model
ASTM E739-80, 168
ASTM STP 313, 168
ASTM STP 731, 267
ASTM STP 738, 522
Asymptotic theory, ML, 235, 291
Average, grand, 174
  life, 55
  sample, 174
Bartlett's test, 184, 436, 490
  generalized, 258
Bartlett's test statistic, 184, 436
Bathtub, curve, 70
  hazard function, 70
Batteries, 10, 75, 282, 407, 524
Bayesian, analysis, 241
  estimate, 241
Bearing capacity, 86
Bearings, 11, 85, 86, 89, 383, 494
  ceramic, 419
  failure, 420
Behrens-Fisher problem, 489
Bernstein distribution, 531

Beta β parameter, 63
Bias, 169
  ML estimate, 297
  sampling, 26
Bias correction, ML estimate, 297
Bibliographies, accelerated testing, 5
Bipolar memories, 381
Birnbaum-Saunders distribution, 70, 533
Birth and death process, 529
Black's formula, 101
Blunders, 183
BMD, 169
BMDP, 169, 240
Boltzmann's constant, 76, 123, 175
Bonds, 10
Bonferroni inequality, 259, 297, 437, 441, 455, 464, 467
Bootstrapping, 236, 438
Boundary points, parameter space, 471
Breakdown, 522
  insulating fluid, 387
Breakdown stress distribution, 548
Building materials, 8, 533
Burn-in, 23, 43, 70

Cable, 383, 385
Calculations, ML fit, 284
  numerical, 48
Capacitors, 9, 282, 385, 494
Cause and effect, 33, 209
Cause of failure, 378, see Failure mode
CCD memories, 381
Cells, 10, 75, 282, 407, 524
Cement, 8
CENSOR, 237, 302
Censored data, 1, 13, 233
  ML comparisons, 451ff
  ML fit, 265ff
  multiply, 15, 139ff, 233ff
  on the left, 13
  on the right, 13, 134ff, 233ff
  singly, 13, 134ff, 233ff
Censoring, 17, 234, 285
  artificial, 135, 234, 273
  choice of, 234
  experimental design, 351
  failure, 235
  failure modes, 408
  informative, 235
  noninformative, 235
  random, 235
  Type II, 235
  value of, 234
Centered variable, 172
Centering, 215
Central limit theorems, 291

Ceramics, 7, 87, 533
Checks, *see* Assess
    data, 189, 209
    linear-lognormal model, 182
    linear-Weibull model, 203
Chemical acceleration, 6
Chemical kinetics, 529, 548
Chemical reaction, 75, 548
Circuit boards, 9
Class-F insulation, 386
Coatings, 6, 8
Cochran's test, 185
CODEX, 120, 132, 133, 188
Coefficient, 99
    graphical estimate, 123, 537
    interaction, 461
    intercept, 103
    LS estimate, 172, 175, 192
    ML estimate, 243*ff*
    relationship, 286
    slope specified, 250
    specified, 216, 255, 282
    value assumed, 216, 255, 282
Coefficients, random, 530
    uncorrelated, 530, 532
Coffin-Manson relationship, 86, 97, 251
Common parameter values, 460
Comparable estimates, 426
Comparison, 154, 425*ff*, 451*ff*, *see* Hypothesis
        test
    coefficients, 278
    complete data, 425*ff*
    confidence limits, 256, 257, 426
    designs, 425
    difference of two ML estimates, 458
    estimate with a specified value, 255
    graphical, 114, 429*ff*, 451
    hypothesis test, 426
    independent samples, 460
    intercepts, 278, 431, 443
    K coefficients, 447
    K estimates for equality, 466
    K means, 439
    K multivariable means, 446
    K multivariable relationships, 447
    K samples, 465*ff*
    K slope coefficients, 442
    K standard deviations, 436
    least-squares, 425*ff*
    lognormal and Weibull, 263
    lognormal σ's, 257
    LR test for a specified value, 454
    LR test for scale parameters, 258
    materials, 425
    means, 437*ff*
    means pairwise, 441

ML, 451*ff*
ML censored data, 451*ff*
ML confidence interval, 452
ML estimate with a specified value, 452
ML estimates for equality, 456
ML interval data, 451*ff*
ML properties, 451
ML quantal-response data, 451*ff*
ML two-sample, 458*ff*
models, 273, 426
multivariable means, 446
multivariable relationships, 217, 445*ff*
one coefficient, 446
one mean, 438
one multivariable mean, 446
one slope coefficient, 441
one standard deviation, 435
one-sample, 452
pairwise, 437
pairwise means, 441
percentiles, 430
probability plots, 431
production periods, 425
Q ML estimates with specified values,
        454
Q sets of K coefficients, 469
ratio of two ML estimates, 460
reasons for, 426
relationships, 431
robust for means, 438
σ's, 278
scale parameters, 431
simple relationships, 441
simultaneous Q ML estimates, 455
simultaneous Q pairs of coefficients, 462*ff*
slope coefficients, 431, 441
slopes and intercepts simultaneously,
        443*ff*
spreads, 431
standard deviations, 434, 445*ff*
step-stress, 505
step-stress models, 505
suppliers, 425
test plans, 324*ff*
Tukey's for means, 441
two coefficients, 447
two means, 438
two ML estimates, 458
two multivariable means, 446
two slope coefficients, 441
two standard deviations, 436
varying stress data, 513
Weibull and lognormal, 211, 263
Weibull β's, 257, 259
Weibull shape to β_0, 208
Comparisons, 42, 43, 47

582   INDEX

Competing failure modes, 15, 377
Competing risks, 378
Complete data, 13
    avoid, 168
    graphical analysis, 114, 128
    LS analyses, 167*ff*
Components of variance, 107
Compressed time test, 15
Computation, statistical, 240
Computer, least squares output, 172
Computer plot, 113
Computer programs, 151, 168
    least squares, 169
    ML, 237*ff*
    user written, 170
Concrete, 8
Condition, single test, 40
Conditions, test, 27
    uncontrolled, 511
    use, 27
Confidence, 168
    demonstration test, 257
Confidence interval, 46, 47, 48, *see*
        Confidence limits
    advantage of, 113
Confidence limits, binomial, 149
    compare a standard deviation, 435
    comparisons with, 256, 426
    consistent with a specified value, 428
    difference of coefficients, 441
    difference of two means, 438
    difference of two ML estimates, 458
    exponential distribution, 248
    hypothesis test, 428
    improved ML, 296
    inspection data, 266
    least squares, 173,178*ff*
    linear-Weibull model, 197*ff*
    LR, 236, 236, 297*ff*
    LS design stress, 182, 199
    LS fraction failing, 181
    LS intercept coefficient, 180, 200
    LS linear-exponential, 202
    LS ln mean, 197
    LS log mean, 178
    LS percentile, 179, 198
    LS shape parameter, 200
    LS slope coefficient, 180, 201
    LS standard deviation, 180
    ML, 235, 245
    ML improved, 296
    ML normal approximation, 236, 245, 295
    ML positive, 296
    ML simultaneous, 297
    ML too short, 236
    ML transformed, 296

nonparametric, 149
one-sided, 296, 297
Poisson approximation to binomial, 149
power-exponential, 236
power-Weibull, 236
simultaneous pairwise, 467
unbounded, 295
Confidence region, joint, 300
    LR simultaneous, 463
    rectangular, 300
    simultaneous rectangular, 464
Connections, electronics, 10
Considerations, engineering, 22*ff*
    statistical, 43*ff*
Consistent hypothesis test, 485
Constant degradation rate, 540
Constant hypothesis, 475
Constrained maximum log likelihood, 457
Constraint hypothesis, 475
Constraints, 36
    model, 487
    test plan, 363
    null hypothesis, 476
Contacts, electrical, 10
Contour plot, 101
Convergence, in distribution, 291
    in law, 291
    iteration, 245, 290
Convincing, 214
    evidence, 429
Correlated failure modes, 382, 412
Correlated lives, positively, 387
Correlation, 514
    engineering, 33
    matrix, 243
    near ±1, 269
    statistical, 33
Corrosion, 6, 12, 522
Covariance, LS coefficient estimates, 172
    ML estimates, 243
    ML estimates of functions, 370
Covariance matrix, failure modes, 415
    local estimate of, 293
    ML estimate of, 369
    of ML estimates, 243, 293
    true theoretical, 369
Covariates, 33, 99
Cox model, 71, 104, 242
    reliability function, 105
Crack initiation, 6, 310, 522
Crack propagation, 6, 522
Cracking, 12
Creep, 6, 12, 522
Creep-rupture, 6, 383
Critical value, equivalent, 480
    test statistic, 480, 481

Cross terms, 276, 99
Crossplot, failure modes, 406
    residuals, 209, 218, 265, 276, 505, 543
    trend, 209
Cumulative damage, cycling effect, 498
    deterministic, 495
    model, 494*ff*
    theory, 494*ff*
Cumulative distribution function, 53
Cumulative distribution plot, 74, 79, 116*ff*
Cumulative exposure, 507, *see* Cumulative
        damage
    basic model, 498, 507
    definition, 501
    model, 497
    model inadequacy, 498
    residual, 512
Cumulative hazard, definition, 57
    value, 140
Cutting tools, 87
Cutting velocity, 87
Cycle length, 397
Cycling damage, 498
Cycling rate, 397

Damage theory, 494*ff*
Data, *see* Example
    all-or-nothing response, 15
    assess degradation, 540
    autocorrelated, 536
    binary, 15
    case, 285
    censored, 1, 13, 233*ff*, 451*ff*
    checks, 189, 209, 264
    cloud, 213, 222
    collection, 46
    complete, 1, 13, 167
    equivalent, 251
    erroneous, 209
    failure modes, 392*ff*
    field, 44
    grouped, 15, 145*ff*
    inspection, 15, 145*ff*
    interval, 15, 312, 371, 451*ff*
    interval analysis, 145
    laboratory, 44
    left censored, 267
    matrix, 284
    missing, 222
    multiply censored, 139, 395
    peculiar, 127
    performance, 13, 521*ff*
    plots, 46, 113*ff*
    probit, 15
    quantal-response, 15, 95, 104, 310, 371,
        451*ff*

read-out, 15, 145*ff*, 518
right censored, 242
sensitivity, 15
singly censored graphical analysis, 134*ff*
step-stress, 495*ff*
suspect, 189, 209
transformed, 283
types of, 12
valid, 127, 44
varying stress, 506
Data analysis, 45
    exploratory, 46
    graphical, 113*ff*
    least squares, 165*ff*, 425*ff*
    maximum likelihood, 233*ff*, 407*ff*, 451*ff*
    methods, 47
Data banks, 5
Data handbooks, 5
Data scale, 116
Dead on arrival, 68
Defectives, 43
Defects, interior, 389
    surface, 389
Degradation, 13, 17, 75, 97, 211
    accelerated, 521*ff*
    adhesive, 521
    applications, 521
    Arrhenius, 524*ff*
    Arrhenius data analysis, 534*ff*
    Arrhenius rate, 528
    assumptions, 523
    bibliographies, 522
    competing modes, 529
    constant rate, 527
    dielectrics, 522
    drugs, 522
    exponential rate, 528
    Eyring rate, 529
    failure, 528
    food, 522
    graphical analysis, 537
    initial, 525
    insulation, 522
    Larsen-Miller generalizations, 533
    Larsen-Miller polynomial rate function,
        533
    Larsen-Miller rate function, 533
    life distribution, 526, 539
    linear, 525, 527
    literature, 544
    mathematical rate function, 533
    mechanisms, 11
    median failure time, 526
    metals, 522
    microelectronics, 522
    ML fit, 538

model assumptions, 523
models, 529
models needed, 523
non-constant stress, 524
percentiles, 525
performance, 521*ff*
pharmaceuticals, 522
plastics, 522
polymers, 522
power function rate, 528
processes, 522
random coefficients model, 530
random increments model, 531
rate, 528
rate depends on stress, 541
rate is constant, 540
rate plot, 541
residuals, 542
semicondutors, 522
temperature-accelerated, 524
test advantages, 521
test plans, 534
wear-in, 527
wear-out, 527
Weibull rate function, 533
Degradation test, advantages, 521*ff*
Degrees of freedom, number of, 174
Demonstrate reliability, 23, 428
Demonstration test, 255, 426, 428, 452, 482, 483
    confidence of, 257
    fail, 257
    for a mean, 438
    GaAs FET, 375
    pass, 257
Density, probability, 55, 285
Dependent, component lives, 497
    failure modes, 406
    failures, 381
    observations, 536
Derating curves, 84, 98, 210
Derivations, 3
Design, experimental, 45
Design life, 152
Design of experiments, 361
Design principles, 32
Design stress, 89
    confidence limits LS, 182, 199
    estimate, 134
    graphical estimate, 123
    LS estimate, 177
    ML estimate, 247
Design temperature, 81, 83, 525
    Arrhenius-lognormal, 81
    Arrhenius-Weibull, 83
Dielectrics, 7, 75, 85, 89, 385, 522, 524

Diesel engines, 532, 533
Difference of coefficients, confidence limits, 441
Difference of means, confidence limits, 438
Difference of ML estimates, confidence limits, 458
Diffusion, 530
Dissipation factor, 98
Distribution, a priori, 241
    all failure modes act, 399*ff*
    assess, 268
    Bernstein, 531
    Birnbaum-Saunders, 70, 533, 545
    bivariate lognormal, 382
    breakdown stress, 548
    χ^2 percentiles table, 554
    curved, 410
    degradation, 531
    estimate for degradation, 539
    exponential, 53 *ff*
    extreme value, 65*ff*, 206
    F percentiles table, 556
    failure modes eliminated, 301*ff*, 402
    fatigue-limit, 94
    function, 53
    gamma, 70
    generalized gamma, 70, 263, *see* log-gamma
    largest extreme value, 206
    line slope, 122
    lines, 117
    log gamma, 278, 412, *see* generalized gamma
    lognormal, 115
    lognormal versus Weibull, 263
    mixture, 69, 379
    multivariate, 382
    multivariate exponential, 382
    multivariate lognormal, 382
    multivariate normal, 291
    multivariate Weibull, 382
    non-central t, 181
    normal, 58*ff*
    normal cumulative distribution table, 550
    normal percentiles table, 552
    normal two-sided factors table, 552
    others, 250
    parameter, 286
    plot, 73
    posterior, 241
    power lognormal, 422
    single, 251
    size effect, 386
    skewed, 305
    smallest extreme value, 65*ff*, 206
    statistical, 73, 285

strength, 94, 269
system of identical parts, 383
t percentiles table, 553
test of fit, 264
under step-stress, 498
valid, 121
volume effect, 387
Weibull, 63*ff*
Weibull versus lognormal, 263
with eternal survivors, 69
with failure at time zero, 68
Distributions, cross, 107
    parallel lines, 74
Drugs, 8, 522, *see* Pharmaceuticals

Effect of area, 386
Effect of size, 385*ff*
Effect of volume, 387
Elastic term, 97
Elastic-plastic relationship, 97
Elastics, 7
Electrical insulation, *see* Insulation
Electromigration, 101
Electron-volts, 76
Electronic components, 73
Electronics, 10, 101, 494
Elephant test, 493*ff*
Eliminate failure modes, 398, 401*ff*
Empirical model, 211
Encapsulants, 9, *see* Example, Au-Al bonds
Endurance limit, 93, 269*ff*
    LR test, 270
    ML fit, 269
    sharp, 269
Engineering considerations, 22
Environmental stress screening, 39, 43, 70
Epoxy, 101
Equal allocation, 321
Equality hypothesis, 466, 475
Equivalent hypothesis test, 482
Equivalent life, 251
Equivalent size, 391
Equivalent start time, 500
Equivalent time, 251, 423
Error, mean squared, 172, 240
    random, 172, 191, 523
    root mean squared, 172
    standard, 172, 293
Errors, independent, 172
ESS, *see* Environmental stress screening
Estimate, accuracy by simulation, 353*ff*
    activation energy, 123
    Bayesian, 241
    coefficient, 175, 537
    confidence interval, 47, *see* Confidence
        limits

degradation life distribution, 539
distribution line, 122
distribution with all failure modes, 399
failure mode distribution, 398
failure mode model, 398
failure modes eliminated, 411
graphical, 118*ff*
Greenwood's, 151
Kaplan-Meier, 146, 151
least squares, 172*ff*
LS design stress, 177
LS fraction failing, 177
LS mean (log) life, 176
LS percentile, 177
life at a stress, 121
linear, 240
linear unbiased, 240
linearly pooled, 466
minimum variance, 240
ML, 245, 290*ff*, 366
ML all failure modes act, 410
ML bias correction, 297
ML existence, 291
ML exponential mean, 249
ML failure modes eliminated, 411
ML for a function, 417
ML invariance, 236
ML of function, 294
ML standard error, 235
ML uniqueness, 291
ML variance, 294
percentage failing, 142
percentile, 141
percentile for degradation, 539
Peto, 151
robust, 121
standard deviation, 124, 538
standard error of, 245
Turnbull, 151
Weibull shape parameter, 130
Estimates, dependent, 461
    independent, 460
    ML for failure modes, 414
Estimation, methods, 240
Estimator, ML best asymptotically normal
        (BAN), 235
    ML sampling distribution, 235
    ML standard error, 235
Eternal survivors, 530
Euler's constant, 66, 191
Example, $1,000,000 discovery, 144, 261, 398,
        158, 314, 489
    $1,000,000 experiment failure modes, 421
    Au-Al bonds, 232, 448
    Au-Al bonds comparison, 489
    battery cells, 99

bearing ball failures, 419
bearings, 157, 230, 305, 384
bolthole cracks, 381
cable, 312
cable insulation, 503ff, 513
capacitor, 63, 100, 302, 387
capacitor test plan, 375
Class-B failure modes, 418
Class-B insulation, 108, 109, 135, 157, 158, 242, 257ff, 266, 285ff, 330
Class-B test plan, 332, 337, 343, 372
Class-H failure modes, 379, 382, 393ff, 403, 408ff, 417
Class-H insulation, 32, 61, 77, 80ff, 115, 121, 123, 126, 144, 156, 171ff, 230, 231
Class-H test plan, 321, 324, 325, 327
CMOS RAM, 164
coffee maker, 25
compressor, 31, 252
cookware, 38
corrosion, 162
cryogenic cable, 387, 495ff, 503ff
degradation of insulation, 534ff, 538
electric cord, 28
electrical machine insulation, 388
encapsulant, 163
engine fan, 54
equality of Poisson λ_k, 486
exponential comparison, 471ff
fatigue limit, 313
fatigue limits comparison, 489
GaAs FET demonstration, 491
GaAs FET test plan, 375
Ground insulation, 395ff
heater, 155, 230
heater failure modes, 420
heater test plan, 319, 323
insulating fluid, 104, 107, 128, 159, 300, 304, 387, see Example, oil
insulating fluid test plan, 372
insulating oil, 190ff, 506ff
insulating oil comparison, 452ff, 488
insulating tape, 32, 33, 98
insulation, 32, 59
insulation degradation, 534ff, 538
insulation endurance limit, 269
left censored data, 159
linear-exponential comparison, 473ff
linear-normal comparison, 473ff
lost specimens, 372
low cycle fatigue, 272
lubricant, 252, 310
lubricating oil, 252
machine insulation, 211ff
machine insulation test plan, 350ff

machine insulations comparison, 458ff
material strength, 66
microprocessor, 145, 149, 152, 164
motor insulation, 26, 28, 386
oil, 231, 315
oil breakdown, 513
oil comparison, 513
oil residuals, 384
other heater failure modes, 421
permalloy corrosion, 162, 231, 548
Phase insulation, 395ff
production shifts, 102
relay, 110, 160, 310
relays comparison, 489
satellite amplifier test plan, 376
snubber comparison, 488
steel fatigue, 312
superalloy fatigue, 110
tandem specimens, 417
tape insulation, 98
taping experiment, 99
three cable insulations, 516, 154, 229, 465ff
three motor insulations comparison, 430ff, 448
toaster, 28
toaster comparison, 488
transformer oil, 86, 90, 92
transformer oil comparison, 449, 488
transformer turn, 161, 310
transistor, 308
turbine disk, 310
Turn failure eliminated, 417, 423
Turn failures, 139, 309, 395ff
TV transformer, 38
two exponentials, 473ff
two insulations, 276, 281
wire varnish, 163
wire varnish test plan, 326, 371, 372
Existence, ML estimate, 291
Expectation, 55, 367
Expected life, 55
Experimental design, 32, 99, 536
    effect of censoring, 351
    principles, 45
Experimental principles, 32, 45
Experimental procedures, 210
Exponent, in power law, 86
Exponential distribution, 53ff, 107
    degradation rate, 528
    hazard function, 57
    identical parts, 384
    likelihood, 289
    literature, 248
    mean, 56
    misuse, 53, 248

ML fit, 248, 374
ML theory, 373
paper, 132
percentile, 55
probability density, 55
relationship to Weibull, 254
size effect, 386
standard deviation, 57
variance, 56
Exponential relationship, 96
Exponential-power relationship, 96
Exposure, 25
cumulative, 494*ff*
Extrapolation, 318
polynomial, 318
Extrapolation factor, 320
Extreme value distribution, 65*ff*
cumulative distribution, 66
density, 66
hazard function, 67
identical parts, 384
likelihood, 289
mean, 66
percentile, 66
probability papers, 206
relationship to Weibull, 67
reliability function, 66
standard deviation, 66
standard percentile, 66
Extreme value theory, 63
Eyring, degradation rate, 529
generalized relationship, 100
model, 109
relationship, 97, 230

F ratio, maximum, 437
F statistic, 183
for equality of *K* relationships, 444, 448
for equality of means, 440
for equality of *K* slope coefficients, 442
incremental, 217,
ratio of exponential means, 483
F test, fit of a multivariable relationship, 217
for linearity, 183
for linearity, 204
incremental, 217, 218
Failure, bearing, 420
catastrophic, 25
cause, 378
component, 378
customer-defined, 25
defined, 25
defined degradation, 526
definition, 25, 272
definition for degradation, 521
degradation, 526, 528, 539

distribution of degradation, 527
first mode, 392
mode, 378
on stress step, 498
single cause of, 114
subassembly, 378
temperature, 381
vibration, 381
Failure analysis, 5
Failure censored, 13
Failure censoring, 235
Failure mechanism, 377
Failure mode, eliminated, 398
estimation of, 382
extraneous, 377
graphical analysis of, 395*ff*
graphically estimate distribution, 398
likelihood equations, 415
partly eliminated, 405
point of view, 395
residuals, 412
separate analysis, 415
single, 75, 493
Failure modes, 30, 31, 40, 41, 43, 76, 183, 204, 251, 326, 377*ff*
censoring, 408
competing, 377*ff*, 413
correlated, 382, 412
covariance matrix, 415
crossplot, 406
dependent, 382, 406
distinct models, 415
Fisher matrix, 415
graphical analysis, 392*ff*
identified, 407
independent, 378
life-stress relationship, 399
literature, 392, 407
local covariance matrix, 416
log likelihood, 413
ML estimate of a function, 417
ML estimates, 414
ML fit, 407*ff*
ML theory, 413*ff*
models, 393
not identified, 407
positively correlated, 382
two, 423
unidentified, 423
Failure rate, 53, 110
bathtub, 70
decreasing, 65
design value, 257
increasing, 64
instantaneous, 57
ML estimate, 247

size effect, 386
specified, 257
Failure rates, addition law, 379
Failures, dependent, 381, 406
Failures are informative, 357
Faith, leap of, 28
Fatigue, 6, 11, 26, 73, 85, 89, 97, 110, 386, 388, 389, 392, 494
Fatigue limit, 92, 269, 312
Fatigue limit relationship, ML fit, 269
Final equation, 223
Finite-element analysis, 388
First failure, 393
Fisher information matrix, 243, 292, 367
    failure modes, 415
    local estimate, 292, 366
    true theoretical, 367
    Weibull, 369
FITs, 53
Fitting, LS, 167ff
    ML, 240ff
    nonparametric, 241
Foods, 8, 522
Force of mortality, 57
Fraction failing, Arrhenius-exponential, 84
    Arrhenius-lognormal, 80
    Arrhenius-Weibull, 83
    confidence limits LS, 181
    LS estimate, 177
    power-exponential, 92
    power-lognormal, 88
    power-Weibull, 90
Freaks, 43, 69
Frequency, stress cycle, 21
Function, ML estimate of, 294, 417

GaAs FET, 375, 383, 491
Gamma distribution, 70
Gaussian distribution, 58
Generalized Eyring relationship, 100
Generalized gamma distribution, 70, 263, see Distribution, log gamma
Generator insulation, 386
Geometric average, sample, 174
Geometric mean, sample, 174
Geometry, 18, 76, 386, 530
GLIM, 240
Global maximum log likelihood, 457
GRAFSTAT, 237, 265
Grand average, 174
Graphical analysis, 113ff, 168
    advantages, 113
    Arrhenius-lognormal, 143
    complete data, 113ff, 128ff
    degradation data, 537ff
    disadvantages, 113

failure modes, 392ff
interval data, 145ff
multiply censored data, 139
power-Weibull, 128ff
read-out data, 145ff
singly censored data, 134ff
Greenwood's variance, 151

Handbooks, acceleration factors, 254
Haphazard sampling, 44
Hazard calculations, 140, 395
Hazard function, 57, 105
    addition law, 379
    base, 105
    bathtub, 70
    size effect, 386
    system of identical parts, 383
Hazard papers, 141
Hazard plot, 140ff
    all failure modes, 400
    failure modes eliminated, 405
    interpretation, 141, 395
    lognormal, 142
    residuals, 262, 265
Hazard rate, 57, see Hazard function
Hazard scale, 141
Hazard value, 140
    cumulative, 140
    modified cumulative, 141
Heteroscedastic model, 105ff, 530, 532
Higher-order terms, ML theory, 236
Homogeneity hypothesis, 466
Homogeneity of variance, Bartlett's test for, 437
Humidity, 102, 162
Hypothesis, 426
    actions, 427
    alternative, 427, 466, 473
    constraint, 475
    equality, 427, 466, 475
    homogeneity, 466
    null, 427, 473
    one-sided, 427
    parameters have specified values, 475
    two-sided, 427
    ways to specify, 475
Hypothesis test, see LR test and Comparison, 47, 428, 482
    comparisons, 426
    confidence interval, 428
    consistent, 485
    homogeneity of σ, 309
    level of, 481
    locally most powerful ML, 235, 451
    ML, 235
    normality, 188

other parameters assumed equal, 467
outlier, 264
performance, 429
Rao's, 486
sample size, 429
statistical, 428
theory, 426
uniformly most powerful, 485
Wald's, 487

Identical parts, dependent, 385
    exponential life, 384
    extreme value life, 384
    other distributions, 385
    series system, 383*ff*
    system hazard function, 383
    system life distribution, 383
    Weibull life, 384
IEEE Std 101, 168
IEEE Std 117, 159, 397
Incremental F statistic, 217
Incremental F test, 217, 218
Independence, 536
    assess failure modes, 406
Independent, errors, 172
    failure modes, 412
    observations, 471, 536
    statistically, 288, 536
Infant mortality, 57
Information, needed, 45
    numerical, 45
Inspection, periodic, 34
Inspection data, 145
Insulating oil, 385
Insulation, 7, 73, 75, 79, 85, 89, 282, 385, 522,
    524
    thickness, 386
Integrated circuit, 9, 383
Interaction, 276
    coefficient, 461
    term, 100, 103, 461
Intercept coefficient, 103
    confidence limits LS, 180, 200
    LS estimate, 172
Interpretation of results, 47
Interval data, computer packages, 151
    confidence limits, 146, 149
    graphical analysis, 145
    literature, 266
    ML comparisons, 451*ff*
    ML estimate, 266
    ML fit, 265*ff*
Invariance property, ML estimate, 236
Inverse absolute temperature, 80
Inverse power, law, 85
    relationship, 85*ff*

relationship plot, 132
Iterative least squares, 241

Journals with accelerated testing, 4

K + E, 120, 122, 133, 188
Kaplan-Meier estimate, 146
Kelvin temperature, 76
Kinetic theory, 548

Lack of fit, 511
Lamps, 10, 75, 85
Larsen-Miller, parameter, 77
    relationship, 77
Least squares, advantages, 168
    analyses, 167*ff*
    compare intercepts, 443
    compare K coefficients, 447
    compare K multivariable means, 446
    compare means, 438*ff*
    compare multivariable relationships,
        445*ff*
    compare one coefficient, 446
    compare one mean, 438
    compare one multivariable mean, 446
    compare one standard deviation, 435
    compare percentiles, 438
    compare relationships, 441
    compare slope coefficients, 441
    compare standard deviations, 434, 445*ff*
    compare two coefficients, 447
    compare two multivariable means, 446
    comparisons, 425*ff*
    complete data, 167*ff*
    confidence limits, 173, 178*ff*
    disadvantages, 168
    iterative, 241
    linear-exponential model, 189*ff*,
    linear-lognormal model, 170*ff*
    linear-Weibull model, 189*ff*
    output, 172
    references, 210
    regression, 168*ff*
    stepwise fitting, 220*ff*
    weighted, 241
    why used, 425
Least squares fit, linear-exponential, 201*ff*
    linear-lognormal, 171*ff*
    lognormal life, 170
    multivariable relationship, 210*ff*
    Weibull life, 189*ff*
Left censored data, 159
    ML fit, 265*ff*
Level of a hypothesis test, 481
Life, 12, 121
    equivalent, 251

typical, 121
Life doubles, 110
Life line, 121
Life-stress relationship, 71*ff*, *see* Relation-
    ship
Likelihood, *see* Log likelihood
    basic, 289
    equations, 290
    exponential, 289
    extreme value, 289
    function, 287
    interval data, 287
    left censored data, 287
    log, 288
    lognormal, 289
    normal, 289
    observed data, 287
    right censored data, 287
    sample, 288, 476
    specimen, 287, 365
    Weibull, 289
Likelihood equations, 366, 478
    failure mode, 415
Likelihood ratio, *see* LR test, 470
    confidence limits, 236
    definition, 479
    test, 235, 475*ff*
LINDEP, 237
Linear damage theory, 494
Linear estimate, 240
Linear function, 99
Linear-exponential model, 190, 201, 315, 372
    confidence limits LS, 202
    LS fit, 189*ff*, 201
    ML fit, 249, 473*ff*
    step-stress, 501
Linear-lognormal model, 171
    checks, 182
    estimate of σ, 175
    LS coefficient estimates, 175
    LS confidence limits, 178*ff*
    LS estimate of a fraction failing, 177
    LS estimate of a percentile, 177
    LS estimate of design stress, 177
    LS estimate of mean (log) life, 176
    LS fit, 171*ff*
    ML fit, 249
    step-stress, 501
Linear-normal model, ML theory, 373
Linear-Weibull model, 190, 372
    checks, 203
    confidence intervals, 197*ff*
    data checks, 209
    LS coefficient estimates, 192
    LS estimate mean ln life, 194
    LS estimate of a percentile, 196

LS estimate of design stress, 196
LS estimate of fraction failing, 195
LS fit, 189*ff*
LS parameter estimates, 192
LS σ estimate, 193
ML fit, 249, 315
ML theory, 374
step-stress, 501
Linearity, assess, 183, 204, 268
    F test for, 183, 204
    lack of, 182, 203
    LR test for, 260
Literature, engineering, 5
    statistical, 4
Literature searches, 52
Loading, constant, 18
    cyclic, 20, 507
    progressive, 20
    ramp, 494, 507
    random, 22, 494
    stress, 18*ff*
Local information matrix, failure modes, 416
Location parameter, 66
Log likelihood, 243, 288
    constrained, 476
    constrained maximum, 455, 457, 470
    failure modes, 413
    global maximum, 457, 470
    maximum, 290
    maximum with equality, 467
    maximum without equality, 467
    observed sample, 365
    sample, 288
    theoretical sample, 365
    unconstrained maximum, 455, 457
Log mean, 60
Log standard deviation, 60
Log-log paper, 133
Logistic relationship, 103
Lognormal distribution, assess, 60*ff*, 125, 185
    base 10, 60, 108
    base e, 62, 80, 108
    cumulative distribution, 60
    hazard function, 62
    hazard plot, 140
    likelihood, 289
    mean, 61
    median, 60, 61
    ML fit, 249
    papers, 119
    percentile, 61
    plot of singly censored data, 136
    probability density, 61
    probability plot, 116
    relationship to normal, 62
    reliability function, 61

σ depends on stress, 105
  standard deviation, 61
  versus Weibull, 65, 211, 263
Lognormal versus Weibull, 65, 211, 263
LR limits, 297*ff*
  advantages, 297
  calculation, 299
  for a coefficient, 300
  for a function, 300
LR statistic, 470
  chi-square approximation, 470
  compare ML estimate with a specified
    value, 454
  compare Q sets of K coefficients, 469
  equality of J ML estimates, 457
  equality of K ML estimates, 467
  equality of Q ML estimates, 455
  equality of Q pairs of coefficients, 463
  equality of two ML estimates, 461
LR test, 470, 475*ff*, 480
  assess relationship, 280
  Behrens-Fisher, 490
  compare ML estimate with a specified
    value, 454
  compare Q sets of K coefficients, 469
  consistent, 485
  depicted, 481
  endurance limit, 270
  equal shape parameters, 510
  equality of K ML estimates, 468
  equality of ML estimates, 457
  equality of Q ML estimates, 456
  equality of Q pairs of coefficients, 463
  equality of two ML estimates, 462
  K models are equal, 484
  K parameters are equal, 484
  linearity, 260, 491
  statistic, 480
  theory, 470*ff*
Lubricants, 8, 75
Lubricating oil, 532

Markov property, 498, 524
Materials, 6*ff*
Matrix, correlation, 243
  covariance, 243, 293
  data, 284
  Fisher, 243, 292
  local Fisher, 292
Maximum, global, 291
  local, 291
Maximum F ratio, 185, 437
Maximum likelihood, all failure modes act,
    410
  Arrhenius-lognormal, 243
  assumptions, 234

asymptotic theory, 291
asymptotically optimum estimate, 235
calculations, 284*ff*
censored data, 265*ff*, 451*ff*
censoring, 234
compare K ML estimates, 466*ff*
compare Q pairs of coefficients, 462*ff*
compare Q sets of K coefficients, 469*ff*
comparisons, 451*ff*
computer programs, 235*ff*
confidence limits, 235, 236, 245, 295*ff*
constrained, 479
degradation data, 538
dependent estimates, 461
endurance limit, 316
estimate of a function, 247
estimates, 245, 290*ff*
estimate under the general model, 478
estimate under null hypothesis, 478
estimator, 235
exponential distribution, 248
failure modes, 407*ff*
failure modes eliminated, 411
fitting, 240*ff*
independent estimates, 460
interval data, 265*ff*, 371, 451*ff*
invariance property, 236
left censored data, 265*ff*
linear-exponential fit, 249
linear-lognormal fit, 249
lognormal distribution, 249
LR limits, 236
LR theory, 470*ff*
method, 233*ff*, 235, 284*ff*
multivariable relationship, 276*ff*
nonconstant σ fit, 273
nonlinear relationship, 261
normal confidence limits, 245
numerical methods, 240
one-sample comparisons, 452
other distributions, 250
output, 245
percentile estimate, 245
properties, 235
quantal-response data, 265*ff*, 451*ff*
reliability estimate, 246
right censored data, 242*ff*
σ depends on stress, 316
same model fitted to Q samples, 456
sampling distribution, 235, 245
simple model, 242*ff*
standard error, 245
statistical properties, 235
step-stress, 502*ff*
stepwise fitting, 281
theory, 364*ff*

theory for failure modes, 413*ff*
three simple steps, 284
two-sample comparisons, 458*ff*
unconstrained estimate, 479
variance of a function estimate, 370
varying stress, 509
varying stress data, 506
versatile, 235
Weibull distribution, 249
Maximum log likelihood, 290
  constrained, 455
  unconstrained, 455
Maximum ratio test, scale parameters, 259
McCool's test, 307
Mean, 55
  confidence limits LS, 178, 197
  geometric, 121, 122
  pooled sample, 439
Mean log life, LS estimate, 176
Mean squared error, 172, 240
Mean time to failure, 53, 56, 248
Means, LS compare K, 439
  LS compare two, 438
  LS comparison, 437*ff*
Measurement, destructive, 536
  error, 34, 545
  nondestructive, 536
  repeat, 534
  single, 534
  stress, 35
Mechanical components, 11
Median, 55
Metal diffusion, 75, 76
Metal fatigue, 97, 361, 494
Metals, 6, 21, 73, 85, 89, 389, 522, 533
Method, LR, 297*ff*, 454*ff*, 470*ff*
  LS, 170*ff*
  ML, 233*ff*
  sampling, 44
Methodology, statistical, 4
Methods, analytic, 113, 168
  estimation, 240
  graphical, 113*ff*, 168
  multivariate, 536
  numerical, 290
  optimization, 290
Metrology, 6
Microcircuits, 9
Microelectronics, 9, 101, 385, 388, 522, 529
MIL-HDBK-217, 72, 84, 91, 96, 98, 110, 248,
  254, 381, 383, 385, 387
MIL-STD-883, 38, 152, 282
Miner's rule, deterministic, 494
  probabilistic, 501
Minitab, 169
Missing data, 222

Mixture, 108
Model, 35, 45, 46
  Arrhenius rate, 536
  Arrhenius rate assumptions, 524
  Arrhenius-exponential, 83, 84
  Arrhenius-lognormal, 79
  Arrhenius-Weibull, 82
  assess degradation, 540
  basic cumulative exposure, 498, 507
  bias, 169
  check, 47
  choice of, 35
  coefficient, 286
  combined for independent data sets, 461
  components-of-variance, 107
  constant degradation rate, 527
  constrained, 470, 473
  constraints, 487
  Cox, 71, 104, 242, 282
  cumulative damage, 19, 494, 524
  cumulative exposure, 19, 494, 497
  cumulative exposure assumptions, 498
  definition, 51
  degradation, 13, 523
  distinct for each failure mode, 415
  empirical, 211
  endurance limit, 93, 269*ff*, 316, 489
  error, 169
  Eyring-lognormal, 98, 109
  failure modes, 393, 413
  fatal shock, 382
  for comparisons, 426
  general, 285, 364, 470
  general for hypothesis test, 471
  heteroscedastic, 530, 532
  Kleinbaum, 536
  linear-exponential, 190, 201, 315
  linear-lognormal, 171
  linear-normal, 365
  linear-Weibull, 190, 315
  needed, 35
  nonlinear, 536
  nonuniform stress, 390
  physical, 36
  Pothoff-Roy, 537
  power-exponential, 91
  power-lognormal, 88
  power-Weibull, 89, 102, 128
  proportional hazards, 104, 242
  Rabinowicz, 518
  random coefficients, 530
  random increments, 531
  series-system, 378*ff*, 395
  shock, 382
  simple, 242
  simple exponential, 315

size effect, 382, 386
standard, 52
statistical, 44
step-stress, 495, 496
uncertainties, 170
uniform-stress, 388*ff*
varying-stress, 493*ff*, 506
Models, compare K, 465
LR comparison, 273
Monte Carlo simulation, 240, 241, 352
MOS, 111, 529
MOS memories, 381
Motors, 11, 385, 494
MTTF, 53, 56, 248
Multiply censored data, 15
graphical analysis, 139*ff*
ML analysis, 233*ff*
Multivariable methods, 536
Multivariable relationship, 98, 210*ff*
LS fit, 210*ff*

Nominal life, 77
Non-central t distribution, 181
Non-constant stress, 493*ff*
degradation, 524
Nonconstant spread, 272
Nonlinear relationship, ML fit, 261
Nonlinearity, 204, 261
reasons for, 182, 203
Nonparametric, analysis, 71
fitting, 241
Nonuniform stress, 387*ff*
model, 390
model extensions
Normal distribution, 58*ff*
approximation for confidence limits, 150
cumulative distribution function, 58
deviate, 58
hazard function, 60
likelihood, 289
mean, 60
papers, 188
percentile, 59
percentiles table, 552
probability density, 59
probability papers, 188
standard cumulative distribution, 58, 80
standard deviation, 60
standardized deviate, 58
table of standard cumulative distribution, 550
two-sided factors table, 552
Notation, 52, 286
Nuclear reactor materials, 9
Null hypothesis, 427, 473
constraints, 476

subspace, 473
ways to specify, 475
Numbering of book, 2
Numerical methods, 290

OC function, 429, 485
One-way analysis of variance, 440
Operating Characteristic function, 429, 485
Optimization, convergence, 290
methods, 290
numerical, 290
unconstrained, 286
Optimum plan, allocation, 321, 337
low stress, 337
test plan, 320
σ depends on stress, 276
Order of a chemical reaction, 548
Organization of book, 2
Outlier, 125, 128, 189, 209, 220, 262, 264, 385, 451, 542
test, 264
Overstress testing, 16
Overview of book, 1
Oxidation, 6, 12, 522

p value, 428
Paints, 8
Palmgren's equation, 86, 251
Paper, Arrhenius, 78, 122
exponential, 132
extreme value, 206
hazard, 141
inverse power, 87
log-log, 87, 133
lognormal, 119
normal, 188
Weibull, 131
Parallel distribution lines, 80, 117
Parameter, of distribution, 285
of relationship, 286
slope, 83
transformed, 296
Parameter space, 471*ff*
of the alternative, 473
of the null hypothesis, 473
Parameterization, other, 286
Parameters, 80, 471
assumed equal, 467
common value, 460
compare for equality, 426
Peck's relationship, 102
Percent failed, graphical estimate, 118, 123
LS estimate, 177, 195
ML estimate, 246
Percentile, Arrhenius-Weibull, 83
confidence limits LS, 179, 198

definition, 54
degradation model, 525
graphical estimate, 118, 123, 133, 141
lines, 73, 81, 121, 246
lognormal, 61
LS estimate, 177
ML confidence limits, 245
ML estimate, 245, 539
sample, 121
standard normal, 59
Percentiles, χ^2 table, 554
compare graphically, 430
F table, 556
normal table, 552
t distribution table, 553
Performance, 11, 13, 24
degradation, 211, 521ff
initial, 525
percent remaining, 525
percentiles degradation, 525
Peto estimate, 151
Peto plot, 312
Pharmaceuticals, 8, 494, 526
stability, 524
Pharmacokinetics, 523
Planning, 36
ahead, 37
aids, 37
scientific, 22
Plans, test, 317ff, see Test plan
Plastic term, 97
Plastics, 6, 75, 522, 524
Plot, 46
Arrhenius, 78
computer, 113
contour, 101
data, 113, 114
degradation rate, 541, 545
degradation relationship, 537
hazard, 140ff
lognormal, 116
probability, 114ff
relationship, 114ff
Weibull, 129
Plotting paper, 72, see Paper
Plotting positions, 115, 116, 136, 141, 155
expected, 116
hazard, 141
median, 116
midpoint, 116
modified hazard, 141
table, 560
Poisson approximation to binomial confidence limits, 149
Polymers, 522, 524, 534
Polynomial, degradation rate, 533

Pooled estimate, sample mean, 439
standard deviation, 175
Population, 29, 44
infinite, 44
sampled, 29
sub, 69
target, 29, 44
Posterior distribution, 241
Power law, 85
Power parameter, 86
Power relationship, 85
Power-exponential model, 91
confidence limits, 236
design stress, 92
failure rate, 91
fraction failed, 92
LS fit, 201ff
ML fit, 242ff, 473ff
percentiles, 92
Power-function, degradation rate, 528
Power-lognormal model, 88
assumptions, 88
design stress, 89
LS fit, 170ff
ML fit, 242ff
percentiles, 89
Power-Weibull model, 89, 128
assess, 134
confidence limits, 236
design stress, 91
fraction failed, 90
graphical analysis, 128
graphical estimates, 133
LS fit, 189ff
ML fit, 242ff
percentiles, 91
plots, 128ff
Press type, 120
Probability density, definition, 55
general, 285
joint, 471
standard normal, 285
Probability paper, 74
Probability papers, extreme value, 206
lognormal, 119
normal, 188
Weibull, 131
Probability plot, comparisons, 431
curved, 125, 380
extreme value, 208
interpretation, 125
interval data, 146
linearity of, 127
lognormal, 116, 136
nonparallel lines, 126
nonparametric fit, 125

normal, 187
read-out data, 146*ff*
residuals, 186, 206, 434
Weibull, 129
Probability plots, parallel, 396
Probability plotting positions, 116
  table, 560
Probability scale, 116
Problem, $1,000,000 experiment, 158, 314,
    489
  $1,000,000 experiment failure modes, 421
  adhesive degradation, 545
  Arrhenius-Weibull comparison, 488
  Au-Al bonds, 232, 448
  Au-Al bonds comparison, 489
  Bartlett's test, 490
  bearing, 157, 230, 305
  bearing ball failures, 419
  Behrens-Fisher, 450, 489
  Birnbaum-Saunders distribution, 545
  cable, 312
  cable insulation, 513
  capacitor, 302
  capacitor test plan, 375
  Class-B failure modes, 418
  Class-B insulation, 157, 158, 310
  Class-B test plan, 372
  Class-H failure modes, 417
  Class-H insulation, 156, 230, 231
  CMOS RAM, 164
  compare two means, 450
  corrosion, 162
  degradation rate plot, 545
  degradation test plan, 547
  encapsulant, 163, 232, 448, 489
  Eyring relationship, 163
  fatigue limit, 313
  fatigue limits comparison, 489
  GaAs FET demonstration, 491
  GaAs FET test plan, 375
  hazard rate for nonuniform stress, 424
  heater, 155, 230
  heater failure modes, 420
  insulating fluid (*see* oil *and* insulating
    oil), 159, 304
  insulating fluid test plan, 372
  insulating oil comparison, 488
  insulation degradation, 545
  linear-exponential comparison, 489
  linear-lognormal comparison, 489
  lost specimens, 372
  lubricant, 310
  measurement error, 545
  microprocessor, 164
  oil, 231, 315
  oil breakdown, 513
  oil model assessed, 515
  oils compared, 513
  order of a reaction, 548
  other heater failure modes, 421
  permalloy corrosion, 162, 231, 548
  power-lognormal, 422
  power-lognormal comparison, 488
  random coefficients model, 545
  random increments model, 545
  random stress, 519
  real, 45
  relay, 160, 310
  relays comparison, 489
  sinusoidal stress, 519
  size compensation, 423
  specimen size, 422
  square-wave stress, 518
  statement of, 45
  statistical, 45
  steel fatigue, 312
  tandem specimens, 417
  test plan for degradation, 547
  three cable insulations, 516
  three insulations, 154, 229
  three insulations comparison, 448
  transformer oil comparison, 449, 488
  transformer turn, 161, 310
  transistor, 308
  turbine disk, 310
  Turn failure eliminated, 417, 423
  Turn failures, 309
  two failures, 423
  unidentified failure modes, 423
  Weibull limits exact, 489
  wire varnish, 163
  wire varnish test plan, 371, 372
Problem statement, 45
Procedures, experimental, 210
Product of likelihoods, 288
Product rule, for reliability, 378, 396
Product size, 388
Products, 9*ff*
Progressive stress, literature, 494, 495
Progressive test, 231, 449
Proportional hazards model, 104, 242
Prot test, 20, 494
Pseudo stress, 272
Purpose, engineering, 24
  statement of, 45
  statistical, 24
  test, 23

Quadratic statistic, 466
Quadratic terms, 99
Quality control, 23, 126
Quantal-response data, 15, 267

ML comparisons, 451*ff*
ML fit, 265*ff*

Rabinowicz model, 518
RAM Symposium, 9
Ramp stress, 494
Ramp voltage, 506
Random coefficients model, 530
Random error, 172, 191, 523
Random failures, 57
Random increments model, 531
Random variation, 172, 191
    degradation, 531
Random vibration, 22
Randomization, 34, 45
Range, of a distribution, 53
    of test stress, 21, 320, 329
Rao's hypothesis test, 486
Rate plot, 545
Reaction order, 548
Reaction rate, 75, 97, 100, 525
Read-out data, *see* Interval data
    confidence limits, 146, 149
    graphical analysis, 145
Reciprocal scale, 120
Reduced off time, 16
Redundancy, 379
References, 561-577
Regression, books, 168
    least squares, 168*ff*
    least squares theory, 172
    maximum likelihood, 233*ff*
    methods, 522
    nonlinear, 522
    weighted, 241
Regularity conditions, 291, 364, 414, 481, 485
Relationship, Arrhenius, 75, 76, 80, 115, 524,
    assess with LR test, 280
    Black's, 101
    cause-and-effect, 33
    coefficient, 286
    Coffin-Manson, 86, 97, 251
    curved, 403, 410
    distribution parameters, 286
    elastic-plastic, 97
    eliminated failure mode, 403
    endurance limit, 92
    endurance limit fit, 269
    exponential and Weibull, 254
    exponential function, 96
    exponential-power, 96
    Eyring, 97, 109, 163, 230
    failure modes, 403
    fatigue limit, 92
    fatigue limit fit, 269
    final, 227

for σ, 105, 273
general, 222, 286
generalized Eyring, 100
heteroscedastic, 105, 273
inverse power, 85
known parameters, 41, 251
Larsen-Miller, 77
Larsen-Miller rate function, 533
least squares fit, 168*ff*
life-stress, 71
life-temperature, 75
line, 132
log-linear, 99
log-linear for spread, 106, 276
logistic, 103
main-effects, 103
maximum likelihood fit, 233*ff*
multivariable, 98, 210*ff*, 276*ff*, 282, 351
multivariable depicted, 213
multivariable linear, 211
nonlinear, 104
not Arrhenius, 144
Palmgren's, 86, 251
paper, 72
partially specified, 41, 251
Peck's, 102, 111
plane, 100
plot, 72, 398
plot of degradation, 537
polynomial, 97
power, 85*ff*
quadratic, 96, 99, 261, 273, 352, 358, 374
reparameterized, 286
scale parameter, 259, 278
single-stress, 95
spread, 105
Taylor's, 87
temperature, 76, 80, 109, 115, 144, 163,
    230
test of fit, 216
test of linearity, 260
three-dimensional, 100
with failure modes, 399
working, 227
Zhurkov's, 102, 111
Relationship plot, acceleration factor, 151
    Arrhenius, 120, 138, 142
    inverse power, 132
    nonlinear, 127
Relationships, compare graphically, 431
    LS comparison, 441
Reliability, books, 42
    demonstration, 23, 255, 375, 426, 438, 452,
    482
    estimation, 43
    field, 28

high, 78
measure, 23
ML confidence limits, 246
ML estimate, 246
test, 28
Reliability function, definition, 54
exponential, 54
lognormal, 61
Weibull, 64
Removal, 145, 146
Repairable system, 52
Repeat measurements, 534
Residual, adjusted for ML fit, 263
censored, 262
cumulative exposure, 512
degradation, 542
exponential, 504
interval, 276
raw, 262
step-stress, 503
uniform, 504
unitized, 273, 278
Residuals, 186, 218
adjusted about the fitted line, 206
adjusted about the mean, 186
crossplot, 209, 218, 505
extreme value plot, 206
failure mode, 412
hazard plot, 186, 262
normal plot, 218
plot, 434
pooled, 186
raw about the fitted line, 186, 206
raw about the mean, 186
standardized, 186
Resistors, 10
Risks, competing, 378, see Failure modes
Robust, comparison of means, 438
comparison of standard deviations, 434
numerically, 245
Robust estimate, 121
Root mean squared error, 172
Rounding properly, 245
Rubber, 7
Runout, 95
Rupture of solids, 102
Rusting, 6

Sample, 29, 44
asymptotically large, 235
biased, 26
extreme values, 27
homogeneous, 27, 34
misleading, 44
random, 27, 44
representative, 27

Sample average, 174
Sample likelihood, for model, 476
Sample size, 36, 336, 340
choice, 327
equivalent, 438
for hypothesis test, 429
Sample standard deviation, 174
Samples, independent, 460
Sampling, bias, 26
haphazard, 44
method, 44
simple random, 44
stratified, 44
two-stage, 44
Sampling distribution, ML estimate, 245
normal approximation, 371
null hypothesis, 428
SAS, 151, 169, 172, 237, 412
Scale, activation energy, 122
cumulative hazard, 141
probability, 116
temperature, 78
Scale parameter, 66
constant, 268
nonconstant ML fit, 272
relationship, 259
Scale parameters, graphical comparison, 431
Scatter of data, 73
Screening, environmental stress, 39
Semiconductors, 9, 75, 79, 392, 522
Sensitivity analysis, 252, 357, 372
of test plan, 342
Sequence effect, 501, 517
Series system, 395
definition, 378
identical parts, 383ff
literature, 379
model, 378ff
size effect, 386
with independent failure modes, 378
Shape parameter, Weibull, 63
Shelf life, 522, 526
Shock model, 382
Significance, 429
level, 428
practical, 48, 429
statistical, 48, 429
Significant, 48
highly, 428
practically, 429
statistically, 428, 429
very highly, 428
Significant figures, 218
Simple random sampling, 44
Simulation, for a test plan, 352ff
larger sample for, 357

Simultaneous confidence limits, 297
Simultaneous confidence region, 464
Simultaneous intervals, Q differences, 463
   Q ML estimates, 455
   Q pairs of ML estimates, 463
Singly censored data, 13
   graphical analysis, 134*ff*
   ML analysis, 242*ff*
Size, 17
   definition, 389
   equivalent, 391
Size effect, 377, 385*ff*, 389, 497, 507
   exponential distribution, 386
   failure rate, 386
   model, 382, 386
   pessimistic, 387
   strength, 386
   Weibull distribution, 386
Size of sample, 36, 327, 336, 340, 429
Sizes, several specimens, 423
Slope, common distribution, 117
Slope coefficient, confidence limits LS, 180,
   201
   LS estimate, 172
   specified, 250
Slope parameter, 63, 83
Software, statistical, 237*ff*
Solder, 86
Solid state devices, 76
Specified coefficient, 250
   error in, 252
Specimen, 27
   creep-rupture, 384
   definition, 272
   design, 17
   geometry, 18, 388
   measured once, 534
   measured repeatedly, 534
   realistic, 26
   standard, 34
Specimens, 26
   number of, 36
   tandem, 383, 384, 417
   various sizes, 423
SPREAD, 245
Spread depends on stress, 105, 272
Spreads, compare graphically, 431
   LS comparisons, 434*ff*
   ML comparisons, 451*ff*
SPSS, 240
SPSS^x, 169
Stability, product, 522, 524
Standard deviation, confidence limits LS, 180
   constant, 80
   definition, 56
   depends on age, 530

estimate, 538
   estimate based on lack of fit, 175
   estimate based on replication, 175
   graphical estimate, 124
   heteroscedastic, 530, 532
   LS comparison, 435
   pooled estimate, 175, 438
   pooled sample, 436, 438
   sample, 174
   test for constant, 184*ff*
Standard deviations, LS compare K, 436
   LS compare two, 436
   LS comparison, 434*ff*
Standard error, 168, 172, 293
   accuracy of, 354
   estimate, 245
   log mean, 178
   LS intercept coefficient, 180
   LS slope coefficient, 180
   ML estimate, 235, 245
   more accurate estimate of, 356
   true asymptotic, 370
Standardized censoring time, 331
Standardized deviate, 181, 289
Standardized slope, 331
Standards, engineering, 5, 28, 38, 39
Standards, reliability, 248
STAR, 151, 237, 304, 412
Statistic, hypothesis test, 428
   LR, 470
   quadratic, 466
   Rao's, 486
   Wald's, 487
Statistical tables, 549*ff*
Statistical test, *see* Hypothesis test
Statistically independent, *see* Independent
Statistically significant, 428
STATPAC, 216, 237, 243, 269, 273, 278, 310,
   354, 407, 503, 538
Step-stress, 30, 493*ff*
   advantages, 19
   data analysis, 502*ff*
   disadvantages, 19
   general model, 501
   interval data, 518
   life distribution, 498
   limitations, 493
   linear-exponential, 501
   linear-lognormal, 501
   linear-Weibull, 501
   literature, 494, 495
   metal fatigue, 494
   ML estimates, 503
   ML fit, 502*ff*
   models, 493*ff*, 496
   multivariable, 502

pattern, 498
read-out data, 518
residuals, 503
sequence effect, 501
test plans, 502
Stepwise, logic, 222
    LS fitting, 220
    ML fitting, 281
Stratified sampling, 44
Strength, distribution, 269
    mean, 94
    size effect, 386
    standard deviation, 94
Stress, amplitude, 21
    choice of levels, 36
    constant, 18, 30, 31, 51
    corrected, 35
    cyclic, 20, 507
    frequency, 21
    history, 524
    inaccurate, 183
    loading, 18ff, 30
    lowest test, 332
    maximum, 388
    nonrepeating, 507
    nonuniform, 387ff, 390
    one level, 282
    optimum levels, 320
    pattern, 390
    progressive, 20
    pseudo, 272
    ramp, 494, 507, 508
    random, 22, 494, 507, 519
    range, 21
    real, 30
    single, 30
    single condition, 37, 51, 282
    sinusoidal, 21, 519
    spectrum, 494
    square-wave, 518
    step, 18, 493ff
    test, 31
    two levels, 249
    uniform, 388
    varying, 31, 493ff, 507, 509
    voltage, 35
Stress screening, 39, 43, 70
Stresses, multiple, 29, 100
Structural materials, 524
Subpopulations, 69
Subspace, alternative, 473
    null hypothesis, 473
Successive test, 363
Sudden-death test, 17, 383
Sum of squares, 172, 175
    for means, 440

SURVCALC, 237, 412
Survivor function, 54
Survivorship function, 54
SURVREG, 237, 486
System, identical parts, 383
    redundant, 379
    weakest link, 378

t statistic to compare mean, 483
Table, χ^2 percentiles, 554
    F percentiles, 556
    normal cumulatve distribution, 550
    normal factors two-sided, 552
    normal percentiles, 552
    probability plotting positions, 560
    t percentiles, 553
Tables, statistical, 549ff
Taylor's model, 87
TEAM, 119, 122, 131, 133, 188, 206
Technology Associates, 122, 153
Temperature, absolute, 76, 78
    acceleration, 97
    corrected, 35
    design, 81, 525
    inverse power approximation, 252
    Kelvin, 76
    range, 86
    Rankine, 76
Terminology, engineering, 3
    statistical, 3
Test, see Hypothesis test
    accelerated, 37
    compressed time, 15
    constraints, 36, 320
    cost, 320
    degradation advantages, 521
    demonstration, 255, 426, 428, 452
    elephant, 37, 493
    engineering, 28
    hypothesis (see Hypothesis test and Comparison)
    index, 28
    length, 36, 320
    pilot, 37
    planning, 36
    progressive, 231, 449
    Prot, 494
    purpose, 23
    ramp, 493
    statistical hypothesis, see Hypothesis test
    stopping, 234
    successive, 363
    sudden death, 383
    temperature-humidity, 101
    up-down, 95
    voltage-endurance, 211

Test condition, one, 151
Test conditions, 27
    number of, 43
    uncontrolled, 131
Test equipment, 5
Test labs, 5
Test of fit, distribution, 264
    multivariable relationship, 217
    simple relationship, 183, 260
Test plan, 2 stresses, 323
    2×3 factorial, 506
    3 stresses, 323
    4 stresses, 324
    accuracy, 319
    allocation unequal, 317
    assumptions, 318, 329
    best low stress, 343
    best traditional, 332ff
    censored data, 328ff
    Class-H, 321, 324
    comparison of optimum, best traditional,
        and good compromise, 341
    complete data, 318ff
    compromise, 341
    compromise standard error, 342
    computer program, 328
    constraints, 363
    cost, 363
    degradation, 534, 547
    drawbacks of optimum, 326, 341
    drawbacks of traditional, 341
    efficient, 317
    equal allocation, 321, 332
    equally spaced stresses, 335
    evaluate by simulation, 349ff
    exponential distribution, 328
    for endurance-limit, 271
    good, 326ff
    heater, 319, 323
    inefficient, 212
    inspection data, 362
    literature, 328, 361
    locally optimum, 331
    logistic distribution, 328
    lognormal distribution, 328
    lowest stress, 332
    LS estimates, 318
    Meeker-Hahn, 343ff
    Meeker-Hahn standard error, 348
    minimum variance, 321, 330
    ML estimates, 330
    ML theory, 364ff
    model, 318, 330
    no failures, 326, 340
    notation, 331
    optimization criteria, 330, 363
    optimum, 240, 320ff, 337ff
    optimum allocation, 321, 337
    optimum drawbacks, 326
    optimum low stress, 337
    optimum ramp, 506
    optimum step-stress, 495
    optimum stresses, 320
    optimum variance, 320
    other relationships, 362
    robust, 326
    sample size, 327
    sample size Meeker-Hahn, 348
    sample size optimum, 340
    sample size traditional, 336
    sensitivity analysis, 342
    simple model, 328ff
    simulation to evaluate, 349ff
    standard error best traditional, 335
    standard error LS, 319
    standard error optimum, 340
    step-stress, 20, 502
    successive testing, 363
    survey, 361ff
    test length, 326
    traditional, 36, 116, 323ff
    unequal allocation, 317
    variance, 319
    Weibull distribution, 328
    without failures, 326, 340
Test planning, 22ff, 210
Test plans, 317ff
    comparison, 341
    complete data, 318ff
    general principle, 317
    literature, 318
    quantal-response, 95
Test statistic, 428
    critical value, 480, 481
Tests, multiple, 52
Theory, large-sample, 236
Thermal cycling, 144, 163, 312, 397
Time, 25
    equivalent, 251
Time censored data, 13
Tolerance limit, 179, 199
Tools, cutting, 87
Traditional test plan, 332ff
Transfer lettering, 120
Transformation, parameter, 296
Transformers, 385
Transistors, 9, 383
Trend, 265
Tukey's comparison of means, 441
Two-stage sampling, 44
Type I censored data, 13, see Time censored
Type II censored data, see Failure censored

Typical life, 55

Uncertainty, of estimates, 170
　model, 170
　of ML estimate, 354
Uncorrelated coefficients, 530, 532
Uniformly most powerful hypothesis test, 485
Uniqueness, ML estimate, 291
Unitized residuals, 273
Up-down testing, 95
Usage, 25
Use conditions, 27

Variability, lot-to-lot, 530
Variable, 0-1, 103, 276
　categorical, 32, 102, 223
　centered, 172, 215, 276
　coded, 172, 215, 276
　confounded, 34
　constant, 32
　continuous, 31, 209
　controlled, 32
　dependent, 24
　dummy, 103
　effect of, 189, 208, 226, 264, 268
　engineering, 31
　experimental, 32, 98
　explanatory, 31
　identifying an important, 211
　independent, 31, 471
　indicator, 102, 103
　performance, 24
　predictor, 98
　response, 24
　significant, 213
　standard accelerating, 52
　statistically significant, 213
　transformed, 99
　uncontrolled, 98, 183
　uncontrolled observed, 32
　uncontrolled unobserved, 33
　unobserved, 33
Variance, components of, 107
　definition, 56
　Greenwood's estimate, 151
　ML estimate, 294
　of ML estimate of a function, 370
Variation, uncontrolled, 542
Varying stress, data analysis, 509ff
　literature, 494, 495
　ML fit, 509ff
　model, 493ff, 506
Vibration, 22, 40, 381
VLSI, 9, 101
Volts per mil, 35
Volume effect, 387

Wald's hypothesis test, 487, 490
Wald's statistic, 487
Weakest link product, 63, 65
Weakest link system, 378
Wear, 6, 12, 522
Wear-in, data, 527
　period, 527
Wear-out, 58, 527
Weathering, 8, 12, 533
Weibull degradation rate, 533
Weibull distribution, 63ff
　assess, 205
　β parameter, 63
　characteristic life, 63
　constant shape parameter β, 89
　cumulative distribution, 63
　hazard function, 64
　identical components, 384
　known β, 315, 248
　likelihood, 289
　mean, 64
　ML fit, 249
　nonuniform stress, 391
　papers, 131
　percentile, 64
　probability density, 64
　probability plot, 129
　relationship to exponential, 65, 254
　reliability function, 64
　scale parameter, 63
　shape depends on stress, 105
　shape parameter β, 63
　shape parameter constant, 204
　shape parameter equals 1, 131
　shape parameter estimate, 131, 206
　shape parameter specified, 131, 254
　size effect, 386
　slope parameter, 63
　spread in log life, 64
　standard deviation, 64
　three-parameter, 65
　versus lognormal, 65, 211, 263
Weighted regression fitting, 241
Wire varnish, 386
Wood composites, 102

Zhurkov's relationship, 102

WILEY SERIES IN PROBABILITY AND STATISTICS
ESTABLISHED BY WALTER A. SHEWHART AND SAMUEL S. WILKS

Editors: *David J. Balding, Noel A. C. Cressie, Nicholas I. Fisher,*
Iain M. Johnstone, J. B. Kadane, Geert Molenberghs. Louise M. Ryan,
David W. Scott, Adrian F. M. Smith, Jozef L. Teugels
Editors Emeriti: *Vic Barnett, J. Stuart Hunter, David G. Kendall*

The *Wiley Series in Probability and Statistics* is well established and authoritative. It covers many topics of current research interest in both pure and applied statistics and probability theory. Written by leading statisticians and institutions, the titles span both state-of-the-art developments in the field and classical methods.

Reflecting the wide range of current research in statistics, the series encompasses applied, methodological and theoretical statistics, ranging from applications and new techniques made possible by advances in computerized practice to rigorous treatment of theoretical approaches.

This series provides essential and invaluable reading for all statisticians, whether in academia, industry, government, or research.

ABRAHAM and LEDOLTER · Statistical Methods for Forecasting
AGRESTI · Analysis of Ordinal Categorical Data
AGRESTI · An Introduction to Categorical Data Analysis
AGRESTI · Categorical Data Analysis, *Second Edition*
ALTMAN, GILL, and McDONALD · Numerical Issues in Statistical Computing for the
    Social Scientist
AMARATUNGA and CABRERA · Exploration and Analysis of DNA Microarray and
    Protein Array Data
ANDĚL · Mathematics of Chance
ANDERSON · An Introduction to Multivariate Statistical Analysis, *Third Edition*
*ANDERSON · The Statistical Analysis of Time Series
ANDERSON, AUQUIER, HAUCK, OAKES, VANDAELE, and WEISBERG ·
    Statistical Methods for Comparative Studies
ANDERSON and LOYNES · The Teaching of Practical Statistics
ARMITAGE and DAVID (editors) · Advances in Biometry
ARNOLD, BALAKRISHNAN, and NAGARAJA · Records
*ARTHANARI and DODGE · Mathematical Programming in Statistics
*BAILEY · The Elements of Stochastic Processes with Applications to the Natural
    Sciences
BALAKRISHNAN and KOUTRAS · Runs and Scans with Applications
BARNETT · Comparative Statistical Inference, *Third Edition*
BARNETT and LEWIS · Outliers in Statistical Data, *Third Edition*
BARTOSZYNSKI and NIEWIADOMSKA-BUGAJ · Probability and Statistical Inference
BASILEVSKY · Statistical Factor Analysis and Related Methods: Theory and
    Applications
BASU and RIGDON · Statistical Methods for the Reliability of Repairable Systems
BATES and WATTS · Nonlinear Regression Analysis and Its Applications
BECHHOFER, SANTNER, and GOLDSMAN · Design and Analysis of Experiments for
    Statistical Selection, Screening, and Multiple Comparisons
BELSLEY · Conditioning Diagnostics: Collinearity and Weak Data in Regression

*Now available in a lower priced paperback edition in the Wiley Classics Library.

* BELSLEY, KUH, and WELSCH · Regression Diagnostics: Identifying Influential Data and Sources of Collinearity

BENDAT and PIERSOL · Random Data: Analysis and Measurement Procedures, *Third Edition*

BERRY, CHALONER, and GEWEKE · Bayesian Analysis in Statistics and Econometrics: Essays in Honor of Arnold Zellner

BERNARDO and SMITH · Bayesian Theory

BHAT and MILLER · Elements of Applied Stochastic Processes, *Third Edition*

BHATTACHARYA and WAYMIRE · Stochastic Processes with Applications

BILLINGSLEY · Convergence of Probability Measures, *Second Edition*

BILLINGSLEY · Probability and Measure, *Third Edition*

BIRKES and DODGE · Alternative Methods of Regression

BLISCHKE AND MURTHY (editors) · Case Studies in Reliability and Maintenance

BLISCHKE AND MURTHY · Reliability: Modeling, Prediction, and Optimization

BLOOMFIELD · Fourier Analysis of Time Series: An Introduction, *Second Edition*

BOLLEN · Structural Equations with Latent Variables

BOROVKOV · Ergodicity and Stability of Stochastic Processes

BOULEAU · Numerical Methods for Stochastic Processes

BOX · Bayesian Inference in Statistical Analysis

BOX · R. A. Fisher, the Life of a Scientist

BOX and DRAPER · Empirical Model-Building and Response Surfaces

*BOX and DRAPER · Evolutionary Operation: A Statistical Method for Process Improvement

BOX, HUNTER, and HUNTER · Statistics for Experimenters: An Introduction to Design, Data Analysis, and Model Building

BOX and LUCEÑO · Statistical Control by Monitoring and Feedback Adjustment

BRANDIMARTE · Numerical Methods in Finance: A MATLAB-Based Introduction

BROWN and HOLLANDER · Statistics: A Biomedical Introduction

BRUNNER, DOMHOF, and LANGER · Nonparametric Analysis of Longitudinal Data in Factorial Experiments

BUCKLEW · Large Deviation Techniques in Decision, Simulation, and Estimation

CAIROLI and DALANG · Sequential Stochastic Optimization

CHAN · Time Series: Applications to Finance

CHATTERJEE and HADI · Sensitivity Analysis in Linear Regression

CHATTERJEE and PRICE · Regression Analysis by Example, *Third Edition*

CHERNICK · Bootstrap Methods: A Practitioner's Guide

CHERNICK and FRIIS · Introductory Biostatistics for the Health Sciences

CHILÈS and DELFINER · Geostatistics: Modeling Spatial Uncertainty

CHOW and LIU · Design and Analysis of Clinical Trials: Concepts and Methodologies, *Second Edition*

CLARKE and DISNEY · Probability and Random Processes: A First Course with Applications, *Second Edition*

*COCHRAN and COX · Experimental Designs, *Second Edition*

CONGDON · Applied Bayesian Modelling

CONGDON · Bayesian Statistical Modelling

CONOVER · Practical Nonparametric Statistics, *Third Edition*

COOK · Regression Graphics

COOK and WEISBERG · Applied Regression Including Computing and Graphics

COOK and WEISBERG · An Introduction to Regression Graphics

CORNELL · Experiments with Mixtures, Designs, Models, and the Analysis of Mixture Data, *Third Edition*

COVER and THOMAS · Elements of Information Theory

COX · A Handbook of Introductory Statistical Methods

*Now available in a lower priced paperback edition in the Wiley Classics Library.

*COX · Planning of Experiments

CRESSIE · Statistics for Spatial Data, *Revised Edition*

CSÖRGÖ and HORVÁTH · Limit Theorems in Change Point Analysis

DANIEL · Applications of Statistics to Industrial Experimentation

DANIEL · Biostatistics: A Foundation for Analysis in the Health Sciences, *Eighth Edition*

*DANIEL · Fitting Equations to Data: Computer Analysis of Multifactor Data, *Second Edition*

DASU and JOHNSON · Exploratory Data Mining and Data Cleaning

DAVID and NAGARAJA · Order Statistics, *Third Edition*

*DEGROOT, FIENBERG, and KADANE · Statistics and the Law

DEL CASTILLO · Statistical Process Adjustment for Quality Control

DeMARIS · Regression with Social Data: Modeling Continuous and Limited Response Variables

DEMIDENKO · Mixed Models: Theory and Applications

DENISON, HOLMES, MALLICK and SMITH · Bayesian Methods for Nonlinear Classification and Regression

DETTE and STUDDEN · The Theory of Canonical Moments with Applications in Statistics, Probability, and Analysis

DEY and MUKERJEE · Fractional Factorial Plans

DILLON and GOLDSTEIN · Multivariate Analysis: Methods and Applications

DODGE · Alternative Methods of Regression

*DODGE and ROMIG · Sampling Inspection Tables, *Second Edition*

*DOOB · Stochastic Processes

DOWDY, WEARDEN, and CHILKO · Statistics for Research, *Third Edition*

DRAPER and SMITH · Applied Regression Analysis, *Third Edition*

DRYDEN and MARDIA · Statistical Shape Analysis

DUDEWICZ and MISHRA · Modern Mathematical Statistics

DUNN and CLARK · Basic Statistics: A Primer for the Biomedical Sciences, *Third Edition*

DUPUIS and ELLIS · A Weak Convergence Approach to the Theory of Large Deviations

*ELANDT-JOHNSON and JOHNSON · Survival Models and Data Analysis

ENDERS · Applied Econometric Time Series

ETHIER and KURTZ · Markov Processes: Characterization and Convergence

EVANS, HASTINGS, and PEACOCK · Statistical Distributions, *Third Edition*

FELLER · An Introduction to Probability Theory and Its Applications, Volume I, *Third Edition,* Revised; Volume II, *Second Edition*

FISHER and VAN BELLE · Biostatistics: A Methodology for the Health Sciences

FITZMAURICE, LAIRD, and WARE · Applied Longitudinal Analysis

*FLEISS · The Design and Analysis of Clinical Experiments

FLEISS · Statistical Methods for Rates and Proportions, *Third Edition*

FLEMING and HARRINGTON · Counting Processes and Survival Analysis

FULLER · Introduction to Statistical Time Series, *Second Edition*

FULLER · Measurement Error Models

GALLANT · Nonlinear Statistical Models

GHOSH, MUKHOPADHYAY, and SEN · Sequential Estimation

GIESBRECHT and GUMPERTZ · Planning, Construction, and Statistical Analysis of Comparative Experiments

GIFI · Nonlinear Multivariate Analysis

GLASSERMAN and YAO · Monotone Structure in Discrete-Event Systems

GNANADESIKAN · Methods for Statistical Data Analysis of Multivariate Observations, *Second Edition*

GOLDSTEIN and LEWIS · Assessment: Problems, Development, and Statistical Issues

GREENWOOD and NIKULIN · A Guide to Chi-Squared Testing

*Now available in a lower priced paperback edition in the Wiley Classics Library.

GROSS and HARRIS · Fundamentals of Queueing Theory, *Third Edition*
*HAHN and SHAPIRO · Statistical Models in Engineering
HAHN and MEEKER · Statistical Intervals: A Guide for Practitioners
HALD · A History of Probability and Statistics and their Applications Before 1750
HALD · A History of Mathematical Statistics from 1750 to 1930
HAMPEL · Robust Statistics: The Approach Based on Influence Functions
HANNAN and DEISTLER · The Statistical Theory of Linear Systems
HEIBERGER · Computation for the Analysis of Designed Experiments
HEDAYAT and SINHA · Design and Inference in Finite Population Sampling
HELLER · MACSYMA for Statisticians
HINKELMAN and KEMPTHORNE: · Design and Analysis of Experiments, Volume 1:
    Introduction to Experimental Design
HOAGLIN, MOSTELLER, and TUKEY · Exploratory Approach to Analysis
    of Variance
HOAGLIN, MOSTELLER, and TUKEY · Exploring Data Tables, Trends and Shapes
*HOAGLIN, MOSTELLER, and TUKEY · Understanding Robust and Exploratory
    Data Analysis
HOCHBERG and TAMHANE · Multiple Comparison Procedures
HOCKING · Methods and Applications of Linear Models: Regression and the Analysis
    of Variance, *Second Edition*
HOEL · Introduction to Mathematical Statistics, *Fifth Edition*
HOGG and KLUGMAN · Loss Distributions
HOLLANDER and WOLFE · Nonparametric Statistical Methods, *Second Edition*
HOSMER and LEMESHOW · Applied Logistic Regression, *Second Edition*
HOSMER and LEMESHOW · Applied Survival Analysis: Regression Modeling of
    Time to Event Data
HUBER · Robust Statistics
HUBERTY · Applied Discriminant Analysis
HUNT and KENNEDY · Financial Derivatives in Theory and Practice
HUSKOVA, BERAN, and DUPAC · Collected Works of Jaroslav Hajek—
    with Commentary
HUZURBAZAR · Flowgraph Models for Multistate Time-to-Event Data
IMAN and CONOVER · A Modern Approach to Statistics
JACKSON · A User's Guide to Principle Components
JOHN · Statistical Methods in Engineering and Quality Assurance
JOHNSON · Multivariate Statistical Simulation
JOHNSON and BALAKRISHNAN · Advances in the Theory and Practice of Statistics: A
    Volume in Honor of Samuel Kotz
JOHNSON and BHATTACHARYYA · Statistics: Principles and Methods, *Fifth Edition*
JOHNSON and KOTZ · Distributions in Statistics
JOHNSON and KOTZ (editors) · Leading Personalities in Statistical Sciences: From the
    Seventeenth Century to the Present
JOHNSON, KOTZ, and BALAKRISHNAN · Continuous Univariate Distributions,
    Volume 1, *Second Edition*
JOHNSON, KOTZ, and BALAKRISHNAN · Continuous Univariate Distributions,
    Volume 2, *Second Edition*
JOHNSON, KOTZ, and BALAKRISHNAN · Discrete Multivariate Distributions
JOHNSON, KOTZ, and KEMP · Univariate Discrete Distributions, *Second Edition*
JUDGE, GRIFFITHS, HILL, LÜTKEPOHL, and LEE · The Theory and Practice of
    Econometrics, *Second Edition*
JUREČKOVÁ and SEN · Robust Statistical Procedures: Aymptotics and Interrelations
JUREK and MASON · Operator-Limit Distributions in Probability Theory
KADANE · Bayesian Methods and Ethics in a Clinical Trial Design

KADANE AND SCHUM · A Probabilistic Analysis of the Sacco and Vanzetti Evidence
KALBFLEISCH and PRENTICE · The Statistical Analysis of Failure Time Data, *Second Edition*
KASS and VOS · Geometrical Foundations of Asymptotic Inference
KAUFMAN and ROUSSEEUW · Finding Groups in Data: An Introduction to Cluster Analysis
KEDEM and FOKIANOS · Regression Models for Time Series Analysis
KENDALL, BARDEN, CARNE, and LE · Shape and Shape Theory
KHURI · Advanced Calculus with Applications in Statistics, *Second Edition*
KHURI, MATHEW, and SINHA · Statistical Tests for Mixed Linear Models
*KISH · Statistical Design for Research
KLEIBER and KOTZ · Statistical Size Distributions in Economics and Actuarial Sciences
KLUGMAN, PANJER, and WILLMOT · Loss Models: From Data to Decisions
KLUGMAN, PANJER, and WILLMOT · Solutions Manual to Accompany Loss Models: From Data to Decisions
KOTZ, BALAKRISHNAN, and JOHNSON · Continuous Multivariate Distributions, Volume 1, *Second Edition*
KOTZ and JOHNSON (editors) · Encyclopedia of Statistical Sciences: Volumes 1 to 9 with Index
KOTZ and JOHNSON (editors) · Encyclopedia of Statistical Sciences: Supplement Volume
KOTZ, READ, and BANKS (editors) · Encyclopedia of Statistical Sciences: Update Volume 1
KOTZ, READ, and BANKS (editors) · Encyclopedia of Statistical Sciences: Update Volume 2
KOVALENKO, KUZNETZOV, and PEGG · Mathematical Theory of Reliability of Time-Dependent Systems with Practical Applications
LACHIN · Biostatistical Methods: The Assessment of Relative Risks
LAD · Operational Subjective Statistical Methods: A Mathematical, Philosophical, and Historical Introduction
LAMPERTI · Probability: A Survey of the Mathematical Theory, *Second Edition*
LANGE, RYAN, BILLARD, BRILLINGER, CONQUEST, and GREENHOUSE · Case Studies in Biometry
LARSON · Introduction to Probability Theory and Statistical Inference, *Third Edition*
LAWLESS · Statistical Models and Methods for Lifetime Data, *Second Edition*
LAWSON · Statistical Methods in Spatial Epidemiology
LE · Applied Categorical Data Analysis
LE · Applied Survival Analysis
LEE and WANG · Statistical Methods for Survival Data Analysis, *Third Edition*
LePAGE and BILLARD · Exploring the Limits of Bootstrap
LEYLAND and GOLDSTEIN (editors) · Multilevel Modelling of Health Statistics
LIAO · Statistical Group Comparison
LINDVALL · Lectures on the Coupling Method
LINHART and ZUCCHINI · Model Selection
LITTLE and RUBIN · Statistical Analysis with Missing Data, *Second Edition*
LLOYD · The Statistical Analysis of Categorical Data
MAGNUS and NEUDECKER · Matrix Differential Calculus with Applications in Statistics and Econometrics, *Revised Edition*
MALLER and ZHOU · Survival Analysis with Long Term Survivors
MALLOWS · Design, Data, and Analysis by Some Friends of Cuthbert Daniel
MANN, SCHAFER, and SINGPURWALLA · Methods for Statistical Analysis of Reliability and Life Data
MANTON, WOODBURY, and TOLLEY · Statistical Applications Using Fuzzy Sets
MARCHETTE · Random Graphs for Statistical Pattern Recognition

*Now available in a lower priced paperback edition in the Wiley Classics Library.

MARDIA and JUPP · Directional Statistics

MASON, GUNST, and HESS · Statistical Design and Analysis of Experiments with Applications to Engineering and Science, *Second Edition*

McCULLOCH and SEARLE · Generalized, Linear, and Mixed Models

McFADDEN · Management of Data in Clinical Trials

* McLACHLAN · Discriminant Analysis and Statistical Pattern Recognition

McLACHLAN, DO, and AMBROISE · Analyzing Microarray Gene Expression Data

McLACHLAN and KRISHNAN · The EM Algorithm and Extensions

McLACHLAN and PEEL · Finite Mixture Models

McNEIL · Epidemiological Research Methods

MEEKER and ESCOBAR · Statistical Methods for Reliability Data

MEERSCHAERT and SCHEFFLER · Limit Distributions for Sums of Independent Random Vectors: Heavy Tails in Theory and Practice

MICKEY, DUNN, and CLARK · Applied Statistics: Analysis of Variance and Regression, *Third Edition*

*MILLER · Survival Analysis, *Second Edition*

MONTGOMERY, PECK, and VINING · Introduction to Linear Regression Analysis, *Third Edition*

MORGENTHALER and TUKEY · Configural Polysampling: A Route to Practical Robustness

MUIRHEAD · Aspects of Multivariate Statistical Theory

MULLER and STOYAN · Comparison Methods for Stochastic Models and Risks

MURRAY · X-STAT 2.0 Statistical Experimentation, Design Data Analysis, and Nonlinear Optimization

MURTHY, XIE, and JIANG · Weibull Models

MYERS and MONTGOMERY · Response Surface Methodology: Process and Product Optimization Using Designed Experiments, *Second Edition*

MYERS, MONTGOMERY, and VINING · Generalized Linear Models. With Applications in Engineering and the Sciences

*NELSON · Accelerated Testing, Statistical Models, Test Plans, and Data Analyses

NELSON · Applied Life Data Analysis

NEWMAN · Biostatistical Methods in Epidemiology

OCHI · Applied Probability and Stochastic Processes in Engineering and Physical Sciences

OKABE, BOOTS, SUGIHARA, and CHIU · Spatial Tesselations: Concepts and Applications of Voronoi Diagrams, *Second Edition*

OLIVER and SMITH · Influence Diagrams, Belief Nets and Decision Analysis

PALTA · Quantitative Methods in Population Health: Extensions of Ordinary Regressions

PANKRATZ · Forecasting with Dynamic Regression Models

PANKRATZ · Forecasting with Univariate Box-Jenkins Models: Concepts and Cases

*PARZEN · Modern Probability Theory and Its Applications

PEÑA, TIAO, and TSAY · A Course in Time Series Analysis

PIANTADOSI · Clinical Trials: A Methodologic Perspective

PORT · Theoretical Probability for Applications

POURAHMADI · Foundations of Time Series Analysis and Prediction Theory

PRESS · Bayesian Statistics: Principles, Models, and Applications

PRESS · Subjective and Objective Bayesian Statistics, *Second Edition*

PRESS and TANUR · The Subjectivity of Scientists and the Bayesian Approach

PUKELSHEIM · Optimal Experimental Design

PURI, VILAPLANA, and WERTZ · New Perspectives in Theoretical and Applied Statistics

PUTERMAN · Markov Decision Processes: Discrete Stochastic Dynamic Programming

*RAO · Linear Statistical Inference and Its Applications, *Second Edition*

RAUSAND and HØYLAND · System Reliability Theory: Models, Statistical Methods, and Applications, *Second Edition*

*Now available in a lower priced paperback edition in the Wiley Classics Library.

RENCHER · Linear Models in Statistics
RENCHER · Methods of Multivariate Analysis, *Second Edition*
RENCHER · Multivariate Statistical Inference with Applications
* RIPLEY · Spatial Statistics
RIPLEY · Stochastic Simulation
ROBINSON · Practical Strategies for Experimenting
ROHATGI and SALEH · An Introduction to Probability and Statistics, *Second Edition*
ROLSKI, SCHMIDLI, SCHMIDT, and TEUGELS · Stochastic Processes for Insurance
    and Finance
ROSENBERGER and LACHIN · Randomization in Clinical Trials: Theory and Practice
ROSS · Introduction to Probability and Statistics for Engineers and Scientists
ROUSSEEUW and LEROY · Robust Regression and Outlier Detection
RUBIN · Multiple Imputation for Nonresponse in Surveys
RUBINSTEIN · Simulation and the Monte Carlo Method
RUBINSTEIN and MELAMED · Modern Simulation and Modeling
RYAN · Modern Regression Methods
RYAN · Statistical Methods for Quality Improvement, *Second Edition*
SALTELLI, CHAN, and SCOTT (editors) · Sensitivity Analysis
*SCHEFFE · The Analysis of Variance
SCHIMEK · Smoothing and Regression: Approaches, Computation, and Application
SCHOTT · Matrix Analysis for Statistics
SCHOUTENS · Levy Processes in Finance: Pricing Financial Derivatives
SCHUSS · Theory and Applications of Stochastic Differential Equations
SCOTT · Multivariate Density Estimation: Theory, Practice, and Visualization
*SEARLE · Linear Models
SEARLE · Linear Models for Unbalanced Data
SEARLE · Matrix Algebra Useful for Statistics
SEARLE, CASELLA, and McCULLOCH · Variance Components
SEARLE and WILLETT · Matrix Algebra for Applied Economics
SEBER and LEE · Linear Regression Analysis, *Second Edition*
*SEBER · Multivariate Observations
SEBER and WILD · Nonlinear Regression
SENNOTT · Stochastic Dynamic Programming and the Control of Queueing Systems
*SERFLING · Approximation Theorems of Mathematical Statistics
SHAFER and VOVK · Probability and Finance: It's Only a Game!
SILVAPULLE and SEN · Constrained Statistical Inference: Order, Inequality and Shape
    Constraints
SMALL and McLEISH · Hilbert Space Methods in Probability and Statistical Inference
SRIVASTAVA · Methods of Multivariate Statistics
STAPLETON · Linear Statistical Models
STAUDTE and SHEATHER · Robust Estimation and Testing
STOYAN, KENDALL, and MECKE · Stochastic Geometry and Its Applications, *Second
    Edition*
STOYAN and STOYAN · Fractals, Random Shapes and Point Fields: Methods of
    Geometrical Statistics
STYAN · The Collected Papers of T. W. Anderson: 1943–1985
SUTTON, ABRAMS, JONES, SHELDON, and SONG · Methods for Meta-Analysis in
    Medical Research
TANAKA · Time Series Analysis: Nonstationary and Noninvertible Distribution Theory
THOMPSON · Empirical Model Building
THOMPSON · Sampling, *Second Edition*
THOMPSON · Simulation: A Modeler's Approach
THOMPSON and SEBER · Adaptive Sampling
THOMPSON, WILLIAMS, and FINDLAY · Models for Investors in Real World Markets

*Now available in a lower priced paperback edition in the Wiley Classics Library.

TIAO, BISGAARD, HILL, PEÑA, and STIGLER (editors) · Box on Quality and Discovery: with Design, Control, and Robustness

TIERNEY · LISP-STAT: An Object-Oriented Environment for Statistical Computing and Dynamic Graphics

TSAY · Analysis of Financial Time Series

UPTON and FINGLETON · Spatial Data Analysis by Example, Volume II: Categorical and Directional Data

VAN BELLE · Statistical Rules of Thumb

VAN BELLE, FISHER, HEAGERTY, and LUMLEY · Biostatistics: A Methodology for the Health Sciences, *Second Edition*

VESTRUP · The Theory of Measures and Integration

VIDAKOVIC · Statistical Modeling by Wavelets

VINOD and REAGLE · Preparing for the Worst: Incorporating Downside Risk in Stock Market Investments

WALLER and GOTWAY · Applied Spatial Statistics for Public Health Data

WEERAHANDI · Generalized Inference in Repeated Measures: Exact MANOVA and Mixed Models

WEISBERG · Applied Linear Regression, *Third Edition*

WELSH · Aspects of Statistical Inference

WESTFALL and YOUNG · Resampling-Based Multiple Testing: Examples and Methods for p-Value Adjustment

WHITTAKER · Graphical Models in Applied Multivariate Statistics

WINKER · Optimization Heuristics in Economics: Applications of Threshold Accepting

WONNACOTT and WONNACOTT · Econometrics, *Second Edition*

WOODING · Planning Pharmaceutical Clinical Trials: Basic Statistical Principles

WOOLSON and CLARKE · Statistical Methods for the Analysis of Biomedical Data, *Second Edition*

WU and HAMADA · Experiments: Planning, Analysis, and Parameter Design Optimization

YANG · The Construction Theory of Denumerable Markov Processes

*ZELLNER · An Introduction to Bayesian Inference in Econometrics

ZHOU, OBUCHOWSKI, and McCLISH · Statistical Methods in Diagnostic Medicine

*Now available in a lower priced paperback edition in the Wiley Classics Library.

Printed in the United States of America
ED-09-19-12